Handbook of

PHYSICS in
MEDICINE and
BIOLOGY

Handbook of
PHYSICS in
MEDICINE and
BIOLOGY

Edited by
Robert Splinter

CRC Press
Taylor & Francis Group
Boca Raton London New York

CRC Press is an imprint of the
Taylor & Francis Group, an **informa** business

CRC Press
Taylor & Francis Group
6000 Broken Sound Parkway NW, Suite 300
Boca Raton, FL 33487-2742

First issued in paperback 2019

ISBN-13: 978-1-4200-7524-3 (hbk)
ISBN-13: 978-0-367-38431-9 (pbk)

Library of Congress Cataloging-in-Publication Data

Handbook of physics in medicine and biology / editor, Robert Splinter.
    p. ; cm.
  Includes bibliographical references and index.
  ISBN 978-1-4200-7524-3 (hardcover : alk. paper)
  1. Biophysics. 2. Biomedical engineering. I. Splinter, Robert.
  [DNLM: 1. Biophysical Phenomena. 2. Biophysics--methods. 3. Biomedical Engineering. QT 34 H2365 2010]
  QH505.H28 2010
  571.4--dc22

2009051347

To my dad Hans, for supporting my goals and believing in my potential.

Contents

Section I Anatomical Physics

Section II Physics of Perception

Section III   Biomechanics

Section IV   Bioelectrical Physics

Section V   Diagnostic Physics

Section VI Physics of Accessory Medicine

Section VII Physics of Bioengineering

Preface

Biology: The New Physics

The focus of this book is the use of physics and physical chemistry in biological organisms, in addition to the physical aspects of the tools used to investigate biological organisms. The term physics is used in the broadest sense of the word, extending to fluid dynamics, mechanics, and electronics. This book provides the physical and technological foundation for understanding biological functioning in general, as well as offering technical and theoretical support for biomedical research.

A frequently cited phrase coined for the new millennium is that biology is the new physics. This concept holds true in bioengineering and diagnostic tool development as well as in general biological research. Both biology and medicine have had a close collaboration with physics over centuries, at both the basic and applied levels. Major new developments in biology are motivated by broad aspects of physics, including electronics and mechanics. In addition to physicists developing instruments for biologists to perform research and create methods to analyze and reproduce biological archetypes, biologists are more and more likely to be the creative minds behind biomedical physics concept development. This book covers the engineering and physics that can help explain clinical conditions and also describes the technology used to examine clinical conditions. In addition, it discusses emerging technologies that will advance the role of medicine and assist in improving diagnostic and therapeutic medical applications.

About the Book

Handbook of Physics in Medicine and Biology is designed as a reference manual for professionals and students in the biomedical profession. It is also a resource on the technological aspects of the broad range of scientific topics for biomedical physics taught in a university setting. The book provides an introduction to engineering technology as applied to biological systems. The scientific background offered will aid in the evaluation and design of devices and techniques that can be used to acquire pertinent clinical information and develop an understanding of pertinent biological processes that will assist in the general understanding of how biology works and how biological entities function.

Each chapter contains references for additional information on specific topics that cannot be discussed in detail due to the focus of the book or based on the exhaustive background available in that particular field.

The goal of this work is to provide a professionally relevant handbook to further the education for students and professionals in biology, medicine, and biomedical physics. The material offers the fundamental knowledge and skill sets required for anyone pursuing a career in the medical and biomedical field. In this book, the reader is given an opportunity to learn about the rapidly developing and innovative techniques for understanding biological processes from an engineering perspective, as well as the science behind diagnostic methods and imaging techniques. Ultimately, we seek to improve our understanding of diseases and our ability to treat them.

Interdisciplinary Treatment

The interaction of physics and living or biological systems starts at the cellular and molecular levels, extending to the description of the operation of the organism in addition to the diagnostic methods of identifying potential problems or gaining a detailed understanding of the operation of the organism. Biomedical physics is often described as biomedical engineering, in contrast to biotechnology, which addresses the biochemistry. All aspects of diagnostic and therapeutic modalities are treated in their relation to normal physiological functions.

The use of physics in medicine and biology will assist in increasing an understanding of the molecular and cellular principles that create the macro diagnostic values measured on the outside with current diagnostic and physical devices.

The objectives of the book are to provide an insight into the elementary laws of physics and the roles they play in the biological organism, in addition to the application of the basic laws of physics applied to the measurement of biological phenomena and the interpretation of these data. Additionally, the authors describe the biological operation by means of physics theory wherever applicable. The various aspects of physics involved in biology are the following: thermodynamics: cellular metabolism; energy: conservation of energy in biological processes; electricity: cellular depolarization; kinetics/mechanics: muscle contraction and skeletal motion, and waves and sound;

radiation: vision; and last but not least fluid dynamics: breathing and blood flow.

Furthermore, the chapters discuss in detail the physics involved in developing biologically alternative detection and signal processing techniques.

Organization

The organization of this text follows the hierarchy of the physical description of the operation of biological building blocks (cells and organs) to the functioning of various organs. The biological function is described at the electric, mechanical, electromagnetic, thermodynamic, and hydrodynamic levels for the respective metric and nonmetric characteristics of the organs. Following the physical description of the biological constituents, the book addresses the principles of a range of clinical diagnostic methods. It also looks at the technological aspects of specific therapeutic applications.

Brief Description of the Chapters

The book begins with a basic background description of specific biological features and delves into the physics of explicit anatomical structures such as the basic building block: the cell. Cellular metabolism is analyzed in terms of cellular thermodynamics, with the ionic mechanism for the transmembrane potential, after which the book describes the concepts of specific sensory functions. Next, the chapters explain more specific biological functions from a physics perspective—starting with electrophysiology, followed by fluid dynamics for blood and air function. After the biological description, the book outlines certain analytical modalities such as imaging and supporting diagnostic methods. A final section turns to future perspectives related to the new field of tissue engineering, including the biophysics of prostheses and the physics concepts in regenerative medicine.

The outline of the book has the following basic structure.

Section I of the book is a detailed description of atomic, molecular, and cellular biological physics. Chapters 1 through 5 describe cellular activity, including action potential generation and conduction.

Section II describes the physics of perception as well as sensory alternatives based on biology developed with physics principles. Chapters 6 through 11 cover the basic senses as well

as additional means of perception and arriving at a diagnostic conclusion.

Section III focuses on the mechanics in biology. Chapters 12 through 19 give a detailed description of the mechanical engineering of motion, and liquid and gas fluid dynamics.

Section IV is an overview of the electrical engineering aspects of medical device design and electrical events in biology. Chapters 20 and 21 delve into electrode design and electrical detection, respectively.

Section V is the diagnostic portion of the book describing imaging techniques and alternate diagnostic methods. Chapters 22 through 40 give the scientific background of several conventional imaging methods and the mechanism of action on which their operation is based, in addition to several novel and upcoming sensing and imaging techniques.

Section VI describes some affiliate biology and physics, with a brief review of the science behind nuclear medicine and some engineering aspects of upcoming diagnostic methods. Chapters 41 through 43 introduce some new technology as well as state-of-the-art technology that has its own place in medicine and biology.

Section VII is the conclusion that delves into the physics behind regenerative medicine aspects in Chapter 44.

Audience

This book targets medical professionals and students involved in physical anthropology, forensic pathology, medical practice, medical device development, and diagnostic design. It is also appropriate for graduate level education in (bio)medical physics as well as in biomedical engineering and related disciplines.

Related Textbooks

1. Cameron JR, Skofronick JG, and Grant RM, *Physics of the Body*, 2nd edition, Medical Physics Publishing, Madison, WI, 1999.
2. Kane SA, *Introduction to Physics in Modern Medicine*, Taylor & Francis Publishing, London, 2003.
3. Davidovits P, *Physics in Biology and Medicine*, 3rd edition, Academic Press, Burlington, MA, 2008.
4. Tuszynski JA and Dixon JM, *Biomedical Applications of Introductory Physics*, John Wiley and Sons, New York, NY, 2001.

Acknowledgments

The authors acknowledge the contributions from various authors as well as the inspiration and contributions provided by students and professionals in the various fields of biophysical sciences. The detailed critique by each respective individual significantly enhanced the value of the biological, physical, and mathematical contents of this book.

The authors also acknowledge the assistance by both illustrative and theoretical means from all hospitals and industry, and the numerous individuals who are not mentioned individually but are nonetheless equally valuable for sharing their biomedical and physics information and illustrations with the reader and the authors, respectively. Great care has been taken to ensure that the source for each illustration is provided and to acknowledge the contributions of all institutions and individuals with proper tribute.

Editor

Robert Splinter received his master's degree from the Eindhoven University of Technology, Eindhoven, the Netherlands and his doctoral degree from the VU University of Amsterdam, Amsterdam, the Netherlands. He has worked in biomedical engineering for over 20 years, both in a clinical setting and in medical device development. He has been affiliated with the physics department of the University of North Carolina at Charlotte for over 20 years as an adjunct professor and was a core team member for the establishment of the interdisciplinary biology graduate program at UNCC. In addition, he served on the graduate boards of the University of Alabama at Birmingham, Birmingham, Alabama and the University of Tennessee Space Institute, Tulahoma, Tennessee. Dr. Splinter has cofounded several biomedical startup companies and served as technology manager in various capacities. He is the co-owner of several US and international patents and patent applications on methodologies using applied physics in the biomedical field. He has published more than 100 peer-reviewed papers in international journals and conference proceedings, as well as several books.

Contributors

Jean-Marc Aimonetti
Laboratory of Integrative and Adaptative
    Neurobiology
CNRS—Université de Provence
Aix-Marseille Université, Marseille, France

Beth J. Allison
Department of Physiology
Monash University
Victoria, Australia

David Baldwin
University of Western Australia
Perth, Western Australia, Australia

Manish Bothara
Portland State University
Portland, Oregon

Sonja Braun-Sand
Department of Chemistry
University of Colorado at Colorado
    Springs
Colorado Springs, Colorado

Hélio Chiarini-Garcia
Department of Morphology
Federal University of Minas Gerais
Belo Horizonte, Minas Gerais, Brazil

Ravishankar Chityala
Image processing and visualization
    consultant
Minnesota Supercomputing Institute
University of Minnesota, Minneapolis

David T. Corr
Rensselaer Polytechnic Institute
Troy, New York

Kevin A. Croussore
Fujitsu Laboratories of America
Fujitsu Network Communications
Richardson, Texas

Ryan C.N. D'Arcy
Institute for Biodiagnostics (Atlantic)
National Research Council Canada
Dalhousie University
Halifax, Nova Scotia, Canada

Mari Dezawa
Department of Stem Cell Biology and
    Histology
Tohoku University
Aoba-ku, Sendai, Japan

Didier Dréau
Department of Biology
University of North Carolina at
    Charlotte
Charlotte, North Carolina

Kert Edward
Department of Physics and Optical
    Sciences
University of North Carolina at
    Charlotte
Charlotte, North Carolina

Paul Elliot
Tessera, Inc.
Charlotte, North Carolina

Martin F. Finlan
Physical Sciences Consultant
Doncaster, United Kingdom

Gábor Fülöp
Heart Center
Semmelweis University
Budapest, Hungary

Paola Garcia
University of Colorado at Colorado
    Springs
Colorado Springs, Colorado

Jodie R. Gawryluk
Institute for Biodiagnostics
    (Atlantic)
National Research Council and
    Neuroscience Institute
Dalhousie University
Halifax, Nova Scotia, Canada

Cynthia Gibas
Department of Bioinformatics and
    Genomics
University of North Carolina at
    Charlotte
Charlotte, North Carolina

Frank Gijsen
Department of Biomedical
    Engineering
Biomechanics Laboratory
Rotterdam, the Netherlands

Mark S. Hedrick
Department of Audiology and
    Speech Pathology
The University of Tennessee
Knoxville, Tennessee

Myriam C. Herrera
Departamento de Biongeniería
Universidad Nacional de Tucumán and
    Instituto Superior de Investigaciones
    Biológicas
Tucumán, Argentina

El-Sayed H. Ibrahim
Department of Radiology
University of Florida
Jacksonville, Florida

Naohiro Ikeda
Department of Ophthalmology
Hyogo College of Medicine
Nishinomiya, Hyogo, Japan

Tomohiro Ikeda
Department of Ophthalmology
Hyogo College of Medicine
Nishinomiya, Hyogo, Japan

Hiroto Ishikawa
Department of Ophthalmology
Hyogo College of Medicine
Nishinomiya, Hyogo, Japan

Takuji Ishikawa
Department of Bioengineering and
    Robotics
Tohoku University
Aoba-ku, Sendai, Japan

Sanae Kanno
Department of Ophthalmology
Hyogo College of Medicine
Nishinomiya, Hyogo, Japan

Orsolya Kiss
Semmelweis University Heart Center
Semmelweis University of Medicine
Budapest, Hungary

Vindhya Kunduru
Department of Electrical and
    Computer Engineering
Portland State University
Portland, Oregon

Amanda W. Lund
Rensselaer Polytechnic Institute
Troy, New York

Rossana E. Madrid
Departamento de Biongenería
Universidad Nacional de Tucumán and
    Instituto Superior de Investigaciones
    Biológicas
Tucumán, Argentina

Michael Markl
Department of Diagnostic Radiology,
    Medical Physics
University of Hospital Freiburg
Freiburg, Germany

Carmen C. Mayorga Martinez
Departamento de Biongenería
Universidad Nacional de Tucumán and
    Instituto Superior de Investigaciones
    Biológicas
Tucumán, Argentina

Pál Maurovich-Horvat
Massachusetts General Hospital
Harvard Medical School
Boston, Massachusetts

Rossana Correa Netto Melo
Department of Biology
Federal University of Juiz de Fora
Juiz de Fora, Minas Gerais, Brazil

Juan Jiménez Millán
Departamento de Geología
Universidad de Jaén
Jaén, Spain

Osamu Mimura
Department of Biology
Hyogo College of Medicine
Nishinomiya, Hyogo, Japan

Timothy Moss
Department of Physiology
Monash University
Victoria, Australia

Sriram Muthukumar
Intel Corporation
Chandler, Arizona

Fernando Nieto
Departamento de Mineralogía y
    Petrología
Granada, Spain

Dmitriv V. Nikolaev
Scientific Research Centre "MEDASS"
Moscow, Russia

John A. Notte
Director of Research and Development
Carl Zeiss SMT, Inc.
Peabody, Massachusetts

Nael F. Osman
Johns Hopkins University
Baltimore, Maryland

István Osztheimer
Semmelweis University of Medicine
Budapest, Hungary

Gregory M. Palmer
Department of Radiation Oncology
Duke University
Durham, North Carolina

Christian Parigger
The Center for Laser Applications
University of Tennessee Space Institute,
    University of Tennessee
Tullahoma, Tennessee

Gleydes Gambogi Parreira
Department of Morphology
Federal University of Minas Gerais
Belo Horizonte, Minas Gerais, Brazil

John Pearce
Department of Electrical and
    Computer Engineering
University of Texas at Austin
Austin, Texas

Daniel S. Pickard
Department of Electrical Engineering
National University of Singapore
Singapore

Pavlina Pike
Department of Radiology
University of Alabama at Birmingham
Birmingham, Alabama

Jane Pillow
School of Women's and Infants' Health
University of Western Australia
Women's and Newborns' Health Service
Perth, Western Australia, Australia

George E. Plopper
Rensselaer Polytechnic Institute
Troy, New York

Shalini Prasad
Department of Electrical Engineering
Arizona State University
Tempe, Arizona

Sasa Radovanovic
Laboratory for Neurophysiology
Institute for Medical Research
Belgrade, Serbia

Sergey G. Rudnev
Institute of Numerical Mathematics
Russian Academy of Sciences
Moscow, Russia

Alexander V. Smirnov
Scientific Research Centre "MEDASS"
Moscow, Russia

Robert Splinter
Department of Physics and Optical
Sciences
University of North Carolina at
Charlotte
Charlotte, North Carolina

Aurélien F. Stalder
Department of Diagnostic
Radiology, Medical Physics
University of Hospital Freiburg
Freiburg, Germany

Ernesto F. Treo
Departamento de Biongeniería
Universidad Nacional de Tucumán and
Instituto Superior de Investigaciones
Biológicas
Tucumán, Argentina

Rudolf Verdaasdonk
Physics and Medical Technology
VU University Amsterdam
Amsterdam, the Netherlands

Karthik Vishwanath
Department of Biomedical Engineering
Duke University
Durham, North Carolina

Juha Voipio
Department of Biosciences
University of Helsinki
Helsinki, Finland

Shelley J. Wilkins
Department of Physical and Theoretical
Chemistry
Oxford University
Oxford, United Kingdom

and

Orbital Instruments Ltd
Doncaster, United Kingdom

Thomas Wolkow
Department of Biology
University of Colorado at Colorado
Springs
Colorado Springs, Colorado

Yamini Yadav
Department of Electrical and Computer
Engineering
Portland State University
Portland, Oregon

Robert Splinter
Department of Physics and Optical
Sciences
University of North Carolina at
Charlotte
Charlotte, North Carolina

Jonathan G. Stadler
Department of Diagnostic
Radiology, Medical Hospital
University Hospital Freiburg
Freiburg, Germany

Ernesto R. Teran
Departamento de Bioquímica
Universidad Nacional de Tucumán and
Instituto Superior de Investigaciones
Biológicas
Tucumán, Argentina

Rudolf Verdaasdonk
Physics and Medical Technology
VU University Amsterdam
Amsterdam, The Netherlands

Karthik Vishwanath
Department of Biomedical Engineering
Duke University
Durham, North Carolina

Juha Voipio
Department of Biosciences
University of Helsinki
Helsinki, Finland

Shelley J. Williams
Department of Physical and Theoretical
Chemistry
University of Oxford
Oxford, United Kingdom

Clinical Imaging, UK Ltd.
Devon, United Kingdom

Thomas Weber
Department of Biology
University of Colorado at Colorado
Springs
Colorado Springs, Colorado

Samuel Valez
Department of Electrical and Computer
Engineering
Portland State University
Portland, Oregon

I

Anatomical Physics

Physics of the Cell Membrane

Thomas Wolkow

1.1 Introduction

The plasma membrane and those of intracellular organelles are composed of lipids and proteins arranged in a dynamic lipid bilayer. The structures and mechanical features of both components are described herein.

1.2 Cellular Membranes are Elastic Solids

The integrity of cellular membranes is routinely challenged by external and internal forces.[1] External forces include high-frequency vibrations, fluid shear stress, and osmotic and gravitational pressure gradients. Internal forces include hydrostatic pressure as well as those produced by cytoskeletal cables that push outward to orchestrate cellular movements, morphological changes, growth, and adhesion.

The physical characteristics of membranes that allow them to withstand such forces have been described by measuring membrane parameters before and after force is applied.[1] Researchers tend to employ human red blood cells (RBCs) and techniques like micropipette aspiration and patch–clamp devices to obtain these measurements. Overall, their results demonstrate that biological membranes respond like elastic solids when mechanical operations are used to compress, expand, bend, and extend a defined membrane area.

Within a drop of fluid that is not surrounded by a cellular membrane, the relationship between surface tension and pressure is described by the law of Laplace. Given a uniform surface tension (σ), internal pressure (P), and radius (R) of the drop, the law states that

$$P = \frac{2\sigma}{R}. \tag{1.1}$$

When modeling this relationship in a cell, one might think that the density of the membrane reacts to pressure differences between the external and internal environments. However, density, which describes the compressibility of lipids within the bilayer, remains constant under physiologically relevant pressures (100 atm).[1] Surface area displays somewhat weaker resistance and does undergo some change, but only 2–4% before rupturing. The tensile force (F_t) on the membrane is expressed in Equation 1.2 for this situation:

$$F_t = K_A \frac{\Delta A}{A_0}, \tag{1.2}$$

where ΔA is the increase in bilayer surface area from the original area A_0, K_A is the area expansion constant (between 10^2 and 10^3 mN/m), and F_t is tension (between 3 and 30 mN/m). And while surface area expands, membrane thickness changes proportionally so that

$$\frac{\Delta A}{A_0} = \frac{\Delta h}{h_0}, \tag{1.3}$$

where h_0 represents original membrane thickness. But the membrane response to shear stress is what clearly describes it as an elastic solid. Using two silica beads and optical traps to exert shear stress across an RBC[2] (Figure 1.1), elasticity can be seen

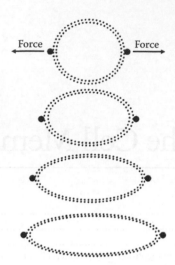

FIGURE 1.1 Membrane response to shear stress has been measured using optical traps and silica beads (black ovals) to apply force across RBCs.[2]

as this membrane elongates in the direction of applied force *F*. It can be shown that the diameter of the RBC obeys Equation 1.4:

$$D = D_0 - \frac{F}{2\pi\mu}, \qquad (1.4)$$

where *D* is the diameter of the RBC, D_0 is the original diameter, and μ is the shear stress applied by the optical trap.

Importantly, this elastic behavior seems to be almost completely dependent upon a flexible network of cytoskeletal elements beneath the membrane surface.[3,4] Thus it is the underlying cytoskeleton that allows membranes to bend, flex, and resist rupture under physiologically stressful conditions.

1.3 Lipids

The major constituents of biological membranes are lipids, which are amphipathic molecules composed of a hydrophobic fatty acid chain and a hydrophilic head group (Figure 1.2).

Fatty acids are hydrocarbon chains (typically 16–18 carbons long) containing a terminal carboxyl group, and are synthesized in the cytoplasm from acetyl-CoA molecules produced in mitochondria. Fatty acids include saturated forms (e.g., palmitate and stearate) wherein carbon atoms are bonded with a maximum number of hydrogens, and the more common unsaturated forms (e.g., oleate) containing at least one kink-inducing double bond. When in excess, free fatty acids are transported

to the space between the leaflets of the endoplasmic reticulum (ER) and the Golgi, where they are transformed into triacylglycerols or cholesteryl esters and stockpiled within membrane-bound organelles called lipid droplets.[5] These lipids will later be used for purposes of adenosine triphosphate (ATP) production, membrane synthesis, and steroid synthesis. For example, lipids contained within droplets and modified with a polar head group containing phosphate, ethanolamine, choline, serine, inositol, or sphingomyelin are used to synthesize the structural lipids of cellular membranes.

Structural lipids are members of the glycerophospholipid and sphingolipid families.[6] Sphingolipids have a ceramide head group built from an amino acid backbone and longer hydrocarbon chains that are predominantly unsaturated. These unsaturated chains allow sphingolipids to pack together closely and form a more solid gel phase within the liquid-disordered glycerophospholipid environment. Cholesterol commonly exists in sphingolipid domains and causes them to transform into a liquid-ordered phase wherein lipids are tightly packed with freedom to diffuse laterally. These liquid-ordered phases or lipid rafts[7] exist only in the outer membrane leaflet and are connected to uncharacterized domains of the inner leaflet.[8] Within lipid rafts, specific sets of membrane-associated proteins are assembled, which promote interactions among specific groups of proteins and inhibit cross talk with others.[9]

Microdomains, such as lipid rafts, can affect the curvature and thus the energy (E_{sphere}) of biological membranes[10] (Figure 1.3). For a membrane without spontaneous curvature, the energy of the membrane is given by

$$E_{sphere} = \gamma S + 8\pi\kappa, \qquad (1.5)$$

where γ is surface tension, *S* is membrane area, and $\kappa = 20 K_B T$ is bending rigidity; the thermal energy scale at room temperature is $K_B T = 4 \times 10^{-21}$ J. Microdomains that induce small invaginations increase the membrane energy by approximately $8\pi\kappa$. Atomic force microscopy (AFM) measurements suggest that these microdomains are stiff, as their elasticity moduli may be approximately 30% higher than the surrounding membrane area.[11]

1.3.1 Membrane-Associated Proteins

Proteins are gene products that can be distinguished based on location, function, and chemical composition. Two principal types of proteins exist within the membrane environment.

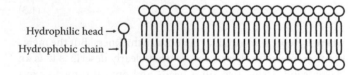

Hydrophilic head →
Hydrophobic chain →

FIGURE 1.2 Lipids are major constituents of biological membranes.

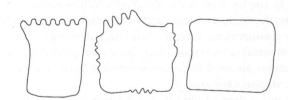

FIGURE 1.3 Cell curvature is a function of structural lipid composition.

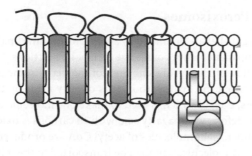

FIGURE 1.4 Transmembrane proteins (left) and peripheral proteins (right) populate biological membranes.

Transmembrane (or intrinsic) proteins cross the bilayer and may function as cellular receptors required for cell-to-cell communication and adhesion, or as transporters that shuttle various particles (including ions, glucose, and proteins) across the membrane. Peripheral (or extrinsic) proteins associate with the internal or external layer of the membrane where they may affect cell shape (BAR domain proteins), or possibly facilitate receptor-dependent signaling pathways (Figure 1.4). Two other groups of proteins (glycoproteins and lipoproteins) are also discussed because of their broad influence on membranes and lipid homeostasis.

1.3.2 Transmembrane Proteins

All transmembrane proteins contain membrane-spanning domains composed primarily of nonpolar residues assembled into secondary structures that neutralize the polarity of peptide bond elements. These membrane-spanning domains typically assume flexible α-helical or rigid β-sheet secondary structures. Ligand binding seems to stimulate downstream receptor signaling events by altering the electrostatic field of the receptor and/or eliciting changes in receptor conformation. Currently, the technology to record measurements of single receptors at the surface of living cells is absent. However, AFM studies of the α-amino-3-hydroxy-5-methylisoxazole-4-propionic acid-type glutamate receptors (AMPAR) demonstrate that upon ligand binding, the elastic modulus of the receptor nanodomains permanently decreases by 20–30% relative to the surrounding membrane area.[12] This probably occurs because activated receptors are sometimes internalized and removed from the cell surface by the endocytotic pathways discussed below.

1.3.3 Peripheral Proteins

Peripheral proteins may associate with membranes through physical interactions with transmembrane proteins or because of post-translational modifications. These modifications result in the addition of a hydrophobic fatty acid, such as palmitate, or a hydrophobic glycolipid group. Glycolipid modifications include glycosylphosphatidylinositol (GPI) modifications that contain two fatty acid chains, an inositol in addition to other sugars, and ethanolamine. By nature of their hydrophobicities, these fatty

acid and GPI groups become inserted into the bilayer and thereby serve to anchor the protein to the membrane.

1.3.4 Glycoproteins

Glycoproteins are carbohydrate-modified proteins. These carbohydrate groups not only assist in protein folding, stability, and aggregation, but when present on transmembrane proteins may affect cell–cell adhesion via electrostatic, hydrophobic, and covalent interactions. Adhesion forces can also result from surface tension that arises in the fluid interface formed with glycoprotein-decorated cellular surfaces. For example, glycoprotein-decorated stigmas and lipoprotein-decorated pollen grains interact with an adhesion force of approximately 2×10^{-8} N due to surface tension and a fluid interface.[13] Four types of glycosylations are known: GPI anchors, C-glycosylation, N-glycosylation, and O-glycosylation. As discussed above, GPI anchors contain sugar moieties that participate in the production of peripheral proteins.

C-glycoproteins are produced when sugars are directly linked to the protein by carbon–carbon bonds, hence the name C-glycoprotein.[14,15] To date, C-mannosylation of human RNase 2 is the most well-characterized example of C-glycosylation. It occurs on the first Try residue in the Trp–X–X–Trp motif of RNase 2 and influences the specific orientation of the modified Trp in the tertiary structure.

N-glycosylation occurs in the ER and begins when a 14-sugar precursor containing three glucose, nine mannose, and two N-acetylglucosamine molecules is transferred to an asparagine (N) residue in one of the following tripeptide motifs, Asn–X–Ser, Asn–X–Thr, or Asn–X–Cys.[14] This precursor complex is then modified to produce mature N-glycosylation patterns that contain, for example, more mannose or different types of saccharides or more than two N-acetylglucosamines.

O-glycosylation occurs in both the ER and the Golgi as glycosyl transferases modify the OH group of amino acids like serine or threonine with one sugar at a time, typically beginning with N-acetylgalactosamine.[14] To date, roughly 100 O-glycosylated proteins are known.[16]

1.3.5 Lipoproteins

Lipids are hydrophobic insoluble molecules and, therefore, travel through the aqueous blood in lipoprotein vesicles composed of a phospholipid monolayer and proteins called apoproteins.[17] Apoproteins wrap around the outside of the phospholipid monolayer using α-helices with hydrophobic residues oriented inside the hydrophobic lipid monolayer and hydrophilic residues interacting with the polar lipid head groups. Apoproteins bind cell surface receptors and allow direct tissue-specific transport of the lipids contained within the lipoprotein vesicles. Two types of lipoproteins are known to transport triacylglycerols. Very low-density lipoproteins (LDL) transport newly synthesized triacylglycerol from the liver to adipose tissue, and chy-

FIGURE 1.5 Lipoproteins transport lipids within vesicles composed of a phospholipid monolayer and apoproteins.

lomicrons transport triacylglycerol to the liver, skeletal muscle, and adipose tissue. Cholesterol is transported inside two other types of lipoproteins. LDL transport cholesterol from the liver to different cells of the body, and high-density lipoproteins (HDL) recycle this cholesterol back to the liver (Figure 1.5).

1.4 Organelle Cell Membranes

Like the cell itself, cytoplasmic organelles are surrounded by complex mixtures of lipids and proteins. Importantly, these organelles compartmentalize a great number of reactions that affect bioenergetics and lipid homeostasis.

1.4.1 Mitochondria

Mitochondria have an outer membrane and an infolded inner membrane with roughly five times the surface area of the outer membrane.[19] The inner membrane space exists between the outer and inner membranes and the matrix exists within the inner membrane. Cristae are the small regions formed within the folds of the inner membrane. The outer membrane contains porin channels composed of the β-sheet secondary structure of integral membrane proteins. These porins are relatively large and allow the passage of molecules smaller than ~5000 Da. In contrast, the inner membrane is highly impermeable, possibly because of the presence of cardiolipin (bisphosphatidyl glycerol). Cardiolipin is a lipid with four fatty acid tails and is only found within the membranes of mitochondria and some prokaryotes.[18] Cardiolipin comprises approximately 20% of the inner membrane of mitochondria and, through its four fatty acid tails, creates a highly impermeable barrier that facilitates proton-gradient formation. Mitochondria also have a small circular genome that encodes a little less than 40 genes.[19] These genes code for ribosomal RNA (rRNA), transfer RNA (tRNA), and protein components of the electron transport system. Within the matrix, the electron transport proteins couple the oxidation of carbohydrates and fatty acids to proton-gradient formation and ATP generation.

1.4.2 Peroxisomes

These organelles are delimited by a single membrane lipid bilayer that compartmentalizes oxidative metabolic reactions.[19] During β-oxidation of fatty acids, enzymes of the peroxisome progressively shorten the hydrocarbon tails of fatty acids by two carbon units, each of which is used to make one molecule of acetyl CoA. Therefore, when a fatty acid with 16 carbons is oxidized in this manner, eight molecules of acetyl CoA are produced. These acetyl-CoA molecules can then be transported to the mitochondria and used to fuel ATP production. Peroxisomes also contain the enzyme catalase, which converts hydrogen peroxide, a by-product of these β-oxidation reactions, into water and molecular oxygen. Catalase also detoxifies alcohol, phenols, formic acid, and formaldehyde.

1.4.3 Lysosomes

These membrane-bound organelles contain different types of acid hydrolases, including lipases, phospholipases, phosphatases, and sulfatases, that use hydrolysis to sever the bonds of numerous cellular materials.[19] Catalytic activity of these enzymes requires an acidic environment of about 5.0, which is produced by membrane-bound H+ ATPase that pumps protons from the cytosol into the lumen of the lysosome. These hydrolases serve to digest the macromolecules that are delivered to lysosomes during endocytic processes (see Section 1.5).

1.4.4 ER and Golgi

Translation of all proteins begins on free ribosomes in the cytoplasm.[19] Those destined for secretion, the plasma membrane, lysosome, ER, or Golgi must first enter the ER. These proteins are either deposited entirely within the lumen of the ER or inserted as transmembrane proteins, depending on the function that each serves in its final destination. Within the ER environment, proteins can be glycosylated and modified with glycolipid anchors and/or disulfide bonds. Chaperone-assisted protein folding and assembly into multisubunit complexes also occur within the ER. Following those modifications, these proteins bud from the ER within vesicles that then fuse with the membrane-enclosed sacs of the Golgi. Further rounds of glycosylation occur within the Golgi stacks before the protein is packaged into vesicles that fuse with the plasma membrane, lysosome, or ER. These vesicles travel along microtubules to their final destination, which is determined by receptor-mediated interactions at vesicle/membrane interfaces. For example, vesicles destined for the plasma membrane have receptors that only interact with those on the cytoplasmic surface of the plasma membrane.

Lipid synthesis also occurs inside the ER and the Golgi.[19] Phospholipids, cholesterol, and ceramide are synthesized in the ER, while enzymes within the Golgi utilize ceramide to produce sphingolipids and glycolipids. Transport of these lipids will depend on the current demands of the cell or organism.

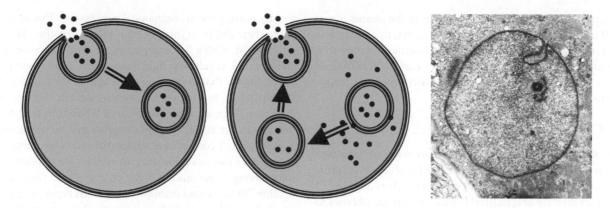

FIGURE 1.6 Endocytosis (left), exocytosis (middle), and phagocytosis. (TEM courtesy of Winston Wiggins, Daisy Ridings, and Alicia Roh, Carolinas HealthCare System, 1000 Blythe Blvd. Charlotte NC.)

1.5 Vesicle Transport

A highly interconnected series of vesicle transport routes mediates the exchange of materials among organelles, as well as between the cell and its environment. Phagocytosis, pinocytosis, clathrin-mediated endocytosis, and caveolae-dependent uptake are mechanisms of endocytosis used during vesicular internalization.[20] Conversely, the process by which materials are delivered from within the cell to the plasma membrane occurs during exocytosis.[21] Together, endocytosis and exocytosis influence membrane surface area homeostasis and the recycling of cellular materials that includes transmembrane receptors and lipids (Figure 1.6).

1.5.1 Phagocytosis

Phagocytosis occurs when large particles, like bacteria or dead cells, are internalized after binding the cellular receptors of phagocytes (e.g., macrophages and neutrophils).[20] Binding results in actin-dependent clustering of the receptors and pseudopodia extension of actin filaments, so that the object becomes surrounded within a large phagosome derived of plasma membrane. During fusion with the lysosome, ligand/receptor interactions are broken within the acidic environment. The receptors can then be placed in new vesicles that return and fuse with the plasma membrane in an effort to recycle these surface receptors. Interestingly, some pathogenic bacteria are able to escape before the phagosome fuses with the lysosome, while others can survive the acidic environment of the lysosome.

1.5.2 Pinocytosis

Pinocytosis, or cell drinking, allows cells to internalize extracellular fluid that may contain nutrients in bulk.[20] It is an actin-dependent process that does not follow a receptor-mediated mechanism. However, pinocytosis may be initiated by upstream pathways that respond to growth factor or hormone signaling. The internalized extracellular fluid is typically transported to the lysosome, where it is metabolized.

1.5.3 Clathrin-Mediated Endocytosis

Clathrin-mediated endocytosis permits the uptake of essential nutrients and the internalization of activated receptors from the cell surface. With regard to receptors, such internalization may down-regulate or modify their activity at the plasma membrane, or allow them to participate in signal transduction cascades within the cell.[22] Clathrin forms a hexameric complex of three 190-kDa heavy chains and two 30-kDa light chains that assemble on adapter molecules located at the cytoplasmic surface of the plasma membrane.[23] Assembly results in the production of a cage (50–100 nm in diameter) that engulfs and coats the membrane invagination. Dynamin is a large 100-kDa guanosine triphosphate (GTPase) that assembles around the neck of the clathrin-coated pit and, upon GTP hydrolysis, constricts the membrane to release the clathrin-coated vesicle. This clathrin coat is later removed from the vesicle and recycled to other adapter molecules as the vesicle is transported to the lysosome.

1.5.4 Caveolae-Dependent Endocytosis

Caveolae-dependent endocytosis[24,25] also facilitates the uptake of essential nutrients and receptors. Caveolins are proteins that localize to lipid rafts, which are small regions of the outer plasma membrane particularly enriched in cholesterol and sphingolipids. Many caveolins weave through the lipid bilayer of these rafts, forming loops as both N- and C-termini remain on the cytoplasmic surface of the membrane. This caveolin scaffold may influence the size and shape of the membrane region to be internalized, a region that has a diameter of 50–80 nm. Like clathrin-dependent endocytosis, caveolae are internalized by a mechanism that depends on dynamin. These caveolin-decorated vesicles are delivered to lysosomes or possibly a different place within the cell or plasma membrane.

1.5.5 Exocytosis

Exocytosis describes the vesicular transport of cellular materials to the plasma membrane.[19] These vesicles may originate from the lysosome, ER, Golgi, or another region of the plasma membrane. If transported in the lumen of the vesicles, this material is secreted from the cell and used to build the extracellular matrix, communicate with other cells, or remove waste. If transported within vesicle membranes, this material will become incorporated into the plasma membrane.

1.5.6 Cytoskeleton

The cytoskeleton establishes and maintains many of the structural characteristics and mechanical activities of cells.[19] In terms of structure, proteins of the cytoskeleton form cables (Figure 1.7) that interact directly and indirectly with cellular membranes to establish and maintain a diverse array of cellular architectures. Meanwhile, mechanical activities of the cytoskeleton usually depend on their interactions with a variety of different motor proteins.

1.5.6.1 Assembly of the Cytoskeleton

The cytoskeleton is composed of filamentous actin (F-actin), microtubules, and intermediate filaments (Ifs), all of which are composed of repeating protein subunits. ATP-bound globular actin monomers (G-actin) stack end-on-end to produce F-actin cables, while GTP-bound tubulin dimers (α and β tubulin) stack end-on-end to produce microtubule cables. The assembly of each cable is regulated by a large number of proteins; some of these serve as nucleation platforms and others regulate the state of the nucleotide cofactor. The state of the nucleotide cofactor is important because nucleotide triphosphate-bound states are required for assembly while diphosphate-bound states tend to promote disassembly of these cables. IFs are composed of repeating protein subunits with coiled-coil structure. Unlike F-actin and microtubules, IF assembly does not appear to be regulated by a nucleotide cofactor.

1.5.6.2 Cytoskeletal Functions

All three filaments support the morphological architecture of cells, but only actin and microtubule filaments are known to

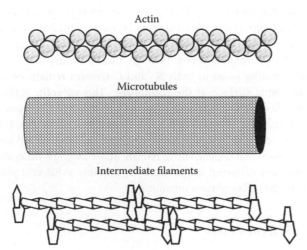

Actin

Microtubules

Intermediate filaments

FIGURE 1.7 The cellular cytoskeleton.

directly participate in mechanical processes.[19] This seems to be due to their ability to interact with motor proteins. Motor proteins consist, with a few exceptions, of a head domain, a heavy chain, and a regulatory light chain. The head domain is an ATPase motor that steps along the length of the filament during ATP hydrolysis by a not yet fully resolved molecular mechanism. The heavy chain associates with cargo while both the heavy and light chains are subject to post-translational modifications that exert regulatory control over movement and cargo affinity. Some motors have two head domains, in which case physical interactions among heavy and/or light chains tether these domains together. These motors transport numerous types of cargo along cytoskeletal cables, including mitochondria and both endocytotic and exocytotic vesicles.

1.6 Motor Proteins

1.6.1 Kinesins, Dyneins, and Myosins

Kinesins and dyneins are microtubule-dependent motor proteins that step toward opposite ends of the microtubule cables. Both these mechanochemical enzymes have a pair of motor domains that take overlapping, hand-over-hand steps along the microtubule surface. *In vitro* measurements suggest that these head domains take 8 nm steps and generate ~6 pN of peak force during one ATP hydrolysis cycle.[26,27] ATP-dependent velocities of single kinesin molecules follow the Michaelis–Menten relationship shown in Equation 1.6:

$$v(c) = \frac{v_{max}c}{K_m} + c, \tag{1.6}$$

where K_m is the mechanochemical Michaelis–Menten constant, c is the ATP concentration, and v_{max} is the velocity at saturating ATP concentration.[28]

Myosins are one- and two-headed actin-dependent motors with step sizes between 5 nm and >40 nm.[29] Two-headed isoforms like Myosin II participate in contractile assemblies that influence sarcomere length in muscle cells and actomyosin ring constriction during cytokinesis of dividing cells.[19] Within these contractile assemblies, Myosin II tails physically associate with each other to form myosin filaments. The head domains of these filaments associate with actin filaments of different polarity. Because the orientation of these myosins reverses along the myosin filaments, ATP hydrolysis by the motor domains causes the actin filaments to slide toward each other. The force generated by each myosin molecule adheres to the following relationship, where force (F) relates to the bending stiffness (Ei) and length (L) of the tail domains that is expressed in Equation 1.4[30]:

$$F = \frac{3Ei}{L^2}. \tag{1.7}$$

In contrast to Myosin II, one-headed myosins do not have tails that form coiled-coils and therefore do not participate in contractile mechanisms. Instead, the tails of these myosins bind vesicles and organelles in order to transport them along actin filaments.

References

1. Hamill OP and Martinac B. Molecular basis of mechanotransduction in living cells. *Physiol Rev* 2001, 81: 685–740.

2. Henon S, Lenormand G, Richert A, and Gallet F. A new determination of the shear modulus of the human erythrocyte membrane using optical tweezers. *Biophys J* 1999, 76: 1145–1151.

3. Lenormand G, Henon S, Richert A, Simeon J, and Gallet F. Direct measurement of the area expansion and shear moduli of the human red blood cell membrane skeleton. *Biophys J* 2001, 81: 43–56.

4. Lenormand G, Henon S, Richert A, Simeon J, and Gallet F. Elasticity of the human red blood cell skeleton. *Biorheology* 2003, 40: 247–251.

5. Martin S and Parton RG. Lipid droplets: A unified view of a dynamic organelle. *Nat Rev Mol Cell Biol* 2006, 7: 373–378.

6. van Meer G, Voelker DR, and Feigenson GW. Membrane lipids: Where they are and how they behave. *Nat Rev Mol Cell Biol* 2008, 9: 112–124.

7. Barnett-Norris J, Lynch D, and Reggio PH. Lipids, lipid rafts and caveolae: Their importance for GPCR signaling and their centrality to the endocannabinoid system. *Life Sci* 2005, 77: 1625–1639.

8. Simons K and Vaz WL. Model systems, lipid rafts, and cell membranes. *Annu Rev Biophys Biomol Struct* 2004, 33: 269–295.

9. Moffett S, Brown DA, and Linder ME. Lipid-dependent targeting of G proteins into rafts. *J Biol Chem* 2000, 275: 2191–2198.

10. Sens P and Turner MS. The forces that shape caveolae. In J. Fielding (Ed.), *Lipid Rafts and Caveolae*. Wiley-VCH Verlag GmbH & Co, 2006.

11. Roduit C, van der Goot FG, De Los Rios P, Yersin A, Steiner P, Dietler G, Catsicas S, Lafont F, and Kasas S. Elastic membrane heterogeneity of living cells revealed by stiff nanoscale membrane domains. *Biophys J* 2008, 94: 1521–1532.

12. Yersin A, Hirling H, Kasas S, Roduit C, Kulangara K, Dietler G, Lafont F, Catsicas S, and Steiner P. Elastic properties of the cell surface and trafficking of single AMPA receptors in living hippocampal neurons. *Biophys J* 2007, 92: 4482–4489.

13. Luu DT, Marty-Mazars D, Trick M, Dumas C, and Heizmann P. Pollen-stigma adhesion in *Brassica* spp involves SLG and SLR1 glycoproteins. *Plant Cell* 1999, 11: 251–262.

14. Vliegenthart JF and Casset F. Novel forms of protein glycosylation. *Curr Opin Struct Biol* 1998, 8: 565–571.

15. Wei X and Li L. Comparative glycoproteomics: Approaches and applications. *Brief Funct Genomic Proteomic* 2009, 8(2): 104–113.

16. Wells L and Hart GW. O-GlcNAc turns twenty: Functional implications for post-translational modification of nuclear and cytosolic proteins with a sugar. *FEBS Lett* 2003, 546: 154–158.

17. Zubay GI. *Biochemistry*, 4th edition. Dubuque, IA: Win C. Brown, 1998.

18. Schlame M. Cardiolipin synthesis for the assembly of bacterial and mitochondrial membranes. *J Lipid Res* 2008, 49: 1607–1620.

19. Cooper GM and Hausman RE. *The Cell: A Molecular Approach*, 4th edition. Sunderland, MA: Sinauer Associates, Inc, 2007.

20. Doherty GJ and McMahon HT. Mechanisms of Endocytosis. *Annu Rev Biochem* 2009, 78: 857–902.

21. Lin RC and Scheller RH. Mechanisms of synaptic vesicle exocytosis. *Annu Rev Cell Dev Biol* 2000, 16: 19–49.

22. Felberbaum-Corti M, Van Der Goot FG, and Gruenberg J. Sliding doors: Clathrin-coated pits or caveolae? *Nat Cell Biol* 2003, 5: 382–384.

23. Ungewickell EJ and Hinrichsen L. Endocytosis: Clathrin-mediated membrane budding. *Curr Opin Cell Biol* 2007, 19: 417–425.

24. Nichols BJ and Lippincott-Schwartz J. Endocytosis without clathrin coats. *Trends Cell Biol* 2001, 11: 406–412.

25. Parton RG. Caveolae and caveolins. *Curr Opin Cell Biol* 1996, 8: 542–548.

26. Svoboda K, Mitra PP, and Block SM. Fluctuation analysis of motor protein movement and single enzyme kinetics. *Proc Natl Acad Sci U S A* 1994, 91: 11782–11786.

27. Svoboda K, Schmidt CF, Schnapp BJ, and Block SM. Direct observation of kinesin stepping by optical trapping interferometry. *Nature* 1993, 365: 721–727.

28. Howard J, Hudspeth AJ, and Vale RD. Movement of microtubules by single kinesin molecules. *Nature* 1989, 342: 154–158.

29. Yanagida T and Iwane AH. A large step for myosin. *Proc Natl Acad Sci U S A* 2000, 97: 9357–9359.

30. Tyska MJ and Warshaw DM. The myosin power stroke. *Cell Motil Cytoskeleton* 2002, 51: 1–15.

Protein Signaling

Sonja Braun-Sand

2.1 Modeling Electrostatics in Proteins

Maxwell's equations describe the properties of electric and magnetic fields, and have been analytically solved for some simple cases. However, calculation of electrostatics in proteins is not a trivial problem, and several models have been used to try to describe these interactions. The question is of extreme interest, however, because electrostatic energies are one of the best correlators between structure and function in biochemical systems.[1-4] There are two main problems in formulating a model to describe protein electrostatics.[5] First, the electrostatic interactions within proteins are occurring at microscopically small distances, which make the dielectric constant ambiguous. In addition, the protein environments are irregularly shaped, such that the analytical models are impractical to use.

Microscopic studies of electrostatics in proteins have emerged with the increase in availability of x-ray crystal structures of proteins and the realization that electrostatic energies provide one of the best ways to correlate structure with function. Many mathematical descriptions of electrostatics in proteins have been proposed, and all are based on Coulomb's law, which gives the reversible work of bringing two charges close together as represented by Equation 2.1:

$$\Delta W = 332 \frac{Q_i Q_j}{r_{ij}}, \tag{2.1}$$

where the distance, r, is given in Å, the charge Q is in atomic units, and the free energy, W, is in kcal/mol. Manipulations of this equation[5] lead to the Poisson equation, Equation 2.2:

$$\nabla^2 U(r) = -4\pi\rho(r), \tag{2.2}$$

where the electric field, $\mathbf{E}$, is expressed as a gradient of the scalar potential, U, and ρ is the charge density. By assuming that a dielectric constant can be used to express effects not treated explicitly, the following equation is reached (Equation 2.3):

$$\nabla\varepsilon(\mathbf{r})\nabla^2 U(\mathbf{r}) = -4\pi\rho(\mathbf{r}). \tag{2.3}$$

If it is assumed that the ion distribution follows the Boltzmann distribution, the linearized Poisson–Boltzmann equation is reached, given by Equation 2.4:

$$\nabla\varepsilon(\mathbf{r})\nabla^2 U(\mathbf{r}) = -4\pi\rho(\mathbf{r}) + \kappa^2 U, \tag{2.4}$$

where κ is the Debye–Huckel screening parameter.

Interesting questions arise when these equations are applied to biological systems. For example, can a dielectric constant be applied to heterogeneous systems, such as an enzyme active site? In addition, are continuum assumptions valid on a molecular level? An early work, the Tanford Kirkwood[6] model, described a protein as a sphere with a uniform dielectric constant. This model was proposed before protein structures were known, and it was thought that ionizable residues were located solely on the protein exterior. Later studies conducted after many protein structures had been elucidated found that the simplification of a protein as a uniform dielectric constant missed some important aspects of the physics of charged residues in a protein interior.[7] Discussed below are alternative approaches to describing the electrostatics of protein interiors.

2.1.1 Theory and Models

Electrostatic models span a wide range of possibilities from continuum dielectric approaches[1,8,9] to all-atom models that explicitly represent the biological molecule and solvent.[10,11] Each model has its own advantages and disadvantages.[5,12-15] There are three primary ways of describing solvent in a system, shown in Figure 2.1.[16] Simulation time decreases from the microscopic all-atom model to the macroscopic model.

2.1.1.1 Protein Dipoles Langevin Dipoles Model

A microscopic dipolar model that is often used in simulations of biological molecules is the protein dipoles Langevin dipoles (PDLD) model.[17,18] This model does not assume a dielectric constant for the solvent molecules; rather, the time-averaged

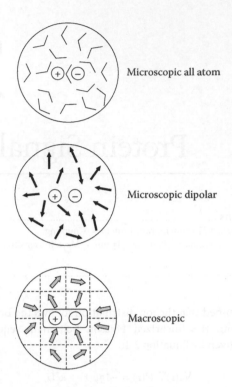

FIGURE 2.1 Three approaches to solvent representation. All atoms are represented in the microscopic approach. A point dipole is used to represent a solvent molecule in the dipolar approach, and several solvent molecules within a certain volume are represented as a polarization vector in the macroscopic approach.

polarization of each solvent molecule is represented as a Langevin dipole.[17,18] The dipoles are placed in a spherical grid, rather than attempting to reproduce the exact locations of the solvent molecules. The net polarization of a thermally fluctuating dipole in response to an electric field is described by the following equations.[16]

$$(\mu_i^L)^{n+1} = e_i^n \mu_0 \left(\coth(x)_i^n - \frac{1}{x_i^n} \right), \qquad (2.5)$$

where x_i^n is described by

$$x_i^n = \frac{C' \mu_0}{k_B T} |\xi_i^n|, \qquad (2.6)$$

and where ξ_i^n is the local field, μ_0 is 1.8 debye, e_i^n is a unit vector in the direction of the local field, C' is a parameter, and the superscript $(n + 1)$ indicates that the equation is solved iteratively.

2.1.1.2 Protein Dipoles Langevin Dipoles/ Semimacroscopic-Linear Response Approximation Model

The microscopic PDLD model discussed above does not use any dielectric constants. A potential problem with these types

of methods is poor convergence of results. One solution to this problem is to scale the dipole contributions of the PDLD model with an assumed protein dielectric constant, ε_p, leading to a semimacroscopic PDLD (PDLD/S) model.[16,19] (For discussions of dielectric constants in proteins, see Refs. [20–22]). To aid in the description of ε_p, the linear response approximation (LRA) is used. In order to apply the LRA approximation, a molecular dynamics (MD) simulation is done to generate a number of protein configurations for the charged and uncharged states of the solute of interest. The LRA approximation uses a thermodynamic cycle to evaluate the PDLD/S energy by averaging over the configurations (with solute charged and uncharged) generated by the MD simulations.[18] This approach is summarized in Figure 2.2, taken from the *Molaris Manual and User Guide*.[16] It is difficult to go from A to B directly; hence a thermodynamic cycle from A → D → C → B is used.

The thermodynamic cycle in Figure 2.2 leads to the following equation, which gives the difference in solvation energy when moving a charge from water to a protein active site shown in Equation 2.7:

$$\Delta\Delta G_{sol}^{w \to p} = \tfrac{1}{2}[\langle \Delta U^{w \to p}(Q = 0 \to Q = Q_0) \rangle]_{Q_0}$$
$$+ \langle \Delta U^{w \to p}(Q = Q_0) \rangle_{Q=0}, \qquad (2.7)$$

where

$$\Delta U^{w \to p} = \left[\Delta\Delta G_p^w(Q = 0 \to Q = Q_0) - \Delta G_Q^w \right] \left(\frac{1}{\varepsilon_p} - \frac{1}{\varepsilon_w} \right)$$
$$+ \Delta U_{Q\mu}^p(Q = Q_0) \frac{1}{\varepsilon_p}, \qquad (2.8)$$

where the subscripts and superscripts p and w refer to protein and water, respectively, ΔG_Q^w is the solvation energy of the charge, Q, in water, $\Delta\Delta G_p^w$ is the change in solvation energy of the protein and bound charge upon changing Q from Q_0 to 0, and $\Delta U_{Q\mu}^p$ is the electrostatic interaction between polar protein groups and the charge.

2.1.1.3 Interactions of Ions with Membrane Proteins

An application of the protein dipoles Langevin dipoles/semi-macroscopic-linear response approximation (PDLD/S-LRA) model (in conjunction with a microscopic determination of the free energy profile for ion transport) is demonstrated in a study of the ion selectivity of the KcsA potassium channel.[23] Remarkably, this ion channel has the ability to discriminate between K^+ and Na^+ ions, allowing K^+ ions to pass through ~1000 times more readily.[24,25] Crystallographic structures are available for this transmembrane protein from *Streptomyces lividans*,[26,27] enabling many computational estimates, using a variety of methods, of the selectivity barrier.[28–33] With the present computing power, the system is too large to allow direct MD simulations of ion transport in a reasonable time. Use of the PDLD/S-LRA

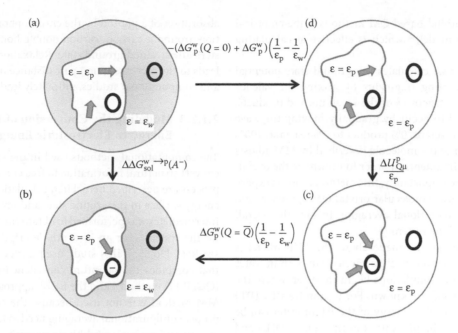

FIGURE 2.2 The PDLD/S-LRA thermodynamic cycle for evaluation of $\Delta\Delta G_{sol}^{w \to p}(A^-)$.

model enables simulation of the K^+ and Na^+ ion currents in a reasonable time in this large system. The simulation system was essentially cylindrical, with a diameter of ~40 Å, which included the channel protein and a small portion of the membrane, and a length of ~89 Å, which included the length of the protein, 30 Å intracellular space, and 25 Å extracellular space. The study was able to examine the effects of several variables on ion current, including the effective dielectric constant used for the protein (which varied from 15 to 30), and the friction for moving the ion through the channel.

What were the factors found by this study to be most important in determining ion selectivity through the channel? First, the study found differences in channel radius at the initial ion loading sites based on the presence of K^+ or Na^+. This led to different geometries, and thus different channel reorganization energies for the ions. This reorganization energy can be captured with the PDLD/S-LRA model because it effectively evaluates the steric interactions between the ion and the channel. Second, it was found that the calculated free energy profiles were very sensitive to the protein dielectric constant, ε_p, used. In this case a large effective dielectric constant for the protein is necessary to properly represent the charge–charge interactions, if the LRA approximation is not used. A typical protein dielectric constant, ε_p, used in many studies without the LRA approximation is less than 6, but this low ε_p may not correctly evaluate charge–charge interactions ($\Delta U_{Q\mu}^p$ in Equation 2.8). The strong dependence of the energy profile on the protein dielectric used may be due in part to the narrow protein channel through which the ions pass. Because the KcsA channel is so narrow, the ion is probably not solvated on all sides by water molecules, as it would be in wider channels, such as porins. This may make interactions with the protein, and thus the value of ε_p, much more important in the calculations.

2.1.2 Bacteriorhodopsin, a Transmembrane Protein

Transmembrane proteins span the width of a lipid bilayer, and have many uses within an organism. For example, transmembrane proteins may be used as ion or water channels, transporters, receptors in signal transduction pathways, proton pumps in the electron transport chain (ETC), adenosine triphosphate (ATP) synthases, and many others. The largest class of cell surface receptors is the seven-transmembrane helix (7TM) family. It is estimated that ~50% of therapeutic drugs on the market today target members of this family. A well-known example of this family is bacteriorhodopsin (bR). bR is present in the membranes of some halobacteria. It harvests light energy and then converts it to electrostatic energy in the form of proton transport (PTR) out of the cell. This creates a large pH gradient across the cell membrane, which is then used to create ATP when the cell allows a proton to move back into the cell, through another transmembrane protein, ATP synthase.

2.1.2.1 Empirical Valence Bond Model

Often scientists want to study biochemical bond breaking and forming processes, such as the breaking of a covalent bond to H, and the forming of a new covalent bond to H. Modeling these types of chemical reactions requires the use of a quantum mechanical (QM) treatment to describe the reacting fragments, coupled with a classical, molecular mechanics (MM) simulation of the surrounding biological molecule (where no reactions are taking place), resulting in a quantum mechanics/molecular mechanics (QM/MM) method. Frequently, a molecular orbital-based QM treatment, which is, for example, very effective for calculating spectroscopic properties of molecules, is prohibitively expensive when studying biochemical systems. An alternative to using molecular orbital-based QM methods is the empirical valence bond (EVB) model,[34] which is effective at calculating bond rearrangements.

The EVB method first calculates the ground state potential energy surface of reacting fragments by mixing the valence bond (VB) resonance structures of the reactants and products. The reaction is forced to occur by gradually moving the wave function from 100% reactant, 0% product to 0% reactant, 100% product. The portion of the molecule described by MM adjusts itself to the reacting fragments in order to minimize the overall energy of the reaction. Importantly, this method is not as expensive as methods that use a molecular orbital QM approach, and allows sufficient configurational averaging to give the overall free energy change for the system.[34,35]

An important point to be made about the EVB model is that the reaction simulation must be carried out both in water and in the protein. The simulations begin in water, where the energetics of the reaction are known. For proton transfer (PT) reactions, the change in free energy of the PT in water can be obtained if one knows the pK_a values of the donor (DH$^+$) and the protonated acceptor (AH) in water. This is given in kcal/mol by Equation 2.9:

$$\Delta G_{PT}^w(DH^+ + A^- \rightarrow D + AH)_\infty = 1.38(\Delta pK_a^w), \quad (2.9)$$

where ∞ indicates that the donor and acceptor are at large separation. All of the adjustable parameters used in the EVB model of the reaction are adjusted so that the water simulation of the PT between donor and acceptor at large separation in water reproduces the experimentally known difference in pK_a values. The parameters are then fixed, and not allowed to vary when moving to simulations of the reaction in the biochemical system. An example where the EVB method was used in bR is given below.

2.1.2.2 Conversion of Light Energy to Electrostatic Energy

The mechanism of light-induced PTR across a membrane and against a proton gradient is of significant interest in bioenergetics.[36-47] bR is a well-characterized model system for this process, with many structures available of the ground state and several intermediates, and many kinetic studies having been performed with it.[36,48-50] This structural information was recently used to propose a detailed molecular picture of how bR converts light energy to electrostatic energy. In the active site, the chromophore is covalently attached to an arginine side chain through a protonated Schiff base (SB), and it forms an ion pair with the negatively charged side chain of Asp85.[51-53] It is known that bR absorbs light, and then a sequence of relaxation and PTR processes occurs, ultimately resulting in a proton being transported outside of the cell. However, the driving force for the initial PT from the SB to the Asp85 was unclear. A previous study[54] and a more recent study[47] suggest that the PT is driven by a light-induced charge separation of the ion pair. Apparently the

absorption of a photon by the chromophore leads to isomerization around a carbon–carbon double bond and a high energy, sterically strained ground state. Relaxation of this steric energy leads to the increase in ion pair distance observed in structural and computational studies, ultimately leading to the PT.

2.1.2.3 Modeling the Conversion of Light Energy to Electrostatic Energy

The computational methods used in the more recent study[47] to convert structural information to free energy barriers for the PT process are described here. First, calculation of the PT potential energy surface in the ground state and several of the photocycle intermediates was performed to obtain the free energy landscape for the system. (For an in-depth description of the photocycle, see Refs. [36,55].) This study used a type of QM/MM method that combines the quantum consistent force field/pi electrons (QCFF/PI) method[56] and the EVB[35] approach. An *ab initio* QM/MM method was not used because the calculations to do the proper configurational averaging needed to provide reliable free energy surfaces are prohibitively expensive. The energetics of the PT are modeled by considering two VB resonance structures of the form

$$\Psi_1 = R\text{---}C(H)\text{==}N^+(R')H\ ^-A$$
$$\quad (2.10)$$
$$\Psi_2 = R\text{---}C(H)\text{==}N^+(R')H\text{---}A$$

where the wave function Ψ_1 represents the protonated SB (SBH$^+$) and deprotonated acceptor, A$^-$ (Asp85), and Ψ_2 represents the deprotonated SB and protonated acceptor (Asp-H). Because the chromophore contains many pi electrons, the SB is represented by the QCFF/PI potential surface, and the acceptor is represented by an empirical potential function (as in the normal EVB). Water and protein surrounding the reacting system are coupled to the QCFF/PI Hamiltonian through a standard QM/MM treatment. This analysis allows the system to be represented by two diabatic or pure states, with the conditions outlined by Equation 2.11:

$$\bar{\varepsilon}^{(1)} = \varepsilon_{QCFF/PI}(SBH^+) + \varepsilon'_{Asp85} + \varepsilon_{SS'}^{(1)} + \varepsilon_{Ss}^{(1)} + \varepsilon_{ss} + \alpha^{(1)},$$
$$\quad (2.11)$$
$$\bar{\varepsilon}^{(2)} = \varepsilon_{QCFF/PI}(SB) + \varepsilon'_{AspH85} + \varepsilon_{SS'}^{(2)} + \varepsilon_{Ss}^{(2)} + \varepsilon_{ss} + \alpha^{(2)},$$

where the $\varepsilon_{QCFF/PI}$ of the indicated form of the chromophore includes the potential from the solvent/protein system, ε' is an EVB description of Asp85 or AspH85, and $\varepsilon_{SS'}$ is the interaction between the classical (EVB) and π-electron systems, which is being treated as in the regular EVB treatment by considering the classical electrostatic and steric interactions between the two fragments. Finally, $\varepsilon_{Ss}^{(i)}$ represents the interaction between the solute (S) and the solvent (s) in the given state, while $\alpha^{(i)}$ is the "gas phase shift." The gas phase shift is a parameter that is adjusted by requiring that the PT at large separation in water between SBH$^+$ ($pK_a = 7.0$) and Asp (Asp-H $pK_a = 4.0$) reproduce the experimental pK_a difference of 3.0 pH units (corresponding to 4.1 kcal/mol). The pure diabatic states shown above do not reflect the actual ground state potential energy surface of the system, which is represented by mixing the two diabatic states according to Equation 2.12:

$$E_g = \frac{1}{2}\left[\left(\varepsilon^{(1)} + \varepsilon^{(2)}\right) - \left(\left(\varepsilon^{(1)} + \varepsilon^{(2)}\right)^2 + 4H_{12}^2\right)^{1/2}\right], \quad (2.12)$$

where H_{12} is the off-diagonal coupling element between Ψ_1 and Ψ_2.

It should be emphasized that all of the required parameters, such as the gas phase shift, are adjusted during the simulation of PT from SBH$^+$ to Asp in water, requiring that the values used reproduce experimental results. The parameters are then fixed, and are not adjusted in protein simulations.

In moving to the protein simulations, it was found that the driving force for the initial PT from SBH$^+$ to Asp is primarily based on electrostatic effects. The absorption of a photon by the chromophore leads to a sterically strained structure. Relaxation of this strain leads to destabilization of the ion pair by increasing the ion pair (SBH$^+$Asp) distance, ultimately driving the PT process. The difference in energy between the ion pair and the neutral state seemed to determine the majority of the barrier to the PT process.

2.1.3 Coupling between Electron and Proton Transport in Oxidative Phosphorylation

Peter Mitchell was awarded the 1978 Nobel Prize for Chemistry for his revolutionary chemiosmotic hypothesis (see Refs. [39,57]). This theory stated that ATP synthesis by the oxidative phosphorylation pathway is coupled to PTR from the cytoplasmic side to the matrix side of the inner mitochondrial membrane. It is now known that transfer of high-energy electrons from the citric acid cycle (from NADH and FADH$_2$) to complexes of the ETC (which are embedded in the inner mitochondrial matrix) results in reduction of an electron acceptor, O$_2$, to H$_2$O. In eukaryotes, there are four transmembrane complexes involved in the ETC, three of which are also proton pumps. Complex I (NADH-Q oxidoreductase), Complex III (Q-cytochrome c oxidoreductase), and Complex IV (cytochrome c oxidase) all pump protons, with a stoichiometry of 2, 2, and 1H$^+$ per high-energy electron.[58] Only Complex II (succinate-Q reductase) does not. This electron transfer (ET) also leads to the creation of a proton gradient across the membrane, creating a proton motive force (*PMF*), which consists of two parts, a chemical gradient and a charge gradient as expressed by Equation 2.13[58]:

$$PMF(\Delta p) = \text{chemical gradient}(\Delta pH)$$
$$+ \text{ charge gradient}(\Delta\Psi). \quad (2.13)$$

The PMF can be used to drive the synthesis of ATP when protons are allowed to move back into the matrix through ATP synthase (sometimes called Complex V). The typical pH difference across the membrane is around 1.4 pH units (outside is more acidic), and the membrane potential is about 0.14 V (outside is positive), leading to a free energy of 5.2 kcal/mol of electrons.[58]

2.1.3.1 Electron Transport Chain Complexes as Proton Pumps

Similar to the KcsA ion channel and bR proteins discussed previously, complexes of the ETC are embedded in a membrane. While there have been several experimental and theoretical studies of ET[59–63] and PT[64–67] individually in proteins, there have been fewer studies of coupled ET/PT reactions.[68–71]

An extensive theoretical analysis of coupled ET and PT in Complex IV of the ETC, cytochrome c oxidase, has been published.[72] Many of the ETC complexes have large, multisubunit structures, and Complex IV is an example. It has a molecular mass of 204 kDa, contains 13 subunits, and has several redox centers, making its mechanism difficult to elucidate. High-resolution structures of Complex IV have been published,[73–75] enabling theoretical studies to probe the coupled ET/PT mechanism of this complex. The author of this analysis postulates that Complex IV couples electron tunneling between redox centers with a proton moving along a conduction channel in a "classical, diffusion-like random walk fashion,"[72] but concludes that much more work is needed to fully understand the coupled ET/PT reactions occurring in the ETC.

2.1.4 Mechanosensitive Ion Channels

Mechanosensitive ion channels (MSCs) are transmembrane proteins that are important in helping cells respond to a variety of mechanical stimuli, such as sound, gravity, and osmotic pressure gradients. They are found in a variety of tissues and are thought to be important for the senses of touch, hearing, and balance.[76] It currently is believed that two types of MSCs exist. The first type is found in specialized sensory cells; forces are applied to these channels through fibrous proteins.[77,78] The second type responds to stresses in the lipid bilayer.[77,79,80] The lipid membrane is often what the initial stress acts upon, and the lipid bilayer must somehow respond to these stimuli (see Chapter 1). A known function of MSCs includes regulating cell volume in response to osmotic pressure gradients, to prevent the cell from bursting. Somehow these transmembrane channels must detect extracellular forces, and transmit the information inside the cell as electrical or chemical signals.[81]

2.1.4.1 Gate Mechanisms

A defining characteristic of MSCs represents large conformational changes between the open and closed forms. For example, the bacterial large conductance mechanosensitive (MscL) channel undergoes a radius change of 5–6 nm between the open and closed forms.[82] An initial descriptor of the energy difference, ΔG, between open and closed forms based on the bilayer tension, T, is given by

$$\Delta G = T \times \Delta A, \quad (2.14)$$

where ΔA is the in-plane area change between open and closed states.[77,83] However, there are likely more conformational changes, or deformations, possible for these MSCs, as the in-plane area does not take account of other channel conformational changes (see below), membrane stiffness, and membrane thickness, all of which can affect channel opening and closing. It is believed that there are three types of changes that channels can undergo.[83] The first is as mentioned above, where a channel can change its in-plane area, A. The second type occurs when a channel actually changes its shape within the membrane. Finally, a channel can also undergo a change in length without changing its shape or in-plane area. See Figure 1 in Markin and Sachs[83] for a nice illustration of the three types of conformational changes.

This system presents another situation where modeling studies can complement experimental studies. In the last several years there have been a few computational studies that look at gating mechanisms using a variety of models.[84-87] A recent computational study of the *Escherichia coli* MscL channel investigated the gating pathways when various conformational deformations are simulated.[87] It is known that, in general, the channel protein and the membrane undergo a deformation leading to the opening of the channel, but a molecular understanding of the process has not yet been achieved. The authors of the recent computational study examined the mechanical roles of structural features, such as transmembrane helices and loops, and found that many of these strongly affect gating ability. However, many questions remain in the drive to understand these interesting and complex channels on a molecular level.

References

1. Sharp, K. A. and Honig, B., Electrostatic interactions in macromolecules: Theory and applications, *Annu. Rev. Biophys. Biophys. Chem.* 1990, 19, 301–332.
2. Perutz, M. F., Electrostatic effects in proteins, *Science* 1978, 4362, 1187–1191.
3. Hol, W. G. J., The role of the α-helix dipole in protein structure and function, *Prog. Biophys. Mol. Biol.* 1985, 45 (3), 149–195.
4. Warshel, A., Energetics of enzyme catalysis, *Proc. Natl. Acad. Sci. U.S.A.* 1978, 75 (11), 5250–5254.
5. Braun-Sand, S. and Warshel, A., *Protein Folding Handbook.* 2005, 1, 163–200, Wiley, Hoboken, NJ.
6. Tanford, C., J. Theory of protein titration curves. II. Calculations for simple models at low ionic strength, *Am. Chem. Soc.* 1957, 79, 5340–5347.
7. Warshel, A., Russell, S. T., and Churg, A. K., Macroscopic models for studies of electrostatic interactions in proteins: Limitations and applicability, *Proc. Natl. Acad. Sci. U.S.A.* 1984, 81 (15), 4785–4789.
8. Georgescu, R. E., Alexov, E. G., and Gunner, M. R., Combining conformational flexibility and continuum electrostatics for calculating pKa's in proteins, *Biophys. J* 2002, 83 (4), 1731–1748.
9. Warwicker, J. and Watson, H. C., Calculation of the electric potential in the active site cleft due to alpha-helix dipoles, *J Mol. Biol.* 1982, 157 (4), 671–679.
10. Lee, F. S., Chu, Z. T., Bolger, M. B., and Warshel, A., Calculations of antibody-antigen interactions: Microscopic and semi-microscopic evaluation of the free energies of binding of phosphorylcholine analogs to McPC603, *Protein Eng.* 1992, 5 (3), 215–228.
11. Warshel, A., Sussman, F., and King, G., Free energy of charges in solvated proteins: Microscopic calculations using a reversible charging process, *Biochemistry* 1986, 25 (26), 8368–8372.
12. Klamt, A., Mennucci, B., Tomasi, J., Barone, V., Curutchet, C., Orozco, M., and Luque, F. J., On the performance of continuum solvation methods. A comment on "Universal Approaches to Solvation Modeling," *Acc. Chem. Res.* 2009, 42 (4), 489–492.
13. Hummer, G., Pratt, L. R., and Garcia, A. F., On the free energy of ionic hydration, *J Phys. Chem.* 1996, 100 (4), 1206–1215.
14. Ha-Duong, T., Phan, S., Marchi, M., and Borgis, D., Electrostatics on particles: Phenomenological and orientational density functional theory approach, *J Chem. Phys.* 2002, 117 (2), 541–556.
15. Parson, W. W. and Warshel, A., 2008. Calculations of electrostatic energies in proteins using microscopic, semi-microscopic and macroscopic models and free-energy perturbation approaches. In: Aartsma, T. J. and Matysik, J. (eds), *Biophysical Techniques in Photosynthesis*, vol 2, Series Advances in Photosynthesis and Respiration, vol 26. Springer, Dordrecht, pp. 401–420.
16. Molaris Manual and User Guide, 2009 (Theoretical Background). USC: Los Angeles, CA, 2009. http://laetro.usc.edu/programs/doc/molaris_manual.pdf
17. Warshel, A. and Russell, S. T., Calculations of electrostatic interactions in biological systems and in solutions, *Q. Rev. Biophys.* 1984, 17 (3), 283–422.
18. Lee, F. S., Chu, Z. T., and Warshel, A., Microscopic and semi-microscopic calculations of electrostatic energies in proteins by the POLARIS and Enzymix programs, *J Comput. Chem.* 1993, 14 (2), 161–185.
19. Warshel, A., Naray-Szabo, G., Sussman, F., and Hwang, J. K., How do serine proteases really work?, *Biochemistry* 1989, 28 (9), 3629–3637.
20. De Pelichy, L. D. G., Eidsness, M. K., Kurtz, D. M., Jr., Scott, R. A., and Smith, E. T., Pressure-controlled voltammetry of a redox protein: An experimental approach to probing the internal protein dielectric constant?, *Curr. Sep.* 1998, 17 (3), 79–82.
21. Schutz, C. N. and Warshel, A., What are the dielectric "constants" of proteins and how to validate electrostatic models?, *Proteins: Struct., Funct., Genet.* 2001, 44 (4), 400–417.
22. Ng Jin, A., Vora, T., Krishnamurthy, V., and Chung, S.-H., Estimating the dielectric constant of the channel protein and pore, *Eur. Biophys. J.* 2008, 37 (2), 213–222.

23. Burykin, A., Kato, M., and Warshel, A., Exploring the origin of the ion selectivity of the KcsA potassium channel, *Proteins: Struct., Funct., Genet.* 2003, 52 (3), 412–426.

24. LeMasurier, M., Heginbotham, L., and Miller, C., KcsA: it's a potassium channel, *J Gen. Physiol.* 2001, 118 (3), 303–313.

25. Cuello, L. G., Romero, J. G., Cortes, D. M., and Perozo, E., pH-dependent gating in the Streptomyces lividans K+ channel, *Biochemistry* 1998, 37 (10), 3229–3236.

26. Zhou, Y., Morals-Cabral, J. H., Kaufman, A., and MacKinnon, R., Chemistry of ion coordination and hydration revealed by a K+ channel-Fab complex at 2.0 A resolution, *Nature* 2001, 414 (6859), 43–48.

27. Doyle, D. A., Cabral, J. M., Pfuetzner, R. A., Kuo, A., Gulbis, J. M., Cohen, S. L., Chait, B. T., and MacKinnon, R., The structure of the potassium channel: molecular basis of K+ conduction and selectivity, *Science* 1998, 280 (5360), 69–77.

28. Aqvist, J. and Luzhkov, V., Ion permeation mechanism of the potassium channel, *Nature* 2000, 404 (6780), 881–884.

29. Luzhkov, V. B. and Aqvist, J., K+/Na+ selectivity of the KcsA potassium channel from microscopic free energy perturbation calculations, *Biochim. Biophys. Acta, Protein Struct. Mol. Enzymol.* 2001, 1548 (2), 194–202.

30. Allen, T. W., Bliznyuk, A., Rendell, A. P., Kuyucak, S., and Chung, S. H., Molecular Dynamics Estimates of Ion Diffusion in Model Hydrophobic and the KcsA Potassium Channel, *J Chem. Phys.* 2000, 112 (18), 8191–8204.

31. Allen, T. W., Hoyles, M., Kuyucak, S., and Chung, S. H., Molecular and Brownian dynamics studies of the potassium channel, *Chem. Phys. Lett.* 1999, 313 (1,2), 358–365.

32. Berneche, S. and Roux, B., Energetics of ion conduction through the K+ channel, *Nature (London)* 2001, 414 (6859), 73–77.

33. Biggin, P. C., Smith, G. R., Shrivastava, I., Choe, S., and Sansom, M. S. P., Potassium and sodium ions in a potassium channel studied by molecular dynamics simulations, *Biochim. Biophys. Acta, Biomembr.* 2001, 1510 (1–2), 1–9.

34. Warshel, A. and Weiss, R. M., An empirical valence bond approach for comparing reactions in solutions and in enzymes, *J. Am. Chem. Soc.* 1980, 102 (20), 6218–6226.

35. Warshel, A., *Computer modeling of Chemical Reactions in Enzymes and Solutions.* Wiley: New York, 1991.

36. Lanyi, J. K., Bacteriorhodopsin, *Annu. Rev. Physiol.* 2004, 66, 665–688.

37. Wikstrom, M. K. F., Proton pump coupled to cytochrome c oxidase in mitochondria, *Nature* 1977, 266 (5599), 271–273.

38. Warshel, A., Electrostatic basis of structure-function correlation in protein, *Acc. Chem. Res.* 1981, 14 (9), 284–290.

39. Mitchell, P., Coupling of phosphorylation to electron and hydrogen transfer by a chemi-osmotic type of mechanism, *Nature* 1961, 191, 144–148.

40. Mathies, R. A., Lin, S. W., Ames, J. B., and Pollard, W. T., From femtoseconds to biology: Mechanism of bacteriorhodopsin's light-driven proton pump, *Annu. Rev. Biophys. Biophys. Chem.* 1991, 20, 491–518.

41. Subramaniam, S. and Henderson, R., Molecular mechanism of vectorial proton translocation by bacteriorhodopsin, *Nature* 2000, 406 (6796), 653–657.

42. Henderson, R., The purple membrane from Halobacterium halobium, *Annu. Rev. Biophys. Bioeng.* 1977, 6, 87–109.

43. Oesterhelt, D. and Tittor, J., Two pumps, one principle: Light-driven ion transport in halobacteria, *Trends Biochem. Sci.* 1989, 14 (2), 57–61.

44. Michel, H., Behr, J., Harrenga, A., and Kannt, A., Cytochrome c-oxidase: Structure and spectroscopy, *Annu. Rev. Biophys. Biomol. Struct.* 1998, 27, 329–356.

45. Stoeckenius, W. and Bogomolni, R. A., Bacteriorhodopsin and related pigments of halobacteria, *Annu. Rev. Biochem.* 1982, 51, 587–616.

46. Warshel, A. and Parson, W. W., Dynamics of biochemical and biophysical reactions: insight from computer simulations, *Q. Rev. Biophys.* 2001, 34 (4), 563–679.

47. Braun-Sand, S., Sharma, P. K., Chu, Z. T., Pisliakov, A. V., and Warshel, A., The energetics of the primary proton transfer in bacteriorhodopsin revisited: It is a sequential light-induced charge separation after all, *Biochim. Biophys. Acta, Bioenerg.* 2008, 1777 (5), 441–452.

48. Luecke, H., Atomic resolution structures of bacteriorhodopsin photocycle intermediates: the role of discrete water molecules in the function of this light-driven ion pump, *Biochim. Biophys. Acta, Bioenerg.* 2000, 1460 (1), 133–156.

49. Herzfeld, J. and Lansing, J. C., Magnetic resonance studies of the bacteriorhodopsin pump cycle, *Annu. Rev. Biophys. Biomol. Struct.* 2002, 31, 73–95.

50. Royant, A., Edman, K., Ursby, T., Pebay-Peyroula, E., Landau, E. M., and Neutze, R., Helix deformation is coupled to vectorial proton transport in the photocycle of bacteriorhodopsin, *Nature* 2000, 406 (6796), 645–648.

51. Song, Y., Mao, J., and Gunner, M. R., Calculation of proton transfers in bacteriorhodopsin bR and M intermediates, *Biochemistry* 2003, 42 (33), 9875–9888.

52. Spassov, V. Z., Luecke, H., Gerwert, K., and Bashford, D., pK_a calculations suggest storage of an excess proton in a hydrogen-bonded water network in Bacteriorhodopsin, *J Mol. Biol.* 2001, 312 (1), 203–219.

53. Sampogna, R. V. and Honig, B., Environmental effects on the protonation states of active site residues in bacteriorhodopsin, *Biophys. J* 1994, 66 (5), 1341–1352.

54. Warshel, A., Conversion of light energy to electrostatic energy in the proton pump of *Halobacterium halobium*, *Photochem. Photobiol.* 1979, 30 (2), 285–290.

55. Lanyi, J. K., X-ray diffraction of bacteriorhodopsin photocycle intermediates, *Mol. Membr. Biol.* 2004, 21 (3), 143–150.

56. Warshel, A., Semiempirical methods of electronic structure calculation. In: G. A. Segal (Ed.) *Modern Theoretical Chemistry*, Vol. 7, Part A, p. 133, 1977, Plenum Press: New York.

57. Mitchell, P. and Moyle, J., Chemiosmotic hypothesis of oxidative phosphorylation, *Nature* 1967, 213 (5072), 137–139.

58. Jeremy, M. Berg, J. L. T., and Lubert S., *Biochemistry*. Sixth edition, W. H. Freeman and Company, New York, 2006.

59. Bollinger, J. M., Jr., Biochemistry. Electron relay in proteins, *Science* 2008, 320 (5884), 1730–1731.

60. Hurley, J. K., Tollin, G., Medina, M., and Gomez-Moreno, C., Electron transfer from ferredoxin and flavodoxin to ferredoxin-NADP+ reductase, In: Golbeck, J. H. (Ed.), *Photosystem I, the Light-Driven Platocyanin:Ferredoxin Oxireductase*, pp. 455–476, 2006, Springer, Dordrecht.

61. Northrup, S. H., Kollipara, S., and Pathirathne, T., Simulation of rates of electron transfer between proteins cytochrome b5 reductase and cytochrome b5, *60th Southeast Regional Meeting of the American Chemical Society*, Nashville, TN, November 12–15 2008, SERM-098.

62. Shao, C., Zhao, C., Wang, Q., and Jiao, K., Direct electron transfer of redox proteins in sodium alginate solution, *Qingdao Keji, Daxue Xuebao (Ziran Kexueban)/ Journal of Qingdao University of Science and Technology* 2008, 29 (6), 474–479.

63. Brittain, T., Intra-molecular electron transfer in proteins, *Protein Pept. Lett.* 2008, 15 (6), 556–561.

64. Braun-Sand S., Olsson M. H. M., Mavri J., and Warshel A., Computer simulations of proton transfer in proteins and solutions, In: Hynes, J. T., Klinman, J. P., Limbach, H.-H., and Schowen, R. L. (Eds.), *Hydrogen Transfer Reactions*, 2007, Wiley-VCH, Hoboken, NJ.

65. Kandori, H., Yamazaki, Y., Sasaki, J., Needleman, R., Lanyi, J. K., and Maeda, A., Water-mediated proton transfer in proteins: An FTIR study of bacteriorhodopsin, *J Am. Chem. Soc.* 1995, 117 (7), 2118–2119.

66. Kim, S. Y. and Hammes-Schiffer, S., Hybrid quantum/classical molecular dynamics for a proton transfer reaction coupled to a dissipative bath, *J Chem. Phys.* 2006, 124 (24), 244102/1–244102/12.

67. Nie, B., Probing hydrogen bonding interactions and proton transfer in proteins. Oklahoma State University, Oklahoma, 2006. http://gradworks.umi.com/32/22/3222751.html

68. Gunner, M. R., Mao, J., Song, Y., and Kim, J., Factors influencing the energetics of electron and proton transfers in proteins. What can be learned from calculations, *Biochim. Biophys. Acta, Bioenerg.* 2006, 1757 (8), 942–968.

69. Camba, R., Jung, Y.-S., Hunsicker-Wang, L. M., Burgess, B. K., Stout, C. D., Hirst, J., and Armstrong, F. A., The role of the proximal proline in reactions of the [3Fe-4S] cluster in *Azotobacter vinelandii* ferredoxin I, *Biochemistry* 2003, 42 (36), 10589–10599.

70. Cukier, R. I. and Nocera, D. G., Proton-coupled electron transfer, *Annu. Rev. Phys. Chem.* 1998, 49, 337–369.

71. Hammes-Schiffer, S., Theoretical perspectives on proton-coupled electron transfer reactions, *Acc. Chem. Res.* 2001, 34 (4), 273–281.

72. Stuchebrukhov, A. A., Coupled electron and proton transfer reactions. Proton pumps, cytochrome oxidase, and biological energy transduction, *J Theor. Comput. Chem.* 2003, 2 (1), 91–118.

73. Yoshikawa, S., Shinzawa-itoh, K., Nakashima, R., Yaono, R., Yamashita, E., Inoue, N., Yao, M., et al., Redox-coupled crystal structural changes in bovine heart cytochrome c-oxidase, *Science* 1998, 280 (5370), 1723–1729.

74. Iwata, S., Ostermeier, C., Ludwig, B., and Michel, H., Structure at 2.8 Å resolution of cytochrome c oxidase from *Paracoccus denitrificans*, *Nature* 1995, 376 (6542), 660–669.

75. Abramson, J., Riistama, S., Larsson, G., Jasaitis, A., Svensson-Ek, M., Laakkonen, L., Puustinen, A., Iwata, S., and Wikstrom, M., The structure of the ubiquinol oxidase from Escherichia coli and its ubiquinone binding site, *Nat. Struct. Biol.* 2000, 7 (10), 910–917.

76. Pivetti, C. D., Yen, M.-R., Miller, S., Busch, W., Tseng, Y.-H., Booth, I. R., and Saier, M. H., Jr., Two families of mechanosensitive channel proteins, *Microbiol. Mol. Biol. Rev.* 2003, 67 (1), 66–85.

77. Gottlieb, P. A., Suchyna, T. M., Ostrow, L. W., and Sachs, F., Mechanosensitive ion channels as drug targets, *Curr. Drug Targets: CNS Neurol. Disord.* 2004, 3 (4), 287–295.

78. Ernstrom, G. G. and Chalfie, M., Genetics of sensory mechanotransduction, *Annu. Rev. Genet.* 2002, 36, 411–453.

79. Martinac, B. and Kloda, A., Evolutionary origins of mechanosensitive ion channels, *Prog. Biophys. Mol. Biol.* 2003, 82 (1–3), 11–24.

80. Sachs, F. and Morris, C. E., Mechanosensitive ion channels in nonspecialized cells, *Rev. Physiol., Biochem. Pharmacol.* 1998, 132, 1–77.

81. Martinac, B., Mechanosensitive ion channels: molecules of mechanotransduction, *J Cell Sci.* 2004, 117 (12), 2449–2460.

82. Chiang, C.-S., Anishkin, A., and Sukharev, S., Water dynamics and dewetting transitions in the small mechanosensitive channel MscS, *Biophys. J* 2004, 86 (5), 2846–2861.

83. Markin, V. S. and Sachs, F., Thermodynamics of mechanosensitivity, *Phys. Biol.* 2004, 1 (2), 110–124.

84. Sukharev, S., Durell, S. R., and Guy, H. R., Structural models of the MscL gating mechanism, *Biophys. J* 2001, 81 (2), 917–936.

85. Sukharev, S. and Anishkin, A., Mechanosensitive channels: What can we learn from 'simple' model systems?, *Trends Neurosci.* 2004, 27 (6), 345–351.

86. Gullingsrud, J. and Schulten, K., Gating of MscL studied by steered molecular dynamics, *Biophys. J* 2003, 85 (4), 2087–2099.

87. Tang, Y., Yoo, J., Yethiraj, A., Cui, Q., and Chen, X., Gating mechanisms of mechanosensitive channels of large conductance, II: Systematic study of conformational transitions., *Biophys. J* 2008, 95 (2), 581–596.

Cell Biology and Biophysics of the Cell Membrane

Didier Dréau

3.1 Cell Interactions

In contrast to unicellular organisms, in multicellular organisms cells interact with one another through cell–cell physical chemical (paracrine or hormonal) contacts. The physical cell–cell interactions occur through cell junctions with binding of proteins present within the membranes of each cell. For distant interactions, molecules secreted (ligands) are recognized by specific receptors on the target cell membrane or in the target cell, depending on the hydrophobicity of the ligand.

3.2 Types of Junctions

Three main types of cell–cell junctions define the physical strength of interactions between cells (Figure 3.1): desmosomes (zonulae adherens), tight junctions (zonulae occludens), and gap junctions. These junctions are cell membrane regions enriched with specific proteins interacting between two cells. In particular, the physical strength of the junction allows a resistance to shear stress (a stress applied parallel or tangential to a contact surface) imparted, for example, by vessel endothelium on blood or blood on vessel,

respectively. The shear stress of a fluid or viscosity drag (t) is a function of its viscosity (η) and velocity (v) exerted on the wall of a lumen (inner radius r). Shear stress resistance characterizes the different cell junctions of a given lumen [e.g., blood vessels and the gastrointestinal (GI) tract]. The greater the viscosity [and the liquid flow (Q)], the higher the shear rate and the viscosity drag (t):

$$t = 4\eta Q / \pi r^3. \qquad (3.1)$$

Note: In denominator $\pi = 3.1415926...$

3.2.1 Desmosomes

Desmosomes are found in all tissues subject to *shear stress*, including the skin and the GI tract. Desmosomes generate strong bonds between cells and between cells and the basal lamina. Proteins, mainly desmoglein and desmocollin, are associated in dense plaques, separated by a 30 nm intercellular space, are present in both cell membranes, and are linked to intracellular structures and tonofilaments in the two connected cells (Figure 3.1). Shear stress leads to deformation of the cell membrane and disruption of filament actin (F-actin) and of

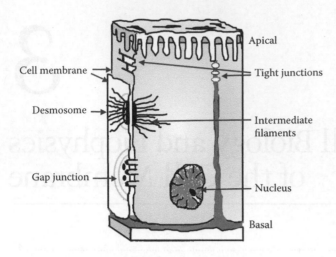

FIGURE 3.1 Type of cell junctions.

cytoskeleton organization. In response to disruption of F-actin, the expression of multiple proteins, including those involved in adherence and cell junctions, is up-regulated, depending on both type of cell and liquid flow. Hemidesmosomes are structures that are present on cells and use similar proteins to anchor cells to the basal lamina through interactions between integrins and proteins in the basal lamina.

3.2.2 Tight Junctions

Tight junctions are common in exchange tissues holding nephron cells in the kidneys, endothelial cells in the vessels, and enterocytes in the GI tract together. Tight junctions in cells are made through the interactions of multiple proteins, including claudins and occludins, both of which are transmembrane proteins embedded in the cytoskeleton. Tight junctions also prevent the movement of proteins with apical functions to the baso-lateral area of the cell membrane (and vice versa), ensuring cell polarity. Cell polarity also defines both chemical interactions and electrical gradients by the types and concentrations of receptors and channels present. As with chemical activity, cell membrane electrical potential can be different between the apical and the baso-lateral area of a cell.

3.2.3 Gap Junctions

Gap junctions are hemichannels or connexons composed of six connexin (Cx) proteins. The alignment of two connexons forms a gap junction between two cells, allowing cytoplasm sharing and the rapid transfer of both electrical and chemical signals (small molecules and ions). The type of connexins associated in the formation of the connexon (homo- and hetero-hexamers) appears to influence the function of the gap junctions (speed of electrical chemical transfers and nature of the chemical molecule transferred). The concentration of multiple (>100) gap junctions forms a complex structure or plaque. Gap junctions allow direct electrical signaling between cells, although differences (10- to 15-fold) in electrical conductance between gap junctions

have been shown. Gap junctions also allow chemical molecules (<1000 Da) to move from one cell to the other and favor chemical communication through the passage of second messengers, including IP3 and Ca^{2+}. The passage of these chemicals is selective, depending on the size and charge of the molecule and the nature of connexin subunits. Although most of the movement of ions through gap junctions does not require energy, the recycling of K^+ in the cochlea, essential to the transduction by auditory hair cells through gap junctions, is facilitated by ATPase activity and connexin conformation changes.

Charge associated with the connexins either repulses or attracts ions, playing a critical role in preventing or allowing for passage. In addition to these chemical and electrical exchange functions, gap junction proteins also promote cell adhesion and have tumor suppressing (C×43, C×36) and cell signaling (C×43) roles. The presence of gap junctions linking multiple cells within a tissue generates a *syncytium*, that is, a cluster of cells with similar response as in the heart muscle, and the smooth muscles of the GI tract.

3.3 Cell Adhesion Molecules

Within tissues, cells interact not only with other cells but also with the extracellular matrix (ECM). Cell attachment to the ECM is a key requirement of multicellular organisms. Produced by multiple cells, including fibroblasts, the ECM is composed of multiple proteins, of which the major ones are collagens, laminin, fibronectin, vitronectin, and vimentin. The ECM constitutes the basal lamina, the basal layer on which cells are anchored by integrins. These cell surface receptors are composed of one α (alpha) and one β (beta) subunit. Each heterodimer binds to a specific molecule of the ECM (e.g., α6β1 binds to laminin) with variable affinities. Integrin expression is cell specific and the strength of the binding to the ECM is variable, depending on the composition of both integrin and ECM. The binding site for the ECM is on the β chain and requires divalent cations to function whereas the α subunit may be involved in protein stabilization.

Integrins attach cells to the ECM through interactions between ECM molecules and microfilaments of the actin cytoskeleton, allowing cells to resist shear stress forces. The intensity of the force needed to deform a cell membrane or dissociate a cell linked by junctions in an epithelium is highly variable and will depend on the force as well as on the characteristics of the specific location of interest.

This cell attachment involves not only integrins but also the formation of cell adhesion complexes consisting of transmembrane integrins and many cytoplasmic proteins, including talin, vinculin, and paxillin. Integrins have a prominent role in regulating cell shape, cell migration, and cell signaling, making them pivotal in multiple cell events (including growth, differentiation, and survival).

3.4 Intracellular Connections

Cell membranes are physically connected with the cellular scaffolding or cytoskeleton. The cytoskeleton, critical in cell shape and motion, intracellular transport (vesicles and organelles), and cell division, is composed of three kinds of filaments: microfilaments, intermediate filaments, and microtubules. Microfilaments are intertwined double-helix actin chains that are concentrated near the cell membrane. Intermediate filaments are very stable and constituted of multiple proteins, including vimentin, keratin, and laminin. Microtubules are composed of tubulin (α and β), which play a major role in intracellular transport and in the formation of mitotic spindles. Connections of the membrane with the cytoskeleton are key in maintaining 3D structures, cell shape and deformation (e.g., generation of processes), and resistance to tension.

3.5 Cell Membranes and Epithelia

As for individual cells, where the membrane is a selective barrier allowing the movement of water and ions through channels and of larger molecules through specific carriers, the epithelium also benefits from the selective permeability of the cells itis made of. The movement of molecules through the semipermeable membrane that is the cell membrane or of cells lining an epithelium relies on various physiological mechanisms. The intrinsic permeability of the cell membrane depends on (1) the presence of a gradient, that is, a difference in the chemical concentration or electrical charge between both sides of the cell membrane, and (2) the movement of molecules through diffusion, leaky channels, or facilitated or active transport. The membrane transport processes are discussed in greater detail in the next section and will be referenced in Chapter 5 for the specific working action.

3.6 Membrane Permeability

3.6.1 Membrane Composition and Structure

Membranes are mostly made up of *hydrophobic* phospholipids (phosphatidylcholine, sphingomyelins, amino phospholipids, phosphatidylglycerol, and phosphatidylinositol), with one polar head and two nonpolar lipid chains. Hydrophobic means that the chemical structure is such that it repels the water dipole, in contrast to hydrophilic, which attracts water due to the inherent polar chemical composition. In an aqueous environment, the nonpolar chains are oriented away from water with the polar head in contact with the water, leading to the spontaneous formation of lipid bilayers. With the exception of the protein anchored internally to the actin or spectrin network or externally to ECM molecules, proteins can move within the lipid bilayer creating a fluid mosaic. The lipid:protein ratio can radically vary between membrane and cell types (Table 3.1).

Membrane composition is also heterogeneous, that is, protein distribution and, to a lesser extent, lipid composition are different throughout the cell membrane. Specifically, the density of a given receptor can be much higher at a specific location, for example, acetylcholine nicotinic receptors concentrated at the motor end plate. This cell polarity, defined by an asymmetry in the protein composition of the baso-lateral and apical membrane areas as

TABLE 3.1 Cell Membrane Composition (%)

Membrane	Carbohydrate	Lipid	Protein
RBC	8	43	49
Myelin	3	79	18
Inner mitochondrial membrane	0	24	76

Source: Silverthorn D.U., *Human Physiology: An Integrated Approach*, 5th edition. Benjamin Cummings, San Francisco, CA, 2010.

delineated by thigh junctions, is critical in the development of epithelium and tissue whose major functions include exchanges.

The heterogeneous and asymmetric lipid bilayer forming the cell membrane through a constant and dynamic redistribution of proteins constitutes a *semipermeable barrier* separating two compartments with different chemical and ionic compositions. These differences in charges and concentrations associated with the asymmetrical membrane proteins generate *electrochemical gradients*, that is, differences in the net electrical charge and concentrations of a given solute inside versus outside the cell (Figure 3.2). Although both gradients are intertwined, each can act independently of the other.

3.6.2 Molecule Movements

Molecule movements between two compartments separated by a plasma membrane or an epithelium use *pericellular transport* (between cells), *transcellular transport* (through the membrane), and *endocytosis* and/or *exocytosis* mechanisms. In pericellular transport, molecules move through an epithelium using spaces between cells. During endocytosis and exocytosis, physical distortions of the cell membrane through vesicle creation or fusion allow the movement of molecules into or out of the cell without transport through the membrane (Figure 3.2).

Transcellular transport depends on multiple parameters, including the hydrophobicity of the molecule and the density of transport proteins. Small molecules use *diffusion* whereas larger molecules require specific *transport proteins*. The cell membrane is highly permeable to most *hydrophobic* molecules or lipid-soluble solutes such as alcohol, vitamins A and E, and steroids. In contrast, the *permeability of water-soluble* or *hydrophilic molecules* is limited to very *small molecules*, including water and hydrophobic molecules with specific carriers. Most membranes are impermeable to water-soluble molecules above 200 Da. Ions are relatively insoluble because of their charge in lipids; therefore, membranes are poorly permeable to ions. Ion diffusion occurs mostly through *ion channels*. Ion channels span the membrane and are specific to an ion or class of ions, mostly depending on size and charge. Amino acids and sugars also require specific *transporters* present in the cell membrane.

3.6.3 Diffusion

Small molecules, gases (O_2, CO_2, and NO), and molecules soluble in polar solvents diffuse through the cell membrane. *Diffusion* is driven by a gradient and continues until equilibrium. The net

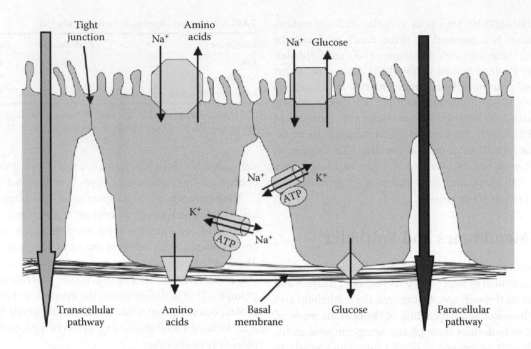

FIGURE 3.2 Transport through the cell membrane and epithelium.

diffusion rate (*J*) is proportional to the coefficient of diffusion (*D*), the surface area (*A*), the thickness of the membrane (Δ*x*), and the gradient or difference in concentration (Δ*C*). The net diffusion rate is defined by Fick's law:

$$J = -DA\frac{\Delta C}{\Delta x}. \quad (3.2)$$

The *diffusion time* is a function of the thickness of the membrane and the permeability coefficient (Einstein relation). The diffusion time (*t*) is a function of average diffusion distance (Δ*x*) and the coefficient of diffusion:

$$t = \frac{(\Delta x)^2}{2D}. \quad (3.3)$$

For small water-soluble molecules with a coefficient of diffusion equal to 10^{-5} cm²/s, the diffusion times are 0.5 m, 50 m, 5 s, 8.3 min, and 14 h for membrane thicknesses of 1, 10, 100, 1000, and 10,000 μm, respectively. The diffusion time (Table 3.2) is proportional to the diffusion coefficient (*D*), which itself is proportional to the speed of movement of the molecule in a given medium: if the molecule is large and the medium is viscous, *D* is small.

For small molecules, *D* is inversely proportional to the molecular weight (MW) in dalton: *D* = 1/MW. For larger spherical molecules, the equation of Stokes–Einstein approximates the

TABLE 3.2 Molecule Size, Coefficient, and Diffusion Time

Molecule	Radius (nm)	*D*	Time (s)
Oxygen	0.2	900	0.001
Sucrose	0.5	400	0.003
Insulin	1.4	160	0.01
Ribosome	10	22	0.06

coefficient of diffusion (Equation 3.4), taking into account the gas constant (*R*), the absolute temperature (*T*), the number π, the Avogadro number (*N*), the solvent viscosity (η), and the radius of the molecule (*r*):

$$D = \frac{RT}{6N\pi r\eta} = 1.38 \times 10^{-23}\frac{T}{\pi r\eta}. \quad (3.4)$$

3.6.4 Protein-Mediated Membrane Transport

Movements of large molecules require intrinsic specific *carriers* or *channels*. Through *conformational changes*, channels form gates, allowing the passage of molecules. Transporter recognition is ligand specific but generally not absolute, and related molecules can compete for or inhibit transport. These channels can be either *voltage gated* or *ligand gated*: the former is activated by difference in the transmembrane voltage difference and the latter by binding to its specific ligand.

Mediated membrane transport is more rapid than simple diffusion, can saturate, and is chemically specific and sensitive to competition. The transport rate (*J*) for a given molecule (S) is defined by its maximum transport rate (*J*ₘ), the Michaelis constant (*K*ₘ) and the concentration of the molecule ([S]), as described by the Michaelis–Menten equation:

$$J = \frac{J_m[S]}{K_m + [S]}. \quad (3.5)$$

The transport through a carrier is limited by the speed and capacity of each carrier with a conformational change of 10^2–10^4 solute molecules/s. For ion channels, through an open channel, ions move at 10^7–10^8 ions/s. In *facilitated transport*, no energy is involved, whereas *active transport* requires energy

(ATP mostly). Facilitated transport benefits from existing charge or concentration gradients to move molecules (e.g., GLU2 transporter and Na$^+$ gradients). Active transports promote the movement against concentration or electrochemical gradients using energy, mostly ATP, to cycle between its conformational states. For example, the Ca^{2+} ATPases move two Ca^{2+} from the lumen to the sarcoplasmic reticulum per ATP and the Na$^+$/K$^+$ ATPase, present in the plasma membrane of the cells, moves three Na$^+$ out of the cells and two K$^+$ into the cell per ATP. Because of the K$^+$ and Na$^+$ concentrations in and out of the cells, these tend to move passively toward equilibrium, and the steady state for these ions is maintained by the constant activity of the Na$^+$/K$^+$ ATPases.

3.7 Osmotic Pressure and Cell Volume

The cell volume is directly related to the internal pressure, and hence the osmotic pressure is also affected by the cell volume.

3.7.1 Osmotic Pressure

Osmosis is defined as the flow of water across a semipermeable membrane (i.e., permeable to water only) from a compartment with a low solute concentration to a compartment with a high solute concentration. *Osmotic pressure* is the pressure that is sufficient to prevent water from entering the cell. Osmotic pressure (Π) is directly associated with the number of ions formed from the dissociation of a solute (i), the molar concentration of the solute (c), and the osmotic coefficient (ϕ) and can be calculated by van't Hoff's law:

$$\Pi = RT(i\phi c). \tag{3.6}$$

Osmotic pressure is a function of the concentration of solute present on either side of the membrane, and the concentration of solute also increases the boiling point and lowers the freezing point. Osmotic pressure (Π) is a function of the concentration of solute present on either side of the membrane. Since the concentration of solute is proportional to the solute freezing point, the osmotic pressure can also be estimated based on the freezing point depression (ΔT_f):

$$\Pi = RT(\Delta T_f/1.86). \tag{3.7}$$

where ΔT_f is the freezing point depression. Two solutions separated by a semipermeable membrane are *isoosmotic* (have equal osmotic pressures), *hyperosmotic* (A hyperosmotic compared to B), or *hypoosmotic* (B compared to A). Osmotic coefficients have been calculated (Table 3.3).

3.8 Tonicity

The plasma membrane of animal cells is relatively impermeable to many solutes but highly permeable to water. Therefore, increase in the osmotic pressure of the extracellular fluid (ECF)

TABLE 3.3 Osmotic Coefficients

Compound	i	MW (Da)	ϕ
NaCl	2	58.3	0.93
KCl	2	74.6	0.92
HCl	2	36.6	0.95
CaCl$_2$	3	111.0	0.86
MgCl$_2$	3	95.2	0.89
Glucose	1	180.0	1.01
Lactose	1	342.0	1.01

Source: Lifson N. and Visscher M.B., In O. Glasser (Ed.), *Medical Physics*, Vol. 1, St. Louis, MO, 1944.

leads to water leaving the cells through osmosis, resulting in cell shrinking. In contrast, if the ECF is diluted, water enters the cells, resulting in cell swelling. Swelling activates channels, increasing efflux of K$^+$, Cl$^-$, and the water that follows by osmosis returns cells to normal size. Both cell shrinking and swelling will continue until the osmotic pressures on both sides are equal or *isoosmotic*.

In vivo, protein concentration is the most important parameter generating oncotic pressure, which contributes to the net flow of a given solute. Both the shrinking and swelling drastically impair cell function and potentially, in extreme cases, its survival. In living organisms, cells are suspended in a mixture of *permeant* and nonpermeant solutes. In those conditions (1) the steady volume of a cell is determined by the concentration of nonpermeant solutes in the ECF, (2) permeant solutes generate only transient alterations of the cell volume, and (3) the greater the cell permeability to a permeant solute, the faster the time course to transient change.

3.9 Electrical Properties of Cell Membranes

The electrical properties of the cell membrane are derived from their insulator potential associated with the composition, especially the amount of lipid present. For example, myelin produced by Schwann cells leads to the insulation of axons, with multiple layers of the cell membrane preventing loss of electrical charges. The electrical properties of the cells are also a function of the constantly maintained disequilibrium of the ions generated by the tight control of ion movements and charges present on either side of the membrane. Additionally, cell transport through channels for ions or carriers for proteins also affects the electrical charges present on each side of the cell membrane, leading to alterations in the local membrane potential.

3.9.1 Forces Acting on Ion Movements

Several forces act on the components surrounding the membrane. The two main categories are electrical forces and chemical gradient forces.

In living animal cells, a comparison of the composition of the cytosol and the ECF underlines the presence of proteins (generally negatively charged) and K$^+$ at high concentrations inside the cell, whereas Ca^{2+}, Na$^+$, and Cl$^-$ concentrations are higher in the ECF. *Permeant* molecules, including some ions, move continuously in or out of the cell through leaky *channels* of the cell membrane following electrical and chemical gradients. The difference in charge between the inside and outside of a cell creates a *membrane potential* or the amount of energy (electrical) associated with the electrochemical gradient present. The net gradient for a given ion and cell remains stable because ATP-dependent ion pumps, especially the Na$^+$/K$^+$ ATPase, continuously and actively maintain these equilibriums.

3.9.2 Distribution of Permeable Ions

Taking into account the ions that cannot diffuse, the distribution of permeable ions is predicted by the Donnan–Gibbs equilibrium: in the presence of a nondiffusible ion (e.g., protein), a diffusible pair of ions of the same valence distributes to generate equal concentration ratios, for example,

$$[K^+]_{in} \times [Cl^-]_{in} = [K^+]_{out} \times [Cl^-]_{out}. \tag{3.8}$$

The overall membrane potential at any time is a function of the distribution (inside versus outside) and membrane permeability to Na$^+$, K$^+$, and Cl$^-$ (Figure 3.3).

The Donnan–Gibbs equilibrium explains the critical role of the Na$^+$/K$^+$ ATPase pump in constantly removing Na$^+$ ions out of the cells to maintain osmotic pressure and cell volume. It also

clarifies the electrical difference generated by the asymmetric distribution of permeable ions between the intracellular and extracellular compartments at equilibrium. Along the membrane on the extracellular side, the charges created by Cl$^-$ are balanced by the K$^+$ ions that are present inside the cell. This effect is also critical in the movement of ions across the capillary wall mostly generated by the higher protein concentration in the plasma compared to the ECF.

3.9.3 Membrane Potential

The relationship between the chemical and electrical forces acting on ions across the plasma membrane and the generation of the resting membrane potential is defined by taking into account the ion valence (Z_{ion}) and ECF ([ion]$_{out}$) and (intracellular fluid) ICF ([ion]$_{in}$) concentrations as described in the Nernst equation:

$$E_{ion} = \frac{RT}{FZ_{ion}} \log_{10}\left(\frac{[ion]_{out}}{[ion]_{in}}\right), \tag{3.9}$$

where R is the gas constant, F is the Faraday constant, and T is the absolute temperature. At 37°C, the equation can be simplified to $E_{ion} = 61.5 \log_{10}([ion]_{out}/[ion]_{in})$. For Cl$^-$, with intra- and extracellular Cl$^-$ concentrations of 9.0 and 125.0 mM, $E_{Cl} = -70$ mV, a value identical to the one measured experimentally. In neurons, calculated E_K^+ (-90 mV) differs from measured E_K^+ (-70 mV). Similarly, the difference between calculated E_{Na}^+

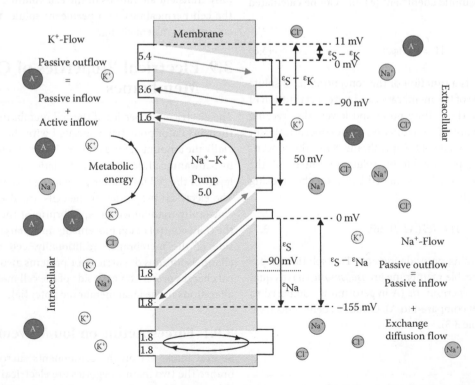

FIGURE 3.3 Ion exchanges and the creation of an electrochemical gradient across the cell membrane. The influence of protein (A$^-$), Cl$^-$, K$^+$, and Na$^+$ extracellular and intracellular concentrations and the role of Na$^+$/K$^+$ ATPase active transport roles in the generation of the electrochemical gradient are represented.

(+60 mV) and measured E_{Na}^+ suggests that other ions may play a role in the equilibrium potential in those cells. These differences from the optimal membrane potentials of each of these ions are actively maintained through the action of the Na$^+$/K$^+$ ATPases pumping three Na$^+$ out and two K$^+$ per ATP molecule.

The K$^+$ efflux is counterbalanced by the electrical gradient of more negative charges (the bulk of which are proteins) inside the cell. The constant activities of the Na$^+$/K$^+$ ATPases pumping K$^+$ inside and removing Na$^+$ prevent these ions from reaching the equilibrium associated with their respective electrochemical gradients.

3.10 ATPases

ATPases are evolutionary conserved proteins with three major types: P, V, and F. The P type involves a phosphorylated intermediate and includes Na$^+$/K$^+$ ATPases and Ca^{2+} ATPases. Mostly present on cell organelles (storage granules and lysosomes), the V type accumulates H$^+$ in vesicle lumen. Most cell membranes also contain Na$^+$/H$^+$ exchangers to prevent the acidification of the cytosol becoming active when the pH of the cytosol decreases, with Na$^+$ moving following its electrochemical gradient in exchange with the movement of H$^+$ out of the cell.

In contrast to P and V ATPases, which consume ATP, the F type represented by ATP synthase of the inner mitochondrial membrane is a major source of ATP. ATP production depends on the oxygen conditions. In anaerobic conditions, one glucose molecule produces two pyruvate molecules transformed in lactate, yielding a net energy of two ATPs. In aerobic conditions, pyruvate molecules enter the citric acid cycle in the mitochondria and through oxidative phosphorylation yield up to 30–32 ATP molecules.

3.10.1 Role of ATPases

In excitable cells, following an action potential in which Na$^+$ and K$^+$ ions move in and out of the cell respectively, the cell repolarizes with an efflux of K$^+$ ions. K$^+$ channels are slow to close and the K$^+$ efflux generates a hyperpolarization of the cell membrane. The equilibrium is re-established by the activity of the sodium/potassium ATPase pump moving three Na$^+$ ions out and allowing two K$^+$ ions in through active transport through a conformational change.

The efflux of Na$^+$ provides the driving force for multiple facilitated transport mechanisms, including glucose, amino acid membrane transport, and creates an osmotic gradient that promotes the absorption of water. In the enterocytes of the GI tract, on the baso-lateral surface of the cell, Na$^+$ is pumped out of the cell through the activity of Na$^+$–K$^+$ ATPase pumps creating a gradient that favors the influx of Na$^+$ from the apical side of the cell.

3.10.2 Regulation of Na$^+$/K$^+$ ATPase Activity

The activity of the ATPase pump is endogenously down-regulated through increases in cyclic adenosine monophosphate (cAMP) associated with G protein coupled receptor activations and up-regulated through decreases in cAMP by ligands leading to G protein coupled receptor inhibition. Thyroid hormones, insulin, and aldosterone increase the expression of Na$^+$/K$^+$ ATPase pumps and therefore the activity. In contrast, in the kidney, dopamine induces the phosphorylation of the Na$^+$/K$^+$ ATPase pump, inhibiting its activity.

3.11 Cell Membrane and pH Regulation

Mechanisms in cells allow the regulation of H$^+$ intracytoplasmic concentrations including through the activity of the Na$^+$/H$^+$ ATPase, which prevents H$^+$ increases into the cytosol by pumping H$^+$ in specific cell compartments. Intracellular H$^+$ concentration is influenced by ECF and plasma H$^+$ concentration.

In the plasma, the H$^+$ concentration is very low compared to other ions (~0.0004 mEq/L) and is expressed as the negative log of the H$^+$ concentration (or pH). The plasma pH ranges from 7.38 to 7.42, with extreme acidosis at 7.0 and extreme alkalosis at 7.7. Throughout the body, pH values are very variable with, for example, gastric HCl (0.8), urine (as low as 4.5), and pancreatic juice (8.0). Because pH homeostasis is essential to organism survival through enzymatic and membrane and capillary exchanges, the pH is tightly monitored through central and peripheral sensors and regulated by buffers, the lung, and kidney activities. Like other chemicals, the organism has sensing mechanisms with properties comparable to those observed in smell and taste senses, including pH sensors, and a constant feedback regulation and monitoring of the ECF and plasma pH.

3.11.1 pH Sensors

Chemoreceptors sensing H$^+$ concentrations in the plasma and cerebrospinal fluid, respectively, are located peripherally in the aortic arch (aortic bodies) and at the bifurcation of the internal and external carotid in the neck (carotid bodies) and centrally in the brain medulla. The blood–brain barrier is poorly permeable to H$^+$ but allows CO$_2$ diffusion. The addition of CO$_2$ displaces the equilibrium bicarbonate hydrogen toward the formation of more H$^+$, increasing the pH of the cerebrospinal fluid (CSF):

$$CO_2 + H_2O \iff H_2CO_3 \iff H^+ + HCO_3^-. \quad (3.10)$$

This decrease in pH is sensed by chemoreceptors that stimulate lung ventilation to bring H$^+$ and CO$_2$ concentrations within range.

3.11.2 pH Regulation

In mammalians, pH regulation is achieved through (1) buffers and the activities of (2) the lung and to a lesser extent (3) the kidneys. Buffers are molecules that combine with H$^+$, neutralizing its effects. The presence of buffers moderates greatly the addition of H$^+$ to a solution. Buffers such as H$_2$PO$_4^-$ and HCO$_3^-$ are present in cells and ECF, respectively. Also, an increase in plasma CO_2 is associated with an increase in H^+ in the CSF sensed in the medulla, leading to a rapid increase in lung ventilation to remove CO_2 and maintain ECF pH. If despite buffer effects and ventilation modulations, pH acidification or alkalination persist, the kidney can secrete or absorb H^+ ion and HCO_3^- ion. Following conversion of CO_2 into HCO_3^- by carbonic anhydrase in proximal tubule cells, HCO_3^- is reabsorbed and H^+ is secreted. Alternatively, H^+ is secreted as ammonium ion NH^{4+}. In the distal nephron, intercalated cells or either type A or B functioning during acidosis or alkalosis excrete H^+ or HCO_3^- and K^+, respectively (see Equation 3.10).

3.11.3 Gas Exchanges and pH Regulation

Cells of the organism receive signaling molecules, nutrients, and O_2 through the cardiovascular system and the ECF. As described above, removal of CO_2 produced by cellular metabolism is critical to the maintenance of a pH compatible with normal cell function. CO_2 is carried in the blood as (about 70%) bicarbonate ions HCO_3^- (see Equation 3.10) by a carbonic anhydrase in the red blood cells (RBCs), 7% is dissolved in the plasma, and 23% is bound to hemoglobin at a site other than O_2. The binding to CO_2, however, decreases hemoglobin O_2 binding, in effect allowing more O_2 release in a region with high CO_2 concentrations. This effect (Bohr effect) is observed when increases in the partial pressure of CO_2 or lower pH values result in the off-loading of oxygen from hemoglobin.

3.12 Summary

In multicellular organisms, cell interactions depending on cell junctions and adhesions to the ECM modulate individual cell function and epithelium membrane permeability. In addition to diffusion, molecules are transported through protein carriers with or without energy requirements. The chemical and electrical inbalance between compartments separated by the lipid bilayer cell membrane is actively maintained by ATPases. These electrical and chemical disequilibria generate electrochemical gradients and the membrane potential critical in cell functions.

Additional Reading

1. Yeagle P.L., *The Structure of Biological Membranes*, 2nd edition. CRC, Boca Raton, FL, 2004.
2. Alberts B., Johnson A., Lewis J., Raff M., Roberts K., and Walter P., *Molecular Biology of the Cell*, 5th edition. Garland Science, Oxford, UK, 2008.
3. Landowne D., *Cell Physiology*, 1st edition. McGraw-Hill Medical, New York, NY, 2006.
4. Ethier C.R. and Simmons C.A., *Introductory Biomechanics: From Cells to Organisms*, 1st edition. Cambridge Texts in Biomedical Engineering, New York, NY, 2007.
5. Barrett K.E., Brooks H., Boitano S., and Barman S.M., *Ganong's Review of Medical Physiology*, 23rd edition. McGraw-Hill Medical, New York, NY, 2009.
6. Silverthorn D.U., *Human Physiology: An Integrated Approach*, 5th edition. Benjamin Cummings, San Francisco, CA, 2010.
7. Koeppen B.M. and Stanton B.M., *Berne and Levy Physiology*, 6th edition. St. Louis, MO, 2008.
8. Lang F., Vallon V., Knipper M., and Wangemann P., Functional significance of channels and transporters expressed in the inner ear and kidney. *Am. J. Physiol. Cell Physiol.*, 293: C1187–208, 2007.
9. Lifson N. and Visscher M.B., Osmosis in living systems in O. Glasser (Ed.), *Medical Physics*, Vol. 1, St. Louis, MO, 1944.

4

Cellular Thermodynamics

Pavlina Pike

4.1 Introduction

Every living organism is built with a network of billions of cells, which communicate among each other and with the surroundings by the controlled exchange of chemical and electrical signals. These signals are molecules and ions that carry information used by the cells to perform certain tasks, such as the activation of enzymes or genes, cellular proliferation, and death. A major role in this exchange is played by the cell membrane and more specifically by structures called ion channels and gap junctions. The purpose of this paragraph is to introduce ion transport in cells from the point of view of thermodynamics. We will start with the basic terms and laws that are the building blocks of thermodynamics. Attention will be given to excitable cells, that is, cells that respond to electrical signals, especially heart muscle cells.

4.2 Energy

Energy is needed to sustain life in all its forms. The most powerful natural source of energy for the Earth is the Sun. The amount of energy that reaches the Earth is 1366 W/m^2, which includes light of all wavelengths irradiated by the Sun.[1] This energy is used in every aspect of life to sustain it. For example, visible light is essential in photosynthesis in plants, activation of photosensitive cells in the eye, regulation of the circadian cycle by the hypothalamus, etc. Energy is also used for electrical signal conduction and neurotransmitter release in nerve cells: importing/exporting and processing ions and molecules in cells. Energy is stored in cells (potential energy) in the form of bonds between the phosphate groups in the adenosine triphosphate (ATP) molecule. If one of these bonds is broken, energy is released and then used for different endergonic (energy gaining) reactions in the cell. But what is energy? So far there is no clear definition. *Energy* is most commonly defined as the ability of an object to do work.

It can be either supplied or taken away from the system in order for the processes (changes) to occur. The energy transfer to or from the system is called *work*.

The sum of the kinetic, potential, chemical, nuclear, and so on energies of particles in a system is called the *internal energy (U)* of a system. For simplicity, we will consider the internal energy to only consist of the potential energy of molecular bonds and the kinetic energy of the microscopic motion of atoms. It is also described as the energy that is required for the system to be created assuming constant temperature and volume.

The laws that govern energy transfer, or the conversion of energy from one form to another, are described by thermodynamics.

4.3 Laws of Thermodynamics

4.3.1 The First Law of Thermodynamics

The first law of thermodynamics states that the change in the internal energy of a system is the net result of the heat (Q) added to the system and the work (W) done by the system on the surroundings, as given by Equation 4.1:

$$\Delta U = \Delta Q - \Delta W. \tag{4.1}$$

The work done can be mechanical (W_M),

$$W_M = P\Delta V + V\Delta P, \tag{4.2}$$

and/or electrical (W_e),

$$W_e = -nF\Delta E, \tag{4.3}$$

where n is the number of transferred charges, F is the Faraday constant equal to 96,485 C/mol, and ΔE is the maximum potential difference due to the motion of charges. In other words, this is a statement of *conservation of energy*.

4.3.1.1 Enthalpy

The sum of the internal energy U and the work done by the reacting molecules to push the surroundings away, that is, to increase the volume of the system at constant pressure, is called *enthalpy* (H), expressed by

$$\Delta H = \Delta U + P\Delta V. \tag{4.4}$$

The change of enthalpy also gives the heat of reactions in chemistry. It is also calculated as the sum of the energies required to break old bonds minus the energies released from the formation of new bonds. The change in enthalpy is negative ($\Delta H < 0$) if the reaction is exothermic and positive ($\Delta H > 0$) if the reaction is endothermic. In the case of constant pressure, the change in enthalpy is simply equal to the heat added to the system (Q).

4.3.1.2 Entropy

Entropy (S) is most commonly described as a measure of the order in a system. Ordered systems have low entropy because the probability that a system is in an ordered state is low. When heat is added to the system it causes particles to move faster, and if the temperature is high enough bonds will be broken and the new state will be less ordered than before. Consider, for example, melting ice. The amount by which entropy has increased is given by

$$\Delta S = \frac{\Delta Q}{T}. \tag{4.5}$$

4.3.2 The Second Law of Thermodynamics

The second law of thermodynamics states that if a system is isolated (no energy is added to it), its entropy will only increase with time. Consider a plant that has been cut and left without water in a closed container. The plant will eventually decay, that is, its entropy has increased. Life requires input of energy to sustain order. Such processes are called *endergonic*. The energy of the final state is higher than the energy of the initial state.

4.3.2.1 Gibbs Free Energy

Gibbs free energy (G) is the energy that determines whether a reaction will be spontaneous or not. It is defined as the change in enthalpy minus the temperature times the entropy, shown in Equation 4.6:

$$\Delta G = \Delta H - T\Delta S = \Delta U + P\Delta V - T\Delta S. \tag{4.6}$$

The free energy for unfavorable (not spontaneous) reactions is positive ($G > 0$). Reactions will be spontaneous if the free energy is negative ($G < 0$). These reactions are classified as endergonic and exergonic, respectively.

During redox reactions, the Gibbs free energy is equal to the maximum electric work:

$$\Delta G^{o} = -nF\Delta E^{o}, \tag{4.7}$$

where the superscript "o" indicates standard conditions (25°C temperature and 1 atm pressure) and ΔE is the electric potential. It can be shown that at any temperature, Equation 4.8 represents the energy balance:

$$\Delta G = \Delta G^{\circ} + RT \ln Q_r, \tag{4.8}$$

where R is the gas constant ($R = 8.314$ J mol^{-1} K^{-1}) and Q_r is the reaction quotient.

As a result of the difference in ionic concentrations in the intra- versus extracellular space in biological cells, a potential difference is generated that can be calculated by combining the previous two equations in the following way:

$$-nF\Delta E = -nF\Delta E^{\circ} + RT \ln Q_r. \tag{4.9}$$

The result is called the Nernst equation, which is also referred to as the Nernst potential:

$$\Delta E = \Delta E^{o} - \frac{RT}{nF} \ln Q_r. \tag{4.10}$$

Therefore, Equation 4.10 gives the potential difference generated in a galvanic cell as a function of the ion concentrations in both compartments. This is called the *resting potential* during equilibrium. In this equation, Q_r represents the ratio of the extracellular to intracellular concentrations. This equation only represents a simple case in which the membrane is only permeable to one type of ion (e.g., only K^+ ions).[2] A more realistic expression for Q_r will be given later.

Cells are surrounded by membranes, which control the flow of ions in and out of the cell by the use of voltage gated ion channels and gap junctions.

4.4 Ion Channels and Gap Junctions

It has been found that there are passages, called *gap junctions*, connecting cardiac cells that are insulated from extracellular space and are wide (with a diameter of about 16 Å).[3] These channels have a very low resistance and are considered to have fixed anions that will select incoming ions based on their electronegativity and size. The other type of structure that is responsible for the flow of charges in the cell is called *ion channels*. The channels are gates that can transmit certain ions or molecules when activated by changes in the potential difference across the membrane or activated by extracellular or intracellular transport molecules (also called ligands) that attach to it.

The membrane potential depends directly on the concentrations of the ions on both sides. For example, let us assume that only K^+, Na^+, and Cl^- are allowed to flow through the membrane.

The expression for the membrane resting potential can be derived from the Nernst equation (Equation 1.10).[4] The resultant equation is also called the Goldman voltage equation, given by[2]

$$V_{\mathrm{m}} = \frac{RT}{F} \ln \frac{P_{\mathrm{Na}}[\mathrm{Na}^+]_{\mathrm{out}} + P_{\mathrm{K}}[\mathrm{K}^+]_{\mathrm{out}} + P_{\mathrm{Cl}}[\mathrm{Cl}^-]_{\mathrm{in}}}{P_{\mathrm{Na}}[\mathrm{Na}^+]_{\mathrm{in}} + P_{\mathrm{K}}[\mathrm{K}^+]_{\mathrm{in}} + P_{\mathrm{Cl}}[\mathrm{Cl}^-]_{\mathrm{out}}}. \tag{4.11}$$

In this equation $P[Na^+]$ is the permeability constant for the Na^+ ion. A typical value for the resting potential is -90 mV.[5] It is negative because cells are more negative relative to the surrounding medium. Cells that are excitable have the ability to rapidly reverse the potential, causing it to be slightly positive. The potential that is generated in this process is known as the *action potential*.

4.5 Action Potential

The resting state of a membrane is the one in which the inside is more negative than the outside. As mentioned in the above section, a typical value is about -90 mV. Excitation from other parts of the membrane can trigger the opening of the sodium ion channels, which will cause Na^+ ions to flow into the cell. As a result the membrane potential will become more positive. This process is called *depolarization*. Once the potential reaches a certain *threshold* value (Figure 4.1), more sodium channels start opening very fast. This corresponds to the part of the potential curve on the picture where the rise of the potential is very steep. The total change in the potential is around 100 mV. When the potential reaches its maximum value, a number of Na^+ channels will begin to close and K^+ channels will open. The potential starts falling back toward its original value—repolarization. It will fall slightly below the resting value (*hyperpolarization*) but it will recover. This is, in short, a description of the action potential in a neuron. It is very similar to the one in a heart muscle cell, except that the action potential of a ventricular myocyte lasts longer and has a "plateau" area in which the potential stabilizes for a short period of time. This is due to the balance between the Ca^{2+} current flowing into the cell and the K^+ current flowing out (Figure 4.3).

4.6 Models of Ionic Current

After performing their famous experiments on a giant *nerve* fiber, A. L. Hodgkin and A. F. Huxley modeled their results comparing the cell membrane as a simple circuit in which there are three conductive elements (ion channels) connected in parallel (Figure 4.2).[6] Sodium, potassium, and "leakage" currents flow through each one of these. The conductance of each channel varies with time, as observed during the experiments, but the potential difference (E_{Na}, E_K, and E_L) for each one stays the same. The ionic currents are expressed in terms of the sodium, potassium, and leakage conductances, as shown by Equation 4.12:

$$I_{Na} = g_{Na}(V - V_{Na}),$$
$$I_K = g_K(V - V_K), \qquad (4.12)$$
$$I_L = g_l(V - V_L),$$

where V_{Na}, V_K, and V_L are the differences between the resting potential and the equilibrium potential for each ion. V is the difference between the measured value of the potential and the absolute value of the resting potential. As previously mentioned,

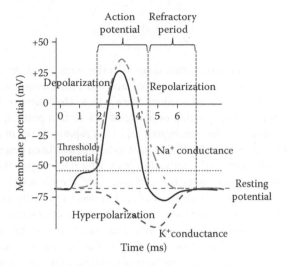

FIGURE 4.1 Action potential of a ventricular myocyte. Action potential on a 100 µm × 100 µm × cardiac tissue.

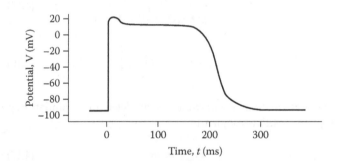

FIGURE 4.2 Action potential of an excitable cell.

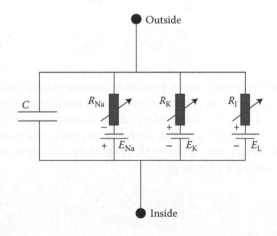

FIGURE 4.3 Circuit model of a squid axon membrane. (Modified from Hodgkin, A. L. and A. F. Huxley, *Journal of Physiology* 117: 500–544, 1952.)

the conductances are functions of both time and voltage. The *conductance of the potassium* ions was modeled by Hodgkin and Huxley, as represented by Equations 4.13 and 4.14:

$$g_K = \bar{g}_K n^4, \qquad (4.13)$$

$$\frac{dn}{dt} = \alpha_n(1-n) - \beta_n n, \tag{4.14}$$

where $\bar{g}_K$ is a constant that is equal to the maximum value of g_K and has units of conductance/cm^2 and n is a dimensionless variable that can only take values from 0 to 1. The latter has been described as the "portion of particles in a certain position (e.g., at the inside of the membrane) and $1 - n$ represents the portion that is somewhere else."[6] This parameter has also been described in the literature as the activation parameter representing the ion channels. It represents the probability of the channels being open if there are a large number of channels.[2] The coefficients α_n and β_n are the rates of ion transfer in both directions: from outside and from inside, respectively. They vary with voltage but not with time and have units of [time]$^{-1}$. The functions α_n and β_n were obtained by fitting the experimental data for n, as shown in Equations 4.15 and 4.16:

$$\alpha_n = 0.01 \frac{(V+10)}{\left[\exp\left((V+10)/10\right) - 1\right]}, \tag{4.15}$$

$$\beta_n = 0.125 \exp\left(\frac{V}{10}\right). \tag{4.16}$$

The *sodium* (Na) *conductance* is modeled in a similar way, Equations 4.17 through 4.19:

$$g_{Na} = m^3 h \bar{g}_{Na}, \tag{4.17}$$

$$\frac{dm}{dt} = \alpha_m(1-m) - \beta_m m, \tag{4.18}$$

$$\frac{dh}{dt} = \alpha_h(1-h) - \beta_h h, \tag{4.19}$$

where m represents the proportion of activating molecules inside the cell, h is the proportion of inactivating molecules on the outside, α_m and β_m are the transfer rates for the activating molecules inside the cell, and α_h and β_h are the same rates for the inactivating molecules. The equations that best fit the data are expressed by Equations 4.20 through 4.23[6]:

$$\alpha_m = 0.1 \frac{(V+25)}{\left[\exp\left((V+25)/10\right) - 1\right]}, \tag{4.20}$$

$$\beta_m = 4 \exp\left(\frac{V}{18}\right), \tag{4.21}$$

$$\alpha_h = 0.07 \exp\left(\frac{V}{20}\right), \tag{4.22}$$

$$\beta_h = \frac{1}{\left[\exp\left((V+30)/10\right) + 1\right]}. \tag{4.23}$$

Therefore, the total ionic current can be expressed as the sum of all the ionic currents shown by Equation 4.24:

$$I = I_K + I_{Na} + I_L,$$

$$I = C_M \frac{dV}{dt} + \bar{g}_K n^4(V - V_K) + \bar{g}_{Na} m^3 h(V - V_{Na})$$

$$+ \bar{g}_L(V - V_L). \tag{4.24}$$

Other models of the action potential include the same type of differential equations but introduced different formulations of the ionic current. The action potential of heart tissue is more complex than that for nerve fibers. That requires the equations and parameters to be adjusted in order to reflect those differences. For example, G. Beeler and H. Reuter modeled the action potential for *mammalian ventricular cardiac* muscles by representing the potassium current as composed of both time-independent outward potassium current, I_{K1}, and a time- and voltage-dependent component, I_{Kx}. The first is also called the "outward rectification" current, and the latter is the inward rectification current.[7] In other words,

$$I = I_{K1} + I_{Kx} + I_{Na} + I_{Ca} - I_S. \tag{4.25}$$

Here I_S is the slow current flowing into the cell that is mainly carried by calcium ions.

Other, more recent models include even more detailed descriptions of the current. When describing action potentials and pacemaker activity, DiFrancesco and Noble suggested that in addition to the currents mentioned above, the total current includes the hyperpolarizing-activated current I_f, the transient outward current I_{to}, the background sodium and calcium currents $I_{b,Na}$ and $I_{b,Ca}$, the Na–K exchange pump current I_p, the Na–Ca exchange current I_{NaCa}, and the second inward current I_{si} given by Equation 4.26[8]:

$$I_{tot} = I_f + I_K + I_{K1} + I_{to} + I_{b,Na} + I_{b,Ca} + I_p$$

$$+ I_{NaCa} + I_{Na} + I_{Ca,f} + I_{Ca,s} + I_{pulse}. \tag{4.26}$$

As more experimental data on the ionic currents became available, the mathematical models became more sophisticated in order to achieve higher accuracy in describing *human atrial* currents. Courtemanche et al.[9] described the membrane currents using the formulations of Luo and Rudy,[10] but adjusting the values for the parameters to fit the action potential for human atrial cells. Their model describes well the variations in Ca^{2+}, Na$^+$, and K$^+$ ion concentrations inside the cell by also including pumps and exchangers. The extracellular concentrations of ions are considered fixed. The total current according to their model is given by

$$I_{tot} = I_{Na} + I_{K1} + I_{to} + I_{Kur} + I_{kr} + I_{ks} + I_{Ca,L} + I_{p,Ca}$$

$$+ I_{NaK} + I_{NaCa} + I_{b,Na} + I_{b,Ca}. \tag{4.27}$$

In this equation I_{Kur} is the ultrarapid delayed rectifier K$^+$ current, I_{kr} is the rapid delayed rectifier K$^+$ current, and I_{Ks} is the slow delayed rectifier K$^+$ current. The model handles the Ca^{2+} ion exchange by representing three calcium ion currents: $I_{Ca,L}$—L-type inward Ca^{2+} current, $I_{p,Ca}$—sarcoplasmic Ca^{2+} pump current, $I_{b,Ca}$—background Ca^{2+} current, and I_{NaCa}—Na$^+$/Ca^{2+} exchange current. Some of the current descriptions are given below.

As in the Luo–Rudy model the expression for the sodium current has an additional parameter, j, called the slow inactivation parameter, as shown in Equation 4.28:

$$I_{Na} = g_{Na}m^3hj(V - V_{Na}). \tag{4.28}$$

The maximum sodium conductance was temperature adjusted ($g_{Na} = 7.8$ nS/pF) to reflect experimental data and also to produce the correct amplitude for the action potential. The expression for the I_{K1} current that best represents current and resistance measurements is given in Equation 4.29 (assuming no temperature dependence):

$$I_{K1} = \frac{g_{K1}(V - V_K)}{1 + \exp[0.07(V + 80)]}. \tag{4.29}$$

Here, the value for g_{K1} was set to be 0.09 nS/pF. The transient outward and ultrarapid are represented in a similar way. The gates o_a and u_a are the activation gates for I_{to} and I_{Kur}, respectively, and o_i and u_i are their inactivation gates, expressed by Equation 4.30:

$$I_{to} = g_{to}o_a^3o_i(V - E_K),$$

$$I_{Kur} = g_{Kur}u_a^3u_i(V - E_K),$$

$$g_{Kur} = 0.005 + \frac{0.05}{1 + \exp\big[(V - 15)/-13\big]}. \tag{4.30}$$

The descriptions of the rest of the currents are given in great detail by the author.[9] The flow of ions across the cell membrane is governed by the Nernst–Planck diffusion equation (Equation 4.31), in which the flux of ions J_K (current/area) is calculated as a function of the concentration of the species of interest C_K and the diffusion constant D_K[11]:

$$\vec{J}_K = -D_K\left(\vec{\nabla}C_K + \frac{C_K}{\alpha_K}\vec{\nabla}V\right), \tag{4.31}$$

where V is the potential due to the electric charge distribution and $\alpha_K = RT/FZ_K$ with Z_K the valence of the ionic species, R the gas constant, F the Faraday constant, and T the absolute temperature.

The diffusion equation can be solved numerically by implementing the Crank–Nicholson scheme expressed by Equation 4.32[12]:

$$\frac{1}{r_a}\left(\frac{\partial^2 V}{\partial x^2} + \frac{\partial^2 V}{\partial y^2} + \frac{\partial^2 V}{\partial z^2}\right) = C_m\frac{\partial V}{\partial t} + I_{Na} + I_{Ca} + I_K + I_{stim}, \tag{4.32}$$

where r_a is the resistance of the intracellular medium, C_m is the capacitance of the cell membrane, I_{stim} is the stimulation current, and I_{Na}, I_{Ca}, and I_K are the sodium, calcium, and potassium currents, respectively. Let V_m^n represent the voltage of the mth spatial element on the grid at the nth iteration (point in time). The second partial derivatives can be rewritten as

$$\frac{\partial^2 V}{\partial x^2} = \frac{1}{2h_x^2}\left[\left(V_{m+1}^{n+1} - 2V_m^{n+1} + V_{m-1}^{n+1}\right) + \left(V_{m+1}^n - 2V_m^n + V_{m-1}^n\right)\right], \tag{4.33}$$

where h_x is the spatial interval along the x-axis. Another approximation has been made for the current as follows:

$$I_m^{n+1} = I_m^n + \frac{dI_m^n}{dV}\left(V^{n+1} - V^n\right) + \cdots. \tag{4.34}$$

After rearranging some terms the diffusion can be factorized, which will help simplify the solution, as shown by Equation 4.35.

$$\frac{1}{(2 + k\Delta t)^2}(2 + k\Delta t - \Delta t\alpha_x)(2 + k\Delta t - \Delta t\alpha_y)$$

$$(2 + k\Delta t - \Delta t\alpha_z) = 2\Delta t\left(\alpha(V)^n - \frac{I_m^n}{C_m}\right), \tag{4.35}$$

where

$$\alpha(V) = \frac{1}{r_a}\left(\frac{\partial^2 V}{\partial x^2} + \frac{\partial^2 V}{\partial y^2} + \frac{\partial^2 V}{\partial z^2}\right), \tag{4.36}$$

$$\alpha_i^n = \frac{1}{h_i}\left(\frac{V_{i+1}^n - V_i^n}{h_iR_{i2}} - \frac{V_i^n - V_{i-1}^n}{h_iR_{i1}}\right), \tag{4.37}$$

$$k = \frac{1}{C_m}\frac{dI_m^n}{dV}. \tag{4.38}$$

In these relations $i = x$, y, or z and h_i is the spatial interval along the x-, y-, or z-axis.

A representative solution to Equations 4.35 through 4.38 for select parameters are presented in Figure 4.4.

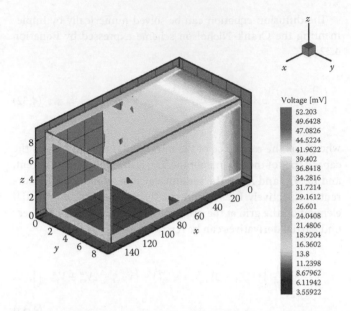

Voltage [mV]

52.203
49.6428
47.0826
44.5224
41.9622
39.402
36.8418
34.2816
31.7214
29.1612
26.601
24.0408
21.4806
18.9204
16.3602
13.8
11.2398
8.67962
6.11942
3.55922

FIGURE 4.4 Numerical solution to Equation 4.35 for a 100 μm × 100 μm × 100 μm dimensional slab of cardiac tissue. Gap junctions have been represented as randomly distributed resistances in the tissue. Parameters used to generate this result are the following: membrane capacitance 6.28 nF/cm; cytoplasmic resistance 79.5 MΩ/cm; gap junction resistance 397.8 MΩ/cm; maximum sodium conductance 15 mS/cm^2; maximum calcium conductance 0.09 mS/cm^2; stimulation current −150 μA/cm^2.[11-14]

References

1. Physikalisch-Meteorologisches Observatorium Davos/ World Radiation Center. 29 Feb. 2008. 10 Apr. 2009; http://www.pmodwrc.ch/

2. Fain, G. L., *Molecular and Cellular Physiology of Neurons.* Cambridge: Harvard, 1999.

3. Zipes, D. P. and M. D. Jose Jalife, Physiology of cardiac gap junction channels. In *Cardiac Electrophysiology*, pp. 62–69. Philadelphia: Saunders, 1990.

4. Hille, B., Ionic selectivity of Na and K channels of nerve membranes. *Membranes* 3: 255–323, 1975.

5. Klabunde, R. E., *Cardiovascular Physiology Concepts.* N.p.: Lippincott, 2004.

6. Hodgkin, A. L. and A. F. Huxley, A quantitative description of membrane current and its application to conduction and excitation in nerve. *Journal of Physiology* 117: 500–544, 1952.

7. Beeler, G. W. and H. Reuter, Reconstruction of the action potential of ventricular myocardial fibers. *Journal of Physiology* 268: 177–210, 1977.

8. DiFrancesco, D. and D. Noble, A model of cardiac electrical activity incorporating ionic pumps and concentration charges. *Philosophical Transactions of the Royal Society of London.* Series B. 307: 353–398, 1985.

9. Courtemanche, M., R. J. Ramirez, and S. Nattel, Ionic mechanisms underlying human atrial action potential properties: Insights from a mathematical model. *Am. J. Physiol. Heart Circ. Physiol.* 275.1: H301–H321, 1998.

10. Luo, C. and Y. Rudy, A model of the ventricular cardiac action potential. *Circulation Research* 68: 1501–1526, 1991.

11. Jack, J. J., D. Noble, and R. W. Tsien, *Electric Current Flow in Excitable Cells.* N.p.: Oxford, 1983.

12. Jacquemet, V. Steady-state solutions in mathematical models of atrial cell electrophysiology and their stability. *Mathematical Biosciences* 208.1: 241–269, 2007.

13. Pergola, N. F., Computer modeling of cardiac electrical activity: Methods for microscopic simulations with subcellular resolution. MS thesis. University of Tennessee, 1994.

14. Buchanan, J. W., et al., A method for measurement of internal longitudinal resistance in papillary muscle. *American Journal of Physiology* 251: H210–H217, 1986.

5

Action Potential Transmission and Volume Conduction

Robert Splinter

5.1 Introduction

The generation of the action potential is described in detail in Chapter 3. The depolarization potential has specific values for specific cell types and is always positive: +30 to +60 mV. The rising potential difference initially causes an avalanche effect due to the influence of the increasing potential on the sodium influx, which is self-maintaining. This process continues until the limit has been reached, after which the cell will actively repolarize the membrane potential locally. Meanwhile this depolarization has initiated the same depolarization process in the neighboring section of the cell, causing the depolarization front to migrate along the cell membrane until its physical end. In principle, the polarization front will obey the telegraph equation since the entire length of the membrane is supposed to follow the repetitive pattern outlined in Section 3.5.

Certain cells that have a different mechanism of propagation of the depolarization wave can be recognized. For instance, the action potential propagates along the length of the axon and dendrites of a myelinated nerve cell differently compared to the unmyelinated nerve cell.

Neurons are cells that specialize in the transfer of information within the nervous system. Neurons are excitable structures that are capable of conducting impulses across long distances in the body. They communicate with other cells by secretion of chemicals (neurotransmitter) to induce an action potential in the neighboring cell structure (e.g., the brain). When depolarization passes down a nerve fiber, there is exchange of ions across the membrane, resulting in changes in membrane potential at each point of the axon. The conduction process of an action potential can be compared to the movement of people in a stadium during a "wave." The motion of one person is induced by its direct neighbor; however, no physical longitudinal transportation of mass takes place. The process of "the wave" is illustrated in Figure 5.1. This is physically different from signal transmission in an electrical cord, where ions are passed through the cord and not exchanged with the environment outside it.

5.2 Components of the Neuron

The general structure of the nerve cell is outlined in Figure 5.2. The description of the nerve is as follows: cell body (perikaryon = a round nucleus); single cell with receiving and transmitting lines: dendrites and axons, respectively. The cell body creates the transmitter molecules (neurotransmitter) for communication with neighboring cells at the synapse at the distal end of the axon. The dendrites are specialized features that receive information from other neurons (conduct the receptor signal to the cell body), and the axon is the cell extremity that has voltage gated channels to facilitate the creation and propagation of an action potential. Axons can be unmyelinated or myelinated. Myelin insulates the nerve cell, creating a leap-frog transmission instead of a cascading depolarization (Figure 5.3). Myelinated nerves have a significantly increased speed of conduction of action potentials compared to unmyelinated axons.

FIGURE 5.1   "The wave" performed by employees of The Spectranetics Corporation and students from UCCS in Colorado Springs, CO. No mass is moving; however, energy is transported from right to left in this picture as neighbors move upwards based on the incentive of the person to the left of each individual (for the viewer: right) as the wave moves to the left on the page. This process is very similar to the migration of turning dipoles on the cell membrane. The figure illustrates a monophasic transfer of potential energy from left to right, which will be described later in this chapter.

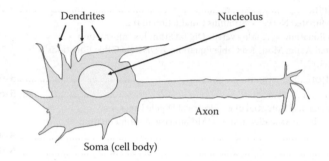

FIGURE 5.2   Schematic description of the construction of a nerve cell.

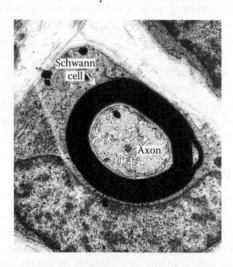

FIGURE 5.3   Electron microscope image of a cross section of a myelinated nerve cell with Schwann cells wrapped around the axon. (Reprinted from CRC-Press: Biomedical Signal and Image Processing, Taylor and Francis Press, 2006. With permission.)

5.3 Operation of a Nerve Cell

Stimulation of the nerve membrane can open ion channels in the membrane. Then Na^+ ions flowing in will depolarize the membrane (movement from -70 mV to say -60 mV) and K^+ ions flowing out of the membrane will hyperpolarize the membrane

(-70 mV to say -90 mV). The spike at one point on an axon causes the adjacent neural membrane to change its permeability so that just beside the area where the spike just took place sodium rushes in and potassium rushes out, resulting in a spike there. This process repeats itself many times, resulting in the propagation of the action potential down the neuron. Since this moving spike is electrochemical rather than strictly electrical, it moves considerably more slowly than electricity in a wire. At body temperature, the depolarization and subsequent repolarization procedure lasts ~0.5 ms at one single point on the membrane.

5.3.1 Unmyelinated Nerve

The transmission of impulses in an unmyelinated membrane propagates as "the wave"; the lag time between the neighboring rise and fall (for the membrane: the transmembrane potential; in the audience: the standing and sitting response) is the refractory period. The refractory period is a function of the initiation of the chemical transfer, which is a voltage gated Na/K pump activation. The chemical migration of the Na and K ions is an active process and hence relatively slow. The propagation of the action potential is shown in Figure 5.4. The accompanying membrane potential gradient resulting from the action potential illustrated in Figure 4.1 is illustrated in Figure 5.5. The expression of the membrane potential at the surface of the body is described later on in this chapter.

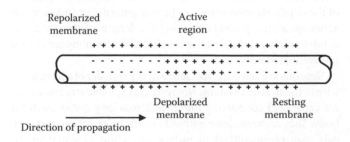

FIGURE 5.4   Schematic description of depolarization propagation in an unmyelinated axon.

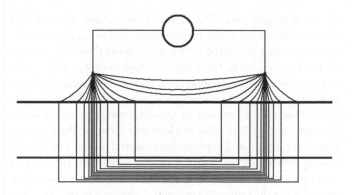

FIGURE 5.5 Electrical potential gradient in a nerve cell surrounding the depolarization region of the unmyelinated axon.

5.3.2 Myelinated Nerve

Certain specialized nerve cells have formed an alliance with a separate cell to create a unique conduction system. In this situation the axon is wrapped in a fatty, white myelin sheath at regular intervals along the length of the axon. The myelin is formed by a specific cell type called glial cells. Oligodendrocytes produce myelin sheaths that insulate certain vertebrate axons in the central nervous system (Figure 5.3). Schwann cells have a similar function in the periphery. Myelin is formed over time and is not fully developed at birth.

5.3.2.1 Transmission of an Action Potential

Electricity in a perfectly conducting wire moves at the speed of light, that is, 2.99792458×10^8 m/s. Neural impulses move at only 1–100 m/s. Impulses travel faster in wide-diameter neurons than in thin neurons, and faster in myelinated neurons (where the impulse skips electrically across the myelin sheath, from one node to another) than in unmyelinated neurons. If you prick your finger with a needle, you will probably feel the pressure before the pain, partly because the pressure signal travels on

neurons that are thicker and more often myelinated compared to the neurons that carry pain information. The membrane underneath the Schwann cell conducts relatively minimally, but if the membrane were a complete insulator, depolarization would not be possible. Due to the low conductivity the myelinated portion of the membrane acts as a capacitor, which will allow a current to flow across the single plane of the capacitor. The basic action potential propagation and electronic equivalent of the myelinated nerve cell are shown in Figure 5.6.

5.3.3 Stimulus Conduction in the Unmyelinated Axon: Cable Equation

All wave propagation phenomena obey the telegraph equation or wave equation. The telegraph equation is the wave equation applied to an electrical system as presented in Equation 5.1:

$$\frac{d^2V}{dt^2} = \frac{1}{v^2}\frac{d^2\varepsilon_m}{dx^2}, \tag{5.1}$$

where v is the speed of propagation of the electrical impulse (membrane potential: ε_m) along the length of the cell membrane over a distance x, all with respect to time t.

Equation 5.1 has solutions of the form

$$\varepsilon_m(x,t) = C_1 e^{ik(x\pm vt)} = C_1 \sin k(x \pm vt), \tag{5.2}$$

where C_1 is a constant and k is the wave number defined as

$$k = \frac{2\pi}{\lambda}. \tag{5.3}$$

Keep in mind that this is the solution for a continuous wave, whereas the action potential is a single pulse and as such needs to be solved for a frequency range under the Fourier principle.

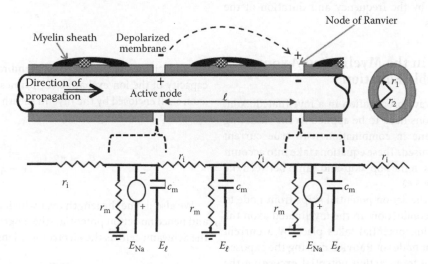

FIGURE 5.6 Representation of the saltatory conduction between two adjacent nodes of Ranvier and the accompanying electronic circuit diagram representing the electrical conductance mechanism.

A repetitive representative irregular pulse, such as the heartbeat, can be solved by using the first eight harmonics of the base frequency (i.e., ~1 Hz, at 60 beats per minute). The action potential is a single pulse only and will theoretically require an infinite number of Fourier components to be solved exactly.

Independent of the presence of a stimulus, the cell will return to its equilibrium state after reaching the depolarization maximum. This process is referred to as repolarization.

After the repolarization process, the cell has a short period in which nothing happens: the absolute refractory period, which lasts for ~1 ms. The cell membrane will not respond to any stimulus during this absolute refractory period. After this absolute refractory period follows a relative refractory period, which will allow a stimulus to initiate a depolarization, however at a higher threshold, that is, the rheobase has been raised during the relative refractory period.

Nerve cells form an exception due to their special construction. In nerve cells there are myelinated and unmyelinated nerve cells. The myelinated nerve cells have Schwann cells wrapped around the nerve axon at regular intervals and the impulse jumps from the open nerve section to the open nerve section (nodes of Ranvier), thus circumventing the entire membrane conduction process and providing a lossless transmitter at very high conduction rates. The larger the diameter of the nerve cell, the greater the depolarization wave conduction. In addition, the greater the conduction rate, the smaller the depolarization threshold and vice versa.

5.3.4 Analog versus Digital Stimulus Information

The occurrence of an action potential alone has no meaning other than that it was present at one time, and it carries no information since it is an all-or-nothing event. This brings the biological signals processing in the digital territory. The action potential is only a yes or no event; however, the intensity of the stimulus is represented by the frequency and duration of the train of action potentials.

5.3.5 Conduction in the Myelinated Axon: Modified Cable Equation

In order to model the wave propagation in a myelinated axon, the passive cable equations need to be applied to the tumbling dipoles on the membrane in combination with true current under the Schwann cells used. These equations take into account the all-or-nothing events in the hop-skip-and-jump between the nodes of Ranvier (Figure 5.6).

In myelinated fibers, the action potential skips from node to node through saltatory conduction. In the myelinated axon the transmission of the action potential takes place by a current flowing between adjacent nodes of Ranvier, causing the exposed membrane to depolarize to an action potential exceeding the action potential required for initiating the Na–K pump process in the adjacent section on the membrane as in unmyelinated

electrical propagation. The current flow between the nodes of Ranvier takes place at the speed of light, and hence signal transmission in the myelinated axon is much faster than in the unmyelinated axon. Due to the limited biological energy driving the current, the high intercellular resistance, and the small electrical potential, the current conduction needs to be reinitiated at discrete intervals, in contrast to the continuous current flowing through an electrical cable in electronic circuits. The propagation of the myelinated axon action potential depends on the following conditions: excitability of the membrane (which is a function of the condition of the membrane) and conductive properties of the intercellular liquid—cable properties. The density and mobility of the ion population in the intercellular fluid are the main contributing factors to current conduction. The threshold potential in myelinated nerves also acts as a filter to eliminate "false positives," and only strong signals will pass through.

The current propagation for conduction between the nodes of Ranvier can be derived using the current balance or the principle of conservation of charge.

5.3.5.1 Current Balance

The membrane current balance is given as follows:

$$2\pi a \Delta x c_m \frac{\partial \varepsilon_m}{\partial t} + 2\pi a \Delta x (i_m - i_e) = -\frac{\pi a^2}{r_L} \left(\frac{\partial \varepsilon_m}{\partial x} \right)_{\text{left}}$$

$$+ \frac{\pi a^2}{r_L} \left(\frac{\partial \varepsilon_m}{\partial x} \right)_{\text{right}} = \frac{\partial}{\partial x} \left(\frac{\pi a^2}{r_L} \frac{\partial \varepsilon_m}{\partial x} \right) \Delta x. \tag{5.4}$$

The internal resistance (R_i) through compartments separated by the myelin sheath is defined by the intrinsic resistance (r_i), the cross-sectional area defined by the radius of the axon (r_1), and the distance between the nodes of Ranvier (ℓ) as follows:

$$R_i = \frac{r_i \ell}{\pi r_1^2}. \tag{5.5}$$

The capacitance across the axon membrane (C_m) is the localized capacity of the ion exchange (C_m) times the surface area of the axon shell enclosed by the myelin sheath ($2\pi r_1 \ell$). The transmembrane capacitance is given by

$$C_m = \pi 2 r_1 \ell c_m \frac{2 r_2}{r_2 - r_1}. \tag{5.6}$$

We also define the length over which there is no ion exchange and hence no action potential (the length of the axon covered by the Schwann cell) as the electrotonic length (λ_l):

$$\lambda_l = \sqrt{\frac{r_1^2 \pi R_m}{2\pi r_1 R_l}} = \sqrt{\frac{r_1 R_m}{2 R_l}}. \tag{5.7}$$

For the depolarization process, the time constant (τ_l) is defined as generally used in electronic circuits:

$$\tau_l = r_m c_m. \tag{5.8}$$

For a large electrotonic length (i.e., the Schwann cell-covered section of the myelinated axon) the depolarization potential needs to be larger than for a short electrotonic length (unmyelinated axon). This phenomenon becomes important when considering multiple sclerosis, which destroys the myelin.

5.3.6 Discrete Cable Equations

In order to illustrate the complexity of the segmented depolarization condition, the wave equation for the myelinated axon is derived briefly without providing the full solution to the wave pattern. The full solution requires advanced mathematical derivation, which falls outside the scope of this book. Additional information on this topic can be found in the article by Joshua Goldwyn.

The wave equation can be shown to obey the following description (Equation 5.9):

$$\lambda_i^2 \frac{\partial^2 \varepsilon_m(x,t)}{\partial x^2} - \varepsilon_m(x,t) - \tau_m \frac{\partial \varepsilon_m(x,t)}{\partial t} = -I_{\text{diff}}(x,t)$$

$$= -\lambda_n^2 \frac{\partial^2 V_S(x,t)}{\partial x^2}, \tag{5.9}$$

where V_S depicts the potential difference between two adjacent nodes of Ranvier.

In order to model the myelinated axon, the variable x can no longer be treated as a continuous variable. By treating it as discretized at nodal points, it is possible to model the myelinated wave propagation in the axon. Each segment is now a step function, comparable to a histogram. The differentiation can hence be replaced by multiplication with a constant factor G. Substitution with discrete segments provides one equation for each node, as presented in Equation 5.10:

$$G_n \delta^2 \varepsilon_m(x,t) - C_m \frac{\partial \varepsilon_m(x,t)}{\partial t} - G_l[\varepsilon_m(x,t) - E_l] = -I_{\text{diff}}(x,t). \tag{5.10}$$

Active axons arise from the dynamics in the nodes of Ranvier due to active ion channels in the membrane. This results in a rewrite for the active wave equation, as expressed by Equation 5.11:

$$G_n \delta^2 \varepsilon_m(x,t) - C_m \frac{\partial \varepsilon_m(x,t)}{\partial t} - \sum_n G_l(\varepsilon_m,t)[\varepsilon_m(x,t) - E_l]$$

$$= -I_{\text{diff}}(x,t) = -G_m \delta^2 \varepsilon_m(x,t). \tag{5.11}$$

Also note that the ion-pump activity is represented by including the variable G as a function of membrane potential.

The conductance across the segment between the two nodes also becomes a function of time and the respective membrane voltages.

5.3.6.1 Electronic Model of the Nerve

The nerve, either unmyelinated or myelinated, has an axial and radial component of an equivalent electronic circuit that governs the propagation of the action potential. Figure 5.6 illustrates the circuit diagram of a myelinated axon.

5.3.7 Optimization of "Conductivity"

It is possible to calculate the optimal conditions for maximum propagation speed in a myelinated nerve axon. In order to approximate this situation, one needs to treat each segment demarcated by two nodes of Ranvier as a multilayer capacitor of thickness $\Delta r = r_2 - r_1$, as shown in Figure 5.6.

The diffusion constant (D) for electrical migration can now be derived as applied to Fick's law for ion movement:

$$D = \frac{\pi r_1^2 \ell}{C_m r_i} = \frac{r_1^2 \log(r_2/r_1)}{2c_m r_i 2r_2} = \text{const}\left[\frac{r_1^2 \log(r_2/r_1)}{r_2^2}\right]. \tag{5.12}$$

Using

$$\gamma_a = \frac{r}{r_2}, \tag{5.13}$$

the optimal thickness for maximal ion diffusion can be derived by partial differentiation with respect to the ratio of the radius of the axon to the outer radius of the myelin wrap: $\gamma_a = r/r_2$, taken at the radius of the bare axon. The maximum value of the diffusion coefficient will occur at an optimal ratio: r_1/r_2 as derived by solving

$$\frac{d}{d\gamma_a}\left(\gamma_a^2 \log\frac{1}{\gamma_a}\right) = 0. \tag{5.14}$$

When solving for the radius ratio, this curve peaks at

$$\log(\gamma_a) = -\frac{1}{2}. \tag{5.15}$$

This yields the optimal ratio for the radii of the inner radius of the myelin sheath at the outer surface of the axon to the outer radius of the myelin sheath as

$$r_1 = \frac{r_2}{\sqrt{e}}. \tag{5.16}$$

Substitution yields

$$D = \frac{r_2^2}{4ec_m r_i d_m}. \tag{5.17}$$

5.4 Electrical Data Acquisition

The depolarization wave fronts can be measured with the help of conductors that are in contact with the cell itself, even inside the cell, or conductors at the surface of the organ that envelops the source of the depolarization potentials. All biological media have an ample supply of ions throughout the entire organism to conduct the currents resulting from electrical potentials anywhere in the organism. The measuring device will refer all potentials against ground or a conveniently chosen reference point on the body itself.

The electrical activity of individual cells can manifest itself on the skin surface as surface potentials. These surface potentials are in fact the culmination of all cellular depolarizations that occur at the exact same time everywhere in the body.

The interpretation of these surface potentials presents a serious challenge for various reasons:

- The biological specimen has finite dimensions. Because the air that surrounds the body is nonconducting, all equipotential lines will be perpendicular to the skin surface.
- The shape of the surface is not a regular geometric form and cannot be described by a simple geometric interpretation; several approximations will need to be made.
- The conductivity of the various tissues is not uniform and ranges from a perfect conductor (e.g., blood) to an almost perfect insulator (e.g., air in lungs).
- Certain tissues have anisotropic conduction, for instance muscle tissue conductors are better in parallel with the muscle strands than perpendicular to it.
- Because of the mathematically unknown shape of conduction distribution inside the body, the reverse problem of finding the source that belongs to a measured signal distribution can be a complicated challenge.

In addition, the skin itself forms a complex electrical impedance (see Chapter 20) that needs to be taken into account for the acquisition of alternating signals in particular, or it needs to be reduced to a simpler mathematical algorithm by sample preparation.

5.5 Volumetric Conduction

In the experimental setup the cellular potential can be measured with a microelectrode, a needle electrode for instance. The relative electrical potential of the inside of the cell with respect to the outside can be measured as negative at first and can be recorded to flip to a positive value during depolarization. In a clinical setting, this is usually not advisable nor technically feasible. In the clinical setting one usually needs to measure the depolarization in close proximity to the source, but still from the outer surface only.

In case of measurement of the cardiac potential, electrodes are usually placed on the arms and legs or on the chest and back. The question is now: How accurate is the information obtained in this way with respect to individual cellular depolarization? In order to compensate for volume conduction, the transfer function can

be modeled and the inverse solution can be used to derive the source of the electric signal from electrode detection at the surface. In order to answer this question, the concept of solid angle needs to be discussed first because the inverse solution calculates back from the surface, under certain approximations of the physical geometry of the outline of the biological specimen.

5.5.1 Concept of Solid Angle

The solid angle from a point O toward a surface area A with curvature C can be described by connecting the edges of the area A to the point of reference O. This concept is illustrated in Figure 5.7.

When a sphere is placed around the reference point O with radius r, this sphere will slice through the cone of lines connecting to the edge of area A. The magnitude of the surface area carved out by the cone has an area A'. In general, the area of a sphere is πr^2. The area A' is also proportional to r^2 in this way, because it is a fraction of the total surface area of the sphere. The solid angle Ω that is enclosed by the cone originating in O is defined as

$$\Omega = \frac{A'}{r^2}. \tag{5.18}$$

The units of the solid angle Ω are in steradians. The solid angle is independent of the shape of the surface, because it only demarcates the outline in space and relief has no impact on that, and it does not depend on the radius to the reference point.

5.5.2 Electrical Potential of an Electrical Dipole Layer

The electrical potential of a charge dipole found in point P at great distance from the dipole, $r \gg \delta$, is defined in Equation 5.19 as

$$V_p = \frac{K(\mathbf{p} \bullet \mathbf{r})}{r^3} = K\frac{p\cos\theta}{r^2}, \tag{5.19}$$

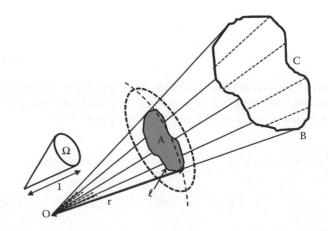

FIGURE 5.7 Definition of the solid angle outlining an area in space with respect to a reference point O.

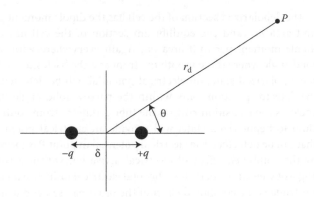

FIGURE 5.8 Electrical dipole resulting from two opposite and equal charges separated by a distance δ.

where *K* is a constant.

The dipole and conditions are outlined in Figure 5.8.

The separation between the two charges is δ, and the dipole moment *p* is the charge times the separation as expressed by Equation 5.20:

$$p = q\delta. \tag{5.20}$$

If the two point charges are replaced by two layers of charge, Equation 5.20 transforms into a situation with a dipole moment per surface area *m*, and the contribution of each area needs to be regarded individually as illustrated in Figure 5.9. The expression for the electrical dipole potential is now Equation 5.21:

$$dV_p = K\frac{m\cos(\theta)\,dA}{r^2} = Km\,d\Omega, \tag{5.21}$$

with θ the angle of the normal to the charge surface with the vector from *O* to *P*. This uses the definition of solid angle, dΩ, for each subsection d*A′* as seen by *P*.

Substitution of Equation 5.18 into Equation 5.21 gives, for the absolute value of the individual surface dipole contributions,

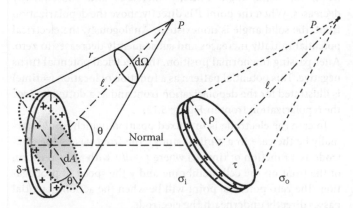

FIGURE 5.9 Diagram of the external observation of the dipole layer at the depolarization cross section seen under an angle θ.

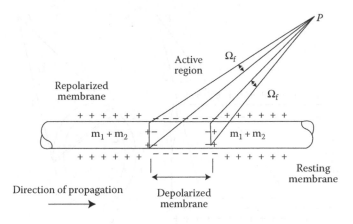

FIGURE 5.10 Diagram illustrating the surface dipole effect represented by dipoles at representative cross sections observed under respective solid angles for the proximal and distal segment of the depolarization wave front.

$$|dV_p| = Km\,d\Omega. \tag{5.22}$$

The sign of the dipole is lost because the solid angle by definition becomes positive. However, if the negative charges are closer to *P*, the contribution d*V_p* is considered negative. The surface dipole is illustrated in Figure 5.10.

The contribution of the entire surface requires the integral over the surface giving the integral version of Equation 5.22 in Equation 5.23:

$$|V_p| = \int_{surface\_area\_angle} Km\,d\Omega = Km\Omega. \tag{5.23}$$

Here point *P* observes the entire surface with the solid angle Ω.

5.5.3 Electrical Potential Perceived by Surface Electrodes

The next step is finding out what the electrode *P* measures in the location on the skin surface an arbitrary distance away from the action potential. The first assumption that needs to be made is that the medium that surrounds the site of the action potential is infinitely widely spread out in all directions. The second assumption is that the cell that produces the action potential is purely cylindrical, with negligible membrane thickness.

These conditions are illustrated in Figure 5.11 for the situation where no depolarization has taken place yet.

The membrane is a dipole layer with dipole moment per unit surface area *m*. The cell is at rest at a potential of -90 mV. Point *P* sees two oppositely charged dipole layers, with equal contributions, since the solid angle and the dipole moments per surface area are identical. As a result, the front and back sides cancel each other out and no potential is registered at *P*. A cell in steady state equilibrium can thus not be detected. In case a depolariza-

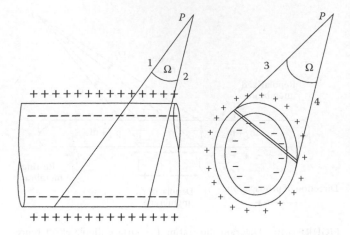

FIGURE 5.11 Diagram of the cell membrane capacitive charge and the observed cancelation of inner and outer charges for the situation where no depolarization has taken place yet. No electrical information will be detected at the surface in this situation.

tion does occur, the initial assumption is made that the action potential is instantaneous, and there is no gradual transfer of Na^+ and subsequent K^+ ions. The resulting depolarization propagation can be seen as two discontinuities, the depolarization and the repolarization front, moving along the length of the cylinder, as shown in Figure 5.12.

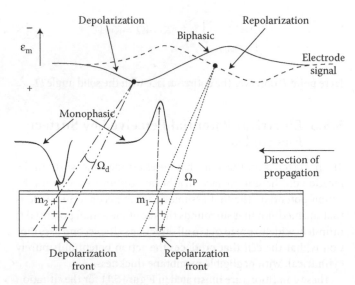

FIGURE 5.12 Registration of dipole movement at the surface of the body during an action potential in a cylindrical cell. Depolarization (solid line) results in monophasic negative deflection, followed by monophasic repolarization (dashed line) with positive deflection on the external electrical potential. The passage of the depolarization dipole will have negative deflection (monophasic) followed by the rear of the dipole which is positive, producing a biphasic wave. Combining depolarization and repolarization creates a triphasic wave form (adding the solid and the dashed line of the electrode signal).

The depolarized section of the cell has the dipole moment per unit area m_2, and the equilibrium section of the cell has the dipole moment per unit area m_1. Again everywhere with one solid angle where there is both the front and the back side in the same polarized state, no electrical potential will be detected at the detector position, only within the narrow solid angle that encloses the transition ring on the tube geometry from equilibrium to depolarized state, will provide an electrical potential that can be detected. The electrical potential in point P is formed by the combined effect of electrical activation within a solid angle of view of the detector. The total electrical influence on the electrode is the combined effect of the proximal, Ω_p, and distal, Ω_d, contributions, which is formulated as

$$V_p = K(m_1 + m_2)\Omega_p - K(m_1 + m_2)\Omega_d. \qquad (5.24)$$

This can also be represented as two cross-sectional charge disks in the cylindrical cell within the same solid angle as shown in Figure 5.12. This view simplifies the mathematical description of a propagating depolarization wave front. The distal section Ω_d will produce a negative potential because a net negative charge is facing the electrode.

The distance separating the depolarization front and the repolarization front is determined from the duration of the action potential and the speed of propagation of the signal along the cell membrane. For instance, if the depolarization takes 1 ms and the propagation velocity is 2 m/s, the distance h equals 2 mm. Because the depolarization front propagates along the cell wall, the angle that the dipoles make with respect to the electrode in point P will continuously change in time.

Dipole migration along the cylindrical cell can be described in a motion algorithm. For convenience, only the depolarization front is discussed. In the discussion it will not make a difference if the depolarization front moves past the *in situ* electrode or if the electrode moves past the motionless front.

As the angle of the normal to the cell wall wave front with the electrode decreases from great distance to directly above the front, the solid angle initially increases and subsequently decreases. When the point P is directly above the depolarization front, the solid angle is nonexistent. Analogously, the electrical potential initially increases and subsequently decreases to zero. After passing the normal position, the electrical potential turns negative. This potential pattern as a function of location (= time) is illustrated for the depolarization front and as a dotted line for the repolarization front in Figure 5.12.

In case the electrode is in a fixed position (which would normally be the case in a clinical study), the front passes the electrode as a function of time (t) where $t = d/v$, with d the position of the front on the cell membrane and v the speed of propagation. The zero potential point will be when the action potential passes directly underneath the electrode.

This type of potential fluctuation is called a biphasic action potential. The combined effect of both depolarization and repolarization can be found by adding the two recordings in Figure 5.11 together. The summation of the two polarization fronts is shown in Figure 5.12. This graphical representation has two crossovers to the opposite sign, making it a triphasic action potential.

5.5.4 Monophasic Action Potential

The monophasic action potential is observed only at the tip of a depolarizing cell, for instance the nerve ending leading away from one of the sensory extremities. The electrode detects the formation of the depolarization, but not the passage, and only the second part of the biphasic phenomenon is revealed. The detected signal will only show the electrical potential returning negative.

At the other end terminal of the cell, the depolarization front is only approaching and will terminate at the junction with the next cell, or the synapse. In this case only the rising part of the biphasic phenomenon is recorded; the electrical potential will register a positive depolarization only.

An example is shown in Figure 8.12.

In Chapters 20 and 21, the mechanism of detecting the electrical potential is described.

5.6 Summary

In this chapter, we discussed the process in which an action potential is transported in cells and, in particular, the modifications to the telegraph equation to accommodate the transmittance of a depolarization wave front in myelinated nerves. The mechanism of detection of distant potentials through volume conduction was also described. The chapter concluded by pointing out specific details of the construction of the perceived action potential at a distal electrode resulting from migrating dipoles.

Additional Reading

Barr RC and Plonsey R, 2007. *Bioelectricity, A Quantitative Approach*, 3rd Edition, Springer, New York, NY, pp. 155–185.

Barrett EF and Barrett JN, 1982. Intracellular recording from vertebrate myelinated axons: Mechanism of the depolarizing after potential. *Journal of Physiology* 323(February): 117–144.

Dayan P and Abbott LF, 2001. *Theoretical Neuroscience, Computational and Mathematical Modeling of Neural Systems*, Chapter 6, The MIT Press, Cambridge, MA.

Georgiev DD, 2004. The nervous principle: Active versus passive electric processes in neurons. *Electroneurobiología* 12(2): 169–230.

Gerstner W and Kistler WM, 2002. *Spiking Neuron Models*, Chapter 6, Cambridge University Press, Cambridge, MA.

Goldwyn J. Analysis of the discrete Nagumo equation: Modeling traveling waves in myelinated neurons. Available at http://www.amath.washington.edu/~jgoldwyn/content/amath504paper.pdf

Hodgkin AL and Huxley AF, 1952. A quantitative description of membrane current and its aplication to conduction and excitation in nerve. *Journal of Physiology* 117(4): 500–544.

Koch C, 1999. *Biophysics of Computation*, Chapters 2 and 6, Oxford University Press, New York, NY.

Loizou PC, 1999. Introduction to cochlear implants. *IEEE Engineering in Medicine and Biology* 18(1): 32–42.

Rall W, 1997. Core conductor theory and cable properties of neurons. In *Handbook of Physiology*, Section 1: The nervous system, Vol.1: Cellular biology of neurons, Kandel ER, (Ed.), pp. 39–97. American Physiological Society, Bethesda, MD.

Schafer DP and Rasband MN, 2006. Glial regulation of the axonal membrane at nodes of Ranvier, *Current Opinion in Neurobiology* 16: 508–514.

II

Physics of Perception

Medical Decision Making

Robert Splinter

6.1 Introduction

Generally the role of physics, engineering, and biological research are all merging in the cumulative knowledge of the clinical details and the expression of the symptoms to form a treatment plan to restore the biological features that are malfunctioning. Additionally, the performance of baseline function assessment provides the platform for diagnostic testing and device development.

The medical process evolves in the efforts associated with the diagnostic methodology as well as creating solutions for treatment. In case the data are not sufficient or certain specific data related to the organ or the function of a region in the biological unit cannot be acquired with the necessary level of detail, this will initiate the process of negotiations, eventually leading to new device development.[1]

6.2 Medical Decision Making

Generally the physician has to rely on the information provided by the patient to decide on the potential disease and course of action to follow. In addition to medical literature supporting every clinical decision made by physicians, there is always the need for a diagnostic procedure to validate the treatment plan. The choice of diagnostic methods varies from tactile and visual examination to imaging and chemical laboratory work. The diagnostic modalities selected by the physician need to be able to corroborate any treatment decision.

6.2.1 Stochastics

Both in physics and medicine, probability indicates the likelihood of success or the accuracy of the branches in the decision tree. In certain cases in physics, probability is the main issue, such as the Heisenberg uncertainty principle.[2–4]

6.2.1.1 Heisenberg Uncertainty

The Heisenberg uncertainty principle resides in the subsection of physics called quantum mechanics. Quantum mechanics refers to the discretization of energy and mass and forms a bridge between classical mechanics and relativistic physics. At a biological level, the quantum mechanics reasoning can provide an insight into the structure of proteins and the operation of smell.

The Heisenberg uncertainty is governed by the realization that one can never establish both the position of an object in motion and the momentum of this object with exact accuracy at the exact same moment in time. This is because the determination of the position reverts back to the energy of the particle and as such a position of a moving object is a wave phenomenon (wave-particle duality).[2–4] The Heisenberg uncertainty principle reads as shown in Equation 6.1:

$$\Delta P \Delta x \geq \frac{\hbar}{2}, \qquad (6.1)$$

where $\hbar$ is Planck's constant ($\hbar = 6.626098 \times 10^{-34}$ m^2 kg/s), ΔP is the uncertainty in the momentum, and Δx is the undefined probability in location. As such the product of the spread in location and the range in momentum has a minimum value.

As a matter of fact, the Heisenberg uncertainty principle claims that the measurement (= the observer) influences the outcome of the measurement. This analogy can be extended to the medical decision tree. The way the question about a patient's condition is posed may affect the answer and as such the treatment. Even the perception by the patient of the physician may affect the outcome of the examination.

The decision tree relies on the physician's ability to ascertain the incidences of the complaint, the prevalence of an ailment, the risks associated with potential medical conditions, and the variability in occurrence within the patient population. Quite frequently, the diversity between patients due to lifestyle, exercise routines, eating habits, and genetic predisposition can create a significant amount of uncertainty. Even within one single patient the left and right legs may not be identical in pathological composition. Chemical tests (i.e., blood work) can be helpful in determining the inclusion and exclusion criteria for the use of certain drugs as well as probability of cause assessment. Additional tests may be performed with various degrees of complication in devices for the sensing and imaging to determine deviations from an accepted control parameter.

To assess the likelihood of making the correct diagnosis, a certain amount of probability theory may be involved.

6.2.1.2 Bayes' Theorem

Bayes' theorem is often quoted in the determination of the conditional probabilities of eliminating a wrong choice in diagnostics when two or more options seem equally probable. Similar to Heisenberg's uncertainty principle, Bayes' theorem determines if a theory is supported by the observations on a probability scale.[5] The more data available to the physician, the greater the likelihood of selecting the proper diagnosis. On the other hand, the Heisenberg uncertainty principle entails that if one would know all the details of the circumstances at one single moment, this would not claim any knowledge of the correlation between preceding states or future states. This is where the analogy breaks up.

Bayes' theorem is used to weigh the marginal probabilities and conditional probabilities of two events ("0" and "1," respectively) against each other where one event is the "null hypothesis" ("0") to be tested with a probability greater than zero by definition. The probability $[P(x)]$ of making the correct diagnosis is performed by testing the null hypothesis against the alternative hypothesis ("1") and can mathematically be formulated as shown in Equation 6.2:

$$P(1\,|\,0) = \frac{P(0\,|\,1)P(1)}{P(0)}, \qquad (6.2)$$

where $P(1)$ is the prior probability that the alternative hypothesis is correct, without taking the conditions of the null hypothesis into account; $P(1\,|\,0)$ is the conditional probability that the alternative hypothesis is correct given the conditions of the null hypothesis, whereas $P(0\,|\,1)$ is the reverse condition probability; and $P(0)$ is the normalization condition prior or marginal probability. There are various derivations in statistics that can be applied to solve for the conditional probability and we would like to refer the reader to a statistics textbook for additional information.

Medical diagnostics is generally a combination of data obtained from medical devices developed by the medical device industry, patient symptom description, and operator experience or research ability (published reference articles and books).

This book focuses on the engineering efforts made in developing devices to assist the medical operator to investigate potential causes and monitor the treatment progress.

The engineering efforts involved in this stage will involve providing the means to monitor the delivery of the treatment as well as tracking the healing process.

6.3 Identifying the Resources

In order to derive a decision on the health status of a patient and make a prognosis on the path to cure the individual, the decision tree will need to be populated. The physician will initially need to identify the information sources available and the mechanisms at her/his disposal to investigate the patient's vital statistics in comparison to certain established baseline parameters.

6.3.1 Vital Signs

Medical health condition is determined by means of measuring specific parameters that are considered to be representative of the baseline health status of the patient. The critical parameters to be measured are defined as the vital signs.

Several vital sign can be identified. The four main vitals are[6,7]

1. Heart rate
2. Body temperature
3. Blood pressure
4. Respiratory rate

An emergency medical technician (EMT) attendant will check for these four parameters when arriving at an emergency scene. The EMT will provide short-term care prior to hospital admittance. The tools of the emergency healthcare provider need to be portable and provide maximum amount of information to indicate the level of immediate life-saving care required. Additional potential vital signs are[6-14]

5. Glucose level
6. End-tidal CO_2 expiration
7. Intracranial pressure
8. Urinary continence
9. Pain level

Glucose level or blood-sugar level is important when dealing with a diabetic, while intracranial pressure will be extremely relevant when dealing with a crash victim.

6.3.2 Medical Device Development

Based on the diagnosis of the patient's ailment there will be a course of action that will involve treatment of various kinds. In certain cases assist devices will be required as part of the rehabilitation process. On the other hand, the devices needed to diagnose are also in a continuous state of flux depending on the engineering knowledge of the biological entity as well as the required detail in the assessment of the biological processes involved. The development of medical devices relies directly on the current state-of-the-art of materials and engineering capabilities. Technological advances directly lead to increasingly sophisticated diagnostic and therapeutic devices. Medical devices are defined with the following purposes in mind:

- Diagnose changes in the status of a disease, injury, or handicap.

- Monitor alleviations in the status of a disease, injury, or handicap (continuously or at discrete intervals; such as doctor's visits).
- Investigate the anatomy and any modifications or replacements associated with the anatomy. This would include structural integrity and physiological performance and integration, respectively.
- Monitor, and when needed moderate, one or more physiological processes.

A medical device is a device designed to diagnose a disease or deficiency. The device can also be designed to support or replace the biological function of an organ or body part. These devices are also geared to treat or mitigate diseases or prevent and cure diseases by providing a plan of action based on the timeline of the progression of the parameters.

A thorough understanding of the biological operating mechanisms allows the device developer as well as the device user to hone in on the critical details necessary to make proper diagnoses. An example of a new era of medical devices is illustrated in Figure 6.1. The wearable cordless monitor can be used to record and remotely store several vital signs on a continuous basis.

The topics covered in the following section describe the basic knowledge needed to develop problem solving techniques for analytical techniques that can be applied to specific cellular processes, organ function analysis, and biological integration. Some of the biomedical engineering efforts are related to prosthetic device development geared to closely match the biological capabilities instead of merely providing a masking tool. Other engineering efforts are in increased resolution, both spatially and temporally as well as source identification.

One specific concern in prosthetic device development is durability and functionality. An example of the quality of medical devices is artificial heart valves; with the thousands of repeat performances each day, the hemodynamic functionality is crucial in addition to being durable.

6.3.3 Medical Device Regulations

Examples of medical devices in common use are blood-sugar meters, blood-pressure meters, insulin pumps, kidney dialysis machines, pacemakers/implantable defibrillators, next to corrective lenses, cochlear implants, and artificial hearts to name but a few. On a more temporary scale there are the following examples: heart–lung bypass machines, external defibrillators, anesthesia ventilators, fever thermometers, internal and external blood pressure units, pulse oximeters, kidney dialysis devices, and fetal monitors. More durable devices would encompass the following: ultrasound imaging units, electrocardiogram (ECG) racks, electroencephalogram (EEG) recorders, endoscopes, positron emission tomography (PET), and magnetic resonance imaging (MRI) machines. A different level of "medical devices" is the use of leeches and maggots.

The requirement of each of these respective devices is based on the established physiological performance and anatomical integrity. The means of establishing the physiological and

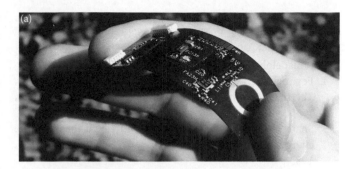

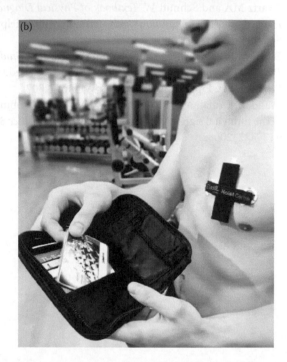

FIGURE 6.1 Wearable vital signs monitor. (a) Monitor for heart rate and other vital parameters. (Courtesy IMEC, Leuven, Belgium and its associated lab at Ghent University.) (b) Patch applied to the body with a wireless data logger kept in the wallet. (Courtesy IMEC, Leuven, Belgium, and the HOLST Centre, Eindhoven, the Netherlands.)

anatomic conditions are some of the features the book will focus on in the following chapters.

The development and performance of medical devices are subject to intense scrutiny by various government bodies, specific for different countries or economic units.

In the United States of America, the medical device industry and the medical treatment practices are strongly regulated by the Food and Drug Administration (FDA); in Japan, this falls under the jurisdiction of TÜV (TÜV SÜD Japan Ltd). In Europe, the European Union has issued guidelines for medical practices but the implementation is left to the organizations in the respective countries.

References

1. Bronzino JD. *Biomedical Engineering Handbook*, 2nd edition. Boca Raton, FL: CRC Press, 2000.

2. Weyl H. *The Theory of Groups and Quantum Mechanics*. New York: Dover Publications, 1950.

3. Ashcroft NW and Mermin ND. *Solid State Physics*. Philadelphia: Saunders College Publishing, 1976.

4. Feynman RP, Leighton RB, and Sands M. *Lectures on Physics, Vol III: Quantum mechancis*. Reading, MA: Addison-Wesley, 1965.

5. Sham LJ and Schlüter M. Density-functional theory of the energy gap. *Physical Review Letters* 51(20): 1888–1891, 1983.

6. Swartz MA and Schmitt W. *Textbook of Physical Diagnosis: History and Examination*, 4th edition. Philadelphia: Saunders, 2001.

7. Bickley LS, Szilagyi PG, and Stackhouse JG. *Bates' Guide to Physical Examination & History Taking*, 8th edition. Philadelphia: Lippincott Williams & Wilkins, 2002.

8. Walid M, Donahue S, Darmohray D, Hyer L, and Robinson J. The fifth vital sign—what does it mean? *Pain Practice* 8(6): 417–422, 2008.

9. Vardi A, Levin I, Paret G, and Barzilay Z. The sixth vital sign: End-tidal CO_2 in pediatric trauma patients during transport. *Harefuah* 139(3): 85–87, 2000.

10. Mower W, Myers G, Nicklin E, Kearin K, Baraff L, and Sachs C. Pulse oximetry as a fifth vital sign in emergency geriatric assessment. *Acad. Emerg. Med.* 6(9): 858–865, 1998.

11. Joseph AC. Continence: The sixth vital sign? *Am. J. Nurs.* 103(7): 11, 2003.

12. Collins SP, Schauer DP, Gupta A, Brunner H, Storrow AB, and Eckman MH. Cost-effectiveness analysis of ED decision making in patients with non-high-risk heart failure. *The American Journal of Emergency Medicine* 27(3): 293–302, 2009.

13. Bultz B and Carlson L. Emotional distress: The sixth vital sign—future directions in cancer care. *Psycho-oncology* 15(2): 93–95, 2006.

14. Bhisitkul DM, Morrow AL, Vinik AI, Shults J, Layland JC, and Rohn R. Prevalence of stress hyperglycemia among patients attending a pediatric emergency department. *The Journal of Pediatrics* 124(4): 547–551, 1994.

7

Senses

Robert Splinter

7.1 Introduction and Overview

In this chapter an overview of the senses is presented. A brief description is given of the senses that have little physics or engineering basis. On the other hand, engineering efforts related to these senses certainly fit the theme of his book.

There are five senses as outlined in traditional classification, each with particular physical attributes. The five senses are hearing, touch, smell, taste, and vision. Additionally, there are derived classifications of senses such as pain, balance (proprioception), direction, and a sense of time. The senses associated with pain, balance temperature, and direction are mostly grouped under somathesis. Generally the senses can be associated with the four mechanisms of perception: chemical (chemoreception); mechanical (mechanoreception); optical (photoreception); and thermal (thermoception), respectively. The senses associated with the chemical detection: taste and smell, technically fall outside the biomedical physics concepts of this book but a brief overview will be presented to place these senses in perspective. Somathesis (often referred to as "touch"), vision and hearing will be discussed in detail in the following chapters.

In the animal world, one specific additional sense is found: electrical perception (electroreception). Electrical sensation as a unique sensation with specific physical applications will be discussed due to the specific physics properties associated with this important mechanism of observation.

Before moving to the specific aspects of the various physical means of sensation, the taste and smell senses will be discussed, respectively, with the limited physical association.

These senses are related to each other in more ways than one.

In order to understand the surrogate methods of both smell and taste, a brief introduction of these two visceral senses is presented followed by an overview of the technological attempts to duplicate these senses.

7.2 Chemoreception: Biochemistry, Biophysics, and Bioengineering Overview

Chemoreception captures the chemical senses grouped as olfaction (smell) and gustation (taste). The general aspect of chemical perception fall outside the engineering aspects of this book; however, a brief introduction of the operations of both chemosenses will provide the basis for engineering efforts surrounding the duplication of these senses. Specific applications include: quality control purposes and remote sensing. Remote sensing is a mechanism to detect poisons or to perform quality assurance for products made for consumption.

Although both smell and taste are chemoreceptor senses, they are anatomically quite different. On the other hand, smell and taste are interrelated; food tastes different when one has a cold. More specifically, this is the case when one has a lessened sense of smell.

7.2.1 Taste

Gustation, the sense of taste, depends on chemical stimuli that are present in food and drink. The stimuli we know as tastes are mixtures of the four elementary taste qualities: sweet, salty, sour, and bitter. These basic tastes are highly sensitive in certain areas of the tongue, but their reception is not restricted only to those locations (Figure 7.1). Some sample taste stimuli that are particularly effective in eliciting these six sensations include sucrose, sodium chloride, hydrochloric acid, quinine, monosodium glutamate, and fats. The sensation of these stimuli depends on chemoreceptors located in taste buds (Berne et al., 1998). There are two additional taste stimuli that are not as familiar, but they are identified as potential nutrients: Umami monosodium glutamate (MSG) and aminoacids.

FIGURE 7.1 Taste sensitive areas of the tongue.

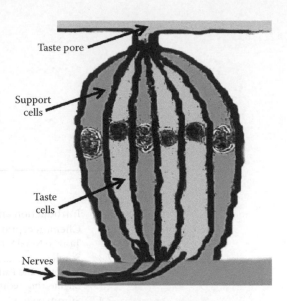

FIGURE 7.3 Oral cavity physiology; detailed diagram of taste cells in taste buds. (From Herness, M.S. and Gilbertson, T.A. 1999. *Annu. Rev. Physiol.* 61: 873–900. With permission.)

7.2.1.1 Receptor Organs

Taste buds are the gustatory receptors. The gustatory receptors form a pore that responds to dissolved chemicals welling the surface of the tongue and upper surface of the oral cavity. The sensation of taste is concentration related; too much sweet can result in a bitter taste. Generally, a change in concentration of less than 30% cannot be detected.

Two-thirds of the taste buds are located on the tongue. The rest of them can be found on the palette and the epiglottis.

7.2.1.2 Function

Taste buds have two distinctive functions during ingestion: (1) nutrient detection, which distinguishes the nutritive and beneficial compounds, and (2) toxin avoidance, which determines toxic or harmful compounds.

Just as the tongue has certain taste-sensitive regions, its taste buds are geographically segregated by their three specialized structures: fungiform, foliate, and circumvallate papillae (Figures 7.2 and 7.3). Regardless of their location, they are similar in their appearance, bulb-like and onion-shaped (Herness and

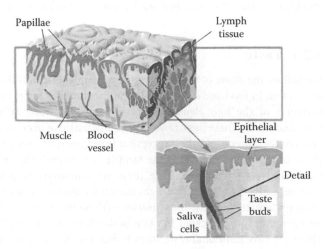

FIGURE 7.2 Papillae on the tongue. (Courtesy of Michael Chinery, from *Het Menselijk Lichaam*, published in Dutch by De Nationale Uitgeverij, 1996.)

Gilbertson, 1999). Fungiform papillae are located on the anterior part of the tongue and contain taste buds imbedded on the apical surface of the papillae.

7.2.1.3 Mechanism of Action

The sense of taste is not based on highly specific receptors, but is based on pattern recognition. Specifically, action potentials are produced from the following chemical properties: pH (acidic or basic), metal ion content (+2 or +3 oxidation state), chemical compound functional groups, and sugar content (Lindemann, 1996). Bioengineering research groups are now trying to blend this knowledge with new engineering technologies to build an "electronic tongue."

Taste receptor cells (TRCs) within the taste buds detect a wide range of chemicals, from small ions or protons to very complex macromolecules. Their major role is to recognize the various chemical signals and translate this information by membrane action potentials or intracellular free calcium concentration. Both translation pathways alter the release of neurotransmitters onto gustatory afferent nerve fibers that relay the information to the brain. Each taste bud comprises 50–150 individual cells that are collected together in a spherical structure (Figure 7.3). Individual taste cells within the taste bud are diverse. There have been many classification schemes for these cells: light and dark cells; S, N, B, and L cells; and I, II, III, and IV cells (using Roman numerals) (reference). Electron microscopy led to the latter scheme by being able to apply ultrastructural criteria to cell types. Type IV (B or dark cells) are basal cells and do not contact taste stimuli directly. Nontaste receptors type II (L or light cells) contain an electron-lucent cytoplasm, large oval nuclei, and a smooth endoplasmic reticulum. The function of these cells is still not fully understood. Cell types I and III (S and N cells) are considered taste cells because they have processes extending into the taste pore (Herness and Gilbertson, 1999). Saliva containing the solubilized food compounds enters the pores and binds to receptors on the surface of the TRCs.

The mechanism of action for sour and acid lies in the detection of protons (H^+). Protons are the primary stimuli and they easily permeate amiloride-sensitive Na^+ channels (ASSCs) and lingual epithelia tight junctions. Protons directly affect a variety of cellular targets and alter intracellular pH. It is therefore not surprising that a number of transduction mechanisms have been proposed for acids. As a matter of concentration, protons may inhibit outward potassium currents, resulting in cell depolarization. In both cases, an action potential is generated, activating Ca^{2+} influx currents. Neurotransmitter release from the Ca^{2+} influx excites the afferent nerve (Herness and Gilbertson, 1999).

Quinine, which produces a bitter taste, inhibits outward potassium currents similarly to protons from acids. Certain taste cells have receptors on the membrane surface that are specific for individual amino acids. The amino acid binds to its receptor type, generating cell depolarization. Intracellular Ca^{2+} is released, initiating an action potential that opens Ca^{2+} influx channels, allowing the increased Ca^{2+} concentration to release neurotransmitter for afferent nerve excitation. Free fatty acids utilize a somewhat different nerve excitation pathway. They diffuse and/or transport through the membrane releasing intracellular Ca^{2+}. The calcium ion concentration is enough to release neurotransmitter and, once again, the pathway culminates in afferent nerve excitation (Murphy, 1996).

One can begin to realize that the various pathways of the basic elementary taste stimuli present a highly sophisticated and complex information translation/transduction system.

7.2.2 Novel Sensing Technology

What can be learned from the biochemistry and biophysics of the gustatory system?

The sense of taste is not based on highly specific receptors, but in based on pattern recognition and the concomitant action potential produced primarily from the following chemical properties:

- pH (acidic or basic)
- Metal ion content (+2 or +3 oxidation state)
- Chemical compound functional groups
- Sugar content

Bioengineering research groups are now trying to blend this knowledge with new engineering technologies to build an "electronic tongue" or an "optical tongue."

7.2.2.1 Taste Sensing Engineering Technology

The Dean Neikirk research group at the University of Texas has built an "electronic tongue" using a combination of new technologies. A silicon chip was designed with an array of microbeads chemically formulated to respond to different analytes similar to the way the tongue responds to salty, sweet, sour, and bitter taste stimuli.

The "electronic tongue" can integrate fluid flow channels, microvalves, and micropumps to access bead storage wells. The technology available has the potential to make "taste buds" for any analyte. Even though the development is still in its early stages, the food and drug industries believe that a commercially developed "electronic tongue" will replace human taste testers. In addition, the machine will be able to create an archive of successful taste patterns based on RGB (red, green, and blue pixel) data files recorded with a charged-coupled device (CCD) color camera. A comparison of the engineering processes with the biochemical pathways of the gustatory system demonstrates how ingenious this invention really is (Colin, 1998).

7.2.2.2 Combinatorial Libraries

Artificial chemoreceptors representing human taste buds made from polymeric microspheres are placed in "buckets" on an integrated circuit. The microspheres had to contain light sensitive ligands that change color in the presence of specific food substances. Since foods contain multiple taste stimuli, it was necessary to form microspheres from a combinatorial library of polymeric monomers so that RGB data files can accurately represent a pattern of the taste stimuli composition (Figure 7.4). First, the polymeric monomers chosen for the microspheres had to contain ligands that would covalently bond the desired optically active reporter molecules responsible for analyte detection. Second, the combination of monomers had to be random to generate reproducible results. A "split lot" technique was chosen to produce the wide diversity necessary for the microspheres. The "split lot" mechanism is outlined in Figure 7.4. The "split lot" is a simple but highly effective method of combining three different polymeric monomers, dividing them in half, and recombining them in all possible combinations to form nine dimers (Figure 7.4c). The nine dimers are divided and coupled to form 33 trimers. Repeating this step again achieves 34 or 81 tetramers. This process is repeated a minimum of 10 times forming a library of

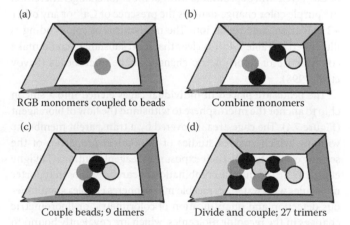

(a) RGB monomers coupled to beads
(b) Combine monomers
(c) Couple beads; 9 dimers
(d) Divide and couple; 27 trimers

FIGURE 7.4 Formation of combinatorial libraries, (a) monomers coupled to beads in pyramid troughs, (b) combined monomers, (c) beads coupled, forming nine dimers, and (d) divided and coupled beads forming 27 trimers. (After Dean and John McDevitt, McDevitt Research Laboratory, Austin, TX [Savoy et al. 1998].)

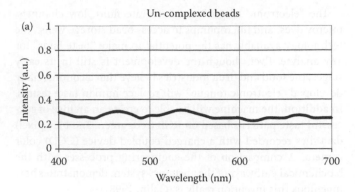

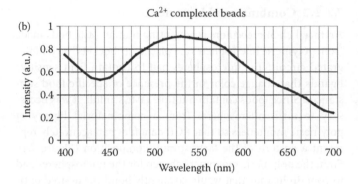

FIGURE 7.5 Spectral response of *o*-cresolphthalein complexone colorimetric to the presence of calcium ions, going from colorless (a), to "purple" (b). (After Dean Neikirk and John McDevitt, McDevitt Research Laboratory, Austin, TX [Savoy et al. 1998].)

over 59,000 different monomers. The libraries combine target binding with an optically active structure that can change color and/or fluorescence, creating a definite RGB detection pattern.

In Figure 7.5, orange carotenoid protein (OCP) in the presence of Ca^{2+} deprotonates at a pKa of 10.1, changing from colorless to a royal purple color. If the pH is raised to 10 without the presence of Ca^{2+}, OCP will deprotonate without the color change. Therefore, the purple color change denotes the presence of Ca^{2+} or any other +2 oxidation-state metal ion. The mechanism of color coding is illustrated in Table 7.1. It is clear that food substances could make colorimetric or fluorometric changes in the microbeads (Savoy et al., 1998).

The beads are placed in individual cages etched into a silicon chip to anchor the microsphere to withstand the flow of bioreagent (Figure 7.4). The cage area is covered by a transparent membrane window, which enables studies of the optical properties of the sensing elements upon their exposure to reagent solutions (Lavigne et al., 1998). The unique combination of carefully chosen reporter molecules with water permeable microspheres enables simultaneous detection and quantification of colorimetric or fluorometric changes in the receptor molecules, which are covalently bound to amine termination sites on the polymeric microspheres.

7.2.2.3 CCD Detection

With the "taste bud" beads mounted on transparent silicon wafers using microelectro-mechanical systems (MEMS)

technology, tracking the bead's color can be accomplished by using a CCD color video camera. The cameras use a prism block incorporating dichroic filters. These filters are coatings that are only a fraction of a light wavelength thick. The thickness determines which colors are passed or reflected. Each camera has three identical chips attached to the prism block and aligned. Since each CCD cell produces only one color, all components of the signal are at full system resolution. Illuminescence is an analog output. Individual samples from each bead produce an analog voltage output (pixel by pixel from the CCD) that is immediately converted to a digital signal. The camera RGB data files obtained from bead illumination are then analyzed by a pattern-recognition software to reveal the elements of the substance being tested. An illustration of the influence of Ca^{2+} (a molecule representative of one specific taste sensation) binding to the ligand on the color spectrum observed by the camera is shown in Figure 7.5.

The "electronic tongue" is just one isolated case where man is advancing engineering technology based on the biochemical and biophysical pioneering research (Savoy et al., 1998).

7.3 Smell

The sensation of smell basically relies on similar chemoreceptor principles to taste.

Some identified primary odors are: floral, etheric, musky, sweaty, rotten, stinging, and camphoraceous. The primary limitation to smell detection is that the reagent needs to be dissolved in liquid before it can be detected.

7.3.1 Receptors and Pathways

Smell relies on the olfactory mucous membrane, located in the nasal mucosa. The sense of smell is directly proportional to the area occupied by the olfactory mucous membrane. In humans the area is approximately 5 cm², containing approximately 10–15 million olfactory receptors. In comparison dogs have a mucous area approximating the surface area of the skin, containing in excess of 220 million receptors.

7.3.2 Stimulation of Receptors

The sense of smell is in actuality perceived from a solution of chemicals. Only substances in direct contact with the olfactory epithelium are detected. In general the olfactory system is relatively impervious to change; a minimum change in concentration of approximately 30% is detected. In comparison the eye can detect a mere 1% change in light intensity. The human sense of small can differentiate between 2000 and 4000 odors that may be considered significantly different from each other. The discrimination is directly related to the level of concern for our health, seeking out nutrition, and our development of other senses related to procreation and other activities. There are certain thresholds for detection for

TABLE 7.1 Colorimetry and Fluorometry of Reporter Molecules

Reagent	pH			Analyte
	3	7	11	
O R H₂C	Gray	Gray	Light gray	H^+
Fluorescein	Pale red	Medium red	Red	H^+
Alizarin complexone	Orange	Dark red	Brown red	H^+
	Red	Red-brown	Dark red	H^+, Ce^{2+}
OCP	Pink	Pink	Dark pink	H^+
	Dark pink	Pink	Purple	H^+, Ca^{2+}

Source: After Dean and John McDevitt, McDevitt Research Laboratory, Austin, TX [Savoy et al. 1998].

different chemical compositions and the thresholds are generally more acute for women than for men. Another feature of smell is saturation, the diminishing sensation when exposed to an odor for an extended period of time, also referred to as olfactory adaptation.

7.3.3 Smell Sensing Engineering Technology

One mechanism used to detect gasses is the gas chromatograph. Another example is the time-of-flight spectrometer. Both of these devices can detect chemicals in the air at minute amounts; however there is no correlation between the detection and the sense of smell. Examples of more smell representative detection are discussed below.

A polyvinyl chloride (PVC) blended lipid membrane converted into a quartz crystal microbalance (QCM) sensor is capable of detecting less than 1 ppm amylacetate gas and other odors. Additional odor sensation was accomplished by various lipid compositions. The crystal oscillates with a Q-factor that is affected by the absorbed odorant mass. The lipid membrane is designed to adhere to a specific molecular size and shape, and hence is very "odor" selective. The range of smells is limited by the choice of lipid at this point.

The change in oscillation frequency (Δf) of the crystal in response to the incremental mass density of adhered gas molecules ($\Delta m/A$) obeys the expression in Equation 7.1:

$$\Delta f = 2.3 \times 10^{-7} f^2 \frac{\Delta m}{A},$$

(7.1)

where f is the resonant frequency of the crystal, Δm the incremental molecular mass, and A the surface area of the lipid. The correlation between the concentrations of odor molecules in air: $[C_a]$, to that on the membrane: $[C_m]$, is established by the relationship expressed in Equation 7.2:

$$C_m = K_D C_a,$$

(7.2)

where K_D is a distribution coefficient (very similar to a diffusion coefficient) that is independent of the concentration and the membrane thickness.

Substitution of Equation 7.2 into Equation 7.1 provides a means of determining odor intensity (= concentration) in air by means of the frequency spectrum as shown by Equation 7.3:

$$\Delta f = 2.3 \times 10^{-7} f^2 \, dK_D C_a,$$

(7.3)

where d is the membrane thickness.

IBM has developed a cantilever gas detection system. The principle is very similar to the mechanism of operation of the atomic force microscope described in Chapter 34. A cantilever has several silicone strings attached, which are coated with a membrane that allows the string to contract under absorption of gas molecules. The direction of deflection indicates the "odor," since each string is sensitive to one chemical only.

The University of Illinois developed a mechanism similar to that shown for taste. The smell imaging technique uses specific dyes. Metalloporphyrins are vapor-sensitive dyes that experience a change in color under exposure to specific chemicals. Signal processing is needed to determine the odor perception.

The artificial tongue and nose technology have a wide range of potential applications:

- Food and water monitoring
- "Real-time" blood analysis

These applications will fall under another concept as well, the "lab-on-a-chip," which will be covered later in the book.

7.4 Summary

The sensory detection of chemicals by means of taste and smell uses the electrical stimulus of specific but largely unidentified receptors. Specific chemical groups can be detected to support the qualification of the nutritional source based on nutritional value and potential risk analysis.

Various technologies are currently being developed to perform an analysis of chemicals using similar principles to those used in taste transduction or smell; however the discovery and recognition mechanism uses well defined physical detection methods using electronic signal processing and data analysis.

The remaining engineering-based senses will be discussed in detail in several subsequent chapters.

Additional Reading

Abramson, D.H. 2005. Retinoblastoma in the 20th century, past success and future challenges. The Weisenfeld lecture, invest. *Ophthalmol. Vis. Sci.* 46: 2684–2691.

Alberts, B., Bray, D., Lewis, J., Raff, M., Roberts, K., and Watson, J.D. 1994. *Molecular Biology of the Cell*, 3rd edition, Chap. 15. New York, NY: Garland Publishing.

Baller, M.K., Lang, H.P., Fritz, J., Gerber, Ch., Gimzewski, J.K., Drechsler, U., Rothuizen, H., et al. Cantilever array-based artificial nose. *Ultramicroscopy*. 82(1–4): 1–9.

Berne, R.M., Levy, M.N., Koeppen, B.M., and Stanton, B.A. 1998. *Physiology*, 4th edition, Chap. 11. St. Louis: Mosby.

Colin, R.C. 1998. Artificial tongue learns to "taste" chemicals. *Electron. Engrg. Times.* 1035: 67–69.

Ganong, W.F. 2005. *Review of Medical Physiology*, 22nd edition. New York, NY: McGraw-Hill Professional.

Herness, M.S. and Gilbertson, T.A. 1999. Cellular mechanisms of taste transduction. *Annu. Rev. Physiol.* 61: 873–900.

Kaiser, L., Graveland-Bikker, J., Steuerwald, D., Vanberghem, M., Herlihy, K., and Zhang, S. 2008. Efficient cell-free production of olfactory receptors: Detergent optimization, structure, and ligand binding analyses. *PNAS*. October 14, 105(41), 15726–15731.

Lang, H.P., Baller, M.K. Berger, R., Gerber, Ch., Gimzewski, J.K., Battiston, F.M., Fornaro, P., et al. 1999. An artificial nose based on a micromechanical cantilever array. *Anal. Chim. Acta*. 393: 59.

Lavigne, J.J., Savoy, S.M., Clevenger, M.B., Ritchie, J.E., McDoniel, B., Yoo, J.S., Anslyn, E.V., et al. 1998. Solution-based analysis of multiple analytes by a sensor array: Toward the development of an electronic tongue. *JACS*. 120: 6429–6430.

Lindemann, B. 1996. Taste reception. *Physiol. Rev.* 76: 719–766.

Marieb, E.N. and Hoehn, K. 2008. *Anatomy and Physiology*. Boston, MA: Benjamin Cummings.

Matsuno, G., Yamazaki, D., Ogita, E., Mikuriya, K., and Ueda, T. 1995. A quartz crystal microbalance-type odor sensor using PVC-blended lipid membrane. *IEEE Trans. Instrum. Meas.*, 44(3), 739.

McKinley, M. and O'Loughlin, V.D. 2007. *Human Anatomy*, 2nd edition. New York, NY: McGraw Hill.

Murphy, K.M. 1996. Taste and Smell Lecture Notes. http://www.science.mcmaster.ca/Psychology/psych2e03/lecture8/taste-smell.htm

Rakow, N.A. and Suslick, K.S. 2000. A colorimetric sensor array for odor visualization. *Nature*. 406: 710–713.

Reece, W.O. 2004. *Dukes' Physiology of Domestic Animals*, 12th edition. Ithaca, NY: Cornell University Press.

Savoy, S.M., Lavigne, J.J., Yoo, J.S., Wright, J., Rodriquez, M., Goodey, A., McDoniel, B., et al. 1998. Solution-based analysis of multiple analytes by sensor array: Toward the development of an electronic tongue. *SPIE Conference on Chemical Microsensors and Applications*. 3539: 17–26; Slides used in conjunction with the presentation at the SPIE Conference can be found at: http://weewave.mer.utexas.edu/MED_files/MED_research/MEMS_chem_snsr/SPIE_tng_98/HTML_Presentation/index.htm.

Suslick, K.S. and Rakow, N.A. 2001. A Colorimetric nose: "Smell-seeing." In *Artificial Chemical Sensing: Olfaction and the Electronic Nose*, Stetter, J.R. and Pensrose, W.R., eds., pp. 8–14. Pennington, NJ: Electrochemical Society.

8

Somesthesis

Jean-Marc Aimonetti and
Sasa Radovanovic

8.1 Introduction

Somesthesis refers to the ability to detect external and internal stimuli, the nature of which could be mechanical, thermal, or even painful, the latter being sufficiently strong to damage tissues. This ability relies upon the activation of populations of different receptors located at the body surface and in deeper structures like muscles or joints. Somatosensory inputs are carried to the brain by different ascending pathways running through the spinal cord, brainstem, and thalamus. These sensory messages reach the primary somatosensory cortex before being integrated in higher-order cortices and subcortical areas.

Although it concerns touch, the sense of body movement and position, and pain, somesthesis remains poorly studied.[1] Everyone can experience the importance of vision only by closing the eyes but, even if less easily experienced, a life deprived of touch is hardly imaginable. Oliver Sacks has popularized the case of a woman who lost all large-diameter peripheral fibers.[2] This patient still experiences pain and thermal sensitivity, as these modalities are provided by smaller diameter fibers, but she has to visually monitor all her movements. This really impairs her daily motor activities although most of these sensory regulations are outside of consciousness. It is also hardly imaginable to experience a life without pain. This is nevertheless what patients suffering from hereditary sensory and autonomic neuropathies experience. Their physical state deteriorates the whole life due to inadvertent self-mutilations and life expectancy may be compromised.[3]

Similar to most conventional textbooks, we will distinguish mechanoreceptors that encode innocuous mechanical deformation of the skin, muscle spindles that encode muscle length and lengthening, thermoreceptors that are sensitive to changes in temperature, and finally nociceptors, rather unspecific receptors specialized in encoding stimuli of higher intensity liable to damage tissues. Although this distinction is somewhat helpful for students and their teachers, this does not reflect any physiological truth. In several instances, mechanoreceptors can be activated by thermal stimuli and the theory of specificity defended by von Frey in 1895 can no longer be supported.[4]

8.2 Touch

Much more than body envelope or barrier, skin protects the organism from physical, microbiological, or chemical attacks. Skin has a key role in thermal regulation, fluid and electrolyte balance and is becoming recognized as a major junction element between the endocrine, immune, and nervous systems.[5] Skin is, moreover, the largest sensory receptor of the body. In an adult, the area surface of the skin takes approximately 2 m^2 compared to only a few square millimeters for the retina.

Touch is a major sense that contributes to cognitive development at the very early stages of life. The famous and controversial experiments of Harlow in the 1960s confirm that early touch privation in the baby monkey causes irreversible psychological troubles, which can be partly overcome using a surrogate mother if it provides tactile comfort.[6] In his short life, David Vetter had to suffer the deprivation of intimate and care touches. Known as the "bubble boy," he suffered from an innate immune depression. He spent his 12 years of life in a sterile confined bubble and was only touched by his mother after death.[7] Although David was curious and intelligent in his younger years, he became depressive around ten and his mood deteriorated until his death. His tragic case illustrates the psychoanalytic theory of Danzieu, the *Moi-peau*, where the skin collects experiences helpful to build the feeling of me.[8]

In its most narrow definition, touch refers to the sensation caused by skin displacements, which occur each time we grasp an object or we kiss our partner. This acceptation is somewhat restrictive because cutaneous sense contributes not only to proprioception[9] but also to kinesthesia since illusory sensations of limb movements can be induced by stretching the skin.[10]

8.2.1 Peripheral Afferents and Mechanoreception

Tactile sensitivity relies upon populations of low-threshold mechanoreceptive units. Thanks to histological studies, the structure of the mammalian somatosensory receptors was known at the end of the nineteenth century. Later, Adrian and Zottermann[11] recorded single units from primary afferents and they emphasized the concept of specific receptors for encoding low threshold mechanical stimuli. In the early 1970s, microneurographic studies conducted in humans have extended the former observations from animal studies.[12] The glabrous skin (i.e., from the Latin "glaber" meaning bald, hairless) of the human hand contains about 17,000 tactile units, as estimated from counts of fibers in the median nerve at the wrist.[13] All their afferent types are large myelinated fibers with conduction velocities in the Aα–β range, that is to say from 26 to 91 m/s[14] (Table 8.1).

Despite some differences, all these mechanoreceptors work in the same way. A stimulus applied to the skin deforms the nerve endings. This mechanical energy alters the spatial conformation of mechanosensitive channels and this affects the ionic permeability of the receptor cell membrane. Changes in sodium permeability, at least, generate depolarizing current in the nerve ending. At this stage, the receptor potential is graded in amplitude and duration, proportional to the amplitude and duration of the stimulus. If the mechanical stimulation is sufficient, the receptor potentials thus generated are greater than the spike threshold. This leads to action potentials getting generated in the trigger zone, close to the initial segment of the axon. These action potentials are propagated along the axon. Details of the action potential are similar to the action potential generation described in Chapters 4 and 5. Sensory transduction refers to this mechanism that converts the mechanical energy of a stimulus into an electrical signal conducted by a first-order sensory neuron.[15] The anatomical details of nerves are described in Chapter 5. Once the receptor potential surpasses the spike threshold, any change in a given stimulus parameter will affect the discharge frequency. In other words, once the spike threshold is passed, sensory transduction shifts from amplitude to frequency encoding.

Four types of mechanoreceptors are differentiated on the basis of two pairs of features. First, we can consider the response adaptation to a sustained indentation: a rapid-adapting (RA) receptor responds with bursts of action potentials, notably at the onset and offset of stimulus application, and is called phasic receptor; a slowly adapting (SA) receptor remains responsive as long as the stimulus is applied upon the receptive field and is called tonic receptor. The phasic or tonic nature of the receptor may depend upon calcium conductance at the transduction site level.[16]

TABLE 8.1 Classification of Somatic Receptors

Receptor Type	Anatomical Characteristics	Location	Sensitive to ...	Adaptation Rate	Activation Threshold	Associated Afferent	Axon Diameter (μm)	Axonal Conduction Velocities (m/s)
Meissner's corpuscles	Encapsulated	Glabrous skin, between dermal papillae	Touch and pressure	Rapid	Low	Aβ	6–12	35–75
Pacinian corpuscles	Encapsulated; onion like	Subcutaneous tissues	Deep pressure and vibration	Rapid	Low	Aβ	6–12	35–75
Merkel's cells	Encapsulated	Skin and hair follicles	Light touch and pressure	Slow	Very low	Aβ	6–12	35–75
Ruffini's corpuscles	Encapsulated	Skin	Stretching of the skin	Slow	Low	Aβ	6–12	35–75
Muscle spindles	Encapsulated	In parallel with extrafusal fibers in skeletal muscles	Stretching of the parent muscle	Rapid and slow	Low	Ia and II	6–20	35–120
Golgi tendon organs	Encapsulated	At the junction between extrafusal fibers and the tendon of skeletal muscles	Increased muscle tension	Slow	Low	Ib	13–20	60
Joint receptors	Encapsulated	Ligaments, junction of synovial and fibrosum of capsule	Extreme joint motions	Slow and rapid	Low and medium	Aβ	8–17	?
Free nerve endings	Minimally specialized nerve endings	Skin and all deeper tissues	Pain, temperature, crude touch	Slow	High	C–Aδ	0.2–5	0.5–30

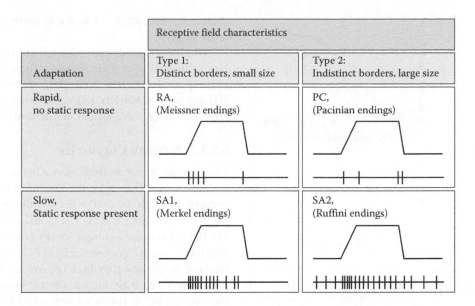

Adaptation	Receptive field characteristics	
	Type 1: Distinct borders, small size	Type 2: Indistinct borders, large size
Rapid, no static response	RA, (Meissner endings)	PC, (Pacinian endings)
Slow, Static response present	SA1, (Merkel endings)	SA2, (Ruffini endings)

FIGURE 8.1 Types of mechanoreceptive units in the glabrous skin of the human hand. (Adapted from Vallbo AB and Johansson RS. *Human Neurobiol.*, 1984, 3: 3–14.)

Secondly, we have to take into account the characteristics of the receptive fields. Type 1 units have small and well-demarked receptive fields, whereas type 2 units have larger receptive fields with confused borders[17] (Figure 8.1).

The related afferents comprise rapid- or fast-adapting type 1 (FA1) afferents that end in Meissner corpuscles, fast-adapting type 2 Pacinian (FA2) afferents that end in Pacinian corpuscles (PCs), slowly adapting type 1 (SA1) afferents that end in Merkel cells, and finally slowly adapting type 2 afferents (SA2) that end in Ruffini corpuscles (Figure 8.2).

In addition to this fast-conducting system, the human skin possesses also slow-conducting unmyelinated (C) afferents that respond to light touch.[18] The related receptors are nonencapsulated endings around hair bulbs and probably in the epidermis. This so-called C tactile system found to be spared in a de-afferented patient may underlie emotional and hormonal responses to caress-like, skin-to-skin contact between individuals.[19] Interestingly, so far, no C tactile afferents have been found in the glabrous skin.[20]

8.2.1.1 Meissner Corpuscles

These large receptors are located into dermal papillae between the epidermal limiting and intermediate ridges.[21] In humans, their receptive field sizes of 10–30 mm^2 with sharply contrasting borders and the highest density are encountered at the fingertip

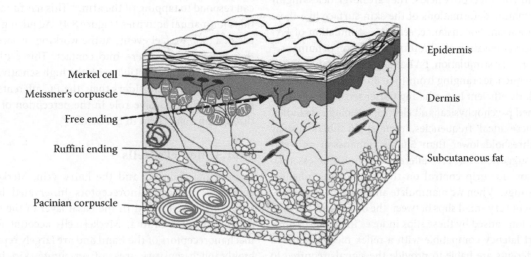

FIGURE 8.2 Skin layer. Skin contains different mechanoreceptors and free nerve endings, each of them liable to be activated by a distinct stimulus. (Redrawn from Irvin Darian-Smith. In: J.M. Brookhart and V.B. Mountcastle (eds), *Handbook of Physiology, Section 1: The Nervous System, Volume III: Sensory Processes.* Bethesda, MD, 1984.)

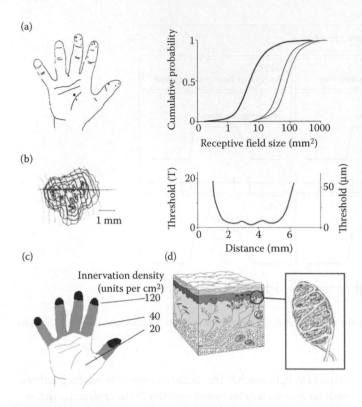

FIGURE 8.3 FA1 units have small receptive fields (a) with distinct borders (b). They are highly represented in the very finger tip (c) and probably connected to Meissner corpuscles located close to the skin surface in the papillary ridges of the dermis (d). (After Vallbo AB and Johansson RS. *Human Neurobiol.*, 1984, 3: 3–14.)

of digits, up to 33 corpuscles per square millimeter,[22] where tactile acuity is greatest[23] (Figure 8.3). They are innervated by two or even three axons[24] that terminate between lamellated Schwann cells into a fluid-filled corpuscle.[25]

FA1 afferents have a poor spatial acuity as they respond uniformly to stimuli over receptive fields. They are nevertheless highly sensitive to dynamic deformations of the skin surface like those induced by vibration. For instance, selective stimulation of FA1 afferents induces a sensation of vibratory stimulation in humans.[26] As regards vibratory stimulation, FA1 afferents respond preferentially to low frequencies ranging from 10 to 80 Hz,[27] which makes them particularly efficient in mediating flutter sensations according to combined psychophysical and electrophysiological studies. Indeed, at these ideal frequencies, certain FA fibers display mechanical threshold lower than 5 µm.[28] Johansson and colleagues have suggested that FA1 afferents may also be essential feedback sensors for grip control on the basis of microneurographic recordings. When we manipulate an object, there are several and frequent very small slips between the object and the skin. Each skin motion caused by these slips induces increases in force grip at a short latency compatible with a reflex mechanism[29] in which FA1 afferents are liable to provide the signals required to initiate and scale grip adjustments.[30] With its high sensitivity and poor spatial resolution, the FA system is often compared to the scotopic system in vision. Meissner corpuscles' density

decreases in the elderly,[31] which may partly account for the lower tactile acuity.[32]

In the hairy skin, dynamic sensitive tactile afferents have endings associated with hair follicles. These hair follicle afferent fibers appear equivalent to the glabrous FA1 afferents. They display the same sensitivity to middle range frequency vibrations around 80 Hz, although their mechanical threshold is higher.[33]

8.2.1.2 Pacinian Corpuscles

These are the largest mammalian mechanoreceptors, three times less represented than FA1 units in humans.[12] PC receptors are encapsulated with one central nerve ending surrounded by lamellae in an onion-like layout. The lamellae are coupled with gap junctions that cause a unique ionic environment partly responsible for the remarkable sensitivity of PCs.[34] Located in the deeper layers of the dermis, they have large receptive fields with ill-defined borders. At least one RA afferent axon lies at the center of the receptor. PC receptors are well suited for signaling vibratory stimulation, especially vibrations in the high frequency range from 80 to 200 Hz,[35] although they can potentially respond until 1000 Hz. They are the most sensitive cutaneous mechanoreceptors since they can respond to indentation of less than 1 µm in the finger pad of the monkey.[28] In humans, the mechanical threshold for detecting vibration falls from 5.0 to 0.03 µm with frequencies increasing from 20 to 150 Hz.[36] This larger sensitivity reflects the high-pass filter function of the PC receptors probably due to the fluid-filled space between the lamellae. Transduction of transient disturbances is thus facilitated at higher frequencies. Selective stimulation of FA2 afferents induces in fact a clear-cut sensation of vibratory stimulation.[26] Because of such an extreme sensitivity and their deep location, FA2 afferents do not have spatial resolution.[37] They may, however, contribute to osseoperception (i.e., static jaw position sensation and velocity of jaw movement as well as forces generated during contractions of the jaw muscles). Indeed, a PC, the receptive field of which is located above a digit, can respond to tapping of the arm.[38] This has functional implications for manual activities (Figure 8.4). Acquiring practice in the use of a tool, we feel events at the working surface of the tool as though our fingers were into contact. This critical perceptual capacity is fully compatible with the high sensitivity of PC afferents and Johnson considers that "the PC afferents play a principal, if not the exclusive role in the perception of distant events through an object held in the hand."[39]

8.2.1.3 Merkel's Cells

In both the glabrous and the hairy skin, Merkel cell–neurite complexes are mechanoreceptors innervated by SA1 tactile nerve. They are located in the basal layer of the epidermis, just above the basal lamina. Merkel cells account for 25% of the mechanoreceptors of the hand and are largely represented in the highly touch-sensitive areas of vertebrate skin, including palatine ridges, touch domes, and finger tips. Each individual SA1 afferent branches over an area of about 5 mm² and innervates on average 28 Merkel receptors in the monkey digit.[40] The receptive

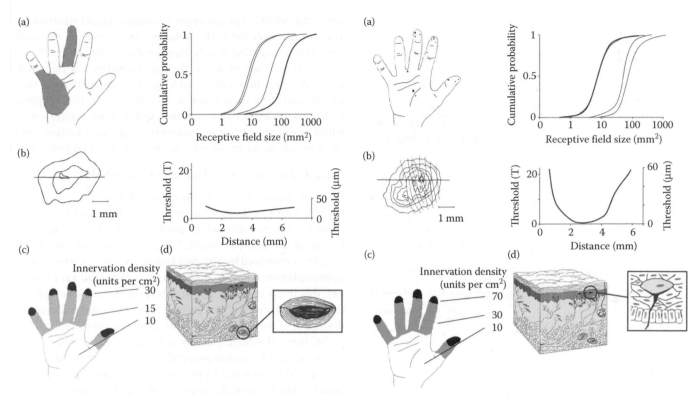

FIGURE 8.4 FA2 units have large receptive fields (a) with obscure borders (b). They are uniformly distributed in the glabrous skin areas (c) and connected to Pacinian corpuscles located in the subcutaneous tissue (d). (After Vallbo AB and Johansson RS. *Human Neurobiol.*, 1984, 3: 3–14.)

FIGURE 8.5 SA1 units have small receptive fields (a) with distinct borders (b). They are highly distributed in the very finger tip (c) and connected to Merkel's cells located at the tip of the intermediate epidermal ridges (d). (After Vallbo AB and Johansson RS. *Human Neurobiol.*, 1984, 3: 3–14.)

field of a SA1 afferent has several hot spots, each of them corresponding to individual branches of the afferent axon. This anatomical organization explains why SA1 afferents can resolve spatial details smaller than their receptive field diameters.[37]

Selective stimulation of SA1 afferents induces a sensation of light pressure although a larger number of impulses is required to elicit a percept in the attending subject than that required for FA1 units.[26] SA1 fibers are completely silent without mechanical stimulation of the touch dome. Sustained indentation evokes a burst of action potentials within an irregular firing pattern followed by a SA response which might continue for more than 30 min.[41] As mentioned above, SA1 afferents have a high spatial resolution and they respond to stimulus features such as curvature, edges, and points rather than indentation itself. SA1 afferents represent very accurately object curvature because they respond linearly to skin indentation to a depth of approximately 1.5 mm, independently of the force of application[42] (Figure 8.5). This makes them particularly suited for representing surface or object form. Two other facts support the major involvement of SA1 afferents in tactile exploration: (1) these afferents are 10 times more sensitive to dynamic than to static stimuli,[43] which makes them a key actor in haptic perception; (2) their responses to repeated identical skin indentations are almost invariant.[42]

8.2.1.4 Ruffini's Corpuscles

A distinct group of SA afferents thought to innervate mechanoreceptor was studied in the 1960s. After being called the "touch fields" in opposition to the "touch spots" standing for SA1 afferents, these afferents were finally called slowly adapting type 2.[44]

Spindle-shaped, the receptor is located in the dermis in hairy skin, but seems rather uncommon in the human glabrous skin.[45] Ruffini endings are small connective tissue cylinders of about 0.5–2 mm length supplied by one to three myelinated nerve fibers of 4–6 μm diameter. SA2 afferents are less represented than SA1 or FA1 afferents; they innervate the human hand more sparsely. Their receptive fields are about six times larger than those of SA1 afferents with ill-defined boundaries.[46] SA2 afferents exhibit sometimes a spontaneous tonic activity. Note that another type of SA afferents has been recently reported in the human hairy skin and named SA3 units. These so-called units share some features with both SA1 and SA2 units and are highly responsive to skin stretches[47] (Figure 8.6).

Although they are much more less sensitive to skin indentation than SA1 afferents,[13] SA2 units are four times more sensitive to skin stretches[48] and exhibit a dynamic sensitivity similar to that of muscle spindle afferents.[49,50] Stretch sensitivity of SA2 units exhibits directional responsiveness both in the glabrous[51] and in the hairy skin. This directional sensitivity makes SA2

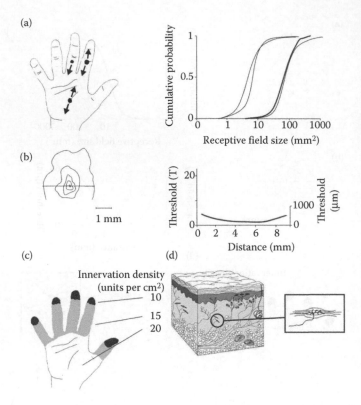

FIGURE 8.6 SA2 units have large receptive fields (a) with confused borders (b). They are widely and uniformly distributed in the skin areas (c) and probably connected to Ruffini endings hooked up to dermal structures (d). (After Vallbo AB and Johansson RS. *Human Neurobiol.*, 1984, 3: 3–14.)

afferents able to send a neural image of skin stretches useful for defining the movement direction of any object moving over the surface of the skin.[52] As they are very responsive to joint movement despite considerable distance from the receptive field[47] and they encode the direction of limb movements following a population encoding,[11] as do muscle afferents,[53] the proprioceptive responsiveness of SA2 afferents makes them key actors in kinesthesia although their selective stimulation does not evoke any sensation[26] or light sensations of movement or pressure occasionally.[54] As previously mentioned, kinesthesia relies not only upon muscle feedback but also upon cutaneous messages. Illusory sensations of movements are induced after activating cutaneous afferents not only at the hand[55] but also at other joints.[56] As regards the congruency between cutaneous and muscle feedbacks, cutaneous inputs and notably those coming from SA2 units may be used to overcome the ambiguities of muscle feedback affected by changes in muscle length and fusimotor drive.[39]

8.2.2 Mechanosensory Discrimination

Tactile acuity, that is, the accuracy with which a tactile stimulus can be sensed, is unequal from one part of the body to another one. It changes also along the life and, surprisingly, the larger reduction occurs in the midlife, between 30 and 40 years, and not in the elderly.[57] The age-related decline in tactile sensitivity is also unequal over the body as the feet and hands are the most impaired, a result that may be due to reduced circulation.[32]

As it is not easy to define the touch functions of the hand, many tests have been developed to quantify sensibility and they can be classified throughout a hierarchy from the simplest, which assess detection of stimulus, to the more sophisticated, which require identification of objects, shapes, or textures and often require active exploration to resolve the spatial features.

8.2.2.1 Test to Assess Detection Threshold

In 1906, Maximilian von Frey showed that the perceptual threshold for touch can be determined in humans by applying hairs of different diameters until the hair that evokes the requested sensation is found.[58] The so-called von Frey hairs have been improved by Semmes and Weinstein into the form of a monofilament kit.[59] Thanks to its low price and very simple use, this test is very popular to detect the degree of severity of the sensory disturbance in given areas of skin. It is indicated in diagnoses of nerve compression syndromes, peripheral neuropathy, thermal injuries, and postoperative nerve repairs. The nylon filament is pressed perpendicularly against the skin area to test and is expected to produce a calibrated force proportional to the filament diameter, as soon as the filament bends. The experimenter holds it in place for 1.5 s and then removes it. In peripheral-nerve compression syndromes at the carpal tunnel, the monofilament test of sensibility correlates accurately with symptoms of nerve compression.[60] A threshold greater than 0.07 g, which corresponds to the 2.83 monofilament size liable, is abnormal and reflects a "diminished light touch." As regards foot sole sensibility, the International Diabetes Federation and the World Health Organization considers as insensate the feet of people who do not detect the 5.07/10-g monofilament applied.

The validity of the monofilament test for evaluating touch threshold is well documented but none of the studies succeeded in correlating test score with actual function.[61] Much more problematic is the fact that the monofilaments seem simple to use, but they are not. They are fragile and can be fatigued after multiple uses. Besides these instrumental troubles, there is no guideline on how the monofilament is to be used or the results interpreted.[62] In healthy subjects, the results of testing with the five smallest monofilaments in the full Semmes–Weinstein kit are not reliable.[63] These are 1.65 (0.008 g), 2.36 (0.02 g), 2.44 (0.04 g), 2.83 (0.07 g), and 3.22 (0.16 g). As previously mentioned, this range includes the limit for considering impaired tactile function. This means that physicians should not be so confident in descriptors such as "diminished light touch" to decide for surgical or therapeutic intervention. As elegantly mentioned, "the probes are simple to use but easy to misinterpret."[64]

8.2.2.2 Test to Assess Spatial Discrimination

All the handbooks copy the same picture of a naked woman, throwing up her arms, adapted from the study of Weinstein (Figure 8.7).[65] The dotted and continuous lines refer to the

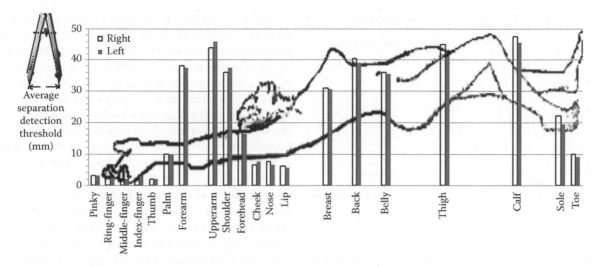

FIGURE 8.7 Two-point discrimination. Tactile acuity can be estimated by measuring the minimal distance required to perceive two stimuli simultaneously applied as distinct. This distance expressed in millimeters is reported in the ordinate axis, labeled here as mean threshold. (After Weinstein, S. In: DR Kenshalo, (ed.), *Skin Senses*. Vol. I11, pp. 195–222. Springfield, IL: Charles C. Thomas, 1968.)

smallest distance at which two points simultaneously applied can be discriminated by the subject. In the latter study as in the original publication of Weber from 1834, the same device was used to investigate the tactile acuity, the two-point discrimination test. This widely used test is a psychophysical evaluation of hand sensibility by quantifying the threshold at which distinction of different spatial properties of stimuli can occur. The classical values for healthy subjects are 2 mm at the fingertips but at least 40 mm for stimuli applied at the forearm.[59]

The two-point discrimination test is very popular and is often used to compare the effectiveness of different repair techniques although the data obtained are extremely variable, which raises the question of the validity of the test.[66] Originally, the test was proposed to assess the innervation density of afferent fibers[46] and this is still the explanation given in some handbooks. The spatial resolution explored by this test depends indeed upon the mean spacing between the receptive field centers of the SA1 afferents, which are highly represented at the fingertips. Changes in receptor density cannot, however, support on its own the extreme variations within subjects, between subjects and between studies.[67] The results of the test depend also upon the way the test is performed, as for monofilaments. For instance, the amount of pressure towards the skin affects considerably the results. Most importantly, the test is sensitive to cognitive functions, which explains very short term changes in two-point discrimination results reported after tactile stimulation[68] or de-afferentation of the contralateral hand.[66]

8.2.2.3 Tests for Tactile Gnosis

As previously mentioned, the former tests are lacking in functional significance and tactile gnosis tests have been thus developed to overcome these shortcomings. The earlier versions of tactile gnosis tests rely on fine motor control of the thumb and index and therefore are useful only for median nerve injuries. These tests require identifying everyday objects placed in a container but they suffer from the same defaults as the formers: there is a lack of standardization and protocols are insufficiently described.[61]

8.2.3 Proprioception

If we close our eyes, we are able to touch our chin even if we do not know *a priori* the precise location of our index finger before the movement has started. This subconscious ability refers to kinesthesia, that is, the sense of limb movement and position. Kinesthesia relies upon the sensory messages coming from a specific class of receptors called proprioceptors meaning "receptors for self." These receptors provide information about the mechanical forces arising from the musculoskeletal system that is essential for controlling limb movements, manipulating objects, and maintaining an upright posture.

Three types of mechanoreceptors are mainly involved in signaling the direction and speed of a movement or the stationary position of a joint. These are muscle spindles that are stretch receptors located in muscle belly, Golgi tendon organs that are force receptors located in the tendons, and finally receptors located in joint capsules that signal extreme flexion or extension of a given joint. As previously mentioned, cutaneous afferents also provide proprioceptive information that can be crucial, for instance, for controlling lip movements in speech.[69] One more time, the reader has to keep in mind that the academic subdivision of the sensory afferents that we make below does not reflect any functional relevance. The sensory endings presented below can respond to several and different stimuli. The labeled line concept, an assumption of specific information transmission in so specific ascending pathways, does not conform to known convergence and divergence patterns in central pathways.[70]

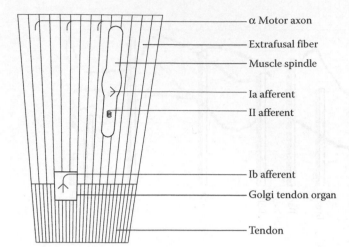

FIGURE 8.8 Location of muscle spindles and Golgi tendon organs in skeletal muscle. (Adapted from Houk, JC, Crago PW, and Rymer WZ. In: J.E. Desmedt (ed.), *Spinal and Supraspinal Mechanisms of Voluntary Motor Control and Locomotion, Volume 8: Progress in Clinical Neurophysiology.* Basel: Karger, 1980.)

8.2.4 Muscle Spindles

In the belly of skeletal muscles, muscle spindles run in parallel with the extrafusal fibers. They consist of specific muscle fibers, the intrafusal fibers, protected by a capsule of connective tissue (Figure 8.8).

Depending on the arrangement of their central nuclei, two types of intrafusal fibers are distinguished. In the few largest fibers, the nuclei are accumulated in a central bag and they are the so-called bag fibers. The nuclei in the smallest fibers are lined up along the central row, so these intrafusal fibers are called chain fibers.[71] The spindle density changes among muscles and this reflects functional differences. Muscles initiating fine movements or maintaining posture exhibit high spindle density, as extraocular, neck, and hand muscles. By contrast, muscles initiating coarse movements as the biceps brachii have a poor density of muscle spindles.[72] Muscle spindles are sensory receptors that signal muscle length and changes in muscle length to the central nervous system (CNS). They receive both sensory and motor innervation. The sensory innervation is distributed along the equatorial regions. Large-diameter Ia axons terminate in primary endings that encircle the central area of both types of intrafusal fibers. As they encode changes in muscle length, Ia afferents have a dynamic sensitivity. Secondary afferents made of lower conduction velocity II fibers innervate the nuclear chain fibers and terminate in secondary endings that exhibit a static sensitivity by encoding sustained muscle lengthening (Figure 8.9).

The motor innervation is distributed along the polar regions and we can distinguish gamma motoneurons that are specific for intrafusal fibers from beta motoneurons that innervate not only the former muscle fibers but also extrafusal fibers.[73] Fusimotor neurons can be furthermore subdivided between dynamic and static fibers.[74] The static fusimotor system may be

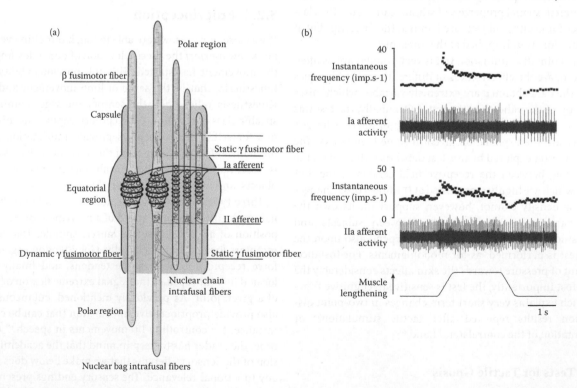

FIGURE 8.9 (a) Simplified diagram of muscle spindle structure in the cat (After Vedel, personal communication). (b) Typical response of Ia and II afferents to ramp-and-hold movements imposed in humans. These movements are used to classify the afferent as primary (upper part) or secondary (lower part). In each diagram are illustrated from the bottom to top the imposed movement, the unitary activity and the corresponding instantaneous discharge. (After Hospod, V, et al., *J. Neurosci.*, 2007, 27(19): 5172–5178. Reprinted with permission of the Society for Neuroscience.)

helpful to prevent spindle silencing during muscle shortening, allowing the CNS to receive feedback throughout the movement, whereas the dynamic system may be concerned with gain control during small perturbations or may fit the spindle sensitivity to the performance of large movements.[75] In line with these assumptions, task-related changes in muscle spindle sensitivity have been recently reported to occur in human muscles (Figure 8.10).[76]

Ia afferents are at the origin of the neural circuit responsible for the stretch reflex. When a muscle is lengthened, the Ia afferents from the primary spindle endings in the parent muscle are activated and their firing rate increases (see Figure 8.10b). The Ia inputs induce a monosynaptic excitation of the motoneurons innervating that muscle, the consequence of which is its reflex contraction. This monosynaptic reflex activation spreads also to the motoneurons innervating the synergistic muscles.[77] At the same time, Ia afferents exert strong inhibitory effects upon the motoneurons of the antagonistic muscles through the activation of Ia inhibitory interneurons. This disynaptic pathway is the neural circuit responsible for reciprocal inhibition, another reflex mechanism that prevents contraction of the antagonistic muscle during the voluntary or the reflex activation of a given muscle.[78] These coordinated motor responses, although involuntary, are highly task-dependent and their efficiency is regulated by local and descending signals liable to affect either presynaptically or postsynaptically the excitability of motoneuronal pools. These regulatory processes explain why a lesion to the spinal cord affects spinal reflexes. Immediately after a complete transection, spinal reflexes can be suppressed. Following this period of spinal shock, spinal reflexes gradually recover by one or two weeks and can sometimes return abnormally enlarged. This hyperreflexia is often seen with spasticity (Figure 8.11).[79]

Beyond their reflex effects, muscle spindles are widely assumed to play a major role in kinesthesia since the 1970s. This assumption is strengthened by the results of famous psychophysical experiments. Applying vibratory stimulation to a muscle tendon induces a powerful illusory sensation of limb movements.[80] More precisely, tendon vibration causes the sensation of an illusory movement, the direction of which is that of the real movement that would have stretched the receptor-bearing muscle. Microneurographic recordings in humans have shown that tendon vibration strongly activates primary muscle spindle endings.[81] Since these pioneer studies, it is known that kinesthesia relies upon a population encoding. All the muscles surrounding a joint provide the muscle spindle information required for the coding of movement parameters.[82]

8.2.4.1 Golgi Tendon Organs

These sensory receptors are located at the musculo-tendinous junctions and connected in series with the extrafusal muscle fibers. The encapsulated structure measures about 1 mm long and 0.1 mm in diameter. Each tendon organ is innervated by a single sensory afferent, a Ib axon, that divides into many fine endings intertwined among the collagen fascicles.[83] When a muscle contracts, the force developed acts upon the tendon and tension increases. This mechanical stress distorts the collagen fascicles and compresses the intertwined sensory endings, which causes them to fire. Golgi tendon organs are thus liable to encode changes in muscle tension, whereas muscle spindles encode rather changes in muscle length. As muscle spindles, Golgi tendon organs encode muscle force in keeping with a population encoding since the averaged discharge of a population of Ib afferents reflects accurately the total force developed in a contracting muscle.[84] Starting from data collected during normal stepping in cats analyzed with the help of mathematical models, it has been shown indeed that firing profiles of only four Ib afferents perfectly predict the directly measured muscle force.[85]

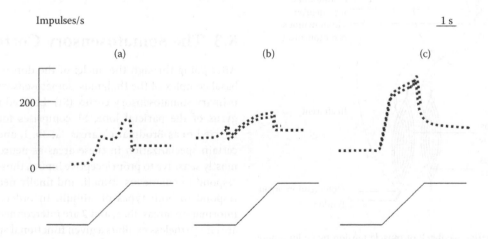

FIGURE 8.10 Effects of static and dynamic fusimotor activation upon the discharge of a single afferent from a spindle primary ending. (a) Applying 6 mm stretch to the parent muscle induces a typical peak of activity in the Ia afferent. (b) Repetitive stimulation of a single static fusimotor axon induces a biasing effect and attenuates the Ia stretch sensitivity; the afferent becomes more sensitive to sustained muscle lengthening. (c) Repetitive stimulation of a single dynamic fusimotor axon, by contrast, strongly increases the stretch sensitivity and the same afferent becomes more sensitive to changes in muscle length. (Adapted from Matthews PBC. *Mammilian Muscle Receptors and Their Central Action.* London: Arnolds, 1972.)

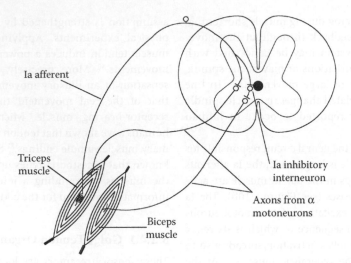

FIGURE 8.11 The Ia monosynaptic reflex loop. Stretching the biceps leads to increased activity in Ia afferents that in turns reflexively induces an increased activity in the α motoneurons innervating the same muscle. At the same time, Ia inhibitory interneurons are activated to reciprocally inhibit the α motoneurons that innervate the antagonistic muscle.

At the spinal level, Golgi tendon organs project upon Ib inhibitory interneurons before reaching the homonymous alpha motoneurons. This spinal pathway, so-called the autogenic inhibition, is viewed as a protective reflex mechanism liable to prevent muscle damage due to an excessive muscle tension although the functional significance of this autogenic inhibition is discussed (Figure 8.12).[86]

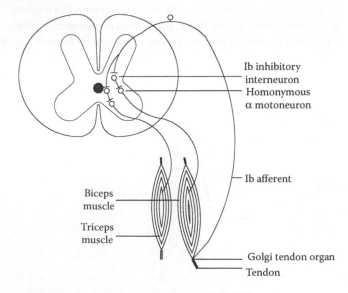

FIGURE 8.12 Negative feedback of muscle tension by Golgi tendon organs. When a muscle contracts, tension increases in the collagen fibrils in the parent tendon. This increased tension activates the endings from Golgi tendon organs. The Ib volley induced reaches Ib inhibitory interneurons in the spinal cord that, in turns, inhibit the α motoneurons from the same muscle to prevent excessive tension.

8.2.5 Joint Receptors

Synovial joints receive two types of sensory afferents: the primary articular nerves specifically supply the joint capsule and ligaments, whereas the accessory articular nerves reach the joint after innervating muscular or cutaneous tissues. These two types of afferents consisting of Aβ fibers terminate in several and different encapsulated and unencapsulated nerve endings sensitive to mechanical, chemical, and thermal stimuli. Among the different types of afferent nerve endings reported in joints, only the encapsulated endings are thought to encode extreme joint motion, rather than the midrange. Beyond their involvement in kinesthesia, these mechanoreceptors may also contribute to protective reflexes.[87]

8.3 The Somatosensory Cortex

After going through the nuclei of the dorsal columns and the basal complex of the thalamus, somatosensory inputs reach the primary somatosensory cortex (S1). Located in the postcentral gyrus of the parietal lobe, S1 comprises four different fields referred to as Brodmann's areas 3a, 3b, 1, and 2. There exists a certain specialization in these areas as neurons in area 3a are mostly sensitive to proprioceptive inputs, those in areas 3b and 1 respond to cutaneous stimuli, and finally neurons from area 2 respond to both types of stimuli. In order to process tactile information, areas 3b, 1, and 2 are interconnected. Each of these area nevertheless exhibits a given functional specialization since induction of selective lesions in S1 induces area-specific deficits in tactile discrimination. Lesion of area 1 impairs texture tasks, roughness and line discriminations, whereas lesion of area 2 affects finger coordination and angle tasks, size, and curve discriminations. As regards area 3b, lesion severely disturbs all

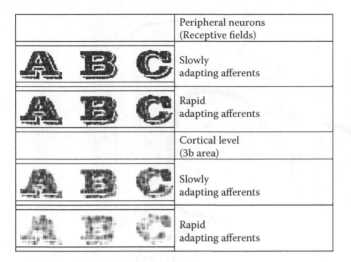

	Peripheral neurons (Receptive fields)
A B C	Slowly adapting afferents
A B C	Rapid adapting afferents
	Cortical level (3b area)
A B C	Slowly adapting afferents
A B C	Rapid adapting afferents

FIGURE 8.13 Spatial event plots reconstructed from discharge of peripheral and central neurons recorded in rhesus monkeys. (a) Both SA and RA afferents (upper and lower traces, respectively) transmit isomorphic neural images of embossed letters applied to their receptive field. (b) At the cortical level in 3b area, SA neurons (upper trace) still exhibit isomorphic responses, whereas the responses of RA neurons have less spatiotemporal structure. (After Phillips JR, Johnson KO, and Hsiao SS, *Proc. Natl. Acad. Sci. USA*, 1988, 85(4): 1317–1321.)

types of tactile discrimination abilities apart from coarse size discrimination.[88]

In the early stages of cortical processing, central neurons exhibit similar encoding properties as peripheral neurons. This is notably the case for SA neurons recorded in area 3b. They exhibit indeed isomorphic responses as acute as those obtained from peripheral afferents, suggesting that they play a major role in processing information underlying tactual pattern recognition (Figure 8.13). This is less true for RA neurons of area 3b; their responses exhibit a smaller spatiotemporal structure.[89] After S1, the higher processing of tactile information may involve the second somatosensory area in keeping with a serial cortical processing similar to that found earlier for vision.[90]

8.3.1 From Columns to Homunculus

By moving microelectrodes step by step in monkeys' brain, Vernon Mountcastle and his colleagues have discovered the repetitive columnar organization of the somatosensory cortex. Cortical columns vary from 300 to 600 μm in diameter and they run vertically along the different layers of the cortex. Neurons of the same column share several properties: they all receive inputs from the same skin area and respond to the same type of receptors. In area 3b for instance, some columns process light touch, while some other ones take care of pressure information. As in the case of peripheral receptors, cortical neurons and columns are either FA or SA. This modality clustering becomes weaker in area 2 and higher cortical areas.[91] Obviously, this columnar organization is an anatomo-functional correlate of the preservation of the topographic arrangement of the receptive fields from the skin to the brain.

Tactile acuity does not only rely upon a high density of receptors and related neural networks. At the dorsal column nuclei, spatial resolution is sharpened by inhibitory networks. For instance, a tactile stimulation applied to the skin area surrounding the receptive field of a cortical neuron may be hardly perceived because the inputs coming from this surrounding area are conveyed through inhibitory interneurons located in the dorsal column nuclei, the ventral posterior lateral nucleus of the thalamus, and even in the cortex. This so-called surround inhibition gradually increases from the periphery to area 3b and then to area 1.[92]

As previously mentioned, from the spinal cord to cortical areas, there exists a mapping of the body. Maps of the brain activity and brain volume involved in the signal processing have been created over the years dating back to the early 1900s. The so-called homunculus has been historically discovered in humans in the early 1930s, although Hughlings Jackson made the assumption of a cortical somatopic organization in 1870 by observing epileptic patients. An electrical stimulation applied to discrete areas of human cortex produces stereotyped motor or sensory phenomena that affects a given part of the body.[93] Some years later, this original study was fully confirmed and the famous cortical map was published[94] (Figure 8.14).

The cortical maps do not represent the body in its actual proportions. The homunculus, that is, little man, exhibits enlarged hands and lips related to the manipulatory and speaking skills in humans. The shape of the cortical maps changes across species. For instance, whiskers are greatly represented in the somatosensory cortex of the rat and we can imagine that trunk is largely represented in the elephant S1, although this has never been studied, to our knowledge. The original map of Penfield and Rasmussen has been so many times reproduced that one can really wonder what is actually represented on these maps.

The homunculus exhibits discontinuity. For instance, the cortical representation of the human penis is located in the mesial wall, that is, to say below the toes. Until recently, we thought that discontinuation resulted from the repetitive stimulation from the feet tucked in by the genitals during the fetal life. Although this may illustrate the phenomenal of plasticity,[118] recent neuroimaging data suggest that the genitals may rather be represented in the logical place at the cortical level, that is, to say between the leg and the trunk.

8.4 Thermoception

The sense of temperature has been studied for a long time in keeping with the theory of specificity of von Frey. Although specific receptors have been reported to encode innocuous temperatures, that is, cold and warmth, and noxious temperatures, that is, cold and heat, cold sensitivity can be also found in large myelinated afferents, which respond rather to mechanical stimuli. SA afferents coming from Merkel or Ruffini corpuscles respond to cooling thermal gradients from normal skin temperature (34°C) to 14.5°C. The temperature sensitivity of these mechanoreceptors

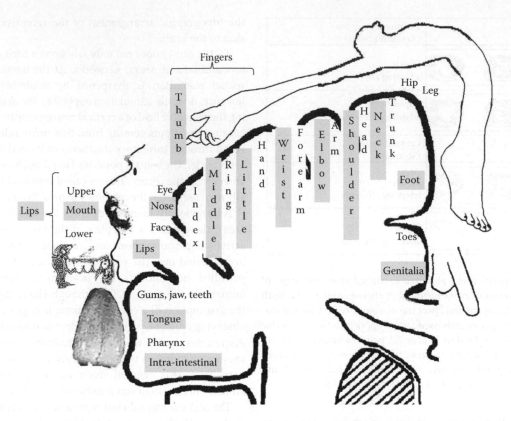

FIGURE 8.14 Homunculus from human S1 (postcentral gyrus) shown on a cross section through the hemisphere. (After Penfield W and Boldrey E. *Brain*, 1937, 60: 389–443.)

provides an explanation for the Thaler illusion, the perception that cold objects are heavier than warm ones.[95]

Thermal sensations result from the analysis of the difference between the temperature of the external stimulus and the normal skin temperature. Thermal responses are elicited from free nerve endings located in both the epidermis and dermis. Thermoreceptor endings are connected to Aδ myelinated as regards cold receptors (sensitive to temperatures ranging from 25°C to 34°C) and heat nociceptors (responsive to temperatures above 45°C) or to C unmyelinated fibers as regards cold nociceptors (sensitive to temperatures below 5°C) and warm receptors (from 34°C to 45°C).[96]

Thanks to the increased popularity of spicy food, the mechanisms of temperature transduction have been explored for the last 10 years. Capsaicin, the principal active agent of hot peppers, activates polymodal nociceptive C fibers by opening transient receptor potential (TRP) channels through which Na+ and Ca²+ ions can enter and thus change the membrane potential. Among the other stimuli that activate the cloned capsaicin receptor, anandamide (an endogenous cannabis-like compound), acid, and heat above 43°C were found efficient.[97] In humans, 28 different TRP channels have been identified and grouped into six families. Each of these receptors operates over a specific temperature range. For instance, seven TRP channels are sensitive to heat, whereas two are more liable to encode cold temperature. Although there is still debate, the transduction

mechanism may involve allosteric interactions between voltage and thermal sensors.[98]

When the skin temperature is set at 34°C, its usual value, both cold and warm receptors exhibit a tonic discharge at low rate although that of the cold receptor is higher.[99] Cold receptors exhibit the highest responsiveness at 25°C, whereas warm receptors are the most responsive at 45°C. For each type of receptor, temperatures departing from these critical values evoke weaker responses. This means functionally that a single receptor cannot precisely encode temperature. To clarify, a population encoding is required where each population of thermoreceptors has a preferred temperature, that is, a temperature value where its sensitivity peaks, and the perceived temperature reflects the activity of pooled responses. This population encoding also explains why we should go naked at the balcony every morning to accurately evaluate air temperature! The larger the skin area exposed, the greater temperature acuity (Figure 8.15).

As regards the ascending pathways of thermal messages, the thermal grill illusion clearly illustrates the strong interactions between thermal and nociceptive systems. The Danish scientist Thunberg was the first to describe this illusory sensation in 1896. Touching interlaced warm and cool bars elicits a sensation of strong, often painful heat, although each single stimulus is innocuous. The mechanism responsible for this illusory sensation has been recently explained and is view as an unmasking phenomenon. The pain sensation may result from the central

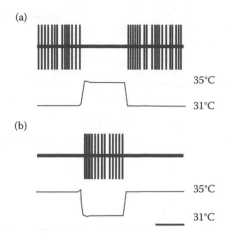

FIGURE 8.15 Resting discharge of a cold C fiber at room temperature. (a) The resting discharge is temporarily suppressed by sudden warming of the receptive field (RF) from 31 to 35°C. (b) From a holding temperature of 35°C, at which the unit is silent, activity is initiated by cooling the RF to 31°C. Time bar: 5 s. (After Campero M, Serra J, Bostock H, and Ochoa JL. *J Physiol.*, 2001, 535: 855–865.)

disinhibition of the polymodal nociceptive spinothalamic channel due to a lower activity in the innocuous thermoreceptive spinothalamic channel.[100] Three types of thermal afferents are activated: those providing warm sensation, others for cool sensation and those connected to nociceptors. When warm and cool bars are interlaced, the warm stimulus may inhibit the ongoing discharge of the cells selectively sensitive to cold stimulus and rather activate the C polymodal nociceptive channel. The thermal grill produces the same pattern of cerebral activation as that produced by the noxious cold stimulus, that is, an increase in blood flow in the mid/anterior insula, in S2 and in the anterior cingulate, and even weaker activation in S1.[101] This study elegantly confirms that illusions can lead to useful insight into the basic mechanisms that support the running of the CNS, for instance, the generation of central pain syndromes. Finally, the studies of Craig and colleagues support evidence that innocuous cold inhibits central pain processing, a finding that may have a clinical relevance.

8.5 Pain

Pain is a submodality of somesthesis. Unlike other modalities, pain is a warning sensation that indicates impending or actual tissue damage. As it is a subjective experience involving awareness, memory and affect, influenced by emotional state and environment, pain shows a large inter-individual variability.[102] In order to survive, animals and humans must be able to "remember" particular noxious stimuli and what they imply. In thousands of years of mankind, pain has been a major concern. People in ancient times easily understood pain as a consequence of injury, but pain caused by disease was linked to evil spirit intrusion into the body and was treated by fighting away the demons or satisfying the gods. Ancient Egyptians and Babylonians believed that the origin of pain were God's

will or the spirit of the dead, while Chinese medicine 2500 years BC attributed pain to an imbalance of Yin and Yang. The ancient Greeks were interested in speculating about body physiology, and Pythagoras's (566–497 BC) thoughts instigated his pupil Alemaeon to study the senses. He was the first to express the idea that the brain was the center for sensation.[103] In his studies, Erasistratus attributed feelings to nerves and the brain. His work was lost until Galen (130–201 AD) recovered it and then provided anatomical studies about the nervous system. The Dark Ages in Europe resulted in suppressing those studies and medical science moved to Arabia, where Avicenna was the first to suggest that pain is a separate sense. The Renaissance brought science back, and Descartes (1596–1650) propounded that pain is conducted via "delicate threads" in nerves that connect the body tissues to the brain,[104,105] and that was the starting point for the knowledge we have today.

Pain, we know, is such an immense subject that even a large book cannot explain all we know so far. Researchers every year, even month, uncover new details about the secret life of pain. "Every living being from its moment of birth seeks pleasure, enjoying it, while rejecting pain, doing its best to avoid it" wrote the seventeenth century poet Racine. Even that is not completely true—it is known that pain could be experienced as pleasure, in the conditions medically known as algolagnia (from the Greek—meaning "lust for pain") and sadomasochisms.

8.5.1 Nociceptors

Pain sensations originate from the activation of specialized sensory receptors called nociceptors, the sensory endings of small-diameter afferent nerve fibers classified into groups III and IV, or Aδ and C, respectively. Unlike the specialized somatosensory receptors for touch or pressure, nociceptors are free nerve endings. Numerous receptors in skin, subcutaneous tissue, muscles and joints are sensitive to a variety of potentially harmful mechanical, thermal and chemical stimuli, and their afferent nerve fibers are responsible for conveying sensory information that is perceived as pain by the CNS. They respond directly to noxious stimuli and indirectly to other stimuli such as chemical stimuli released from injured tissue. Several substances have been proposed to act as chemical pain inducers- histamine, K^+, bradykinin, substance P, serotonin, capsaicin and synthetic derivatives, such as resiniferatoxin, and so on.[106–109] Bradykinin is one of the most potent pain-producing endogenous substances.[110–112] It vigorously activates muscle nociceptors and induces longer-lasting effects than, for example, serotonin or KCl in experimentally induced pain.[113,114] Also, bradykinin acts more selectively on receptors in muscles than in skin.[115]

Most nociceptors are sensitive to the concentration of irritant chemicals released by noxious thermal or mechanical stimuli or to exogenous chemicals that may penetrate the skin and come into contact with sensory endings. Using this approach, three classes of nociceptors can be distinguished on the basis of the type of the activating stimulus: mechanical and thermal, activated by a

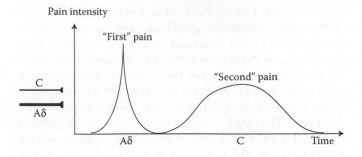

FIGURE 8.16 Schematic representation of sharp and aching pain. "First" (sharp) and "second" (aching) pain are carried by two different primary afferent axons: Aδ and C fibers.

particular form of stimuli (pinch, squeeze of the skin, and extreme temperature, both heat above 45°C and cold below 5°C), and polymodal, sensitive to destructive effects of stimuli (mechanical, thermal, or chemical) rather than to its physical properties.[116] These classes of nociceptors are distributed in skin and deep tissues and often respond together—Aδ mechanical transmit sharp pain, Aδ thermal-mechanical transmit burning pain, and C thermal–mechanical transmit freezing pain, while C polymodal transmit slow, burning pain (see Table 8.1). When one hits a finger with a hammer, or one is hit by a ball in a certain region of the body, first a sharp pain is felt, followed later by prolonged aching pain, the pain that comes second. The fast sharp pain is transmitted by Aδ fibers, while the later dull, annoying pain is transmitted by C fibers (Figure 8.16).

Pain is a complex perception, a threat of potential or actual tissue damage associated with an unpleasant emotional experience. A distinction has to be made between pain as a perception and the neural mechanisms of nociception—nociception does not necessarily lead to the experience of pain, and this relationship describes a concept where perception is a sum of processed sensory inputs by the brain. Many factors in addition to the firing of the nociceptors determine the location, intensity or quality of experienced pain (Figure 8.17). This is because action potentials in the population of slowly conducting nonmyelinated axons are dispersed in time, and the current generated by an action potentials in those axons is smaller than the current in myelinated axons.

We all know that different forms of pain sensations can be experienced. Categories of pain can be established in various ways, stressing different aspects of pain generation, duration, maintenance, and subjective characteristics[117]:

- Nociceptive pain—normal acute sensation evoked by a potentially damaging stimulus, generated as a result of physiological activity in pain receptors.
- Clinical pain—manifests as *spontaneous pain* without evident origin or noxious stimulation, and pain *hypersensitivity*, which appears as a result of activity-dependent neuroplastic changes in nociceptors and spinal cord neurons.

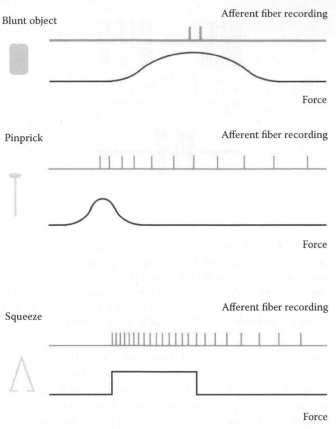

FIGURE 8.17 Schematic representation of nociceptor response based on the shape of an object, for example, blunt, pinprick, and squeeze. Mechanical nociceptors have a differential response (against a "reference"), as compared to magnitude only.

- Chronic pain—involves a multitude of mechanisms by which the sensory function of nociceptive afferents and their central connections can be altered, initiating and maintaining plastic changes in nociceptive transmission and processing.[118] Chronic pain can be generally divided into subcategories: *inflammatory pain*—as a consequence of tissue inflammation, and *neuropathic pain*, which results from some form of nerve injury or dysfunction of the central or peripheral nervous system, and often manifests itself as burning sensation. Neuropathic pain may be further divided into spontaneous or stimulus-induced. Furthermore, stimulus-induced neuropathic pain is divided into *allodynia* (pain induced by a stimulus that is normally not painful) and *hyperalgesia* (increased pain level induced by a stimulus that normally causes pain).

Hyperalgesia can appear in and around the site of injury. The hyperalgesia occurring at the site of injury is defined as *primary* and is responsive to heat and mechanical stimuli,[119] while the hyperalgesia that occurs in a wider area of undamaged surrounding skin is defined as *secondary* and is responsive mostly to mechanical stimuli. Primary hyperalgesia results from the sensitization of the nociceptive receptors at the injury site, while secondary hyperalgesia is due to plastic changes in the CNS.[120] Peripheral mechanisms may also contribute to hyperalgesia. Nociceptor sensitization denotes a greater sensitivity of nociceptors in skin and deep tissues, and is manifested by an increase in spontaneous discharge rate, a lowered threshold, prolonged responses to stimuli, a sensitization to chemical stimuli and heat, an expansion of receptive fields, and so on.[121,122]

Chronic muscle pain is a growing problem, causing not only large sufferings of those affected, but also large costs to the modern society, due to increased sick leave, early retirement, and production losses. Chronic work-related musculoskeletal disorders account for the majority of all occupational illnesses. Common symptoms in patients suffering from these disorders include pain and discomfort in muscles and a feeling of stiffness and fatigue of the painful muscles. The pain is likely to originate in the activation of sensory receptors, but within the CNS, these signals will evoke the sensation of pain and at the same time initiate and maintain a complex assembly of parallel and serial processes that ultimately influence many body systems. The body systems modulated by nociceptive signals include the emotional–affective system, the neuroendocrine system, the autonomic nervous system and the motor control system. The activation of these processes may in turn augment the peripheral processes underlying the activation of nociceptors, thus leading to feedback actions.[123] Furthermore, prolonged nociceptive activation may permanently change the ways in which nociceptive signals are processed and thus induce and sustain chronic pain. Research findings suggest complex multiple interactive pathophysiological mechanisms, which may work at different times in the disease process. Related pathophysiological mechanisms might be prolonged, sustained or repetitive muscle activation,[124,125] since muscle fibers belonging to low-threshold motor units are the first to be recruited and the last to be de-recruited, according to the ordered recruitment principle.[126] Also, muscle blood flow dysregulation may increase the risk of accumulation of metabolites and inflammatory substances and augment the activation of group III and IV muscle afferents, thus potentially giving rise to a vicious circle, maintaining the pain.[127]

8.5.2 Evaluating Pain

Pain comes in many flavors—causalgia, migraine, muscular and visceral pain, central pain, dental pain, referred pain, and so on. Pain from infectious, inflammatory or neoplastic processes is a feature of many visceral diseases. In evaluating patients with pain, it is important to determine the level of the nervous system at which the pain arises. Attention of the physician should be focused on the duration, nature, severity, mode of onset, location of the pain and any associated symptoms. In terms of the underlying cause and clinical context, pain can be divided into:

- Peripheral nerve pain—arising from peripheral nerve lesions. When mixed nerves are involved, there may be accompanying motor deficits. The *causalgia* is the term to describe severe persistent, burning pain resulting from nerve trauma, and such pain often radiates to a wider area. *Complex regional pain syndrome type I*, or reflex sympathetic dystrophy, denotes sympathetically mediated pain, precipitated by a tissue or bone injuries.
- Radicular pain—it is localized to the distribution of nerve roots and it is exacerbated by coughing, sneezing, or other maneuvers that increase intraspinal pressure or stretch the affected roots. Additionally, paresthesia, numbness in a dermatomial distribution, muscle weakness, and reflex changes could be seen.
- Thalamic pain—thalamic lesions may lead to pain in all or part of the contralateral half of the body.
- Low back and/or neck pain—spinal disease can cause pain, referred to other parts of the involved dermatomes. Local pain may lead to reflex muscle spasm, which in turn causes further pain, and may result in posture changes and limitation of movements. Causes of low back or neck pain could be trauma, prolapsed lumbar intervertebral or cervical disk, osteoarthropathy, spondylosis, osteoporosis, disease of the hips, cervical injuries such as whiplash, and so on.
- Herpes zoster pain—burning pain in involved dermatome caused by reactivated latent viral infection, with inflammatory reaction in the dorsal root or cranial nerve ganglia, root, or nerve.[128]

The measurement of human pain involves both voluntary and involuntary response activity. Involuntary responses, comprising autonomic nervous system activity such as galvanic skin response, heart rate or muscle tension, are less stable and reliable, and they might represent arousal rather than pain.[129] Also, for example, skin response may disappear, while pain is still present. Voluntary response is probably a better tool to measure pain, even though critics of that method claim that it is under the direct control of the subject or patient and therefore could be distorted or falsified. Categories used in voluntary estimation of pain are pain threshold, defined as the point at which pain is perceived 50% of the time, then pain tolerance, regarded as an upper threshold, pain sensitivity range, just-noticeable-difference, and so on.[130] A test commonly used for the clinical evaluation of small-caliber afferent pathways is the quantitative somatosensory thermotest (QST),[131] where a ramp of ascending and descending stimulating temperatures is applied to the skin. This test does not have very precise diagnostic value in pain evaluation but could point toward certain abnormalities. However, a simple, widely accepted and clinically useful method is to rate pain intensity and measure changes in pain over time or response to medical procedure or treatment by using verbal or visual analog scales (VAS). VAS rates pain subjectively between 0 and 10, zero being "no pain" and 10 being "the worst pain one can imagine."[132-134]

8.5.3 Pain Modulation

Pain, especially muscle pain, has functional implications on our motor activity and affects movements and performances of our daily life. Painful limb draw our attention and we try to protect it by limiting its usage, employing other nonpainful muscles, by mechanisms that we have just briefly mentioned in this Chapter. Also, we are developing different facial expressions, and pain can be visualized by our surroundings. However, it is evolutionarily useful that in animals pain might be modulated and alleviated, even in the case of serious injury, due to the serious threat of predator attack. Modulation refers to behavioral and neuronal changes mediated by neuronal systems that participate in control of transmission and processing of pain. The activity of those systems is context dependent, the factors being task, arousal, attention and motivation.[135,136]

There are a number of pain-modulating systems, both ascending and descending, described as gate control theory, placebo analgesia or placebo effect, endogenous opioides, peptidergic modulation, modulation by a system of brainstem regions (periaqueductal gray matter, nucleus cuniformis, nucleus raphe magnus and nucleus reticularis gigantocellularis, locus coeruleus, etc.). Activation of ascending tracts, which may be perceived as painful, can be modulated by descending pathways from the brain. The CNS pain-modulating systems can either suppress or enhance the transmission of nociceptive information. The suppression of nociception engages several levels of the CNS, involving supraspinal descending inhibition, propriospinal heterosegmental inhibition and segmental spinal inhibition.[137] The latter two concern spinal inhibitory systems referred to as *Gate control* and *Diffuse Noxious Inhibitory Control*, while the former involves several *descending pain modulating systems* distributed throughout the forebrain, diencephalon, brainstem and spinal cord.[136]

8.5.3.1 The Gate Control Theory

It can happen to all of us that we hit the head against the door of a car or a house, or to get a finger hit by some piece of furniture. In that case we alleviate the pain by rubbing the surrounding skin. As an explanation of this phenomenon, it has been proposed that activity in large afferent fibers may modulate activity in the pain pathways. Melzack and Wall[138] proposed the gate control theory of pain suggesting that activity in low-threshold mechanoreceptors inhibits the discharge in second-order nociceptive afferents through inhibitory interneurons. Wide-dynamic range neurons (WDRNs) receive excitatory input from both group IV fibers, transmitting the nociceptive information, and large diameter afferents, and without additional activation of mechanoceptors, nociceptive information is transmitted unaffected. A study of segmental interaction between electrically evoked sensitive cutaneous mechanoreceptors with thick myelinated axons and nociceptor afferents resulted in suppression of nociceptive responses of dorsal horn neurons (DHNs).[139] De Koninck and Henry[140] investigated the effects of skin vibration on nociceptive DHNs, a stimulation procedure that could have induced activity in deep receptors. Studies concluded that the effects resulted from spinal cord neuronal circuitry. Recently, gate control theory was questioned with experiments showing sensitization and expansion of receptive areas of WDRNs following a transient nociceptive input to the spinal cord.[141] These plastic changes would allow input from low-

threshold mechanoreceptors to excite second-order neurons, signaling pain.

However, based on the gate-theory assumption, it is proved clinically useful to stimulate large diameter afferents via transcutaneous electrical stimulation (TENS) to alleviate some forms of chronic pain. TENS at low intensity activate group I and II fibers and elicits short-lasting analgesia, confirming practically the gate-control theory.[142] The study of Tardy-Gervet and colleagues[143] showed that combined stimulation of vibration application and TENS provided pain relief in a larger number of pain patients than either vibration or TENS used alone, and the relief was greater and longer-lasting.

Another form of nociceptive modification might be the muscle stretching. In the treatment and rehabilitation of chronic painful musculoskeletal disorders, stretching of the stiff and tender muscles is a commonly used method to alleviate the pain.[112-114] Hypothetical mechanisms underlying the effects of stretching usually implicate peripheral events, such as changes in the viscoelastic properties of muscle-tendon elements.[143] However, the mechanism might also reside along the pain pathways, where nociceptive processing is modifiable by afferent inputs originating from various peripheral receptors and tissues, in this case muscle.[144] Muscle afferent inputs might exert modulatory effects on muscle nociception at levels of overlapping between noxious and nonnoxious inputs, in Rexed's laminae I and II.[122,136,145]

Diffuse noxious inhibitory control[146] is based on a propriospinal heterosegmental inhibitory system, by which "counterirritation" on a distant part of the body may inhibit the pain evoked by nociceptive stimuli in another part. Those prolonged, distant noxious stimuli may underlie treatment methods such as acupuncture, mustard plaster and deep tissue massage.

8.5.4 The Endogenous Opioids

The role of endogenous opioids in the modulation of nociception and pain perception is hinted by findings that morphine exerts pharmacological action by binding to specific receptors and that opioid antagonist naloxone blocks morphine-induced analgesia.[147] The endogenous opioids regulate nociception by two inhibitory mechanisms: a postsynaptic and presynaptic inhibition. The effect of postsynaptic inhibition is produced mainly by increasing K^+ conductance, which in turn decreases Ca^{2+} entry into the sensory terminals and decreases transmitter release from primary afferents. The presynaptic inhibition of release of glutamate, substance P and other transmitters from the terminals of the sensory neurons is conducted directly through a decrease in Ca^{2+} conductance.[148]

Three classes of opioid receptors have been described: μ, δ, and κ. All three classes of opioid receptors are located on terminals of the nociceptive afferents and the dendrites of postsynaptic dorsal horn neurons.[149] Also, three classes of endogenous opioid peptides have been identified: enkephalins, β-endorphin and dynorphins. Enkephalins bind to μ and δ receptors, β-endorphin is released in response to stress, while dynorphin is selective for κ

receptors. The μ receptors, where morphine and naloxone bind, are located in the periaqueductal gray matter, the ventral medulla and the superficial dorsal horn, as well as along the nervous system in many other structures.[150] Similar localization is also found for cells and axons releasing enkephalins and dynorphins, while β-endorphin cells are found in hypothalamus. Each of the peptides is definitively located in areas of nociceptive processing and modulation.

The use of opioids to manage moderate and severe pain in patients with cancer is well accepted. In contrast to the local anesthetics, which act at the spinal nerve root level, opioids act selectively at receptor sites. However, due to the widespread distribution of opioid receptors, not only in areas responsible for pain mediation, but also in the brain stem, responsible for respiration and cardiovascular regulation, use of opioids might cause significant side effects. An overdosage or overstrength of opioids such as morphine has life-threatening consequences, including respiratory depression (difficulty in or lack of breathing) and low blood pressure. Therefore, instead of systemic injection of, for example, morphine for the treatment of patients' pain, epidural injection produces sufficient analgesia, with minimal side effects because the injected substance does not diffuse into the systemic circulation. For continuous chronic pain, opioids should be administered around-the-clock. Slow dose titration also helps to reduce the incidence of events such as nausea, vomiting, constipation and sedation, thus maintaining tolerability to chemotherapy.[151,152]

An example of peptidergic modulation is neuropeptide Y (NPY). NPY receptors are distributed in nerve terminals throughout the dorsal horn, suggesting a role in the processing of sensory inputs. NPY receptors could distribute to include the number of cells after nerve injury, and may also mediate some analgesic effects in inflammation-induced pain by acting either on primary afferent terminals or on second-order neurons to inhibit sensory inputs.[153,154]

8.5.5 Functional Implications

The nociceptive activity modulates firing of the motor neurons, confirmed by changes in electromyographic (EMG) activity after painful stimuli of the muscle.[130,155] The level of pain-modulated muscle activity depends upon the specific muscle function.[156] As a result, decreased muscle activity is found in the muscle producing the movement,[155] and reduced movement amplitudes have been found in low-back pain patients.[130,157] This is probably a functional adaptation in order to limit movements employing painful muscles. These changes have a strong influence on reflex pathways[158] without involvement of supraspinal mechanisms. Also, activation of group III and IV muscle afferents reduces the efficacy of proprioceptive input from Ia afferents, thus reducing motor resolution in response to input, which reduces the accuracy of motor action. Furthermore, supraspinal influences can participate in modulation of muscle activity and motor function in the condition of pain.

Pain-related changes in cortical activity were also described, being the reduced cortical excitability after both muscle and cutaneous tonic pain induction,[148,159] in patients with fibromyalgia,[160] and also decreased cerebral activation during pain episodes in Complex Regional Pain Syndrome type I, a.k.a. reflex sympathetic dystrophy.[161] That again could lead to deranged motor control action in the pain condition.

References

1. Craig JC and Rollman GB. Somesthesis. *Annu. Rev. Psychol.*, 1999, 50: 305–331.
2. Sacks O. *The Man Who Mistook His Wife for a Hat*, pp. 43–55. New York: Touchstone, 1998.
3. Axelrod FB and Gold-von Simson G. Hereditary sensory and autonomic neuropathies: types II, III, and IV. *Orphanet J. Rare Dis.*, 2007, 2: 39.
4. Melzack R and Wall PD. Gate control theory of pain. In: A. Soulairac, J. Cahn, and J. Charpentier (eds), *Pain*. pp. 11–31. London: Academic Press, 1968.
5. Chuong CM, Nickoloff BJ, Elias PM, Goldsmith LA, Macher E, Maderson PA, et al. What is the 'true' function of skin? *Exp. Dermatol.*, 2002, 11(2): 159–187.
6. Harlow HF. The nature of love. *Am. Psychol.*, 1958, 13: 673–685.
7. Lawrence RJ. David the 'Bubble Boy' and the boundaries of the human. *JAMA*, 1985, 253(1): 74–76.
8. Anzieu D. *Le moi-peau*. 254p. Paris: Dunod, 1985.
9. Aimonetti JM, Hospod V, Roll JP, and Ribot-Ciscar E. Cutaneous afferents provide a neuronal population vector that encodes the orientation of human ankle movements. *J. Physiol.*, 2007, 580(2): 649–658.
10. Collins DF, Refshauge KM, Todd G, and Gandevia SC. Cutaneous receptors contribute to kinesthesia at the index finger, elbow, and knee. *J. Neurophysiol.*, 2005, 94(3): 1699–1706.
11. Adrian ED and Zottermann Y. The impulses produced by sensory endings, part II. The response of a single end organ. *J. Physiol.*, 1926, 61: 151–171.
12. Iggo A and Andres KH. Morphology of cutaneous receptors. *Annu. Rev. Neurosci.*, 1982, 5: 1–31.
13. Johansson RS and Vallbo AB. Tactile sensibility in the human hand: Relative and absolute densities of four types of mechanoreceptive units in glabrous skin. *J. Physiol.*, 1979, 286: 283–300.
14. Knibestöl M. Stimulus-response functions of rapidly adapting mechanoreceptors in human glabrous skin area. *J. Physiol.*, 1973, 232(3): 427–452.
15. Lumpkin EA and Bautista DM. Feeling the pressure in mammalian somatosensation. *Curr. Opin. Neurobiol.*, 2005, 15(4): 382–388.
16. Bensmaïa SJ, Leung YY, Hsiao SS, and Johnson KO. Vibratory adaptation of cutaneous mechanoreceptive afferents. *J. Neurophysiol.*, 2005, 94(5): 3023–3036.

17. Knibestöl M and Vallbo AB. Single unit analysis of mechanoreceptor activity from the human glabrous skin. *Acta Physiol. Scand.*, 1970, 80(2): 178–195.

18. Kumazawa T and Perl ER. Primate cutaneous sensory units with unmyelinated (C) afferent fibers. *J. Neurophysiol.*, 1977, 40(6): 1325–1338.

19. Olausson H, Lamarre Y, Backlund H, Morin C, Wallin BG, Starck G, Ekholm S, et al. Unmyelinated tactile afferents signal touch and project to insular cortex. *Nat. Neurosci.*, 2002, 5(9): 900–904.

20. McGlone F, Vallbo AB, Olausson H, Loken L, and Wessberg J. Discriminative touch and emotional touch. *Can. J. Exp. Psychol.*, 2007, 61(3): 173–183.

21. Cauna N. Nerve supply and nerve endings in Meissner's corpuscles. *Am. J. Anat.*, 1956, 99(2): 315–350.

22. Nolano M, Provitera V, Crisci C, Stancanelli A, Wendelschafer-Crabb G, Kennedy WR, and Santoro L. Quantification of myelinated endings and mechanoreceptors in human digital skin. *Ann. Neurol.*, 2003, 54(2): 197–205.

23. Caruso G, Nolano M, Lullo F, Crisci C, Nilsson J, and Massini R. Median nerve sensory responses evoked by tactile stimulation of the finger proximal and distal phalanx in normal subjects. *Muscle Nerve*, 1994, 17(3): 269–275.

24. Takahashi-Iwanaga H and Shimoda H. The three-dimensional microanatomy of Meissner corpuscles in monkey palmar skin. *J. Neurocytol.*, 2003, 32(4): 363–371.

25. Paré M, Smith AM, and Rice FL. Distribution and terminal arborizations of cutaneous mechanoreceptors in the glabrous finger pads of the monkey. *J. Comp. Neurol.*, 2002, 445(4): 347–459.

26. Vallbo AB, Olsson KA, Westberg KG, and Clark FJ. Microstimulation of single tactile afferents from the human hand. Sensory attributes related to unit type and properties of receptive fields. *Brain*, 1984, 107(3): 727–749.

27. Ribot-Ciscar E, Roll JP, Tardy-Gervet MF, and Harlay F. Alteration of human cutaneous afferent discharges as the result of long-lasting vibration. *J. Appl. Physiol.*, 1996, 80(5): 1708–1715.

28. Talbot WH, Darian-Smith I, Kornhuber HH, and Mountcastle VB. The sense of flutter-vibration: Comparison of the human capacity with response patterns of mechanoreceptive afferents from the monkey hand. *J. Neurophysiol.*, 1968, 31(2): 301–334.

29. Johansson RS and Westling G. Roles of glabrous skin receptors and sensorimotor memory in automatic control of precision grip when lifting rougher or more slippery objects. *Exp. Brain Res.*, 1984, 56(3): 550–564.

30. Macefield VG, Häger-Ross C, and Johansson RS. Control of grip force during restraint of an object held between finger and thumb: responses of cutaneous afferents from the digits. *Exp. Brain Res.*, 1996, 108(1): 155–171.

31. Bolton CF, Winkelmann RK, and Dyck PJ. A quantitative study of Meisner's corpuscles in man. *Neurology*, 1965, 16: 1–9.

32. Stevens JC and Choo KK. Spatial acuity of the body surface over the life span. *Somatosens. Mot. Res.*, 1996, 13(2): 153–166.

33. Mahns DA, Perkins NM, Sahai V, Robinson L, and Rowe MJ. Vibrotactile frequency discrimination in human hairy skin. *J. Neurophysiol.*, 2006, 95(3): 1442–1450.

34. Munger BL and Ide C. The enigma of sensitivity in Pacinian corpuscles: A critical review and hypothesis of mechano-electric transduction. *Neurosci Res.*, 1987, 5(1): 1–15.

35. Ribot-Ciscar E, Vedel JP, and Roll JP. Vibration sensitivity of slowly and rapidly adapting cutaneous mechanoreceptors in the human foot and leg. *Neurosci Lett.*, 1989, 104(1–2): 130–135.

36. Brisben AJ, Hsiao SS, and Johnson KO. Detection of vibration transmitted through an object grasped in the hand. *J. Neurophysiol.*, 1999, 81(4): 1548–1558.

37. Phillips JR and Johnson KO. Tactile spatial resolution. II. Neural representation of bars, edges, and gratings in monkey primary afferents. *J. Neurophysiol.*, 1981, 46(6): 1192–1203.

38. Macefield VG. Physiological characteristics of low-threshold mechanoreceptors in joints, muscle and skin in human subjects. *Clin. Exp. Pharmacol. Physiol.*, 2005, 32(1–2): 135–144.

39. Johnson KO. The roles and functions of cutaneous mechanoreceptors. *Curr. Opin. Neurobiol.*, 2001, 11(4): 455–461.

40. Güçlü B, Mahoney GK, Pawson LJ, Pack AK, Smith RL, and Bolanowski SJ. Localization of Merkel cells in the monkey skin: an anatomical model. *Somatosens. Mot. Res.*, 2008, 25(2): 123–138.

41. Iggo A and Muir AR. The structure and function of a slowly adapting touch corpuscle in hairy skin. *J. Physiol.*, 1969, 200(3): 763–796.

42. Vega-Bermudez F and Johnson KO. SA1 and RA receptive fields, response variability, and population responses mapped with a probe array. *J. Neurophysiol.*, 1999, 81(6): 2701–2710.

43. Johnson KO and Lamb GD. Neural mechanisms of spatial tactile discrimination: Neural patterns evoked by braille-like dot patterns in the monkey. *J. Physiol.*, 1981, 310: 117–144.

44. Chambers MR, Andres KH, von Duering M, and Iggo A. The structure and function of the slowly adapting type II mechanoreceptor in hairy skin. *Q. J. Exp. Physiol. Cogn. Med. Sci.*, 1972, 57(4): 417–445.

45. Paré M, Behets C, and Cornu O. Paucity of presumptive ruffini corpuscles in the index finger pad of humans. *J. Comp. Neurol.*, 2003, 456(3): 260–266.

46. Johansson RS and Vallbo AB. Spatial properties of the population of mechanoreceptive units in the glabrous skin of the human hand. *Brain Res.*, 1980, 184(2): 353–366.

47. Edin BB. Cutaneous afferents provide information about knee joint movements in humans. *J Physiol.*, 2001, 531(1): 289–297.

48. Edin BB. Quantitative analysis of static strain sensitivity in human mechanoreceptors from hairy skin. *J. Neurophysiol.*, 1992, 67(5): 1105–1113.

49. Grill SE and Hallett M. Velocity sensitivity of human muscle spindle afferents and slowly adapting type II cutaneous mechanoreceptors. *J. Physiol.*, 1995, 489(2): 593–602.

50. Edin BB. Quantitative analyses of dynamic strain sensitivity in human skin mechanoreceptors. *J. Neurophysiol.*, 2004, 92(6): 3233–3243.

51. Johansson RS. Tactile sensibility in the human hand: Receptive field characteristics of mechanoreceptive units in the glabrous skin area. *J. Physiol.*, 1978, 281: 101–125.

52. Olausson H, Wessberg J, and Kakuda N. Tactile directional sensibility: Peripheral neural mechanisms in man. *Brain Res.*, 2000, 866(1–2): 178–187.

53. Bergenheim M, Ribot-Ciscar E, and Roll JP. Propioceptive population coding of two-dimensional limb movements in humans: I. Muscle spindle feedback during spatially oriented movements. *Exp. Brain Res.*, 2000, 134(3): 301–310.

54. Macefield G, Gandevia SC, and Burke D. Perceptual responses to microstimulation of single afferents innervating joints, muscles and skin of the human hand. *J. Physiol.*, 1990, 429: 113–129.

55. Collins DF and Prochazka A. Movement illusions evoked by ensemble cutaneous input from the dorsum of the human hand. *J. Physiol.*, 1996, 496(3): 857–871.

56. Collins DF, Refshauge KM, Todd G, and Gandevia SC. Cutaneous receptors contribute to kinesthesia at the index finger, elbow, and knee. *J. Neurophysiol.*, 2005, 94(3): 1699–1706.

57. Low Choy NL, Brauer SG, and Nitz JC. Age-related changes in strength and somatosensation during midlife: Rationale for targeted preventive intervention programs. *Ann. N. Y. Acad. Sci.*, 2007, 1114: 180–193.

58. von Frey M. The distribution of afferent nerves in the skin. *JAMA*, 1906, 47: 645.

59. Weinstein S. Fifty years of somatosensory research: From the Semmes-Weinstein monofilaments to the Weinstein Enhanced Sensory Test. *J. Hand Ther.*, 1993, 6(1): 11–22.

60. Gelberman RH, Szabo RM, Williamson RV, and Dimick MP. Sensibility testing in peripheral-nerve compression syndromes. An experimental study in humans. *J. Bone Joint Surg. Am.*, 1983, 65(5): 632–638.

61. Jerosch-Herold C. Assessment of sensibility after nerve injury and repair: A systematic review of evidence for validity, reliability and responsiveness of tests. *J. Hand Surg. [Br].*, 2005, 30(3): 252–264.

62. McGill M, Molyneaux L, Spencer R, Heng LF, and Yue DK. Possible sources of discrepancies in the use of the Semmes-Weinstein monofilament. Impact on prevalence of insensate foot and workload requirements. *Diabetes Care*, 1999, 22(4): 598–602.

63. Massy-Westropp N. The effects of normal human variability and hand activity on sensory testing with the full Semmes-Weinstein monofilaments kit. *J. Hand Ther.* 2002, 15(1): 48–52.

64. Levin S, Pearsall G, and Ruderman RJ. Von Frey's method of measuring pressure sensibility in the hand: An engineering analysis of the Weinstein-Semmes pressure aesthesiometer. *J. Hand Surg. [Am]*, 1978, 3(3): 211–216.

65. Weinstein S. Intensive and extensive aspects of tactile sensitivity as a function of body part, sex and laterality. In: DR Kenshalo, (ed.), *Skin Senses*. Vol. I11, pp. 195–222. Springfield, IL: Charles C. Thomas, 1968.

66. Lundborg G and Rosén B. The two-point discrimination test–time for a re-appraisal? *J. Hand Surg. [Br].*, 2004, 29(5): 418–422.

67. Johnson K, Van Bowen R, and Hsiao S. The perception of two points is not the spatial resolution threshold. In: J Boivie, P Hansson, and U Lindblom (eds), *Touch, temperature and pain in health and disease: mechanisms and assessments, progress in pain research and management*. Vol. 2, pp. 389–404. Seattle: IASP Press, 1994.

68. Godde B, Stauffenberg B, Spengler F, and Dinse HR. Tactile coactivation-induced changes in spatial discrimination performance. *J. Neurosci.*, 2000, 20(4): 1597–1604.

69. Connor NP and Abbs JH. Orofacial proprioception: Analyses of cutaneous mechanoreceptor population properties using artificial neural networks. *J. Commun. Disord.*, 1998, 31(6): 535–542.

70. Baldissera F, Hultborn H, and Illert MI. Integration in the spinal neuronal system. In: V Brooks, (ed.), *Handbook of physiology. The nervous system II*. pp. 509–595. Bethesda: American Physiological Society, 1981.

71. Barker D and Banks RW. The muscle spindle. In: AG Engel, C Franzini-Armstrong, (eds), *Myology*. pp. 333–360. New York: McGraw-Hill, 1994.

72. Liu JX. Human muscle spindles. Complex morphology and structural organization. Umeå: Umeå University Thesis, 2004, pp. 9–11.

73. Boyd IA. The isolated mammalian muscle spindle. *TINS*, 1980, 3: 258–265.

74. Matthews PBC. The differentiation of two types of fusimotor fibre by their effects on the dynamic response of muscle spindle primary endings. *Q. J. Exp. Physiol. Cogn. Med. Sci.*, 1962, 47: 324–333.

75. Matthews PBC. *Mammilian Muscle Receptors and Their Central Action*. London: Arnolds, 1972.

76. Hospod V, Aimonetti JM, Roll JP, and Ribot-Ciscar E. Changes in human muscle spindle sensitivity during a proprioceptive attention task. *J. Neurosci.*, 2007, 27(19): 5172–5178.

77. Baldissera F, Hultborn H, and Illert M. Integration in spinal neuronal systems. In: VB Brooks (ed.), *Handbook of Physiology, section 1, The Nervous System, vol. II, Motor Control*. pp. 509–595. Bethesda, MD, USA: American Physiological Society,. 1981.

78. Jankowska E. Interneuronal relay in spinal pathways from proprioceptors. *Prog. Neurobiol.*, 1992, 38(4): 335–378.

79. Stein RB, Yang JF, Bélanger M, and Pearson KG. Modification of reflexes in normal and abnormal movements. *Prog. Brain Res.*, 1993, 97: 189–196.

80. Goodwin GM, McCloskey DI, and Matthews PBC. Proprioceptive illusions induced by muscle vibration: Contribution by muscle spindles to perception? *Science*, 1972, 175(28): 1382–1384.

81. Roll JP and Vedel JP. Kinaesthetic role of muscle afferents in man, studied by tendon vibration and microneurography. *Exp. Brain Res.*, 1982, 47(2): 177–190.

82. Roll JP, Albert F, Ribot-Ciscar E, and Bergenheim M. "Proprioceptive signature" of cursive writing in humans: A multi-population coding. *Exp. Brain Res.*, 2004, 157(3): 359–368.

83. Jami L. Golgi tendon organs in mammalian skeletal muscle: functional properties and central actions. *Physiol. Rev.*, 1992, 72(3): 623–66.

84. Crago PE, Houk JC, and Rymer WZ. Sampling of total muscle force by tendon organs. *J. Neurophysiol.*, 1982, 47(6): 1069–1083.

85. Prochazka A and Gorassini M. Ensemble firing of muscle afferents recorded during normal locomotion in cats. *J. Physiol.*, 1998, 507: 293–304.

86. Chalmers G. Do Golgi tendon organs really inhibit muscle activity at high force levels to save muscles from injury, and adapt with strength training? *Sports Biomech.*, 2002, 1(2): 239–249.

87. Sjölander P and Johansson H. Sensory nerve endings in ligaments: response properties and effects on proprioception and motor control. In: L Yahia, (ed.), *Ligaments and igamentoplasties*. pp. 39–83. Berlin: Springer-Verlag, 1997.

88. Carlson M. Characteristics of sensory deficits following lesions of Brodmann's areas 1 and 2 in the postcentral gyrus of Macaca mulatta. *Brain Res.*, 1981, 204(2): 424–430.

89. Phillips JR, Johnson KO, and Hsiao SS. Spatial pattern representation and transformation in monkey somatosensory cortex. *Proc. Natl. Acad. Sci. USA*, 1988, 85(4): 1317–1321.

90. Pons TP, Garraghty PE, Friedman DP, and Mishkin M. Physiological evidence for serial processing in somatosensory cortex. *Science*, 1987, 237(4813): 417–420.

91. Mountcastle VB. The columnar organization of the neocortex. *Brain*, 1997, 120(4): 701–722.

92. Sripati AP, Yoshioka T, Denchev P, Hsiao SS, and Johnson KO. Spatiotemporal receptive fields of peripheral afferents and cortical area 3b and 1 neurons in the primate somatosensory system. *J. Neurosci.*, 2006, 26(7): 2101–2114.

93. Penfield W and Boldrey E. Somatic motor and sensory representation in the cerebral cortex of man as studied by electrical stimulation. *Brain*, 1937, 60: 389–443.

94. Penfield W and Rasmussen T. *The Cerebral Cortex of Man*, pp. 214–215. New York: Macmillan, 1950.

95. Schepers RJ and Ringkamp M. Thermoreceptors and thermosensitive afferents. *Neurosci. Biobehav. Rev.*, 2009, 33: 205–212.

96. Spray DC. Cutaneous temperature receptors. *Annu. Rev. Physiol.*, 1986, 48: 625–638.

97. Caterina MJ, Schumacher MA, Tominaga M, Rosen TA, Levine JD, and Julius D. The capsaicin receptor: A heat-activated ion channel in the pain pathway. *Nature*, 1997, 389(6653): 816–824.

98. Latorre R, Brauchi S, Orta G, Zaelzer C, and Vargas G. ThermoTRP channels as modular proteins with allosteric gating. *Cell Calcium*, 2007, 42(4–5): 427–438.

99. Darian-Smith I, Johnson KO, and Dykes R. "Cold" fiber population innervating palmar and digital skin of the monkey: Responses to cooling pulses. *J. Neurophysiol.*, 1973, 36(2): 325–346.

100. Craig AD and Bushnell MC. The thermal grill illusion: Unmasking the burn of cold pain. *Science*, 1994, 265(5169): 252–255.

101. Craig AD, Reiman EM, Evans A, and Bushnell MC. Functional imaging of an illusion of pain. *Nature*, 1996, 384(6606): 258–260.

102. Mogil JS. The genetic mediation of individual differences in sensitivity to pain and its inhibition. *Proc. Natl. Acad. Sci. USA*, 1999, 96: 7744–7751.

103. Keele KD. *Anatomies of Pain*. Springfield, IL, USA: Charles C Thomas, 1957.

104. Descartes R. L'Homme, English Translation in: *Lectures on the History of Physiology During the 16th, 17th and 18th Century*. Cambridge, UK: Cambridge University Press, 1664.

105. Bonica JJ. *Pain. Research Publications-Association for Research in Nervous and Mental Disease*. Vol. 58, New York, USA: Raven Press Books, 1980.

106. Belcher G. The effects of intra-arterial bradykinin, histamine, acetylcholine and prostaglandin E1 on nociceptive and non-nociceptive dorsal horn neurones of the cat. *Eur. J. Pharmacol.*, 1979, 56: 385–395.

107. Fock S and Mense S. Excitatory effects of 5-hydroxytryptamine, histamine and potassium ions on muscular group IV afferent units: A comparison with bradykinin. *Brain Res.*, 1976, 105: 459–469.

108. Foreman RD, Schmidt RF, Willis WD. Effects of mechanical and chemical stimulation of fine muscle afferents upon primate spinothalamic tract cells. *J. Physiol. Lond.*, 1979, 286: 215–231.

109. Kniffki KD, Mense S, and Schmidt RF. Responses of group IV afferent units from skeletal muscle to stretch, contraction and chemical stimulation. *Exp. Brain Res.*, 1978, 31: 511–522.

110. Besson JM, Conseiller C, Hamann KF, and Maillard MC. Modifications of dorsal horn cell activities in the spinal cord, after intra-arterial injection of bradykinin. *J. Physiol. Lond.*, 1972, 221: 189–205.

111. Dray A and Perkins M. Bradykinin and inflammatory pain. *Trends Neurosci.*, 1993, 16: 99–104.

112. Graff-Radford SB, Reeves JL, and Jaeger B. Management of chronic head and neck pain: Effectiveness of altering factors perpetuating myofascial pain. *Headache*, 1987, 27: 186–190.

113. Pfeifer MA, Ross DR, Schrage JP, Gelber DA, Schumer MP, Crain GM, Markwell SJ, and Jung S. A highly successful and novel model for treatment of chronic painful diabetic peripheral neuropathy. *Diabetes Care*, 1993, 16: 1103–1115.

114. Solveborn SA. Radial epicondylalgia ("tennis elbow"): treatment with stretching or forearm band. A prospective study with long-term follow-up including range-of-motion measurements. *Scand. J. Med. Sci. Sports*, 1997, 7: 229–237.

115. Tardy-Gervet MF, Guieu R, Ribot-Ciscar E, Gantou B, and Roll JP. Two methods for the sensory control of pain: Transcutaneous mechanical vibration, applied either alone or associated with TENS. *Eur. J. Pain*, 1994, 15: 13–21.

116. Fields HL. *Pain*. New York, USA: McGraw-Hill, 1987.

117. Windhorst U. Neuroplasticity and modulation of chronic pain. In: H Johansson, U Windhorst, M Djupsjobacka, M Passatore (eds), *Chronic Work-Related Myalgia*. pp. 207–224. Gavle: Gavle University Press, 2003.

118. Koltzenburg M. The stability and plasticity of the encoding properties of peripheral nerve fibres and their relationship to provoked and ongoing pain. *Semin. Neurosci.*, 1995, 7: 199–210.

119. Hardy JD, Wolff HG, and Goodell H. The nature of cutaneous hyperalgesia. In: *Pain Sensations and Reactions*. pp. 173–215. Baltimore: Williams and Wilkins, 1952.

120. LaMotte RH, Shain CN, and Simone DA. Neurogenic hyperalgesia: Psychophysical studies of underlying mechanisms. *J. Neurophysiol.*, 1991, 66: 190–211.

121. Levine JD, Fields HL, and Basbaum AI. Peptides and the primary afferent nociceptor. *J. Neurosci.*, 1993, 13: 2273–2286.

122. Mense S. Nociception from skeletal muscle in relation to clinical muscle pain. *Pain*, 1993, 54: 241–289.

123. Johansson H, Windhorst U, Djupsjöbacka M, and Passatore M. Chronic work-related myalgia. In: H Johansson, U Windhorst, M Djupsjöbacka, and M Passatore (eds), *Neuromuscular Mechanisms Behind Work-Related Chronic Muscle Pain Syndromes*. pp. 1–310. Gävle, Sweden: Gävle University Press, 2003.

124. Ariens GA, van Mechelen W, Bongers PM, Bouter LM, and van der Wal G. Physical risk factors for neck pain. *Scand. J. Work Environ. Health*, 2000, 26: 7–19.

125. van Dieen JH, Visser B, and Hermans V. The contribution of task-related biomechanical constraints to the development of work-related myalgia. In: H Johansson, U Windhorst, M Djupsjöbacka, and M Passatore, (eds), *Chronic Work-Related Myalgia. Neuromuscular Mechanisms Behind Work-Related Chronic Muscle Pain Syndromes*. pp. 83–93. Gävle, Sweden: Gävle University Press, 2003.

126. Henneman E, Somjen G, Carpenter DO. Excitability and inhibitability of motoneurons of different sizes. *J. Neurophysiol.*, 1965, 28: 599–620.

127. Hansen J, Thomas GD, Harris SA, Parsons WJ, Victor RG. Differential sympathetic neural control of oxygenation in resting and exercising human skeletal muscle. *J. Clin. Invest.*, 1996, 98: 584–596.

128. Greenberg DA, Aminoff MJ, Simon RP. *Clinical Neurology*. USA: Lange Medical Books/McGraw-Hill, 2002.

129. Wolff BB. Measurement of human pain. In: JJ Bonica (ed.), *Pain*. New York, USA: Raven Press, 1980.

130. Arendt-Nielsen L, Graven-Nielsen T, Svarrer H, and Svensson P. The influence of low back pain on muscle activity and coordination during gait: A clinical and experimental study. *Pain*, 1996, 64: 231–240.

131. Verdugo RJ, Ochoa JL. Quantitative somatosensory thermotest: A key method for functional evaluation of small caliber afferent channels. *Brain*, 1992, 115: 893–913.

132. Huskisson EC. Measurement of pain. *J. Rheumatol.*, 1982, 9: 768–769.

133. Graven-Nielsen T, McArdle A, Phoenix J, Arendt-Nielsen L, Jensen TS, Jackson MJ, and Edwards RH. In vivo model of muscle pain: Quantification of intramuscular chemical, electrical, and pressure changes associated with saline-induced muscle pain in humans. *Pain*, 1997, 69: 137–143.

134. Breivik EK, Bjornsson GA, and Skovlund E. A comparison of pain rating scales by sampling from clinical trial data. *Clin. J. Pain* 2000, 16: 22–28.

135. Millan MJ. Descending control of pain. *Prog. Neurobiol.*, 2002, 66: 355–474.

136. Jankowska E. Interneuronal relay in spinal pathways from proprioceptors. *Prog. Neurobiol.*, 1992, 38: 335–378.

137. Sandkuhler J. The organization and function of endogenous antinociceptive systems. *Prog. Neurobiol.*, 1996, 50: 49–81.

138. Melzack R and Wall PD. Pain mechanisms: A new theory. *Science*, 1965, 150: 971–978.

139. Handwerker HO, Iggo A, and Zimmermann M. Segmental and supraspinal actions on dorsal horn neurons responding to noxious and non-noxious skin stimuli. *Pain*, 1975, 1: 147–165.

140. De Koninck Y and Henry JL. Peripheral vibration causes an adenosine-mediated postsynaptic inhibitory potential in dorsal horn neurons of the cat spinal cord. *Neuroscience*, 1992, 50: 435–443.

141. Dubner R. Neuronal plasticity and pain following peripheral tissue inflammation or nerve injury. In: MR Bond, JE Charlton, and CJ Woolf (eds). Proceedings of the 6th World Congress of Pain. pp. 263–276. Amsterdam: Elsevier Science Publications, 1991.

142. Deyo RA, Walsh NE, Martin DC, Schoenfeld LS, and Ramamurthy S. A controlled trial of transcutaneous electrical nerve stimulation (TENS) and exercise for chronic low back pain. *N. Engl. J. Med.*, 1990, 322: 1627–1634.

143. Taylor DC, Dalton JDJ, Seaber AV, and Garrett WEJ. Viscoelastic properties of muscle-tendon units. The biomechanical effects of stretching. *Am. J. Sports Med.*, 1990, 18: 300–309.

144. Bjorklund M, Radovanovic S, Ljubisavljevic M, Windhorst U, and Johansson H. Muscle stretch-induced modulation of noxiously activated dorsal horn neurons of feline spinal cord. *Neurosci. Res.*, 2004, 48: 175–184.

145. Hoheisel U and Mense S. Response behaviour of cat dorsal horn neurones receiving input from skeletal muscle and other deep somatic tissues. *J. Physiol. Lond.*, 1990, 426: 265–280.

146. Le Bars D, Dickenson AH, and Besson JM. Diffuse noxious inhibitory controls (DNIC). I. Effects on dorsal horn convergent neurones in the rat. *Pain*, 1979, 6: 283–304.

147. Akil H, Mayer DJ, and Liebeskind JC. Antagonism of stimulation-produced analgesia by naloxone, a narcotic antagonist. *Science*, 1976, 191: 961–962.

148. La Pera D, Graven-Nielsen T, Valeriani M, Oliviero A, Di Lazzaro V, Tonali PA, and Arendt-Nielsen L. Inhibition of motor system excitability at cortical and spinal level by tonic muscle pain. *Clin. Neurophysiol.*, 2001, 112: 1633–1641.

149. Mansour A, Watson SJ, and Akil H. Opioid receptors: Past, present and future. *Trends Neurosci.*, 1995, 18: 69–70.

150. Morgan MM, Heinricher MM, and Fields HL. Circuitry linking opioid-sensitive nociceptive modulatory systems in periaqueductal gray and spinal cord with rostral venromedial medulla. *Neuroscience*, 1992, 47: 863–871.

151. Nicholson B. Responsible prescribing of opioids for the management of chronic pain. *Drugs*, 2003, 63: 17–32.

152. Pergolizzi J, Boger RH, Budd K, Dahan A, Erdine S, Hans G, Kress HG, et al. Opioids and the management of chronic severe pain in the elderly: Consensus statement of an International Expert Panel with focus on the six clinically most often used World Health Organization step III opioids (buprenorphine, fentanyl, hydromorphone, methadone, morphine, oxycodone). *Pain Pract.*, 2008, 8: 287–313.

153. Mark MA, Colvin LA, and Duggan AW. Spontaneous release of immunoreactive neuropeptide Y from the central terminals of large diameter primary afferents of rats with peripheral nerve injury. *Neuroscience*, 1998, 83: 581–589.

154. Polgar E, Shehab SA, Watt C, and Todd AJ. GABAergic neurons that contain neuropeptide Y selectively target cells with the neurokinin 1 receptor in laminae III and IV of the rat spinal cord. *J. Neurosci.*, 1999, 19: 2637–2646.

155. Graven-Nielsen T, Arendt-Nielsen L, Svensson P, Jensen TS. Effect of experimental muscle pain on muscle activity and co-ordination during static and dynamic motor function. *EEG Clin. Neurophysiol.*, 1997, 105: 156–164.

156. Lund JP, Stohler CS, and Widmer CG. The relationship between pain and muscle activity in fibromyalgia and similar conditions. In: H Vaeroy and H Merskey (eds), *Progress in Fibromyalgia and Myofascial Pain*. pp. 311–327. Amsterdam: Elsevier Science Publications, 1993.

157. Ahern DK, Follick MJ, Council JR, Laser-Wolston N, and Litchman H. Comparison of lumbar paravertebral EMG patterns in chronic low back pain patients and non-patient controls. *Pain*, 1988, 34: 153–160.

158. Wang K, Svensson P, and Arendt-Nielsen L, Modulation of exteroceptive suppression periods in human jaw-closing muscles by local and remote experimental muscle pain. *Pain*, 1999, 82: 253–262.

159. Farina S, Valeriani M, Rosso T, Aglioti S, Tamburin S, Fiaschi A, and Tinazzi M., Transient inhibition of the human motor cortex by capsaicin-induced pain. A study with TMS. *Neurosci. Lett.*, 2001, 314: 97–101.

160. Salerno A, Thomas E, Olive P, Blotman F, Picot MC, and Georgesco M. Motor cortical dysfunction disclosed by single and double magnetic stimulation in patients with fibromyalgia. *Clin. Neurophysiol.* 2000, 6: 994–1001.

161. Apkarian AV. Functional imaging of pain: New insight regarding the role of the cerebral cortex in human pain perception. *Semin. Neurosci.*, 1995, 7: 279–293.

162 Irvin Darian-Smith. The sense of touch: Performance and peripheral neural processes. In: J.M. Brookhart and V.B. Mountcastle (eds), *Handbook of Physiology, section 1: The Nervous System, Volume III: Sensory Processes*. Bethesda, MD, 1984.

163 Vallbo AB and Johansson RS. Properties of cutaneous mechanoreceptors in the human hand related to touch sensation. *Human Neurobiol.*, 1984, 3: 3–14.

164 Houk, JC, Crago PW, and Rymer WZ. Functional properties of the Golgi tendon organs. In: J.E. Desmedt (ed.), *Spinal and Supraspinal Mechanisms of Voluntary Motor Control and Locomotion, Volume 8: Progress in Clinical Neurophysiology*. Basel: Karger, 1980.

165 Campero M, Serra J, Bostock H, and Ochoa JL. Slowly conducting afferents activated by innocuous low temperature in human skin. *J Physiol.*, 2001, 535: 855–865.

9

Hearing

Mark S. Hedrick

While the study of the acoustics of sound is relatively well known, being a part of most physics textbooks, the study of the function of the auditory system, and our perception of sound, is still evolving. The purpose of this chapter is to review some basic acoustics, and then use this information as a springboard for studies of the structure and function of the auditory system, how this may lend itself to our perception of the world of sound, and how we may remediate this loss of perception in the case of hearing loss.

9.1 Acoustics

9.1.1 How a Sound is Produced

Nearly all objects in our environment have properties of inertia (Newton's principle of an object preserving its present state) and elasticity (the ability of an object to resist deformation or return to its exact original shape following stress). When the object is impinged by energy, the object's properties of inertia and elasticity operate reciprocally, resulting in the object moving back and forth, or vibrating. When this vibration disturbs the medium surrounding the object, a sound is created. The medium's particles, of course, must also have properties of inertia and elasticity to be set into vibration. This disturbance may then be propagated through the medium, taking a pattern similar to the original back and forth motion of the object. This disturbance may then be detected by a person or by a recording instrument.

Return for a moment to the back and forth vibration of the object. To get the object to move, the force or energy applied to the object must move the mass (an index of inertia); this motion is then opposed by the elasticity (or restoring force) of the object. This is frequently illustrated by the motion of a mass (inertia) attached to a spring (elasticity); the motion of the mass–spring system is described as the force needed to move the mass:

$$F = ma \quad \text{or} \quad F = sa, \tag{9.1}$$

where F represents force, m mass, a acceleration, and s elasticity. Because the inertial and elastic effects are equal and opposite, a description of the object's displacement from its resting state at an instant in time can be given by the following equation:

$$d(t) = A\sin\left(2\pi\theta\left(\frac{s}{m}\right)^{1/2}t + \theta\right) \text{ or as } d(t) = A\sin(2\pi ft + \theta), \tag{9.2}$$

where $d(t)$ represents instantaneous displacement, A amplitude, s restoring force, m mass, f frequency, t time, and θ starting phase or the point along the vibratory cycle at which the object began to vibrate. The unit of measurement would be the distance of the object's displacement. This equation shows that instantaneous displacement can be defined by amplitude, starting phase, and frequency—in essence, by a sinusoidal function. Frequency represents the number of cycles of back and forth vibration per unit time, and the time taken to complete one cycle of vibration is the period. Frequency and period are reciprocals, that is, $f = 1/p$.

9.1.2 How the Frequency Components of Sound are Determined

Although the sinusoidal function may not be the only waveform representation of sound, it does lend itself to a coherent mathematical description of more complex vibrations. Most sounds in our environment are created by complex vibrations, and thus contain more than one frequency. According to the Fourier theorem, any complex periodic vibration may be shown to be a summation of sinusoidal vibrations, termed a Fourier series. For a periodic, time-domain waveform of period T, the Fourier series equation is

$$x(t) = A_o + \sum_{n-1}^{\infty} \left[A_n \cos(\omega_n t) + B_n \sin(\omega_n t) \right], \qquad (9.3)$$

where $x(t)$ represents the time-domain waveform, n are integers representing the different sinusoidal frequencies in the frequency domain, $\omega_n = 2\pi n/T$, and A_n and B_n represent coefficients that can be determined by projecting (or multiplying) the function $x(t)$ on respective trigonometric functions with T/n as their period. Thus,

$$A_n = \frac{2}{T} \int_{-T/2}^{T/2} d(t) x(t) \cos(\omega_n t) \qquad (9.4)$$

and

$$B_n = \frac{2}{T} \int_{-T/2}^{T/2} dt\, x(t) \sin(\omega_n t). \qquad (9.5)$$

Finally, to find A_o, the average value of x is computed:

$$A_o = \frac{1}{T} \int_{-T/2}^{T/2} dt\, x(t). \qquad (9.6)$$

Thus, one may say that the Fourier theorem allows a transformation of the signal from the time domain to the frequency domain. In other words, the Fourier theorem allows us to determine the frequencies within a complex sound. The Fourier series is most useful for describing a periodic signal, but cannot be used to describe a nonperiodic signal, that is, a signal that does not have a repetitive pattern. For a frequency analysis of a nonperiodic longitudinal signal, a Fourier transform is employed:

$$X(\omega) = \int_{-\infty}^{\infty} dt\, e^{-i\omega t} x(t), \qquad (9.7)$$

where i represents the imaginary value of the complex numbers needed to solve the equation.

Thus, the Fourier transform is a function of angular frequency ω, and involves both real and imaginary numbers. The Fourier transform can also be used to analyze a periodic signal, in which case it simply resolves into a Fourier series. The Fourier transform allows a change in representation of a signal from the time domain to the frequency domain; to go from the frequency to the time domain, there is the inverse Fourier transform:

$$x(t) = \frac{1}{2}\pi \int_{-\infty}^{\infty} d\omega\, e^{-i\omega t} X(\omega). \qquad (9.8)$$

The Fourier theorem for analyzing a signal into its component frequencies is very important to a study not only for acoustics of vibration but also for hearing. It is thought that the auditory systems of mammals, and most nonmammalian species, perform something like a Fourier analysis of the acoustic signal to develop a neurological representation of the acoustic signal. Such a representation may be considered an intermediate step between acoustic sound and perception of the sound.

9.1.3 Propagation of Sound

Before beginning our discussion of the auditory system, however, we must still discuss how the vibration of an object reaches the ear. As mentioned above, a vibrating object must disturb a surrounding medium. Sound waves may travel through gases, liquids, and solids, although not always in the same way. Sound waves through gases and liquids are longitudinal waves in that the direction of wave propagation is parallel to that of the displacement of particles in the medium. Sound waves through solids may be longitudinal or transverse. In a transverse wave, the direction of wave propagation is perpendicular to that of medium displacement. Transverse waves may propagate along the surface of a liquid, and also occur on a string. Transverse sound waves do not occur in gases or liquids because of lack of a mechanism to displace particles perpendicular to the wave propagation. For human hearing, longitudinal sound waves propagating through the air are most typical of hearing and will thus be emphasized in this chapter. When the object moves, air molecules are compressed together. This follows from the fact that air molecules are in constant motion (Brownian motion) and are continually transmitting kinetic energy from one to another via collisions. From the ideal gas law, that pressure is proportional to density,

$$P = \rho \frac{R}{M} T, \qquad (9.9)$$

where P represents pressure, ρ density, R a constant, M molar mass, and T temperature. Thus, as air molecules move together, the density and pressure are increased above the static air pressure—this is referred to as condensation. When the vibrating object then moves in the opposite direction to complete its vibratory cycle, air molecules then diffuse to occupy the space now momentarily vacated. This decrease in molecular density and pressure of air relative to static air pressure is referred to as rarefaction. This pattern of condensations and rarefactions

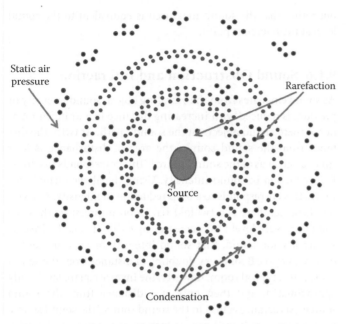

FIGURE 9.1 Graphical representation of the molecular distribution in air resulting from a sound source captured at a single moment in time. The pressure wave with condensation (compression) and rarefaction (expansion) moves outward from the source in a circular pattern (actually the propagation is in 3D, thus spherical wave propagation). As the pressure wave moves outward, the density distribution is spread over a larger spherical surface area, hence thinning out the density of condensation and rarefaction which equals diminished sound.

emanates in a wave pattern from the vibrating object through the air (Figure 9.1). The air molecule motion occurs parallel to the wave of condensations and rarefactions, which means that sound waves in air are longitudinal waves.

If the Newtonian laws of motion and the gas laws are combined, with the addition of the concept of the conservation of mass, a general wave equation for a plane wave can be formulated:

$$\frac{\partial^2 y}{\partial x^2} = \frac{1}{v^2}\frac{\partial^2 y}{\partial t^2}, \tag{9.10}$$

where v represents phase velocity, x direction, and y the variable. This general, one-dimensional wave equation can demonstrate that particle velocity, particle displacement, pressure, and density are connected in space/time by calculus derivatives and integrals. One can determine particle velocity, particle displacement, pressure, or density by substituting any of these for y in the wave equation. Note that the particle velocity discussed here is different from the speed of sound through a medium, which will be discussed next.

9.1.4 Wavelength and Speed of Sound

As propagation continues through the medium, there is a distance between successive condensations. This distance is referred to as wavelength. The size of the wavelength is dependent on the speed of propagation (or speed of sound) through a medium and the frequency of sound:

$$\lambda = \frac{c}{f}, \tag{9.11}$$

where c represents the speed of sound and f is the frequency. Lower frequencies have longer wavelengths, which is why lower frequencies can "bend" or diffract more easily around objects. Having shorter wavelengths, higher frequencies are more likely to be reduced in sound level by objects in the sound field.

The speed of sound through an ideal gas may be described as the ratio of gas elasticity to its density, or as

$$v = \sqrt{\frac{\theta\gamma RT}{M}}, \tag{9.12}$$

where γ represents the adiabatic constant, R the gas constant, T the absolute temperature, and M the molecular mass of the gas. Thus the molecular mass of different gases will result in different speeds. In terms of speed of sound through our day-to-day atmosphere, the equation resolves to approximately

$$v = 331.4 + 0.6T \,\text{m/s}, \tag{9.13}$$

where T represents temperature in Celsius. Note that the speed of sound is not dependent on the sound's frequency, intensity, or wavelength. Recall that speed is actually a scalar quantity, defined as distance per unit time. Velocity (v in the above formulas) is a vector quantity measuring the rate and direction of movement and is the first derivative of motion (v = distance/time). Acceleration is the change of the rate and is the second derivative of motion (a = distance/time/time).

9.1.5 Sound Intensity and Pressure

As noted above, a vibrating object will cause air molecules to move together, increasing both density and pressure. The instantaneous pressure of air molecules [$P(t)$] may be described by the following equation:

$$P(t) = \frac{mv}{tA}, \tag{9.14}$$

where m is the mass of the vibrating object, v the velocity of the vibrating object, t the time, and A the area. Recall that velocity is distance/time and is therefore proportional to displacement. Since pressure is proportional to velocity, it is also proportional to displacement. Thus, pressure measured by an instrument such as a sound level meter can represent the displacement pattern of the vibrating object. Thus, one can determine the sound pressure that the vibrating object creates by calculating the root-mean-square (rms) of the instantaneous pressures. In terms of amplitude, the equation is expressed as

$$A_{rms} = \sqrt{\frac{1}{T}\int_0^T d^2(t)\,dt}, \qquad (9.15)$$

where T is the period of the signal. Amplitude often refers to measures of pressure or displacement. Pressure is typically defined in units of micropascals (μPa).

The vibrating object accomplishes work by moving air molecules a given distance. Power is the rate at which work is done, and intensity is a measure of a sound's power. Sound intensity is typically defined in units of watts per area^2, where the area may be in either cm or m. The relation of pressure and intensity is described by the following formula:

$$I = \frac{P^2}{\rho c}, \qquad (9.16)$$

where I is the intensity, P the pressure, ρ the density of the medium, and c the speed of sound.

According to Yost, the exquisite range of our hearing ability encompasses a range of 10^{14} intensity units, making for cumbersome calculations. Therefore, to make calculations easier, sound intensity or pressure is usually expressed in terms of logarithmic ratios. A *bel* is the log of an intensity ratio I_a/I_b. For convenience, these log ratios are usually expressed in tenths of a bel, or in decibels (dB). Expressed in dB, humans have a pressure dynamic range of 140 dB. The sound intensity in dB is

$$dB\ IL = 10\log\left(\frac{I^o}{I^r}\right), \qquad (9.17)$$

where I^o refers to the observed or measured intensity of sound in watts per area^2 and I^r is the reference intensity. For Watt/cm^2, the reference intensity is 10^{-16} W/cm^2, and for W/m^2, the reference intensity is 10^{-12} W/m^2. Measures of sound level, in pressure, are most frequently used to describe human hearing. The sound pressure in dB is

$$dB\ SPL = 20\log\left(\frac{P^o}{P^r}\right), \qquad (9.18)$$

where P^o refers to the observed or measured pressure of sound in μPa and P^r is the reference pressure of 20 μPa. The pressure reference was determined to be the softest level at which a tone in the 1–4 kHz range could be detected by an average young adult. Note that in both the formulas the reference for dB is specified. This is necessary, since dB is not a unit of measurement *per se*, but only a ratio. Further, 0 dB does not mean absence of sound,

but rather that the sound measured is equivalent to the sound level at the reference point.

9.1.6 Sound Obstruction and Interaction

As sound emanates from the vibrating object, sound intensity or pressure is reduced with increasing distance in a uniform manner, if there are no objects in the sound's path. At twice the distance from the sound source, the area of disturbance is four times as great as at the sound source. This corresponds to a four-fold reduction in sound intensity. Using the above formula for dB IL, this corresponds to a 6 dB reduction in intensity. At twice the distance, there is a twofold reduction in pressure; thus the sound pressure level also decreases by 6 dB. This is the inverse square law for sound, that sound intensity or pressure ideally decreases by 6 dB for every doubling of distance from the sound source. In a typical room, however, the inverse square law is only approximated, and then only at a distance from the sound source. At distances close to the sound source, the sound source may act more as a wall source instead of a point source, with large nonuniform changes. This near-field condition makes for less than a 6 dB decrease in sound (Figure 9.2). Other physical objects in the room, and the size of the objects relative to the wavelength of frequency components of the sound, may act to impede the sound by more than 6 dB. For instance, high frequencies have short wavelengths; if the size of the object is small relative to the wavelength, the sound will pass around the object with little or no sound level reduction. If the object is large relative to the wavelength, and the acoustic impedance mismatch is large, then the sound will be reflected off the object (this fact is used in acoustical imaging). If the object is of "moderate" size relative to the wavelength, there will be a marked reduction in sound level just on the other side of the object; this reduction in sound level is called a sound shadow. Sound reflections from physical objects and the ceiling/walls/floor, termed reverberation, may actually increase the sound level such that at distances relatively far from the sound source, the sound level may be higher than that predicted from the inverse square law.

When a sound encounters an object (or a change of media), the sound may be reflected off the object, transmitted through the object or media, absorbed by the object or media, or converted to heat energy. To what degree each of these four options is realized depends on the impedance mismatch between the media. Acoustical impedance (Z) to transmission of energy is described by the formula

$$Z = \sqrt{R^2 + (X_m - X_s)^2}, \qquad (9.19)$$

where R is resistance, X_m is mass reactance, and X_s is stiffness reactance (Figure 9.3). Friction is a type of resistive force that impedes vibration and is not dependent on frequency. Mass reactance is the impedance to vibration caused by mass (the inertial force mentioned at the beginning of this section); if the frequency of vibration increases, the degree to which mass

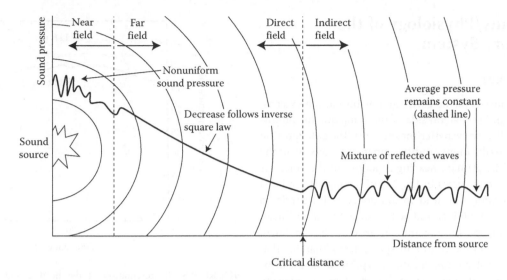

FIGURE 9.2 Graphical representation of the near- and far-field phenomena, as well as direct and diffuse field relationships with respect to the sound source in free space in a closed environment. (Reproduced from Emanuel D.C. and Letowiski T., 2009, *Hearing Science*, Philadelphia, PA: Lippincott Williams & Wilkins. With permission.)

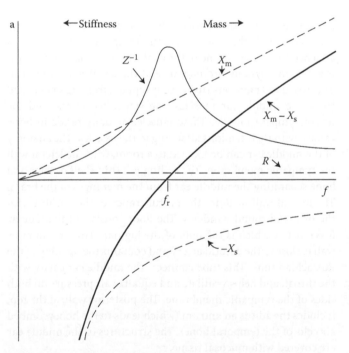

FIGURE 9.3 Illustration of the magnitude of reactance and resistance of a forced oscillation of a mass–spring system as a function of frequency. The power transfer is proportional to inverse impedance (i.e., admittance: Z^{-1}). The resonance frequency f_r occurs where mass inertia and stiffness cancel each other ($X_M - X_S = 0$) leaving only frictional dampening (R) yielding maximal vibrational response. The graph exemplifies that a mass–spring system is stiffness dominated at low frequencies and inertia dominates at high frequency. (From Durrant J. D. and Lovrinic J. H., 1995. *Bases of Hearing Science*, 3rd edition, Baltimore, MD: Williams & Wilkins. With permission.)

reactance impedes vibration also increases. Stiffness reactance is the impedance to vibration caused by stiffness (the elastic force); if the frequency of vibration is lowered, the degree to which stiffness reactance impedes vibration increases. The resonant frequency of an object is the point at which the mass and stiffness reactances cancel each other. The greater the similarity between impedances of the media, the more the sound energy transmitted through the new media. The larger the impedance mismatch between media, the more the sound energy reflected from the encountered media. For instance, the impedance mismatch between air and water results in a reflection of 99.9% of the acoustic energy (a 30 dB reduction).

In the case of reflected sound, the reflected wave may interact with a wave from the sound source that has not yet reached the reflecting object. In this case, the sound waves will interact. Extreme changes in level may occur if the waves interact constructively, resulting in reinforcement and an increase in sound level, or if the waves interact destructively, resulting in cancelation and a decrease in sound level. Related to reflected sound is the measurement of reverberation time (RT) or the time for a sound to decrease by 60 dB SPL in a room. RT is calculated using the Sabine equation:

$$RT = k\left(\frac{V}{A}\right), \tag{9.20}$$

where V represents the room volume in m^3, A is the total sound absorption by individual surfaces in the room, and k is a temperature-dependent constant (for 22°C, $k = 0.161$).

This brief section on acoustics has shown how a sound is created, how to determine the frequency and intensity/pressure of a sound, how a sound is propagated through the common medium (air), and how sound may generally be obstructed or interact. What happens when the sound reaches the ear?

9.2 Anatomy/Physiology of the Auditory System

9.2.1 Outer Ear

The outer ear consists of the cartilaginous pinna, the external auditory canal, and the lateral side of the tympanic membrane (Figure 9.4 as well as www.netterimages.com). The outer portion of the canal is cartilaginous, and the inner part consists of the temporal bone. The primary hearing functions of the outer ear are resonance of frequencies in the 2–5 kHz range (for adults) and sound localization. The head, in addition, acts as a barrier to high frequencies (>1500 Hz) and will distort the sound field, resulting in greater pressure at the entrance of the ear canal. Sound striking the pinna results in spectral modulations that help the listener resolve sound location. Typical sound localization cues are interaural time differences and interaural intensity differences. However, localization is possible even in cases where there are no interaural differences (labeled cones of confusion). In these cases, a slight head movement will result in spectral modulations arising from the pinna, causing changes in the head-related transfer functions (HRTFs) and enabling localization. Examples of HRTFs combining ear canal resonance, pinna, and head effects are shown in Figure 9.5. The pinna also aids in front-back localization of sounds in that frequencies in the 3–6 kHz region arising from the back of the head are diffracted by the pinna relative to those arising from directly in front.

9.2.2 Middle Ear

The middle ear consists of the middle ear cavity and the bones or ossicles of the middle ear [the hammer (malleus), anvil (incus), and stirrup (stapes)], as depicted in Figure 9.4. The malleus is attached on one side to the medial part of the tympanic

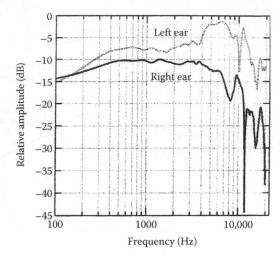

FIGURE 9.5 Representation of the head-related transfer functions (HRTFs) of the adult human ear in response to a momentary transient source delivered directly at the left ear (Transfer function, see Chapter 22). The HRTF measured within the ear canal for both the left and right ear is presented in a relative scale (arbitrary decibel scale). The attenuation of higher frequencies perceived at the right ear is more pronounced, specifically at higher frequencies. (From Yost, W. A., 2007. *Fundamentals of Hearing*, 5th edition, San Diego, CA: Academic Press. With permission.)

membrane, and on the other side to the incus. The incus, in turn, articulates with the stapes, and the footplate of the stapes is attached to the membrane of the oval window. The oval window is a hole in the bony protuberance of the cochlea, or inner ear. The ossicular chain gets suspension support from ligaments and from two middle ear muscles, the stapedius muscle and the tensor tympani muscle. These muscles actually reside in bony canals, with their tendons attaching to the ossicles. The anatomy of the middle ear can be likened to a room, with the lateral wall the tympanic membrane. The ceiling is a thin shelf of temporal bone separating the middle ear from the meninges of the brain. The medial wall includes the protuberance of the cochlea with the oval and round windows. The floor consists of the jugular fossa, below which is the bulb of the jugular vein. The anterior wall is close to the carotid artery and contains the opening of the Eustachian tube. This tube connects the middle ear cavity with the throat, and helps ventilate and equalize air pressure on both sides of the tympanic membrane. The posterior wall, at the top, includes the aditus ad antrum (which leads to the honeycombed air cells of the temporal bone). The structures of the middle ear are covered with mucosal tissue.

What happens when a sound wave encounters the outer and middle ear? The outer ear, as stated above, will help localize and amplify the incoming sound wave. An incoming sound wave in the external ear canal will cause the tympanic membrane to vibrate, which in turn causes the ossicular chain to move back and forth, which in turn causes the footplate of the stapes to move in and out of the oval window (more like a door than a plunger). The stapes movement then causes pressure waves in the fluids of the cochlea. The main function of the middle ear is

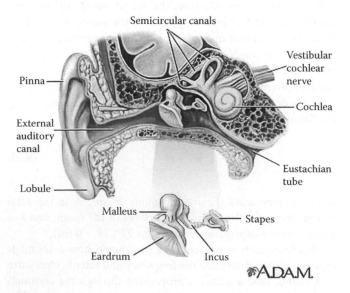

FIGURE 9.4 From left to right: the outer, middle, and inner ear. (From A.D.A.M. Inc. With permission.)

to act as an impedance transformer, to help match the approximately 30 dB of energy lost because of the impedance mismatch between air and the fluids of the cochlea. There are three mechanisms by which the middle ear acts as an impedance transformer. The first is the area difference between the tympanic membrane and the oval window. Recall that pressure may be described as

$$P(t) = \frac{mv}{tA} \quad \text{or as} \quad P = \frac{F}{A}, \tag{9.21}$$

where *F* represents force and *A* the area. Not all the surface area of the tympanic membrane vibrates for all acoustic frequency stimulation: the functional area of the tympanic membrane in adults is about 55 mm². The area of the oval window is about 3.2 mm², about a 17:1 area difference. The area ratio is the most significant mechanism by far for the impedance matching action of the middle ear. The second mechanism is the lever action of the ossicles, which appears to increase pressure by a ratio of 1.3:1. The third is the buckling action of the tympanic membrane, which appears to increase pressure by a ratio of 2:1. Because of the mass and stiffness reactances of the middle ear, the actual transfer function is frequency dependent, with maximum gain at around 1 kHz.

Contraction of the stapedius muscle will stiffen the ossicular chain and thereby attenuate sound energy transmitted to the cochlea. The stapedius muscle is the effector muscle in a feedback mechanism known as the acoustic reflex arc, a pathway coursing from the VIIIth cranial nerve to the brainstem and back to the stapedius muscle. Environmental sounds 85 dB above the hearing threshold will elicit contraction of the stapedius muscle, thus providing some protection for intense, low-frequency, nontransient sounds. Contraction of the stapedius and tensor tympani may prevent distortion caused by slight separation of the ossicles during vibration, and may help preserve our ability to understand speech amid background noise.

9.2.3 Cochlea

The inner ear is a relatively inaccessible, complex structure anatomically and physiologically (Figure 9.6, panel B; also see www. netterimages.com/image/7306.htm; http://147.162.36.50/cochlea/index.htm and http://www.iurc.montp.inserm.fr/cric/audition for anatomical details). The inner ear, housed within the temporal bone, consists of a network of canals in the bone referred to as the bony labyrinth. Within the bony labyrinth are tissues of the inner ear, referred to as the membranous labyrinth. These tissues are suspended by fluids called perilymph and endolymph. Perilymph has a high sodium concentration and is similar to a saline solution. The labyrinth of the inner ear can be divided into three parts: the semicircular canals, the vestibule, and the cochlea. The semicircular canals and the vestibule (consisting of the utricle and saccule) together form the vestibular system of the inner ear, and help determine angular and linear acceleration/deceleration of the head. These are balance functions, and will not be discussed in this chapter. The auditory function of the inner ear occurs in the cochlea.

The bony labyrinth of the cochlea makes approximately $2\frac{2}{3}$ turns in the human. Recall that the footplate of the stapes fits into the oval window. As seen in the figure, the oval window opens into a canal filled with a fluid similar to cerebrospinal

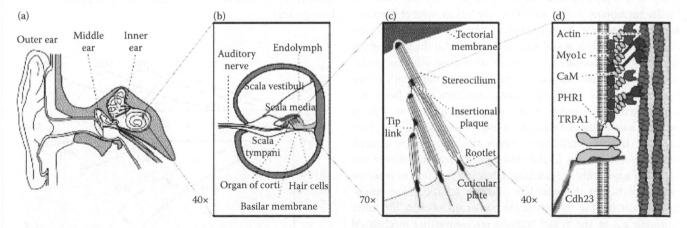

FIGURE 9.6 Graphical representation of the anatomy of the inner ear. (a) From left to right diagram of outer, middle, and inner ear relative positions within the human auditory system. (b) Cross-sectional view of the cochlear duct, illustrating the three scala and the auditory nerve; the scala media contains extracellular fluid called endolymph, and the scala media and scala tympani are separated by a membrane containing the organ of Corti that has hair cells protruding in the scala media. (c) Outer hair cell within the organ of Corti hair-bundle detail. The three stereocilia of the hair bundle are each connected to their own respective actin filament. The stereocilia are imbedded in the cuticular plate on the proximal end; the abundant other stereocilia have been omitted for clarity. The shorter stereocilia are connected at the tip to the adjacent taller neighbors. The tallest stereocilium is connected permanently to the overlying tectorial membrane. (d) Mechanotransduction mechanism on a molecular level with a single stereocilium of an outer hair cell. Key molecules in this diagram are Cdh23 on the tip of the hair providing a connective link, the transduction channel is formed by TRPA1, and Myo1c is the motor for adaptation, along with the crosslinking protein PHR1 in addition to calmodulin light chains (CaM). (Adapted from LeMasurier M. and Gillespie P.G., 2005, *Neuron* 48: 403–415. With permission.)

fluid called perilymph (the space above the cochlear duct). This canal is called the scala vestibuli. This location is referred to as the base of the cochlea. Following the scala vestibuli to the top of the turns (the apex), the canal is now below the cochlear duct. This canal is now called the scala tympani. The scala tympani ends at the round window, which is located below the oval window, in the medial wall of the middle ear cavity. The passage point between the scala at the apex is the helicotrema.

A cross section through the cochlear duct is shown in Figure 9.6, panel B. Here again are the scala vestibuli and scala tympani, with the cochlear duct the triangle-shaped tissue in the middle. The fluid-filled space in the middle of this triangle is the scala media, which is filled with endolymph. Endolymph has a high concentration of potassium, which will become important when discussing cochlear function. The membrane separating the scala vestibuli from the scala media is Reissner's membrane. The tissue located at the open end of the scala media (the right side in the figure) is the stria vascularis, which contains the blood supply to the cochlea. The stria also helps maintain the endolymph and the unusual high positive endolymphatic potential (+80 mV). The structure at the bottom of the scala media is the organ of Corti. The organ of Corti is a complex structure, consisting of the tectorial membrane (at the top), outer and inner hair cells, support cells such as Deiter and Hansen's cells, the arches and tunnel of Corti, synaptic connections between neurons and the hair cells, and afferent and efferent neural elements. These neural elements will course out of the cochlea through holes in the temporal bone (the habenula perforata) to become the auditory portion of the VIIIth cranial nerve. On the bottom of the organ of Corti is the basilar membrane.

The mass and stiffness characteristics of the basilar membrane allow for tonotopicity, or frequency tuning, of the membrane. The basilar membrane is stiff and relatively less massive at the base, which is ideal for high-frequency vibration. The membrane becomes relatively less stiff and more massive at the apex, which is ideal for low-frequency vibration. This mass–stiffness gradation changes in an exponential pattern along the length of the basilar membrane. Upon pure-tone stimulation, a high frequency would then have a point of maximum displacement at a point near the base of the cochlea, and a low frequency at a point near the apex. In some vertebrate animals, the hair cells are actually tuned for frequencies. In the mammalian cochlea, however, the hair cells apparently are not tuned, but still function as receptor cells. In particular, the outer hair cells may act as cochlear amplifiers, enhancing the motion of the basilar membrane. The inner hair cells act as the actual transducers, converting mechanical energy to neural/chemical energy at the hair cell/neural synapse.

What happens when a pure tone sound wave encounters the cochlea? Recall above that the middle ear has transformed acoustic energy into mechanical energy, and that the stapes moves in and out of the oval window. This creates a near-instantaneous pressure wave throughout the cochlea, since the cochlear fluids are essentially incompressible. Then, a "slow" traveling wave begins at the base of the cochlea and moves to the apex. At the point of maximum displacement, outer hair cells would have provided amplification

for the motion of the basilar membrane. Inner hair cells would then be stimulated to release a neurotransmitter (probably the excitatory amino acid glutamate), which would then stimulate afferent auditory nerve fibers. These fibers will eventually fire an action potential in the hearing portion of the VIIIth nerve.

9.2.3.1 Motion of the Basilar Membrane

Many of the mechanisms involved in the cochlear activity just described are not definitively known and have been the subject of intense study for the past 60 years. Some of these issues, and their complexity, follow. One of these is the motion of the basilar membrane. Studies by Georg von Bekesy using cadaver cochleas and cochlear models showed that the basilar membrane, upon stimulation by a pure tone, responded in a series of displacements known as a traveling wave, always moving from base to apex. The envelope of these displacement patterns and their phase delays are depicted in Figure 9.7, adapted from von Bekesy's work. These motion patterns can be modeled in terms of acoustic impedance for sections of the cochlear partition by the following equation:

$$Z = \frac{1}{A^2} \sqrt{\left[\left(M\omega_o - \frac{K}{\omega_o} \right)^2 + D^2 \right]}, \qquad (9.22)$$

where M represents the mass of a cochlear partition, K the stiffness of the partition, D the damping of the partition, and ω_o the radian frequency of the stimulus. There will be a section of the cochlear partition in which the frequency of the stimulus will be the resonant frequency, or where the mass and stiffness reactances cancel each other. At this resonant point, the traveling wave envelope will be at its greatest, since impedance is at its lowest value.

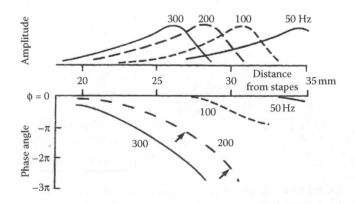

FIGURE 9.7 Diagram of phase and amplitude response for four different frequencies with respect to the distance to the stapes: (a) amplitude envelop of traveling waves, (b) angular phase shift between a fixed location on the basilar membrane and the relative motion of the stapes. When the phase shift and wave period are known, the time to travel from one point to the next on the membrane can be calculated. (From Yost, W. A., 2007, *Fundamentals of Hearing*, 5th edition, San Diego, CA: Academic Press. With permission.)

Thus, as a mid-frequency stimulus wave propagates along the basilar membrane, it will disturb sections along the base, but no resonance will occur because of high impedance; little energy will be absorbed by the basilar membrane. As the wave moves along the length of the basilar membrane, impedance to the wave will lessen, and finally reach its lowest point for that frequency. At this resonance point, large displacement of the membrane will occur, and most of the wave's energy is absorbed. There is little energy left to stimulate more apical regions of the membrane, and the wave dies out. In fact, this description of motion holds true for dead or badly damaged cochleas, but not for a living, healthy mammalian cochlea.

Using the Mossbauer technique of measuring radioactivity from an emitting source placed on the basilar membrane, Rhode (1971) found that at low stimulus intensity levels the displacement of the basilar membrane was sharper than at higher levels. This meant that the movement of the basilar membrane could be nonlinear with respect to stimulus level. Rhode also subsequently showed that this nonlinearity was dependent on a living animal. This finding has been replicated several times in various animals. It has also been replicated by measuring laser reflectance from surfaces placed on the basilar membrane, a technique called laser interferometry. More precise measurements have been determined using laser vibrometry, which does not require placement of a reflective surface upon the basilar membrane. In this technique, monochromatic light is directed at the membrane; the light scattered by the membrane motion will have a frequency shift proportional to the velocity of membrane motion (the Doppler effect). Tracking the Doppler shift can then provide information about basilar membrane motion. The physiologically vulnerable nonlinearity or compression of basilar membrane motion occurs for a given frequency at that frequency's point of maximum displacement only (Figure 9.8). An interesting note about the nonlinearity of basilar membrane motion is that it produces new tones, tones that were not in the original acoustic stimulus. This is referred to as intermodulation distortion.

Accurate descriptions of basilar membrane motion patterns have been presented by Robles and Ruggero (2001; Figures 9.8 and 9.9). Figure 9.9 shows a set of isointensity curves measuring basilar membrane velocity as a function of frequency. Figure 9.8 shows the sharp tuning and sensitivity of basilar membrane motion, and nonlinearity of response as the intensity of the stimulus increases. A few points should be noted here regarding basilar membrane motion. First, Figure 9.9 is a plot of basilar membrane velocity rather than displacement; research has suggested that stimulation of afferent VIIIth nerve auditory fibers is a function of both basilar membrane displacement and velocity (Ruggero et al., 2000). Second, because this is nonlinear vibration, the motion could be analyzed using cross-correlation of the basilar membrane's response to noise. The first-order cross-correlation provides information regarding odd-numbered nonlinearities, and the second-order cross-correlation provides information regarding quadratic and other even-numbered nonlinearities. These first- and second-order Wiener kernels of basilar membrane motion, particularly at low stimulus

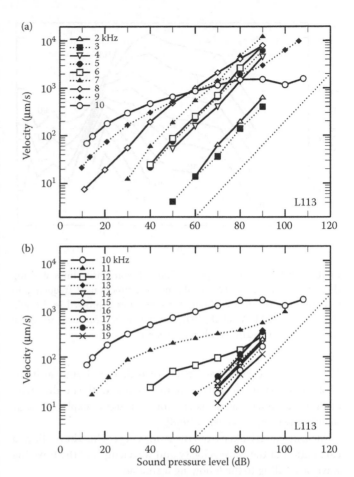

FIGURE 9.8 Graphical representation of the velocity–intensity function response recordings in cochlea L113 for the basilar membrane when excited by frequencies lower or higher than 10 kHz (i.e., characteristic cochlear tuning frequency or characteristic frequency, CF). A reference line, dashed, has a linear slope of 1 dB/dB. (From Ruggero M. A., et al., 1997, *Journal of the Acoustical Society of America* 101: 2151–2163. With permission.)

intensities, show sharp tuning and relatively low harmonic distortion. As mentioned above, however, the nonlinearities of the motion result in intermodulation distortion, which is discussed below. Third, there appear to be differences in basilar membrane motion between the base and the apex of the cochlea. At the apex, characteristic or best frequency (CF) octaves are more compressed than at the base, similar to the idea of smaller physical keys along a keyboard at one end of a piano compared to the other. Apical regions are not as sharply tuned as basal regions, and nonlinearities at the apex, when present, may exist throughout the frequency range. In addition, the spiral shape of the cochlea will cause the basilar membrane to tilt to the side, giving more motion to hair cell cilia at the center of the spiral (the apex) and thereby improving our sensitivity to low frequencies (Manoussaki et al., 2006). Fourth, because of the nonlinearities, the physical extent of basilar membrane motion required at threshold is larger than the linear extrapolations predicted from cadaver cochleas: basilar membrane motion ranging from 0.3 to

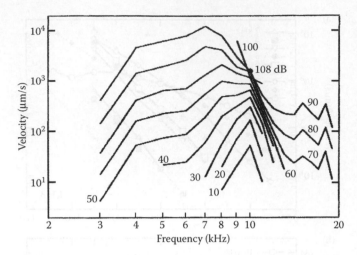

FIGURE 9.9 Graphical representation of isointensity lines showing the velocity of the basilar membrane as a function of frequency in response to tone pipes at a range of intensities expressed in dB SPL. (From Ruggero M. A., et al., 1997, *Journal of the Acoustical Society of America* 101: 2151–2163. With permission.)

1–2 nm at threshold have been recorded in living animals (Robles and Ruggero, 2001). Finally, because of the nonlinearity of motion, Fourier analysis cannot thoroughly explain basilar membrane motion (Geisler, 1998).

Measurements taken from the guinea pig cochlea have allowed calculation of the wave velocity and wavelength of the traveling wave, according to the following formulae:

$$\delta t = \delta\varphi(2\pi f), \qquad (9.23)$$

$$\text{wave velocity} = \frac{\delta x}{\delta t}, \qquad (9.24)$$

$$\text{wavelength} = \frac{2\pi\delta x}{\delta\varphi}, \qquad (9.25)$$

where δt represents travel time, $\delta\varphi$ the phase difference between two locations along the basilar membrane, δx the distance between the two locations, and f the stimulus frequency. At the base of a guinea pig cochlea, the wavelength was measured to be about 0.67 mm and the wave velocity was 10 m/s (Robles and Ruggero, 2001).

For illustrative purposes, further explanation of basilar membrane motion in this section will be guided by a theoretical framework and four equations (Equations 9.26, 9.27, 9.31, 9.36) from Neely et al. (2003). Even with the nonlinearities, basilar membrane motion may be represented by the following linear equation:

$$Z_b = Z_p + \gamma Z_a, \qquad (9.26)$$

where Z_a is active impedance (which plays an important role at low levels), Z_p is passive impedance, γ is a parameter ranging from 1 at low levels to 0 at high stimulus levels, and Z_b is the

combined basilar membrane impedance (de Boer et al., 2007; Neely et al., 2003). As will be discussed below, outer hair cells are thought to be responsible for the physiologically vulnerable nonlinearity of basilar membrane motion. In the above equation, "γ" represents outer hair cell mechanoelectrical transduction, and simulates compression of basilar membrane motion and intermodulation distortion. To write a time-domain representation of the above equation for points along the cochlear partition, one modeling equation that may be used is

$$M_b\xi_b + R_b\xi_b + K_b\xi_b + g_cR_c\xi_c + g_cK_c\xi_c = -A_pP_f, \qquad (9.27)$$

where M is the mass, ξ_b the basilar membrane velocity, R the resistance, K the stiffness, g the basilar membrane to inner hair cell lever gain, ξ_c the shear displacement between the tectorial membrane and the reticular lamina, and P_f the fluid pressure (Neely et al., 2003; Neely and Kim, 1986). The reticular lamina is the tough border between the endolymph of the scala media and the hair cells, with the hair of the hair cells (the stereocilia) sticking out from the reticular lamina.

The above equations prompt discussion of three topic areas before further modeling equations can be mentioned. These topics include intermodulation distortion, cochlear micromechanics (vibration of other structures in the cochlear partition, such as the tectorial membrane), and how the outer hair cells serve as the source of nonlinear motion of the basilar membrane. Each of these topics will be discussed below.

9.2.3.2 Intermodulation Distortion

Nonlinear systems will create frequency components that were not present in the input signal. Similarly, certain tone combinations cause human listeners to hear additional components that were not in the original acoustic stimulus. This is termed intermodulation distortion. The most prominent of these combination tones psychophysically is the cubic difference tone, or $2f_1 - f_2$, where f_1 and f_2 represent the two tones in the original acoustic stimulus.

Kemp (1978) was the first to record the ear making sound as it was listening to sounds. In addition to distortion of an acoustic stimulus, the cochlea may also produce sounds without any obvious acoustic input. Sounds produced by the ear, whether evoked by an acoustic stimulus or not, are termed otoacoustic emissions (OAEs). For a number of years following Kemp's discovery, OAEs have been labeled based on whether a stimulus was necessary to record them present in the ear canal, and, if so, what type of stimulus was used. Thus, spontaneous OAEs are produced by the ear without any stimulus present, and evoked OAEs are produced in response to an acoustic stimulus. Evoked OAEs are produced in response to a click (transient OAEs), a two-tone complex (distortion product OAEs), or a pure tone (stimulus frequency OAEs). Classifying OAEs based on the underlying mechanism of their generation, however, is a better way of defining these phenomena (Shera, 2004).

One mechanism by which OAEs are produced is cochlear nonlinearity. The traveling wave causes local mechanical distortions near the maximum displacement of the basilar membrane. These distortions are likely caused by the force of outer hair cells feeding back onto the basilar membrane, effectively changing the membrane's stiffness. This distortion may result from, say, a two-tone acoustic stimulus that produces a cubic distortion tone. de Boer et al. (2007) modeled this situation as follows. The pressure generated by the outer hair cells for each tone separately could be expressed by

$$P_{\text{act}}(x; f) = -\frac{1}{2} v_{\text{prim}}(x; f) Z_{\text{bm}}^{\text{act}}(x; f), \quad (9.28)$$

where $v_{\text{prim}}(x; f)$ represents a template of excitation produced by each tone and $Z_{\text{bm}}^{\text{act}}(x; f)$ represents the nonlinearity the tone undergoes. Then, the amplitude of the created cubic distortion tone could be represented by

$$A_{\text{DP}}(x) = \frac{3}{4} A \psi_1^2(x; f_1) \psi_2^*(x; f_2). \quad (9.29)$$

This can result in waves traveling in two directions from the point of maximum displacement. The reverse wave would be the wave then recorded in the external ear canal.

The second mechanism by which OAEs are produced is energy reflection. Disparities in cochlear anatomy (such as number or geometric shape of hair cells) or in physiology (such as variations in the distribution of muscle fibers or proteins in the outer hair cells) may result in micromechanical impedance perturbations that cause reflection of the energy of the traveling wave. Reflectance could be modeled by the following equation:

$$R(f) = \int_0^\infty \delta(\chi, f) T^2(\chi, f) \, d\chi, \quad (9.30)$$

where $\delta(\chi, f)$ represents the spatial arrangement of impedance perturbations and $T^2(\chi, f)$ represents the basilar membrane wave. Wavelets here that interact constructively will compose the net reflected wave (Shera, 2003). As an example, spontaneous OAEs may actually be reflections of cochlear standing waves produced in response to physiological noise or background environmental noise (Shera, 2004).

9.2.3.3 Micromechanics

The reflectance mechanism for generating OAEs provides a segue into cochlear micromechanics. It is presumed that the primary event in the cochlea is the motion of the basilar membrane (cochlear macromechanics); motions of parts of the organ of Corti and the tectorial membrane are referred to as cochlear micromechanics. An early model of cochlear mechanics by Neely and Kim (1986) represented the tectorial membrane as a mass that could oscillate in the radial direction and rotate around its hinge. The activity of the tectorial membrane was

then modeled as an additional degree of freedom coupled to the basilar membrane (which involves radial motion of the tectorial membrane in conjunction with the reticular lamina and outer hair cells):

$$m_r \xi_r + R_r \xi_r + K_r \xi_r - g_c R_c \xi_c = g_c K_c \xi_c = -g_r V_m, \quad (9.31)$$

where parameters listed in Equation 9.27 are reproduced and ξ_c represents the difference between the two mechanical degrees of freedom (Neely et al., 2003). In the model, the outer hair cells were thought to exert a force onto the tectorial membrane. Indeed, research has shown that outer hair cell motility can drive radial tectorial membrane motion (e.g., Jia and He, 2005).

A recent study has demonstrated the existence of longitudinal traveling waves along the tectorial membrane. The motion of the tectorial membrane is represented by a distributed impedance model (Ghaffari et al., 2007; Figure 9.10). These longitudinal waves of the tectorial membrane may interact with and/or stimulate hair cells, and interact with the basilar membrane traveling wave in fundamental ways that are yet to be explored (Ghaffari et al., 2007). In fact, a recent study combining atomic force microscopy, fluorescence microscopy, and modeling has suggested that collagen fibers in the tectorial membrane are paired to outer hair cells as hammers are to strings on a piano, and that this pairing may allow the tectorial membrane to influence and increase the amplification provided by the outer hair cells (Gavara and Chadwick, 2009).

9.2.3.4 Outer Hair Cell Activity

The above discussions have frequently mentioned the activity or motility of outer hair cells. Gold (1948) was the first to suggest that the cochlea required an electromechanical feedback source to counter the viscous drag on basilar membrane motion caused by the cochlear fluids. The source of this electromechanical feedback, and the source of the physiologically vulnerable nonlinearity of basilar membrane motion, is thought to be tied to the outer hair cells. Just over 20 years ago it was discovered that the outer hair cells move in response to stimulation; this movement or motility may act to feed energy back onto the basilar membrane, and thereby improve the sensitivity and tuning of the basilar membrane motion that has been alluded to earlier. There are still details to be learned on how this may happen, as the following discussion will reveal.

But first, a bit of anatomy. In humans there are roughly 11,000 outer hair cells, situated in 3–4 rows. The hair cells are referred to as "outer" because they are positioned on the lateral side of the arches of the organ of Corti. In this position, they are relatively farther from the center of the cochlea (the modiolus) than the inner hair cells, which are on the medial side of the arches of the organ of Corti (Figure 9.11). The outer hair cells are not the primary sensory transducer of hearing, synapsing with only about 5% of afferent (sending information from the cochlea to the brain) auditory neurons forming the hearing portion of the VIIIth cranial nerve. These are type II afferent auditory fibers, with each

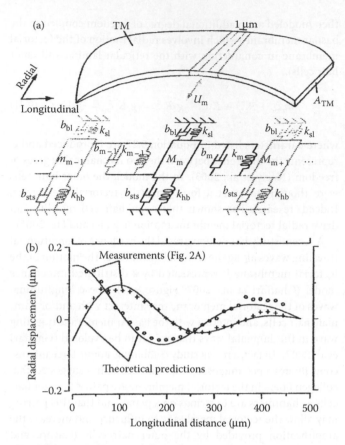

FIGURE 9.10 Schematic representation of the mechanical equivalence of the temporal membrane (TM). (a) A 1-μm-wide section is selected as a basis with a rectangular area: A_{TM} (upper). Radial motion can be induced in this section of the TM through vibration support, providing longitudinal coupling with propagation velocity U_M. The lower part illustrates the mechanical circuit diagram consisting of a central portion with mass M_n beaconed off by neighbors M_{n-1} and M_{n+1}, linked though viscous friction expressed by b_m and b_{m-1} as well as elastic coupling expressed by coefficients k_m and k_{m-1}. The effective mass for each section includes the mass of the lymph liquid above and below the TM. Other mechanical effects such as the influence of the liquid layers is represented as a dashpot between ground and the respective mass, as well as dashpot–spring combinations for viscous dampening from subtectorial space (b_m), limbal attachment (k_{sl}), and hair bundle rigidity (k_{hb}). (b) Theoretical predictions based on the mechanical circuit model described in (a) for wave propagation (solid lines represent the least-square fit to theory) are compared against physical TM wave measurements illustrated by + and 0 symbols. Theoretical analysis against the measured curves yielded the following values for TM with respect to shear storage modulus and shear velocity, respectively: $G\boxtimes$ = 30 kPa and h = 0.13 Pa s. Boundary conditions accounted for the physical constraints of TM due to adhesion to the cochlea. (From Ghaffari R., et al., 2007, *Proceedings of the National Academy of Sciences USA* 104(42): 16510–16515. With permission.)

fiber synapsing with 10 or so outer hair cells. There is little information regarding their activity. Rather, outer hair cells are primarily innvervated by efferent (sending information from the brain to the cochlea) fibers coursing from the superior olivary complex in the brainstem. Some efferent fibers run from the

medial superior olive, cross the midline of the brainstem, and innervate outer hair cells of the contralateral cochlea, with one fiber innervating several outer hair cells. This fiber bundle is called the crossed olivocochlear bundle and uses acetylcholine as its primary neurotransmitter. Other fibers from the medial superior olive innervate outer hair cells in the cochlea on the same side of the body (ipsilateral). Other efferent fibers run ipsilaterally from the lateral superior olive and synapse onto type I afferent fibers just below the inner hair cell–afferent fiber synapse. These fibers may act to modulate the output of the inner hair cell–afferent fiber synapse, and include dopamine and enkephalin neurotransmitters. It should be noted that the tallest of the stereocilia of the outer hair cells are likely embedded in the tectorial membrane, which, along with the above discussion of the tectorial membrane, may suggest an important interaction between outer hair cells and the tectorial membrane that is not yet fully explored.

Exactly how outer hair cell movement may effect cochlear amplification is not precisely known. Two general hypotheses are (1) motion of the stereocilia atop the outer hair cell creates the amplification (e.g., Choe et al., 1998) and (2) a change in length of the outer hair cell creates the amplification (e.g., Brownell et al., 1985). Let us consider the latter hypothesis first. Mammalian outer hair cells change length in response to electrical stimulation, and this is likely a novel type of cell motility. The motility is likely from a two-state area motor located along the lateral membrane of the cell. This motor can be switched to lengthen or compact via a membrane polarization change. The motor is not driven by an intracellular force, but rather is like a piezoelectric element, switching states upon a supply of external energy. The electrically induced change in length of the cell is proportional to charge displacement across the membrane and can be written as

$$L(V) = L_{max} + \frac{(L_{min} - L_{max})}{\{1 + \exp[\beta(V - V_o)]\}}, \qquad (9.32)$$

where Q represents charge movement. Charge displacement of ion channels is too slow to represent hair cell motion in the range of 80 kHz (Frank et al., 1999). Membrane transport proteins may operate at such a high rate; one such protein found along the membrane of outer hair cells is prestin. A strain of mouse in which the gene for prestin was partially deleted shows loss of outer hair cell motility and a hearing loss of 40–60 dB as measured by distortion-product OAEs and auditory brainstem responses (ABRs). The most straightforward account of charge movement is that chloride ions move partway in a membranous pore consisting of prestin (Figure 9.12). This is theory at this point, however, and the actual structure of the prestin molecule is yet to be accurately described (Ashmore, 2008).

The second hypothesis for generation of cochlear amplification is movement of the stereocilia atop the outer hair cell. Deflection of the stereocilia toward the tallest stereocilia results in hair cell depolarization, and deflection away from the tallest stereocilia results in hair cell hyperpolarization. As seen from

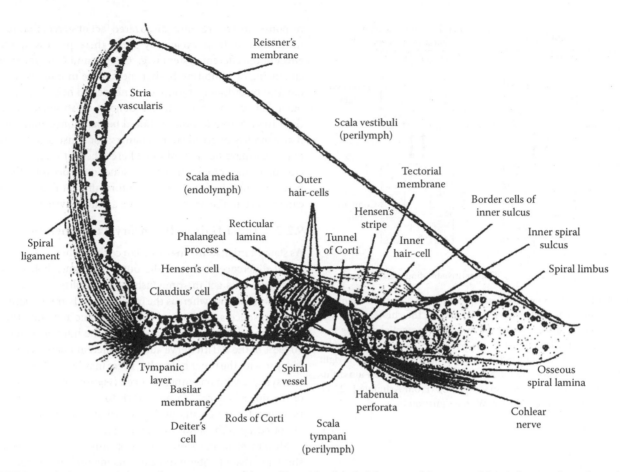

FIGURE 9.11 Artist representation of a cross section of the cochlea with a detailed drawing of the organ of Corti, showing the scala media with respect to the basilar membrane. (Drawing by Sara Crenshaw McQueen, Henry Ford Hospital, Detroit, from Yost Y.A., 2007, *Fundamentals of Hearing*, 5th edition, San Diego, CA: Academic Press. With permission.)

Figure 9.6, panels A–D, deflection toward the tallest stereocilia causes tip links to open potassium and calcium ion channels, thereby depolarizing the cell. The tip links are formed by the proteins cadherin 23 and protocadherin 15 (Kazmierczak et al., 2007). There is constant tension on the stereocilia of the outer hair cell. There are two general explanations of how movement of the stereociliar bundle may account for amplification. One is that the motor protein myosin, found in the stereocilia, may act to generate movement. This may be possible for frequencies below 10 kHz, but is unlikely for frequencies of movement higher than that. The second general explanation is that calcium interacts with the transduction channel to induce movement. In this explanation, the stereocilia transduction channel may be opened by stimulus force; the binding of calcium to the channel will close the channel. The transduction channel may be closed (1), forced open (2), having calcium attached at two binding sites (3 and 4), closed (5), and having calcium disattached (5 and 6). The six states of the channel system $S(t)$ may be represented by the following equation:

$$S(t) = [X, V, p_1, p_2, p_3, p_4, p_5, p_6]^T, \qquad (9.35)$$

where X is the displacement of the stereociliar bundle, V the bundle velocity, T the time, and p_1–p_6 the stages of transduction channel opening or closing (Choe et al., 1998). Further refinement of this model suggests that calcium may not directly cause channel closure, but indirectly by causing the release of a mechanical element of the transduction machinery, which then makes the gating spring slack, resulting in transduction channel closure (LeMasurier and Gillespie, 2005).

The debate over how the outer hair cells effect cochlear amplification, or even if their role may be overemphasized relative to the tectorial membrane, is as yet unresolved. It may be that both outer hair cell motility and stereociliar bundle displacement work together to create electromechanical feedback and cochlear amplification. Both mechanisms may be coupled together (e.g., Zenner et al., 1988).

To suggest how overall outer hair cell activity may be mathematically modeled, we return to the cochlear modeling equations by Neely and colleagues. According to their model, outer hair cell motility can be described as

$$C_m V_m + G_m V_m = \gamma g_f \xi_b, \qquad (9.36)$$

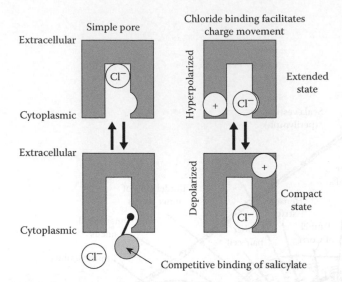

FIGURE 9.12 Chemomechanical representation of Prestin as an area motor. The dot represents chloride. On the left, the chloride is driven into a cytoplasm-facing pore by the membrane potential, leading to a shape change of the lateral membrane. Salicylate may act as a chloride-binding inhibitor when adhered to a cytoplasmic binding site. The right side illustrates allosterically linked charge motion based on ion diffusion and protein kinetics (see also Chapter 2). (From Ashmore J., 2008, *Physiological Reviews* 88: 173–210. With permission.)

where

$$\gamma = \gamma_\circ \left(1+\left|\frac{\xi_b}{d_h}\right|^{3/2}\right)^{-2/3} \tag{9.37}$$

and represents outer hair cell force generation gain, and other parameters as listed from Equation 9.27 and Equation 9.31. Equation 9.36 can account for much data, but has yet to explain some aspects of intermodulation distortion product growth (Neely et al., 2003). When the coefficients of Equations 9.27 through 9.36 are constant, the ratio of fluid pressure to basilar membrane velocity is proportional to the combined basilar membrane impedance (Z_b) and can then be expressed in Equation 9.26 ($Z_b = Z_p + \gamma Z_a$). This mathematical model then helps describe basilar membrane motion (Equation 9.27), activity of a coupled mechanical second degree of freedom suggested to be tectorial membrane motion (Equation 9.31), and outer hair cell activity (Equation 9.36) in a unified fashion.

It is hoped that the reader has gained an appreciation for the complexity and the not-completely-understood nature of cochlear mechanics. There are a plethora of unresolved issues, including (but not limited to) the details of micromechanics, activity of the tectorial membrane, the actual mechanism of cochlear amplification, the apparent lack of adaptation of cochlear mechanics, how support cells in the cochlea (e.g., Deiter cells) affect the mechanical response, how coupling of the outer hair cells via the support cells may affect the mechanical

response, how the recently discovered fact of stereociliar replacement on a 48-hour cycle may affect these processes, and the effect of the efferent system (e.g., Cooper and Guinan, 2006) on enhancing or inhibiting basilar membrane motion. New physiological recording techniques are needed to help determine the outer hair cell response to frequencies of stimulation greater than 10 kHz. In addition, it should be mentioned that there are many models of cochlear mechanics; for illustrative purposes, only a few have been mentioned here. There still remains a discussion of the actual sensory transduction process by the inner hair cells, and the function of auditory afferent fibers, before we can proceed to the next levels of the auditory system.

9.2.3.5 Inner Hair Cell Sensory Transduction

As stated earlier, in the mammalian cochlea, there are clear functional differences between the outer and inner hair cells. The outer hair cells act in some fashion to amplify the mechanical response to sound, whereas the inner hair cells are the actual sensory transducers. Again, we begin with some anatomy. There are approximately 3500 inner hair cells in the human cochlea, and 90–95% of the VIIIth nerve auditory afferent fibers (type I fibers) synapse with the inner hair cells. There may be 10–30 type I fibers that synapse with one inner hair cell (Figure 9.13). The stereocilia of the inner hair cell do not attach to the tectorial membrane; therefore, the stereocilia of the inner hair cell are moved by the viscous drag of the fluid surrounding the hair cell.

Movement of the stereocilia atop inner hair cells would be similar to that of outer hair cells, causing tip links to open large-conductance calcium-activated potassium channels and delayed rectifier potassium channels and generate a receptor potential. This receptor potential opens voltage-sensitive calcium channels, which in turn modulates the secretion of vesicles containing glutamate into the synaptic cleft. In the mammalian cochlea, each synapse between an inner hair cell and the dendrite of an afferent nerve fiber is called a "ribbon" synapse because of its similarity to ribbon synapses found in retinal photoreceptor and bipolar cells. Although these synapses are so sensitive that a single vesicle of glutamate may be able to cause a postsynaptic potential in the dendrite, it is likely that temporally precise coding arises from the synchronous release of many vesicles (Glowatzki and Fuchs, 2002; Moser et al., 2006b). For precise temporal coding, modulation of vesicular release by calcium likely occurs at the nanodomain; that is, stochastic gating of one or only a few calcium channels modulates vesicular release (Moser et al., 2006b). Glutamate transporters, such as GLAST, help clear the synaptic cleft of the transmitter once the postsynaptic receptor has been activated. On the dendritic postsynaptic membrane, there are likely three glutamatergic receptors: α-amino-3-hydroxy-5-methyl-isoxazole-4-propionic-acid(AMPA), *N*-methyl–D–aspartate (NMDA), and kainate binding proteins. It is likely, however, that it is the AMPA receptor that mediates fast synaptic transmission. Activation of the postsynaptic receptor creates a graded potential in the dendrite that may later cause an action potential along the myelinated portion of the fiber (located in the modiolus). It should also be noted that efferent

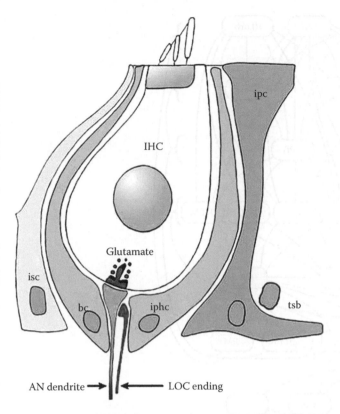

FIGURE 9.13 Graphical illustration of the signal transfer mechanism for an inner hair cell (IHC) synaptic complex. The schematic representation of IHC innervation by a Type I afferent dendrite shows a synaptic ribbon which has itself surrounded by glutamate (Glu) containing vesicles along with a lateral olivocochlear efferent fiber (LOC). The IHC is supported by border cells (bc) and inner phalangeal cells (iphc). Other anatomical features include the inner sulcus and inner pillar cell (isc and ipc, respectively). (Reprinted from Ruel J., et al., 2007, *Hearing Research* 227: 19–27. With permission.).

fibers from the lateral superior olive, which synapse onto the afferent fibers just below the ribbon synapse, act to modulate the output of the afferent fiber in ways not yet fully explored (e.g., Ruel et al., 2007).

There are still a number of details to be worked out to fully explain this sensory transduction process. For instance, electron micrographs show that the anatomy of the inner hair cell–VIIIth nerve fiber synapse is somewhat different in detail from one species to another, so a "typical" synapse is a misnomer (e.g., Moser et al., 2006a). Use of electron tomography and immunoelectron microscopy will hopefully provide more detail into the anatomy at the molecular level. To better understand the temporal coding ability of the synapse, studies of neurotransmitter release rate, calcium channel gating, and synaptic delay should be done, preferably with paired pre- and postsynaptic recordings (Moser et al., 2006b).

9.2.3.6 Inner Hair Cell and VIIIth Nerve Responses

Because of the relative inaccessibility of the cochlea, numerous studies examining auditory physiology were performed from recordings of VIIIth nerve auditory afferent fibers. Much of the activity reported from these fibers is now thought to arise from the cochlea. For instance, these fibers are typically organized according to their spontaneous action potential firing rates, or their activity sans obvious auditory stimulation. Spontaneous firing of these fibers seems to rely entirely on inner hair cell neurotransmitter release. Similarly, neural adaptation to sound stimulation may arise from partial depletion of the readily releasable neurotransmitter vesicle pool (RPP) in the inner hair cell. One of the prime responses of VIIIth nerve afferent fibers to sound is phase-locking; that is, the neuron fires an action potential during one part of the stimulus cycle. This behavior is actually mediated by the inner hair cell synapse, perhaps reflecting the back and forth movement of the stereocilia.

Before accurate measures of basilar membrane motion were made, the broad frequency tuning of the cadaver basilar membrane compared with the sharp frequency tuning of VIIIth nerve afferent fibers suggested the existence of a "second" filter between the basilar membrane and the VIIIth nerve to explain the improved tuning. Subsequent basilar membrane and neural measurements at the 9 kHz place along the chinchilla cochlea has definitively shown that there is no need for such a second filter (Narayan et al., 1998; Ruggero et al., 2000). In addition, many of the phenomena first recorded in VIIIth nerve fibers (e.g., frequency dependence of slopes of rate-intensity functions, changes in frequency tuning with increasing stimulus intensity, two-tone distortion, two-tone suppression) actually arise from basilar membrane motion (Robles and Ruggero, 2001).

9.2.4 Central Auditory Nervous System

When an action potential is fired along the VIIIth nerve, information may arrive in the primary auditory cortex of both hemispheres of the brain by a number of pathways. The most common pathway is contralateral, for example from the right ear to the left hemisphere primary auditory cortex. One pathway that would include all the major auditory midbrain nuclei would first include the VIIIth nerve synapse in the cochlear nucleus, then crossing over the midline of the brainstem via the trapezoid body to the superior olivary complex, then to the lateral lemniscus, the inferior colliculus, the medial geniculate in the thalamus, and then to the primary auditory cortex by way of auditory radiations. From there, information would proceed to intermediate cortical areas, such as the language area, the frontal cortex for decision making, and even sensorimotor and motor cortices and spinal neurons if a behavioral response is made to the sound.

9.2.4.1 Auditory Brainstem

The anatomical interconnections of the auditory midbrain are exceedingly complex; much work has been expended to determine both structure and function of these areas—and there is much yet to be learned (Figure 9.14). Introductory anatomical texts usually illustrate a simplification of these pathways. General principles of the central auditory nervous system include tonotopicity (neurons with similar best frequencies are grouped together), morphological and response pattern variability as

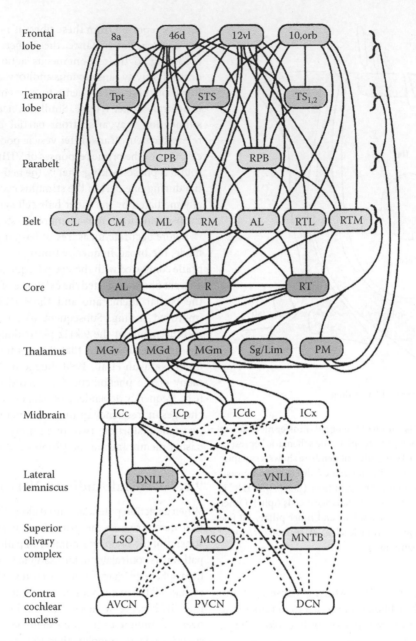

FIGURE 9.14 Diagram of the cortical and subcortical connections composing the auditory system of the primate. All major cortical and subcortical regions are illustrated in horizontal rows, with solid lines indicating established connections. The bracketed joined lines indicate connections to all fields with that horizontal region. The dashed lines represent hypothetical connections derived based on observations in various mammals. The "belt" region may also include an additional field: MM. The following nomenclature was used: anteroventral cochlear nucleus: AVCN; posteroventral cochlear nucleus: PVCN; dorsal cochlear nucleus: DCN; lateral superior olivary nucleus: LSO; medial superior olivary nucleus: MSO; medial nucleus of the trapezoid body: MNTB; dorsal nucleus of the lateral lemniscus: DNLL; ventral nucleus of the lateral lemniscus: VNLL; central nucleus of the inferior colliculus: ICc; pericentral nucleus of the inferior colliculus: ICp; dorsal cortex of the inferior colliculus: ICdc; external nucleus of the inferior colliculus: ICx; ventral nucleus of the medial geniculate complex: MGv; dorsal nucleus of the medial geniculate complex: MGd; medial/magnocellular nucleus of the medial geniculate complex: MGm; suprageniculate nucleus: Sg; limitans nucleus: Lim, medial pulvinar nucleus: PM; auditory area 1: A1; rostral area R; rostrotemporal area: RT; caudolateral area: CL; caudomedial area: CM; middle lateral area: ML; rostromedial area: RM; anterolateral area: Al; lateral rostrotemporal area: RTL; medial rostrotemporal area: RTM; caudal parabelt: CPB; rostral parabelt: RPB; temporoparietal area: Tpt; superior temporal area 1 and 2: $TS_{1,2}$; frontal lobe areas numbered according to the Brodmann convention and Preuss and Goldmann-Rakic as periarcuate: 8a; dorsal principal sulcus: 46d; ventrolateral area: 12vl; frontal pole: 10; orbitofrontal areas: orb. (From Kaas J.H., and Hackett T.A., 2000, *Proceedings of the National Academy of Sciences* 97: 11793–11799. With permission.)

compared to afferent fibers of the VIIIth nerve, and interneurons within the brainstem nuclei, often forming such combinations as excitatory–excitatory neurons (in which a target neuron adds the input from two other neurons) and excitatory–inhibitory neurons (in which a target neuron differentiates the input from two other neurons). Some neurons may form an array called a lateral inhibitory network (similar to those in the visual system), in which changes or contrasts in a stimulus are enhanced.

The first auditory nucleus in the brainstem is the cochlear nucleus. Cochlear nucleus neurons may be involved in neural suppression, in which activating one neuron may drive the firing rate of another neuron down to the spontaneous rate, even though the second neuron is being stimulated by its CF. Computation across neurons with different CFs in the cochlear nucleus may help determine how complex stimuli differ in their frequency spectra. Lateral inhibitory networks may also function in the cochlear nucleus (Yost, 2007). In addition, the dorsal cochlear nucleus has recently been implicated as the source of tinnitus (ringing in the ears) following noise damage inflicted on the outer hair cells (e.g., Schaette and Kempter, 2008).

The second major auditory nucleus is the superior olivary complex. This is the first point in the central auditory nervous system that receives binaural input. The lateral superior olivary area contains primarily neurons sensitive to high frequencies as well as excitatory–inhibitory circuits, which increase the firing rate as interaural level differences are increased. Interaural level differences are the primary cue for localizing high frequencies. The medial superior olivary area contains primarily neurons sensitive to low frequencies, and has excitatory–excitatory circuits, which increase firing when there is an interaural time delay. Interaural time delays are the primary cue for localizing low frequencies, and for complex stimuli with low-frequency repetition patterns (Yost, 2007). There is considerable evidence, however, that cortical processing is also involved in sound localization (e.g., see Recanzone and Sutter, 2008).

The third major auditory nucleus is the inferior colliculus. This nucleus has combinations of the previous two major nuclei. Spectral processing and binaural processing combine, perhaps allowing for multidimensional representation of sound. There are excitatory–inhibitory and excitatory–excitatory circuits. Some cells fire upon monaural stimulation, some upon binaural stimulation. Some respond differentially to interaural time cues, and some respond differentially to changes in the vertical location of a sound (Yost, 2007).

9.2.4.2 Auditory Cortex

The last few years have shown a veritable explosion of studies into the workings of auditory cortical areas, largely guided by the brain imaging techniques mentioned in other parts of this text. But first, a bit of anatomy and introductory physiology. In recent years, the macaque monkey has been used as an animal model for the study of the auditory cortex. The auditory cortex of the macaque can be divided into three main areas (core, belt, and parabelt), each of which is subdivided into sections (Figure 9.15). Each of the main areas is defined by histology,

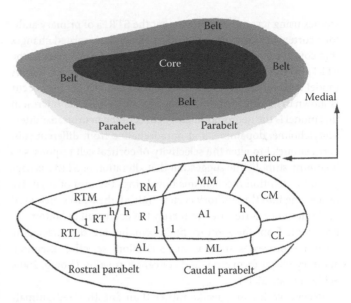

FIGURE 9.15 Schematic representation of the auditory complex of the primate. Top: general organization into core (black) enclosed by the belt and parabelt region, respectively. Bottom: dissection of the cortical areas into representative functional sections, nomenclature as in Figure 9.14, not to scale. The parabelt is subdivided into caudal and rostral areas. (From Recanzone G.H. and Sutter M.L., 2008, *Annual Review of Psychology* 59: 119–142. With permission.)

corticocortical and thalamocortical connections, and functional activity. The primary auditory cortex is labeled A1 and, along with the rostral (*R*) field, is tonotopically arranged. Human studies have shown wide variability in the anatomical markers of the auditory cortex, with several tonotopic fields. As in the macaque, there appear to be three main areas, with structure and function comparable to that in macaques (Recanzone and Sutter, 2008). In humans, there is yet to be an exhaustive, definitive description of the auditory cortex.

Common themes in studies of the auditory cortex now include comparisons to the visual cortex, adaptive tuning, and cognition. A common finding is that spatial and nonspatial features of sounds are processed in parallel streams. This is similar to the ventral "what" and dorsal "where" processing segregation in the visual cortex. Studies in the auditory cortex suggest that spatial features are analyzed in caudal areas and nonspatial features are analyzed in ventral auditory cortical areas (Recanzone and Sutter, 2008). Surprisingly, the frequency tuning of cortical cells can be changed by training; this change may persist and is not caused by the state of arousal of the animal (Bakin and Weinberger, 1990). Tuning changes in associative cortical areas were also observed in humans (Molchan et al., 1994; Morris et al., 1998). These experiments suggest that the response field of cortical neurons is plastic. One measure that is widely used in the study of all sensory cortical cells is the spectro-temporal response field (STRF) (de Boer, 1967; see Shamma, 2008). In terms of auditory cortical cells, the STRF may be defined as the best linear fit between the attributes of a sound (frequency and intensity over time) and the neural firing rate the sound evokes (Christianson et al., 2008). Recent studies using ferrets have shown that the STRFs of primary auditory cortex (A1) neurons showed dynamic, task-related changes that could fade after the task or persist for hours. Changes in the STRFs were likely triggered by attention to the specific task. A model describing how these attention-driven changes may occur is shown in Figure 9.16 (Fritz et al., 2007). Of particular interest in this model is the neuromodulatory effect of the neurotransmitters acetylcholine, dopamine, and noradrenaline from different subcortical nuclei to alter the selectivity of cortical cell responses to incoming stimuli. The suggestion that alteration of STRFs is triggered by attention mechanisms now brings the idea of cognitive processing into the auditory cortex. Indeed, the STRFs measured in the auditory cortex are not merely the end result of bottom-up processing from the auditory brainstem, but are also affected by cognitive processes such as categorization, selective attention, auditory object formation, and concept formation (Konig, 2005; Schiech et al., 2005).

Recent work using awake rather than anesthetized animals has shown that cortical cells do exhibit sustained firing to acoustic stimuli (e.g., Bieser and Muller-Preuss, 1996; Recanzone, 2000). Contrary to previous work, the sustained responses by cortical cells may actually provide a platform for representing time-varying signals in the auditory cortex (Lu et al., 2001; Liang et al., 2002; Wang et al., 2003). These studies revealed two types of cortical cells in A1: those that encode the acoustic stimulus by spike timing and those that encode by spike firing rate. The spike timing population could represent slowly changing long-duration stimuli in a manner that is a replica of the acoustic stimulus (or an isomorphic representation). The nonsynchronous spike firing rate cells could encode rapidly changing acoustic stimuli in a manner that is not a replica of the acoustic stimulus, but rather a conversion from a stimulus parameter to an internal representation (or a nonisomorphic representation). Thus, the cortex may have both unprocessed (isomorphic) and processed (nonisomorphic) representations of sounds (Wang, 2007).

A very generalized tentative model, then, of auditory function to this point could be the following. Acoustic features related to percepts are extracted in the periphery and auditory brainstem, with isomorphic moment-by-moment and perhaps some nonisomorphic processing occurring, perhaps controlled by neurons projecting from the auditory cortex to the midbrain superior olive and inferior colliculus (Coomes Peterson and Schofield, 2007). At the level of the medial geniculate (in the thalamus) and A1, there would be the beginnings of auditory object formation. Different time scales might be evidenced by the synchronized and nonsynchronized cell populations in A1, and the nonsynchronized cell output may allow for integration of information, or segment-by-segment processing. In addition, the output of these cells may initiate the beginning stages of multisensory integration. The nonisomorphic representations would then be an intermediate transition stage from acoustic to perceptual identification of the auditory objects, with later processing occurring via distributed networks throughout the temporal, parietal, and frontal lobes (Nelken et al., 2005; Poremba and Mishkin, 2007; Wang, 2007; Winer and Lee, 2007).

9.3 Psychoacoustics

9.3.1 Threshold

Given the previous background of auditory anatomy and physiology, we will now discuss briefly how anatomy and physiology may lead to perception of sound. The most basic perception is that of detection of a sound. The lowest sound pressure level of a sound that results in its detection is referred to as the absolute threshold for that sound. A quantum of sound is called a phonon, with energy described by $h\nu$ (h is Planck's constant and ν is frequency). The ratio $h\nu/k_BT$ for the auditory system is about 10^{-10} (where k_B is Boltzmann's constant and T is absolute temperature). Auditory detection is limited by the presence of thermal noise; a single phonon is

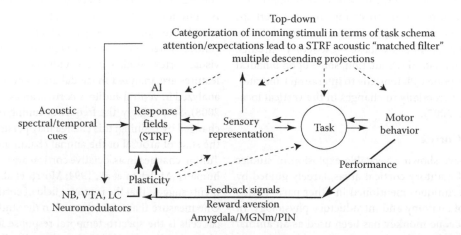

FIGURE 9.16 Schematic interpretation of the network model pertaining to rapid STRF plasticity in the primary auditory cortex (A1). While performing tasks the neuromodulator activity will selectively increase based on reference (suppress) and target (enhance) stimuli, thus mitigating receptive field plasticity in A1. The signal migrates top-down into respective subcortical nuclei: nucleus basalis (NB), ventral segmental area (VTA), and locus coeruleus (LC). (Reprinted from Fritz J.B., et al., 2007, _Hearing Research_ 229: 186–203. With permission.)

not likely to be detected by itself, unlike a photon in the visual system (Ashmore, 2008). Thus, it is unlikely that a single neurotransmitter vesicle released at the inner hair cell–VIIIth nerve afferent synapse would result in perceptual detection of a sound, but if extra spikes were triggered in a small number of neurons, coincidence detector circuits in the cochlear nucleus may begin the process that would lead to perceptual detection (Heil and Neubauer, 2005).

Although the preceding paragraph provides the basis for perception, it does not define perception. In what follows, we must be careful not to equate the ear with listening; a given sound pressure variation will not always result in one and only one given perception. If so, listening and perception would be nothing more than memorizing physical signatures of sound. Perception is rather the combining of sensations into entities or objects. Please keep this in mind as the following paragraphs examine the behavioral sensitivity and accuracy of the auditory system to physical changes in frequency, intensity, and time. These behavioral measures are akin to the working parts for perception of complex acoustic events such as music or speech (Handel, 1989).

The detection of sounds by humans varies with frequency; we are most sensitive to mid-frequency tones (in the 1–4 kHz region) and less sensitive to lower and higher frequencies. This pattern of absolute thresholds probably arises from the transfer functions of the outer and middle ear. Less sensitivity to frequencies less than 500 Hz may arise from a less active cochlear amplifier at the apex of the cochlea. Absolute threshold may also depend on the duration of the tone. For instance, for tones between 10 and 300 ms, threshold detection reflects constant energy. That is, in this duration range, if the stimulus duration is reduced by half, the power of the tone must be increased by 3 dB to maintain detection. This is referred to as the 3 dB per doubling rule, and implies some form of temporal integration. The actual physiological process here is likely some form of sampling rate employed by the auditory system, with the sampling rate reset by any acoustic discontinuity (White and Plack, 1998).

In addition to absolute sensitivity, sensitivity to change in frequency or intensity has been measured. These just noticeable differences (jnds) are usually expressed as a Weber fraction, in which the perceived difference between two stimuli is proportional to the value of the smaller number ($\Delta S/S$). For mid-frequency stimuli (0.5–2 kHz) the Weber fraction for frequency discrimination is constant, at around 0.002 (about 1 Hz for 0.5 kHz). For pure tones, the Weber fraction for intensity discrimination is a "near miss," slightly improving with increases in overall SPL from about 0.5 to 0.2 (in the 1–0.5 dB range).

9.3.2 Frequency Selectivity

Frequency selectivity is the ability to resolve the different frequency components in a complex sound, and is most important for determining the tuning of the auditory system. To perceive and understand complex signals such as music and speech, we rely on the tuning properties of the auditory system, which was discussed in the physiology section as tonotopicity and as basilar membrane and neural tuning curves. An important concept here is that of critical bandwidth, which is directly associated with internal filtering. Masking experiments, in which an acoustic signal is detected in the presence of a background noise, showed that the ear operates as a set of level-dependent, overlapping bandpass filters. The motion of the basilar membrane is a large factor in the shape of these bandpass filters. The excitation pattern is the combined output from a number of filters, and thus the excitation pattern presents an estimation of the isomorphic internal representation of an acoustic stimulus in the auditory periphery (inner ear and VIIIth nerve).

9.3.3 Pitch

In the acoustics section of this chapter, much attention was given to frequency, which is a physical phenomenon. Pitch is basically the psychological attribute of frequency; although frequency is the main cue for pitch, the intensity of the sound will also affect the pitch. Because the auditory system processes frequency in a logarithmic fashion and because other attributes of sound (such as intensity) affect pitch, there is no one-to-one correspondence between frequency and pitch. Measurement of pitch is done by presenting a standard 1 kHz tone at a given level (itself having a reference pitch of 1000 mels), presenting a stimulus tone, and then having a listener judge the stimulus tone as having n times the pitch of the 1 kHz standard tone. Testing a number of tones and levels in this way produces the Mel pitch scale. The two main theories for frequency encoding in the auditory system are place (each frequency has its own point of maximum excitation along the basilar membrane) and volley (neurons may phase-lock to the stimulus waveform, and a central mechanism can then compute the reciprocal of the periodic firings to compute the frequency of the waveform). While most of the pitch data from listeners can be described using either place or volley theories (or some combination thereof), there are still a few experimental conditions (complex pitch, pitch shift of the residue) that are not adequately explained by either place or volley theories at this time.

9.3.4 Loudness

Loudness is the psychological attribute of intensity or sound pressure level. Loudness is measured by matching two tones to equal sensation (the basis of phons and the equal loudness contours) or by scaling (the basis of sones). The scaling procedure for sones is similar to that of the Mel scale for pitch, described above, with one sone defined as the reference loudness for a 1 kHz tone presented at a 40 phon level. Loudness can be described by the following power law relation: $L = kI^{0.3}$, where k is a constant and I is stimulus intensity. Loudness is related to the extent of basilar membrane motion; that is, loudness is proportional to the square of basilar membrane velocity. This basilar membrane motion may be combined to yield the loudness of a broadband or complex acoustic signal. The more critical bandwidths a sound traverses, the louder the sound will seem to be, even if the sound intensity remains constant. This is referred to as loudness summation.

9.3.5 Temporal Perception

In addition to temporal integration (discussed above), another aspect of temporal coding is temporal resolution, or the ability to detect changes in the acoustic stimulus over time. Experiments in which a stimulus is turned off and then quickly back on produce a silent gap. Gap detection is thus a measure of temporal resolution, and can be smaller than 5 ms. In addition, listeners can detect amplitude modulation rates of nearly 1000 Hz when the modulation depth is 100%, which makes the auditory system much more temporally acute than the visual system. Such a function showing modulation depth as a function of modulation rate is called a temporal modulation transfer function.

Masking experiments have also examined the effects of narrow bands of noise spectrally removed from the critical band of a signal. When a narrow band of noise centered on the critical band of the signal is modulated at the same rate as a band of noise spectrally removed from the critical band, the signal is much more easily detectable than when the maskers are present but not modulated. This effect is known as comodulation masking release (CMR). Other experiments have tested listeners' ability to detect a modulated signal; when a masker having the same modulation as the signal is presented, the detection of the signal is poor. However, if the masker has a different modulation rate from that of the signal, the detection of the signal is not changed by much. This effect is known as modulation detection interference (MDI). Both CMR and MDI suggest that the auditory system has a set of modulation filterbanks; indeed there are cells in the auditory cortex that fire at a best modulation rate, analogous to a CF for frequency tuning.

Temporal masking is also demonstrable, with either the signal or the masker preceding the other in time. When the masker is presented first and then the signal, the condition is known as forward masking. When the signal is presented first and then the masker, the condition is known as backward masking. As one might surmise, backward masking is likely a central phenomenon, is highly variable, and can be markedly reduced by training. Forward masking, however, is quite consistent within a subject regardless of extensive training. Forward masking may reflect persistence of neural activity in response to the masker, and may be modeled by the following circuit: a set of overlapping bandpass filters (critical bands), a nonlinear device (half-wave rectification caused by phase-locking or basilar membrane motion), a sliding temporal integrator or window of about 7 ms that averages or smooths intensity during this time period, and a decision device for whatever task is being performed.

9.3.6 Binaural Hearing

The duplex theory of localization was mentioned (not by name) in the above discussions of the function of the outer ear and the superior olivary complex. Interaural intensity cues help localize high frequencies, and interaural time or phase cues help localize low frequencies or complex stimuli with a low-frequency repetition. The interaural time cues that we can detect may be as small as 150 μs. Operation of EE and EI interneuron networks in the superior olive is likely the physiological basis for localization in mammalian auditory systems rather than coincidence detectors. Because both interaural cues are weakest for the mid-frequency range (around 1500 Hz), most localization errors occur for midfrequencies. The minimum audible angle is a measure of how well listeners can detect a shift in the location of a sound source in the horizontal plane. Localization in the vertical plane is usually made accurate by small head movements.

These small head movements cause changes in the HRTFs, which then enable localization. HRTFs can be recorded from microphones placed in the ear canals of listeners, and then these HRTFs can modify sounds presented via headphones to give a listener a "virtual" three-dimensional listening experience. A unique form of binaural masking is the masking level difference (MLD). When a noise is correlated in both ears, and a signal presented to both ears is flipped 180° in phase in one ear relative to the other, then the signal is released from masking and is easier to hear. MLDs have been measured using speech signals, and are similar to the cocktail party effect in which a listener is able to separate out a signal in a complex listening environment using spatial separation cues.

9.3.7 Auditory Scene Analysis

When you are in almost any listening environment, you hear sounds from several sources. As you sit at your desk in the library, you may hear the sound of heating/cooling fans, voices across the next desk, paper shuffling, the elevator bell, and the trundle of a library cart. All these sounds are impinging onto the basilar membrane simultaneously, and some have overlapping spectral content, yet you are able to parse out and identify the various sound sources. How the auditory system does this is the process of auditory scene analysis. Some cues that may aid this process include temporal onsets and offsets, spectral separation, harmonicity, spectral profile, and temporal modulation. Auditory scene analysis may lead to auditory object formation, a concept of the "what" stream of processing in the auditory cortex (Kubovy and van Valkenburg, 2001).

9.4 Hearing Loss

9.4.1 Types of Hearing Loss

To determine whether an individual has hearing loss, thresholds are obtained for tones from 250 to 8000 Hz in octave steps, and these thresholds are then compared to the modal (most frequently occurring) thresholds obtained from listeners aged 18–30 with no otologic pathology (see Figure 9.17). Thresholds obtained in such a manner are referenced as dB HL (hearing level), and are plotted on a chart called an audiogram (e.g., ANSI,

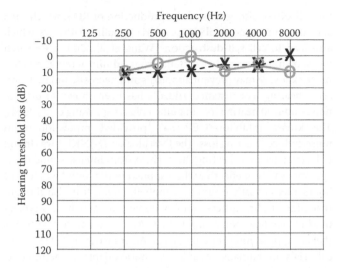

Frequency (Hz)

FIGURE 9.17 Representative audiogram illustrating the hearing response of an individual with normal hearing; the responses for the right and left ear are represented by o and x, respectively.

1996; Bess and Humes, 2008; Martin and Clark, 2007). Thresholds greater than 25 dB HL represent hearing loss, and may range from mild (26–40 dB HL) to profound (>90 dB HL). Hearing loss is categorized according to where a blockage or insult occurs in the auditory system. If the blockage or disorder occurs in the outer or middle ear, the hearing loss is conductive and is at most about 60 dB HL, if the disorder or insult occurs in the cochlea and/or VIIIth nerve, the hearing loss is sensorineural, and (extremely rare) if the disorder occurs centrally because of lesion or arterial blockage, the hearing loss is central deafness (Martin and Clark, 2007). Clinical assessment of hearing can include not only behavioral threshold and speech recognition testing, but also measures of the impedance of the middle ear system (tympanometry), contraction of the stapedius muscle in the middle ear (the acoustic reflex test), measures of OAEs, and electrophysiological tests of the central auditory pathways' neural activity in response to sound. The electrophysiological tests include the ABR, the auditory steady-state response (ASSR), the auditory middle latency response (AMLR), and late latency cortical responses (ALR).

Disorders in the outer ear resulting in hearing loss include earwax impaction, infection, tympanic membrane perforation, and narrowing or absence (atresia) of the external ear canal, with the latter often having a genetic origin. All except atresia may be treated medically; atresia may be corrected by either surgery or bone-conduction hearing aids. Disorders in the middle ear resulting in hearing loss include infection (otitis media), ossicular chain discontinuity, congenital disorders, and a genetic disorder called otosclerosis, in which the stapes becomes fixed in the oval window. These conditions may be treated medically or surgically, in some cases requiring prostheses to replace the ossicles and/or middle ear reconstructive surgery (tympanoplasty).

Disorders in the cochlea resulting in hearing loss include anoxia, prenatal Rh factor incompatibility, trauma or toxemia

prenatally or at birth, viral infections, ototoxic drugs, fevers and systemic illnesses, middle ear infections or disorders that may involve the cochlea, noise, head trauma, autoimmune responses in the cochlea, disruption of blood flow to the cochlea, labyrinthitis, Meniere's disease, and aging (presbycusis). These disorders may also secondarily affect VIIIth nerve fibers. In addition, there are many genetically based etiologies of cochlear hearing loss. VIIIth nerve disorders affecting hearing include tumors in the nerve or near the cerebellopontine angle (CPA) and dys-synchrony of fiber firing. Fiber dys-synchrony, also called auditory neuropathy, may also have a genetic etiology (e.g., Varga et al., 2003). Effective medical or surgical treatment for cochlear hearing loss is quite limited; in the case of VIIIth nerve tumors, excision of the tumor may or may not allow for preservation of hearing.

9.4.2 Genetics, the Cochlea, and Hearing Loss

In the United States, approximately one child in a thousand has hearing loss from birth; this hearing loss may prevent the normal development of language. About 50–60% of these children have a genetic etiology for their hearing loss. For the genetic loss group, about 70% have a form of nonsyndromic hearing loss (hearing loss with no associated clinical manifestations), and about 30% have a genetic syndrome, in which hearing loss is a primary characteristic (Battey, 2003). More than 100 gene loci have been identified in nonsyndromic hearing loss (Petersen et al., 2008), and there are hundreds of genetic syndromes that include hearing loss (Friedman et al., 2003). Up-to-date online catalogs of genetic hearing loss may be found at many websites, including the Hereditary Hearing Loss Homepage http://webh01. ua.ac.be/hhh/ and the Online Mendelian Inheritance in Man http://www.ncbi.nlm.nih.gov/omim. As suggested by the latter website, most of the inheritance patterns of hearing loss follow Mendelian patterns (e.g., autosomal dominant DFNA, autosomal recessive DFNB, sex-linked DFN), although a few may involve gene–gene interactions (e.g., connexin 30).

Studies of genetic hearing loss have illuminated how the cochlea functions at the cellular level (Battey, 2003). Knowing what gene product arises from a mutated gene known to cause a hearing loss provides insight into the cellular basis of hearing. For instance, the *KCNQ4* gene, located on chromosome 1, has as its product a protein that functions as a potassium ion channel in cells (including hair cells in the cochlea). A mutation of this gene results in a non- or dysfunctional potassium channel, and in sensorineural hearing loss that progresses in severity with age. In these examples, knowledge of a genetic basis of the hearing loss, that is, knowing what gene was involved and its product, helped clarify the vital importance of potassium recycling for normal cochlear function (e.g., Van Laer et al., 2003).

Errors that may occur during transcription (making a copy of the gene from DNA into mRNA) and translation (modifying and converting the mRNA into a gene product such as a protein) may also result in hearing loss. For instance, mutation of the *POU3F4* gene results in a dysfunctional protein that normally acts as a transcription factor; transcription factors attach to the DNA and, when functioning correctly, aid in correct "reading" of the DNA during transcription. Another example involves mRNA splicing. The mRNA molecule is spliced to arrive at a final copy for translation; improper splicing may result in a dysfunctional cadherin 23 protein that results in stereociliar splaying of hair cells and subsequent hearing loss in Usher's syndrome (Siemens et al., 2002).

Genetic hearing loss may also arise from mutations that occur in genes located in the mitochondria of cochlear cells. There are several hundred mitochondria in each cell and they are involved in producing energy via oxidative phosphorylation, maintaining reactive oxygen species (ROS) levels, and cell death pathways. A given mitochondrion may have 2–10 chromosomes. Mitochondrial DNA is transmitted matrilineally. Hearing loss associated with mitochondrial DNA may be syndromic [e.g., with neuromuscular syndromes such as mitochondrial encephalomyopathy; lactic acidosis; stroke (MELAS) and myoclonic epilepsy; ragged red fibers (MERRF)] or nonsyndromic. In certain ethnic populations, mitochondrial mutations can result in susceptibility to hearing loss arising from ototoxic medications. Finally, hearing loss in old age (presbycusis), once thought solely to be a result of the cochlea wearing out or of environmental noise exposure over many years, may also stem from a reduction of energy production due to mutations in mitochondrial DNA (Fischel-Ghodsian, 2003).

The mouse auditory system is frequently used as a model for the human system. One possible disorder of human hearing is operationally defined as an auditory processing disorder (APD), in which an individual has difficulty interpreting auditory information yet does not have hearing loss (Martin and Clark, 2009). Mice exhibit genetic mutations that affect their central auditory nervous systems, affecting the formation of VIIIth nerve fibers and the efferent system, causing defects of specific cell types in auditory brainstem nuclei with corresponding abnormal electrophysiological measures, and causing enzyme deficiencies. Specific central auditory system anomalies are rare in humans (Steel et al., 2002). However, with the finding of a genetic component to auditory dys-synchrony, could there also be a genetic basis (perhaps similar to those found in mice) for some APDs? It may be that APDs result from an interaction of genetic predisposition and environmental factors.

9.4.3 Noise-Induced Hearing Loss

Intense continuous noise over 115 dBA, or impulse noise at this or higher levels may result in physical damage to hair cell stereocilia, alteration in blood flow to the cochlea, detachment of pillar cells, and physical damage to inner hair cell–VIIIth nerve synapses. Additional effects may include overstimulation of inner hair cells, resulting in levels of glutamate transmitter that are toxic to afferent fibers (Henderson et al., 2007). Noise levels lower than this, however, can still stress the cochlea and result in hearing loss via apoptosis or hair cell death. Under conditions of high-level noise exposure, more pro-apoptotic molecules enter the mitochondria within the hair cell; these molecules initiate caspase enzyme activity, effectively signaling a self-destruct pathway for the cell. Noise also results in the production of ROS, which then activate c-Jun N-terminal kinase (JNK) signaling pathways, which also lead to cell self-destruction (Wang et al., 2007). Research has shown, however, that certain drugs administered at or after noise exposure may prevent apoptosis and necrosis, and may preserve hearing. For instance, riluzole has been shown to prevent noise-induced hearing loss in guinea pigs (e.g., Wang et al., 2002). The kinase inhibitor CEP-1347 provided protection from noise-induced hearing loss. The JNK blocker D-JNKI-1 was found to be effective in protecting the cochlea from intense sound damage, and was even effective when applied to the round window 12 hours after acoustic trauma (Campbell, 2007). It is also thought that efferent transmitters, notably dopamine, may have a protective function (e.g., Niu et al., 2007). Currently, otoprotective agents that are approaching Food and Drug Administration (FDA) clinical trials in humans include the antioxidants D-methionine, ebselen, and N-acetylcysteine (NAC). In future, pharmacogenetics and pharmacogenomics may be used to provide an individually tailored drug to, say, help prevent noise-induced hearing loss or hearing loss arising from life-saving medications.

9.4.4 Hair Cell Regeneration

Since the discovery that avians can regrow functional hair cells following noise-induced hearing loss (Corwin and Cotanche, 1988; Ryals and Rubel, 1988), much effort has gone into regenerating hair cells in the mammalian cochlea. Investigators have attempted to manipulate the cell cycle, cell differentiation, and stem cells to produce new hair cells in the mammalian cochlea. Control of the cell cycle has attempted to remove tumor suppressor genes (via transgenic and/or knockout genetic manipulation) to allow new cell growth, albeit with little success. New cell differentiation may be controlled by genes producing transcription factors; indeed, transfecting the transcription factor gene *Atoh1* into the cochlea of deafened guinea pigs led to structural and functional recovery of the inner ear (e.g., Izumikawa et al., 2005; Kawamoto et al., 2003). Consistency in where new cells are produced, however, has proven problematic. Introducing stem cells into the mammalian cochlea, or activating stem cells that remain in the cochlea, has shown some promise but is still in the initial stages of study (e.g., Ito et al., 2001; Parker et al., 2007; see also Cotanche, 2008; Ryals et al., 2007). In certain genetic, anatomical, or biochemical disorders, however, hair cell regeneration may not be a viable treatment for hearing loss. It may be that future treatment for certain sensorineural hearing loss etiologies may involve a combination of hair cell regeneration, drug treatment, and either hearing aids or cochlear implants (Li et al., 2004; Ryals et al., 2007).

9.4.5 Hearing Aids

(Re)habilitation of hearing loss has largely focused on devices such as hearing aids. At present, hearing aids, via automatic gain control (AGC) compressor circuits, multiband compression applied across different frequency bands, or adaptive dynamic range optimization, can help manage the reduced dynamic range of listeners with sensorineural loss, but cannot restore "normal" loudness perception. Likewise, hearing aids cannot restore normal frequency selectivity or temporal fine structure resolution. Advances have been made to help listeners wearing hearing aids understand speech in background noise via the use of directional microphones and binaural hearing aids; work with noise-cancelation algorithms has shown promise, but limited success to date (Moore, 2007). The introduction of digital signal processing in hearing aids has provided flexibility for the addition of processing algorithms tailored for particular listening conditions, which can be selected by the wearer (e.g., Staab et al., 1997; Staab, 2002), as well as opportunities for block processing of the signal (e.g., Levitt, 2004). Some recent hearing aids and cochlear implants perform processing akin to auditory scene analysis. Advances have also been made with bone-conduction middle-ear implantable hearing devices (see http://www.cochlearamericas.com/index.asp or http://www.medel.com).

9.4.6 Cochlear Implants

The functional goal of the cochlear implant is to bypass the basilar membrane and hair cells of the cochlea and to instead electrically stimulate the remaining afferent VIIIth nerve fibers. Basic components of a cochlear implant include a microphone, a digitizing speech processor, a radio-frequency transmitter, an implanted radio receiver, circuits that convert the signal into an electrical signal, and a stimulator that activates an electrode array inserted into the inner ear (Wilson, 2004; Zeng, 2004). Cochlear implants attempt to replicate the frequency selectivity and compression features of the cochlea. The implant separates the acoustic stimulus into 8–20 narrowband signals by either analog or digital filters. However, postfiltering may be quite different among implants. The speech processing strategies to date include compressed analog (CA) or simultaneous analog stimulation (SAS), in which the narrowband analog waveform is delivered to a tonotopically appropriate electrode that directly stimulates the auditory nerve. Another strategy extracts the temporal envelope of the narrowband signal (via rectification and low-pass filtering); the envelope then amplitude-modulates a biphasic pulse carrier. The pulses between electrodes are interleaved to prevent simultaneous stimulation between different electrodes. If the number of analysis bands is the same as the number of electrodes, this is called continuous interleaved sampling (CIS). When only some of the bands are activated, the strategy is called the n-of-m or SPEAK strategy. There is usually a different amplitude compression scheme for each strategy: CA employs a gain control mechanism for compression, CIS uses a logarithmic compression function, and SPEAK uses a power function. Many commercially available devices have combinations of the above strategies (Zeng, 2004; see also the above websites and http://www.advancedbionics.com). New developments include presenting the electrical signal at high rates (up to 90,000 pulses per second) and increasing the number of available processing channels via current steering.

Part of the variability seen across listeners wearing cochlear implants does stem from the device itself. Improvements in implant performance in the future will likely come from a better understanding of the electrical–neural interface (Abbas and Miller, 2004); use of bilateral cochlear implants, combining electrical and acoustic hearing by using electrode arrays that are introduced only in the base of the cochlea (e.g., von Ilberg et al., 1999); delivery of neurotrophic drugs through the implant to promote neural growth and health (e.g., Qun et al., 1999); new processing strategies that increase the number of frequency bands or that increase the pulse rate of stimulation, electrode placement and design, and new processing strategies that copy the transforms created by the basilar membrane and hair cell–fiber synapse discussed earlier (Wilson, 2004). In addition, knowledge gained from auditory brainstem implants and from central auditory neuron processing may also help improve implant performance (Abbas and Miller, 2004; Schwartz et al., 2008).

Through this chapter, it is hoped that the reader has gained an appreciation for, and a humble acknowledgment of, the exquisite and somewhat still mysterious human auditory system. There is much that needs to be done in the future to help optimize individuals' hearing, especially in modern society's complex information lifestyle.

Additional Reading

Abbas, P. J. and Miller, C. A., 2004. Biophysics and physiology. In F.-G. Zeng, A. N. Popper, and R. R. Fay (Eds), *Cochlear Implants*. New York: Springer-Verlag.

Advanced Bionics, Available at http://www.advancedbionics.com

American National Standards Institute, 1996. *Specification for audiometers* (ANSI S3.6-1996). New York: Author.

Ashmore, J., 2008. Cochlear outer hair cell motility. *Physiological Reviews* 88: 173–210.

Atencio, C. A., Sharpee, T. O., and Schreiner, C. E., 2008. Cooperative nonlinearities in auditory cortical neurons. *Neuron* 58: 956–966.

Bakin, J. S. and Weinberger, N. M., 1990. Classical conditioning induces CS-specific receptive field plasticity in the auditory cortex of the guinea pig. *Brain Research* 536(1–2): 271–286.

Battey, J. F., 2003. Using genetics to understand auditory function and improve diagnosis. *Ear and Hearing* 24(4): 266–269.

Beranek, L. L. (Ed.), 1986. *Acoustics*. New York: American Institute of Physics.

Bess, F. H. and Humes, L., 2008. *Audiology: The Fundamentals*. Philadelphia: Lippincott Williams & Wilkins.

Bieser, A. and Muller-Preuss, P., 1996. Auditory responsive cortex in the squirrel monkey: Neural responses to amplitude-modulated sounds. *Experimental Brain Research* 108: 273–284.

Brownell, W. E., Bader, C. R., Bertrand, D., and De, R. Y., 1985. Evoked mechanical responses of isolated cochlear outer hair cells. *Science* 227: 194–196.

Bruel and Kjaer, Available at http://www.bksv.com/Products/LaserVibrometry.aspx

Choe, Y., Magnasco, M. O., and Hudspeth, A. J., 1998. A model for amplification of hair-bundle motion by cyclical binding of Ca^{2+} to mechanoelectrical-transduction channels. *Proceedings of the National Academy of Sciences USA* 95: 15321–15326.

Christianson, G. B., Sahani, M., and Linden, J. F., 2008. The consequences of response nonlinearities for interpretation of spectrotemporal receptive fields. *Journal of Neuroscience* 28(2): 446–455.

Cochlear Corporation, Available at http://www.cochlearamericas.com/index.asp

Coomes Peterson, D. and Schofield, B. R., 2007. Projections from auditory cortex contact ascending pathways that originate in the superior olive and inferior colliculus. *Hearing Research* 232(1–2): 67–77.

Cooper, N. P. and Guinan, J. J., 2006. Efferent-mediated control of basilar membrane motion. *Journal of Physiology* 576(1): 49–54.

Corwin, J. T. and Cotanche, D. A., 1988. Regeneration of sensory hair cells after acoustic trauma. *Science* 240: 1772–1774.

Cotanche, D. A., 2008. Genetic and pharmacological intervention for treatment/prevention of hearing loss. *Journal of Communication Disorders* 41: 421–443.

Dallos, P., 2003. Some pending problems in cochlear mechanics. In A. W. Gummer (Ed.), *Biophysics of the Cochlea*. River Edge, NJ: World Scientific Publishing.

de Boer, E., 1967. Correlation studies applied to the frequency resolution of the cochlea. *Journal of Auditory Research* 7: 209–217.

de Boer, E., 1996. Mechanics of the cochlea: Modeling efforts. In P. Dallos, A. N. Popper, and R. R. Fay (Eds), *The Cochlea*. New York: Springer-Verlag.

de Boer, E., Nuttall, A. L., and Shera, C. A., 2007. Wave propagation patterns in a "classical" three-dimensional model of the cochlea. *Journal of the Acoustical Society of America*, 121: 352–362.

Durrant, J. D. and Lovrinic, J. H., 1995. *Bases of Hearing Science*, 3rd edition, Baltimore: Williams & Wilkins.

Edeline, J.-M., 1999. Learning-induced physiological plasticity in the thalamo-cortical sensory systems: A critical evaluation of receptive field plasticity, map changes and their potential mechanisms. *Progress in Neurobiology* 57: 165–224.

Eggermont, J. J., 2005. Correlated neural activity: Epiphenomenon or part of the neural code? In R. Konig, P. Heil, E. Budinger, and H. Scheich (Eds), *The Auditory Cortex: A Synthesis of Human and Animal Research*. Mahwah, NJ: Lawrence Erlbaum.

Emanuel, D. C. and Letowski, T., 2009. *Hearing Science*. Philadelphia: Lippincott Williams & Wilkins.

Fischel-Ghodsian, N., 2003. Mitochondrial deafness. *Ear and Hearing* 24(4): 303–313.

Frank, G., Hemmert, W., and Gummer, A. W., 1999. Limiting dynamics of high-frequency electromechanical transduction of outer hair cells. *Proceedings of the National Academy of Sciences USA* 96: 4420–4425.

Friedman, T. B., Schultz, J. M., Ben-Yosef, T., Pryor, S. P., Lagziel, A., Fisher, R. A., Wilcox, E. R., et al., 2003. Recent advances in the understanding of syndromic forms of hearing loss. *Ear and Hearing* 24(4): 289–302.

Fritz, J. B., Elhilali, M., David, S. V. and Shamma, S. A., 2007. Does attention play a role in dynamic receptive field adaptation to changing acoustic salience in A1? *Hearing Research* 229: 186–203.

Gavara, N. and Chadwick, R. S., 2009. Collagen-based mechanical anisotropy of the tectorial membrane: Implications for inter-row coupling of outer hair cell bundles. *PLoS One* 4(3): e4877, doi:10.1371/journal.pone.0004877.

Geers, A., Brenner, C., Nicholas, J., Uchanski, R., Tye-Murray, N., and Tobey, E., 2002. Rehabilitation factors contributing to implant benefit in children. *The Annals of Otology, Rhinology, and Otolaryngology* 189(Suppl.), 127–130.

Geisler, C. D., 1998. *From Sound to Synapse*. New York: Oxford University Press.

Ghaffari, R., Aranyosi, A. J., and Freeman, D. M., 2007. Longitudinally propagating traveling waves of the mammalian tectorial membrane. *Proceedings of the National Academy of Sciences USA*, 104(42): 16510–16515.

Glowatzki, E. and Fuchs, P. A., 2002. Transmitter release at the hair cell ribbon synapse. *Nature Neuroscience* 5: 147–154.

Gold, T., 1948. Hearing II: The physical basis of the action of the cochlea. *Proceedings of the Royal Society of London B: Biological Sciences* 135: 492–498.

Grant, L. and Fuchs, P. A., 2007. Auditory transduction in the mouse. *Pflugers Archiv European Journal of Physiology* 454: 793–804.

Handel, S., 1989. *Listening*. Cambridge, MA: MIT Press.

Hartmann, W. M., 1997. *Signals, Sound, and Sensation*. New York: American Institute of Physics.

Hedrick, M. S. and Younger, M. S., 2007. Perceptual weighting of stop consonant cues by normal and impaired listeners in reverberation versus noise. *Journal of Speech, Language, and Hearing Research* 50(2): 1–16.

Heil, P. and Neubauer, H., 2005. Toward a unifying basis of auditory thresholds. In R. Konig, P. Heil, E. Budinger, and H. Scheich (Eds), *The Auditory Cortex: A Synthesis of Human and Animal Research*. Mahwah, NJ: Lawrence Erlbaum.

Henderson, D., Hu, B., Bielefeld, E., and Nicotera, T., 2007. Cellular mechanisms of noise-induced hearing loss. In K. C. M. Campbell (Ed.), *Pharmacology and Ototoxicity for Audiologists*. Clifton Park, NY: Thomson Delmar.

Hereditary Hearing Loss Homepage, Available at http://webh01.ua.ac.be/hhh/

Hromadka, T. and Zador, A. M., 2007. Toward the mechanisms of auditory attention. *Hearing Research* 229: 180–185.

Irvine, D. R. F., 2007., Auditory cortical plasticity: Does it provide evidence for cognitive processing in the auditory cortex? *Hearing Research* 229: 158–170.

Ito, J., Kojima, K., and Kawaguchi, S., 2001. Survival of neural stem cells in the cochlea. *Acta Oto-laryngologica* 121(2): 140–142.

Izumikawa, M., Minoda, R., Kawamoto, K., Abrashkin, K. A., Swiderski, D. L., Dolan, D. F., Brough, D. E., and Raphael, Y., 2005. Auditory hair cell replacement and hearing improvement by *Atoh1* gene therapy in deaf mammals. *Nature Medicine* 11: 271–276.

Jia, S. and He, D. Z., 2005. Motility-associated hair-bundle motion in mammalian outer hair cells. *Nature Neuroscience* 8: 1028–1034.

Kaas J. H. and Hackett T. A., 2000. Subdivisions of auditory cortex and processing streams in primates. *Proceedings of the National Academy of Sciences* 97: 11793–11799.

Kawamoto, K., Ishimoto, S., Minoda, R., Brough, D. E., and Raphael, Y., 2003. *Math1* gene transfer generates new cochlear hair cells in mature guinea pigs in vivo. *Journal of Neuroscience* 23(11): 4395–4400.

Kazmierczak, P., Sakaguchi, H., Tokita, J., Wilson-Kubalek, E. M., Milligan, R. A., Muller, U., and Kachar, B., 2007. Cadherin 23 and protocadherin 15 interact to form tip-link filaments in sensory hair cells. *Nature* 449(7158): 87–91.

Kemp, D. T. 1978. Stimulated acoustic emissions from within the human auditory system. *Journal of the Acoustical Society of America* 64: 1386–1391.

Konig, R., 2005. Plasticity, learning, and cognition. In R. Konig, P. Heil, E. Budinger, and H. Scheich (Eds), *The Auditory Cortex: A Synthesis of Human and Animal Research.* Mahwah, NJ: Lawrence Erlbaum.

Kubovy, M. and Van Valkenburg, D., 2001. Auditory and visual objects. *Cognition* 80(1–2): 97–126.

LeMasurier, M. and Gillespie, P. G., 2005. Hair-cell mechanotransduction and cochlear amplification. *Neuron* 48: 403–415.

Levitt, H., 2004. Compression amplification. In S. P. Bacon, R. R. Fay, and A. N. Popper (Eds), *Compression: From Cochlea to Cochlear Implants.* New York: Springer-Verlag.

Li, H., Corrales, C. E., Edge, A., and Heller, S., 2004. Stem cells as therapy for hearing loss. *Trends in Molecular Medicine* 10(7): 309–315.

Liang, L., Lu, T., and Wang, X., 2002. Neural representations of sinusoidal amplitude and frequency modulations in the primary auditory cortex of awake primates. *Journal of Neurophysiology* 87(5): 2237–2261.

Lu, T., Liang, L., and Wang, X., 2001. Temporal and rate representations of time-varying signals in the auditory cortex of awake primates. *Nature Neuroscience* 4: 1131–1138.

Manoussaki, D., Dimitriadis, E. K., and Chadwick, R. S., 2006. Cochlea's graded curvature effect on low frequency waves. *Physical Review Letters* 96: 088701.

Martin, F. N. and Clark, J. G., 2006. *Introduction to Audiology*, 9th edition, Boston: Allyn & Bacon.

Martin, F. N. and Clark, J. G., 2009. *Introduction to Audiology*, 10th edition, New York: Allyn & Bacon.

McKay, C., 2004. Psychophysics and electrical stimulation. In F.-G. Zeng, A. N. Popper, and R. R. Fay (Eds), *Cochlear Implants.* New York: Springer-Verlag.

Med-El Corporation, Available at http://www.medel.com

Miller, J. M. and Towe, A. L., 1979. Audition: Structural and acoustical properties. In T. Ruch and H. D. Patton (Eds), *Physiology and Biophysics, Volume I: The Brain and Neural Function.* Philadelphia: W. B. Saunders.

Molchan, S. E., Sunderland, T., McIntosh, A. R., Herscovitch, P., and Schreurs, B. G., 1994. A functional anatomical study of associative learning in humans. *Proceedings of the National Academy of Sciences USA* 91(17): 8122–8126.

Moller, A. R., 2003. *Sensory Systems.* Boston: Academic Press.

Moore, B. C. J., 2003. *Introduction to the Psychology of Hearing*, 5th edition. London: Academic Press.

Moore, B. C. J., 2007. *Cochlear Hearing Loss*, 2nd edition. Chicester, West Sussex: Wiley.

Morris, J. S., Friston, K. J., and Dolan, R. J., 1998. Experience-dependent modulation of tonotopic neural responses in human auditory cortex. *Proceedings of the Royal Society of London B: Biological Sciences* 265: 649–657.

Moser, T., Brandt, A., and Lysankowski, A., 2006a. Hair cell ribbon synapses. *Cell Tissue Research* 326: 347–359.

Moser, T., Neef, A., and Khimich, D., 2006b. Mechanisms underlying the temporal precision of sound coding at the inner hair cell ribbon synapse. *Journal of Physiology* 576(1): 55–62.

Nagel, K. I. and Doupe, A. J., 2008. Organizing principles of spectro-temporal encoding in the avian primary auditory area field L. *Neuron* 58: 938–955.

Narayan, S. S., Temchin, A. N., Recio, A., and Ruggero, M. A., 1998. Frequency tuning of basilar membrane and auditory nerve fibers in the same cochleae. *Science* 282: 1882–1884.

National Institute on Deafness and Other Communication Disorders, Available at http://www.nidcd.nih.gov/

Nave, C. R., 2006. Hyper Physics, Available at http://hyperphysics.phy-astr.gsu.edu/hbase/HFrame.html

Nedzelnitsky, V., 1980. Sound pressures in the basal turn of the cat cochlea. *Journal of the Acoustical Society of America* 68: 1676–1689.

Neely, S. T. and Kim, D. O., 1986. A model for active elements in cochlear biomechanics. *Journal of the Acoustical Society of America* 79: 1472–1480.

Neely, S. T., Gorga, M. P., and Dorn, P. A., 2003. Growth of distortion-product otoacoustic emissions in a nonlinear, active model of cochlear mechanics. In A. W. Gummer (Ed.), *Biophysics of the Cochlea.* River Edge, NJ: World Scientific Publishing.

Nelken, I., Las, L., Ulanovsky, N., and Farkas, D., 2005. Levels of auditory processing: The subcortical auditory system, primary auditory cortex, and the hard problems of auditory perception. In R. Konig, P. Heil, E. Budinger, and H. Scheich (Eds), *The Auditory Cortex: A Synthesis of Human and Animal Research.* Mahwah, NJ: Lawrence Erlbaum.

Niu, X., Tahera, Y., and Canlon, B., 2007. Environmental enrichment to sound activates dopanimergic pathways in the auditory system. *Physiology and Behavior* 10(92, 1–2): 34–39.

Online Mendelian Inheritance in Man (OMIM), Available at http://www.ncbi.nlm.nih.gov/omim

Parker, M. A., Corliss, D. A., Gray, B., Roy, N., Anderson, J. K., Bobbin, R. P., Snyder, E. Y., and Cotanche, D. A., 2007. Neural stem cells injected into the sound-damaged cochlea migrate throughout the cochlea and express markers of hair cells, supporting cells, and spiral ganglion cells. *Hearing Research* 232(1–2): 29–43.

Petersen, M. B., Wang, Q., and Willems, P. J., 2008. Sex-linked deafness. *Clinical Genetics* 73: 14–23.

Pickles, J. O., 1988. *An Introduction to the Physiology of Hearing*, 2nd edition, London: Academic Press.

Plack, C. J., 2005. *The Sense of Hearing*. Mahwah, NJ: Lawrence Erlbaum.

Poremba, A. and Mishkin, M., 2007. Exploring the extent and function of higher-order auditory cortex in rhesus monkeys. *Hearing Research* 229: 14–23.

Qun, L. X., Pirvola, U., Saarma, M., and Ylikoski, J., 1999. Neurotrophic factors in the auditory periphery. *Annals of the New York Academy of Sciences* 884: 292–304.

Recanzone, G. H. and Sutter, M. L., 2008. The biological basis of audition. *Annual Review of Psychology* 59: 119–142.

Recanzone, G. H., 2000. Response profiles of auditory cortical neurons to tones and noise in behaving macaque monkeys. *Hearing Research* 150: 104–118.

Rhode, W. S., 1971. Observations of the vibration of the basilar membrane in squirrel monkeys using the Mossbauer technique. *Journal of the Acoustical Society of America* 49: 1218–1231.

Robles, L. and Ruggero, M. A., 2001. Mechanics of the mammalian cochlea. *Physiological Reviews* 81(3): 1305–1352.

Ruel, J., Wang, J., Rebillard, G., Eybalin, M., Lloyd, R., Pujol, R., and Puel, J.-L., 2007. Physiology, pharmacology and plasticity at the inner hair cell synaptic complex. *Hearing Research* 227: 19–27.

Ruggero, M. A., Narayan, S. S., Temchin, A. N., and Recio, A., 2000. Mechanical bases of frequency tuning and neural excitation at the base of the cochlea: Comparison of basilar-membrane vibrations and auditory-nerve fiber responses in chinchilla. *Proceedings of the National Academy of Sciences USA* 97: 11744–11750.

Ruggero, M. A., Rich, N. C., Recio, A., Narayan, S. S., and Robles, L., 1997. Basilar membrane responses to tones at the base of the chinchilla cochlea. *Journal of the Acoustical Society of America* 101: 2151–2163.

Ryals, B. M. and Rubel, E. W., 1988. Hair cell regeneration after acoustic trauma in adult Cotumix quail. *Science* 240: 1774–1776.

Ryals, B. M., Matsui, J. I., and Cotanche, D. A., 2007. Regeneration of hair cells. In K. C. M. Campbell (Ed.), *Pharmacology and Ototoxicity for Audiologists*. Clifton Park, NY: Thomson Delmar.

Schaette, R. and Kempter, R., 2008. Development of hyperactivity after hearing loss in a computational model of the dorsal cochlear nucleus depends on neuron response type. *Hearing Research* 240(1–2): 57–72.

Scheich, H., Brechmann, A., Brosch, M., Budinger, E., and Ohl, F. W., 2007. The cognitive auditory cortex: Task-specificity of stimulus representations. *Hearing Research* 229: 213–224.

Schiech, H., Ohl, F. W., Schulze, H., Hess, A., and Brechmann, A., 2005. What is reflected in auditory cortex activity: Properties of sound stimuli or what the brain does with them? In R. Konig, P. Heil, E. Budinger, and H. Scheich (Eds), *The Auditory Cortex: A Synthesis of Human and Animal Research*. Mahwah, NJ: Lawrence Erlbaum.

Schwartz, M. S., Otto, S. R., Shannon, R. V., Hitselberger, W. E., and Brackmann, D. E., 2008. Auditory brainstem implants. *Neurotherapeutics* 5(1): 128–136.

Shamma, S. A., 2008. Characterizing auditory receptive fields. *Neuron* 58: 829–831.

Shera, C. A., 2003. Wave interference in the generation of reflection- and distortion-source emissions. In A. W. Gummer (Ed.), *Biophysics of the Cochlea*. River Edge, NJ: World Scientific Publishing.

Shera, C. A., 2004. Mechanisms of mammalian otoacoustic emission and their implications for the clinical utility of otoacoustic emissions. *Ear and Hearing* 25: 86–97.

Siemens, J., Kazmierczak, P., Reynolds, A., Sticker, M., Littlewood-Evans, A., and Muller, U., 2002. The Usher syndrome proteins cadherin 23 and harmonin form a complex by means of PDZ-domain interactions. *Proceedings of the National Academy of Sciences* 99: 14946–14951.

Speaks, C. E., 1999. *Introduction to Sound*. San Diego: Singular.

Staab, W., 2002. Characteristics and use of hearing aids. In J. Katz (Ed.), *Handbook of Clinical Audiology*, 5th edition, Philadelphia: Lippincott Williams & Wilkins.

Staab, W., Preves, D., Yanz, J., and Edmonds, J., 1997. Open platform DSP hearing instruments: A move toward software-controlled devices. *Hearing Review* 4(12): 22–24.

Steel, K. P., Erven, A., and Kiernan, A. E., 2002. Mice as models for human hereditary deafness. In B. J. B. Keats, A. N. Popper, and R. R. Fay (Eds), *Genetics and Auditory Disorders*. New York: Springer-Verlag.

Valencia, D. M., Rimell, F. L., Friedman, B. J., Oblander, M. R., and Helmbrecht, J., 2008. Cochlear implantation in infants less than 12 months of age. *International Journal of Pediatric Otorhinolaryngology* 72(6): 767–773.

Van Laer, L., Cryns, K., Smith, R. J. H., and Van Camp, G., 2003. Nonsyndromic hearing loss. *Ear and Hearing* 24(4): 275–288.

Varga, R., Kelly, P. M., Keats, B. J., Starr, A., Leal, S. M., Cohn, E., and Kimberling, W. J., 2003. Non-syndromic recessive auditory neuropathy is the result of mutations in the otoferlin (OTOF) gene. *Journal of Medical Genetics* 40: 45–50.

von Bekesy, G., 1947. The variation of phase along the basilar membrane with sinusoidal vibrations. *Journal of the Acoustical Society of America* 19: 452–460.

von Bekesy, G., 1960. *Experiments in Hearing*. New York: McGraw-Hill.

von Ilberg, C., Kiefer, J., Tillein, J. Pfennigdorff, T., Hartmann, R., Sturzebecher, E., and Klinke, R., 1999. Electric-acoustic stimulation of the auditory system. *ORL Journal for Otorhinolaryngology and its Related Specialties* 61: 334–340.

Wang, J., Puel, J.-L., and Bobbin, R., 2007. Mechanisms of toxicity in the cochlea. In K. C. M. Campbell (Ed.), *Pharmacology and Ototoxicity for Audiologists*. Clifton Park, NY: Thomson Delmar.

Wang, X., 2007. Neural coding strategies in auditory cortex. *Hearing Research* 229: 81–93.

Wang, X., Lu, T., and Liang, L., 2003. Cortical processing of temporal modulations. *Speech Communication* 41: 107–121.

Weinberger, N. M., 2007a. Auditory associative memory and representational plasticity in the primary auditory cortex. *Hearing Research* 229: 54–68.

Weinberger, N. M., 2007b. Associative representational plasticity in the auditory cortex: A synthesis of two disciplines. *Learning and Memory* 14(1): 1–16.

White, L. J. and Plack, C. J., 1998. Temporal processing of the pitch of complex tones. *Journal of the Acoustical Society of America* 103: 2051–2063.

Wilson, B., 2004. Engineering design of cochlear implants. In F.-G. Zeng, A. N. Popper, and R. R. Fay (Eds), *Cochlear Implants*. New York: Springer-Verlag.

Winer, J. A. and Lee, C. C., 2007. The distributed auditory cortex. *Hearing Research* 229: 3–13.

Yost, W. A., 2007. *Fundamentals of Hearing*, 5th edition, San Diego: Academic Press.

Zeng, F.-G., 2004. Compression and cochlear implants. In S. P. Bacon, R. R. Fay, and A. N. Popper (Eds), *Compression: From Cochlea to Cochlear Implants*. New York: Springer-Verlag.

Zenner, H. P., Zimmermann, R., and Gitter, A. H. 1988. Active movements of the cuticular plate induce sensory hair motion in mammalian outer hair cells. *Hearing Research* 34: 233–239.

10

Vision

Hiroto Ishikawa,
Naohiro Ikeda,
Sanae Kanno,
Tomohiro Ikeda,
Osamu Mimura, and
Mari Dezawa

10.1 Introduction

The human eye is mainly composed of a cornea, conjunctiva, sclera, ciliary body, lens, vitreous body, retina, and anterior and posterior chambers filled mainly with aqueous humor.[1-3]

Light entering the eye passes through the cornea and the lens to form an image on the retina. If images are formed only on the retina, vision is correct. However if images are formed behind the retina, the eyes are hyperopic, and if images are formed in front of the retina, the eyes are myopic.

10.2 Lens

The lens is located behind the iris; its diameter is about 10 mm and its thickness is 3–4 mm, showing a convex lens shape (Figure 10.1).[2-5] Around the equator plane and the ciliary body, there are numerous muscles of the zonule of Zinn, which hold the lens in front of the vitreous body. The zonule of Zinn is a tissue without color, blood vessels, and nerve fibers.

Histologically, a lens is enveloped with a lens capsule (the anterior part by the anterior capsule and the posterior part by

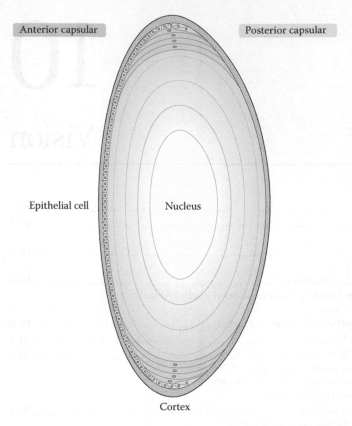

Anterior capsular   Posterior capsular

Epithelial cell   Nucleus

Cortex

FIGURE 10.1   Histological anatomy of a lens.

the posterior capsule), and under the anterior capsule, a monolayer of lens epithelial cells is recognized. Numerous hexagonal lens fibers are regularly and tightly arranged. They form a corneal substance. The central part of the lens becomes hard after 25 years of age, and the lens forms a nucleus. As aging progresses, the lens nucleus becomes harder, which results in a deterioration of transparency and finally in nuclear cataract (Figure 10.2). A lens cortex localizes around the nucleus of a lens, and is the portion between the nucleus and the lens capsule. The Cataract with opacity mainly in the cortex is called cortex cataract. According to the turbidities and shapes, the cataract shows many symptoms. The causes of cataract are varied, such as aging, diabetes, atopic dermatitis, dosage of steroid drugs, high myopia, external injuries, and so on.

Emery-Little's Classification		
Grade	**Hardness**	**Color**
1	Soft	Clear, milky-white
2	Semisoft	Yellowish white
3	Medium	Yellow (no brown)
4	Hard	Brownish yellow
5	Rock-hard	Brown, black

FIGURE 10.2   Emery's classification.

Zinn's zonule is composed of collagen fibers resembling elastic fibers with a diameter of 1–10 mm. Zinn's zonule arises from both the ciliary body and the flattened part of the ciliary body, and most of its fibrous strands attach to the capsule of the lens. By the attached portions of collagen fibers to the lens capsule, it is divided into an anterior zonule, an equator one, and a posterior one. The attached portions of the anterior zonule to the lens capsule are different according to age, namely, at 17 years the attached portion from the lens equator is 0.25 mm, at 43 years it is 0.4 mm, at 57 years it is 0.86 mm, at 85 years it is 1.2 mm, and in the aged man the attached portions gradually move to the center of the lens capsule. The fine fibers attached to the lens capsule protrude to the thinner fine fibers; they go to the central part of the lens and finally become the fine fiber layer of the lens capsule.

The Lens has a very important role in the refraction of light together with the cornea. An image on the retina is controlled by the change in thickness of the lens. When looking at an object close to the eyes, zonule ciliaris muscles contract and the ciliary process rotates to the frontal direction. The zonule ciliaris becomes loose and the lens increases its thickness by its own elasticity, which causes the refractory power of the lens to increase. On the other hand, if the eyes look at objects that are fartheraway, the ciliary muscles are relaxed and the frontal rotation of the ciliary process decreases. If the zonule ciliaris is pulled, the lens is also pulled and flattened. According to age, the elasticity of the lens decreases, which results in presbyopia.

10.2.1 Diseases of the Lens

A. *Cataract*
  1. Nuclear cataract: The central part of the lens gradually hardens with aging. Emery's classification is mostly used for classification of the grades of hardness of the nucleus. In this classification, slit lamp examinations are conducted.
  2. Cortex cataract: The main cause of this disease is opacity in the lens cortex around the lens nucleus. Other causes are aging diabetes.
B. *Ectopic lens*
  1. Traumatic lens luxation: Not only is Zinn's zonule disrupted by a bruise to the eye(s), there are several other causes for the lens to drop down into the vitreous body.
  2. Congenital ectopic lens: Abnormal development of Zinn's zonule is thought to be the cause of this disease. Except for the ectopic lens, other parts of the eye are normal.
  3. Marfan syndrome: this is a typical disease attributed to genetic mutation.[6,7] The lens is small and less developed, about 30% of the normal size. The shapes of the ciliary processes and their distribution are not symmetrical. The fibers attached to the lens capsule are very thin and their number is very small. Therefore, keeping the lens at the normal position is very difficult. This disease is attributed to dominant inheritance and the amino acid sequence of Zinn's zonule is

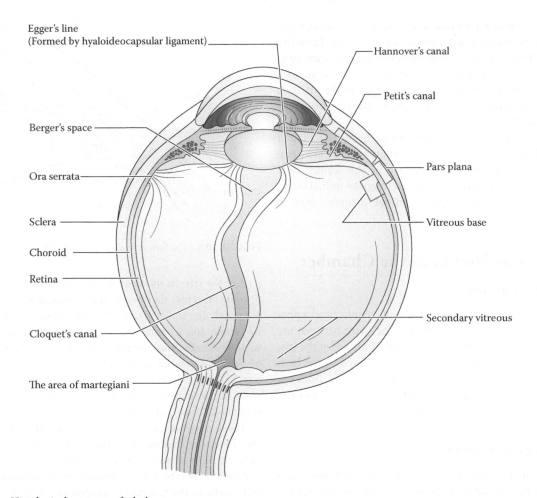

Egger's line
(Formed by hyaloideocapsular ligament)

Hannover's canal

Petit's canal

Berger's space

Pars plana

Ora serrata

Sclera

Vitreous base

Choroid

Retina

Cloquet's canal

Secondary vitreous

The area of martegiani

FIGURE 10.3 Histological anatomy of whole eye.

abnormal. Lens dislocation is found mostly to the upper part, particularly to the upper ear side.

10.3 Vitreous Body

The vitreous body is the largest part of the eye and occupies about 80% of the total eye volume (4 cc) (Figure 10.3).[2,4,5] It is composed of collagen fibers with transparent gel tissue, and is localized at the space behind the lens. A vitreous body consists of a fiber-like matrix and fluid that contains hyaluronic acid. Adherence between the vitreous body and retina is strongest at the basal part of the saw-like margin of the retina and the optic disk.[1,8-11] The bottom region of the vitreous body contains much more collagen fibers and hyaluronic acid, and thus its viscosity is high. Therefore, this region is called the vitreous body cortex. On the contrary, the central part of the vitreous body contains fewer amounts of these substances. Therefore, the viscosity is low. In the bottom part of the vitreous body and the cortex of the optic disk margin, there are vitreous body cells, macrophages, and fibroblasts. There is no blood vessel, but in the viviparous stage of embryonic development branches of the central artery extend from the retina. The artery disappears soon after birth. In a few cases, however, they remain in the adult.[12]

According to aging, the vitreous body liquefies more and more, and also the posterior part of the vitreous body gradually exfoliates at 40–60 years of age. The vitreous body attaching to the optic disc of nerve fibers detaches like a ring, which is called the Weis ring. When light illuminates the retina near the ring, we perceive that flies are flying. This phenomenon is called "flying flies." There are many causes for this morbid phenomenon other than the Weis ring, for example, cell penetration into the transparent vitreous body, opacity, bleeding, and so on.

The vitreous body is the transparent intermediate structure that lets in light to the retina, and maintains eye pressure in order to maintain eye shape. It has another important role as a reserve of various substances. For example, it has functional roles of storage of glucose and amino acids and storage of waste substances such as lactic acid. In addition, it maintains the ionic balance and supplies oxygen to the retina and other parts of the eye.

10.3.1 Diseases of the Vitreous Body

A. *Vitreous opacity:* This disease is related to physiological opacity, inflammation of the zonule ciliaris, endophthalmitis, obsolete vitreous body bleeding, vitreous body denaturation (asteroid hyalosis, synchsis scintillans, vitreous body amyloidosis), eye tumors, and so on. Turbidity of the vitreous body is caused by change in the quality of fibers in the vitreous body, invasion of inflammatory cells, hardness of blood, quality changes in the vitreous body by diseases of the whole body, invasion of tumor cells, and so on.

B. *Vitreous hemorrhage:* Vitreous hemorrhage is attributed to bleeding in the retina or to the rupture of retinovitreal neovascularization. A typical case of the former type includes macular degeneration and tumor of retinal blood vessels. The second type is retinal disorder due to diabetes.

10.4 Aqueous Humor in the Chamber

10.4.1 Eye Chamber

The eye chamber is divided into an anterior chamber and a posterior chamber. The anterior chamber is the space between the posterior surface of the cornea and the frontal surface of the iris, and the posterior chamber is the space surrounded by the lens, the vitreous body, and the posterior surface of the iris. Both eye chambers are filled with aqueous humor. Aqueous humor is divided into anterior and posterior humor.

10.4.2 Aqueous Humor

The chamber aqueous humor is produced at the ciliary body, and from the posterior chamber aqueous humor enters through the space between the lens and iris and reaches the anterior chamber. It is discharged from the eye, and the main outflow route is Schlemm's canal at the corner of the anterior chamber and the sinus in the sclera. In human eyes, 80–90% of the total aqueous humor flows out through Schlemm's canal and 5–20% of the rest of the aqueous humor flows out from the uveal sclera.

Aqueous humor in the chamber is composed of both anterior chamber aqueous humor (which contains approximately 0.25 mL) and posterior chamber aqueous humor (which contains approximately 0.06 mL). The osmotic pressure is slightly higher than that of blood plasma.

10.4.3 Anterior Chamber Angle; Trabecular Meshwork

In order to understand the outflow of aqueous humor into a chamber (Figure 10.4), the anterior chamber angle has a very important role. The area of the chamber angle is from the terminal end of the corneal Descemet's membrane (Schwalbe line) to the root part of the iris. At the corneal side of the angle, the trabecular meshwork is composed of anterior chamber nets, angle sclera, endothelial nets, and endothelial nets.

The uveal membrane is located at the region closest to the anterior chamber, and is tissue of 2–3 layers with string-like trabecular meshwork showing wire netting that adheres to the root

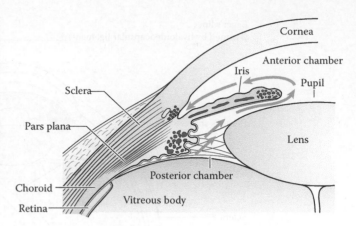

FIGURE 10.4 Outflow of the aqueous humor.

part of the iris. In uveal membrane nets, large holes (spaces of fibrous trabecular meshwork) with diameters of 30–100 mm are observed. The trabecular meshwork has extracellular matrix at its center, and its surface is covered with flattened trabecular meshwork cells. The corneal–scleral plexus is located at the side of Schlemm's canal of the uveal side, and its posterior side attaches to the tip of the sclera. It shows multilayers made up of plate-like structure. Plates of the trabecular meshwork have small holes (spaces among trabecular meshwork) with diameters of 10–30 mm, which are routes of aqueous outflow. Plates of the trabecular meshwork of the corneal–scleral plexus show a sandwich-like structure. Namely, both sides of the extracellular matrix are covered with a monolayer of trabecular meshwork cells. Cells of the trabecular meshwork are thin and flattened, but its nuclear region is thick. The trabecular meshwork contains connective tissue embedded in a matrix of 2–4 layers.

The extracellular matrix of the trabecular meshwork consists of fibrous components as well as extracellular macromolecules. Fibrous components are made up of type I and type III collagen fibers and elastic fibers. Elastic fibers are composed of microfibril, elastin, and insoluble substances without structure. The main component of microfibril is a kind of protein named fibrillin. Extracellular macromolecules are gel substances without structure such as glycosaminoglycan. Glycosaminoglycan can be divided into two types: proteoglycan (which is combined with nuclear protein) and hyaluronic acid. In Molecules such as glycoprotein, fibronectin, laminin, and nonfibrous collagen type IV and type VI are present. These extracellular matrixes do not exist in intercellular space, but assemble and adhere to other substances without any structure.

10.4.4 Schlemm's Canal

Schlemm's canal is approximately 36 mm in circumference and the width is 300–500 mm, showing a flattened elliptic structure. Schlemm's canal is not a clear ring but it shows a meandering structure. It can be divided into 2–3 branches and again the branches meet to form a single duct. Inside the duct, there are pores opening to the outside called Sonderman's inner duct.

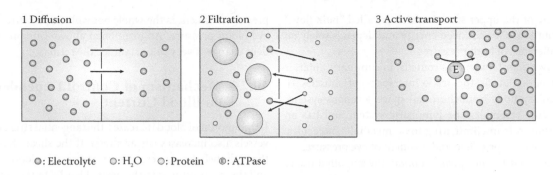

1 Diffusion   2 Filtration   3 Active transport

○: Electrolyte   ○: H₂O   ○: Protein   ⊛: ATPase

FIGURE 10.5 Mechanism of aqueous humor.

Outside the duct, there is an outer assembling duct, and through the ducts of aqueous humor vein and upper sclera vein, aqueous solution discharges into the vein system of the whole body.

Aqueous humor has an important role in the control of eye pressure and is essential for maintaining normal eye shape. In addition, it sustains tissues such as the cornea and lens, which have no blood circulation, and has an important role as an optical part of the eyes.

Aqueous humor is produced in the epithelium of the ciliary body. The following three theories represent the mechanism of aqueous humor: diffusion, ultrafiltration, and active transport (Figure 10.5).

1. *Diffusion:* Diffusion is the flow of a substance from a high concentration to a low concentration through a membrane. Lipid-soluble molecules, however, flow freely into both sides. In fact, the concentration of ascorbic acid in aqueous humor is much higher than that of blood plasma; therefore, ascorbic acid flows into blood plasma.

2. *Ultrafiltration:* In the case of molecules in solution on one side that are permeable to the membrane and other molecules in solution on the other side that are not permeable, and when they are separated unevenly by the membrane, only distributions of membrane-permeable molecules alter the concentration on both sides. This phenomenon is called ultrafiltration. In the eye, the flow of blood plasma from endothelial tissue in the ciliary body matrix ciliary body is one example of this case. The higher the aqueous humor pressure, the more the filtered amount.

3. *Active transport:* Using energy adenosine triphosphate (ATP), even molecules with a low concentration in solution can be transported through the membrane against reversal concentration to the other side of the solution with higher concentration. This active transportation is the most important for the production of aqueous humor. The nonpigmented epithelium of the ciliary body actively transports Na^+ into the posterior chamber by $Na^+K^+ATPase$; at the same time, in order to maintain electrical balance, Cl^- and HCO_3 are transported. In this way, aqueous humor is produced. At some physiological condition, active transport is involved at 80–90%.

10.5 Mechanism of the Outflow of Aqueous Humor

There are two main routes: the angle trabecular meshwork through Schlemm's canal, namely, the conventional route (trabecular outflow), and from the root of the iris, interspaces of the ciliary body, and subchoroid, namely, the unconventional route (uveoscleral outflow) (Figure 10.6).

1. *Conventional route:* This route is, in summary, anterior chamber → trabecular meshwork → Schlemm's canal → collecting tube → upper sclera vein. In humans and monkeys, the barriers (60–80%) of the outflow have two routes: the interval between the anterior chamber and Schlemm's canal, and the connective tissue of para-Schlemm's canal and the inner wall of Schlemm's canal. The main barrier is the connective tissue of para-Schlemm's canal and the endothelial cells in the inner wall of Schlemm's canal. This region contains macromolecules such as hyaluronic acid and proteoglycan (glycoprotein + glycosaminoglycan), collagen, and fibronectin.

The outflow of aqueous humor is attributed to the difference of aqueous humor pressure (eye pressure—vein

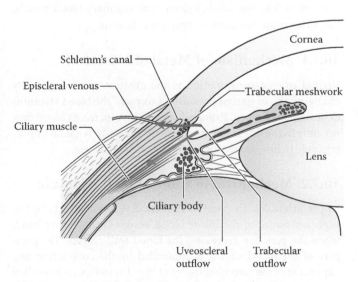

FIGURE 10.6 Trabecular outflow and uveoscleral outflow.

pressure of the upper sclera), which is called "bulk flow." This outflow does not need energy and depends solely on the difference of both pressures.

2. *Unconventional route:* This route is, in summary, anterior chamber → root of the iris → interspace of muscles of the ciliary body → subchoroidal space → venous system. This route resembles the lymphatic system and has an important role in eliminating toxic metabolic molecules. The route is, in principle, independent of eye pressure.

3. *Control of outflow of aqueous humor:* The smooth muscles of the Iris, ciliary body, and trabecular meshwork are innervated by the autonomic nervous system. Namely, drugs activate the sympathetic nerves and accelerate the outflow. In addition, in humans and monkeys, the drugs increase uveoscleral outflow. Activating drugs also accelerate uveoscleral outflow. Parasympathetic nerves contract the muscles of the ciliary body; consequently, when the spaces in the trabecular meshwork are wide open, the resistance for outflow decreases.

10.6 Main Diseases of the Eye

A. *Glaucoma:* When the route of outflow is disturbed, eye pressure abnormally increases to cause glaucoma.[13] This disease damages the optic nerves due to the high eye pressure. Glaucoma is classified into two types according to the opening of the angle. For treatment, the decrease in eye pressure is primary. Recently, however, it has been proposed that the defecting blood circulation around the optic disc may be another mechanism for glaucoma.

10.7 Mechanism of Circulation Control in the Retina

Retinal vessels physiologically construct microcirculation such as arteriole microcirculation, microveins, and capillary vessels. The retina, like the brain, kidney, and coronary blood vessels, has an accurate circulation control mechanism.

10.7.1 Mechanism of Metabolic Control

Retinal oxygen consumption is the greatest in all organs. By changing the of partial pressure of oxygen, the blood stream is delicately influenced. Especially, adenosine increases blood flow not only in the retina but also in the whole body. It has an important role in low oxygen and retinal ischemic disease.

10.7.2 Mechanism of Control by the Muscle

In the arterioles, the wall of the blood vessel contracts by the higher pressure of the inside blood vessel. On the other hand, when the pressure decreases, the blood wall relaxes. The pressure of the blood vessels is controlled by the contraction and expansion of the smooth muscles of blood vessels, which is called myogenic control. This condition, which is dependent on

pressure, influences the whole body blood pressure and also the eye blood pressure. As mentioned above, this pressure-dependent change is very important for the retina.

10.7.3 Mechanism of Control Dependent on Blood Current

If the amount of blood increases, the tangential stress on the blood vessels also increases (shear stress). If the shear stress increases, the endothelial cells perceive this stress, the blood vessels dilate, and the stress returns to the normal level. On the contrary, if the amount of blood decreases, the reverse occurs, namely, the blood vessels contract and the stress returns to the normal level.

10.7.4 Mechanism of Nervous Control

The choroid and ciliary body are controlled by the nervous system. The retinal blood vessels, however, are not controlled by the nervous system. Different from the other organs, the retina is not influenced by sympathetic nerves. That is to say, with respect to blood circulation in the retina, the circulation system may be included in the retina itself. The nervous controls of the eye arteries that are located at the upper retinal artery are much less influenced.

10.8 Diseases of Retinal Circulation

10.8.1 Diabetic Retinopathy

In diabetes mellitus, pathological changes can be observed in the endothelial cells of the capillary at the early stage, such as increase of vesicles in cells, tight junction deterioration, degeneration of cells around the retina blood vessel cells, and disappearance of endothelial cells. Thickening of the basement membrane and denatured retinal pigment epithelium are often recognized. Vascular endothelial growth factor (VEGF) is the main factor that induces the increase in permeability of the blood–retina barrier and promotes neovascularization (Figure 10.7).

10.8.2 Retinal Vascular Disorders

Central retinal artery and vein occlusion (CRAO and CRVO), ocular ischemic syndrome, and macroaneurysm (bleeding from an aneurysm extending from the inner retina to the subretina, and occasionally to the vitreous body) are included in this category.

10.9 Cornea

Together with the sclera, the cornea is a part of the outermost layer of an eyeball, but the greatest difference from the sclera is that the cornea is a transparent tissue that allows light to pass into the eyeball (Figure 10.8). In addition, the cornea is convex shaped so as to play the role of a lens. Therefore, the cornea has the following functions: (1) the outer wall borders between the outside

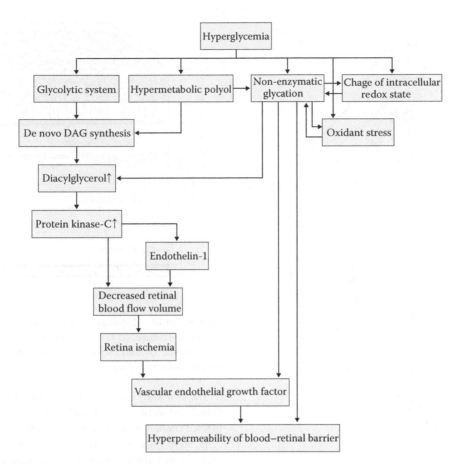

FIGURE 10.7 Mechanism of blood–retinal barrier hyperpermeability in diabetic retinopathy.

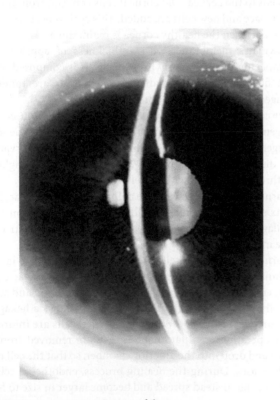

FIGURE 10.8 Anterior segment of the eye.

and the inside of the eyeball, (2) it is a transparent tissue to transilluminate the light, and (3) it is a strong convex lens that refracts of light in order to make a sharp image on the retina.

The cornea is not exactly round, but oval. Its vertical diameter is about 11 mm and its horizontal diameter is about 12 mm. The thickness of a convex-shaped cornea is not constant, namely the central part is thinner and the peripheral region is thicker. Because of this structure, the cornea does not curve uniformly. The peripheral area is flattened, and the cornea shows a gradual rising slope in the peripheral margin. The refractive power that determines visual acuity is dependent on the 3 mm diameter of the central area of the cornea, which shows a convex shape.

Images are focused on the retina by accommodation. Accommodation mainly depends on increased refractivity, which is mostly provided by increase in curvature of the anterior surface of the lens, increase in lens thickness, anterior movement of the lens, and increase in curvature of the posterior surface of the lens. The lens is connected to the ciliary body by the zonule of Zinn. Contraction of ciliary muscle causes relaxation of zonule of the Zinn, which may further cause thickening of the lens to increase the refractive power (Figure 10.9).

As the cornea has no blood vessels, oxygen supply depends on the diffusion of oxygen from air that is dissolved into the tear. Other nutrition supplied to the cornea is through the solution contained in the anterior chamber. But when the cornea is under

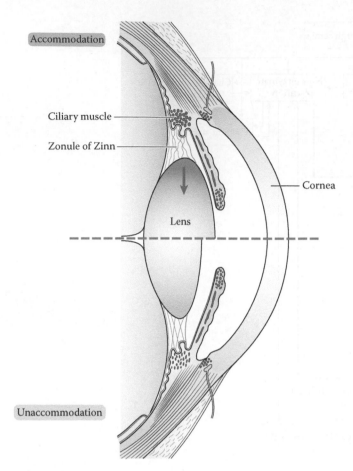

FIGURE 10.9 Schema of accommodation.

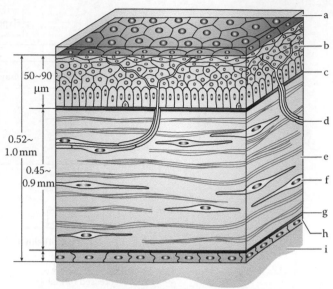

FIGURE 10.10 Histological anatomy of the cornea: (a) tear film, (b) corneal epithelium, (c) Bowman's layer, (d) nerve in the cornea, (e) corneal stroma, (f) keratocyte, (g) Descemet's membrane, (h) corneal endothelium, and (i) aqueous humor.

poor nutritional condition as in the case of severe injury, abnormal angiogenesis into the cornea through the barrier of the corneal–conjunctival junction occurs. From this newly formed blood supply, enough nutrition is supplied; however, unnecessary substances are also conveyed. Such a situation causes sedimentation of the substances, which leads to decreased transparency of the cornea.

From the histological point of view, a cornea is made up of five different layers: from the surface, epithelial layer, Bowman's membrane, substantial layer, Descemet's membrane, and endothelial layer (Figure 10.10).

The epithelial layer, which is the surface of the cornea, is stratified squamous epithelium. Blinking of eyes constantly peels off dust and old epithelial cells from the corneal surface. For protection from such forces, epithelial cells are tightly connected to each other by interdigitation between the cells. In addition, they have tight junctions, desmosomes, and gap junctions between the cells. Due to the structural connections, the epithelial layer is strongly protected from the vertical and horizontal external forces. The anchoring fibril coming out from the cell membrane with many hemi-desmosomes elongates just like a plant complicatedly extends the roots in every direction. Therefore, epithelial cells connect each other quite strongly to the substantial layer (Figure 10.11).

The boundary between the epithelium and the conjunctiva is called the rimbus. Stem cells of the corneal epithelium are located there, which constantly supply new cells to the epithelial cells. Stem cells are found in the palisade of Vogt (POV). Stem cells do not divide normally, but transient amplifying cells derived from stem cells actively divide and constantly supply new cells to the cornea. Old corneal cells come off from the corneal surface and new cells are added. This cycle repeats every 1–2 weeks, and then the healthy cornea is maintained.

Bowman's membrane, with a thickness of approximately 10 μm, is found particularly in primates. It is composed of collagen fibers with uniform diameter, but they are randomly arranged. Its clinical role is not known.

The corneal stroma is composed of approximately 200 sheets of collagen fibers, and corneal stromal cells exist among the fibers. Collagen fibers are of type I, arranged in a constant interval, and their diameter is approximately 22–32 μm. Therefore, light can pass through the fibers without reflection (Figure 10.12).

Corneal stromal cells have the ability to produce extracellular matrix such as collagen. They adhere to mesh sheets consisting of collagen fibers and have a role to maintain the layer structure of the matrix.

Corneal endothelial cells are composed of a single layer of columnar cells in the innermost layer of the cornea. In the adult, endothelial cells lose the ability to proliferate and stay at the cell cycle of stage G1. A columnar cell shows a hexagonal structure (Figure 10.13). When endothelial cells are injured by either operation or inflammation, they are removed from the cornea and drop into the anterior chamber, so that the cell number decreases. During the healing process, endothelial cells do not divide but instead spread and become larger in size to fill up the lost space.

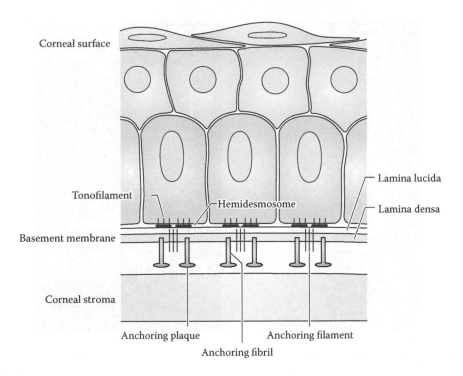

Corneal surface

Tonofilament

Hemidesmosome

Lamina lucida

Lamina densa

Basement membrane

Corneal stroma

Anchoring plaque

Anchoring filament

Anchoring fibril

FIGURE 10.11 Connection of the corneal epithelial cell.

Because of eye pressure, the water solution of the anterior chamber is always ready to enter into corneal substances. Because of the collagen and glycosaminoglycan contained in the corneal matrix, the inside of the cornea is always hypertonic, and water comes into the cornea. The function of endothelial cells is to protect the incoming excessive water by pumping out the excessive water into the anterior chamber. The mechanism of this function is achieved by the existence of ions with different concentrations between the substance and the anterior chamber. To maintain the difference, Na–K ATPase and carbon dioxide dehydrogenase are involved.

10.9.1 Diseases of the Cornea

A. *Bullous keratopathy:* If the corneal endothelial cells decrease in number (less than about 500 cells/mm^2) or the cell functions deteriorate even if the cell number is within

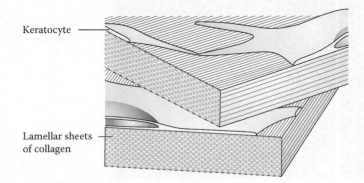

Keratocyte

Lamellar sheets of collagen

FIGURE 10.12 Corneal lamellae.

normal range, water of the anterior chamber flows into the corneal substance. If this situation continues for a long period, the disease is called bullous keratopathy.

B. *Keratoconus:* Here the central part of the cornea protrudes and this part is abnormally thin. The disease arises particularly during adolescence, and gradually becomes worse. The symptom, however, stops at about 30 years of age. The cornea becomes thinner, and Descemet's membrane expands and sometimes degenerates. The cause of keratoconus is not known.

C. *Herpes simplex virus (HSV):* If the eye is infected with HSV, the virus stays in the ganglion cells of ciliary nerves in inactive state. An attack of fever or cold or other stresses induce corneal herpes. The symptom of this disease differs according to the portions of the eyes, but are generally characterized by a decrease in the sensitivities of the cornea and precipitation at the posterior cornea. If the disease arises repeatedly, the necrosis like inflammation occurs in the cornea. In this situation, blood vessels enter into the substance, where an immune complex is formed against the virus antigen.

10.10 Sclera

Sclera is a strong membrane located in the outermost surface of the eye. The surface layer of sclera is derived from the neuroectoderm, and the substance is comprised of condensed mesenchymal cells. Sclera is composed of three layers: the outer upper sclera, the propria layer, and the brown plate.

Type I collagen is the main component of the fibers, and proteoglycan partly contributes. Different from the cornea, the

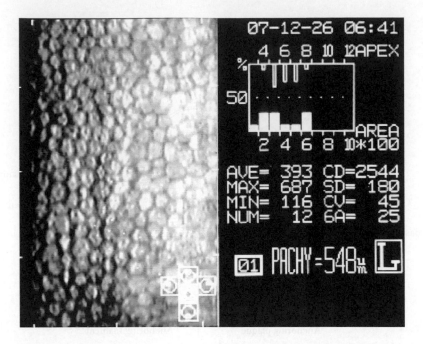

FIGURE 10.13   Structure of the corneal endothelial cell.

collagen fibers are big and are arranged at randomly. Therefore, sclera is very strong and opaque.

In the adult sclera, the most posterior part is the thickest at ~0.6 mm, the ring area is 0.8 mm, and the thinnest part is the portion of four muscle insertions at 0.3 mm. Because a dull outer force, the rupture of the eyeball occurs mostly at the portion attached to the rectus muscles. Particularly when the eyelids are closed, the eyeball rotates upward, and the outer force also works upward, injuries mainly arise in parallel with the ring part attached to the frontal part of the superior rectus muscle.

10.10.1 Diseases of the Sclera

In the sclera, blood vessels are scarce. Hence, inflammation rarely occurs, but patients with autoimmune diseases (chronic rheumatism, systemic lupus erythematosus (SLE), Wegener's

bud of sarcoma, etc.) are susceptible to inflammation of the sclera. Among diseases of the sclera, most patients suffer from anterior scleritis, and the color is salmon pink (Figure 10.14, left). Necrotizing scleritis is the most dangerous disease; in this case, the sclera gradually becomes thinner (Figure 10.14, right). The disease extends to the anterior segment of the eye, and sometimes results in blindness.

The diagnosis of posterior scleritis is in most cases difficult. Patients complain of the visual acuity loss, sharp pain, and double vision caused by paralysis of the lateral rectus muscle.

10.11 Retina

The retina is located at the innermost layer of the eye wall, and is transparent membranous tissue. Its innermost surface faces the vitreous body, and its outermost surface faces the choroidal

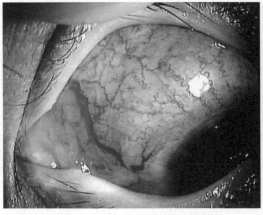
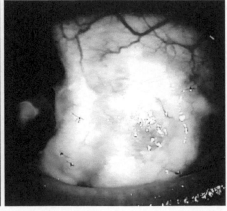

FIGURE 10.14   Anterior scleritis (left) and necrotizing scleritis (right).

layer. From the developmental point of view, the retina is divided into two layers: sensory retina and retinal pigment epithelium. In the sensory retina, visual cells, that is photoreceptor cells, perceive light in the vision. The number of visual cells in a retina is about 100 million. When visual cells perceive light, they transform light energy into electrical signals. The signals are treated and integrated by nerve cells in the retina, and the visual information is sent via the lateral geniculate body to the visual area at the occipital lobe in the cerebrum.

10.11.1 General Structure of the Retina

The retina can be seen through the pupil and such an observed retina is called the fundus of the eye. The central area of the retina contains xynthophyll and the area is called "macula." At the center of the macula, there is a little hollow that is called as fovea centralis. At the center of the macula, about 4 mm to the nasal side, there is an optic disc with a diameter of about 1.5 mm (Figure 10.15). The macula is anatomically divided into more detailed small regions; however, the clinical map of the retina is not clear. Therefore, the anatomical map and the clinical one are not coincident.

A good guideline showing the equator in an eye is the sclera entrance of the vortex vein. By the circle connecting each posterior tip of the vortex vein, the fundus of the eye can be divided into the posterior fundus and a peripheral one. The anatomical equator is located at 3 mm in front of the sclera entrance of the vortex veins.

At the edge of the retina, the thickness of the sensory retina decreases sharply to make the transition to the pars plana ciliaris. In this region, the retinal pigment epithelium of the retina is switched over with ciliary body pigment epithelium of the two layers. Thus, this transitional part from the retina to the pars plana ciliaris is called ora serrata.

The thickest part of the retina is about 0.3 mm at the posterior part of the fundus of the eye, and the thinnest part is in the fovea

centralis 0.13 mm, at the equator region 0.18 mm, and at the ora serrata only 0.1 mm.

10.11.2 Histological Structure of the Retina

The retina is divided into two parts: the sensory retina and retinal pigment epithelium. The sensory retina is composed of six different kinds of nerve cells: visual cells, horizontal cells, bipolar cells, Amacrine cells, reticular interlayer cells, and ganglion cells (Figure 10.16). Nerve cells in the sensory retina are surrounded by Mueller cells, which supply nutrition to the nerve cells.

Light stimulates photoreceptors, and its sensory information is conducted finally to ganglion cells. In addition to conduction along this lengthwise direction, there is another direction, namely, conduction in the breadthwise direction. In the outer plexiform layer, there are horizontal cells, and in the inner plexiform layer, there are Amacrine cells. These cells are involved in the process of inhibition in order to increase visual contrast, and are also related to the central–peripheral mechanism of vision. In addition, the connecting cells, that is the reticular interlayer cells, located between the inner and outer plexiform layers, are related to the function of the feedback mechanism in the retina.

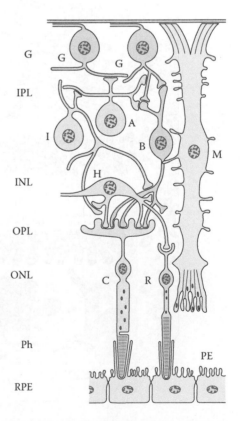

FIGURE 10.16 Schema of retinal neuron and Mueller cell: (A) Amacrine cell, (B) bipolar cell, (C) cone photoreceptor cell, (G) ganglion cell, (H) horizontal cell, (I) inner plexiform cell, (M) Mueller cell, (PE) pigment epithelium, and (R) rod photoreceptor cell.

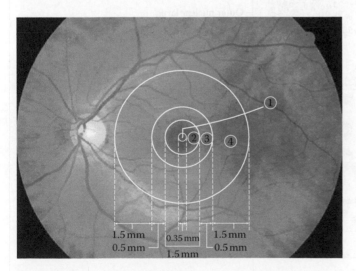

FIGURE 10.15 Anatomical definition of macula area. (1) Foveola, (2) fovea, (3) parafovea, and (4) perifovea.

Except for the macula region, the retina can be divided into the following 10 layers (Figure 10.17): from the outer side (choroidal layer), retinal pigment epithelium layer; visual cell layer, that is photoreceptive layer (rods and cones, and the inner and outer segments of visual cells); outer limiting membrane (adherent belt connecting the inner segments of visual cells and Mueller cells); outer granular cell layer (assembly of the nuclei of visual cells), outer plexiform layer (synapses and processes of visual cells, bipolar cells, and horizontal cells); inner granular layer (assembly of nuclei of bipolar cells, horizontal cells, and Mueller cells); inner plexiform layer (synapses and processes of bipolar cells and ganglion cells); ganglion cell layer (assembly of nuclei of ganglion cells), optic nerve fiber layer (layer of axons of the ganglion cells); and inner limiting layer (basement membrane of Mueller cells).

In the region of the central fovea that is the center of the macula, the retina is the thinnest part, and the region is occupied by the inner and outer segments of visual cells and the outer granular layer. In the surface layer, a small number of fibers from the outer granular layer is observed (Figure 10.17). Axons from visual cells in the central fovea radiate on the surface of the retina to form a Henle fiber layer.

10.11.3 Visual Cells

Visual cells are differentiated nerve cells that perceive light stimulation, and according to its structures of the outer segment that contain visual substances, visual cells are classified into rods and cones (Figure 10.18). Rod cells are about 92 million in number, and they are very sensitive to weak light intensity. Cone cells are about 5 million in number; they respond to strong light, increase visual acuity, and are directly related to color vision. Cone cells are classified into three types: blue cones, green cones, and red cones. Among the total cone cells in the retina, blue ones occupy 10–15%, green ones 50–54%, and red ones 33–35%.

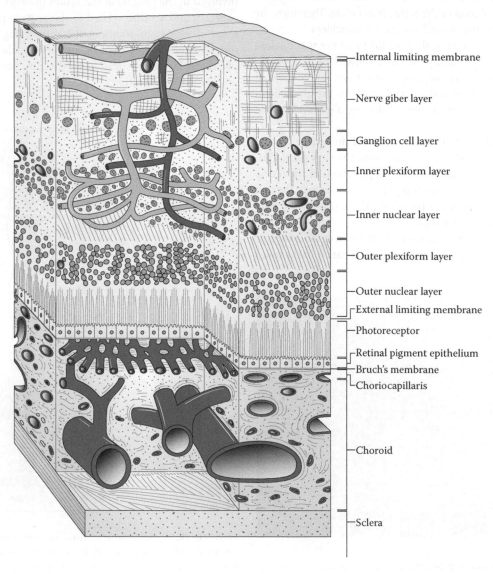

Internal limiting membrane
Nerve giber layer
Ganglion cell layer
Inner plexiform layer
Inner nuclear layer
Outer plexiform layer
Outer nuclear layer
External limiting membrane
Photoreceptor
Retinal pigment epithelium
Bruch's membrane
Choriocapillaris
Choroid
Sclera

FIGURE 10.17 Ten layers of retina.

Vision

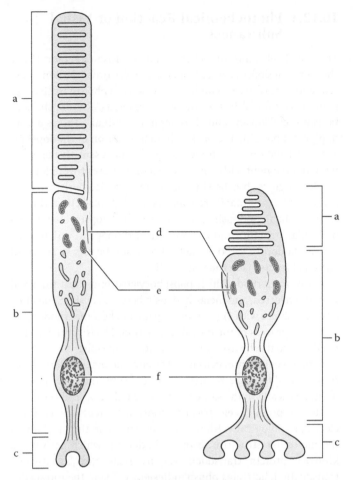

FIGURE 10.18 (Left) Rod photoreceptor cell and (right) cone photoreceptor cell: (a) outer segment, (b) inner segment, (c) synaptic terminal, (d) cilium, (e) mitochondria, and (f) nucleus.

The visual cell is composed of outer segment, inner segment, cell body, and nerve fiber running on the surface of the inner most layer of the retina. The outer segment and the inner segment are located between the retinal pigment epithelium and the outer limiting membrane. The outer segment and the inner segment are connected by a connective cilium.

The outer segment is composed of tight stacks of numerous disk membranes. In the outer segment of the cone cell, a disk membrane is connected to the outer enveloping cell membrane of the outer segment. In the rod cell, however, the disk membranes are not connected to such membranes, and they are independent from the cell membranes. A new disk membrane is formed at the base of the outer segment. Old disk membranes are pushed gradually to the tip of the outer segment and finally are removed from the outer segment, phagocytosed by the pigment epithelial cells.

The outer segment, where the disk membranes are tightly stacked, contains a huge amount of visual substance. Light passes through the numerous disk membranes, and the more the disk membranes that are tightly stacked, the more the probability of light that hits the visual substance. In the rod outer segment, about 90% of incoming light of 500 nm wavelength is absorbed by the visual substance.

The inner segment synthesizes the necessary substance of the outer segment. Mitochondria in the inner segments supplies the necessary energy for conduction of light information.

10.11.4 Bipolar Cells

Bipolar cells as secondary neurons accept light information from the visual cells and send it to Amacrine and ganglion cells. The bipolar cell makes the connections with the rod and cone cells. Bipolar cells are divided into two types: rod bipolar cells and cone bipolar cells. A rod bipolar cell near the fovea centralis makes synapses with 17–18 rod cells. In addition, a few bipolar cells make synapses with a single ganglion cell in this region. Such a convergence of information is more frequently observed in the peripheral area of the retina. For example, information from more than 100 rod cells makes a connection with a single ganglion cell. On the contrary, near the fovea centralis, the cone cell, rod cell, and ganglion cell correspond as 1:1:1, respectively. This means higher visual resolution acuity.

10.11.5 Horizontal Cells and Amacrine Cells

The main route of visual information is as follows: visual cells, bipolar cells, ganglion cells, and, in addition, several other kinds of related interneurons. These interneurons horizontally elongate their processes in the retina.[10] They integrate information from groups of closely located visual cells.

Horizontal cells elongate their processes to visual cells to neighboring cells, and they make synapses together with bipolar cells. Horizontal cells function as the inhibitory feedback system to visual cells, and they also act inhibitorily to bipolar cells.

Most Amacrine cells are localized in the inner side of the inner granular layer, but a few cells locate in the ganglion cell layer, which is called the ectopic Amacrine cells. Amacrine cells function as the interface among ganglion cells.

10.11.6 Cells of Interplexiform Layers

Cell bodies are located in the inner granular layer, and they elongate their processes into outer and inner plexiform layers. In the inner plexiform layer, cell processes have a role as postsynaptic endings. In the outer plexiform layer, however, cells have a role as presynaptic endings. It can be said that the cells are quite unique cells, because they send information ain the ntidromic direction from the inner plexiform layer to the outer plexiform layer.

10.11.7 Ganglion Cells

Ganglion cells are large multipolar neurons, and their total number in the retina is approximately 1 million. The cells are innervated with bipolar cells and Amacrine cells. Retinal ganglion cells send visual information to the central nervous system via the optic nerve fibers. Optic nerve fibers run on the surface of the retina, and they form the optic nerve fiber layer. Optic nerve fibers are, in general, unmyelinated in the retina; sometimes, however, myelinated optic nerve fibers are observed, particularly in the area near the optic disk.

10.11.8 Mueller Cells

Mueller cells are supporting glial cells and they fill the spaces among the neurons that are related to neuroglial interaction. They function in the protection of neurons, supplying nutrition to the neurons, insulation of other signals, and restoration to the injury. Mueller cells themselves are connected to one another with the adherence junction to form the outer limiting membrane. On the other hand, basal surfaces of the cells spread in the innermost retinas to form the inner limiting membranes.

10.11.9 Pigment Epithelial Cells

Retinal pigment epithelial cells, which form a monolayer, are located at the outermost layer of the retina. The cells are connected with one another by well-developed tight junctions, so that they protect to penetrate the blood plasma from the choroid layer (the outer retinal barrier to outer blood).

Facing the tips of visual cells, there are numerous microvilli, and they surround the outer segments of visual cells for protection. A single retinal pigment epithelial cell covers about 20 visual cells. Between the visual cells and pigment epithelial cells, there is no adherent structure. Therefore, retinal detachment occurs in this layer.

The retinal pigment epithelial layer and Bruch membrane are tightly adhered. The Bruch membrane consists of five layers: from the retinal side, basement membrane of the retinal pigment epithelial layer, inner collagen fiber layer, elastic fiber layer, outer collagen fiber layer, and basement membrane of endothelial cells of the blood capillary in the choroid layer.

In the pigment epithelial cell, there are numerous melanin granules and phagosomes derived from disk membranes of the outer segments of visual cells, and lipofuscin is produced by phagocytosis of the outer segment of visual cells and the final products of metabolic processes.

10.12 Function of the Retina

Visual cells are divided into two groups: the first one is related to light and dark vision, and the second one is related to color vision. Visual substances of the rod cells and three kinds of cone cells are different at the molecular level. These substances belong to a kind of G-protein coupling receptor of approximately 40,000 molecular weight. The molecule penetrates the membrane seven times at the disk membrane, and 11-*cis*-retinal is involved in the molecules (Figure 10.19).

10.12.1 Photochemical Reaction of Visual Substances

The visual substance of rod cells shows a rose red color; hence the name rhodopsin comes from the Greek words "rhodon" (rose red) and "optos" (see). Rhodopsin, which is a photosensitive pigment, is contained in the rod outer segments. Rhodopsin contains protein "opsin" and chromophore "retinal." This is a kind of pigment protein. The peak of light absorption of rhodopsin is at around 500 nm, which means it is the most sensitive to green light, but red light with longer wavelength passes through it.

When light is absorbed by rhodopsin, photoisomerism occurs. Namely, 11-*cis*-retinal is transformed to all-*trans*-retinal (Figure 10.19). When rhodopsin is isomerized, rhodopsin changes its molecular structure and becomes meta-rhodopsin, which is the active rhodopsin. Active meta-rhodopsin begins a series of chemical reactions in the visual cell.

Active meta-rhodopsin is phosphorylated in a short time and retinal is released, and meta-rhodopsin becomes opsin. Soon after, opsin combines with 11-*cis*-retinal again and becomes rhodopsin.

In this process of the chemical reaction, 11-*cis*-retinal is supplied to visual cells from pigment epithelium cells.

Pigment epithelium cells have the enzyme that isomerizes all-*trans*-retinal to 11-*cis*-retinal. The visual substance of cone cells occurs basically in the same chemical reactions as rod cells.

The cell produces electrical response with hyperpolarization, electrophysiologically. Rhodopsin absorbs light; then it changes the molecular structure to meta-rhodopsin, which activates a kind of G-protein "transducin" (Figure 10.20). Next, as activated transducin, it activates phosphodiesterase. Then, the intracellular concentration of cGMP decreases. Consequently, Na^+ channels that are dependent on cGMP close. The Na^+ channel is open in the dark. Therefore, the membrane potential recorded intracellularly by electrophysiology becomes more negative. A single meta-rhodopsin molecule activates several hundred molecules of transducin, and then each transducin activates phosphodiesterase, which degrades several hundred molecules of cGMP per second. By this very high amplification of molecular activation, a rod cell can respond by a single photon. Ca^{2+} can pass freely through Na^+ channels. If the Na^+ channels close the gates, the concentration of Ca^{2+} also decreases. Then, the guanylate cyclase that is inhibited by Ca^{2+} is activated, and cGMP synthesis is accelerated. As a result, the channel is opened in the dark before light stimulates the visual cell.

10.12.2 Cone Cells

Because cone cells are smaller in number than rod cells and because they are smaller in volume, the extraction of visual substance is very difficult. Measurement with a microspectral photometer, however, revealed the absorption spectrum of visual substance in the outer segments. The result indicated that peak absorption exists at 430, 530, and 560 nm, respectively, which means there are three different types of cone cells, namely, blue, green, and red.

Cone cells respond to light in the same manner as rod cells, namely, the membrane receptor potential recorded intracellularly becomes more negative to light stimulation. In cone cells, the life span of activated visual substance is too short; therefore, only when groups of many cone cells are simultaneously stimulated with light, a clear hyperpolarization potential can be observed. As a result, the amplification rate is low, but the responses are not sustained longer; hence they can detect the fast change of brightness.

When light enters the eye, nearly 4% of light intensity is lost by reflection on the surface of the cornea. However, the rest of the light intensity goes through the cornea, aqueous fluid, lens, and vitreous body sequentially along with the slight absorption of light in each site and finally reaches the retina. When the retina is activated by light, their activation is conducted (or transmitted) to bipolar cells, retinal ganglion cells and to optic nerve axons. The optic nerve form optic pathway which connect to the pretectal nucleus of the superior colliculus where its activity is finally transmitted to parasympathetic nerve to control the pupillary sphincter muscle that works for myosis.

10.12.3 Integration of Visual Information in the Retina

The visual cell stimulated by light responds to hyperpolarization of the receptor potential, and the cell decreases the release of glutamic acid from the synaptic terminal of the visual cell. Bipolar cells that receive glutamic acid at the synaptic site from visual cells are classified into two types: the first type of bipolar cell responds to the depolarization (on-type) of the receptor potential; the second type responds to hyperpolarization (off-type), which is the reverse direction of the first type. Both types are, therefore, different kinds of receptors to glutamic acid.

The visual cell, bipolar cell, and horizontal cell produce only slow receptor potential at the synaptic site when the eye is stimulated with light. The ganglion cell, however, only responds to impulses in the retina. The ganglion cell is the output of visual information to the central nervous system. The ganglion cell has the long axon sending impulses without any decrease in amplitude. The retinal area in which a single ganglion cell responds with impulses is called the receptive area.

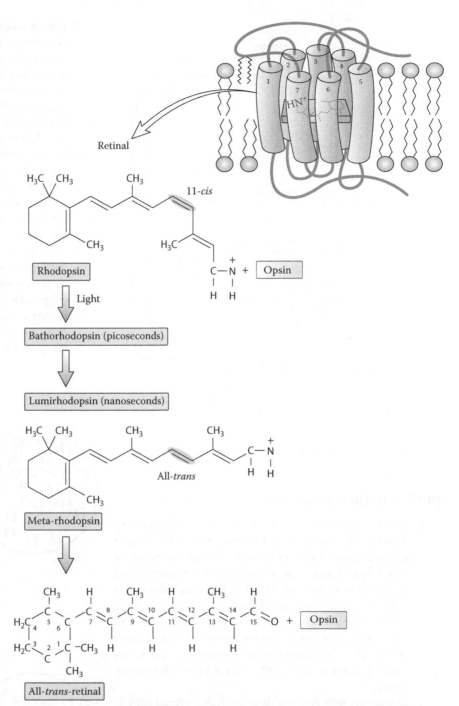

FIGURE 10.19 Photochemical reaction of rhodopsin.

In the receptive area, the responses of the central part and the peripheral region within the receptive area are reversed with respect to each other. If the central part is illuminated, the impulse frequency increases; however, if the peripheral region is illuminated, the impulse frequency decreases. This type of response pattern is called the on-center type (Figure 10.21). In addition, there is another type that shows the reverse response pattern. This is called the off-center type (Figure 10.21). On the formation of such a receptive area, interneurons such as horizontal cells and Amacrine cells have important roles.

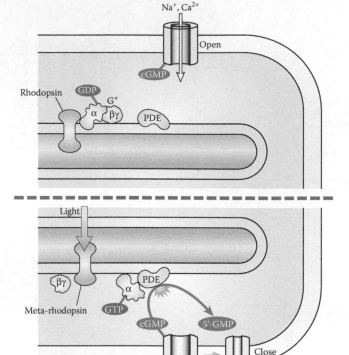

FIGURE 10.20 Activation of transducin.

10.12.4 Retinal Diseases

1. *Retinal congenital disorder:* Retinopathy of prematurity (ROP) is a disease affecting the retina of premature infants and is a major course of childhood blindness, which is caused by proliferative neovascularization.[14] Other retinal congenital disorders are Von Hippel–Lindau disease (a well-known retinal disease characterized by hemangioma derived from retinal blood capillary), Leber disease (one of the major four causes of infantile low vision and based on guanylic acid synthetase gene mutation), and Coats disease (exudative retinopathy caused by retinal vascular ectasia).[15]

2. *Retinopathy with collagen diseases:* SLE:—retinal pathology occurs in 20–30% of SLE patients and is in most cases, binocular. This disease is characterized by cotton wool spots and retinal hemorrhage.

3. *Infectious disease:* Herpes virus, HSV, varicella-zoster virus, and cytomegalovirus are the major causes.

4. *Macular diseases:* Age-related macular degeneration (AMD). This disease is one of the major causes of blindness in middle-aged and aged populations. Age-related degeneration in Bruch's membrane causes choroidal neovascular, exudate hemorrhage. Photodynamic therapy (PDT), intravitreal injection of VEGF antibody, or both are used for the treatment.

5. *Retinal detachment:* This includes rhegmatogenous retinal detachment (tear of the retina is caused by vitreous traction, and results in retinal detachment) and exudative retinal detachment (exudative fluid accumulates in subretinal space because of functional disorders in choroid retinal pigment epithelium and retinal vascular systems). Causative disorders include Harada's disease, and tractional retinal detachment is included.[9]

6. *Retinal tumor:* This includes retinoblastoma (infantile tumor with parenchymal mass with calcification; enucleation of the eye is applied when extraocular invasion is recognized, otherwise conservative treatment is selected), cavernous hemangioma, and tumors of the retinal pigment epithelium.

7. *Cancer-associated retinopathy (CAR):* Autoimmune retinal degeneration is caused by cancers and is characterized by photophobia and visual loss. Autoimmune response is triggered by cross antigenicity of tumor-derived abnormal proteins and normal retinal proteins. In some cases, autoantibodies against recoverins that exist in the photoreceptor cell are detected.

(a)

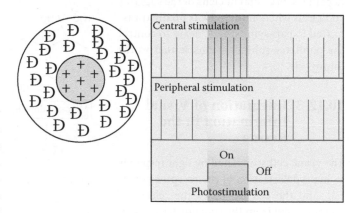

(b)

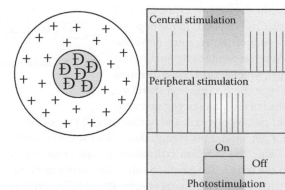

FIGURE 10.21 Response pattern of the retinal ganglion cell. (a) On-center, off-surround type and (b) off-center, on-surround type.

10.13 Uvea and Ciliary Body

10.13.1 Uveal Tract

10.13.1.1 Iris

The iris is located at the position between the cornea and lens, and it separates the anterior and posterior chambers filled with aqueous humor. At the center of the iris, there is a pupil where light passes into the eye.

The size of the pupil changes according to light intensity, namely, in dark conditions, the size becomes larger by the contraction of the sphincter muscle, and in bright conditions, the pupil becomes smaller by the dilator muscle. Changes in pupil size have a functional role like the diaphragm of a camera. The diameters of the pupil are different according to age, but, in general, are 2–4 mm.

The iris shows various colors according to human race, such as bright blue and brown. The colors are dependent on the amount of melanin pigment, which is contained in tissues of the substance and its anterior region. Because albinos completely lack the melanin pigment in melanocyte and because to the hemoglobin in blood is red, the iris and retina are seen as red, and hence their eyes look like as red eyes.

10.13.2 Histological Anatomy of the Iris

The iris is divided into following the parts: anterior border, stroma, and posterior border of the iris (Figure 10.22).

- *Anterior border of the iris:* In the frontal surface, the anterior border of the iris contacts with the aqueous humor and has neither epithelium nor endothelium. Therefore, various kinds of macromolecules can easily pass through and enter the stroma (Figure 10.23).
- *Stroma of the iris:* The stroma has plenty of blood vessels and it is rather sparse and hydrophobic connective tissue. Melanocytes, fibroblasts, macrophages, mast cells, and a clump of pigment cells are observed. Endothelial cells of the blood vessels in the iris, including capillaries, have no fenestration, and cells are connected with tight junction and desmosome; thus they function as a barrier.
- *Posterior border of the iris:* In the posterior part of the iris, there are two layers of pigment epithelia: (1) anterior

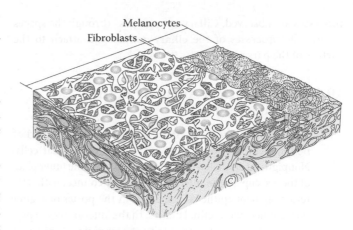

FIGURE 10.23 Anterior border of the iris.

pigment epithelium (derived from muscle) and (2) posterior pigment epithelium and sphincter muscle and dilator muscle (Figure 10.24).

10.13.3 Ciliary Body

10.13.3.1 Anatomy of the Ciliary Body with Naked Eyes

The ciliary body is located between the iris and choroid, and is about 6 mm (ear side 6.5 mm and nasal side 5.5 mm). To the frontal direction, it elongates until the scleral spur, and to the posterior direction, it reaches the ora serrata.

The ciliary body is divided into two parts: the anterior region of 1/3 ciliary body is called pars plicata, and the posterior 2/3 is pars plana. In pars plana, there are a few blood vessels located in the anterior region of the eyes. In pars plicata, 70–80 ciliary

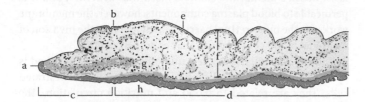

FIGURE 10.22 Axial section of the iris: (a) pigment ruff of pupillary margin, (b) collarette, (c) pupillary portion of the iris, (d) ciliary portion of the iris, (e) anterior border, (f) stroma, (g) sprinter muscle, and (h) posterior border of iris.

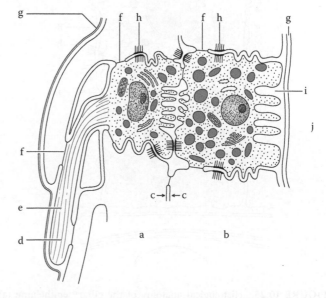

FIGURE 10.24 Posterior border of the iris. (a) Anterior retinal pigment epithelium (RPE), (b) posterior RPE, (c) apex, (d) basal process, (e) dilator muscle, (f) tight junction, (g) basal lamina, (h) desmosome junction, (i) basal infolding, and (j) posterior chamber.

processes are observed. Ciliary zonules pass through the spaces among the processes of the ciliary zonula and attach to the surface of the pars plicata.[16]

10.13.4 Histological Anatomy of the Ciliary Body

1. *Ciliary epithelium:* The ciliary epithelium is composed of nonpigment epithelial cells and pigment epithelial cells. Nonpigment epithelial cells are derived from the inner plate of the eye cup. In the anterior region they connect with posterior pigment epithelial cells, and in the posterior region they connect with retinal nerves. In the anterior region pigment epithelial cells contact the anterior pigment epithelial cells of the iris, and in the posterior regionthey contact the retinal pigment epithelial cells (Figure 10.25).

   The basal lamina of the nonpigment epithelial cells face the posterior chamber, the basal part of the vitreous body, and the basement membrane of the pigment epithelial cells face the parenchyma of the ciliary body.

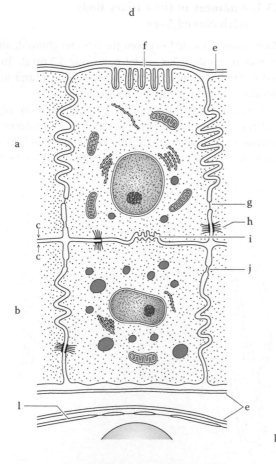

FIGURE 10.25 Histological anatomy of the ciliary epithelium: (a) nonpigment epithelial cell, (b) pigment epithelial cell, (c) apex, (d) posterior chamber, (e) basal lamina, (f) basal infolding, (g) tight junction, (h) desmosome, (i) ciliary canaliculus, (j) gap junction, (k) ciliary stroma, and (l) fenestrated capillary.

Based on the structure mentioned above, the tips of both cells face each other, and there is a small space called the ciliary channel. The nonpigment epithelial cells of pars plicata actively function in the transportation of water and various ions and produce the secretory substance called aqueous humor. In the aqueous humor, the concentration of protein is about 1%, and it is transparent without any color.

Pigment epithelial cells exist between nonpigment epithelial cells and the parenchyma of ciliary body substance. They form a monolayer on the basal lamina, which is connected with Bruch's membrane. The basal lamina has many folds and the cells there are actively related to transportation of ions. Pigment epithelial cells contain a huge number of melanin granules.

2. *Ciliary stroma:* The ciliary stroma is composed of connective tissue, blood vessels, and muscles of the ciliary body. Ciliary muscle is made of smooth muscle and originates from the tendon of the ciliary muscle of scleral spur.

10.13.5 Choroid

10.13.5.1 Anatomy of the Choroid by the Naked Eye

The choroid is located between the retina and the sclera, and it constitutes the middle layer of the eye from 0.1 mm (from the anterior end) and 0.3 mm (from the posterior end); it contains many blood vessels and melanocytes. The choroid is located behind the uveal tract, and in the frontal region it is transitive to the parenchyma of the ciliary body. In the posterior region, it disappears at the optic disk. From the region where there are optic nerves, the choroid attaches firmly to the sclera.

10.13.6 Histological Anatomy of the Choroid

Histologically, choroid is divided into four layers: from the retinal side, Bruch's membrane, the choriocapillaris, vascular layer, and suprachoroid (Figure 10.26).[8]

Bruch's membrane is 2–4 mm thick. It is like a vitreous body with homogeneous appearance and no structure, and it is positive to periodical acid-schiff (PAS) stain. It is composed of five layers: basement membrane of retinal pigment epithelium, inner collagen fiber layer, elastic fiber layer, outer collagen fiber layer, and basement membrane of endothelial cells of the choriocapillaris. Bruch's membrane shows a mesh-like structure, and it is permeable to blood plasma components; however, the membrane works for the outer retinal barrier which prevents the invasion of choroidal blood cells and vascular endothelial cells.

The choriocapillaris is a few micrometers in thickness and is monolayered blood vessel capillaries with flattened fenestrated cells. They are related to blood and nutrition circulation. The choriocapillaris supplies blood from the outer granular layer to the retinal pigment epithelium. The choriocapillaris is the biggest capillary in the whole body (Figure 10.27).

The vascular layer make up the majority of choroid. In order to keep the light coming inside the eye from other stray light, the

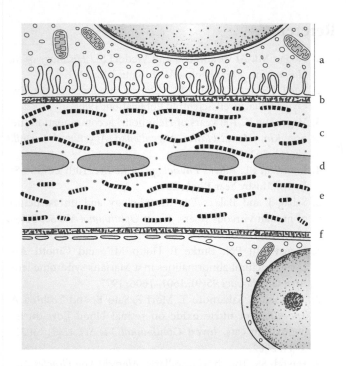

FIGURE 10.26 Layers of Bruch's membrane: (a) retinal pigment epithelium, (b) basal lamina of the retinal pigment epithelium, (c) inner collagenous zone, (d) elastic layer, (e) outer collagenous zone, and (f) basal lamina of the choriocapillaris.

vascular layer functions as a black box. Moreover, many of the middle- and large-sized blood vessels, particularly 2/3 of the outer layer, contain melanocytes.

The suprachoroid is the transitional region between the choroid and the sclera, and collagen fibers and elastic fibers are the matrix components.

10.13.7 Structure of Blood Vessels in the Choroid

The characteristic property of the structure of blood vessels in the choroid is that the inflow route (artery) and the outflow route

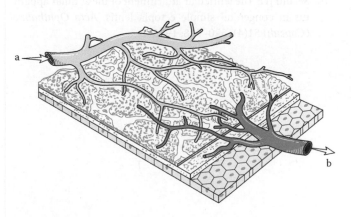

FIGURE 10.27 Lobular structure of the choriocapillaris: (a) choroidal arteriole and (b) choroidal venule.

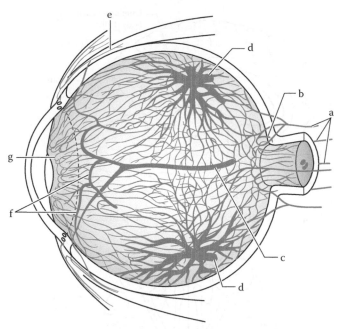

FIGURE 10.28 Schema of uveal vessels: (a) short posterior ciliary arteries, (b) circle of Zinn–Haller, (c) long posterior ciliary arteries, (d) vortex vein, (e) anterior ciliary arteries, (f) major arterial circle of the iris, and (g) minor arterial circle of the iris.

(vein) are separated. The inflow route starts from the most posterior part of the eye, which is called the short posterior ciliary artery (SPCA). This represents the U-turned branches of the long posterior ciliary artery (LPCA) and the anterior ciliary artery (ACA). They are all branches of the eye artery. The outflow route begins from the voltex vein near the equator area (Figure 10.28).

10.13.8 Main Disease

Iritis: collective disease of inflammation in the iris, ciliary body, and choroid. Typical diseases are Behcet's disease, Harada's disease (Vogt–Koyanagi–Harada disease), and sarcoidosis, all of which are accompanied by constitutional symptoms. Definitive diagnosis needs ophthalmological and general examinations.

10.14 Optic Nerve

The optic pathway spans from the optic nerve to the occipital visual center (Figure 10.29). The optic nerve fiber is the axon of the retinal ganglion cell, and goes through the optic disk and lamina cribrosa. There, they leave the eyeball and become optic nerves. The total number of optic nerve fibers is approximately 1–1.2 million, and within the eyeball they are unmyelinated. Once they leave the eye, they all become myelinated so that the nerve signals (impulses) conduct with much higher speed than in the retina. It is known that central nervous system glia, namely, oligodendrocytes and astrocytes, are inhibitory for axonal regeneration, while peripheral nervous system glia Schwann cells are very supportive to the regeneration. Because the optic nerve is

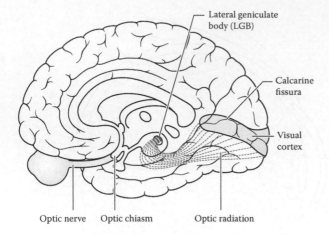

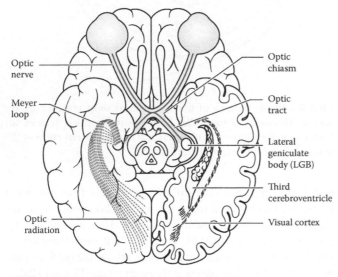

FIGURE 10.29   Optic pathway.

myelinated by oligodendrocytes but not by Schwann cells, RGC axons mostly do not regenerate after damage.

An optic disk is about 1.5 mm in diameter and shows round or standing-oval shapes. In the retrobulbar portion, however, it is about 3 mm in diameter, which increases from that of the retrobulbar portion. This increase is attributed to the myelination of optic nerve fibers and also to the formation of septum, which consists of connective tissue.

References

1. Hogan MJ, Al varado SA, and Weddcl SE. *Histology of the Human Eye*. Philadelphia: W. B. Saunders; 1971.
2. Snell RS and Lemp, MA. *Clinical Anatomy of the Eye*. Oxford: Blackwell Scientific Publication; 1989.
3. Vogelsang K. 100 years of Helmholtz' accommodation theory. *Klin Monatsblatter Augenheilkd Augenarztl Fortbild* 126(6): 762–765; 1955.
4. Kaufman PL. *Adler's Physiology of the Eye*, 10th Edition St. Louis: Mosby Year Book; 2003.
5. Sebag J and Balazs EA. Morphology and ultrastructure of human vitreous fibers. *Invest Ophthalmol Vis Sci* 30(8): 1867–1871; 1989.
6. Farnsworth PN, Burke P, Dotto ME, and Cinotti AA. Ultrastructural abnormalities in a Marfan's syndrome lens. *Arch Ophthalmol* 95(9): 1601–1606; 1977.
7. Nagaoka T, Sakamoto T, Mori F, Sato E, and Yoshida A. The effect of nitric oxide on retinal blood flow during hypoxia in cats. *Invest Ophthalmol Vis Sci* 43(9): 3037–3044; 2002.
8. Hayreh SS. The choriocapillaris. *Albrecht Von Graefes Arch Klin Exp Ophthalmol* 192(3): 165–179; 1974.
9. Hayreh SS, Kolder HE, and Weingeist TA. Central retinal artery occlusion and retinal tolerance time. *Ophthalmology* 87(1): 75–78; 1980.
10. Miller NR. *Walsh and Hoyt's Clinical Neuro-Ophthalmology*. Philadelphia: Lippincott Williams & Wilkins; 2004.
11. Stephen JR. *Retina*. St Louis: Mosby; 2006.
12. Farnsworth PN and Shyne SE. Anterior zonular shifts with age. *Exp Eye Res* 28(3): 291–297; 1979.
13. Ritch R. *The Glaucomas II*. St Louis: Mosby; 1996.
14. Aiello LP, Bursell SE, Clermont A, Duh E, Ishii H, Takagi C, et al. Vascular endothelial growth factor-induced retinal permeability is mediated by protein kinase C *in vivo* and suppressed by an orally effective beta-isoform-selective inhibitor. *Diabetes* 46(9): 1473–1480; 1997.
15. Coats G. Forms of retinal disease with massive exudation. *Roy London Ophth Hosp Rep* 17: 440–525; 1908.
16. Seland JH. The lenticular attachment of the zonular apparatus in congenital simple ectopia lentis. *Acta Ophthalmol (Copenh)* 51(4): 520–528; 1973.

Electroreception

Robert Splinter

11.1 Introduction

It has been known for centuries that rubbing amber will generate an electrical charge. In Greek the word for "amber" is electron, which places the origin of the name for electricity around 600 BC. In addition to the visible impact of electrostatic forces, moving charges are recognized as having secondary effects that can be measured. As such, moving charges and the associated electric and magnetic fields, respectively, can be detected by biological organs that have a structure that accommodates the interaction. The biological detection of electric and magnetic fields is ranked as a separate sense called electroreception, sometimes also called electroception.

Various biological effects involve the exchange of electrical charge and hence produce a current and display electric and magnetic field effects. Examples are the action potential in both nerve signal conduction and muscle contraction (see Chapters 4 and 12). A single electric charge will, by definition, generate an electric field: terminating or originating on the charge (Figure 11.1). On the other hand, a magnetic field can also produce

electrical events. The magnetic induction of current or electrical potential also supports the detection of magnetic fields. The earth's magnetic field for example is used for guidance by several species (Figure 11.2).

Certain species have the ability to sense electrical and/or magnetic activity, respectively. The perception of the electrical activity will be a primary function of the magnitude of the signal and, hence, will be more pronounced in animals that have a conducting environment, such as aquatic creatures. The liquid environment in oceans and other salt water is approximately $2 \times 10^{+10}$ more conductive than air, and as such, air is a highly unlikely medium to accommodate animals that use electrical detection as a means of sensing.

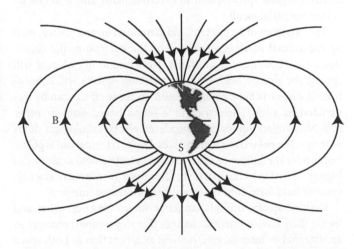

FIGURE 11.2 The Earth's magnetic field lines running from the South Pole to the North Pole. The convention is to call the "North Pole" the direction in which the north of the magnetic dial is directed.

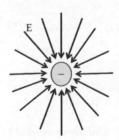

FIGURE 11.1 Electric field lines terminating on negative point charge.

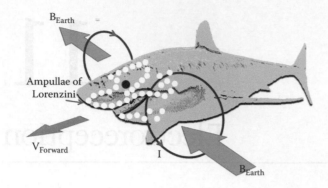

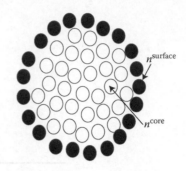

FIGURE 11.3 Sharks sense for the indirect influence of electric and magnetic fields by means of passive perception resulting from electrical induction in dozens of Ampullae of Lorenzini acting as induction coils.

FIGURE 11.4 Aggregation of dielectric particles forming column-like structures in an electrorheological liquid with charged core particles (n^{core}) surrounded by charged surface particles ($n^{surface}$).

On the other hand, magnetic detection does not require a highly conductive medium. Sharks, amongst other animals, use the earth's magnetic field to guide them as they move in the spatial magnitude distribution of the magnetic field. Moving through the changing magnetic density as a function of location, the shark can orient itself (Figure 11.3). The shark also senses prey based on the sensation of both changing magnetic and electric field lines resulting from muscular motion.

11.2 Biological Detection of Electrical and Magnetic Fields

The generation of an action potential is processed within a biological entity by various means. The most well-known effect of an action potential is the release of chemicals, which in turn generate a secondary action potential in a neighboring cell. The electric potential inherently produces an electric field as described later in this chapter. It is not hard to imagine that specialized organs can respond to external fields and generate an action potential locally.

The existence of an electric field generated by any source, even by the animal itself, is influenced by the environmental conditions (i.e., boundary conditions). Nonconducting objects will spread the electric field, while conducting objects will concentrate the electric field; changes based on these effects can be recognized as such. Dielectrics in a blood-based fluid or other non-Newtonian liquids are considered electrorheological fluids with specific polarization properties. Dielectric material is polarized under the influence of an external electric field as shown in Figure 11.4. Dielectric particles dissolved in a nonconducting viscous fluid form column-like aggregation structures.

Certain aquatic animals can sense other living organisms and locate their environmental changes due to perceived changes in an external or internal electric field as a function of both space and time. For instance, the platypus and the electric eel communicate and sense with the use of an electric field that is produced by the animal itself[1-3] (Figure 11.5).

11.2.1 Receptor Mechanism

Two types of electroreception can be distinguished: passive and active.[2, 3]

In the passive detection, the electrical and magnetic fields and impulses are generated by external sources and received by specialized sensory devices. In the active electroreception, either electrical or magnetic fields are generated by specific signaling sources and are monitored for deviations as well as time- and place-dependent variations by sensors specifically adapted to handle the subtle changes.

Both for the passive and active mechanisms of action, specific elementary physics principles will apply.

11.2.1.1 Passive Electroreception

Passive electroreception relies on electrical signals emitted by nearby inanimate and animate sources other than the animal itself.

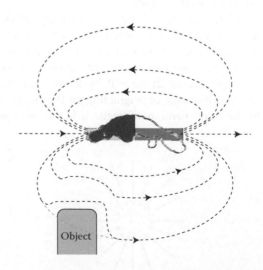

FIGURE 11.5 Electric filed lines generated by the platypus, allowing the animal to sense the environmental geometry due to changes in the field lines induced by conducting and nonconducting objects.

11.3 Sensor Design

The sensors for electroreception can be catagorized into two groups:

- Pulse receptors, which uses frequency as a measure of signal strength
- Burst receptor, which relies on amplitude as an indication of the magnitude of the electric field properties

11.3.1 Active Electroreception

In active electroreception, the animal monitors the electrical discharges emitted by the animal itself. The discharge of electrical energy can be in a continuous fashion or in intermittent sequences.

11.3.2 Animals with Electroreceptor Sense

Various animals use one or more forms of electrical and magnetic sensing to identify their location and the migration pattern and locate prey.[3-5]

Examples of animals with passive and active electroreception, respectively, can be identified as follows.

11.3.2.1 Animals with Passive Electroreception

Examples of animals that have passive electroreception are: catfish, sharks, sting-ray, platypus, lamprey, lungfish, electric fish, and various other animals (Figures 11.3 and 11.5).

The sensor mechanism is primarily epithelial tissue that has been modified to perform a sensory function.

11.4 Electro-Sensory Perception

In sharks the detection of electric and magnetic field changes is made possible by a specific sensory element called the Ampullae of Lorenzini (Figure 11.3). The Ampullae of Lorenzini are jelly-filled organs with alveoli clusters. The Ampullae were discovered in the late eighteenth century by Stephan Lorenzini. All alveoli are lined with nerve cells attached to individual receptor cells. The receptor cells detect time-dependent electric and magnetic fluctuations based on the Faraday principle of electromagnetic induction. The sensing mechanism is highly sensitive due to the positive feedback mechanism that is delicately balanced at threshold, technically operating as a differential amplifier. In fact, there are dozens of differential amplifiers that are self-regulating (they compensate for steady-state conditions) and that work in series with each other. The response time is less than most man-made differential amplifiers, but the cascade effect makes up for this insensitivity. In addition to the changing field resulting from the motion of the prey, conducting objects also produce changes in the earth's magnetic field lines and as such are sometimes mistaken for prey. Sharks have been known to swallow cans and license plates, presumably based on their electric properties. The shark sense is known to be able to detect

changes in electric fields are as small as 1×10^{-10} V/m. This allows sharks to detect muscle activity at a distance of several meters.

The Ampullae of Lorenzini are of the burst receptor category and respond to the dipole potential generated by the activity of the prey in addition to the earth's magnetic field during motion.[6,7] The bio-electric potential (V_d) generated by a dipole moment resulting from a potential gradient across a segment of the body of the fish is expressed as

$$V_d = \frac{p_{prey}\cos(\theta)}{r^2}, \tag{11.1}$$

where θ is the angle between the dipole axis and the detector location, and r the distance between the dipole and the detector. The dipole moment generated by the electrical activity of the prey p_{prey} is defined in vector format as:

$$\overrightarrow{p_{prey}} = Q\vec{d}, \tag{11.2}$$

where Q is the dipole charge and $\vec{d}$ the direction and distance between the charges of the dipole. Note that the charge can be the transmembrane charge, which changes during depolarization.

The electric field (E_{dipole}) generated by a dipole of the prey is expressed as

$$E_{dipole} = \frac{1}{4\pi\varepsilon_0}\frac{p_{prey}}{r^3}, \tag{11.3}$$

where ε_0 is the dielectric permittivity ($\varepsilon_0 = 8.854187817$ A^2 s^4 kg^{-1} m^{-3}).

When the prey moves, both the distance to the source and the direction of the dipole moment would change.

The Ampullae of Lorenzini detect the change in electric field strength and/or direction, respectively, or the associated voltage drop when the prey moves.

In addition to passive electrical sensing, various other animals rely on the earth's magnetic field to orient themselves, such as birds, but humans have also shown rudimentary sensory interaction with the earth's magnetic field.

11.4.1 Animals with Active Electroreception

The electric eel is the primary example in this group; other examples are other marine animals such as electric skates and other weak and strong electric fish. The electric eel uses its electric charge as a weapon in addition to as a means of communication. The sensor technology in this class is frequently designed of some sort of gel-filled tubular organ and has a sensory range from 30 Hz to several kHz. The platypus, skate, and sting-ray each, respectively, senses changes in the contours of its environment due to the induced changes in the magnetic as well as electric field lines surrounding its body (Figure 11.5). The platypus will detect changes in the pattern of the electric field it generates itself. These changes will help identify prey or recognize geometric changes in the direct environment of the platypus.[8]

11.5 Electrosensory Mechanism of Action

Electroreception is based on the detection of electric and magnetic fields, either produced by the animate object (the animal) or an inanimate object by means of specialized sensory organs. The electric and magnetic fields, respectively, are generally the result of moving charges. The electric charge will move under the influence of an external force, which in turn is associated with an electric field. A moving charge by definition provides a current, which will in turn generate a magnetic field.[9-11] The moving charges can be depolarization charges from muscle contraction or polarization of organs designed as capacitor plates.

11.5.1 Electrical Principles

Before moving to the electrical foundations that allow sensors to record the influence of electric and magnetic fluctuations, a few basic principles and definitions in electromagnetics are discussed.

Electric current, denoted as I, is defined as the rate at which an incremental amount of electric charges ($\mathrm{d}q$) pass through a cross-sectional area in a time interval $\mathrm{d}t$. The cross-sectional area can be arbitrary in space or within a confined volume of a material: for example, wire, tissue, or organ. The mathematical definition of current is shown in Equation 11.4:

$$I = \frac{\mathrm{d}q}{\mathrm{d}t}. \tag{11.4}$$

The current originates from the electrical potential difference between two end points resulting in a force on the charges exposed to the electrical filed. The electric field, and hence the electrical potential, is the result of the build-up of electric charges in one location, either artificially or naturally. The electric field ($\mathbf{E}$) is directly proportional to the accumulated charge (q) and inversely proportional to the square of the physical distance from the charge cloud (r) and the point of observation. The squared dependency with respect to distance is the result of the area–surface enclosing the charges.

The electric field is defined in vectorial form as follows:

$$\mathbf{E} = \frac{q}{4\pi\varepsilon_0 r^2}\mathbf{r}, \tag{11.5}$$

where ε_0 is the dielectric constant ($\varepsilon_0 = 8.85 \times 10^{-12}$ C^2 Nm^{-2}) of empty space and is also known as "permittivity of free space." The permittivity is a measure of the ease at which an electric field can penetrate through empty space. The vector $\mathbf{r}$ is a unit vector showing direction only. (Note: if $\mathbf{r}$ were a true vector with

direction and magnitude it would result in a factor $\mathbf{r}^3$ in the denominator, instead of $\mathbf{r}^2$.)

The force experienced by the charge ($\mathbf{F}_E$) is represented by the charge times the electric field resulting from a change in proximity, or as stated by Coulomb's law:

$$\mathbf{F}_E = \frac{q_1 q_2}{4\pi\varepsilon_0 r^2}\mathbf{r} = q_1\mathbf{E}, \tag{11.6}$$

where q_2 is the source of the electric field and q_1 is the test charge. The charge role can be reversed where the net magnitude of the Coulomb force remains the same; however, the direction of the field would change. This force is the driving mechanism of moving a charge particle and hence generates a current.

The voltage or potential difference (V_b) sustaining the charge separation is found from the work performed on a unit charge by the electric field caused by the other charge to separate them from point a to point b, where a and b represent the locations of the accumulated charges in space (better indicated in vector format), or the negative of the work performed on the charges:

$$V_b = -\int_a^b \mathbf{E}\cdot\mathrm{d}\ell = \int E\,\mathrm{d}\ell\cos(\alpha). \tag{11.7}$$

Electric potential is present during signal transmission in biological media as explained in Chapter 5, and can in principle be measured.

11.5.2 Magnetic Detection

Since the mechanism of action of magnetosensory detection usually reverts back to the electricity, the correlation between magnetism and electricity will be described next.

Another electronic phenomenon associated with moving charges (i.e., current) is the force acting on a moving charge when subject to an external magnetic field ($\mathbf{B}$). The moving charges produce the magnetic fields of their own, which in turn interact with the eternal magnetic field resulting from other charges. The net force on the charge then has two origins: electrostatic and magnetic. The electrostatic force is omnipresent between two charges and the magnetic force arises only when the charges are in relative motion. The total force as the vector sum of electrostatic and magnetic influences is known as the Lorentz force. The Lorentz force principle is described in Section 11.5.2.1.

11.5.2.1 Lorentz Force

When a charged particle is moving through an external magnetic field, it will experience a force. This phenomenon also applies to a medium filled with charged particles moving through a magnetic field. The charged particles will experience a force that will force the charges to move in a direction that is described by the Fleming "right-hand rule" (Figure 11.6). The

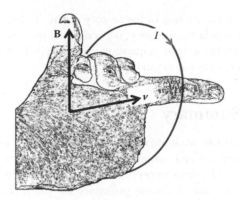

FIGURE 11.6 Fleming's "right-hand rule." An external magnetic field perpendicular to the motion of the medium will result in a current of moving charges that encircles the magnetic field.

resulting magnetic force is always perpendicular to the plane defined by the direction of the magnetic field and the direction of the velocity vector of the moving charge particle. In case the velocity of motion and the magnetic field are parallel or anti-parallel, the magnetic force vanishes, while it is maximal when the two are orthogonal with each other. This may explain why sharks zigzag in their motion. The case where the external magnetic field is perpendicular to the motion of the medium will result in a current of the moving charges that encircles the magnetic field.[9-12] Any angular interaction between the direction of motion ($\vec{v}$) and the external magnetic field ($\vec{B}$) yielding the direction of the force on the charged particle ($\vec{F}$) and hence the direction of the induced current is described in vector format.

The magnetic force ($\mathbf{F_B}$) on an individual moving charge is directly proportional to the magnetic field, which is perpendicular to the motion of the charge, the magnitude of the charge, and the velocity of the charge and is given by

$$\mathbf{F_B} = q\mathbf{v} \times \mathbf{B} = qvB\sin(\theta), \qquad (11.8)$$

where θ is the angle between the velocity and the magnetic field direction, B the magnetic field magnitude, v the speed at which the animal moves or the speed of the source. The Lorentz force ($\mathbf{F_L}$) is the vector sum of the electrostatic force and the magnetic force as expressed by Equation 11.9:

$$\mathbf{F_L} = \mathbf{F_B} + \mathbf{F_E}. \qquad (11.9)$$

The force on the charge will result in a change in direction of the charge and as such may provide a means of detection with specialized sensory equipment to the magnitude and direction of external magnetic fields, for example, the earth's magnetic field.

The consequences of the changing magnetic field are described by Faraday's law of induction.

11.5.2.2 Faraday's Law of Induction

Similarly to Gauss's derivation of the electric flux, Michael Faraday (1791–1867) proposed a definition of the magnetic flux around 1820/1821:

$$\Phi_B = \int \mathbf{B} \cdot d\mathbf{A} = \int B\, d\mathbf{A} \cos(\alpha), \qquad (11.10)$$

where α is the angle between the normal to the area and the magnetic field piercing through the surface and Φ_B the magnetic flux.

When a changing magnetic field is linked to a conducting loop (or for that matter even in free space), an electric current and consequently an electrical potential is generated by magnetic induction. A magnetic flux changing over time is found to induce an electrical potential, or "electromotive force" emf, measured in volts. A "volt" is equal to the work done on a positive test charge of 1 Coulomb magnitude moving from infinity (∞) to a point "A" where the electrical potential is measured.

$$\text{emf} = \frac{d\Phi_B}{dt}. \qquad (11.11)$$

The magnetic induction (resulting from changing magnetic flux) is also used in a generator where a coil is revolving in a static magnetic field, but due to the changing area captured by the coil the magnetic flux changes as a function of time, inducing a voltage that will make a bulb to glow (Figure 11.7). The changing magnetic flux expressed by Equation 11.10 can be accomplished by either a changing area, a changing magnetic field magnitude, or a change in the relative angle between the magnetic field and the (normal to the) surface of the area.

Consider a Cartesian coordinate system. The electric field in the positive y-direction and the magnetic field perpendicular to the electric field, in the z-direction, also propagates in the positive x-direction.

$$\text{emf} = -\frac{\Delta\Phi_B}{\Delta t} = B\frac{\Delta A}{\Delta t} = B\frac{h\Delta x}{\Delta t} = Bhv$$

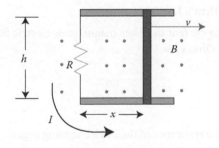

FIGURE 11.7 Diagram of magnetic induction by a magnetic field piercing through a closed loop. Expanding the area of the loop or changes in the magnitude of the magnetic field when a fixed area loop travels through an inhomogeneous magnetic field will induce a current in the loop.

The magnetic induction follows Lenz's law, which is a mere manifestation of conservation of energy. This conservation principle is equally applicable to the biological current used in the senor organ and relies on Lenz's law. Lenz's law is described in Section 11.5.2.3.

11.5.2.3 Lenz's Law

It was recognized in early stages of electricity and magnetism that nature tries to maintain a balance. In the case of induced current resulting from a changing magnetic flux, the current itself will produce a magnetic field according to the Biot–Savart law. The generated magnetic flux is opposite and equal to the magnetic field that induces the εmf creating the current. This phenomenon is referred to as Lenz's law.

One can also consider the generation of a current in the sensor organ without taking into account that an electrical potential is required. The use of either technique will depend on the exact mechanism of sensing that is involved in the inner structure of the sensor organ.

A more general, but special form, of the Biot–Savart law is Ampere's law. Ampere's law describes the relationship between the magnitude of an enclosed magnetic field generated by the respective quantity of closed loop electric current.

11.5.2.4 Ampere's Law

The current produced in specialized sensor elements arranged in a closed-loop configuration produce a magnetic field that can be derived by using Ampere's law. An induced current in the organ will be a function of the magnitude of the magnetic flux change and the motion of the charges that are part of the magnetic sensor organ of the animal. The magnetic field generated in the sensor unit by the current I is described by Ampere's law and is given by Equation 11.12:

$$\mu_0 I = \oint B \cdot d\ell = \oint B \, d\ell \cos(\alpha) \tag{11.12}$$

considering motion over a distance $d\ell$.

An induced current and associated εmf will in principle be able to produce an action potential and will be detected as such.

11.5.2.5 Ohm's Law

The resulting current in either magnetic or electric field induction I obeys Ohm's law:

$$I = \frac{\varepsilon mf}{R}, \tag{11.13}$$

where R is the resistance of the device/sensing organ.

The induced current in the sensory organ can be a true current or a depolarization wave, depending on the mechanism of transfer for the animal in question and the specific organ used, when more than one organ is present.

11.6 Summary

In this chapter, we discussed the ability of certain animals to detect electrical and magnetic influences. Several examples of receptor mechanisms were provided with exhibits of representative species. The electronic principles for the mechanism of detection were introduced along with the basic electric field theory supporting the mechanism of action of detection.

References

1. Evans DH and Claiborne JB. *The Physiology of Fishes*. Boca Raton: CRC Press, 2005.
2. Kalmijn A. Electric and magnetic field detection in elasmobranch fishes. *Science* 218(4575): 916–918, 1982.
3. Snyder J, Nelson M, Burdick J, and MacIver M. Omnidirectional sensory and motor volumes in electric fish. *PLoS Biology* 5(11): 2671–2683, 2007.
4. Bodznick D, Montgomery JC, and Tricas C. *Electroreception: Extracting Behaviorally Important Signals from Noise*. New York: Springer-Verlag, 2003.
5. Lohmann KJ and Johnsen S. The neurobiology of magnetoreception in vertebrate animals. *TINS*, 24: 153–159, 2000.
6. Heyers D, Manns M, Luksch H, Güntürkün O, and Mouritsen H. A visual pathway links brain structures active during magnetic compass orientation in migratory birds. *PLoS One*, 2(9): e937: 1–6, 2007.
7. http://www.lightandmatter.com/area1book4.html, Electricity and Magnetism, An on-line physics textbook.
8. Knight RD, Jones B, and Field S. *College Physics: A Strategic Approach*. Reading, MA: Addison-Wesley Publishing Company, 2003.
9. Serway R and Jewett J. *Principles of Physics: A Calculus Based Text*, 6 Edition. Pacific Grove, CA: Thomson Brooks-Cole, 2004.
10. Maxwell JC. *A Treatise on Electricity and Magnetism*. Oxford: Clarendon Press, 1873.
11. Yang F. Aggregation of dielectric particles in electrorheological fluids. *Journal of Materials Science Letters* 16: 1525–1526, 1997.
12. Jackson JD. *Classical Electrodynamics*. New York: John Wiley & Son, Inc., 1962, Chap. 4.

III

Biomechanics

III

Biomechanics

12

Biomechanics

Robert Splinter

12.1 Introduction

Muscle in humans makes up 40–43% of the total body weight in men and 23–25% in women. For a woman who weights 135 pounds, 15 pounds is composed of bones, 35 pounds of organs and skin, 35 pounds of fat, and 50 pounds of muscle.[1]

Three different types of muscle can be distinguished in the human anatomy: skeletal, smooth, and cardiac muscle. Smooth muscle can be found lining organs in the body, and cardiac muscle pertains to the heart. The movements of the bones at joints and posture maintenance are the main job of skeletal muscle.

12.2 Muscle Tissue

The three categories of muscles are *smooth*, *cardiac*, and *skeletal*.[1,2] Each muscle type comprises contractile fibers, which are actually elongated cells.[3–5] Several nuclei are present in the cells of the cardiac and skeletal muscles, while the cells in the smooth muscles have only one nucleus. Many mitochondria are also present in each cell to provide the required cell energy used during muscle movement. The energy produced is supplied to long, thin threads of protoplasm, known as myofibrils that contract and relax the cell in response to impulses from the autonomic nerve system, from its neighbors, or directly from the brain, respectively. The impulses from the brain are passed through motor nerves that attach to the cell membrane. When a cell receives an impulse, it initiates the release of chemicals, resulting in an electrochemical reaction involving filament proteins, calcium ions, and adenosine triphosphate (ATP).[6–9] A by-product

of the chemical reactions is heat that helps maintain body temperature. Thick and thin filaments make up the myofibrils. The thick filaments are made from protein myosin, while the thin filaments are made from *actin*, *tropomysin*, and *topin proteins*. A magnified view of a skeletal muscle shows a striated pattern with parallel bands of different color tissue. The darker bands are the thick myofibril filaments and the lighter color bands are the thin filaments. Skeletal muscles are made up of millions of discrete contractile fibers that are bound together into bundles by connective tissue. Connective tissue also provides the link of the muscle with the skeleton bones.

Skeletal muscles' mode of control is voluntary, meaning that the user has control over its actions, whereas smooth and cardiac muscles are involuntary. Figure 12.1 show the composition of skeletal muscle observed under scanning electron microscopy.

Smooth muscle is controlled by the autonomic nervous system and is found primarily in the walls of organs and tubes.[1,2,9] The spindle-shaped cells that make up the composition of this type of muscle are arranged in sheets. These cells contain small amounts of sarcoplasmic reticulum but lack transverse tubules. The cells are made up of thick and thin myofilaments but do not contain sarcomeres and therefore are not striated like skeletal muscle. Chemically, calcium binds to a protein and induces contraction.

There are two types of smooth muscle: visceral and multiunit. Visceral is found in the digestive, urinary, and reproductive systems. For this type of smooth muscle, a unit is composed of multiple fibers that contract and in some cases are self-excitable. In multiunit smooth muscle, nervous stimulation activates motor units. This type of muscle is found lining the walls of large

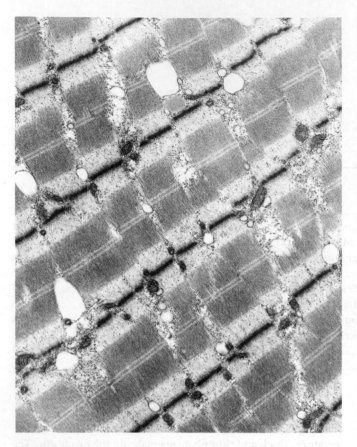

FIGURE 12.1 Electron microscopy image of skeletal muscle. (From Bronzino, J., ed. 2000. *The Biomedical Handbook Volume II*. Boca Raton, FL: CRC Press LLC. With permission.)

blood vessels and in the eye. They are also located at the base of follicles and can produce a "goose bump" effect.[8]

Heart walls have three distinct layers, which are the endocardium, myocardium, and epicardium. The endocardium lines the circulatory system, is the innermost layer, and is composed of epithelial tissue. The myocardium is the thickest layer and consists of cardiac muscle. The epicardium is a thin layer and is the external membrane around the heart.[8]

Cardiac muscle is striated and contains sarcomeres, which makes this muscle similar to skeletal. Adjacent cells are joined end-to-end at structures known as intercalated discs. These discs contain two types of junctions, desmosomes and gap junctions. Desmosomes act like rivets and hold cells tightly together, while gap junctions allow action potentials to spread from one cell to another.[10]

Cardiac muscle forms two functional units called the atria and ventricles. There are two of each whereby both the atria act together and then the ventricles. For contraction, there are two types of cells: contractile cells and autorhythmic. The contractile cells contract when stimulated, while the autorhythmic cells are automatic.[10]

For the focus of discussion of this chapter, only skeletal muscle will be explored. Skeletal muscle has the ability to shorten quickly and recover; this action is defined as contraction. One end of the muscle, origin, is stationary while its other end, insertion, shifts.

A contraction always moves from the insertion to the origin. When a contraction occurs, both the agonist muscle and antagonist muscle work against each other. The agonist muscle or flexion is the muscle that is being acted upon while the antagonist or extension muscle contracts in opposition to the flexion.

A contraction is only the shortening of the muscle and does not describe the actual generation of a force. A force occurs only when there is resistance to the muscle contraction. To understand the nature of contractions, the anatomy of skeletal muscle must be studied. Secondly, the physiology of muscular contraction, meaning the excitation and relaxation will be investigated. The physics of the muscle acting as a lever and measurement techniques will be explored. Lastly, factors that affect strength performance and fatigue will be discussed.[5]

12.3 Anatomy of Skeletal Muscle

Muscle is composed of several kinds of tissue including striated cells, nerve, blood, and various connective tissues. The basic unit of contraction in an intact skeletal muscle is a motor unit, composed of a group of muscle fibers and a motor neuron that controls them (Figure 12.2). When the motor neuron fires an action potential, all muscle fibers in the motor unit contract. The number of muscle fibers in a motor unit varies, although all muscle fibers in a single motor unit are the same fiber type (Figure 12.3). Thus, there are fast-twitch motor units and slow-twitch motor units. The determination of which kind of fibers associated with a particular neuron appears to lie with the neuron itself. The force of contraction within a skeletal muscle can increase by recruiting additional motor units whereby this process is called motor unit recruitment.[1,4,9] An individual muscle is separated

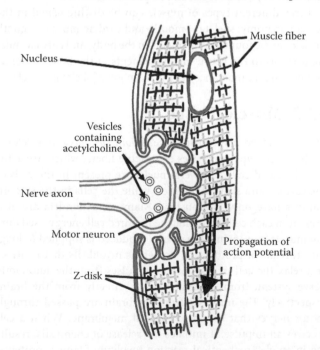

FIGURE 12.2 Diagram of the skeletal muscle with the stimulus provided by the acetylecholine released from the nerve at the motor plate.

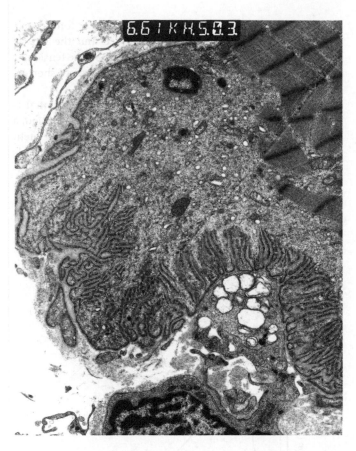

FIGURE 12.3 Electron microscope image of the motor plate at the muscle. (Reprinted from CRC-Press: Biomedical signal and image processing, Taylor and Francis Press, 2006. With permission.)

from adjacent muscles and held into position by layers of fibrous connective tissue called fascia (a bundle). This fascia is part of a network of connective tissue that extends throughout the skeletal muscle system and its attachment with other parts.

These connective tissues become either tendons or aponeuroses. The fibers in the tendon become intertwined with bone, which allows muscle to be connected with bone. Aponeuroses are broad fibrous sheets that may be attached to adjacent muscles. Each muscle fiber within a fascicle is surrounded by a layer of connective tissue in the form of a thin, delicate covering called endomysium. These layers allow for some independent movement as well as allowing blood vessels and nerves to pass through.

A skeletal muscle fiber is a thin elongated cylinder with rounded ends and may extend the full length of the muscle. Within the sarcoplasm are numerous threadlike myofibrils that lie parallel to one another. They play an important role in muscle contractions. The myofibrils contain two kinds of protein filaments, thick ones composed of the protein myosin and thin ones composed of the protein actin. The arrangement of these filaments in repeated units is called sarcomeres, which produce the characteristic alternating light and dark striations of muscles. Light areas are the I-bands and the darker areas are A-bands. Z-lines are where adjacent sarcomeres come together and thin

myofilaments of adjacent sarcomeres overlap slightly. Thus, a sarcomere can be defined as the area between Z-lines.[1]

Myosin filaments are located primarily within the dark portions of the sarcomeres, while actin filaments occur in the light areas.[10]

Within the cytoplasm of a muscle fiber is a network of membranous channels that surrounds each myofibril and runs parallel to it, which is the sarcoplasmic reticulum. Transverse tubules (t-tubules) extend inward from the fiber's membrane. They contain extracellular fluid and have closed ends that terminate within the light areas. Each t-tubule contacts specialized enlarged portions (cisternae) of the sarcoplasmic reticulum near the region where the actin and myosin filaments overlap (Figure 12.4). These parts function in activating the muscle contraction mechanism when the fiber is stimulated.

Each skeletal muscle fiber is connected to a fiber from a nerve cell. Such a nerve fiber is a branch of a motor neuron that extends outward from the brain or spinal cord. The site where the nerve fiber and muscle fiber meet is called a neuromuscular junction (myoneural junction). At this junction the muscle fiber membrane is specialized to form a motor end plate.[1]

The end of the motor nerve fiber is branched, and the ends of these branches project into recesses (synaptic clefts) of the muscle fiber membrane. The cytoplasm at the ends of the nerve fibers is rich in mitochondria and contains many tiny (synaptic) vesicles that store chemicals called neurotransmitters. When a nerve impulse traveling from the brain or spinal cord reaches the end of a motor nerve fiber, some of the vesicles release neurotransmitter into the gap between the nerve and the motor end plate. This action stimulates the muscle fiber to contract.[10]

A muscle fiber contraction is a complex action involving a number of cell parts and chemical substances. The final result is a sliding movement within the myofibrils in which the filaments of actin and myosin merge. When this happens, the muscle fiber

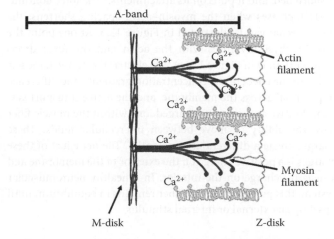

FIGURE 12.4 Schematic representation of the muscle structure. Actin and myosin filaments interact with each other to produce contraction. Polarization of the myosin filament under the influence of Ca^{2+} ions makes the "tentacles" flare out and come in closer proximity to the actin filament. The chemical attraction between the myosin filament and the actin pulls the actin into the myosin, thus shortening the "A-band" and contracting.

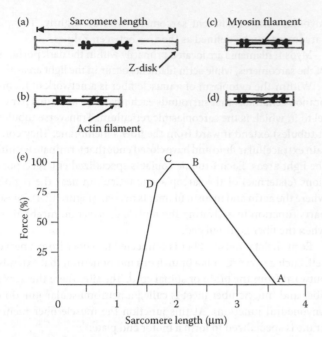

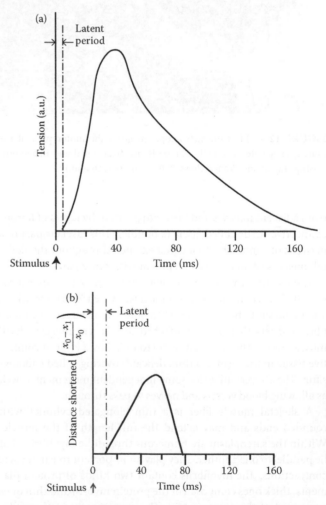

FIGURE 12.5 Schematic representation of muscle sarcomere contraction. (a) Totally distended muscle (relaxed); (b) cross-bridge formation during initial contraction: the actin enters into the H-zone; (c) incremental shortening of the sarcomere results in the crossing of the midline of the sarcomere, and additional cross-bridge formation starts counteracting the contraction and even repelling it; (d) myosin filaments start contacting the Z-line and compressing the actin filaments. At this point permanent chemical bonds can be formed, damaging the muscle, and (e) force diagram associated with the contraction positions shown in this figure. After an increase in force during shortening, the sarcomere length exceeds a minimum distance at which point the force decreases, until the muscle becomes ineffective.

is shortened, and it pulls on its attachments.[1] The force continuously increases when the myosin–actin spacing shortens; the force mechanism is illustrated in Figure 12.5. At one point the myosin physically fuses with the actin and the force drops (shorter than D in Figure 12.5e). The electrical activity in muscle is due to the effect of the concentration gradients, the difference in potential across the membrane, and the active transport system. Positive and negative charged ions within the muscle fiber have the ability to move between intercellular fluids. These charges create a differential in potential. The net effect of these charges is a positive charge on the exterior of the membrane and a negative charge on the interior. In a healthy neuromuscular system, this polarized muscle fiber remains in equilibrium until upset by an external or internal stimulus.

12.4 Physiology of Skeletal Muscular Contraction

The length and tension of muscles vary indirectly. When a muscle contracts because of a load applying a force, the muscle produces one of two types of contraction: isotonic or isometric.

An isotonic contraction is one where the tension or force generated by the muscle is greater than the load and the muscle shortens in length. When the load is greater than the tension or force generated and the muscle contracts but does not shorten, this response is called isometric.[4,8,10,11]

The response to a single stimulation or action potential is called a twitch. A twitch contains three parts, a latent period, a contraction period, and a relaxation period. During the latent period, there is no change in length but the impulse travels along the sarcolemma and down the t-tubules to the sarcoplasmic reticulum. Calcium is released during this period, triggering the muscle to contract. The contraction period is when the tension in the muscle increases, causing a swiveling action of the sacomere components. The muscle tension decreases during the relaxation period and returns to its original shape. Figure 12.6 shows the three components of the twitch and the different types of contraction.

FIGURE 12.6 Muscle contraction motion during isometric and isotonic contraction. (a) Isometric contraction: the muscle length is fixed and tension increases with time as illustrated; and (b) isotonic contraction: the muscle moves freely with constant force, providing a shortening as a function of time as illustrated.

With an increase in the frequency with which a muscle is stimulated, the strength of contraction increases; this is known as wave summation. This occurs by re-stimulation of muscle fibers while there is still muscle contraction activity. With several stimulations in rapid succession, calcium levels in the sarcoplasm increase and this produces more activity and a stronger contraction. When rapid stimulation occurs and the length of time between successive stimulations is not long enough, a buildup of calcium can occur and there will be no relaxation between stimulations. This leads to a smooth sustained contraction known as tetanus.[12]

12.5 Muscle Contractility

The contractile force of skeletal muscle is proportional to the cross-sectional area of the muscle and has been established to fall in the range: 30–40 N/cm^2.

When measuring the contraction velocity of an isotonic contraction, the muscle force can be related to the contraction velocity as derived by A.V. Hill and shown in Equation 12.1[11,12]:

$$(F + a)(v + b) = (F_0 + a)b, \qquad (12.1)$$

where v is the contraction velocity, F is the isotonic load, F_0 is the maximum isometric tension, and a and b are constants. This relationship is exemplified in Figure 12.7 as well.

12.6 Major Skeletal Muscles

The major skeletal muscle groups are facial, mastication, head, pectoral girdle, arms, abdominal, pelvic, and legs. Figure 12.8 shows the anterior view of skeletal muscles within the human body.

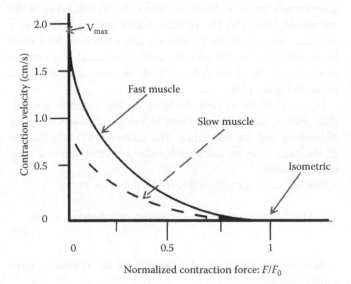

FIGURE 12.7 Contraction velocity of skeletal muscle as a function of force (normalized) for fast and slow contracting muscles.

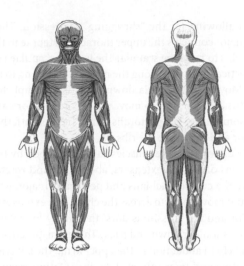

FIGURE 12.8 Diagram of skeletal muscles of the human body. (This figure was produced using Servier Medical Art; http://www.servier.com/Default.aspx)

The facial muscles are epicranius, orbicularis oculi, orbicularis oris, buccanitor, zygomaticus, and platysma. The epicranius covers the upper part of the cranium and has two parts, one that lies over the frontal bone and another that lies over the occipital bone. Lifting the eyebrows and making the forehead wrinkle are its main functions. Multiple stimulations of these muscles can lead to headaches. The orbicularis oculi is a ring of muscle that surrounds the eye and is used for closing or blinking of the eye. It also aids in the flow of tears. The orbicularis oris orbits the mouth and is also know as the kissing muscle since it allows the mouth to pucker. Help keeping food in place is done by the buccanitor muscles. They also aid in blowing air out of the mouth. Zygomaticus' main function is allowing the corner of the mouth to smile or laugh. The platysma extends from the chest to the neck and its function can be seen by the expression of pouting.

Chewing movements are accomplished by four pairs of muscles that are part of mastication. These muscle pairs are masseter, temporalis, medial pterygoid, and lateral pterygoid. Two of these pairs close the mouth while the others allow for side-to-side movements. The masseter is primarily for raising the jaw. A fan-shaped muscle used also for raising the jaw is known as the temporalis. The medial pterygoid is used for closing the jaw and allowing for the sliding movements. The lateral pterygoid allows the jaw to close and move forward.[1]

There are four pairs of muscles that move the head, allowing for rotation. The sternocleidomastoid is located in the side of the neck, and when one side contracts, the head is turned to the opposite side. When both are contracted, the head is bent towards the chest. On the back of the neck, the splenius capitis allows the head to rotate and bend towards one side. For rotating, extending, and bending the head, the semispinalis capitis and longissimus capitis both help with these functions.

Muscles that move the pectoral girdle or chest area include the trapezius, rhomboideus major, levator scapulae, serratus anterior, and pectoralis minor. The trapezius muscle raises the shoulders allowing for the "shrugging" expression. The rhomboideus major connects the upper thoracic vertebrae to the scapula and helps to raise the scapula. The levator scapulae run in an almost vertical direction along the neck, also helping to raise the scapula. Moving the scapula downwards is accomplished with the serratus anterior. It also moves the shoulders forward when pushing something. The pectoralis minor lies beneath the pectoralis major and can move the ribs.

For the upper arm, the muscles can be grouped by their primary actions of flexors, extensors, abductors, and rotators. The flexors are the coracobrachialis and pectoralis major, which flex the arm and move the arm across the chest. The extensors are the teres major and the latissimus dorsi that allow for rotation and moving the shoulder down and back. The supraspinatus and deltoid both help in abduction in the upper arm. The subscapularis, intraspinatus, and teres minor allow for rotation medially and laterally.[1]

In the forearm, there are seven major muscles and these can be categorized by their action. The flexor muscles are the biceps brachii, brachialis, and the brachioradialis all of which help in the rotation and lateral movements about the elbow. The primary extensor of the elbow is the triceps brachii. To rotate the arm, the supinator, pronator teres, and pronator quadratus allow for movements.

In the wrist, hand, and fingers the two major groups are flexors and extensors. Flexing of the wrist is accomplished with the assistance of the flexor carpi radialis and the flexor carpi ulnaris. The palmaris longus connects the humerus to the palm and functions to flex the wrist. To make a fist the flexor digitorum profundus flexes the distal joints of the fingers. All four of these are flexors. The extensors of the wrist and hand are the extensor carpi radialis longus, extensor carpi radialis brevis, extensor carpi ulnaris, and the extensor digitorum. The first three aid in extending the wrist while the last one deals with the functions of the fingers.

Four muscles located in the abdominal wall region help with breathing defecation, urination, and childbirth. These muscles are the external oblique, internal oblique, transversus abdominis, and the rectus abdonimis. All of these muscles function in a similar manner and compress the abdominal cavity.[1]

The pelvis region consists of two major muscle sheets that are the pelvic diaphragm and the urogenital diaphragm. The pelvic diaphragm has two major muscles that are the levator ani and the coccygeus that provide a sphincterlike action in the anal canal and the vagina in women. The three major muscles within the urogenital diaphragm are the superficial transversus perinei, bulbospongiosus, and the ischiocavernosus muscles. The superficial transversus perinei supports the pelvis. The other two help in urination within men and constriction of the vagina from opening in women.

In the thigh, there are two groups, the anterior and posterior groups. The anterior groups' main job is for flexing, while that of the posterior groups is for extending or rotation. The psoas major and the iliacus make up the anterior group and their function is to flex the thighs. The posterior group consists of the gluteus maximus, gluteus medius, gluteus minimus, and the tensor fasciae latae. These muscles help in flexing and rotating the thigh. A third group of muscles, the adductor longus, adductor magnus, and gracilis, adducts, flexes, and/or rotates the thigh.

The muscles of the leg can be separated into two categories that are the flexors and extensors about the knee. The flexors are the biceps femoris, semitendinosus, semimembranosus, and sartorius and all help to flex and rotate the leg. The extensors of the lower legs are the quadriceps femoris, which extends the leg at the knee.

In the ankles, feet, and toes, there are four categories of muscles based on their function. The dorsal flexors, which include the tibialis anterior, peroneus tertius, and extensor digitorum longus, aid in moving the foot upward. The plantar flexors move the foot downward and contain the gastrocnemius, soleus, and flexor digitorum longus muscles. Turning the sole of the foot outward or inward is accomplished by the two groups, the invertor and evertor. The tibialis posterior is within the invertor group and allows for the inversion of the foot. The eversion of the foot is done by the peroneus longus that is within the evertor group.

Although each muscle is specialized within a certain area, all of these muscles help in maintaining posture and movement.[5]

12.7 Physics of Skeletal Muscles

A lever is a simple mechanical machine consisting of a bar on a fixed point that transmits a force. Lever types are defined by three classes according to the axis of rotation, resistive force, and movement force. In a third-class lever, the movement force is applied between the axis of rotation and the resistive force. This lever favors movement speed over movement force. Most muscles in the human body use this type of lever. A second-class lever is one in which the resistance is between the movement of force and the axis of rotation. This type allows for great movement force. When the axis of rotation is between the movement force and the resistance this lever is known as a first-class lever. First-class levers are very similar to third-class levers since they are big on movement speed and lack in movement force.[13] The three different lever arrangements are illustrated in Figure 12.9.

Figure 12.10 shows how the bicep muscle works as a third-class lever where the movement of force is between the axis of rotation and the resistance. The mathematical relationship of the force can be calculated upon knowing some muscle measurements.[3,4]

The lever action can be captured by Equation 12.2:

$$\text{Muscle force} \times \text{force arm} = \text{resistance} \times \text{resistance arm.}$$

$$(12.2)$$

Rearranging Equation 12.2 to solve for the resistance term gives Equation 12.3:

$$\frac{\text{Muscle force} \times \text{force arm}}{\text{Resistance arm}} = \text{Resistance.} \qquad (12.3)$$

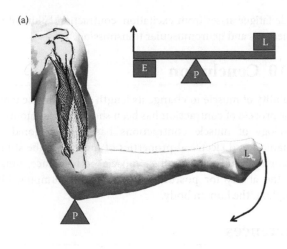

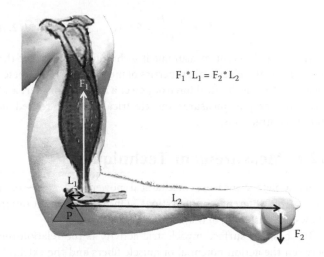

$$F_1 * L_1 = F_2 * L_2$$

FIGURE 12.10 Biceps torque: the biceps is attached to the tibia at a distance L_1 from the elbow joint as the fulcrum. The vertical force applied by the biceps provides a rotation around the elbow joint, forming a torque to counteract and overcome the torque applied by a weight in the hand at a distance L_2 from the fulcrum (elbow). The torque is defined as the magnitude of the force acting perpendicular to the direction of the arm times the length over which the force is moved from a pivot point. In contrast, the chimpanzee has the biceps attached at a point farther removed from the elbow joint providing a greater torque than a human with less muscle mass.

People are built differently and depending on their muscle force, force arm or resistance arm they might be able to hold more or less resistance.[13]

Torque (τ) is the force F that tends to rotate and turn things when applied to a lever arm with length d. Mathematically it is equal to the force multiplied by the distance. Equation 12.4 illustrates the definition of torque:

$$\tau = F \times d. \tag{12.4}$$

This distance is usually radial from a central point or pivot point (the point where the moving section is attached and where it can rotate from). Power is the measure of how fast work can be done. It equals the amount of torque on an object multiplied by the rotational speed. To solve power as a function of human output, the following question can be proposed: How much power can a human output? One way to solve this problem is to see how fast one can run up a flight of stairs. Firstly, measure the height of a set of stairs that reaches three stories. Secondly, record the time to run up the stairs. Lastly, divide the height of the stairs by the time, where this equals the instantaneous speed. The weight of the individual must also be known.[5]

For instance, if it took 15 s to run up 10 m, then the speed equals 0.66 m/s (only the speed in the vertical direction is important). Then one needs to figure out how much force was exerted over 10 m; this force is equal to your weight. The amount of power output (W) equals the weight (= gravitational force: F) multiplied by speed (v).

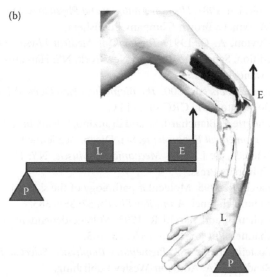

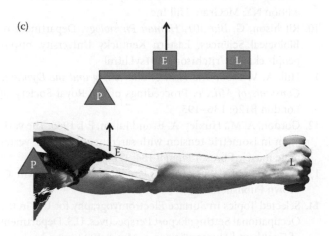

FIGURE 12.9 Classifications of muscle contraction: (a) class 1, the fulcrum lies between the applied muscle force and the resisting force (e.g., scissor action, or the head resting on the spine), (b) class II, the applied force is removed from the fulcrum with the resisting force in between (e.g., standing on toes), and (c) class III, the muscle force is applied between the fulcrum and the counteracting force. This situation is most common for skeletal muscle.

$$W = \frac{h \times F}{t} = F \times v. \qquad (12.5)$$

There are different measurement techniques to measure the mechanical and electrical properties of muscle. A dynamometer measures the mechanical force or power a muscle can exert and electromyography measures the electrical activity related to muscle contractions.[14]

12.8 Measurement Techniques

Dynamometers measure mechanical power and come in a variety of different configurations. One such configuration allows measurement of the clench action of hands.[15]

The basis of surface myoelectric activity is the relationship between the action potential of muscle fibers and the extracellular recording of those action potentials at the skin surface. Electrodes, external to the muscle fiber, can be used to detect action potentials. The most common electrodes are surface and fine wire electrodes. For general muscle groups, surface electrodes are suitable; however, they are limited in detecting small muscles or muscles deep within the body, where fine wire electrodes are recommended.[14]

Detection electrodes are usually used in a bipolar manner. Two electrodes are positioned on the muscles, while a third is a reference (ground) to the body. A difference preamplifier increases the amplitude of the signals between each of the detecting electrodes and the common mode reference. Signals that are nearly zero are deemed common mode signals. The difference preamplifier improves the signal-to-noise ratio allowing the signal of small twitch to be detected. The most common noise problem is 60 Hz interference. This signal distortion occurs when the reference electrode is not applied properly, a loose wire, or when electrical fields persist. The acquisition of the electrical activation data is described in detail in Chapters 20 and 21.

12.9 Factors that Affect Skeletal Muscle Strength Performance

Sex, age, training experience, muscle strength, and fatigue all affect strength performance. Both male and females can develop muscular strength at the same rate, although due to hormones, namely testosterone, male muscles have a greater potential for size and strength. As humans mature, strength increases but by age 30 muscles degenerate. The degeneration is a reversible process but continuous training is required. The length of time devoted to training and the type of training regiment influence muscle strength. The initial length of the muscle fibers and the insertion length affect the amount that muscles can lift and support.[13]

Fatigue relates to the condition that the muscle does not allow for the same amount of power output. Fatigue is highly variable and is influenced by the intensity and duration of the contractile activity, by whether the muscle fiber is using aerobic or anaerobic metabolism, by the composition of the muscle and by the fitness level of the individual. Most experimental evidence suggests that muscle fatigue arises from excitation–contraction failure of control neurons and neuromuscular transmission.[5]

12.10 Conclusion

The ability of muscle to change its length or lengthwise tension by the process of contraction has been shown. The anatomy and physiology of muscle contractions has been explored. The mechanical and electrical properties of muscle can be studied. Although several factors affect muscle performance, muscles have the ability for posture, digestion, and pumping blood throughout the human body.

References

1. Hole, J. Jr. 1981. *Human Anatomy and Physiology*. Dubuque, IA: Wm. C. Brown Company Publishers.
2. Guyton, A. C. 1991. *Textbook of Medical Physiology*, 8th edition. San Diego, CA and New York, NY: Harcourt Brace Jovanovich, Inc.
3. Bronzino, J., ed. 2000. *The Biomedical Handbook Volume II*. Boca Raton, FL: CRC Press LLC.
4. Enderle, J., Blanchard, S., and Branzino, J. 2000. *Introduction to Biomedical Engineering*. San Diego: Academic Press.
5. Schneck, D. J. 1992. *Mechanics of Muscle*. NY: New York University Press.
6. Barchi R. 1995. Molecular pathology of the skeletal muscle sodium channel. *Annu Rev Physiol* 57: 355–385.
7. Cohen S. and Barchi R. 1993. Voltage-dependent sodium channels. *Int Rev Cytol* 137c: 55–103.
8. Epstein, H. T. 1963. *Elementary Biophysics: Selected Topics*. Reading, MA: Addison-Wesley Publishing.
9. Harrison, T. R. 1980. *Principles of Internal Medicine*, 9th edition NY: McGraw-Hill Inc.
10. Ritchison, G. *Bio 301, Human Physiology*. Department of Biological Sciences, Eastern Kentucky University. http://people.eku.edu/ritchisong/301syl.html
11. Hill, A. V. 1938. *The Heat of Shortening and the Dynamic Constants of Muscle*. Proceedings of the Royal Society of London B126: 136–195.
12. Gordon, A. M., Huxley, A. F., and Julian, F. J. 1966. The variation in isometric tension with sarcomere length in vertebrate muscle fibres. *J Physiol* 184: 170–192.
13. Westcott, W. L. 1989. *Strength Fitness*. Dubuque, IA: Wm. C. Brown Publishers.
14. Selected Topics in Surface Electromyography for Use in the Occupational Setting: Expert Perspectives. U.S. Department of Health and Human Services. March 1992.
15. Pflanzer, R. *PhD. Lesson 2: Electromyography II*. Biopac Systems Inc. Version 3.0. http://www.biopac.com/Manuals/bsl_l02dr.pdf
16. Wiggins, W., Ridings, D., and Roh, A. Carolinas HealthCare System, 1000 Blythe Blvd. Charlotte NC.
17. CRC-Press: Biomedical signal and image processing, Taylor and Francis Press, 2006.

13

Artificial Muscle

Robert Splinter and
Paul Elliott

13.1 Introduction

Webster's dictionary simply defines a muscle as something that produces movement by contracting and dilating. Biologically a muscle contracts and dilates due to the release of chemicals throughout the muscle fiber resulting in an action potential. Similarly, research is being done, and has also been done previously, to create a micro-fabricated muscle whose action potential is generated by electrostatic forces using micro-electronic techniques. The technology that is discussed in this chapter will present how Micro Electro-Mechanical Systems (MEMS) are producing similar results in scalable packages.[1-3]

A brief biological description of muscle contraction will help understand its relation to synthetic muscle contraction. To excite the muscle for contraction, a motor neuron cell body located in the spinal cord sends motor neurons to neuromuscular junctions on muscle fibers. The motor neuron sends an action potential that results in the release of a neurotransmitter acetylcholine. The acetylcholine spreads across the entire sarcolemmal surface and diffuses into the muscle fiber's plasma membrane traveling down the transverse tubules. The action potentials cause the endoplasmic reticulum to release Ca^{2+} into the cytoplasm. The calcium triggers binding of myosin to actin, which causes the filament to slide, resulting in contraction of the muscle fiber. The release of the contraction is as simple as removing the action potential thereby removing Ca^{2+} from the cytoplasm.

13.2 Micro Electro-Mechanical Systems

Technology has already proven itself by making functioning prosthetics that emulates muscle action. How can we create a muscle-like structure close to the same scale as just discussed? Micro-fabrication using micro-electronic techniques is used to produce micron-thick muscles.[3-5]

The integrated force array (IFA) and the spiral wound transducer (SWT) are both modeled as muscles. These devices are fabricated using standard micro-electronic techniques and are grouped as MEMS. These muscle-like actuators receive their input by switched voltage, comparable to the spinal cord signaling the muscle with a motor neuron. With an applied voltage, the devices actuate by contracting due to electrostatic forces between two conducting plates. This process is illustrated in Figure 13.1. When the voltage is released, the actuators dilate or relax due to the spring-like nature of the structure. This process

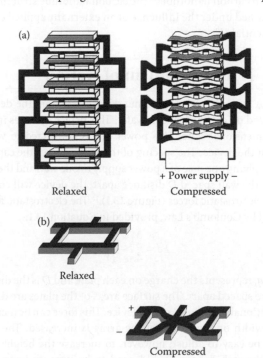

FIGURE 13.1 Integrated force array, (a) operate array elements in flat position, two-dimensional device. Electrostatic force is inversely proportional to the distance "R" apart (Coulombs law): $Fe = (1/4\pi \varepsilon)(q_1 q_2/R^2)$ and (b) perspective of an array element; shaded areas indicate a thin metal layer.

can be compared to the motor neuron releasing acetylcholine into to muscle fiber. The presence of the acetylcholine initiates contraction. Likewise, the cutoff of acetylcholine and the presence of acetylcholinesterase relax the muscle. The ability to move is among the most important processes necessary for the survival of eukaryotic organisms. Several different strategies have been evolved to perform various biological tasks that require directional movement. The current understanding of the molecular mechanism of muscle contraction with a primary focus on the motor protein myosin is discussed in detail.

13.3 Synthetic Molecular Motors and Artificial Muscle

One example of artificial muscles is the synthetic molecular motor. One of the various types of synthetic molecular motors is the poly-azobenzene peptide array. The working action of the poly-azobenzene peptide array results from the rearrangement of the molecular change from a short (*cis*) format to a long (*trans*) state when excited by light of a specific wavelength.[6,7,8]

Several other types of materials that can respond to external stimuli have been developed as artificial muscle structures.[9] One example is the family of electroactive polymers. These elastomer polymers can be polarized by external electrical stimulation and will subsequently constrict. Other alternatives are polymer/metal composite materials. Another solution may be found in carbon nanotubes. The carbon nanotube structure can be modified under the influence of an externally applied electrical potential.[9,10]

13.4 IFA Mechanism of Action

The IFA is a microfabricated muscle-like actuator. The device is fabricated on a 4-inch silicon wafer. This device operates in a flat two-dimensional state. This power bus is connected to vertical walls of the device. The spacing of the walls defines the capacitor gap. Applying one pole of a power supply to one wall and the other pole to the wall spaced *D* distance apart, the device will contract due to electrostatic forces (Figure 13.1).[11] The electrostatic force is defined by Coulomb's Law, provided in Equation 13.1:

$$Fe = \frac{1}{4\pi\varepsilon}\frac{q_1 q_2}{D^2},$$ (13.1)

where q_i represents the charge on each plate and *D* is the distance they are spaced apart. The surface areas of the plates are directly proportional to the force of the device. This force can be increased if the width or the height of the array is increased. The width would be easy to adjust; however, to increase the height of the plate, the etch process would need to be very un-isotropic to result in such deep etching without underetching the structures. Equation 13.2 provides the force as a function of the change in potential energy stored in a mechanism:

$$Fe = -\Delta PE \text{ (Energy}_{\text{stored}}).$$ (13.2)

The potential energy (stored energy) is defined by Equation 13.3

$$\text{Energy}_{\text{stored}} = \frac{CV^2}{2},$$ (13.3)

where the capacitance of the MEMS device acting as an artificial muscle is given by Equation 13.4:

$$C = \frac{\varepsilon A}{D}.$$ (13.4)

The capacitance of the device can likewise be calculated. Note that the capacitance is dependent on the distance between the plates. The thicknesses of the polyimide along with distance in air between the metal plates are both accounted for in Equation 13.5. The dielectric constant of air is represented as ε_0 and the dielectric constant of the polyimide is represented as ε. Therefore, the IFA can be monitored as to how much the muscle has contracted, by monitoring the value of the capacitor. This feedback continues to support the integration of this technology modeling a biological muscle. The equivalent capacitance of the MEMS device is given by Equation 13.5:

$$C_{\text{equiv}} = \frac{1}{D/\varepsilon_{\text{air}}A} + \frac{1}{D/\varepsilon_{\text{poly}}A},$$ (13.5)

which can be rewritten as Equation 13.6:

$$C_{\text{equiv}} = \frac{\varepsilon\varepsilon_o A}{D_{\text{poly}} + \varepsilon D_{\text{air}}}.$$ (13.6)

The IFA can also act as a generator.[12] This can be accomplished by charging the plates at a low voltage. Disconnect the applied voltage and then stretch the device to increase the value of *D* in Equation 13.6. This stretch allows the device to be discharged at a higher potential.

The IFA is fabricated at lengths of 6 cm and thicknesses of 3 μm. These devices were fabricated at the Microelectronic Center of North Carolina (MCNC).[13-15] They require chemical vapor deposition (CVD) of oxide. The structure supporting the metal plates is a polymer named polyimide. This is a very flexible film allowing the electrostatic forces to overcome their opposing spring forces. However, these spring forces provide the relaxation mechanism of the "muscle." The conducting plates are aluminum-deposited at an angle θ as shown in Figure 13.2. Then the device is lifted off from the wafer to function as a three-dimensional muscle that is contracting linearly along one axis.

The IFA is a microfabricated device that has proven itself by functioning successfully. The design of this muscle is scalable. The width and length can be increased or decreased to serve as a muscle of any size. The slide in the presentation titled "Anthropomorphic Hand" illustrates the IFA as being attached to a plastic skeletal structure. This example is using 6 cm by 6 cm by 3 μm so that they are rolled into 6 cm by 0.6 mm (diameter) cylinders. The actuation of this muscle bends the joint it is attached to. The devices are bundled together from 20, 40–60

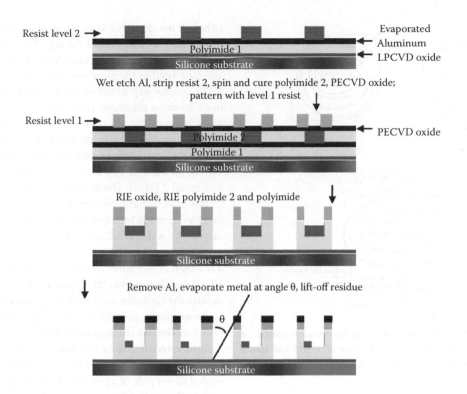

FIGURE 13.2 IFA—fabrication process sequence as performed at MCNC.

devices per muscle to result in a force of 2–24 g with a compression of 2 cm.

The SWT is based on the fundamental operation of the IFA. The SWT is schematically represented in Figure 13.3. This device is fully volumetric and as a final product operates as a three-dimensional device. The SWT is fabricated in a two-dimensional state. After very large scale integrated circuit (VLSI) fabrication, the device is lifted off from the wafer. The flat tape operates as a single layer and can be compressed by 1.4 μm. The overall thickness of one tape is around 6.9 μm. The device compresses 20% of its thickness. Winding it around a preform and attaching it to itself increases this compression. The average number of turns is 40. Each layer, one tape thick, can be compressed by 1.4 μm. Since the device is wound around a form 40 times, with two sides of the form, the overall compression is around 0.11 mm.

Referring to the cross-sectional illustration of the SWT, Figure 13.4, the fabrication process will be explained. This device was created in the University of North Carolina at Charlotte, Cameron Applied Research Center's clean room under the supervision of Dr. Stephen Bobbio. The process sequence of the SWT begins with a 1-micron pad of thermally grown oxide. This pad of oxide is grown in a wet oxidation process with a furnace temperature of 1100°C. This is called the lift off oxide because this allows the latter fabricated device to be lifted from the silicon wafer.

Polyimide is coated over the pad of oxide using a 0.86 μm double spin process. The double spin is performed to lower the probability of shorting between the metal layers. Pinholes may appear using a single spin coating of polyimide. A spinning of polyimide adhesion promoter reduces the occurrence of

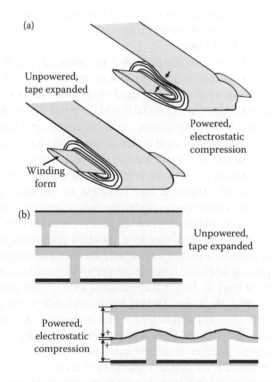

FIGURE 13.3 Spiral wound force transducer (SWT), (a) fully volumetric, three-dimensional device formed by winding a flat structure each turn is glued to the previous over the flats of the winding form all compressions add together and (b) mechanical detail of compression versus relaxed.

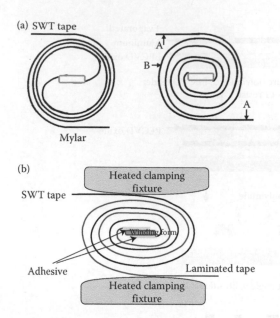

(a) SWT tape

Mylar

(b)

SWT tape

Heated clamping fixture

Winding form

Adhesive

Laminated tape

Heated clamping fixture

FIGURE 13.4 Winding configuration in artificial muscle design. (a) Not the metal but the polyimide causes the tape to be stiff and spring like and (b) description of detail of SWT tape compression.

pinholes. Each layer of polyimide is prebaked in two stages at 90°C and 150°C. It is then cured (hard-baked) in a convection oven at 150°C ramped to 350°C in N_2. The polyimide cures for 30 min and is ramped down to room temperature. To complete the double coating, the same process is performed after the first prebake and hard-bake.[4]

Each metal layer consists of 200 Å of chromium and 2000 Å of gold. Chromium is used to ensure the adhesion of gold. Electron beam evaporation is used to deposit the metal. The smoothness of this film is crucial due to the possibility of electrical shorting. If the evaporation rate is too high, metal splatters (molten metal blobs) deposit on the substrate and burn through the plastic layers. The splattered metal appears rough and scatters light. During the winding process, the splattered metal may be mechanically displaced and complete an electrical short. Therefore much attention is paid to the evaporation process. The best results were obtained when the gold is deposited at 0.5 Å/s. The metal layer is patterned in a standard photolithographic process. Polyimide is then coated over the metal to encapsulate it and provide additional protection against electrical shorts.

The next level in the SWT structure is a chromium etch stop layer of 200 Å thickness. During the plasma etch of the oxide above, the chromium provides a barrier against etching the underlying polyimide. The etch stop layer also provides a means to determine whether or not the oxide has been fully etched by applying Chrome etchant in a selected test area.

Spin-on-glass (SOG) is used to define the capacitor gap regions. A thin film of evaporated silicon, 200 Å, is used to promote the adhesion of the SOG. This oxide film is applied and cured in a double layer process, resulting in a total SOG thickness of 1.43 μm. Some cracking has been observed in the SOG, which,

although not desirable, is acceptable (does not impair functionality) as long as there is no flaking. The remaining layers of the structure (two polyimide, one oxide, and two metal) are applied and patterned in the same way as previously described. The photolithography step to define the etching of the second layer oxide must be done precisely. This oxide via must be centered to the space between the neighboring via's of the first oxide. Totally, six photolithographic levels are used. A seventh and final photolithography step defines the contact holes, which are etched through the device to the two buried metal layers [metal 1 (M1) and metal 2 (M2)]. The uppermost metal layer on the SWT tape (metal 3: M3) is also the top layer of the tape, and electrical contact can be made to it directly. After the VLSI processing is completed, the wafer is cut into strips with a diamond-blade dicing saw. The narrow rectangular wafer pieces, each containing one SWT tape, are then immersed in undiluted aqueous HF to remove all oxides. This allows the tapes to lift off from the silicon and defines the capacitor gaps by etching away both SOG layers.

Immediately after liftoff, the single tape structures may be tested in the flat configuration. Contacts were made using conductive epoxy and gold bonding wire. Metal 1 (M1) was externally shorted to metal 3 (M3), as well as the test fixture, and connected to one pole of a power supply; metal 2 (M2) was connected to the other pole. When powered, the entire tape visibly tautened as the tape compressed across its thickness. This compression could be directly observed using a microscope. Even without winding, analog motion (i.e., proportional response to the applied voltage) could be easily seen, both visually (in the tautening of the tape) and in a direct way with the microscope. Deflection was observed over the range 36–90 V. Full compression occurred at the upper limit. More compliant structures that compress at lower voltages may readily be designed. The compliance could, for example, be increased by a factor of 2 by changing the spacing of the oxide vias from 55 μm to $55 \times (2)^{1/3}$ μm = 69 μm. Likewise, if the thickness of the polyimide is reduced, its corresponding spring force will be reduced to allow the reduction of the electrostatic force.

To create a three-dimensional device, the SWT tape is now wound around a form. The cross-section of the wound device before adhesion shows how the spring forces naturally cause the device to wind non-uniformly (grouping to the outer windings). This is corrected by holding the SWT tape and the Mylar adhesion tape fixed. Then turn the form so that the windings are pulled toward the form. Dupont's metallized Mylar is used because liquid cements will not stay localized to the flat parts of the device, and they bond the device all over. If the wound device is completely bonded, it has nowhere to go when it is energized to compress. For an infinite number of springs, no finite electrostatic force could cause the device to compress. Figure 13.5 presents a schematic representation of the spring stack. The spring force of the Mylar coils is given by Equation 13.7:

$$F_s = kx + 2kx + 3kx + \cdots + nkx = kx\frac{n(n+1)}{2}. \quad (13.7)$$

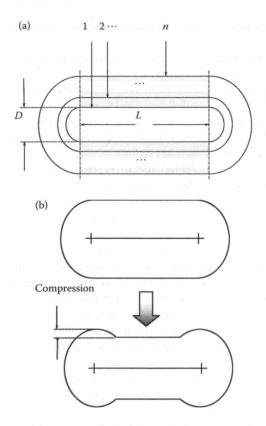

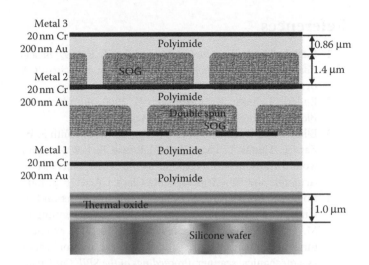

FIGURE 13.6 SWT process sequence as developed in the clean room at the University of North Carolina at Charlotte.

FIGURE 13.5 (a) Diagram of spring equivalent winding structure where each coil acts as composite spring, comprising n-springs and (b) detail of composite spring compression.

Not all the spring constants are equal; they diminish as n becomes larger. The spring forces result from the deformation of the semicircular parts on the sides of each oval and are proportional to t^3/D, where t is the thickness of the spring material and D is the diameter. Using the substitution $f = 2t/D$, Equation 13.7 can be rewritten as Equation 13.8:

$$F_s = kx(1 + 2/(1 + f)^3 + 3/(1 + 2f)^3 + \cdots + n/(1 + (n-1)f)^3,$$

$$F_s = kx\left(1 + \frac{2}{(1+f)^3} + \frac{3}{(1+2f)^3} + \cdots + \frac{n}{(1+(n-1)f)^3}\right).$$

(13.8)

Since f is small, the equation can be approximated to Equation 13.9:

$$F_s = \frac{1}{2}\frac{kx}{f^2}$$

(13.9)

The measured spring force for the coiled artificial muscle MEMS device is approximately 20 dyne/μm.

The SWT is much like that of the IFA in its functionality. However, it is scalable so that it can produce much larger amounts of force. Referring back to Equation 13.2, the potential energy is dependent upon the surface area of the conducting plates. The SWT design has 1 mm and 2 mm wide by 6 cm long devices. Both of these dimensions can be scaled up or down for the needed application. As the length of the IFA increases, so does the compression. The thickness (force) is maintained at 3 μm. However, the SWT can increase its compression by having longer tapes allowing more windings. This results in a much thicker and greater volume device.

As explained for the IFA, the compression of the SWT can be monitored by the capacitive value across either metal 1 (M1) and metal 2 (M2) or metal 2 (M2) and metal 3 (M3) (Figure 13.6). In addition to sharing similarities to an artificial muscle, both of these devices can be mounted on a catheter and used as linear actuators to move ultrasonic scanners for biomedical imaging. Illustrations of both devices are in the accompanying presentation.

13.5 Summary

To conclude, micro-electronic techniques provide a means of fabricating artificial muscles from 3 μm to infinity in thickness. To further explore the integration of these particular devices, several considerations must be made. All material must be biocompatible. The devices must be encapsulated so that it does not act as a pump but acts in a way that it can easily be bonded to other muscles, bones, and so on. The voltage actuation levels must be determined and isolated to prevent any side effects, such as shock, or to avoid stimulating other muscles. Best of all, they provide a means so that motor neurons can form neuromuscular junctions on the synthetic muscle (this number is associated with the number of muscle fibers the synthetic muscle represents), so that the total neuron energy would be enough to stimulate the synthetic muscle.

References

1. Campbell N. A., Lawrence G. M., and Jane B. R. 1997. *Biology Concepts and Connections* 2nd edition. San Francisco, CA: Benjamin Cummings.

2. Enderle J., Susan B., and Joseph B. 2000. *Introduction to Biomedical Engineering*. London: Academic Press.

3. Elliott P. C., Bobbio, S. M., Pennington M. A., Smith S. W., Zara J., Hudak J., Pagan J., and Rouse B. J. 2000. Spiral wound transducer. *Proceedings of the SPIE Conference on Actuators and Flow Sensors*, Santa Clara, CA, 4176, p. 130.

4. Bobbio S. M., Smith S. W., Zara Goodwin-Johansson J. S., Hudak J., DuBois T., Leamy H., Godwin J., and Pennington M. 1999. Microelectromechanical actuator with extended range and enhanced force: Fabrication, test, and application as a mechanical scanner, *Proceedings of the SPIE Conference on Smart Structures and Materials*, Newport Beach, CA, 3673, p. 210.

5. Jacobson J. D., Goodwin-Johansson S., and Bobbio, S. M. 1995. Integrated force arrays I: Theory and modeling of static operation. *J. MEMS*, 43, 139.

6. Hugel T., Holland N. B., Cattani A., Moroder L., Seitz M., and Gaub H. E. 2002. Single-molecule optomechanical cycle. *Science* 296: 1103–1106.

7. Holland N. B., Hugel T., Neuert G., Cattani-Scholz A., Renner C., Oesterhelt D., Moroder L., Seitz M., and Gaub H. E. 2003. Single molecule force spectroscopy of azobenzene polymers: Switching elasticity of single photochromic macromolecules. *Macromolecules* 36: 2015–2023.

8. Chuang C. J., Li W. S., Lai C. C., Lui Y. H., Peng S. M., Chao I., and Chiu S. H. 2008. A molecular cage based [2] rotaxane that behaves as a molecular Muscle. *Org. Lett.* 11(2): 385–388.

9. Baughman R. H. 2005. Playing nature's game with artificial muscles. *Science* 30: 63–65.

10. Aliev A. E., Oh J., Kozlov M. E., Kuznetsov A. A., Fang S., Fonseca A., Ovalle R. et al. 2009. Giant-stroke, superelastic carbon nanotube aerogel muscles. *Science* 323: 1575–1578.

11. Saduki Matthem N. O. 1994. *Elements of Electromagnetics*, 2nd edition. Saunders College.

12. Bobbio S. M., Pennington M. A., Smith S. W., Zara J., Leamy H., Hudak J., Pagan J., and Elliott P. C. 2000. Scalable synthetic muscle actuator. *Proceedings of the SPIE Conference on Smart Structures and Materials*, Newport Beach, CA, 3990, p. 206.

13. Bobbio M. S., Goodwin-Johansson S., DuBois T., Tranjan F., Jacobson J., Bartlett T., and Eleyan N. 1995. Integrated force arrays: Scaling methods for extended range and force. *Proceedings of the SPIE Conference on Smart Structures and Materials*, San Diego, CA, 2448, p. 15.

14. Bobbio S., Goodwin-Johansson S., DuBois T., Tranjan F., Smith S., Fair R., Ball C., Jacobson J., Bartlett C., Eleyan N., Makki H., and Gupta R. 1996. Integrated force arrays: Positioning drive applications. *Proceedings of the SPIE Conference on Smart Structures and Materials*, San Diego, CA, 2722, p. 23.

15. Bobbio S. M., Goodwin-Johansson S., Tranjan F. M., Dubois T. D., Jones S., Kellam M. D., Dudley B. W., Bousaba J. E., and Jacobson J. D. 1995. Integrated force arrays: Loaded motion, fabrication, and applications. *Proceedings of the Symposium on Microstructures and Microfabricated Systems, Electrochemical Society*, 94–14, p. 67.

14

Cardiovascular System

István Osztheimer and
Gábor Fülöp

14.1 General Characteristics of the Cardiovascular System

Besides the endocrine and nervous systems the cardiovascular system constitutes the principal integrating system of the body. The circulatory system serves as a medium for transportation and distribution of essential metabolic substrates to the cells of each organ and it plays an important role in removing the by-products of metabolism. Furthermore, its purpose is to facilitate the exchange of heat between cells and the outside environment, to adjust the nutrient and oxygen supply in different physiological conditions and to serve as a medium for humoral communication throughout the body. All of these functions are prerequisite of maintaining the "consistency," the so-called homeostasis of the internal environment of the human body. Before the detailed functional description of the parts of the cardiovascular system, it is important to consider it as a whole from an anatomical viewpoint. The circulatory system is made up of two primary components: the pump (the heart) and the collecting tubes (blood vessels). A third component, the lymphatic system, has an important exchange function, it does not contain blood.

14.2 Functional Anatomy of the Heart

The heart can be viewed as two pumps in series: the right ventricle to propel the pulmonary circulation, which is involved in the exchange of gases between the blood and alveoli; the left ventricle to propel the systemic circulation, which perfuses all other tissues of the body (only the gas exchange parts of the lungs are not supplied by the systemic circulation).

The right side of the heart consists of two chambers: the right atrium and the right ventricle. Figure 14.1 gives a general overview of the construction of the heart and the flow pattern.

The venous blood returns to the right atrium from the systemic organs and the right ventricle pumps it into the pulmonary circulation, where the gas exchange between the blood and the alveolar gases occurs. The left side of the heart comprises the left atrium and the left ventricle. The blood is ejected across the aortic valve into the aorta. There are two types of valves in the heart—the atrioventricular (AV) valves (tricuspid and mitral) and the semilunar valves (pulmonic and aortic). The valves are endothelium-covered, fibrous tissue flaps that are attached to the fibrous valve rings. The unidirectional blood flow is ensured by the orientation of the heart valves; their movements are essentially passive, regulated by the blood pressure gradients.

The atria function more as reservoirs of blood for their respective ventricles than as important pumps for the circulation. The thin-walled, highly distensible atria can easily accommodate the venous return at a low pressure. The right atrium receives blood from the systemic circulation via the superior and inferior venae cavae. The pressure in the right atrium is between 0 and 4 mm Hg. From the right atrium, the blood flows into the right ventricle through the tricuspid valve. The right ventricle is triangular in form, and it wraps around part of the thicker left ventricle. The blood propels through the right ventricle's outflow

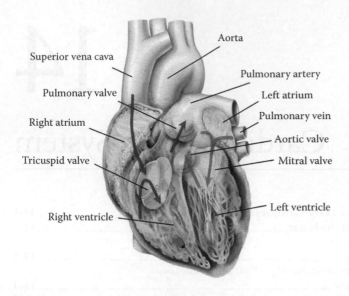

FIGURE 14.1 Anterior view of the heart. The blue arrows represent the directions of the venous blood flow. The red arrows indicate the route of the oxygenated blood flow. (Modified after Patrick J. Lynch, illustrator; C. Carl Jaffe, MD, cardiologist.)

tract and the semilunar pulmonic valve into the pulmonary artery. Following the gas exchange in the lungs, the blood returns to the left atrium via four pulmonary veins. The pressure in the left atrium ranges from 8 to 12 mm Hg. The blood flows through the mitral valves into the left ventricle. The left ventricular wall is about three times thicker than the right ventricular wall. The thick muscular wall allows for high pressure generation during contraction.

14.2.1 The Law of Laplace

The afterload is the "load" against which the heart must contract to eject blood. One way to characterize the afterload is to examine the ventricular wall tension. The relationship that exists between forces within the walls (wall tension) of any curved fluid container and the pressure of its contents is described by the law of Laplace. Within a cylinder, the law of Laplace states that wall tension is equal to the pressure within the cylinder times the radius of curvature of the wall (Equation 14.1):

$$T = P \times R. \tag{14.1}$$

Here the wall tension is expressed in terms of tension (T) per unit of cross-sectional area [dynes per centimeter (dyn/cm)], P is pressure (dyn/cm²), and R is the radius (cm). So, the wall tension is proportional to radius. The wall thickness must be taken into consideration when the equation is applied for the cardiac ventricles. Because of the thick ventricular walls, wall tension is distributed over a large number of muscle fibers, thereby reducing

tension on each. The equation for a thick-walled cylinder (heart) is described by the Equation 14.2:

$$T = \frac{P \times R}{h}, \tag{14.2}$$

where h is the wall thickness.

From the described relationship stems that the tension on the ventricular walls increases as ventricular cavity volume increases, even if intraventricular pressure remains constant (ventricular dilation). Therefore, afterload is increased whenever intraventricular pressures are increased during systole and by ventricular dilation. On the other hand, the tension is reduced, even as pressure rises, as the ventricle empties or the ventricle wall is thickened and hypertrophied. Thus, ventricular wall hypertrophy can be thought of as an adaptive mechanism of the ventricle to offset increase in wall tension caused by increased aortic pressure or aortic valve stenosis.

14.3 The Cardiac Cycle

The sequential contraction and relaxation of the atria and ventricles constitute the cardiac cycle. The mechanical events during the cardiac cycle can be described by the pressure, volume, and flow changes that occur within one cycle (Figure 14.2). Furthermore, it is important to note that to understand the regulation of cardiac contraction and relaxation, one must know how these mechanical events are relate to the electrical activity of the heart. In the following paragraphs the mechanical changes on the left side of the heart during a cardiac cycle will be discussed. Pressure and volume changes and the timing of mechanical events on the right side of the heart are qualitatively similar to those on the left side. The main difference is that the pressures in the right cardiac chambers are much lower compared to the left side of the heart.

14.3.1 Ventricular Systole

The ventricular contraction begins when the action potential propagates over the ventricular muscle; this event coincides with the QRS complex on the electrocardiogram, which represents ventricular depolarization. As the ventricles depolarize, contraction of the ventricular myocytes leads to a rapid rise of the intraventricular pressure. When the intraventricular pressure exceeds the atrial pressure the AV valves close (Figure 14.2).

Closure of the AV valves results in the first heart sound (S1). During the interval between the closure of the AV valves and the opening of the semilunar valves (when the ventricular pressure rises abruptly), the ventricular volume is constant for this brief period and is termed as isovolumic contraction. The ventricular pressure rises steeply as the myocardial contraction intensifies. When the ventricular pressure exceeds the systemic pressure, the aortic valve opens, and the ventricular ejection occurs. The ejection phase may be subdivided into a rapid phase (earlier and shorter) and a reduced phase (later and

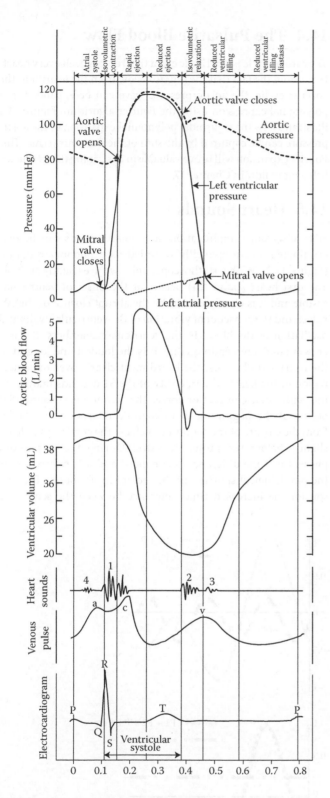

FIGURE 14.2 Pressure, flow, volume, electrocardiographic, and phonocardiographic events constituting the cardiac cycle. (Reprinted from Berne, R.M. (Editor) and Levy, M.N. *Physiology*, 2nd revised edition, p. 444, MO: Mosby, 1988, © Elsevier. With permission.)

longer). The rapid ejection phase has three major characteristics: first, it terminates at the highest ventricular and aortic pressure; second, the decrease in ventricular volume is rapid; and third, the aortic blood flow is greater than during the reduced ejection phase. It is important to note that the peak of the blood flow velocity during the ejection coincides in time with the point when the ventricular pressure curve intersects the aortic pressure curve. At this time the pressure gradient reverses and the blood flow starts to decrease. At the end of the ejection phase, a residual volume of blood remains in the ventricles (end-systolic volume). This blood volume is approximately equal to the ejected blood volume, and this residual volume is fairly constant in healthy hearts.

14.3.2 Ventricular Diastole

At the time point of aortic valve closure the ventricular diastole begins. The abrupt closure of the semilunar valves is associated with an incisura (notch) in the aortic and pulmonary pressure curves and produces the second heart sound (S2). During the period between the closure of the semilunar valves and opening of the AV valves (i.e., all valves are closed) the ventricular pressure falls quickly without a change in ventricular volumes; this first part of the diastole is termed as the isovolumic relaxation. The major part of the ventricular diastole is the filling phase, which is subdivided into the rapid filling phase and diastasis (slow filling phase). The left ventricular rapid filling phase starts, when the ventricular pressure falls below left atrial pressure and the mitral valve opens. Ventricular pressure continues to decrease after the AV valve opening, because of the continued ventricular relaxation, which is facilitated by the elastic recoil of the ventricular wall. Meanwhile, the ventricular volume increases sharply. The rapid filling phase is followed by diastasis, during which atrial and ventricular pressures and volumes increase very gradually. The P wave on the electrocardiogram indicates the atrial depolarization, which is soon followed by the atrial contraction. The atrial systole causes a small increase in atrial, ventricular, and venous (a wave) pressures. The atrial contraction forwards the blood from the atrium to the ventricle and completes the period of ventricular filling with the late rapid-filling phase. The cardiac cycle is completed by the subsequent ventricular depolarization.

14.3.3 Pressure–Volume Loop

The pressure–volume loop illustrates the changes in left ventricular pressure and volume throughout the cardiac cycle independent of the time (Figure 14.3). The diastolic filling phase starts at the point A and finishes at C, when the mitral valve closes. Due to ventricular relaxation and distensibility the pressure decreases initially (from A to B), despite the rapid ventricular filling. During the remainder of the diastole (B–C) the ventricular pressure increases slightly. During the isovolumic ventricular contraction the pressure steeply increases, and the volume

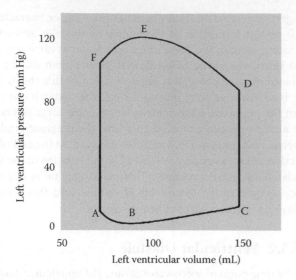

FIGURE 14.3 Pressure–volume loop of the left ventricle. The loop illustrates the changes (ABCDEF) in left ventricular pressure and volume throughout the cardiac cycle.

remains constant (C–D). At D the aortic valve opens, and rapid ejection of blood to the aorta occurs (D–E), during this part of the systole the ventricular volume decreases rapidly with a less steep increase of the ventricular pressure. The rapid ejection is followed by the reduced ejection phase (E–F) and a small decrease in ventricular pressure. At F the aortic valve closes and the isovolumic relaxation starts (F–A). The opening of the mitral valve at point A completes the pressure–volume loop.

14.4 The Pulsatile Blood Flow

The cardiac cycle is a periodic event that repeats itself every heartbeat. From the fact that the pulsatile flow from the heart into the aorta is cyclic, the aortic pressure wave can be expanded in a polynomial expression using the Fourier algorithm. Figure 14.4 illustrates how the sinusoidal polynomial expansion of the aortic pressure can be captured by the sum of the first four terms. This Fourier expansion will prove valuable in describing the frequency behavior of flow in Chapter 17.

14.5 Heart Sounds

With electronic amplification, four heart sounds can be discriminated and graphically recorded as a phonocardiogram (Figure 14.2). With a stethoscope, only two are usually audible. The first heart sound (S1) occurs at the onset of ventricular systole and it is chiefly caused by the abrupt closure of the AV valves and the consequent vibration of the ventricular walls and oscillation of the blood in the ventricular chambers. It has a crescendo–decrescendo quality. It is the loudest and longest of the heart sounds; it can be heard most clearly over the apical region of the heart. The intensity of the S1 depends mainly on the ventricular contraction force. The second heart sound (S2) occurs at the beginning of isovolumic relaxation and arises from the closure of the semilunar valves. The aortic valve closes slightly before the pulmonic valve. During the inspiratory phase, this time difference is enhanced and it is referred to as the physiological splitting of the second heart sound. The reason for the increased time difference between the semilunar

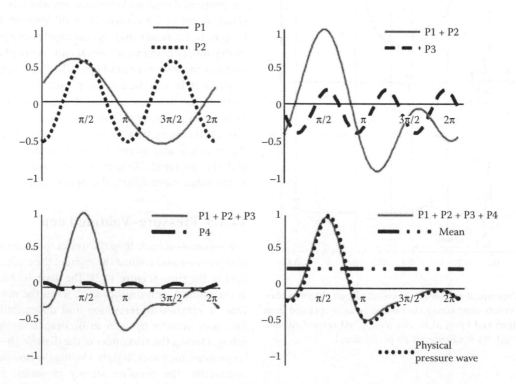

FIGURE 14.4 Fourier transform of a single period of aortic pressure wave: merely four sinusoidal terms will provide a fit.

valves closure is the prolonged ejection of the right ventricle. During inspiration, the intrathoracic pressure decreases and this enhances venous return and the diastolic filling of the right ventricle; consequently, the ejection of the extra blood volume needs a little extra time. The third heart sound (S3) sometimes can be detected in healthy children; however, it is more commonly heard in patients with heart failure. It occurs in early diastole and the possible cause of this low-intensity sound is the vibration of the ventricular wall caused by abrupt cessation of ventricular distension during ventricular filling. The fourth heart sound (S4) is rarely heard in normal individuals. It is created by the atrial contraction, which results in oscillation of blood and cardiac chambers.

14.5.1 Measurement of Arterial Blood Pressure: Korotkoff Sounds

Measuring a patient's systolic and diastolic blood pressures is one the most routine diagnostic tools in the hands of a physician. The most widely used noninvasive method is based on auscultation technique. A right-sized inflatable cuff is wrapped around the patient's upper arm and the pressure within the cuff is monitored with a mercury manometer.

The pneumatic cuff is inflated with air to a pressure (P_{cuff}) well above the normal systolic pressure, which makes the blood vessels in the upper arm to be compressed and then the cuff is slowly deflated. The blood pressure (P_{vessel}) assessment is based on the detection of low-pitched Korotkoff sounds over a peripheral artery with a stethoscope at a point distal to cuff compression of the artery (Figure 14.5). The tapping sounds occur when cuff pressure falls below the peak arterial systolic pressure and results from abrupt arterial opening and turbulent jet of flow in the partially collapsed artery. The Korotkoff sounds are heard when the cuff pressure is between the systolic and diastolic arterial pressures. The highest cuff pressure at which the tapping sounds

are detected defines the systolic arterial blood pressure. The cuff pressure at which the Korotkoff sounds diminish defines the diastolic arterial pressure.

14.6 Arterial System: Arteries and Arterioles

The arterial system can be subdivided into vessel sizes arranged in accordance with their respective function and location in the circulatory system.

14.6.1 Basic Functions of the Systemic Arterial System

The principal function of the circulatory system is to provide an adequate quantity of blood flow to the capillary beds throughout the body. The metabolic rates, and consequently the blood flow requirements in different organs and regions of the body, change continuously as people live their daily life. Thus, the magnitude of the cardiac output and its distribution has to be constantly adjusted to the actual need of the systemic organs.

In the following section the general principles involved in the regulation of blood pressure and blood flow will be discussed.

The highest blood pressure can be measured in the aorta and as the blood flows away from the heart the pressure decreases (Figure 14.6). In normal adult the mean aortic pressure is about 95 mm Hg. As indicated earlier, the aorta and the large arteries have a relatively little resistance to flow; thus the mean blood pressure does not decrease significantly along their lengths. The highest resistance to flow is caused by the small arteries and

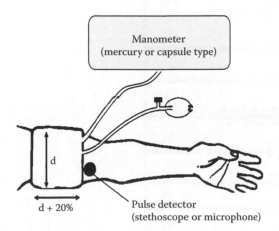

FIGURE 14.5 The pneumatic cuff pressure exceeds the arterial pressure at the upper arm ($P_{cuff} > P_{vessel}$). The cuff is slowly deflated and the blood pressure assessment is based on the detection of Korotkoff sounds over the peripheral artery with a stethoscope.

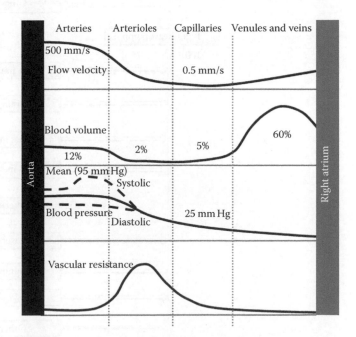

FIGURE 14.6 Schematic illustration of the systemic vasculature from the aorta to the right atrium: flow velocities, blood volumes, blood pressures, and vascular resistance.

arterioles, by the so-called resistance vessels resulting in 50–70% of pressure drop. At the capillary level the mean blood pressure is around 25 mm Hg. As the blood travels back to the heart, its pressure decreases further; at the level of the right atrium, it is very close to 0 mm Hg. The basic flow equation is analogous to Ohm's law (Equation 14.3):

$$I = \frac{U}{R}. \qquad (14.3)$$

which describes the relationship between current (I), voltage (U), and resistance (R). In the basic flow equation (Equation 14.4) the blood flow rate (Q in volume/time) corresponds to the electric current, the pressure difference (ΔP in mm Hg) to the electric potential difference, and the resistance to flow (R in mm Hg × time/volume) to electric resistance.

$$Q = \frac{\Delta P}{R}. \qquad (14.4)$$

Blood flow rate (Q) increases with an increase in the pressure gradient ($\Delta P = P_{outlet} - P_{inlet}$). The pressure gradient gives the driving force for the flow. The vascular resistance indicates how much of pressure gradient is needed to achieve a certain flow.

The vessels network's overall resistance can be calculated analogous to the overall resistance in electrical circuits. When vessels with individual resistances $R_1, R_2, \ldots, R_n$ are connected in series as indicated in Figure 14.7, the total series resistance (R_s) across the network is given by Equation 14.5:

$$R_s = R_1 + R_2 + \cdots + R_n. \qquad (14.5)$$

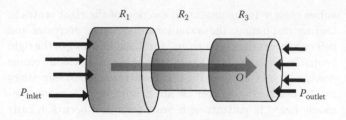

FIGURE 14.7 Illustration of the series resistance network.

When the vessels with individual resistances ($R_1, R_2, R_3, \ldots, R_n$) are arranged to form a parallel network, as illustrated in Figure 14.8 the overall resistance for the parallel network (R_p) can be calculated according to the following formula (Equation 14.6):

$$\frac{1}{R_p} = \frac{1}{R_1} + \frac{1}{R_2} + \cdots + \frac{1}{R_n}. \qquad (14.6)$$

As the equation implies, in general, the more parallel elements constitute the network, the lower the total resistance of the network. This is true for the capillary vascular beds as well, where the individual vessel has a relatively high resistance; however, many capillary vessels in parallel configuration can have a low total resistance. The total blood flow through a series and a parallel resistance network is $Q_s = \Delta P/R_s$ and $Q_p = \Delta P/R_p$, respectively.

The conduits of the circulatory system are arranged in series and in parallel (Figure 14.8). The two sides of the heart are in

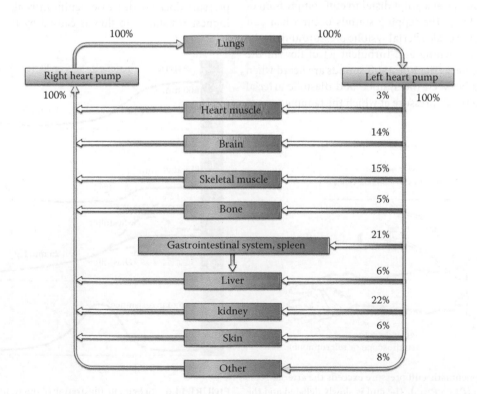

FIGURE 14.8 The schematic arrangement of the cardiovascular system with the relative distribution of cardiac output to systemic organs during rest.

series with each other and are separated by the pulmonary and systemic circulation. Consequently, the right and the left ventricles must each pump the same amount of blood each minute. This blood volume is called the cardiac output. The systemic organs are functionally arranged in parallel (i.e., side by side). One major exception is the liver, which receives blood directly from the intestinal tract via the hepatic portal system. The liver circulation is also supplied via the hepatic artery of the aorta. As a consequence of the latter, the liver circulation has parts that are in series and parts that are in parallel with the intestinal circulation. The hemodynamic implications of the parallel arrangement of the systemic circulation are the following: First, the flow through any one of the systemic organs can be independently regulated; thus, blood flow changes in one organ do not significantly affect the blood flow in other organs. Second, the arterial blood composition is almost identical for all systemic organs.

14.6.2 Components of the Sytematic Arterial System

The arterial system is divided into the systemic and pulmonary circulation. The principal function of the circulation is to carry blood from the heart to the capillary beds and back to the heart. The distribution of blood volume is controlled by the anatomical vessel arrangement and physiologic control of vessel constriction and dilation. The main regulatory vessels are the arterioles, which are the terminal components of the arterial system. Between the arterioles and the heart, the main conduit vessels, namely, the aorta and the pulmonary artery, and their branches transfer the blood toward the periphery. The intermittent output of the heart is converted to a steady flow as the blood reaches the capillary beds. This dampening function of the large vessels can be attributed to the elastic properties of the large arteries.

The entire stroke volume is ejected into the arterial system during a systole. The energy of cardiac contraction is dissipated as a forward capillary flow and as potential energy stored in the stretched arteries. During the diastole, which occupies about two-third of the cardiac cycle, the elastic recoil of the arteries converts the stored potential energy into blood flow. The elastic nature of the vessels is characterized by the parameter called compliance (C) as described by Equation 14.7.

$$C = \frac{\Delta V}{\Delta P}. \qquad (14.7)$$

where compliance (C) describes how much the vessel's volume changes (ΔV) in response to the change in distending pressure (ΔP). The distending pressure is the pressure difference between the intraluminal and outside pressures on the vascular wall. As they are branching off the aorta, the large arteries distribute the blood to specific regions and organs of the body. The consecutive branching of the arteries causes an exponential increase in the number of arteries. While the individual vessel diameter gets smaller and smaller, the total vessel cross-sectional area increases several fold compared to the aortic cross-section. Arteries are often termed as conduit vessels, since they have relatively low

and constant resistance to blood flow. Once the vessel diameter decreases below 200 μm, the arteries are termed as arterioles. The majority of arterioles range in diameter between 5 and 100 μm, and in the proportion of luminal dimensions they have a thicker vessel wall with much more smooth muscle and less elastic fiber than in the arteries. The arterioles regulate the blood flow and blood pressure within organs. Their wall is rich in smooth muscle cells; thus their diameter can be actively changed. Because of these characteristics, they are often termed as resistance vessels. The arterioles give rise directly to the capillaries (5–10 μm). Some capillaries have diameters less than that of the red blood cells (7 μm); it is necessary that the cells must deform to pass through them. The capillary wall is composed of only endothelial cells and a basement membrane. The capillaries are the smallest vessels of the vascular system, although they have the greatest total collective cross-sectional area because they are so numerous.

The combined cross-sectional area of the capillary system is about 1000 times greater than the cross-section of the aortic root; consequently, the mean blood flow velocity is about 1000-fold less than the velocity in the aorta (0.5 mm/s versus 500 mm/s). The capillaries have a cumulative 100 m² of surface area, which serves as the location for exchange of gases and solutes between blood and interstitial fluid. Referring to their function, capillaries are often termed as the primary exchange vessels of the circulatory system.

14.7 Venules and Veins

After leaving capillaries, the blood is collected in highly permeable postcapillary venules, which serves as exchange vessels for fluid and macromolecules. As postcapillary venules converge, they form larger venules with vessel wall containing smooth muscle cells. These vessels like the arterioles are capable of active diameter change. By contracting or dilating they play an important role in intracapillary pressure regulation.

Venules converge and form larger veins. Together venules and veins are the primary site where more than 50% of the total blood volume is found and where the circulating blood volume is regulated. Consequently, they are termed as primary capacitance vessels of the body. The final veins are the inferior and superior venae cavae, which are connected to the right atrium of the heart.

14.7.1 Control of Peripheral Blood Flow Distribution

It is important to note the difference between blood flow (volume/time) and blood flow velocity (distance/time) in the vascular system. Consider that the blood flows rapidly in the peripheral vascular system in the region where the total cross-sectional area is the smallest (the aorta), and the flow is most slow in the region with the largest cross-section (the capillary beds). The blood flow velocity varies inversely with the local cross-sectional area (Figure 14.6).

As was previously described, the major portion of the circulating blood is contained within the veins of the systemic organs. This large blood reservoir is termed as a peripheral venous pool. The inferior and superior venae cavae and the right atrium constitute the central venous pool, which is smaller in volume than the peripheral pool. When the veins of the peripheral pool constrict, the blood enters the central venous pool and consequently the cardiac filling enhances, which in turn increases stroke volume according to the Frank–Starling law, which is described in more detail in Chapter 15. Thus, the peripheral veins have great importance in controlling the cardiac output.

The aortic pressure fluctuates between systolic and diastolic values with each heartbeat. The difference between the systolic and diastolic pressures is the aortic pulse pressure. As blood flows down the aorta and the consecutive segments of the arteries, characteristic changes take place in the shape of the pressure curve. As the pulse wave moves away from the aorta the systolic pressure rises and the diastolic pressure decreases (Figure 14.5). The probable reason for this hemodynamic phenomenon is the decreased compliance of the peripheral vessels and the generated reflective waves, which travel back toward the heart and summate with the pulse wave traveling down the aorta and the arteries. The mean arterial pressure decreases by only a small amount in the arterial system, because the distributing arteries have relatively low resistance. The largest pressure drop occurs in the arterioles, and by the capillary beds the blood pressure declines to 25 mm Hg. In the venules and veins, the blood pressure decreases further, and the filling pressure for the right side of the heart (central venous pressure) is close to 0 mm Hg.

The different organs receive different amounts of blood during rest and exercise. Normally, the kidney, muscle, and abdominal organs receive about 20% each, while the skin receives 10%, the brain 13%, the heart 4%, and other regions about 10%. During exercise, blood flow will be tenfold in the muscles, triples in the heart and skin, while it decreases in the kidney, abdomen, and other regions to about a half. The brain has no change compared to rest. Control of blood flow is regulated at different levels and at different times, both locally and by the autonomic nervous system.

At the tissue level, at the smaller arterioles, mostly local metabolites are active. Within minutes, flow is mostly controlled by the needs of the tissue (for oxygen, nutrients) and also by some special needs (heat loss in the skin, excretion by the kidneys). Blood flow is increased if CO_2 is high, pH is low, or O_2 is decreased. Low O_2 causes production of vasodilator substances; also, an increased metabolic demand (low concentration of O_2 or other nutrients locally) causes vasodilatation. Blood pressure also has a regulating role. If blood pressure is high, nutrient supply will also be high, while the concentration of vasodilating substances decreases (washout), which in turn causes vasoconstriction. The reverse is true if blood pressure is low, which causes ischemia and an elevation in the concentration of vasodilating substances. During *active hyperemia*, increased activity in the tissue causes an elevated local blood flow (like in the muscles during exercise). During *reactive*

hyperemia, transient decrease in blood flow is followed by an increased blood supply.

In most of the organs, blood flow is relatively constant within certain pressure limits, which is caused by *autoregulation*. This is achieved by the local metabolic regulation mentioned above (washout of dilating substances by the initial higher flow) and also by the myogenic response to higher pressures causing elevated arteriolar tone and resistance. For example, in the brain this is because blood flow is very sensitive to concentration of CO_2 and H^+.

Persistent changes in blood flow (like chronic ischemia) cause structural changes on the long term, like the development of collateral vessels as well as dilatation of vessels.

Besides local factors, blood flow is influenced by central mechanisms—the autonomic nervous system. The blood vessels are innervated by the sympathetic nervous system—a constant level of output maintains vessels in a partially contracted state called the vasomotor tone. An increase in the sympathetic activity causes increased resistance in the smaller arterial vessels, while in the larger arteries and veins it influences the distribution of blood volume.

Lastly, blood flow is also regulated by several vasoconstricting and vasodilating humoral factors [vasoconstrictors: adrenaline–noradrenaline, angiotensin, vasoactive hormones (ADH), etc.; vasodilators: histamine, bradykinin, etc.].

14.7.2 Mechanisms of Vasoconstriction

The increased concentration of calcium (Ca^{2+} ions) within vascular smooth muscle leads to vasoconstriction. There are several mechanisms that can elicit such a response. The two most important ones are circulating *epinephrine* and activation of the sympathetic nervous system (through release of *norepinephrine*). These compounds connect with cell surface adrenergic receptors. Circulating catecholamines probably have a role under situations when their levels are much higher than normal (like vigorous exercise or hemorrhagic shock). There is also a vasodilating role of moderate levels of epinephrine via β2 receptors (e.g., in the muscle during exercise). Many other factors are involved in vasoconstriction as mentioned in the previous chapter. *Angiotensin II* is produced from angiotensin I by renin (released from the kidney) in response to decreased blood pressure. *Vasopressin (ADH)* is a hormone produced by the hypothalamus and released from the posterior pituitary gland due to hypovolemia or decreased extracellular fluid osmolality; it has a vasoconstrictive effect besides decreasing water excretion by the kidney. The latter two substances probably have vasoconstrictor roles during hypovolemic shock states. *Endothelin I* (ET-1) is a 21-amino-acid peptide that is produced by the vascular endothelium from a 39-amino-acid precursor, big ET-1. Its level is increased by different stimuli (like angiotensin II, ADH, increased shear stress, etc.). Endothelin has a vasodilating action if connected to the endothelium ET_B receptors through the formation of NO, while it causes vasoconstriction if it is bound to smooth muscle ET_A and ET_B receptors (under normal conditions, ET_A receptors predominate in vessels). Some *prostaglandins* (like thromboxane) are also capable of eliciting a vasoconstrictive response. Besides local metabolic and humoral factors, arterioles also respond to pressure changes. In case pressure increases suddenly, the vessel will dilate first passively and then constrict actively (the opposite happens with a sudden decrease in pressure). This is called the myogenic response.

14.8 Vascular Control in Specific Tissues

Blood flow in some organs is mostly controlled by local metabolic control (like the brain, skeletal muscle, and heart), while in others (like the kidneys, skin, and abdominal organs) it is influenced more by sympathetic nervous activity.

Coronary blood flow is mainly determined by myocardial oxygen requirement. This is met by the heart's ability to change coronary vascular resistance considerably, while the myocardium extracts a high and relatively fixed amount of oxygen.

In the heart, external compressive forces are more pronounced in the endocardial region; under normal situations this is balanced by the longer diastolic filling time. However, in the case of left ventricular hypertrophy or coronary stenoses, decreased blood flow will first cause ischemia in these regions. Increased activity of sympathetic fibers (exercise or stress) causes vasodilatation rather than vasoconstriction in the heart. It is caused by the concurrent elevation in myocardial metabolic needs (increased heart rate and contractility), increased oxygen consumption, and therefore a more pronounced local metabolic vasodilating action that counteracts the sympathetic vasoconstriction.

In the skeletal muscle, blood flow is low under normal circumstances, but is still above the metabolic needs. It is under a very strong metabolic control, and muscle oxygen consumption is the primary determinant of blood flow.

The brain has a nearly constant rate of metabolism and its circulation is autoregulated very tightly by local mechanisms causing an even flow under most circumstances (unless blood pressure is very low or under changes of pCO_2). On the other hand, blood flow to different regions of the brain can change due to increased local activity. Cerebral blood flow increases/decreases with higher/lower partial pressure of CO_2 (pCO_2), respectively. pCO_2 changes influence cellular H^+ concentrations, which has an effect on the arterioles. Cerebral vasodilatation is also caused by a decrease in pO_2. The sympathetic and parasympathetic fibers have little effect on cerebral blood flow during normal conditions, although sympathetic tone may protect cerebral vessels from excessive distention if the arterial pressure rises suddenly to high levels.

Abdominal organs are innervated by sympathetic vasoconstrictor nerves, which play a major role in blood flow regulation. Increased tone causes a drastic decrease in flow and mobilization of large amounts of blood into the venous pool. Food ingestion also increases blood flow in these organs. The kidneys are under strong sympathetic control; under prolonged and marked vasoconstrictor tone, even renal failure can develop.

Blood flow to the skin is altered by metabolic changes and environmental conditions (temperature homeostasis). An extensive venous system of small vessels (the venous plexus) contains much of the blood volume of the skin; increased tone of sympathetic fibers causes vasoconstriction and a decrease in heat loss. The arteriolar, resistance vessels in the skin are also innervated by sympathetic fibers and have a high baseline tone. Heat loss is achieved by a decreased sympathetic tone, but also in some areas cholinergic sympathetic fibers cause vasodilatation and sweating. Not only the body, but also local skin temperature can affect blood flow.

Pulmonary blood flow equals cardiac output. Pulmonary vessels have low resistance and are more compliant than the systemic arteries. A rise in pulmonary arterial pressure causes a decrease in pulmonary vascular resistance. Another major difference is that hypoxia causes vasoconstriction in pulmonary arteries, its mechanism is not clear. It helps to maintain blood flow to areas of the lung with adequate ventilation.

References

Berne, R.M. and Levy, M.N., *Cardiovascular Physiology*. Maryland Heights, MO: Mosby, 2001.

Fuster, V., Walsh, R.A., O'Rourke, R.A., and Poole-Wilson, P., *Hurst's the Heart*, 12th Edition. New York, NY: McGraw-Hill Medical Publishing Division, 2008.

Klabunde, R.E., *Cardiovascular Physiology Concepts*. Philadelphia, PA: Lippincott Williams & Wilkins, 2005.

Mohrman, D.E. and Heller, L.J., *Cardiovascular Physiology*, 6th Edition. New York, NY: McGraw-Hill Medical Publishing Division, 2006.

Control of Cardiac Output and Arterial Blood Pressure Regulation

István Osztheimer and
Pál Maurovich-Horvat

15.1 Introduction

Metabolism in the human body is dependent on blood flow. Just 5 min after cessation of blood flow in the arteries, irreversible changes happen in the most sensitive tissues. Each tissue type needs a more or less continuous blood supply. Oxygen and substances have to be transported to the cells, and carbon dioxide and end product metabolites have to be washed out, taken away, and excreted.

The human heart and vasculature have to meet all these needs life long under all circumstances. Just consider the necessary changes during exercise, digesting a meal, after standing up from a lying position, under stress, during hemorrhage, and so on. Also think of the not so easy anatomy of the human vasculature: the double circuit and double pump system.

Engineering of another system with comparable endurance and control has still not occurred. No wonder that cardiovascular morbidity and mortality are among the major issues of the modern world and health systems.

This section aims to describe the control of this unique system. The first two subsections deal with heart rate (HR) (Section 15.2) and stroke volume (SV) (Section 15.3), the two determinants of cardiac output (CO). Subsection 15.3 deal with neural control of CO and all of its determinants. After a short section about cardiovascular reserve, measurements of the determining factors of circulation are discussed.

15.2 Control of HR

HR is controlled by the sinoatrial (SA) node. The SA node is a pacemaker that is capable of initiating a heartbeat on its own, but has a strong neural control mechanism. The average resting heartbeat is around 70 min^{-1}, which decreases during sleep and gets higher during exercise or stress. Clinically a heartbeat of less than 60 min^{-1} is called bradycardia and a heartbeat of greater than 100 min^{-1} is called tachycardia. Regulation of the heartbeat is under autonomic nervous control. The two main functional components of this system are the sympathetic and parasympathetic pathways. Sympathetic activity enhances HR, whereas parasympathetic activity diminishes HR.

The parasympathetic pathway of the heart originates from the medulla oblongata (nucleus dorsalis nervi vagi or parts of the nucleus ambiguous). As for all efferent autonomic nervous pathways, the first fibers have to synapse with postganglionic cells to reach the desired organ. The first fibers are carried by the vagus nerve, which travels through the neck close to the carotid arteries to the mediastinum and from there to the epicardial surface of the heart. The parasympathetic ganglia lie in the epicardial surface or very close to the sinus and the atrio-ventricular (AV) nodes. The neurotransmitter of the postganglionic neuron is acetylcholine, which acts rapidly on the postganglionic potassium channels. Both the SA and the AV node are rich in cholinesterase, which breaks down acetylcholine very fast. Thus a beat-by-beat regulation can be achieved.

The sympathetic preganglionic fibers originate from the C6-T6 (variations exits) segments of the spinal cord and synapse in the paravertebral chains of ganglia (stellate or middle cervical). The postganglionic sympathetic fibers form an extensive epicardial plexus together with preganglionic parasympathetic fibers. The postsynaptic neurotransmitter is norepinephrine, which activates a relatively slow second messenger system, namely the adenyl cyclase system. Norepinephrine is taken up by nerve terminals and the reminder is washed out slowly by the bloodstream. Thus a beat-by-beat regulation is not possible for the sympathetic nervous system. Sympathetic activity has a delayed and prolonged response on the heartbeat.

β-Adrenergic receptor antagonists (propranolol, metoprolol, bisoprolol, and nebivolol) can prevent sympathetic activity on the sinus node, whereas atropine can block the postsynaptic effect of acetylcholine on muscarinic receptors. Usually, the sinus node is under parasympathetic tone. If atropine is given, the HR quickly increases, while β blockers have just a slight HR-lowering effect.[1]

15.3 Control of SV

The volume of blood ejected from the ventricle is called the stroke volume (SV).[2] SV (whose unit is liter) and HR (whose unit is L/min) are the two determinants of CO. CO is the volume of blood pumped by the heart per minute and its unit is L/min.

$$CO = HR \times SV, \tag{15.1}$$

HR—as described above—is under strong neural control, although the SA node is able to maintain a stable rhythm spontaneously. SV, on the other hand, has strong intrinsic control systems. Denervated hearts can adapt to different needs in CO, mainly by altering SV.

The Frank–Starling (Otto Frank and Ernest Starling) mechanism is a well-studied quality of the heart; nevertheless, the presumed physiological basis has changed a lot since its discovery a century ago. These two physiologists independently discovered the same responses of isolated hearts to changes in preload and afterload.

15.3.1 The Frank–Starling Mechanism

If the myofilaments of the heart are more stretched before contraction (at the end of diastole) they generate a stronger contraction during systole. This effect cannot be described by a precise equation; rather it can be measured and plotted on a graph. This graph is called the ventricular function curve (Figure 15.1).

Preload is the force that dilates the ventricular musculature or myofilaments before the onset of contraction. This can be measured as the ventricular filling pressure, or more accurately the end diastolic pressure. Since it is the left ventricle that pumps blood into the systemic circulation, we analyze the left ventricu-

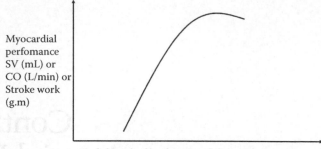

FIGURE 15.1 The Frank–Starling mechanism. Ventricular function curve: increased stretching of myofilaments before contractions results in the increased strength of the contraction and hence results in increased myocardial performance.

lar end diastolic pressure (LVEDP) or volume (LVEDV). LVEDV represents, more precisely, the stretch of myofilaments.

Afterload is the mean aortic pressure that has to be overcome by the ventricular contraction. The Frank–Starling mechanism shows that in case the preload rises (stretching of myofilaments), the heart reacts with a higher performance [elevated SV, CO, or elevated stroke work, which is the product of SV and mean arterial pressure (MAP)].

The physiological basis is a stretch-induced change in the sensitivity of the cardiac myofilaments to calcium. More calcium during contraction causes enhanced myocardial performance.

In case the venous return increases (systemic resistance drops because of the need for a higher CO state), the SV and CO rise because of this intrinsic control mechanism. So, if cardiac performance has to be measured, this should always be made according to preload levels. Moreover, the ventricular function curve (fiber length on the abscissa and ventricular performance on the ordinate axis) has to be plotted (Figure 15.1). This function curve represents the myocardial contractility: ventricular performance that is independent of preload and afterload. The myocardial function curve has a downslope at very high filling pressures, indicating an optimum fiber length for myocardial performance. If the curve moves to the left it means an improvement of contractility (healing of stunned myocardium after an acute myocardial infarction). If the curve moves to the right it means deterioration of contractility [worsening of chronic heart failure (CHF)].

Neural control of SV (more correctly of ventricular contractility) is achieved mainly through sympathetic fibers. Parasympathetic fibers do innervate the ventricular myocardium but their depressing effect on contractility is less pronounced. The sympathetic postganglionic fibers secrete norepinephrine to the myocardial synapses and shift the ventricular function curve to the left by raising intracellular cyclic AMP levels, which activates Ca channels. Although the myocardium has a strong intrinsic control, these effects can modify the SV and performance significantly.

Circulating catecholamines secreted from the medulla of the adrenal gland (epinephrine and norepinephrine) reach the heart through the bloodstream, but their effect on contractility is less than moderate compared with neurally released norepinephrine.

15.3.2 Changes in Ventricular Contractility

In the normal state, ventricular contraction meets the needs of the circulation. Central venous pressure (CVP) representing preload is kept within the normal range during any kind of change in CO. Hence myocardial contractility changes according to the needs of the circulation.

There are pathological states where the contractility of the ventricle diminishes. Acute heart failure is mostly caused by acute myocardial infarction. A part of the myocardium, the one that was supplied with blood by the occluded coronary artery, fails to contract. This prompt drop in myocardial contractility often cannot be compensated for by the circulatory system and results in a subsequent drop of CO. The reduced CO is not enough to meet with the needs of the vascular system, and hypotension and shock arise as a result.[3]

CHF, on the other hand, has a different clinical path. CHF is caused by essential hypertension, ischemic heart disease, and toxic agents like alcohol. Slow onset of the diminished ventricular contractility allows the circulatory system to adapt (Figure 15.2).

Retention of fluid by the kidneys enhances the filling pressure, which keeps the CO at or near the normal level. This effect was achieved not by enhancing the myocardial contractility (which is not possible in the case of heart failure) but by the Frank–Starling mechanism. As it is obvious from the picture this adaptation can only work in the case of moderate CHF. Severe CHF lowers the contractility of the myocardium to some degree, whereas the Frank–Starling mechanism cannot compensate for higher filling pressures.

There are reversible causes of reduced contractility (reperfusion of stunned myocardium in acute myocardial infarction, and recovery after a severe myocarditis). In this case, the temporal high-filling-pressure phase is followed by a normalization of contractility and fluid excretion by the kidneys to a normal CVP.

15.3.3 Ventricular Adaptation during Spontaneous Activity

For didactic reasons, HR and SV have been discussed separately above. However, this does not represent physiological function. By increasing the HR, a number of factors are going to change: diastolic filling time and consequently filling pressure and LVEDV are going to drop. This drop means a shift, on the function curve, toward less stretch of myofilaments and lower SV. The opposite is true for elevated HR. A change in CVP alters not only SV but also HR.

Spontaneous activity also involves movement of the body, which causes flow and pressure changes because of gravity. Although in a closed circuit the counterforce of the arterial side should drive the venous blood back toward the heart, gravity tends to pool blood in the lower extremities. Venous compliance and the insufficiency of venous valves and muscle contraction cause piling up of blood and diminish the filling pressure.

Heat or cold can dilate or constrict the arterial system, thereby causing changes in peripheral vasculature resistance.

Exercise dilates the vasculature of the muscles to enhance blood supply and to meet metabolic requirements. This in turn lowers afterload and enhances venous return and hence elevates preload. This state has to be compensated for by increasing the CO. This can mean a four- to sixfold increase in CO. These factors indicate that knowledge about the vascular part of the circulatory system is essential. It is more about the coupling of the heart as a pump to the vascular system that determines ventricular adaptation.

A vascular function curve can be drawn according to the ventricular function curve (Figure 15.3). If the CO is zero (the heart does not pump) a certain CVP can be measured. By increasing the level of CO, the CVP will drop. This draws a graph showing a first-order connection between CO and CVP. As CO rises beyond a certain limit, CVP drops below zero, which means a reduction in venous return, what keeps the CO level constant.

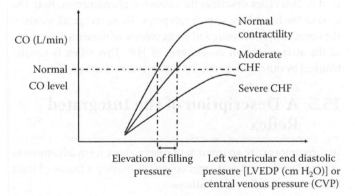

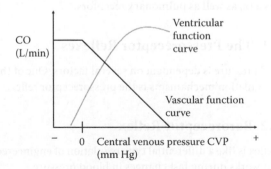

FIGURE 15.2 Changes in contractility during heath failure and compensatory mechanisms. In case of heart failure the ventricular function curve has a right shift compared to the normal state. This means a drop in contractility. Normal CO levels can only be achieved with elevated preload levels. (Adapted from Koeppan, B.M. and Stanton B.A., *Berne and Lewy Physiology*, Sixth Edition, Philadelphia, PA: Elsevier, 2008.)

FIGURE 15.3 Vascular function curve (black) and ventricular function curve (gray) plotted on the same graph. The right intersection of the vascular function curve with the abscissa represents the theoretical point where the heart stops to pump. There is a first-order connection between CO and CVP (Equation 8.4) to the point where CVP is zero. CVP cannot fall below zero; thus on the left of the CVP = 0 point, the vascular function curve is horizontal. (See main text.) (Adapted from Koeppan, B.M. and Stanton B.A., *Berne and Lewy Physiology*, Sixth Edition, Philadelphia, PA: Elsevier, 2008.)

The connection between CO and CVP is the peripheral resistance (R). CVP is inversely proportional to the arteriovenous pressure difference (ΔP).

$$\Delta P_{arteriovenous} = \frac{k}{CVP}, \quad \text{where } k \text{ is a constant.} \quad (15.2)$$

The peripheral resistance is defined by the arteriovenous pressure difference and the CO.

$$R = \frac{\Delta P_{arteriovenous}}{CO}. \quad (15.3)$$

Combining Equations 15.2 and 15.3 leads to the following expression for the CO:

$$CO = \frac{k}{R} \cdot \frac{1}{CVP}, \quad \text{where } \frac{k}{R} \text{ is a constant.} \quad (15.4)$$

This equation represents the first-order connection between CO and CVP.

By plotting the vascular and ventricular function curves on the same graph, an intersection of the two curves is visible (Figure 15.3). This intersection is the equilibrium point. If these curves change (ventricular contractility changes, and vascular filling and vascular resistance changes), the intersection point is going to point out a new equilibrium state.

CO is influenced by HR, contractility, and vascular coupling (preload and afterload).

15.4 Neural Control over Cardiac Performance (Blood Pressure, CO, and HR)

Two main reflexes can be distinguished in the control of arterial pressure: baroreceptor reflex and chemoreceptor reflex. Additional reflexes include the hypothalamus, cerebrum, skin, and viscera, as well as pulmonary receptors.[3]

15.4.1 The Pressoreceptor Reflexes

Arterial pressure is dependent on several factors. One of the primary regulation mechanisms is the pressoreceptor reflex.

15.4.2 Baroreceptor Reflex

This reflex is like a differential type regulation of engineered systems. It works during fast changes in blood pressure.

Stretch receptors are located in the aortic arch and carotid sinuses. Impulses from the carotid sinuses travel through the nerve of Herring to the glossopharyngeal nerve (cranial nerve IX) and impulses from the aortic arch travel through the vagus nerve (cranial nerve X) to end in the nucleus of the tractus solitarius (NTS). NTS is the center of the medulla oblongata where afferent fibers from chemo- and baroreceptors merge. Afferent fiber impulses inhibit sympathetic tone in the peripheral

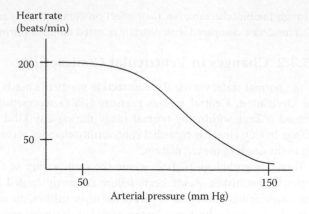

FIGURE 15.4 Baroreceptor response regarding HR. (Adapted from Kollai, M. and Koizumi, K., *Pflügers Arch.* 1989; 413(4): 365–371.)

vasculature by affecting the NTS and enhance parasympathetic tone by the vagus nerve.

The two main effects of the baroreceptors are the following.

A blood pressure rise results in a drop in HR, whereas a pressure drop results in an elevation of HR. This inverse effect can be seen only in case of rapid changes of blood pressure and only within an intermediate range of pressure. Below the cutoff point of 100 mm Hg and above 180 mm Hg, this effect is blunted (Figure 15.4).

Baroreceptors react to sudden changes in blood pressure, or rather to the pulse pressure. During upslope of the pulse amplitude, firing of the afferent receptor fibers increases, whereas during downslope (diastole), the firing gets inhibited. The impulses from the baroreceptors inhibit the cerebral vasoconstrictor areas, resulting in a decrease of the peripheral resistance and lowering of the blood pressure.

15.4.3 Bainbridge Reflex

The Bainbridge reflex is named after the researcher who discovered it. This reflex describes the following phenomenon. Both the atria of the heart have stretch receptors. These receptors measure the venous filling pressure and so the volume of the atria. Stretching of the atria results in an increase of HR. This reflex is usually blanked by other reflexes that elicit opposite effects on HR.

15.5 A Description of the Integrated Reflex

Baroreceptors play an important role in short-term adjustments of blood pressure. Long-term control is mainly a factor of fluid balance regulated by the kidneys.

15.5.1 Chemoreceptor Reflexes

Peripheral chemoreceptors are located in the region of the aortic arch and carotid sinuses. These are well-vascularized bodies measuring the pO_2, pCO_2, and pH of the arterial blood. Afferent fibers influence both the respiratory and the cardiovascular system. Hence their effect in a normal functioning body is diverse.

Chemoreceptors are excited by hypoxia, hypercapnia, and acidosis. Their effects are the greatest in case hypoxia and hypercapnia coexist. The main function of these receptors is to regulate respiration, and cardiovascular effects are just secondary. The primary cardiovascular response is bradycardia in case chemoreceptors get stimulated. Afferent fibers excite the medullary vagal center, which in turn has a negative chronotropic effect on the heart. In a clinical situation bradycardia is only seen in case respiratory effects are blocked (sedation, respiratory insufficiency, airway obstruction, etc.). In a physically normal individual, secondary responses due to the enhanced respiration block bradycardia (hypocapnia and enhanced lung stretch).

Afferent fibers also stimulate the vasoconstrictor regions, thereby increasing peripheral resistance.

15.5.2 Chemoreceptor Effects from the Medullary Centers

pCO_2 is the main stimulus for the chemosensitive regions of the medulla. A raise in pCO_2 raises peripheral resistance by enhancing vascular tone, whereas a fall in pCO_2 diminishes vascular tone and peripheral resistance. pH also affects these receptor areas: lowering of the blood pH elevates the vascular tone and peripheral resistance. Hypoxia has a minimal effect on these medullary receptors. Moderate hypoxia can stimulate the vasomotor region, but severe hypoxia depresses central neural activity.

15.5.3 Other Reflexes

15.5.3.1 Hypothalamus

Behavioral and emotional reactions can have huge effects on blood pressure and HR. Certain areas of this brain region can elevate the blood pressure and HR or decrease it. This is also central to body temperature control. In case temperature drops, the vasculature of the skin gets constricted to prevent heat loss.

15.5.3.2 Cerebrum

The cerebrum can control blood flow to different areas of the body. Motor and premotor areas of the brain mostly elevate blood pressure, whereas fainting represents a blood pressure drop caused by cerebral centers located in the upper part of the brain.

15.5.3.3 Skin and Viscera

Both pressor and depressor reflexes can be elicited by pain or distention.

15.5.3.4 Pulmonary Reflexes

15.5.3.4.1 Respiratory Sinus Arrhythmia

Neural activity in the vagal fibers increases during expiration and decreases during inspiration. The opposite effect is true for sympathetic fibers. As discussed earlier, it is the parasympathetic nervous system that is able to have beat-to-beat influence on the HR. The result is an alteration in HR according to respiration. A very pronounced respiratory sinus arrhythmia can be found in trained subjects. It was proposed that by measuring this phenomenon the vagus tone could be quantified. However, expectations were not met fully.

Inflation of the lungs causes systemic vasodilatation and a fall in blood pressure. This reaction is often seen after intubation and mechanical ventilation. Collapse of the lungs elevates blood pressure. Vagus and sympathetic nerves are the main afferent fibers.

15.5.3.5 Sleep Apnea

It is a complex reflex caused by the chemoreceptors during temporal airway obstruction while sleeping. Blood pressure elevation due to hypoxia is exaggerated, because pulmonary stretch receptors are deactivated.

15.6 Cardiovascular Reserve

As described above, CO can increase up to four- to sixfold during exercise, depending on the level of training and personal abilities. HR is a major determinant of cardiovascular reserve. Increasing HR allows an increase in CO. However, after a certain limit, CO drops because diastolic filling time reaches a point where no effective filling can occur. The maximum HR is often calculated as 220(L/min) + age (years).

15.7 Measurements of Pressure and Flow

Although measurements of pressure and fluid flow appear to be very easy to achieve, they are very hard to accomplish in the living body. The vasculature is not rigid tubing; moreover, it is flexible, distensible, and can be regulated by neural activity. The cardiac pump creates a periodically changing pressure and flow. Blood is not a Newtonian fluid; it consists of cellular elements and many others. Its viscosity changes with variation of the temperature, the size of the vessel, an so on. It is also a question where to measure pressure and flow, since it is different in each segment of the vasculature. These factors make the human circulatory system very complex. Attempts to create a full mathematical model have still been a difficult task even with the use of computing devices.

Nevertheless, blood pressure and flow are the most important clinical measurements. Blood pressure can be measured directly by catheterizing peripheral arteries and connecting them to strain gauges.

Systolic blood pressure is the pressure peak created by the heart during systole (P_s), whereas diastolic blood pressure is the lowest pressure during cardiac relaxation measured in the arterial system (P_d). The difference in these pressures is termed pulse pressure (P_P).[4]

$$P_P = P_s - P_d. \tag{15.5}$$

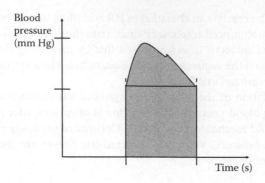

FIGURE 15.5 MAP.

MAP is calculated as the area under the pressure curve divided by the time interval involved (Figure 15.5, Equation 15.6).

$$\overline{P_a} = \frac{\int_{t_2}^{t_1} P_a \, dt}{t_2 - t_1}. \tag{15.6}$$

Under normal conditions and normal HR, this equation can be simplified since MAP is closer to the diastolic pressure; MAP lies approximately one third of the pulse pressure above the diastolic level.

$$\overline{P_a} = P_d + \frac{P_s - P_d}{3}. \tag{15.7}$$

It is a very important phenomenon that the systolic pressure raises with distance from the heart, whereas the diastolic pressure drops with distance from the heart. MAP falls in order to establish blood flow. This phenomenon is created by the interference of pressure impulses coming from the heart and rebounded ones coming from the periphery.

Blood pressure can also be measured noninvasively. For this procedure, we use a sphygmomanometer, which is a cuff connected to a pressure sensor. The classical and most precise way is the so-called Riva–Rocci method (Figure 15.6), which has been named after the inventors. The pressure cuff is placed on the extremity, usually the left arm. A stethoscope is placed distally above the artery. The cuff is inflated to a pressure that is surely above the systolic blood pressure, so there is no flow to the distal part of the artery and no murmur can be heard above it. By slowly lowering the pressure in the cuff, a certain point can be reached where the systolic blood pressure is high enough to pump blood under the cuff to the distal part of the vessel. A sound called Korotkoff can be heard above the distal part of the artery in systole. By further lowering the pressure in the cuff, this sound decreases and stops. At this point the applied pressure is in equilibrium with the arterial diastolic blood pressure, hence, the blood flow is restored during the full heart cycle, so no murmur can be heard. By listening to the Korotkoff sound while watching the manometer, the systolic and diastolic blood pressures can be measured (Figure 15.6).

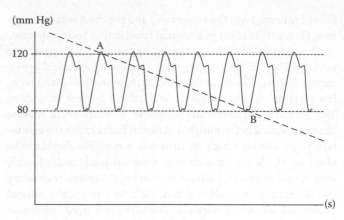

FIGURE 15.6 Blood pressure measurement, Riva–Rocci. A: Systolic blood pressure and B: diastolic blood pressure.

Automatic devices often use a technique called oscillometry. This method just measures the pressure in the cuff. During the gradual lowering of pressure in the cuff, arterial pulsation gets transferred to the cuff filled with air. This pulsation has the highest amplitude somewhere between systolic and diastolic blood pressures. These automatic devices have a mathematical program that calculates systolic and diastolic blood pressures from the point of maximal pressure oscillation amplitude in the cuff.

Flow measurements are far more difficult than pressure measurements. It is also a question what to measure. In a clinical setting, the amount of blood pumped by the heart per minute [CO (L/min)] is an important factor. There are different ways to measure it, although in clinical situations the technique called thermodilution is most often used. We use heat as a tracer substance, since it is easy to measure. Cold saline (usually 0°C) is injected through the proximal port of a central venous or a Schwann–Ganz catheter and temperature is measured distally in the pulmonary artery. CO is then calculated using an equation that considers the temperature and specific gravity of the injectate and the temperature and specific gravity of the blood, along with the injectate volume. Thermodilution can also be used transpulmonarily, where the cold saline is injected into the central venous infusion line and temperature is measured in one of the major arteries.

References

1. Libby P. and Bonow, R.O., *Braunwald's Heart Disease*, Eighth Edition. Philadelphia, PA: W.B. Saunders Company, 2007.
2. Koeppan, B.M. and Stanton B.A., *Berne and Lewy Physiology*, Sixth Edition, Philadelphia, PA: Elsevier, 2008.
3. Standring S., *Gray's Anatomy*, Thirty-Ninth Edition, The Anatomical Basis of Clinical Practice, Philadelphia, PA: Elsevier, 2004.
4. Irwin, R.S. and Rippe, J.M., *Intensive Care Medicine*, Sixth Edition, Philadelphia, PA: Lippincott Williams & Wilkins, 2007.
5. Kollai, M. and Koizumi, K., Cardiac vagal and sympathetic nerve responses to baroreceptor stimulation in the dog, *Pflügers Arch.* 1989; 413(4): 365–371.

16

Fluid Dynamics of the Cardiovascular System

Aurélien F. Stalder and
Michael Markl

16.1 Introduction

According to WHO estimates, cardiovascular disease is the leading cause of death in industrialized nations, accounting for over 16.6 million deaths worldwide each year [1]. In addition, 600 million people with high blood pressure are at risk of heart attack, stroke, and cardiac failure [2].

Clinical care is costly and prolonged and it is therefore crucial to establish and improve efficient methods for diagnosis, treatment, and monitoring of cardiovascular disease. It is therefore important to study and understand local hemodynamics within anatomically complex regions of the human body and it is of high interest since these sites are predisposed to vascular disease.

The pulsatile blood flow through the arterial vasculature generates various types of hemodynamic forces—wall shear stresses (WSSs), hydrostatic pressures, and cyclic strains—that can impact vessel wall biology. Although much of the biological activity in pathophysiology is at the cell and molecular level, the macroscopic flow environment strongly influences biological phenomena at the microscopic level, and is believed to play an important factor in the pathogenesis and progression of vascular disease [3,4].

16.2 The Cardiovascular System

16.2.1 Cardiovascular Circulation

The cardiovascular system consists of blood, blood vessels, and the heart. The blood transports necessary substances such as nutrients and oxygen to the cells of the body and carries back the waste products of those cells. As depicted in Figure 16.1, the

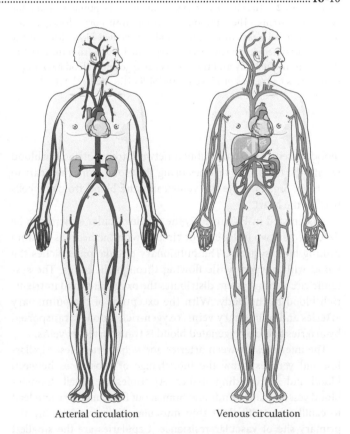

Arterial circulation   Venous circulation

FIGURE 16.1 Arteries and veins of the human body. Arteries (left) carry blood from the heart to the body. Veins (right) transport blood from the body back to the heart. (This figure was produced using SERVIER Medical Art [5].)

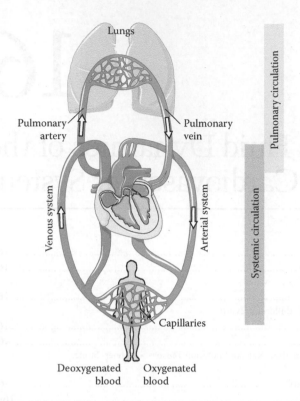

FIGURE 16.2 The circulatory system includes the pulmonary and the systemic systems. The pulmonary system transports deoxygenated blood from the heart to the lungs and oxygenated blood back to the heart. The systemic system carries oxygenated blood from the heart to the organs of the body and transports deoxygenated blood back to the heart. (This figure was produced using SERVIER Medical Art [5].)

16.3 Heart

As depicted in Figure 16.3, the heart is composed of a double circulatory system on its left and right side. The right side of the heart is connected to the superior and inferior vena cava as well as to the pulmonary trunk, both parts of the pulmonary circulation. The more powerful left side of the heart is connected to the pulmonary veins (left and right) as well as to the aorta and is the pump of the systemic circulation. Both sides of the heart are separated by the septum (atrial and ventricular septum) and consist of similar elements: the atria, the atrioventricular valves, the ventricles, and the aortic or pulmonic valves. Blood enters the heart via the atria and is ejected from the ventricles. At the separation between atria and ventricles as well as at the outflow of the ventricles, valves are present and ensure that blood flows only forward. While atria and ventricles are separated by the atrioventricular valves (also called mitral and tricuspid valves for the left and right side of the heart, respectively), the aortic and pulmonic valves are located at the outflow of the left and right ventricles, respectively.

The heart possesses a complex electrical conduction system [6] that allows proper functioning of its muscles. The functioning of the heart throughout the cardiac cycle is explained in Figure 16.4.

The heart cycle can be divided into four phases: isovolumetric relaxation, ventricular filling, atrial systole, and ventricular systole. During diastole (isovolumetric relaxation and ventricular filling) the heart is filled with blood. During the atrial systole and the ventricular systole, the heart contracts and ejects blood. The succession of both diastole and systole composes the cardiac cycle with a typical frequency, the heart rate, of about 70 beats per minute at rest.

blood vessels consist of a tubular network transporting the blood within the body. The arteries bring the blood from the heart to the body organs while the veins carry the blood from the cells back to the heart.

As depicted in Figure 16.2, the cardiovascular system can be described as two independent circuits with the heart as a pump linking both systems. The pulmonary circulation enriches the blood with oxygen while flowing through the lungs. The systemic circulation system distributes the oxygenated and nutrient-rich blood to the body. With the exception of the pulmonary arteries and pulmonary veins, oxygen-rich blood is transported by arteries and deoxygenated blood is transported by veins.

The interfaces between arteries and veins, arterioles, capillaries, and venules allow the interchange of substances between blood and surrounding tissues. Arterioles are small diameter blood vessels that extends and branch out from an artery and lead to capillaries. With their thin muscular walls arterioles are the primary site of vascular resistance. Capillaries are the smallest blood vessels, measuring 5–10 μm in diameter, which connect arterioles and venules, and enable the interchange of water, oxygen, and many other nutrient and waste chemical substances between blood and surrounding tissues.

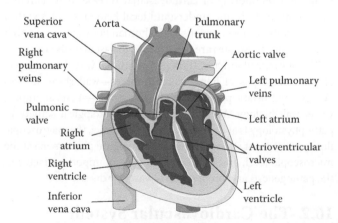

FIGURE 16.3 Anatomy of the human heart. The heart is based on a double circulatory system described as the left and right sides. Both sides contain an atrium, an atrioventricular valve, a ventricle and the aortic (left side) or pulmonic (right side) valve. The left side is connected to the pulmonary veins and the aorta while the right side is linked to the vena cava (superior and inferior) and the pulmonary trunk. (This figure was produced using SERVIER Medical Art [5].)

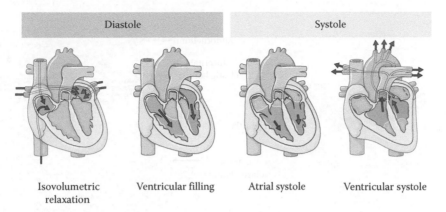

FIGURE 16.4 Heart pump operation. During the isovolumetric relaxation, the atria are filled with blood and the atrial pressure rises gradually. When the atrioventricular valves open, the blood flows into the ventricles; this is the ventricular filling. During the atrial systole, the atria contract to top off the ventricles. Eventually, for the ventricular systole, the pressure in the ventricles rise as they contract, the atrioventricular valves close and the blood is ejected through the aortic and pulmonic valves into the pulmonary trunk and the aorta, respectively. (This figure was produced using SERVIER Medical Art [5].)

16.3.1 Blood Vessels, Pulsatile Flow, and Blood Pressure

As a consequence of the periodic operation of the heart, blood flow shows a pulsatile temporal waveform throughout the body. Blood pressure is the force exerted by circulating blood on the walls of blood vessels, and constitutes one of the principal vital signs. The pressure of the circulating blood decreases as blood moves through arteries, arterioles, capillaries, and veins.

The arterial system is the higher-pressure part of the circulatory system. Arterial pressure varies between the peak pressure during systolic contraction and the minimum during diastole contractions, when the heart expands and refills. This pressure variation within the artery produces the pulse that is observable in any artery, and reflects heart activity.

The blood vessels include arteries, arterioles, capillaries, venules, and veins. While the blood flow can be strongly pulsatile in the arteries, it is relatively constant in the veins. The normal systolic pressure (i.e., maximum pressure during the cardiac cycle) in the arteries is about 120 mm Hg (≈ 16 kPa) for a diastolic pressure (i.e., minimum pressure during the cardiac cycle) of 80 mm Hg (≈ 11 kPa). In contrast, the pressure in the veins is relatively constant and much lower with less than 10 mm Hg (≈ 1.3 kPa). The capillaries, providing an exchange interface for the substances in the blood, are the smallest vessels of the human body. Their cumulated cross-section is yet larger than for arteries or veins and hence the pressure in the capillaries is low.

All vessels, and arteries particularly, can deform or move as a result of the pulsatile flow of blood. Compliance is defined as the ratio of the change in volume over the change in pressure. Due to compliance, arteries expand under high pressure at systole and shrink back at diastole. Compliance is hence directly related to the elasticity of the vessels. Compliance is a buffering element and limits the pressure oscillations during the cardiac cycle.

Compliance generally decreases with age and is one reason why in turn blood pressure increases [7].

Due to the non-negligible elasticity of the arterial vessel walls, the pressure pulse is transmitted through the arterial system 15 or more times more rapidly than the blood flows itself. The aortic pulse wave is initially relatively slow moving (3–6 m/s). As it travels towards the peripheral blood vessels, it gradually becomes faster reaching 15–35 m/s in the small arteries.

Among the blood vessels, the arteries in particular are subject to the development of diseases and are thus of particular interest. As shown in Figure 16.5, the arterial wall can be separated into three layers: the intima, the media, and the adventitia. Among these layers, the intima has a particular role as it is in direct contact with the flowing blood.

The arteries are affected by the blood flow. The blood flowing in the arteries exerts a stress (force per unit surface) on their intima. As illustrated in Figure 16.6, the stress can be decomposed in the hydrostatic stress (pressure $\vec{p}$) and the tangential stress (WSS $\vec{\tau}$). The WSS represents the friction force of the flowing blood on the vessel wall. While the blood pressure in the healthy human body reaches up to 16 kPa, the physiological range of WSS is only 1–2 Pa [7]. Nonetheless, much larger spatial and temporal variations are found for the WSS compared to blood pressure and it has been shown that the WSS is influencing arterial remodeling [8–12].

16.4 Blood

The human blood is a suspension of cells in an aqueous solution and accounts for about 8% of the total body weight [14]. It consists of plasma (55%) and of formed elements (45%). Plasma contains proteins (7%), water (91%), and other solutes (2%). The formed elements (i.e., formed from precursors or stem cells) are the platelets (or thrombocytes), the leukocytes (or white blood cells), and the erythrocytes (or red blood cells).

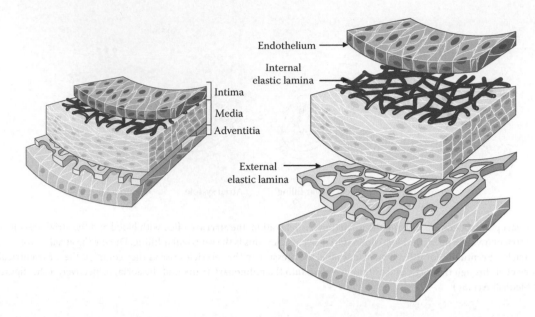

FIGURE 16.5 Anatomy of the arteries. From inside to outside, the arteries are composed of the intima, media, and adventitia. The intima is essentially made of a single layer of endothelial cells and is the thinnest layer of the arteries. The media is made up of smooth muscle cells and elastic tissue and is the largest layer. The outermost layer, the adventitia, is composed of connective tissue. The three layers are separated by the internal and external elastic lamina. (This figure was produced using SERVIER Medical Art [5].)

The red blood cells, transporting the oxygen, are the most abundant in the blood with 5×10^6 cells/mm³. They are disk shaped with a diameter of 7.6 μm and a thickness of 2.8 μm.

Red blood cells are the reason for the non-Newtonian aspect of blood flow: when the shear rate increases, the red blood cells orient themselves with the flow and the viscosity is reduced (the Fahraeus–Lindquist effect). The blood plasma (i.e., blood without its formed elements) is found to behave like a Newtonian fluid [15] (linear dependence of the shear stress on the shear rate). But the blood presents a non-Newtonian character [16,17]

with viscosity depending on the shear rate. Based on the work of Chien et al. [16], the viscosity of blood can be determined for various hematocrit (proportion of blood volume occupied by red blood cells). The Newtonian aspect of blood at normal hematocrit levels and the non-Newtonian aspect of plasma are shown in Figure 16.7.

While the viscosity of plasma remains constant over the range of shear rates, the viscosity of blood changes. The blood presents a strong non-Newtonian behavior at low shear rates, while the

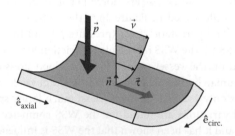

FIGURE 16.6 Stresses acting on the vessel wall. Pressure ($\vec{p}$) and WSS ($\vec{\tau}$). The pressure is acting perpendicular to the vessel wall (i.e., parallel to the normal vector) while the WSS is tangential. WSS arise from the friction of the blood flow on the vessel wall and is related to the blood velocity profile. The direction vectors of the local reference system in the axial and circumferential directions are given by and, respectively. In this illustration, the velocity profile presents only an axial component and consequently is oriented in the axial direction as well. (Adapted from Stalder AF. Quantitative analysis of blood flow and vessel wall parameters using 4D flow-sensitive MRI, PhD Thesis, University of Freiburg, Germany, 2009.)

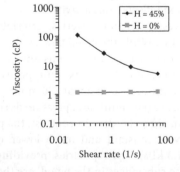

FIGURE 16.7 Non-Newtonian properties of blood according to Chien et al. [16]. Relationship between the logarithm of the viscosity and the logarithm of the shear rate for blood with a physiologically realistic hematocrit of 45% (H = 45%) and plasma (H = 0%). While the plasma presents an almost constant viscosity over the range of shear stresses, the blood at 45% hematocrit presents a non-Newtonian behavior. This non-Newtonian behavior is more pronounced at low shear rates than at higher shear rates where it can be approximated as Newtonian. The viscosity is given in cP: 1 cP = 10^{-3} Pa × s. (Adapted from Stalder AF. Quantitative analysis of blood flow and vessel wall parameters using 4D flow-sensitive MRI, PhD Thesis, University of Freiburg, Germany, 2009.)

behavior is nearly Newtonian at higher shear rates. In large and medium size arteries, shear rates can be higher than 100 s[⊠1] and viscosity can be assumed to be practically constant. As a result, blood flow in medium to large arteries can be reasonably assumed to have a Newtonian behavior.

16.5 Fluid Dynamics

Fluid dynamics is the study of fluids in motion. Fluids include gases and liquids and present the particularities of no or limited resistance to deformation (viscosity) and the ability to flow (i.e., to take on the shape of the container). While solids can be subjected to shear stress and to both compressive and tensile normal stresses, fluids can only be subject to normal compressive stresses (i.e., pressure) and to low levels of shear stress (if they display viscosity).

The behavior of fluids follows specific rules that can be described by the Navier–Stokes equations. In addition, the reaction to the friction stress of a fluid on some boundaries, the shear stress at the wall, is of particular interest.

16.5.1 The Stress Tensor

Stress is by definition a surface force per unit area. The surface of interest need not to be a real surface but a conceptual one, such as that surrounding an infinitesimal fluid particle. In general, the stress vector $\vec{\sigma}$ is a function of both position $\vec{r}$ and surface orientation $\vec{n}$:

$$\vec{\sigma} = \vec{\sigma}(\vec{r}, \vec{n}), \tag{16.1}$$

where $\vec{\sigma}$ is a vector that includes both the hydrostatic pressure (i.e., the part of the stress vector normal to the surface) and the shear stress (i.e., the part of the stress vector tangential to the surface).

According to Newton's third law:

$$\vec{\sigma}(\vec{r},\ \vec{n}) = \vec{\sigma} = (\vec{r},\ \vec{n}), \tag{16.2}$$

that is, for every action (force or stress) there is an equal but opposite reaction.

It is possible to show that the stress vector $\vec{\sigma}$ can be related to a second-order tensor $\ddot{\sigma}$ that depends on the position $\vec{r}$ but not on the surface orientation $\vec{n}$ [18]:

$$\vec{\sigma}(\vec{r},\vec{n}) = \vec{n} \cdot \ddot{\sigma}(\vec{r}). \tag{16.3}$$

There is thus a relation between the three-component vector $\vec{\sigma}$, the nine-component second-order tensor $\ddot{\sigma}$, and the normal $\vec{n}$. Not all the components are yet independent of each other and it is possible to show that $\ddot{\sigma}$ must be symmetric, that is, $\ddot{\sigma} = (\ddot{\sigma})^{\mathrm{T}}$

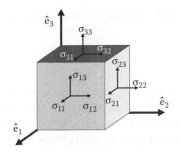

FIGURE 16.8 Components of the stress tensor on an infinitesimal fluid control volume. (Adapted from Stalder AF. Quantitative analysis of blood flow and vessel wall parameters using 4D flow-sensitive MRI, PhD Thesis, University of Freiburg, Germany, 2009.)

and hence $\vec{\sigma} = \vec{n} \cdot \ddot{\sigma} = \ddot{\sigma} \cdot \vec{n}$. The elements of the stress tensor $\ddot{\sigma}$ are shown in Figure 16.8.

If a fluid with constant velocity field is assumed and the gravity ignored, $\vec{\sigma}$ and $\ddot{\sigma}$ have no dependence on $\vec{r}$ and the fluid possess no shearing motion and no shear stress. As a consequence, $\ddot{\sigma}$ depends only on the hydrostatic pressure p:

$$\ddot{\sigma}_p = -p\vec{I}, \tag{16.4}$$

where p is the hydrostatic pressure (where by convention a compressive stress is negative) and $\vec{I}$ is the identity matrix.

If the hydrostatic pressure ($-p\vec{I}$) is subtracted from the general form of the shear tensor, the viscous stress tensor $\ddot{\tau}$ is obtained:

$$\ddot{\tau} = \ddot{\sigma} + p\vec{I}. \tag{16.5}$$

16.5.2 The Deformation-Rate and Rotation Tensors

Considering, a fluid velocity $\vec{v}(\vec{r})$, the gradient of $\vec{v}$ results in a second-order tensor that can be decomposed in a sum of a symmetric and antisymmetric tensors:

$$\nabla \vec{v} = \ddot{\varepsilon} + \ddot{\omega}, \tag{16.6}$$

where the symmetric tensor $\ddot{\varepsilon}$, called the rate-of-deformation tensor, is given by

$$\ddot{\varepsilon} = \frac{1}{2}\left(\nabla \vec{v} + (\nabla \vec{v})^T\right), \tag{16.7}$$

where $(\cdot)^{\mathrm{T}}$ denotes the transposition operation or in index notation:

$$\varepsilon_{ij} = \frac{1}{2}\left(\frac{\partial v_i}{\partial x_j} + \frac{\partial v_j}{\partial x_i}\right), \tag{16.8}$$

with the three orthogonal coordinates $i, j = [1, 2, 3]$, the velocity components v_i and the spatial dimensions x_i.

The antisymmetric part of $\Delta\vec{v}$ is called the rotation tensor:

$$\bar{\bar{\omega}} = \frac{1}{2}\left(\nabla\vec{v} - (\nabla\vec{v})^T\right). \tag{16.9}$$

The symmetric tensor $\bar{\bar{\varepsilon}}$ contains six independent components. While the diagonal elements of $\bar{\bar{\varepsilon}}$ are related to the linear strain, the nondiagonal elements are related to the shear strain. The antisymmetric tensor $\bar{\bar{\omega}}$ possesses only three independent components and is related to the solid body rotation of the fluid.

16.5.2.1 Newtonian Fluid

In solids the shear stress is a function of the strain, but in a fluid the shear stress is a function of the rate of strain. Yet, there are different ways to relate shear stress to the deformation-rate tensor. A fluid is called Newtonian if it presents a linear relation between the shear stress $\bar{\tau}$ and the shear rate $\bar{\bar{\varepsilon}}$. In addition, it needs to be isotropic in which a coordinate rotation or interchange of the axis leaves the stress, rate-of-deformation relation unaltered. Under these assumptions, the viscous stress tensor is given by [18]

$$\bar{\tau} = 2\eta\bar{\bar{\varepsilon}} + \lambda(\nabla\cdot\vec{v})\bar{I} \tag{16.10}$$

where η refers to the shear viscosity (first coefficient viscosity) and λ is the second coefficient of viscosity. $\bar{I}$ is the identity matrix.

16.5.3 Navier–Stokes Equations

Based on the conservation laws applied to mass and momentum, it is possible to derive the basic Navier–Stokes equations [18,19]. The equations can be simplified using assumptions on the fluid such as incompressibility (i.e., constant density), linear dependency of the stress on the strain rate (Newtonian fluid), or constant viscosity.

16.5.3.1 Conservation Laws

According to the Reynolds transport theorem, the change of some intensive property (L) defined over a control volume Ω must be equal to what is lost or gained through the boundaries of the volume ($\partial\Omega$) plus what is created/consumed by sources and sinks inside the control volume:

$$\frac{d}{dt}\int_\Omega L\,dV = -\int_{\partial\Omega} L\vec{v}\cdot\vec{n} - \int_\Omega Q\,dV, \tag{16.11}$$

where $\vec{v}$ is the velocity of the fluid and Q represents the sources and sinks in the fluid.

Using Gauss's theorem [20], it is possible to rewrite Equation 16.11:

$$\frac{d}{dt}\int_\Omega L\,dV = -\int_\Omega \nabla\cdot(L\vec{v}) - \int_\Omega Q\,dV. \tag{16.12}$$

Applying Leibniz's rule to the integral on the left:

$$\int_\Omega \left(\frac{dL}{dt} + \nabla\cdot(L\vec{v}) + Q\right)dV = 0. \tag{16.13}$$

It is now possible to combine the three integrals:

$$\int_\Omega \left(\frac{dL}{dt} + \nabla\cdot(L\vec{v}) + Q\right)dV = 0. \tag{16.14}$$

Equation 16.14 must be true for any control volume; this implies that the integrand itself is zero:

$$\frac{dL}{dt} + \nabla\cdot(L\vec{v}) + Q = 0. \tag{16.15}$$

Equation 16.15 represents the conservation law of a general intensive property L and can now be applied for different concepts represented by different L.

16.5.3.2 Conservation of Mass

The law of conservation given in the general form in Equation 16.15 is now applied to the conservation of mass. Assuming no sources or sinks of mass (i.e., $Q = 0$) and considering the density ($L = \rho$), it is possible to rewrite in Equation 16.15 as

$$\frac{\partial\rho}{\partial t} + \nabla\cdot(\rho\vec{v}) = 0, \tag{16.16}$$

where ρ is the mass density.

This equation is called the mass continuity equation. In the case of an incompressible fluid, ρ is a constant and the equation reduces to

$$\Delta\cdot\vec{v} = 0. \tag{16.17}$$

In other words, an incompressible fluid with no sources or sinks of mass (conservation of mass) is divergence free.

16.5.3.3 Conservation of Momentum

The law of conservation given in the general form in Equation 16.15 is now applied to the conservation of momentum. If the momentum $\vec{L} = \rho\vec{v}$ is considered:

$$\frac{\partial}{\partial t}(\rho\vec{v}) + \nabla\cdot(\rho\vec{v}\vec{v}) + \vec{Q} = 0. \tag{16.18}$$

Note that $\vec{v}\vec{v}$ is a dyad, a second rank tensor. It is possible to show that Equation 16.18 can simplify to [18]

$$\rho\left(\frac{\partial\vec{v}}{\partial t} + \vec{v}\cdot\nabla\vec{v}\right) = \vec{b}. \tag{16.19}$$

where $\vec{b}$ is a body force representing the source or sink of momentum per volume. This results can yet be obtained based on Newton's second law. Recall, Newton's second law in terms of body force, it is possible to write

$$\delta\vec{F} = \delta m \cdot \frac{d}{dt}(\vec{v}(x_1, x_2, x_3, t))$$

$$= \rho\delta V\left(\frac{\partial\vec{v}}{\partial t} + \frac{\partial\vec{v}}{\partial x_1}\frac{dx_1}{dt} + \frac{\partial\vec{v}}{\partial x_2}\frac{dx_2}{dt} + \frac{\partial\vec{v}}{\partial x_3}\frac{dx_3}{dt}\right), \qquad (16.20)$$

where ρ represents the density of the fluid element. Assuming $\vec{v} = (v_1, v_2, v_3) = (dx_1/dt, dx_2/dt, dx_3/dt)$, Equation 16.20 can be simplified to

$$\delta\vec{F} = \rho\delta V\left(\frac{\partial\vec{v}}{\partial t} + \frac{\partial\vec{v}}{\partial x_1}v_1 + \frac{\partial\vec{v}}{\partial x_2}v_2 + \frac{\partial\vec{v}}{\partial x_3}v_3\right) = \rho\delta V\left(\frac{\partial\vec{v}}{\partial t} + \vec{v}\cdot\nabla\vec{v}\right).$$
$$(16.21)$$

Considering that $\vec{b} = (\delta\vec{F}/\delta V)$, Equation 16.21 is equivalent to Equation 16.19. The conservation of momentum allows relating an external force to some partial derivatives on the velocity field.

16.5.3.4 General Form of the Navier–Stokes Equations

The body force $\vec{b} = (\delta\vec{F}/\delta V)$ in Equation 16.19 is now broken into two terms representing stresses and other forces such as gravity [18]:

$$\rho\left(\frac{\partial\vec{v}}{\partial t} + \vec{v}\cdot\nabla\vec{v}\right) = \nabla\cdot\vec{\sigma} + \vec{f}, \qquad (16.22)$$

where $\vec{\sigma}$ is the stress tensor and f represents external forces. By decomposing the stress tensor in the viscous stress tensor and the hydrostatic pressure: $\vec{\sigma} = \vec{\tau} - p\vec{I}$, Equation 16.5, it is possible to rewrite Equation 16.22 as

$$\rho\left(\frac{\partial\vec{v}}{\partial t} + \vec{v}\cdot\nabla\vec{v}\right) = -\nabla\cdot(p\vec{I}) + \nabla\cdot\vec{\tau} + \vec{f}, \qquad (16.23)$$

$$= -\nabla p\cdot\vec{I} - p\nabla\vec{I} + \nabla\cdot\vec{\tau} + \vec{f}, \qquad (16.24)$$

$$= -\nabla p + \nabla\cdot\vec{\tau} + \vec{f}, \qquad (16.25)$$

since $\vec{A}\cdot\vec{I} = \vec{A}$ and $\nabla\cdot\vec{I} = 0\cdot\vec{A}$. Equations 16.16 and 16.25 represent the general form of the Navier–Stokes equations.

16.5.3.5 Newtonian and Incompressible Fluid

Recalling the conservation of mass for incompressible fluids ($\Delta\cdot\vec{v} = 0$, Equation 16.17), the second coefficient of viscosity in the equation of the viscous stress tensor for a Newtonian fluid (Equation 16.10) vanishes:

$$\vec{\tau} = 2\eta\vec{\varepsilon}. \qquad (16.26)$$

And the Navier–Stokes equation for a Newtonian incompressible fluid simplifies to

$$\rho\left(\frac{\partial\vec{v}}{\partial t} + \vec{v}\cdot\nabla\vec{v}\right) = -\nabla p + \nabla\cdot[2\eta\vec{\varepsilon}] + \vec{f}, \qquad (16.27)$$

$$= -\nabla p + \nabla\cdot[\eta(\nabla\vec{v} + (\nabla\vec{v})^T)] + \vec{f}. \qquad (16.28)$$

Furthermore, by assuming that the viscosity η is constant, it is possible to simplify Equation 16.28 to [18]

$$\rho\left(\frac{\partial\vec{v}}{\partial t} + \vec{v}\cdot\nabla\vec{v}\right) = -\nabla p + \eta\nabla^2\vec{v} + \vec{f} \qquad (16.29)$$

This is the standard formulation of the Navier–Stokes equation for Newtonian and incompressible fluids.

16.5.4 Wall Shear Stress

The spatial velocity gradient at the vessel wall can be calculated from a known velocity field to derive the tangential stress. Considering the vectorial nature of the blood flowing in the arteries, the WSS $\vec{\tau}$ is a vector quantity as well. It is tangential to the vessel wall and can hence be separated in its axial and circumferential components as illustrated in Figure 16.9.

Recalling the formulation for the stress tensor (Equation 16.5):

$$\vec{\sigma} = \vec{\tau} - p\vec{I}, \qquad (16.30)$$

with $\vec{\tau}$ being the viscous stress tensor, p the pressure, and $\vec{I}$ the identity matrix. The stress vector is given by

$$\vec{\sigma} = \vec{\sigma}\cdot\vec{n} = \underbrace{\vec{\tau}\cdot\vec{n}}_{\text{viscous stress}:\vec{\tau}} - \underbrace{p\vec{I}\cdot\vec{n}}_{\text{pressure}} \qquad (16.31)$$

with $\vec{n}$ the inward unit normal vector at the boundary. Fluids can exert two kinds of stresses on surfaces: viscous stress and

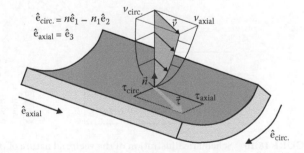

FIGURE 16.9 Illustration of the vectorial nature of the WSS presenting axial and circumferential components. (Adapted from Stalder AF. Quantitative analysis of blood flow and vessel wall parameters using 4D flow-sensitive MRI, PhD Thesis, University of Freiburg, Germany, 2009.)

hydrostatic stress (pressure). While pressure is normal to the surface, viscous stress is tangential to the surface (as depicted in Figure 16.9). Recalling the relation between the viscous stress tensor and the deformation-rate tensor for a Newtonian and incompressible fluid (Equation 16.26), the viscous shear stress vector becomes

$$\vec{\tau} = \vec{\tau} \cdot \vec{n} = 2\eta\vec{\varepsilon} \cdot \vec{n}, \qquad (16.32)$$

with the viscosity η, the deformation-rate tensor $\vec{\varepsilon}$, and the unit normal to the surface of interest $\vec{n}$.

16.5.4.1 Simplified WSS Estimation

Estimations of WSS are often simply calculated assuming a global axial parabolic velocity profile. The WSS ($\vec{\tau}$) is then oriented along the axial direction only and it simplifies to

$$\tau = |\vec{\tau}| = -\frac{4\eta Q}{\pi a^3}, \qquad (16.33)$$

with the vessel radius a, the flow Q, and the dynamic viscosity η.

For a parabolic velocity profile, it is thus possible to calculate WSS based on the flow, the vessel radius, and the viscosity. However, Equation 16.30 follows the law of Poiseuille and is only valid for steady laminar flow in a straight stiff and uniform tube. Furthermore, it assumes a Newtonian fluid and a constant viscosity as well as zero velocity at the wall.

16.5.4.2 Vectorial WSS

Based on Equation 16.32, the WSS vector $\vec{\tau}$ for a Newtonian incompressible fluid can be derived from the velocity field based on the deformation tensor ($\vec{\varepsilon}$) at the vessel wall: The relation between the WSS vector and the three-directional velocity field is then given by

$$\vec{\tau} = \eta \cdot \begin{pmatrix} 2n_1\dfrac{\partial v_1}{\partial x_1} + n_2\left(\dfrac{\partial v_1}{\partial x_2} + \dfrac{\partial v_2}{\partial x_1}\right) + n_3\left(\dfrac{\partial v_1}{\partial x_3} + \dfrac{\partial v_3}{\partial x_1}\right) \\[2ex] n_1\left(\dfrac{\partial v_1}{\partial x_2} + \dfrac{\partial v_2}{\partial x_1}\right) + 2n_2\dfrac{\partial v_2}{\partial x_2} + n_3\left(\dfrac{\partial v_2}{\partial x_3} + \dfrac{\partial v_3}{\partial x_2}\right) \\[2ex] n_1\left(\dfrac{\partial v_1}{\partial x_3} + \dfrac{\partial v_3}{\partial x_1}\right) + n_2\left(\dfrac{\partial v_2}{\partial x_3} + \dfrac{\partial v_3}{\partial x_2}\right) + 2n_3\dfrac{\partial v_3}{\partial x_3} \end{pmatrix}. \qquad (16.34)$$

Calculation of the deformation tensor (Equation 16.7) requires the three-dimensional (3D) differentiation of the velocity vector field. The resulting shear stress vector at the boundary can be separated into its axial and circumferential components as schematically illustrated in Figure 16.10 for nonparabolic flow within an artery.

In practice, the full 3D velocity information is often not available and blood flow velocities are measured in two-dimensional (2D) analysis planes transecting the vascular lumen. If the analysis plane was normal to the vessel surface, it is possible to calculate $\vec{\tau}$ from 2D data with three-directional velocity information by enforcing a no flow condition at the vessel surface: $\vec{n} \cdot \vec{v} = 0$ such that

$$\vec{n} \cdot \frac{\partial \vec{v}}{\partial x_3} = n_1\frac{\partial v_1}{\partial x_3} + n_2\frac{\partial v_2}{\partial x_3} = 0 \qquad (16.35)$$

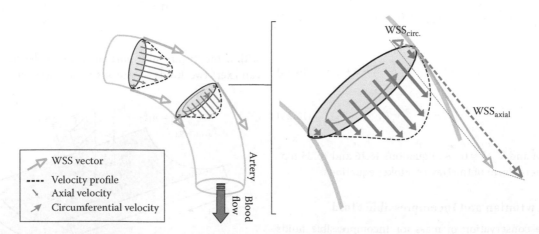

FIGURE 16.10 Schematic illustration of the vectorial nature of the WSS. Reduced flow along the inner curvature side of the artery results in asymmetric velocity profiles (dashed curves) and consequently different WSS vectors (arrows with open arrowhead) at the vessel wall. Components of secondary flow (circumferential velocity) can induce shear stresses along the lumen circumference ($WSS_{circ.}$) in addition to stresses in the main flow direction (WSS_{axial}). (Adapted from Stalder AF. Quantitative analysis of blood flow and vessel wall parameters using 4D flow-sensitive MRI, PhD Thesis, University of Freiburg, Germany, 2009.)

and Equation 16.34 simplifies to

$$\vec{\tau} = \eta \cdot \begin{pmatrix} 2n_1 \dfrac{\partial v_1}{\partial x_1} + n_2 \left(\dfrac{\partial v_1}{\partial x_2} + \dfrac{\partial v_2}{\partial x_1} \right) \\[2mm] n_1 \left(\dfrac{\partial v_1}{\partial x_2} + \dfrac{\partial v_2}{\partial x_1} \right) + 2n_2 \dfrac{\partial v_2}{\partial x_2} \\[2mm] n_1 \dfrac{\partial v_3}{\partial x_1} + n_2 \dfrac{\partial v_3}{\partial x_2} \end{pmatrix}. \tag{16.36}$$

Note that if a one-dimensional problem (e.g., axial flow) is considered, $\vec{\tau}$ simplifies to

$$\vec{\tau}_{1D} = \eta \frac{\partial v}{\partial h}, \tag{16.37}$$

with the axial velocity v and the distance to the boundary h.

16.5.4.3 Oscillatory Shear Index

The oscillatory shear index (OSI) represents the temporal oscillation of WSS during the cardiac cycle. It was originally defined by Ku et al. [21]. For a purely axial WSS, it is given by:

$$OSI_{axial} = \frac{1}{2} \left(1 - \frac{\left| \int_0^T \tau_{axial} \cdot dt \right|}{\int_0^T |\tau_{axial}| \cdot dt} \right) \tag{16.38}$$

$$= \frac{1}{2} \left(1 - \frac{A_{pos} - A_{neg}}{A_{pos} + A_{neg}} \right) = \frac{A_{neg}}{A_{pos} + A_{neg}}, \tag{16.39}$$

where τ_{axial} is the axial component of the WSS vector. T is the duration of the cardiac cycle. A_{pos} and A_{neg} represent respectively

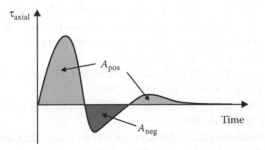

FIGURE 16.11 τ_{axial} as a function of time for OSI calculation. The positive area of the τ_{axial}-time curve is given by A_{pos} (light gray). The negative area is represented by A_{neg} (dark gray). A_{pos} and A_{neg} can be used to calculate OSI as in Equation 16.39. The axial OSI (OSI_{axial}) represents the relative inversion of the axial WSS (τ_{axial}) during the cardiac cycle ($0 \le OSI_{axial} \le 0.5$ with $OSI_{axial} = 0$ if τ_{axial} is always positive). (Adapted from Stalder AF. Quantitative analysis of blood flow and vessel wall parameters using 4D flow-sensitive MRI, PhD Thesis, University of Freiburg, Germany, 2009.)

the positive and negative area of the τ_{axial}—time curve as illustrated in Figure 16.11.

The original definition of OSI [21] was dependent on a predominant axial direction and for that reason difficult to extend to a general 3D case. An alternative general definition was thus introduced by He and Ku [22]:

$$OSI = \frac{1}{2} \left(1 - \frac{\left| \int_0^T \vec{\tau} \cdot dt \right|}{\int_0^T |\vec{\tau}| \cdot dt} \right), \tag{16.40}$$

where $\vec{\tau}$ is the instantaneous WSS vector. From Equation 16.40, it is clear that OSI is bound between 0 and 0.5 as well. Compared to OSI_{axial}, OSI considers not only the amplitude variations but the changes of direction as well.

16.5.4.4 Role of *In Vivo* WSS

Complex vascular geometry and pulsatile flow in the human arterial system lead to regionally different flow characteristics and thus spatial and temporal changes in shear forces acting on the vessel wall [9,11,23]. Recent reports stressed the importance of WSS and OSI with respect to the formation and stability of atherosclerotic plaques [24]. A number of studies have shown that low WSS and high OSI represent sensitive markers for formation of plaques in the aorta, carotid, or coronary arteries [25,26]. Particularly, the assessment of both WSS and OSI can help to determine the complexity of the lesions. A recent study with animal models and deliberately altered flow characteristics in the carotid arteries demonstrated the close correlation of low WSS with the development of vulnerable high-risk plaques whereas high OSI induce stable lesions [24]. In addition, the effects of selected pathologies on regionally varying WSS and OSI values have been reported [27,28]. Furthermore, multiple applications based on computational fluid dynamics for the numerical simulation of human flow characteristics have been reported in the past years including, for example, the simulation of time-resolved 3D flow patterns in the left ventricle, in geometrically complex aneurysms, or the carotid artery bifurcation [29–33].

16.6 Laminar and Turbulent Flow

Laminar flow occurs when a fluid flows in parallel layers, with no disruption between the layers and is the normal condition for blood flow throughout most of the circulatory system. It is characterized by concentric layers of blood moving in parallel down the length of a blood vessel. The flow profile is parabolic once laminar flow is fully developed. However, under conditions of high flow, particularly in the ascending aorta, laminar flow may be disrupted and become turbulent.

Turbulence and velocity fluctuations of the blood flow are believed to play a role in hemolysis, platelet activation, and thrombus formation [34]. Previous invasive studies have already assessed

TABLE 16.1   Dimensionless Numbers for Turbulence Estimation in Pulsatile Blood Flow

Reynolds Number	Womersley Number	Strouhal Number
Inertial forces/viscous forces	Unsteady forces/viscous forces	Dimensionless stroke volume
$Re(t) = \dfrac{\rho v(t) D(t)}{\mu}$	$\alpha = \dfrac{D_m}{2}\sqrt{\dfrac{\rho \cdot 2\pi f}{\mu}}$	$St = \dfrac{D_m}{2}\dfrac{f}{V_p - V_m}$

With density: ρ; dynamic viscosity: μ; mean cross-sectional velocity: $v(t)$; diameter: $D(t)$; $D_m = $ mean $(D(t))$; $v_p = \max(v(t))$; $v_m = \text{mean}(v(t))$; heart rate: f; and $Re_{mean} = \text{mean}(Re(t))$; $Re_{peak} = \max(Re(t))$.

turbulence *in vivo* based on catheter hot-film anemometry or perivascular Doppler ultrasound in animals [35,36] and in humans [37,38].

The Reynolds number (Table 16.1) has long been used to define critical values for turbulence. Nerem et al. [35] defined *in vivo* a critical peak Reynolds number proportional to the Womersley number. More recently, those results were amended by a systematic analysis of the influence of Reynolds, Womersley, and Strouhal numbers on the development of turbulence *in vitro* for physiologically realistic pulsatile flow [39]. Based on the work of Peacock et al. [39], the critical peak Reynolds number, indicating the transition to turbulence, was derived as

$$Re_{peak}^c = 169\alpha^{0.83}St^{-0.27}. \qquad (16.41)$$

The transition to turbulence is thus related to the ratio inertial forces/viscous forces as well as to the ratio unsteady forces/viscous forces and to the dimensionless stroke volume. Furthermore, the oscillatory component of the blood flow (high Womersley number and low Strouhal number) has a stabilizing effect such that high transient Reynolds number (beyond 4000 in the aorta) can be observed without presence of turbulence.

References

1. Murray C, Lopez A, Rodgers A, Vaughan P, Prentice T, Edejer TTT, Evans D, and Lowe J, The world health report 2002—reducing risks promoting healthy life, *World Health Organization*, 2002.
2. Puska P, Mendis S, and Porter D, Fact Sheet—Chronic Disease, *World Health Organization*, 2003.
3. Friedman MH and Giddens DP. Blood flow in major blood vessels—modeling and experiments. *Annals of Biomedical Engineering* 2005; 33(12): 1710–1713.
4. Gimbrone MA, Jr, Topper JN, Nagel T, Anderson KR, and Garcia-Cardena G. Endothelial dysfunction, hemodynamic forces, and atherogenesis. *Annals of the New York Academy of Sciences* 2000; 902: 230–239; discussion 239–240.
5. SERVIER Medical Art.
6. Beers MH and Berkow R. *The Merck Manual of Diagnosis and Therapy*. 17th edition, Whitehouse Station, NJ: Merck Research Laboratories, 1999.
7. Westerhof N, Stergiopulos N, and Noble MIM. *Snapshots of Hemodynamics*. New York: Springer, 2005.
8. Chien S, Li S, and Shyy YJ. Effects of mechanical forces on signal transduction and gene expression in endothelial cells. *Hypertension* 1998; 31(1): 162–169.
9. Davies PF. Flow-mediated endothelial mechanotransduction. *Physiological Reviews* 1995; 75(3): 519–560.
10. Davies PF. Haemodynamic influences on vascular remodelling. *Transplant Immunology* 1997; 5(4): 243–245.
11. Glagov S, Weisenberg E, Zarins CK, Stankunavicius R, and Kolettis GJ. Compensatory enlargement of human atherosclerotic coronary arteries. *The New England Journal of Medicine* 1987; 316(22): 1371–1375.
12. Malek AM, Alper SL, and Izumo S. Hemodynamic shear stress and its role in atherosclerosis. *JAMA* 1999; 282(21): 2035–2042.
13. Gijsen FJ. *Modeling of Wall Shear Stress in Large Arteries*. Eindhoven, NL: Eindhoven University of Technology, 1998.
14. Pschyrembel W, Dornblueth O, and Amberger S. *Pschyrembel Klinisches Woerterbuch*. Berlin: de Gruyter, 2004.
15. Merrill EW, Benis AM, Gilliland ER, Sherwood TK, and Salzman EW. Pressure-flow relations of human blood in hollow fibers at low flow rates. *Journal of Applied Physiology* 1965; 20(5): 954–967.
16. Chien S, Usami S, Taylor HM, Lundberg JL, and Gregersen MI. Effects of hematocrit and plasma proteins on human blood rheology at low shear rates. *Journal of Applied Physiology* 1966; 21(1): 81–87.
17. Fung YC. *Biomechanics: Mechanical Properties of Living Tissues*. New York: Springer, 1993.
18. Emanuel G. *Analytical Fluid Dynamics*. Boca Raton, FL: CRC Press, 2001.
19. Batchelor GK. *An Introduction to Fluid Dynamics*. Cambridge: Cambridge University Press, 2005.
20. Gustafson GB and Wilcox CH. *Analytical and Computational Methods of Advanced Engineering Mathematics*. New York: Springer, 1998, 729pp.
21. Ku DN, Giddens DP, Zarins CK, and Glagov S. Pulsatile flow and atherosclerosis in the human carotid bifurcation. Positive correlation between plaque location and low oscillating shear stress. *Arteriosclerosis* 1985; 5(3): 293–302.

22. He X and Ku DN. Pulsatile flow in the human left coronary artery bifurcation: Average conditions. *Journal of biomechanical Engineering* 1996; 118(1): 74–82.

23. Langille BL and O'Donnell F. Reductions in arterial diameter produced by chronic decreases in blood flow are endothelium-dependent. *Science* 1986; 231(4736): 405–407.

24. Cheng C, Tempel D, van Haperen R, van der Baan A, Grosveld F, Daemen MJAP, Krams R, and de Crom R. Atherosclerotic lesion size and vulnerability are determined by patterns of fluid shear stress. *Circulation* 2006; 113(23): 2744–2753.

25. Chatzizisis YS, Jonas M, Coskun AU, Beigel R, Stone BV, Maynard C, Gerrity RG, Daley W, Rogers C, Edelman ER, Feldman CL, and Stone PH. Prediction of the localization of high-risk coronary atherosclerotic plaques on the basis of low endothelial shear stress: An intravascular ultrasound and histopathology natural history study. *Circulation* 2008; 117(8): 993–1002.

26. Irace C, Cortese C, Fiaschi E, Carallo C, Farinaro E, and Gnasso A. Wall shear stress is associated with intima-media thickness and carotid atherosclerosis in subjects at low coronary heart disease risk. *Stroke; a Journal of Cerebral Circulation* 2004; 35(2): 464–468.

27. Frydrychowicz A, Arnold R, Hirtler D, Schlensak C, Stalder AF, Hennig J, Langer M, and Markl M. Multidirectional flow analysis by cardiovascular magnetic resonance in aneurysm development following repair of aortic coarctation. *Journal of Cardiovascular Magnetic Resonance* 2008; 10(1): 30.

28. Frydrychowicz A, Berger A, Russe MF, Stalder AF, Harloff A, Dittrich S, Hennig J, Langer M, and Markl M. Time-resolved magnetic resonance angiography and flow-sensitive 4-dimensional magnetic resonance imaging at 3 Tesla for blood flow and wall shear stress analysis. *The Journal of Thoracic and Cardiovascular Surgery* 2008; 136(2): 400–407.

29. Taylor CA, Hughes TJ, and Zarins CK. Finite element modeling of three-dimensional pulsatile flow in the abdominal aorta: Relevance to atherosclerosis. *Annals of Biomedical Engineering* 1998; 26(6): 975–987.

30. Boussel L, Rayz V, McCulloch C, Martin A, Acevedo-Bolton G, Lawton M, Higashida R, Smith WS, Young WL, and Saloner D. Aneurysm growth occurs at region of low wall shear stress: Patient-specific correlation of hemodynamics and growth in a longitudinal study. *Stroke; A Journal of Cerebral Circulation* 2008; 39(11): 2997–3002.

31. Lee SW, Antiga L, Spence JD, and Steinman DA. Geometry of the carotid bifurcation predicts its exposure to disturbed flow. *Stroke; A Journal of Cerebral Circulation* 2008; 39(8): 2341–2347.

32. Steinman DA, Milner JS, Norley CJ, Lownie SP, and Holdsworth DW. Image-based computational simulation of flow dynamics in a giant intracranial aneurysm. *AJNR American Journal of Neuroradiology* 2003; 24(4): 559–566.

33. Steinman DA and Taylor CA. Flow imaging and computing: Large artery hemodynamics. *Annals of Biomedical Engineering* 2005; 33(12): 1704–1709.

34. Gnasso A, Carallo C, Irace C, Spagnuolo V, De Novara G, Mattioli PL, and Pujia A. Association between intima-media thickness and wall shear stress in common carotid arteries in healthy male subjects. *Circulation* 1996; 94(12): 3257–3262.

35. Nerem RM and Seed WA. An *in vivo* study of aortic flow disturbances. *Cardiovascular Research* 1972; 6(1): 1–14.

36. Stein PD and Sabbah HN. Measured turbulence and its effect on thrombus formation. *Circulation Research* 1974; 35(4): 608–614.

37. Nygaard H, Paulsen PK, Hasenkam JM, Pedersen EM, and Rovsing PE. Turbulent stresses downstream of three mechanical aortic valve prostheses in human beings. *The Journal of Thoracic and Cardiovascular Surgery* 1994; 107(2): 438–446.

38. Stein PD and Sabbah HN. Turbulent blood flow in the ascending aorta of humans with normal and diseased aortic valves. *Circulation Research* 1976; 39(1): 58–65.

39. Peacock J, Jones T, Tock C, and Lutz R. The onset of turbulence in physiological pulsatile flow in a straight tube. *Experiments in Fluids* 1998; 24(1): 1–9.

40. Stalder AF. Quantitative analysis of blood flow and vessel wall parameters using 4D flow-sensitive MRI, PhD Thesis, University of Freiburg, Germany, 2009.

17

Fluid Dynamics

Takuji Ishikawa

17.1 Introduction

Hydrostatics or fluid-statics describes the condition of fluids in static equilibrium, or fluids at rest. Fluid dynamics describes the process of fluids in motion. Hydrostatics and fluid dynamics are part of the scientific focal point of fluid mechanics. When the fluid is water the flow is captured under hydrodynamics, and when it concerns the flow of blood the nomenclature becomes hemodynamics. Fluids are described by Webster's dictionary as follows: "having particles that easily move and change their relative position without a separation of the mass and that easily yield to pressure: capable of flowing." Fluids are substances that, under pressure or stress, deform and as such flow. The definition of fluids in this manner covers both gases and liquids. Fluid dynamics, and generally fluid mechanics, is governed by a set of basic physics principles. The physics principles of fluid mechanics can be summarized with the use of the conservation laws. The conservations laws are: conservation of mass, conservation of energy (or the First Law of Thermodynamics), and conservation of linear momentum (Newton's Second Law of Motion). These principles are all derived from classical mechanics. These laws applied to fluid mechanics are captured in the Reynolds transport theorem. For example, the conservation of energy under the Reynolds transport theorem converts into Bernoulli's equation. Under fluid mechanics Bernoulli's equation is more equivalent to a conservation of energy density, taking the fluid compressibility into account.

This chapter will primarily focus on liquids, not including gasses, which is covered in Chapter 19.

The general conservation laws and the specific applications are described next, with particular attention being given to pulsatile flow and the consequences to the boundary conditions.

17.2 Hydrostatics

17.2.1 Pressure and Density

Pressure is the force per unit area and has the units Pa ($=N/m^2$). Since pressure always acts perpendicular to the plane of interest, the pressure has no specific direction. Thus, pressure at one point can be expressed only by magnitude, which is called a scalar quantity. For example, density, temperature, and concentration are scalar quantities, but velocity, force, and momentum are not.

Density is the mass per unit volume and has the units kg/m^3. The density of a gas, such as air or oxygen, varies with pressure, but that of a liquid, such as water or blood, does not vary considerably. In fluid dynamics, fluids with constant density are referred to as incompressible fluids, whereas those with variable density are compressible fluids. The effect of fluid compressibility can be negligible even for a gas if the fluid velocity is less than about $0.2v$, where v is the speed of sound in the gas.

17.2.2 Pressure in the Gravity Field

When an incompressible liquid with density ρ is quiescent in the gravity field, the pressure P at z below the surface is given by

$$P = P_0 + \rho g z, \tag{17.1}$$

where P_0 is the pressure on the liquid surface, as depicted in Figure 17.1. In measuring blood pressure, the units of mm Hg are often used, where 1 mm Hg refers to the pressure induced by mercury of 1 mm in height. The pressure can be converted as 1 mm Hg = 133.3 Pa.

17.2.3 Viscosity

17.2.3.1 Stress

Stress is the force per unit area and has the units Pa, similar to pressure. However, stress and pressure differ in two ways. First, pressure is positive when it acts in the direction to compress the volume, whereas stress is positive when it acts in the direction to expand the volume. Second, pressure is a scalar quantity, but stress is a second-order tensor with nine components given by

$$\tau = \begin{pmatrix} \tau_{xx} & \tau_{xy} & \tau_{xz} \\ \tau_{yx} & \tau_{yy} & \tau_{yz} \\ \tau_{zx} & \tau_{zy} & \tau_{zz} \end{pmatrix}, \tag{17.2}$$

where τ_{ij} is the force in the direction j acting on a plane perpendicular to axis i. Now let us assume a cubic control volume, as depicted in Figure 17.2. When $i = x$, for example, forces in the x-, y-, and z-directions are expressed as τ_{xx}, τ_{xy}, and τ_{xz}, respectively. Stresses acting normal to the surface are called normal stresses (i.e., τ_{xx}, τ_{yy}, and τ_{zz}), whereas stresses acting tangential to the surface are called shear stresses. When the surface is a wall boundary, the shear stress is called wall shear stress.

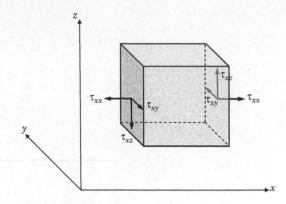

FIGURE 17.2 Stress components acting on the plane perpendicular to the x-axis.

By considering the balance of moments acting on the control volume, the following conditions can be derived:

$$\tau_{xy} = \tau_{yx}, \tau_{xz} = \tau_{zx}, \tau_{yz} = \tau_{zy}. \tag{17.3}$$

Thus, the stress tensor is a symmetric tensor.

17.2.3.2 Rate of Strain

In the case of a solid, such as an elastic material, it deforms when stress is exerted, but returns to its original shape if the stress is removed. In contrast, fluid flows when stress is exerted and does not return to its original shape even if the stress is removed. This is because an elastic material generates stress against the strain, whereas fluid generates stress against the rate of strain. The rate of strain is also called the velocity gradient and has the units 1/s.

Let fluid be filled between flat plates with gap h [m], and let the upper wall be moved in the x-direction with speed U [m/s], while the lower wall is fixed, as illustrated in Figure 17.3. After a sufficiently long time, the velocity field between the flat plates becomes linear in the y-direction. This type of flow is called Couette flow or simple shear flow. The velocity gradient in the y-direction is given by

$$\frac{\partial u_x}{\partial y} = \frac{U}{h}, \tag{17.4}$$

where u_x is the velocity component in the x-direction.

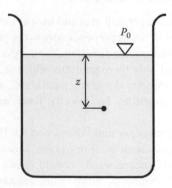

FIGURE 17.1 A liquid is quiescent in the gravity field when the pressure P_0 is exerted on the surface.

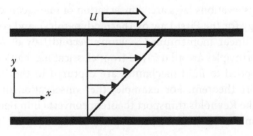

FIGURE 17.3 Fluid flow between flat plates.

In the case of a three-dimensional flow field, the rate of strain becomes a second-order tensor with nine components. For incompressible fluids, the rate of strain e_{ij} is given by

$$e_{ij} = \frac{1}{2}\left(\frac{\partial u_i}{\partial j} + \frac{\partial u_j}{\partial i}\right), \tag{17.5}$$

where i and j can be replaced by x-, y-, or z-coordinates.

17.2.3.3 Viscosity

In the case of Couette flow, as shown in Figure 17.3, the force F [N] is necessary to move the upper plate in the x-direction with speed U. When the area of the upper plate is A [m²], a relationship exists such that

$$\frac{F}{A} = \mu\frac{U}{h}, \tag{17.6}$$

where μ is the viscosity, which has the units Pa·s. The left-hand side of Equation 17.6 is the shear stress τ_{xy}, and U/h on the right-hand side can be written as du_x/dy. Thus, Equation 17.6 can be rewritten as

$$\tau_{xy} = -\mu\frac{du_x}{dy}. \tag{17.7}$$

The minus sign in the right-hand side appears due to the direction of shear stress. This equation is called Newton's law of viscosity.

In the case of a three-dimensional flow field, Newton's law of viscosity for incompressible fluids is given by

$$\tau_{ij} = \tau_{ji} = -\mu\frac{1}{2}\left[\frac{du_i}{dj} + \frac{du_j}{di}\right], \tag{17.8}$$

where i and j can be replaced by x-, y-, or z-coordinates. This equation gives the relationship between the stress and the rate of strain for a Newtonian fluid.

The viscosity of a Newtonian fluid will be constant regardless of the flow field if the temperature and the density are invariant. Fluids that consist of small molecules, such as air and water, are usually Newtonian fluids. The viscosities of air and water are

TABLE 17.1 Density and Viscosity of Water and Air

Temperature (°C)	Water		Air	
	Density (kg/m³)	Viscosity (Pa s)	Density (kg/m³)	Viscosity (Pa s)
0	999.84	1.792×10^{-3}	1.293	1.724×10^{-5}
10	999.70	1.307×10^{-3}	1.247	1.773×10^{-5}
20	998.20	1.002×10^{-3}	1.205	1.822×10^{-5}
30	995.65	0.797×10^{-3}	1.165	1.869×10^{-5}
40	992.22	0.653×10^{-3}	1.127	1.915×10^{-5}

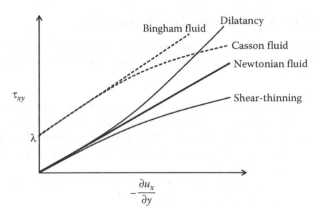

FIGURE 17.4 Laminar flow in a straight tube.

given as examples in Table 17.1. If a fluid contains large molecules or particles, it often does not obey Newton's law of viscosity; such fluids are called non-Newtonian fluids. Many biofluids, such as blood, saliva, synovial fluid, and cytoplasm, are non-Newtonian fluids and exhibit complex relationships between stress and rate of strain.

17.2.4 Non-Newtonian Fluids

A non-Newtonian fluid is one that does not obey Newton's law of viscosity (Equation 17.7). The relationship between τ_{xy} and du_x/dy for various non-Newtonian fluids is illustrated in Figure 17.4. The thick solid line with constant slope crossing the origin of the figure represents a Newtonian fluid. The two curves crossing the origin, shown by thin solid lines, signify non-Newtonian fluids. When the apparent viscosity decreases as the velocity gradient increases, this is called shear-thinning, and when the apparent viscosity increases as the velocity gradient increases, this is called dilatancy.

The broken line and the broken curve in the figure do not cross the origin, but cross the y-axis with an intercept coefficient of λ. This means that the fluids do not deform until the exerted stress exceeds the plasticity λ. The broken line represents a Bingham fluid and the broken curve, a Casson fluid. The property of blood is well expressed by the Casson model.

A second non-Newtonian phenomenon is related to the narrowing of vessels downstream. When the vessel diameter falls below a critical value the viscosity becomes a function of the vessel diameter. The critical diameter has been documented as approximately 2 mm. The dependence of the viscosity on the vessel radius is known as the Fahraeus–Lindqvist effect. This phenomenon is illustrated in Figure 17.4.

17.3 Conservation Laws in Fluid Dynamics

Fluid dynamics is governed by the same laws that apply in other fields of physics: laws of conservation, thermodynamics, and Newton's laws. Additionally, the conservation of momentum also applies.

17.3.1 Conservation of Momentum

Based on Newton's second law the next conservation law for fluid dynamics can be derived. The three laws of Newton are as follows:

- *Newton's First Law of Motion:* Every object in motion remains in motion unless an external force is applied.
- *Newton's Second Law of Motion:* The force applied on an object is directly proportional to the acceleration and the mass of the object in motion.
- *Newton's Third Law of Motion:* For every action there is an equal but opposite reaction.

Newton's second law is expressed as follows:

$$F = ma = m\frac{dV}{dt} = \frac{d(mV)}{dt}, \quad (17.9)$$

where the momentum is expressed by Equation 17.10:

$$p^* = mV \quad (17.10)$$

In equilibrium the pressure applied to the liquid will balance the shear forces on the liquid in cylindrical coordinates, assuming the direction of flow to be in the z-direction:

$$\sum F_i = \sum PA - \sum \tau\,(2\pi r\,dz). \quad (17.11)$$

In equilibrium, the sum of the forces is zero and Equation 17.11 becomes

$$\sum PA = \sum \tau\,(2\pi r\,dz). \quad (17.12)$$

Applying Equation 17.12 on an infinitesimally small volume and rearranging the infinitesimal radius and length to the respective sides of the equal sign, the equation can be shown to yield

$$\frac{d(r\tau)}{dr} = \frac{r\,dP}{dz}. \quad (17.13)$$

Substitution of the shear rate with Equation 17.7 and rearranging all equal terms to the respective sides of the equal sign will yield

$$\frac{1}{r}\frac{d(r(du/dr))}{dr} = \frac{1}{\mu}\frac{dP}{dz}. \quad (17.14)$$

Integrating Equation 17.14 over the radius of the lumen of flow will yield the Hagen–Poiseuille equation, shown later in Equation 17.28.

For the sum of forces acting on a liquid in a compliant volume, Equation 17.9 translates into the following volumetric deformation description:

$$\sum F = \sum \frac{d(mV)}{dt} = \oint_{surface\_of\_volume} \rho V(V \cdot n)dA + \frac{\partial}{\partial t}\oint_{volume}\rho V\,dV. \quad (17.15)$$

In one-dimensional flow, Equation 17.15 transforms into

$$\sum F = \frac{\partial}{\partial t}\oint_{volume}\rho V\,dV + \left[\sum\left(\frac{\partial m}{\partial t}V\right)_{out} - \sum\left(\frac{\partial m}{\partial t}V\right)_{in}\right]. \quad (17.16)$$

Moving on from the momentum, the next step is the change of momentum, the impulse i, as illustrated by

$$i = d\,(mV) = Ft. \quad (17.17)$$

The energy (E) applied to achieve the motion described by the momentum is the change in energy between the before and after situation and is called "work" and can be captured by the change in impulse over time integrated over the distance the motion takes place, or the force integrated over the infinitesimal distance of motion (ds). The work (W) is expressed as

$$W = \Delta E = \int F\,ds. \quad (17.18)$$

This brings us to the next set of conservation principles: Conservation of Energy.

17.3.2 Conservation of Energy

There are two basic forms of energy: potential energy and kinetic energy. Potential energy describes the potential for releasing energy due to the boundary conditions. One example of potential energy is storing an object at a raised altitude with respect to a reference level; this object can fall and release the potential energy as kinetic energy during its fall. The potential energy (U) in this case is described by the gravitational energy expressed as

$$U = mgh, \quad (17.19)$$

which equals the work performed to raise the object with mass m to the elevated level.

The kinetic energy released during the fall is the force acting on the mass over the distance traveled. The distance traveled can be expressed as the speed during the motion times the duration the motion takes place. Since the speed is not constant, integration will be required for yielding the average speed times the duration of fall for the distance traversed:

$$s = \overline{V}t = \frac{V_{max}}{2}t. \quad (17.20)$$

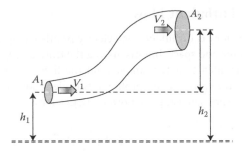

FIGURE 17.5 Bernouilli flow diagram.

This will give the kinetic energy (E) as

$$E = \frac{1}{2}mV^2. \tag{17.21}$$

Combining the work, kinetic energy, and potential energy with any form of internal energy (i.e., chemical or radioactive) will give the total energy of the system, which is interchanged between sources of energy but is never lost. Consider the flow through the pipe in Figure 17.5 with input cross-sectional area A_1, output cross-sectional area A_2, flow velocity at entrance V_1, flow velocity at exit point V_2, and elevation $h_2 - h_1$ between input and output. The conservation of energy equation reads this as follows:

$$E_{\text{exit}}^{\text{total}} = E_{\text{entry}}^{\text{total}} = \left[mgh_2 + \frac{1}{2}mV_2^2 + E_{\text{internal}} - P_2 As \right]_{\text{exit}}$$

$$= \left[mgh_1 + \frac{1}{2}mV_1^2 + E_{\text{internal}} + P_1 As \right]_{\text{entry}}. \tag{17.22}$$

Note that the work done by pressure on the liquid as well as the friction is also included in this conservation law. Friction converts to internal energy and raises the temperature by increasing the internal kinetic energy since temperature is the average internal kinetic energy of the medium.

When dividing Equation 17.22 by the volume of the liquid and neglecting E_{internal}, the conservation of energy converts into conservation of energy density, also known as Bernoulli's equation. The potential energy in this case will involve any change in height, such as between the head and the heart. This is expressed as

$$\rho gh_2 + \frac{1}{2}\rho V_2^2 + P_2 = \rho gh_1 + \frac{1}{2}\rho V_1^2 + P_1. \tag{17.23}$$

17.4 Osmotic Pressure

In a vascular circulation the internal energy involves the chemical solution of all ions, minerals, and colloid materials such as proteins. The solution of chemicals is described by the osmotic pressure Π. The colloid osmotic pressure is indicated as Π_c in this way. Including the osmotic pressure in the Bernoulli equation gives a more representative description of the flow dynamics in organs,

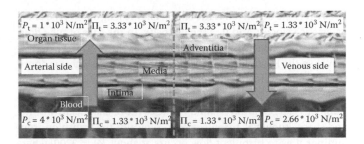

FIGURE 17.6 Arterial/venous pressure diagram showing tissue pressures (P_t), osmotic pressure (Π_t), and luminal pressures: colloid osmotic pressure (Π_c) and luminal hydraulic pressure (P_c) illustrating the combined pressure gradient facilitating the exchange of dissolved gasses.

since the osmotic pressure is a real part of the material exchange between the blood and the organ under dynamic conditions.

The osmotic pressure of a solution has the same value as what the particle solution would have as a gas with the same volume and temperature as the solution. The osmotic pressure can in this way be expressed in the form used by the Ideal Gas Law:

$$\Pi = \frac{nRT}{V} = CRT, \tag{17.24}$$

where R is the gas constant (8.135 J/mol K), T is the local temperature in Kelvin, n is the number of moles of solution, and C is the concentration. Note that the gas laws apply in more ways than one; all solutions will add to the total osmotic pressure of the liquid.

The use of the Van't Hoff equation becomes evident when considering a physiological salt solution (saline) that is in osmotic equilibrium with blood plasma at 0.9% weight NaCl, or 300 mOsm/L concentration.

The analogy between gas pressure and dissolved particle pressure was identified by Van't Hoff (1887) and Equation 17.24 is sometimes referred to as the Van't Hoff Law.

In the same way as the blood has osmotic pressure, the vessel wall or lung alveoli has an osmotic pressure that forms a mechanism of exchange for dissolved materials in combination with the other terms in the Bernoulli equation that are attributed to the exchange process

$$\rho gh_2 + \frac{1}{2}\rho V_2^2 + P_2 + \Pi_2 = \rho gh_1 + \frac{1}{2}\rho V_1^2 + P_1 + \Pi_1. \tag{17.25}$$

Figure 17.6 illustrates the pressure gradient across the membrane wall at the boundary of the blood flow over a length of circulation.

17.5 Flow in a Straight Tube

17.5.1 Reynolds Number

The Reynolds number is a dimensionless parameter that indicates the ratio of the momentum force to the viscous force in the flow field, defined as

$$Re = \frac{\rho LU}{\mu}, \tag{17.26}$$

where ρ is the density, L is the characteristic length, U is the characteristic velocity, and μ is the viscosity. In the case of tube flow, L is usually taken as the diameter and U as the average velocity.

When the Reynolds number is small, the viscous effect is dominant in the flow field. Thus, fluid particles move regularly with time and space in the flow, known as laminar flow. When the Reynolds number is large, however, fluid particles move chaotically in time and space, known as turbulent flow. Typically, the transition from laminar flow to turbulent flow occurs at about $Re_c = 2200$, where Re_c is the critical Reynolds number. For example, when ink is injected to a flow in a straight pipe, the ink flows as a straight line if $Re < Re_c$, whereas the ink becomes mixed if $Re < Re_c$.

17.5.2 Laminar Flow

When the flow in a tube is fully developed and steady, the velocity profile of laminar flow is given by

$$u_x = -\frac{R^2}{4\mu}\frac{dp}{dz}\left[1-\left(\frac{r}{R}\right)^2\right], \qquad (17.27)$$

where z- and r-coordinates are taken in the axial and radial directions, respectively, R is the radius of the tube, and dp/dz is the pressure gradient, as illustrated in Figure 17.7. The velocity is zero on the wall and has a maximum value on the axis. In the case of laminar flow, the velocity profile is parabolic.

The flow rate Q can be calculated by integrating the velocity in the cross-section as

$$Q(t) = 4\int_0^{R(z)} u(z,r,t)r\,dr = \frac{\pi R^4}{8\mu}\frac{dp(t)}{dz}, \qquad (17.28)$$

where $R(z)$ is the axial-dependent radius of the vessel. This equation is known as the Hagen–Poiseuille Law. The flow rate is proportional to R^4, and is therefore very sensitive to the tube radius when the pressure gradient is fixed. The average velocity u_a is calculated by dividing the flow rate by the cross-section as

$$u_a = -\frac{R^2}{8\mu}\frac{dp}{dz}, \qquad (17.29)$$

which is half of the maximum velocity on the axis.

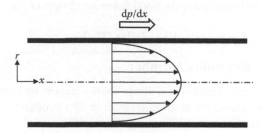

FIGURE 17.7 Representative Poiseuille flow diagram. The flow velocity decreases with the square of the radius.

17.5.3 Turbulent Flow

When the flow is turbulent, the velocity profile changes chaotically in time and space, but discussing the time-averaged velocity is warranted to better understand the flow characteristics. The time-averaged velocity profile of turbulent flow in a tube is typically expressed using the power law, given by

$$\frac{u}{u_{\max}} = \left(\frac{y}{R}\right)^{1/n}, \qquad (17.30)$$

where $u_{\max}$ is the maximum velocity, y is the distance from the wall, and n is a constant that varies with Re as described in Table 17.2. The time-averaged velocity profile of turbulent flow is more flat than that of laminar flow.

17.6 Factors Influencing Flow-Velocity Profile

Blood is a non-Newtonian fluid as mentioned earlier; in addition, blood has a particle solution that creates additional flow conditions that are causing deviations from the theoretical description provided. On top of the particulate factor, the flow through vessels encounters a compliance of the vessel wall, which also contributes to the need for adjustments to the standard flow equations. Last but not least, all mechanical factors of the composition of the blood itself and the vessel wall elasticity introduce mechanical impedance similar to electrical impedance. The mechanical impedance has similar effects on the mechanical flow as in electrical conduction, especially when applied to a pulsatile flow. In alternating electric current, the frequency response at the distal end of an electronic circuit is described by the "Telegraph Equation," taking into account phase retardations resulting from inductance and capacitance. The equivalent consequences applied to vascular flow, and in particular pulsatile flow, are described in the next sections.

17.6.1 Particulate Flow Pattern

Owing to the approximately 45% particulate concentration, RBCs, white blood cells, platelets, and other dissolved components, the particles tend to separate out from the liquid in the blood flow. This results in a two-layer model describing the flow in the vessels. The boundary layer will be formed with a low concentration of particles. Beyond the boundary layer the flow can be described using the Poiseuille flow. The thickness of the boundary layer is generally a function of vessel diameter, flow

TABLE 17.2 Relationship between Re and n in the Power Law

Re	4×10^3	10^4–1.2×10^5	2×10^5–4×10^5	7×10^5–1.3×10^6	1.5×10^6–3.2×10^6
N	6	7	8	9	10

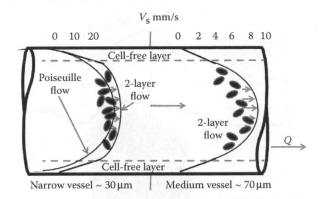

FIGURE 17.8 Representation of two-layer flow for narrow (left) and wide (right) vessels. The cell-free layer on the wall creates a deviation from the Poiseuille flow. In narrow vessels, the RBCs tend to group together (Rouleaux formation), which results in a flattening of the velocity profile.

velocity and particle concentration. The two-layer model is illustrated in Figure 17.8.

17.6.2 Vascular Compliance

The innate mechanical stretch of the vascular wall under pressure is necessary to accommodate the rapid influx of massive amounts of blood with a downstream restrictive flow matrix. Especially in the arterial system, the vessels become smaller but more plentiful moving distally. The mechanical impedance (primarily resistance) of the smaller vessels in the flow path will require an alternative solution to accept the capacity of the inflow locally. The electronic equivalent to this condition will be capacitance and to a degree inductance, which translate to mechanical compliance of the vessel wall. The arterial wall stretches under the increased pressure from the inflow of blood, providing a temporary holding of the transient flow.

17.6.3 Pulsatile Flow

Since the heart beats periodically, the output of the heart has a pulsatile character. The pulsatile behavior has all the characteristics of a wave. Since the wave-like flow phenomenon is periodic, Fourier analysis can be applied to dissect the wave pattern into sinusoidal components. Standard wave propagation theory can thus also be applied to the pulsatile blood flow. The mechanical "Telegraph equation" applied to blood flow shows frequency and phase effects on the transmission pattern of flow. The main issue in blood flow is the mechanical wall compliance.

When the flow in a tube is pulsatile, like blood flow in a large artery, the velocity profile is no longer parabolic, even though the flow is laminar. In the accelerated period, the velocity profile tends to be flatter than that of a parabolic profile. In the decelerated period, however, fluid in the center region tends to flow downstream because of considerable momentum, but fluid near

the wall may flow upstream due to the large negative pressure gradient. The wavy velocity profile was derived for various flow conditions by Womersley [1] using Bessel functions. The effect of pulsation can be discussed in terms of the Womersley number, defined as

$$Wo = \sqrt{\frac{\rho a^2 \omega}{\mu}}, \qquad (17.31)$$

where a is the radius and ω is the angular velocity of the pulsation. Wo is a dimensionless constant to express the ratio of the momentum force due to the pulsation to the viscous force. When Wo is small, the flow is quasi-steady and we can assume that the velocity profile is parabolic. When Wo is large, however, we need to consider the oscillatory boundary layer developing near the wall.

In order to describe the pulsatile flow in the human vasculature, the Poiseuille equation needs to be applied to a frequency-dependent flow instead of a steady-state flow.

17.7 Blood Flow in Large and Small Vessels

17.7.1 Flow Conditions in Various Vessels

The diameter of vessels varies from 20 to 30 mm in large arteries to between 5 and 10 μm in capillaries. In large arteries, the value of Re becomes several thousands and inertia plays an important role in the flow. Stream lines often separate from the wall downstream of a stenosis, at an aneurysm or after a bifurcation, and a secondary flow is generated in a curved artery. Most flows in arteries are laminar, except for the very large arteries that exit the left ventricle. In capillaries, however, the value of Re is less than 1, and viscous force plays a major role. Pulsation of the flow can be found in all vessels, even in capillaries, but the flow becomes quasi-steady in the capillaries. The variation of Re and Wo in different vessels is listed in Table 17.3 [2].

TABLE 17.3 Flow Conditions in Different Vessels [2]

Site	Internal Diameter (cm)	Peak Velocity (m/s)	Peak Reynolds Number	Womersley Number
Ascending aorta	1.5	1.2	4500	13.2
Descending aorta	1.3	1.05	3400	11.5
Abdominal aorta	0.9	0.55	1250	8
Femoral artery	0.4	1.0	1000	3.5
Arteriole	0.005	0.75	0.09	0.04
Capillary	0.0006	0.07	0.001	0.005
Venule	0.004	0.35	0.035	0.035

17.7.2 Blood Flow in Large Vessels

The blood flow in large arteries involves several complicated processes and concepts, such as the complex geometry of vessels, deformation of vessel walls, pulsatile flow due to the heartbeat, and the non-Newtonian property of blood. Recent advances have considerably improved the clinical images obtained by computed tomography (CT) and magnetic resonance imaging (MRI), and patient-specific geometries are often employed for computational fluid dynamics (CFD) analysis. In a large artery, blood can be assumed to be a homogeneous fluid because the size of blood cells is much smaller than the scale of the flow field. The non-Newtonian property of blood is sometimes neglected for the sake of mathematical simplicity. The pulsation of flow is important not only in large arteries, but also in large veins, because the flow in large veins also oscillates due to the heartbeat. Vessel wall deformation in large arteries is about 5–10%. Deformation in veins is usually greater than that in arteries, and veins can even collapse if the outer pressure is greater than the inner pressure. Due to vessel wall deformation and pulsatile flow, pressure waves propagate on the wall. The pressure wave velocity (PWV) is given by the Moens–Korteweg equation:

$$PWV = \sqrt{\frac{hE}{2\rho a}}, \tag{17.32}$$

where h is the thickness of the wall, E is the Young's modulus of the wall, ρ is the density of the fluid, and a is the radius.

17.7.3 Blood Flow in Small Vessels and Capillaries

In a small vessel, blood is no longer assumed to be homogeneous. This is because the flow field is affected by the individual motions of blood cells, and blood cells are not distributed evenly in a vessel. Typically, a cell-free layer that is several microns thick forms in the vicinity of endothelium cells. The apparent viscosity of blood changes with vessel diameter due to the heterogeneity of the blood cells, which is known as the Fahraeus–Lindqvist effect [3]. The hematocrit also tends to decrease in small vessels because the plasma layer flows into smaller, daughter vessels.

The diameter of capillaries is about 5–10 μm, which is comparable to the 8 μm size of RBCs. An RBC can pass through a narrow constriction by deforming to a parachute- or slipper-like shape. The deformation of an RBC is defined by capillary number, which is the ratio of the viscous force to the elastic force, defined as

$$Ca = \frac{\mu \dot{\gamma} a}{hE}, \tag{17.33}$$

where μ is the viscosity, $\dot{\gamma}$ is the shear rate, a is the radius, h is the membrane thickness, and E is the Young's modulus of the membrane. Even in a large vessel, the shear rate can be large near the wall, which leads to the deformation of RBCs. Figure 17.9

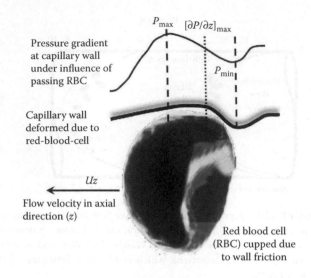

FIGURE 17.9 Capillary wall deformation (compliance) and pressure gradient discontinuity under capillary flow conditions.

illustrates the wall compliance in the capillaries. Wall deformation takes place under the rubbing of the RBCs. The rubbing of the RBCs against the capillary wall also creates a discontinuity in the pressure gradient of the flow. At small shear rates, RBCs in plasma rotate like a rigid disk. At large shear rates, on the other hand, the RBCs orient themselves at a constant angle to the flow and their membrane appears to rotate about the interior, referred to as a tank-treading motion.

17.7.4 Sinusoidal Fluctuation in Flow

The heartbeat, and the resulting flow, generates a complex flow rate pattern. Fourier analysis of this signal will show that the wave pattern can be approximated by the first eight sinusoidal harmonics, with the base wave being the heart rate. The frequency-dependent flow requires the introduction of a frequency component in the flow parameter. Using the axial flow $u(r)$ as the base for steady-state flow, the frequency-dependent flow can be developed in a harmonic expansion with the use of a Poisson series and can be described as

$$u(r, t) = u'(r)e^{i\omega_n t}, \tag{17.34}$$

where ω_n represents the frequency harmonics of the pulsatile flow.

Substitution of the time-dependent flow in the Poiseuille equation yields the foundation for the flow dynamic equivalent of the Telegraph equation. The solution is presented without derivation, which can be found in Uri Dinnar's "Cardiovascular Fluid Dynamics":

$$u(r, t) = Re\left\{-\frac{ia_n}{\omega_n}e^{i\omega_n t}\left[1 - \frac{j_0\left(\sqrt{-i}\alpha(r/R)\right)}{j_0\left(\sqrt{-i}\alpha\right)}\right]\right\}, \tag{17.35}$$

where a_n represents the coefficients of the Fourier series of the frequency polynomial series, $\infty^2 = Re * S_r$, and R represents the radius of the vessel.

Here, Re is the Reynolds numbers and S_r is the Strouhal number, and J_0 is the first-order Bessel function. The frequency parameters of blood flow are determined by the Strouhal number. The Strouhal number quantifies the ratio of fluidic forces due to turbulence phenomena with respect to the internal forces resulting from localized flow accelerations. The Strouhal number can be used to estimate the time scale for vortex formation. The Strouhal number is defined as

$$S_r = \frac{\omega D}{V}, \tag{17.36}$$

where D is the vessel diameter, V is the average flow velocity, and ω is the angular velocity of the periodic phenomenon of flow.

This means that the flow characteristics are inherently tied to the flow rate, the vessel diameter, and the frequency.

Equation 17.35 can be solved for small μ yielding a fluctuating parabolic flow profile in Poiseuille format, expressed as

$$u(r,t) = a_n \left[\left(1 - \frac{r^2}{R^2} \right) \cos(\omega_n t) \right]. \tag{17.37}$$

The shear stress resulting from blood flow through a vessel can be described using the Herschel–Bulkley approximation to model blood as a flow medium. This gives the shear stress as a function of flow as

$$\tau = \mu'^{1/n} \left(\frac{-\partial \overline{(u(r,t))}}{\partial r} \right)^{1/n} + \mu', \quad \text{when } \tau \geq \tau', \tag{17.38}$$

$$\frac{-\partial \overline{(u(r,t))}}{\partial r} = 0 \quad \tau \geq \tau', \quad \text{when } \tau < \tau', \tag{17.39}$$

where τ' is the Herschel–Bulkley yield-stress and μ' is the Herschel–Bulkley coefficient of friction.

Additionally, changes in the quantitative values of parameters such as vessel diameter, flow velocity, and pulsation frequency among other parameters will affect the shear rate definition and can be captured by the mechanical compliance of the vessel or more generally the mechanical impedance of the time-dependent flow through a compliant vessel.

The mechanical compliance of the vessel can be separated in a static compliance and dynamic compliance. The static compliance (C_V^{Stat}) is the result of the combined effect of circumferential compliance (C_C^{Stat}) and axial compliance (C_a^{Stat}) as described in the equations given below:

$$C_V^{Stat} = dV/V\, dP = \frac{d(\ln(V))}{dP}, \tag{17.40}$$

$$C_C^{Stat} = dD/V\, dP = \frac{d(\ln(D))}{dP}, \tag{17.41}$$

$$C_L^{Stat} = dL/V\, dP = \frac{(d\ln(L))}{dP}, \tag{17.42}$$

where V is the volume of the lumen of the vessel, D is the diameter of the lumen of the vessel, and D is the length of the vessel over which the pressure gradient is considered. It is given without proof that the total volumetric compliance satisfies the equation

$$C_V^{Stat} = C_L^{Stat} + 2C_d^{Stat}. \tag{17.43}$$

Equation 17.43 relies on the assumption that the vessel is circular and remains circular throughout the flow.

The dynamic values of the compliance can be split in a similar fashion. In this case, the time-dependent volume of the lumen of the vessel is considered. Additionally, the pressure gradient is replaced by the pressure difference between maximum (systolic: P_{Sys}) and minimum (diastolic: $P_{diastol}$) pressures: $\hat{P}$ is described as

$$\hat{P} = P_{Sys} - P_{diastol}. \tag{17.44}$$

The time-dependent compliance in this case is considered to be a function of the mean pressure and time gradient of the volume of the vessel segment. Describing the volumetric, circumferential, and axial dynamic compliance will look like the following:

$$C_V^{Dynamic} = \frac{1}{V_0\, \hat{P}} \Delta V, \tag{17.45}$$

$$C_C^{Dynamic} = \frac{1}{D_0\, \hat{P}} \Delta V, \tag{17.46}$$

$$C_a^{Dynamic} = \frac{1}{L_0\, \hat{P}} \Delta V, \tag{17.47}$$

where V_0 is the volume of the vessel lumen, ΔV is the time-dependent volume change, D_0 is the luminal diameter, ΔD is the luminal expansion diameter, L_0 is the length of the vessel under consideration at rest, and ΔL is the axial stretch.

The total compliance is the sum of the static and dynamic compliances, respectively, for each component:

$$C_V = C_V^{Dynamic} + C_V^{Stat}, \tag{17.48}$$

$$C_C = C_C^{Dynamic} + C_C^{Stat}, \tag{17.49}$$

$$C_a = C_a^{Dynamic} + C_a^{Stat}. \tag{17.50}$$

Additionally, the mechanical behavior is described by the material properties of the vessel such as the elastic modulus of the vessel (E), or Young's modulus and Poisson's ratio (υ). The

Young's modulus is defined by the ratio of the tensile stress over the tensile strain:

$$E = \frac{\tau}{e_{ij}}. \tag{17.51}$$

In the case of the expanding vessel the change will be both in the radial and axial directions, yielding two independent moduli. This translates into the Young's modulus for the circumferential (E_c) and axial (E_a) directions as follows:

$$E_c = (1-v^2)\frac{d\tau}{dr/R} \equiv (1-v^2)\frac{dP}{dr/R} \approx (1-v^2)\left(\frac{R}{d_w C_c}\right), \tag{17.52}$$

$$E_a = (1-v^2)\frac{d\tau}{dl/L} \equiv (1-v^2)\frac{dP}{dl/L}\left(\frac{R}{2d_w}\right) \approx (1-v^2)\left(\frac{R}{2d_w C_a}\right). \tag{17.53}$$

Generally, for compliant materials, the Poisson ratio equals 0.5; however, elastic biological materials do not expand isometrically due to the fiber/muscle structure. As a result, the expansion has a time lag, which results in a phase retardation of the periodic expansion and contraction.

The impedance for flow ($Z(r,t)$) is defined as the ratio of the pressure gradient over a length of vessel ($\Delta P(z,t)$) divided by the flow through that vessel ($Q(r,t)$), as shown by the equation

$$Z(r,t) = \frac{\Delta P(z,t)}{Q(r,t)} = \frac{(-dP(z,t)/dz)f(t)}{Q(t)}, \tag{17.54}$$

where $f(t)$ describes the frequency profile of the flow.

Combining the compliance and impedance and applying this to pulsatile flow of the beating heart embedded in the flow equivalent "Telegraph" equation or Cable equation, it can be shown that the flow in the radial distribution as well as in the axial direction has phase retardation resulting in turbulence, especially when considering curved vessels.

The Cable equation for current (I) in a circuit with capacitance (C), inductance (L), and resistance (R) is formulated as follows:

$$I_m = I_{in} - I_{out} = \frac{1}{R_a}\partial^2 V/\partial x^2, \tag{17.55}$$

which can describe the current passing through the membrane channel conductance. Here, I_{rm} is the resistive current through the membrane flowing in the axial direction in an electronic equivalent model and the amount stored on the membrane capacitance, I_{cm}; R_a is the axial resistance, C_m is the membrane capacitance, and V is the transmembrane potential, leading to

$$I_m = I_{in} - I_{out} = \frac{1}{R_a}\partial^2 V/\partial x^2 = I_{rm} + I_{cm} = \frac{1}{R_m}V + C_m\partial V/\partial t. \tag{17.56}$$

Solving this partial differential equation and rearranging to solve for the partial with respect to time will yield

$$C_m\partial V/\partial t = \frac{1}{R_a}\partial^2 V/\partial x^2 - \frac{1}{R_m}V, \tag{17.57}$$

which is referred to as the Telegraph equation or the Cable equation.

Multiplying both sides by r_m/r_m yields

$$\partial V/\partial t = \frac{1}{C_m}\left(\frac{1}{R_a}\partial^2 V/\partial x^2 - \frac{1}{R_m}V\right) = \frac{1}{R_m C_m}\left(\frac{R_m}{R_a}\partial^2 V/\partial x^2 - V\right). \tag{17.58}$$

The Telegraph equation for flow can now be formulated by the Moens–Korteweg equation, which gives the fluidic pulsatile velocity [$v(r,t)$] as a function of the fluid characteristics as well as the vessel compliance:

$$v(r,t) = \sqrt{\frac{Eh}{2\rho R}} = \sqrt{\frac{A}{\rho C_u}} = \sqrt{\frac{1}{L_u C_u}}, \tag{17.59}$$

where the wall thickness is given by h and the elastic modulus is E; ρ is the density of blood, R is the radius of the blood vessel, L_u is the inertance per unit length, and C_u is the compliance per unit length; $v(r,t)$ is the speed of blood flow in the vessel as a function of radial location and as a function of time. In all this descriptive analysis, the pulsatile pressure as a function of axial and radial location and z, the axial location, is represented by $P(r,z,t)$. Numerical examples of compliance are presented in Table 17.4.

An indication of the radial flow dynamics of a pulsatile transportation of blood through a rigid vessel is illustrated by the work of Uri Dinnar in Figure 17.10. The calculated flow patterns show back-flow due to the phase retardation resulting from the compliance as well as the non-Newtonian behavior of the blood.

17.8 CFD for Blood Flow Analysis

17.8.1 CFD Methods

The isothermal laminar flow of an incompressible Newtonian fluid can be solved by coupling the conservation of mass (i.e., the continuity equation) with the conservation of momentum (i.e., the Navier–Stokes equation). Many algorithms to couple the two

TABLE 17.4 Examples of Specific Tissue Compliance for Selected Biological Materials

Tissue Type	Compliance (%mm Hg)
Artery (average)	590 ± 50
Bovine heterograft	260 ± 30
Saphenous vein	440 ± 80
Umbilical vein	370 ± 50

Source: Salacinski, H. J. 2001. *J. Biomater. Appl.* 15(3), 241–248.

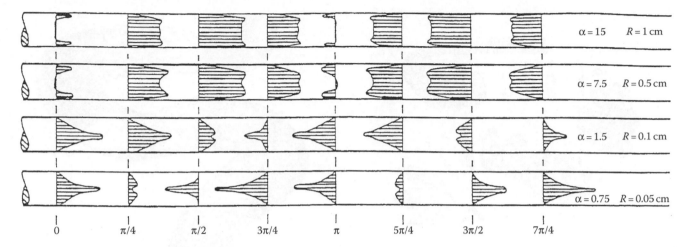

FIGURE 17.10 Flow pattern in relative vessel diameter under the influence of pulsatile non-Newtonian liquid flow. Flow conditions are presented for different values of $\alpha = \omega_n R^2 \nu$, where ω_n is the nth harmonic of the pulsatile flow frequency $\omega_n = 2\pi\nu n$, and ν is the heart pump frequency. The upper graph is for aortic flow, while the bottom three graphs are for smaller vessels. Note the flow reversal and the differences in this phenomenon with respect to vessel diameter and flow velocity. The arterial vessel diameter in these graphs ranges from 100 mm to 1.5 cm. (Reproduced from Dinnar U. 1981. *Cardiovascular Fluid Dynamics*. CRC Press, Boca Raton. With permission.)

equations, as well as discretization methods, have been proposed. Since these methodologies are beyond the scope of this book, please refer to standard CFD textbooks for further details. To solve the continuity equation and the Navier–Stokes equation simultaneously, boundary conditions are necessary. Usually, the initial condition of the flow field, inlet, and outlet flow conditions, and wall boundary conditions are required to solve the problem. These equations are discretized into each computational mesh and solved by massive computational power.

Over the past few decades, numerical methods for CFD have advanced considerably. Based on this progress, current trends in computational hemodynamics favor the more practical applications for diagnosing cardiovascular diseases and planning any related surgery. For example, CFD analysis on blood flow at a cerebral aneurysm has been reported by many researchers. Cerebral aneurysm is an extremely important disease in clinical medicine, since the rupture of a cerebral aneurysm is the most common cause of subarachnoid hemorrhage, which is well known for its very high mortality. Although how cerebral aneurysms originate is still unclear, hemodynamics is believed to play a vital role. Thus, many researchers have investigated the blood flow around an aneurysm. Currently, an intravascular stent is often used to treat a cerebral aneurysm. CFD analysis is also used to develop an effective stent to prevent the inflow to an aneurysm. CFD is now one of the fastest growing fields in medical engineering.

17.8.2 Example of CFD Analysis in a Large Artery

In this section, an example of CFD analysis in a large artery is presented. A patient-specific geometry of the human internal carotid artery is used. For the calculation of blood flow, blood is assumed to be an incompressible and Newtonian fluid with density $\rho = 1.05 \times 10^3$ kg/m^3 and viscosity $\mu = 3.5 \times 10^{-3}$ Pa · s. The

governing equations for such a blood flow are the continuity and Navier–Stokes equations. Pulsatile flow with the maximum Reynolds number of 200 is solved. In this section, the vessel wall deformation is neglected for numerical simplicity. Boundary conditions include a parabolic velocity profile at the inlet, zero pressure at the outlet, and the no-slip condition on the wall. The blood flow calculation is accomplished through an in-house three-dimensional flow solver based on a MAC algorithm. The total number of grid points is 52,065.

Figure 17.11a illustrates the magnitude of axial velocity at each cross-section at peak velocity. Since the arterial shape has high curvature, the velocity profile is no longer parabolic, as analytically derived for a straight tube. The secondary flow at each cross-section is shown in Figure 17.11b. When an artery has a bend, the high-velocity fluid in the center region experiences a

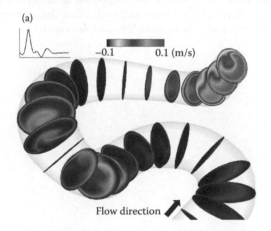

FIGURE 17.11 Velocity field of blood flow in an internal carotid artery at peak velocity. (a) Magnitude of the axial velocity. (b) Velocity vector of secondary flow. The velocity profile is also shown in the upper left of (a) and (b). (Image provided by Y. Shimogonya and T. Yamaguchi, Tohoku University.)

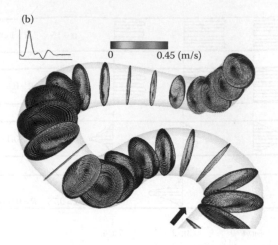

FIGURE 17.11 (Continued).

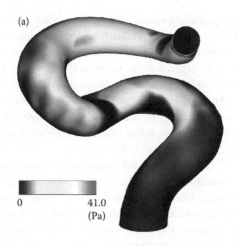

FIGURE 17.12 Wall shear stress distribution in an internal carotid artery at peak velocity. (a) Front and (b) rear views. (Image provided by Y. Shimogonya and T. Yamaguchi, Tohoku University.)

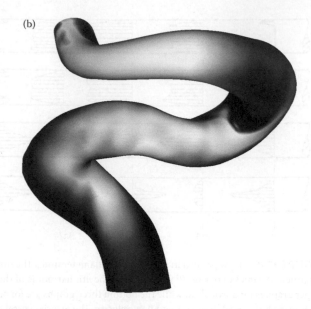

FIGURE 17.12 (Continued).

large centrifugal force, which leads to the secondary flow in the cross-section. The magnitude of the secondary flow is about half that of the axial flow; therefore, the flow field is fully three-dimensional in this case. Figure 17.12 depicts the wall shear stress. By solving the flow field, one can easily calculate the wall shear stress. Since CFD analysis gives us detailed information on the hemodynamics in a cardiovascular system, it is currently an area of much research attention.

References

1. Womersley, J. R., 1955. Method for the calculation of velocity, rate flow, and viscous drag in arteries when the pressure gradient is known. *J. Physiol.*, **127**, 553–563.
2. Pedley, T. J., 1980. *The Fluid Mechanics of Large Blood Vessels.* Cambridge, UK: Cambridge University Press.
3. Fahraeus, R. and Lindqvist, T., 1931. The viscosity of blood in narrow capillary tubes. *Am. J. Physiol.*, **96**, 562–568.

18

Modeling and Simulation of the Cardiovascular System to Determine Work Using Bond Graphs

Paola Garcia

18.1 Introduction

As demonstrated in previous chapters, the heart is a complex organ that pumps oxygenated blood throughout the body via a closed loop called the circulatory system. This chapter discusses the mechanics of the heart and circulatory system that requires a non-steady-state-compliant algorithm. A modeling method is introduced to account for the staged time-dependent boundary conditions, in order to solve for the fluid dynamics characteristics. Since the heart is much too complex to analyze in its entirety, it is modeled in this chapter as a two-cycle, four-cylinder engine. The purpose of representing the heart as a mechanical system is not only to simplify it, but to calculate the total work of the heart as well. This four-cylinder engine is represented in two halves: the arterial chambers and the coronary chambers. The two arterial chambers (i.e., ventricles) are represented mathematically by a pump, while the two venous chambers (i.e., atria) are represented as a reservoir. The reasoning behind this is that the arterial chambers push blood forward and therefore can be represented as a force acting on a fluid (i.e., a pump), while the venous chambers "draw" blood into it and act as a pressure difference where one pressure limit is theoretically zero (i.e., a reservoir). The engine analogy is to combine these two systems into one to allow for the total work of the system to be considered. The rest of the cardiovascular system (i.e., the arteries, coronaries, and veins) is represented by a thin-walled tube with a constant diameter taken from the average of the diameters of the major arteries and the major veins. This system has a finite length with no "branching" or breaks. The "tube" has the compliance and wall friction characteristics of an actual artery.

The overall model is a hypothetical statistical construct of a standard person's cardiovascular system, which allows the use of average values of physiological parameters, for example, 120/80 mmHg for systolic/diastolic blood pressure. (See Chapter 14 for details of the systolic and diastolic blood pressure.) This reduces the degrees of freedom and therefore the number of equations needed. The entire chapter is approached from an engineering analysis standpoint and the examination of the physiological process can take place at different levels. For example, from a thermodynamic point of view, the circulatory system can be approximated as a closed system that is in quasi-steady state with its surrounding environment. The analysis done in this chapter is reduced to a fluid dynamics and a mechanics point of view.

Given all of these assumptions, simplifications, and reductions, the amount of work (W) the heart does is empirically analyzed from an engineering standpoint, rather than a biological one. The standard definition for work is given by the equation:

$$W(\text{work done}) = Q(\text{flow rate}) \times (\text{height or length traveled}). \quad (18.1)$$

However, even with all the aforementioned simplifications and assumptions, it is difficult to assess the amount of work the heart does, because it is a compliant system with a time derivative. In order to model the entire system and solve for total work done by the heart, bond graphs must be used to formulate a finite series of first-order differential equations to solve for work.

18.2 Background

The cardiovascular system of vessels and organs consists of two separate flow paths, namely the systematic circulation and the pulmonary circulation, as explained in Chapter 19. These two paths will be modeled in series, which means they are flowing and operating simultaneously, as well as in conjunction with one another. Those flow paths combined have the following sequence of actions. Pulmonary circulation flows to the left side of the heart with freshly oxygenated blood and is then shot-gunned from the cardiac output (left ventricle) and into the arterial system to be circulated among all the major organs. Once the organs have had their fill of oxygen, the deoxygenated blood is then returned to the right side of the heart via the venous vessels. Since this an extremely turbulent ride for the blood inside and near the heart (think, stop and go roller coaster), there has to be some sort of commanding order to ensure that the blood cannot accidentally flow backwards into another blood rollercoaster car. A series of four valves (one per chamber) constrains the blood flow to an irreversible path through the circulatory system. Flow traveling from the cardiac output through the systemic circulation averages 4–6 L/min and gets its supply from the left ventricle, whereas the pulmonary circulation is supplied by the right ventricle. The left and right ventricles expand and contract in parallel in order to ensure that the cardiac output and venous return are equal.

The circulation in this system does not work too much since the normal maximum pressure in the pulmonary artery is 25–28 mmHg (a relatively low pressure), while the normal maximum pressure in the aorta happens to be around 4–5 times greater at 100–120 mmHg.[1] This is due to the fact that the left ventricle must pump against a pressure in order to open the aortic valve, which is 4–5 times greater than the pressure that the right ventricle must pump against to open the pulmonic valve. Because of this, the right ventricular wall is like a nerd (working smarter, not harder) and therefore has less muscular build and is considerably thinner than the left ventricular wall, which resembles a linebacker.

The pressure driving the flow causes the vessels to behave in a certain way based on their mechanical properties, which in turn acts upon the blood to cause it to flow. It is the blood flow interacting with the vessels that causes a pressure. The relationships among these pressures, flows, and fluid resistance in arterial vessels are dynamic processes. These processes depend on factors such as pulsation, blood viscosity, mechanical properties of the vessel wall (i.e., compliance), the connections of the circulatory vessels (i.e., series versus parallel), and the regulatory mechanisms that stabilize the arterial blood pressure and volume (mechanisms that prevents one from exploding when changing altitude). This chapter will take the aforementioned relationships used in applied mathematics to model the larger vessels such as arteries.

18.3 Inertia, Capacitance, and Resistance

There is an entire branch of fluid dynamics dedicated to the processes mentioned above and the flow of blood called hemodynamics. The reason is that blood in itself is so complex that it does not fall under any of the normal fluid dynamic assumptions. For example, the blood flowing through the circulatory system is not at a steady state; it is pulsating at the frequency of the heart. The pressure, velocity, and volumetric flow rate in the arteries and ventricles vary with time over a single cardiac cycle. However, if the time increments are small enough and normal resting conditions are considered, the variations can be approximated as periodic so that the cycle repeats itself. In first-order approximation, using a Fourier expansion (see Chapter 14), this allows for the analysis of a single pulse at a time. For the purposes of this chapter, the time-constrained values of pressure, velocity, and flow are to be considered as steady state.

Now with this in mind, the flow can be examined, starting with an electrical analogy. Most people know that electricity flows through a wire because the voltage is pushing it at a rate known as current, it is hindered by something called resistance, and what they get on the other side is the time derivative of work or power. The equation relating all of these things is called Ohm's law for the rate of transport of a quantity between any two points. This seems to be a good equation to use. This concept can be applied to the problem and thus the equation can be used to combine pressure, velocity, and flow, in order to give the amount of work done by the heart (Figure 18.1).

Rather than using electrical terms, Equation 18.1 has been modified to show the basic relation of the variables:

$$\text{Rate of flow} = \frac{\text{Driving force}}{\text{Resistance to flow}}. \qquad (18.2)$$

By analogy, this can be rewritten for steady laminar volumetric flow of blood between any two points in the circulation as

$$Q\left(\frac{L}{\min}\right) = \frac{\Delta P(\text{mmHg})}{R_{eq}\left[\dfrac{\text{mmHg}}{\text{L/min}}\right]}, \qquad (18.3)$$

where Q is the volumetric flow rate, P is the pressure (ΔP being the change in pressure), and R_{eq} is the total amount of resistance the blood flow encounters, which is nondimensionalized

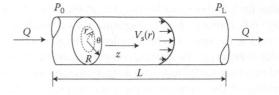

FIGURE 18.1 Laminar flow through a tube.

viscosity. The viscosity is nondimensionalized because it is taken relative to the viscosity of water.[2] What this implies is that the rate of volumetric flow between two points equals the driving force for flow (which happens to be the difference in pressure between the two points or ΔP) divided by the equivalent hemodynamic resistance (friction against the vessel walls, i.e., shear stress: Chapters 14 and 15). This is just one piece of the model. Flow and pressure are both time varying, and the vessels have resistive, capacitive, and inductive (inertia) elements.

While Equation 18.3 can be used to find resistance, there is still an issue of representing the inductance and capacitance elements (i.e., compliance) involved in the heart model. Using the change in pressure as the moving force again and taking the time derivative of volumetric flow rate, we can obtain the inductance elemental equation:

$$\Delta P = \frac{\rho \Delta L}{A} \dot{Q}, \tag{18.4}$$

where ρ (rho) is the density of blood, ΔL is the length of the vessel, A is the cross-sectional area of the vessel, and $\dot{Q}$ is the time derivative of the volumetric flow rate. The inductive element is represented by the $\rho \Delta L/A$ term. That leaves the capacitive element, which describes the compliance of the vessel walls and how it affects the change in pressure and volume.

$$\Delta P = \frac{1}{e} \int V, \tag{18.5}$$

where e is the elastance of the vessel, P is the pressure, and V is the volume. The capacitive element is represented by the e term.

18.4 Bond Graphs

In order to model the circulatory system with all of its elements using differential equations, bond graphs must be used to connect the elements to each other to make up the entire system. For the longest time, mechanical engineers used the Lagrangian or Hamiltonian equations to model mechanical systems. However, there are two problems in pursuing this route: first, if nonconservative forces are present, the equations are no longer based solely on the energy of the system but are additionally based on the virtual work of the nonconservative forces, and second, it is not possible to use the energy of subsystems (or elements) and then connect these subsystems together in a modular fashion. Bond graphs are a graphical representation of a physical dynamic system and a technique for building mathematical models of those systems.[3] By computing the time derivative of energy and using the power instead of energy E for modeling, both of the aforementioned difficulties vanish as shown in Equation 18.6.

$$P(t) = \frac{dE(t)}{dt} \quad \text{or} \quad P(t) = \frac{dW(t)}{dt}. \tag{18.6}$$

By using bond graphs, power continuity equations are formulated instead of energy conservation laws. The modeler can express power balance equations for each subsystem separately and then connect all the subsystems, as long as he or she ensures that the power is also balanced at all the interfaces between elements.

Now for a brief overview of bond graphs; the symbols used are defined and what bond graphs can do is discussed. In a bond graph, energy flow from one point of a system is denoted by a harpoon (an arrow with half its pointer missing). Power is the product of effort and flow (Equation 18.7).

$$P = e \times f. \tag{18.7}$$

In electrical systems, it is polite to select the voltage as the effort variable and current as the flow. If not, one may suffer the wrath of an electrical engineer named Mackie. In mechanical systems on the other hand, the force will be treated as effort and the velocity as flow. However, the assignment is subjective. Effort and flow are dual variables and the assignment could just as well be done the other way around.

In a so-called junction, where the power is either split or combined, the power continuity equation dictates that the sum of the incoming power streams must equal the sum of the outgoing power. This can be satisfied in more ways than one, but the simplest ones are to keep either effort or flow constant around the junction and formulate the balance equation for the other. For bond graphs these two situations are called 0-junctions and 1-junctions, respectively.

As displayed in Figure 18.2, the 0-junction is a common effort junction (where all of the efforts are equal to each other) and the 1-junction is a common flow junction (where all of the flows are equal). This is useful in solving for either flow or effort, and then using that to balance the equation and find the other.

The most important features of a bond graph are that it operates on energy flows rather than individual signals and that they are inherently acausal. This is where the causality strokes come into play in the methodology of bond graphs, turning the formerly acausal bonds into those of a causal variety. Causal bonds have a cause and effect relationship with the elements attached to them, whereas acausal bonds do not. It is the element themselves

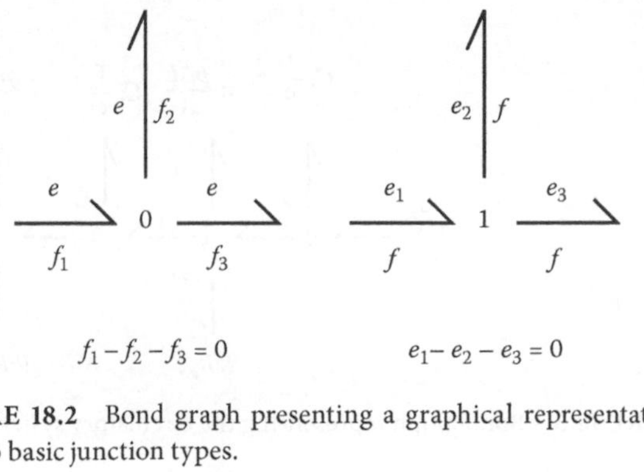

$$f_1 - f_2 - f_3 = 0 \qquad e_1 - e_2 - e_3 = 0$$

FIGURE 18.2 Bond graph presenting a graphical representation of the two basic junction types.

that determine on which end of the "bond" the causal stroke will fall upon. For example, an effort source or element will have a causal stroke on the broken arrow part of the bond, while a flow will display it at the butt of the arrow. The purpose of this is to display the relationship between the elements themselves, as well as between the elements and the junctions.

Each element has its own model symbol, as well as a representative, general differential equation. Capacitor-like elements will fall under the symbol "C" and are generally representatives of elements that store power. Differential equations representing a capacitor will have the inverse of a constant (i.e., $1/C$). Similarly, resistive elements fall under the symbol "R" and generally hinder the power's flow. The equation will have a constant in front of a variable. Inertial or inductive elements are ones that describe the amount of resistance to change in velocity in a system. This differential equation usually contains the second time derivative of a variable with a constant before it. The constant is the inductive element. The last two are relatively straightforward. Flow and effort sources are denoted by "S_f" and "S_e," respectively, and generally have a first derivative. So, with all of this in mind, we begin to introduce the circulatory system to the methodology of bond graph modeling.

First and foremost, the heart in all of its entirety is to be considered a flow source. The arterial and ventricular systems are represented by capacitors, resistors, and inertial elements.

By extracting the fluid dynamics from the beginning of the chapter, a differential equation can govern every element and relate to the entire system via the bonds. Figure 18.3 displays a very simplistic bond graph of a circulatory system; however, for the purposes of this chapter it is sufficient. Moving from left to right, the first term in $S_f = QS_f$ dictates a flow source and Q (flow rate) is the flow source identified in the system. The series of 0-junctions and 1-junctions indicates the various sections of the body where hemodynamical properties differ from other sections. The very first element represented is the flow source (Q), which, as mentioned previously, is the heart. The flow leaving the heart (Q) enters a 0-junction, where there is common effort. This states that the compliance of the arterial walls $1/e$ has the same amount of force being applied to it as the flow source, which also supplies the 1-junction with the same amount of force. However, that force is no longer the same since it is split between the

friction against the wall R_{eq} and the sudden change in velocity ($\rho \Delta L/A$). The elements do in fact have the same blood flow in their equations as denoted by the 1-junction. This proceeds the same way up until the 1-junction containing the effort source (ΔP). The purpose of this effort source is to model the muscles "pushing" the blood back up to the heart from the lower extremities. Lastly, since this is a closed-loop system, the force being exerted by both the muscles pushing the blood forward, as well as the actual blood pressure itself, is what goes back into the flow source. This completes one cycle and the cycle is then repeated, similar to the circle of life, in a hemodynamic sort of way.

18.5 Differential Equations

The model is only beginning, as the equations used to represent each of the elements has to be tied into the system via the power continuity equations of each junction. Linear time-invariant systems may be described by either a scalar nth order differential equation with constant coefficients or a coupled set of n first-order differential equations with constant coefficients using state variables. For the purposes of this chapter, the analysis is done using the state variable and the forced response. The forced response is the part of the response of the form of the forcing function and the state variables are determined by the system.

The equation (Equation 18.3) relating pressure, velocity, and fluid resistance will be used in addition to two other equations. Equation 18.5 will be used by the capacitance elements, which relate the change in pressure to the change in velocity by Bulk's modulus of an arterial wall. This demonstrates how the vessel expands and collapses, thus changing the flow of the system. Equation 18.4 relates density, a section of length of the artery, and the area, which is representative of how easily the blood can flow and the resistance to change in speed of that flow. The following definitions are introduced:

Equation 18.8 is the representation of the hemodynamic resistance (i.e., shear stress) that hinders flow rate. The resistance (R) is being defined as the equivalent resistance of the whole system, which was defined earlier as relative viscosity (or viscosity of blood taken relative to the viscosity of water).

$$R := R_{eq}. \tag{18.8}$$

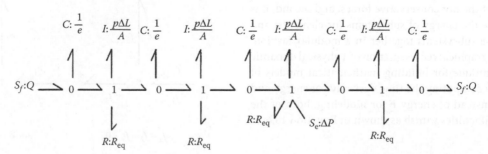

FIGURE 18.3 Bond graph representing the circulatory system.

Equation 18.9 is the representation of the elastance of the vessel wall (i.e., capacitance) or the change in volume during a pulsation. In other words, the capacitance (C) is defined as the inverse of the elastance of the vessel walls.

$$C := 1/e. \tag{18.9}$$

Equation 18.10 represents the inductance of the vessel, also known as fluid inertia. What is seen below is that the inertia (I) is defined as the density of blood multiplied by the length of travel, all divided by the cross-sectional area of the vessel:

$$I := \rho \Delta L / A. \tag{18.10}$$

By using the equations at their respective locations in the system model, more information can be obtained about the system as a whole using operational values for the heart. The values being used are the ones mentioned previously in the Introduction.

With this completed bond graph, the state variables are chosen and the mathematical models are manipulated until there is a system of first-order differential equations. The first step in getting to this system of differential equations is to select inputs and state variables, where the inputs are the sources or forcing functions and the state variables are the variables of energy storage (i.e., I and C). Bond graphs use inputs and state variables, which are sources or forcing functions and variables of energy storage, respectively, that define the elements of the system and control how they operate. For this model, the inputs are the flow and effort sources (see Equation 18.11). Since state variables are variables of energy storage, state variables for this model are I and C elements, which are represented by p and q, respectively (see Equation 18.12).

$$\text{Inputs: } S_f = Q; \; S_e = \Delta P, \tag{18.11}$$

State variables: $\dot{p}_2 = f_2$; $\dot{p}_4 = e_4$; $\dot{p}_7 = f_7$; $\dot{p}_9 = e_9$; $\dot{p}_{12} = f_{12}$;

$$\dot{p}_{14} = e_{14} \; \dot{q}_{18} = f_{18}; \; \dot{p}_{20} = e_{20}; \; \dot{q}_{23} = f_{23}. \tag{18.12}$$

The next step is to formulate the system of equations by first getting the derivative of state variables equal to either an effort or flow. Then use causality to track through the bond graph and eliminate nonstate variables. Equations 18.13 and 18.14 show the relationships involving common effort and flow (respectively) as caused by the 0-junctions and 1-junctions.

$$e_1 = e_2; \; e_7 = e_8; \; e_{11} = e_{12} = e_{13};$$

$$e_{17} = e_{18} = e_{19}; \; e_{22} = e_{23} = e_1, \tag{18.13}$$

$$f_4 = f_5; \; f_9 = f_{10}; \; f_{14} = f_{15} = f_{16}; \; f_{20} = f_{21}. \tag{18.14}$$

Equations 18.15 and 18.16 are the balance equations for these very same junctions, so that the power continuity equation is satisfied. With the balance equations, common effort and flow relationships, and the state variable equations, the elimination of all nonstate variables can take place. From this, the power can be determined and the integral can be taken in order to find the work done by the heart.

$$f_1 + f_2 + f_3 = 0; \; f_6 + f_7 + f_8 = 0; \; f_{11} + f_{12} + f_{13} = 0;$$

$$f_{17} + f_{18} + f_{19} = 0; \; f_{22} + f_{23} + f_{24} = 0, \tag{18.15}$$

$$e_3 + e_4 + e_5 + e_6 = 0; \; e_8 + e_9 + e_{10} + e_{11} = 0;$$

$$e_{13} + e_{14} + e_{15} + e_{16} = 0; \; e_{19} + e_{20} + e_{21} + e_{22} = 0. \tag{18.16}$$

Equation 18.17 shows that the velocity of the change in volume ($\dot{q}_2$) is directly equivalent to the flow at that point (f_2). Essentially, all 0-junctions represent this at different points throughout the body and it is only the constants that vary. The following equation represents the first 0-junction in Figure 18.4:

$$\dot{q}_2 = f_2 = -(f_1 + f_3); \; f_1 = Q = \frac{p_1}{q_1}; \; f_3 = f_4 = \frac{\Delta L \rho}{A} \dot{Q}. \tag{18.17}$$

As shown in Equation 18.18, the speed of the blood ($\dot{p}_4$) at any given point is directly related to the force being applied to it

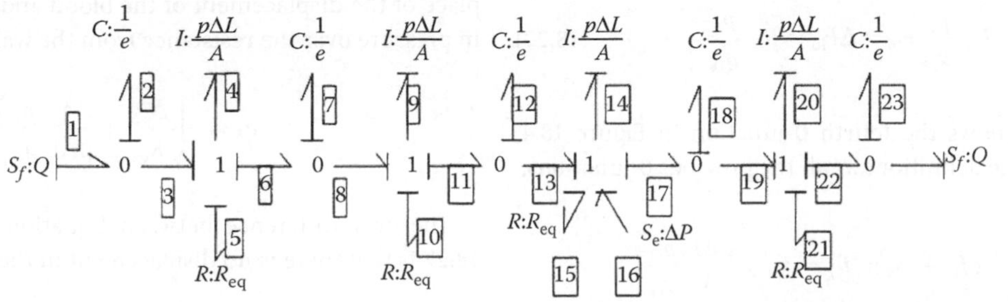

FIGURE 18.4 Bond graph with causality strokes and bond number. Note: the bond numbers are in the orange boxes and any variables relating to that bond have that number as a subscript.

(i.e., the change in pressure, represented by e_4 or effort). The following equation represents the first 1-junction in Figure 18.4:

$$\dot{q}_4 = e_4 = e_3 - e_5 - e_6; \ e_3 = e_2 = \frac{\Delta V_2}{e};$$

$$e_5 = R_{eq}Q_5; \ e_6 = e_7 = \frac{\Delta V_7}{e}. \tag{18.18}$$

Equation 18.19 represents the second 0-junction in Figure 18.4 and has the same definition as the first 0-junction (Equation 18.17).

$$\dot{q}_7 = f_7 = -(f_6 + f_8); \ f_6 = f_8 = \frac{\Delta P_5}{R_{eq}}; \ f_8 = f_9 = \frac{\rho \Delta L}{A} \dot{Q}. \tag{18.19}$$

Equation 18.20 shows the representation of the second 1-junction in Figure 18.4 and has the same definition as seen in Equation 18.18.

$$\dot{p}_9 = e_9 = e_8 - e_{10} - e_{11}; \ e_8 = e_7 = \frac{\Delta V_7}{e} q_7;$$

$$e_{10} = R_{eq} \frac{p_{10}}{q_{10}}; \ e_{11} = e_{12} = \frac{\Delta V_{12}}{e}. \tag{18.20}$$

The third 0-junction is represented by Equation 18.21 and has the same definition as the first and second 0-junctions (Equations 18.17 and 18.19).

$$\dot{q}_{12} = f_{12}; \ f_{12} = -(f_{11} + f_{13}); \ f_{11} = f_9 = \frac{\rho \Delta L}{A} \frac{\dot{p}_9}{\dot{q}_9};$$

$$f_{13} = f_{14} = \frac{\rho \Delta L}{A} \frac{\dot{p}_{14}}{\dot{q}_{14}}. \tag{18.21}$$

What differentiates the third 1-junction (Equation 18.22) from Equations 18.20 and 18.18 1-junction equations is that it has an extra source putting effort into the system. This is where the muscles pushing the blood toward the heart are represented.

$$\dot{p}_{14} = e_{14}; \ e_{14} = e_{13} + e_{16} - e_{15} - e_{17};$$

$$e_{13} = e_{12} = \frac{\Delta V_{12}}{e}; \ e_{17} = e_{18} = \frac{\Delta V_{18}}{e};$$

$$e_{15} = R_{eq} \frac{p_{15}}{q_{15}} e_{16} = \Delta P_{16} = R_{eq} \frac{p_{16}}{q_{16}}. \tag{18.22}$$

Equation 18.23 shows the fourth 0-junction in Figure 18.4, which bears the same definitions as all the previous 0-junctions.

$$\dot{q}_{18} = f_{18}; \ f_{18} = -(f_{17} + f_{19}); \ f_{17} = f_{14} = \frac{\rho \Delta L}{A} \frac{\dot{p}_{14}}{\dot{q}_{14}};$$

$$f_{19} = f_{20} = \frac{\rho \Delta L}{A} \frac{\dot{p}_{20}}{\dot{q}_{20}}. \tag{18.23}$$

The fourth 1-junction is represented by Equation 18.24 and follows the same definition as the first two 1-junctions (Equations 18.20 and 18.18).

$$\dot{p}_{20} = e_{20}; \ e_{20} = e_{19} - e_{22} - e_{21}; \ e_{19} = e_{18} = \frac{\Delta V_{18}}{e};$$

$$e_{22} = e_{23} = \frac{\Delta V_{23}}{e}; \ e_{21} = R_{eq} \frac{p_{22}}{q_{22}}. \tag{18.24}$$

The fifth 0-junction, the last junction in Figure 18.4, is represented in Equation 18.25 and resembles Equation 18.17 in the fact that the effort being put on that last section of the vessel wall is equivalent to the effort being exerted onto the heart.

$$\dot{q}_{23} = f_{23}; \ f_{23} = -(f_{22} + f_{24}); \ f_{22} = f_{20} = \frac{\rho \Delta L}{A} \frac{\dot{p}_{20}}{\dot{q}_{20}};$$

$$f_{24} = Q_{24} = \frac{p_{24}}{q_{24}}. \tag{18.25}$$

Stepping down e_4 the sequential bonds yield to conditions of each respective bond as shown in Equations 18.17–18.25, stopping at bond number 5 (box 5 in Figure 18.4). A series of first-order differential equation is formulated, as shown in Equations 18.26–18.34.

The velocity of change in volume of the vessels can be equated (as shown in Equation 18.26) using the displacement of the wall and its speed at the point before it, the displacement and speed of the blood at the point before it, and the inductance (which is the area, length, and density of the vessel in question).

$$\dot{q}_2 = -\left(\frac{p_1}{q_1} + \frac{A}{\Delta L \rho} \frac{\dot{p}_1}{\dot{q}_1} \right). \tag{18.26}$$

The speed of the blood (Equation 18.27) in the upper section of the body can be calculated by using elastance of the artery wall, the friction against it, the change in volume, as well as the displacement of the blood and vessel wall at two different points:

$$\dot{p}_4 = \frac{\Delta V_2}{e} q_2 - R_{eq} \frac{P_s}{q_s} - q_7 \frac{\Delta V_7}{e}. \tag{18.27}$$

Equation 18.28 is similar to Equation 18.26 except that in place of the displacement of the blood and the wall, the change in pressure over the resistance from the wall is used:

$$\dot{q}_7 = -\left(\frac{\Delta P_s}{R_{eq}} + \frac{A}{\rho \Delta L} \frac{\dot{p}_s}{\dot{q}_s} \right). \tag{18.28}$$

The only difference between Equation 18.29 and Equation 18.27 is that there is no displacement in the second section.

$$\dot{p}_9 = \frac{\Delta V_7}{e} q_7 - R_{eq} \frac{p_{10}}{q_{10}} - \frac{\Delta V_{12}}{e}. \tag{18.29}$$

Since this section is the return path of the lower part of the body, it has some added effort into it (as shown in Equation 18.30) and is therefore the flow that is affected by this. There is only inductance and the time derivative of the flow rate in this equation.

$$\dot{q}_{12} = -\left(\frac{A}{\rho \Delta L} \frac{\dot{p}_{14}}{\dot{q}_{14}} + \frac{A}{\rho \Delta L} \frac{\dot{p}_{20}}{\dot{q}_{20}} \right). \tag{18.30}$$

It is in Equation 18.31 that the added source is taken into account with the added displacements and resistance and the removal of the displacements.

$$\dot{p}_{14} = \frac{\Delta V_{12}}{e} + R_{eq} \frac{p_{16}}{q_{16}} - R_{eq} \frac{p_{15}}{q_{15}} - \frac{\Delta V_{13}}{e}. \tag{18.31}$$

Equations 18.32 through 18.34 follow the original patterns of Equations 18.26 through 18.28.

$$\dot{q}_{18} = -\left(\frac{\rho \Delta L}{A} \frac{\dot{p}_{14}}{\dot{q}_{14}} + \frac{\rho \Delta L}{A} \frac{\dot{p}_{20}}{\dot{q}_{20}} \right), \tag{18.32}$$

$$\dot{p}_{20} = \frac{\Delta V_{18}}{e} - \frac{\Delta V_{23}}{e} - R_{eq} \frac{p_{21}}{q_{21}}, \tag{18.33}$$

$$\dot{q}_{23} = -\left(\frac{\rho \Delta L}{A} \frac{\dot{p}_{20}}{\dot{q}_{20}} + \frac{p_{24}}{q_{24}} \right). \tag{18.34}$$

Using these in conjunction with the parameters laid out in the Introduction, the work can be solved for in terms of speed and displacement. Since $Q \times h = W$ and $Q = p/q$, the work can be determined by taking the integral of $\dot{Q}$. Because all of the state variables are time derivatives, performing integration is necessary to find Q. Using Equations 18.26 through 18.34, each section will contribute a $\dot{Q}$ and by summing all of the sections, the total $\dot{Q}$ can be obtained.

Some unmentioned constants that are used for calculating the total $\dot{Q}$ include: A is set to be 3.1415 mm², ΔL is approximately 41 mm per section, ρ is 1.06e-6 kg/m³, R_{eq} is 3.59, e (elastance) is 1 kPa, ΔV is 0.1162 mm², and ΔP is an average of 0.01599 kPa. Since they are constants, they will remain the same value throughout the entire system. With these constants and the aforementioned equations, the $\dot{q}$ and $\dot{q}$ can be solved for

$$\dot{Q}_1 = \frac{\dot{p}_4}{\dot{q}_2} = \frac{\left((\Delta V_2/e)q_2 - R_{eq}(p_5/q_5) - q_7(\Delta V_7/e) \right)}{-\left((p_1/q_1) + (A/\Delta L \rho)(\dot{p}_1/\dot{q}_1) \right)} =$$

$$= \frac{-0.359}{-[(0.1) + 72284.85(\dot{p}_1/\dot{q}_1)]}, \tag{18.35}$$

$$\dot{Q}_2 = \frac{\dot{p}_9}{\dot{q}_7} = \frac{(\Delta V_2/e)q_7 - R_{eq}(p_{10}/q_{10}) - \Delta V_{12}/e}{-\left((\Delta P_5/R_{eq}) + (A/\rho \Delta L)(\dot{p}_8/\dot{q}_8) \right)}$$

$$= \frac{-0.1162q - 0.4752}{-[0.00445 + (72284.85)(\dot{p}_8/\dot{q}_8)]}, \tag{18.36}$$

$$\dot{Q}_3 = \frac{\dot{p}_{14}}{\dot{q}_{12}} = \frac{(\Delta V_{12}/e) + R_{eq}(p_{16}/q_{16}) - R_{eq}(p_{15}/q_{15}) - \Delta V_{18}/e}{-\left((A/\rho \Delta L)(\dot{p}_{14}/\dot{q}_{14}) + (A/\rho \Delta L)(\dot{p}_{20}/\dot{q}_{20}) \right)}$$

$$= \frac{0}{-72284.85((\dot{p}_{14}/\dot{q}_{14}) + (\dot{p}_{20}/\dot{q}_{20}))}, \tag{18.37}$$

$$\dot{Q}_4 = \frac{\dot{p}_{20}}{\dot{q}_{18}} = \frac{(\Delta V_{18}/e) - (\Delta V_{23}/e) - R_{eq}(p_{21}/q_{21})}{-\left((\rho \Delta L/A)(\dot{p}_{14}/\dot{q}_{14}) + (\rho \Delta L/A)(\dot{p}_{20}/\dot{q}_{20}) \right)}$$

$$= \frac{-0.359}{-(72284.85)[(\dot{p}_{14}/\dot{q}_{14}) + (\dot{p}_{20}/\dot{q}_{20})]}, \tag{18.38}$$

$$\dot{Q}_{\text{Total}} = \dot{Q}_1 + \dot{Q}_2 + \dot{Q}_3 + \dot{Q}_4$$

$$= \frac{-0.359}{[(0.1) + 72284.85(\dot{p}_1/\dot{q}_1)]}$$

$$+ \frac{0.1162q - 0.4752}{-[0.00445 + (72284.85)(\dot{p}_8/\dot{q}_8)]}$$

$$+ \frac{-0.359}{-(72284.85)[(\dot{p}_{14}/\dot{q}_{14}) + (\dot{q}_{20}/\dot{q}_{20})]}. \tag{18.39}$$

Taking the integral of $\dot{Q}_{\text{Total}}$, Equation 18.40 yields the total flow rate Q_{Total}:

$$Q_{\text{Total}} = \left[\frac{(72284.85)}{Q} + \left(0.363455 + \frac{4.9664}{Q} \right) + \left(\frac{2.48323 \times 10^{-6}}{Q} \right) \right]$$

$$= 12047.475 \text{kg/s}. \tag{18.40}$$

The work now follows by multiplying Equation 18.40 with the distance traveled (i.e., Equation 18.1).

$$W = Q \times h = (12047.475) \times (1.654) = 19926.523 \text{ kN} \times \text{m}. \tag{18.41}$$

It is also possible to use the differential equations to find the frequency response of the system. In Figure 18.5, it is shown that the frequency response of the heart as calculated by the differential equations resembles that of an electrocardiogram (ECG or EKG) (see Chapter 23 for details), which validates that the system

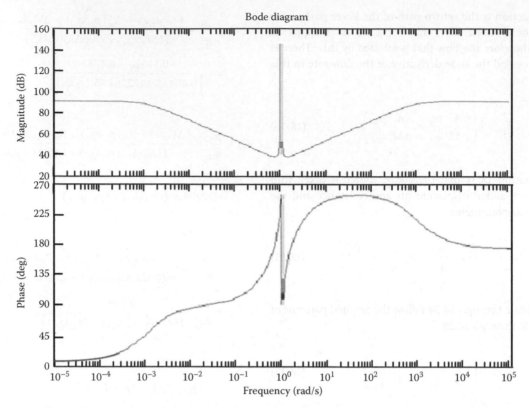

FIGURE 18.5 Bond diagram of the frequency response using the differential equations.

being analyzed is behaving as it should, even with all of the assumptions in place (i.e., Newtonian, steady state, average blood pressure, neglecting branching, and smaller vessels).

18.6 Conclusion

The heart and circulatory system is an extremely complex system with multiple variables, inputs, outputs, and elements. Luckily, there is not a high level of accuracy needed to analyze the heart; most of the time, it is considered in a much simpler context. Given this, the heart was modeled using a pump, reservoir, capacitors (C), inductors (I), and resistors (R). The cardiovascular system was modeled as a whole using bond graphs and differential equations. Using these tools, the power of the system (rather than the energy) was analyzed in order to determine the work of the system. The cardiovascular system does about 19926.523 kN m of

work per cycle (i.e., leaving the aorta to flow through the body, returning to the heart, and going through the lungs to be sent back through the aorta). Using the same differential equations, the frequency response was determined and it was that the actual behavior of the system was not compromised, even though numerous assumptions were made.

References

1. Wilnor, W. R. (1982). *Hemodynamics*. Unknown: Williams and Wilkins.
2. White, F. (2006). *Fluid Mechanics*, 6th Edition. Columbus: McGraw-Hill.
3. Karnopp, D. C., Margolis, D. L., and Rosenberg, R. C. (2006). *System Dynamics: Modeling and Simulation of Mechatronic Systems*. New York: Wiley.

19

Anatomy and Physics of Respiration

Timothy Moss,
David Baldwin,
Beth J. Allison, and
Jane Pillow

19.1 Introduction

The provision of oxygen to the body's cells for oxidation of substrate to generate energy and the removal of the gaseous waste product, carbon dioxide, is critical for the survival of most organisms. In the animal kingdom, there are systems of varying complexity for transport of oxygen from the external environment to the body's tissues. The principal focus of this chapter is the human respiratory system, but knowledge derived from other relevant species has been used in order to describe the biophysical properties of this critical organ system. We consider also human disease states, developmental stages, and other variations that impact on the biophysical properties of the respiratory system. The primary purpose of the respiratory system is to exchange the respiratory gases, namely oxygen and carbon dioxide, but it performs many other tasks, including thermoregulation (e.g., panting in dogs), host defense against pathogens, and metabolic functions.

19.2 Anatomic and Physical Relationships

The respiratory system consists of the lungs, conducting airways, pulmonary circulation, diaphragm and chest wall, and the associated blood vessels, muscles, and nerves.

The fundamental requirement of any exchange organ is a large surface area over which exchange occurs, sufficient supply of this surface with the substances to be exchanged, and a thin barrier across which exchange can easily occur.

The lungs are paired organs; the left consists of two lobes (an upper and a lower) and the right consists of three (upper, middle, and lower). The lung tissue is sponge-like in appearance with 300–700 million small air sacs, termed alveoli.[1] These millions of small airspaces result in a large surface area for gas exchange (50–100 m²).[2] The pulmonary interstitial tissue contains fibroblasts, elastin and collagen fibers, proteoglycans and other elements such as smooth muscle. These cells and tissues within the interstitium contribute to the structural integrity and mechanical properties of the lung.

The lung tissue is elastic and, *in situ*, is held in a slightly stretched state as a result of indirect coupling to the chest wall and diaphragm. The inner chest wall and the upper surface of the diaphragm, and the outer surface of the lungs are lined with, respectively, parietal and visceral pleural membranes. Surface tension within the intrapleural space creates a subatmospheric (negative) intrapleural pressure that holds the lungs slightly open at rest, against the chest wall and diaphragm.

During normal inspiration, when the diaphragm and external intercostal muscles of the ribcage contract, expansion of the internal dimensions of the thorax draws the visceral pleural membrane outwards, thus stretching open the lungs. This results in the creation of subatmospheric pressure within the chest; air moves into the lungs to equilibrate pressure. Effective inspiration is thus reliant on the integrity of neural control mechanisms and the structural characteristics of the chest wall. Musculoskeletal disease and other factors that affect chest wall mechanics can have a large impact on the function of the respiratory system (e.g., preterm neonates have compliant chest walls that tend to retract, rather than expand, during diaphragmatic contraction). Relaxation of the inspiratory muscles results in passive recoil of the lungs and chest wall, and causes expiration. During exercise or forced expiration, contraction of the expiratory muscles (the internal intercostals of the ribcage and the muscles of the abdominal wall) generates a positive intrathoracic pressure, thus forcing air from the lungs.

19.2.1 The Airways

The alveoli within the lungs communicate with the external environment via a system of branching airways. The complicated airway branching pattern appears to be a result of the coordination of three distinct simple modes of branching that operate during development.[3] The airway system can be divided into two sections: the conducting zone, which is essentially a series of progressively narrowing and branching tubes that carry air from the external environment into the lungs; and the respiratory zone, where gas exchange occurs.

19.2.1.1 The Conducting Zone

During inspiration, when the glottis opens and the inspiratory muscles contract, air entry into the respiratory system occurs through the mouth and/or nose. Warming and humidification of inspired air occur here and throughout the conducting airways, of which there are 15–16 generations.[4] The trachea communicates between the upper airway and the lungs. Its walls are rigid as a result of cartilage rings embedded within its wall. Within the chest, the trachea divides into the left and right main stem bronchi. From this major division, the airways branch further and progressively narrow. With progressive branching, the proportion of cartilage surrounding the airways decreases and the proportion of smooth muscle found in the walls increases. When cartilage is no longer present the airways are termed bronchioles.

The absence of cartilage rings from bronchioles influences their mechanical properties. They are compliant, relative to the bronchi, and since they are tethered to the interstitial tissue framework of the lung, transpulmonary pressure (i.e., the pressure difference across the airway wall) influences their caliber. Disease states that influence the mechanical properties of the lung tissue can therefore influence airflow through these small airways.

The conducting zone of the lungs and the upper airway together form the *anatomic dead space* of the lungs. The volume of air contained within these structures is exchanged with the external environment during breathing but it does not contribute to gas exchange. The volume of the anatomical dead space is largely independent of changes in lung volume.

Movement of air through the conducting zone of the lungs appears to occur largely as a result of bulk transport.[5]

19.2.1.2 Respiratory Zone

Gas exchange occurs within the airway generations contained within the respiratory zone. Respiratory bronchioles have alveoli budding from their walls and further divide into terminal alveolar ducts. Alveolar ducts end with blind alveolar sacs containing only alveoli separated by septal walls. There are small holes in the walls of the interalveolar septa, termed the pores of Kohn, which permit gas exchange between adjacent alveoli. Thin interalveolar septae increase the surface area for gas exchange in this region.

Upon inspiration, the air that moves into the lungs is not distributed equally throughout the available space. Similarly, expiration does not empty gas equally from all regions of the lungs. This *unequal distribution of respired air* is dependent on differences in compliance, airway resistance, and balance between gravitational and intrapleural forces on the lungs. In a seated or standing subject, the weight of the lungs results in the intrapleural pressure being higher (i.e., less negative) at the base of the lungs than at the apices. This results in relative closure of the lower regions of the lung relative to the upper regions, which are held more open by the greater negative intrapleural pressure. During inspiration, the lower regions of the lungs are opened more than the upper regions, as a result of volume-related differences in compliance and greater expansion of the internal thoracic dimensions in the lower chest. Thus, ventilation (i.e., the amount of air moved into the lung) is relatively greater in the lower regions of the lungs than in the upper regions.

Within the respiratory zone of the lungs, there is chaotic mixing of air[6] and gas movement occurs largely by diffusive transport.[5]

19.2.2 Airway Resistance and Flow

The complex branching pattern of the airways renders application of standard flow concepts such as laminar and turbulent flows difficult. Mathematical modeling of airflows based on data obtained from anatomical specimens, bronchoscopy, computed tomography, and x-ray has demonstrated the complex flow configurations that result from the branching pattern of the lungs.[7] Throughout much of the lung, airflow is likely transitional between laminar and turbulent states, with laminar flow restricted to the very small airways in the periphery, and turbulent flow occurring through much of the conducting airway system.[8]

Respiratory system resistance is contributed to by both airway and tissue components of the lungs. Resistance of the entire respiratory system can be made by *plethysmography*, using measurements of airflow and pressure change[9] and using the *forced oscillation technique* (see below). The *interrupter technique* or *end-inspiratory interruption* can be used, respectively, for measurement of airway or tissue components of resistance.[8]

19.2.3 Lung Compliance

Examination of the pressure–volume relationship of the lungs during inflation and deflation demonstrates a number of important mechanical characteristics of the lungs. Figure 19.1 shows the *pressure–volume curve* when a positive pressure is delivered to the airways and is then removed. During inflation from zero volume, pressure must increase until the *opening pressure* of the lung is reached, at which point the small airways and alveoli begin opening. Inflation of the lung is nonuniform at this point because of regional differences in opening pressures. Once the opening pressure is reached, lung volume increases as pressure is increased until *total lung capacity* is reached. During deflation, as airway pressure falls, lung volume decreases; but the pressure–volume relationship during deflation is different from that during inspiration.

The lung volume change per unit pressure change (i.e., the gradient of the relationship) is known as the *compliance* of the lungs, as illustrated by Equation 19.1.

$$C = \frac{\Delta V}{\Delta P}. \tag{19.1}$$

Lung compliance is influenced by the stiffness of the lung tissue, which can be markedly affected by disease. For example, pulmonary fibrosis makes the lung tissue stiffen, thus decreasing compliance; emphysema causes the lung tissue to become less stiff, thus increasing compliance.

The idealized *pressure–volume curve* illustrated in Figure 19.1 is a simplification of the relationship between pressure and volume in the lung *in situ*. Technically, the compliance can be derived from the slope of the curve. In reality, elastic and resistive characteristics of the chest wall (which is pulled inward by the negative intrapleural pressure, just as the lungs are pulled outward) influence this relationship. Thus diseases that affect these structures can affect compliance of the respiratory system.

The *hysteresis* in the inflation/deflation pressure volume curve is a result of the viscoelastic properties of the lung tissue. Thus, the mechanical characteristics of the lung parenchymal tissue are large contributors to lung function and its perturbation in disease. In recent years, experimentation and mathematical modeling have contributed substantially to the understanding of the mechanical properties of the lung tissue in healthy subjects and in various disease states.[10,11]

19.2.4 The Blood–Gas Barrier

The blood–gas barrier consists only of the cytoplasmic extensions of Type I alveolar epithelial cells, the endothelial wall of the pulmonary capillary, and a thin layer of pulmonary interstitial tissue.[12] The thickness of the barrier varies throughout regions of the lung;[13] at its thinnest it is 0.2 μm in width,[2] with a mean thickness of 1.25 μm.[14]

Type I alveolar epithelial cells have long and thin cytoplasmic extensions and make up the majority of the surface area of the lung. Indeed, although Type I cells account for ~10% of total lung cell population, they cover more than 95% of the lung surface area.[15] Type II cells constitute ~12% of the overall lung cell population; they are cuboidal in shape and contain many organelles, including lamellar bodies, which are the sites of surfactant production, storage, and recycling within the lungs.

19.2.5 Surface Tension and Surfactant

The fluid lining the internal surface of the lungs creates surface tension as a result of attractive forces between molecules in the fluid. At the alveolar surface, the effect of this surface tension is to encourage airspace collapse. However, the alveoli and small airways are lined by pulmonary surfactant, which lowers the surface tension at the air liquid interface. The surface-active properties of surfactant are due mainly to phospholipids, which are the major constituents of surfactant. Surfactant proteins (SP-A, -B, -C, and -D) contribute to surfactant storage and secretion (SP-B), adsorption of surfactant on the alveolar surface (SP-B and -C), and to host defense within the lung (SP-A and -D).[16] Diseases that affect the function of surfactant thereby have deleterious effects on lung compliance. The surface tension, alveolar pressure, and volume of the alveoli will be described later with the help of the law of Laplace.

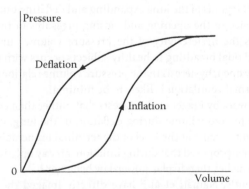

FIGURE 19.1 A schematic of the relationship between volume and pressure in an excised human lung. (After Salmon RB, Primiano FP Jr, Saidel GM and Niewoeher DE. *J Appl Physiol* 1981; 51(2): 353–362.)

19.2.6 Pulmonary Circulation

The blood supply to the lungs leaves the right side of the heart through the pulmonary arteries, which branch to eventually provide a capillary surface area of ~70 m² for gas exchange.[4] The pulmonary vasculature is maintained within the lung interstitial (parenchymal) tissue at a much lower pressure (mean pulmonary arterial pressure is ~15 mm Hg) than the systemic circulation (mean pressure in the aorta is ~100 mm Hg). This low pressure provides substantial reserve to pulmonary blood flow.

Pulmonary vascular resistance can be greatly reduced from its already-low value at rest, by only small increases in pressure (e.g., when cardiac contractility increases during exercise). This occurs initially by recruitment of capillaries with no flow, but at higher pressures, distension of the capillary bed occurs.

The pulmonary capillaries and the alveolar epithelial cells are fused over a large proportion of the alveolar surface area (8). As a result, the alveolar and capillary walls are mechanically coupled such that forces exerted in one compartment will influence the other, resulting in a capillary–alveolar transmural relationship. Thus, capillary–interstitial tissue transmural pressure is an important determinant of capillary caliber and therefore pulmonary blood flow and resistance. At very low lung volumes, smooth muscle tone and the elasticity of the blood vessel walls keep the caliber of the pulmonary vessels small. Increases in lung volume from this low point open the pulmonary vasculature as the airways open. If airway and interstitial tissue pressures rise, relative to capillary pressure (e.g., at high lung volumes), compression of the vascular network can result in increases in pulmonary vascular pressure sufficient to reduce pulmonary blood flow.[17]

The caliber of the pulmonary vessels, unlike those in the periphery, is decreased by hypoxia (low oxygen levels).[18] This characteristic is a result of the direct effects of low oxygen levels in the airspace, not the pulmonary circulation itself, and is mediated by direct effects on the smooth muscle and endothelial cells lining the pulmonary vessels. Hypoxic pulmonary vasoconstriction has the effect of directing pulmonary blood flow away from poorly ventilated regions of the lung, in which the oxygen level is low. In cases of generalized hypoxia (such as at altitude or in diseases that cause reductions in ventilation), this characteristic of the pulmonary circulation can lead to pulmonary hypertension.

19.2.7 Lung Volumes

The *total lung capacity* is the maximal absolute volume of air that can be contained within the lungs. By convention the total lung capacity is subdivided into three different overlapping capacities and their four constituent nonoverlapping volumes (Figure 19.2). Conventional equations for the estimation of normal lung volumes include sex, body size, and age as contributing factors.[19]

Tidal volume is the volume of air that enters (and leaves) the lungs during a breath. At rest, tidal volume of a normal adult male is approximately 500 mL. The *minute ventilation* is the amount of air moved into (and out of) the lungs in 1 min; it is the product of tidal volume and breathing rate.

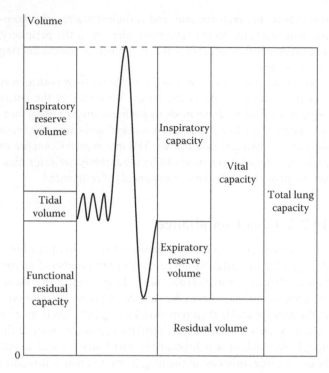

FIGURE 19.2 A schematic spirometry recording showing tidal breathing, followed by a maximal inspiration and expiration.

The volume of air remaining within the lungs after a normal expiration is the *functional residual capacity* (FRC), which is made up of the *expiratory reserve volume* (the volume of air that could be forcibly exhaled beyond a normal expiration) and the *residual volume* of the lungs, which cannot be exhaled. The *vital capacity* is the sum of the *inspiratory reserve volume* (the volume of air that can be inhaled above that resulting from a normal inspiration), *tidal volume* (which together constitute the *inspiratory capacity*), and the *expiratory reserve volume*.

19.2.8 Lung Recruitment

Work by Frazer et al. suggests that during inflation and deflation of the lung, individual lung "units" are open and closed, rather than all regions of the lung expanding and deflating uniformly.[20] Difference in the opening and closing pressures of these units confers the hysteresivity of the pressure volume curve. Since normal tidal breathing in healthy subjects occurs with the lungs already open, hysteresis in the pressure volume relationship during normal ventilation is likely to be minimal.

The work by Frazer et al. suggests that, rather than collapse of alveoli to zero volume during deflation of the lung, closure of lung units occurs at the level of the terminal bronchioles.[20] Suki et al. have proposed that during inflation, airway openings occur in avalanches that display power-law behavior.[21]

Recently, Namati et al.[22] have directly imaged the opening of individual alveoli during inflation of excised mouse lungs using scanning laser confocal microscopy, and observed alveolar opening at relatively high pressures (~25 cm H₂O). During inflation, at lung volumes close to total lung capacity, they observe the appearance of increasing numbers of open alveoli and a corresponding decrease in alveolar size. To explain their observations, they propose the presence within an alveolar duct of two types of alveoli: "mother" alveoli and several "daughter" alveoli. These two types of alveoli are connected by pores of Kohn, which provide a duct via which the mother alveoli may recruit the daughter alveoli. It is postulated that as pressure increases throughout inspiration the mother alveoli increase in size, stretching the alveolar walls. This stretch increases the size of the pores of Kohn and thins the surfactant layer lining the alveoli, which forms a film that occludes the pores. Once a threshold pressure is reached, blockage of the pore by surfactant is overcome and air enters the daughter alveoli. It is proposed that during deflation, equilibration of pressure between mother and daughter alveoli results in deflation of them all, until pore size reduces sufficiently to re-establish blockage by a film of surfactant. This hypothesis provides a theoretical mechanism whereby closed alveoli are recruited during inspiration and remain open during subsequent expiration.

19.2.9 Measurement of Lung Volumes

Some lung volumes can be measured directly, by spirometry, but FRC and residual volume cannot because a maximal exhalation cannot empty the lung. The two commonly used techniques for measurement of FRC are body plethysmography and gas dilution techniques the physical principles of which are summarized below (for a more detailed review of these techniques, and critical appraisal of their application to studies of infants, see Ref. [23]).

Body plethysmography was first described by DuBois et al. in 1956.[24] This technique is based on Boyle's law, which states that for a gas at constant temperature the product of pressure (P) and volume (V) remains constant:

$$P \times V = k. \tag{19.2}$$

Body plethysmography involves the subject sitting (or lying) in a large airtight "box" (the plethysmograph), wearing a nose-clip, and breathing through a mouthpiece that can be remotely closed at any stage of the respiratory cycle. At the end of a normal expiration, when the lung is at FRC, the mouthpiece is closed and the subject makes respiratory movements against this obstruction. These obstructed movements result in compression of the air within the respiratory system: that is, there is a rise in pressure (to P_{obst}), which is measured, and a fall in volume (ΔV_{lung}) within the respiratory system from their initial values, before the obstructed movement (P_0 and FRC, respectively). Within the plesythmograph there is a corresponding change in air pressure, which is measured, and a corresponding change in volume. The volume change in the plethsymograph can be calculated from Boyle's law, since the pressure change in the plethysmograph is known. The change in air volume within the plethysmograph corresponds to the change in lung volume

(ΔV_{lung}) during the obstructed respiratory movement. Hence, from Boyle's law:

$$P_0 \times FRC = (FRC - \Delta V_{lung}) \times P_{obst}. \tag{19.3}$$

Thus,

$$FRC = \Delta V_{lung} \times \left(\frac{P_{obst}}{P_{obst} - P_0} \right). \tag{19.4}$$

Gas dilution methods to determine FRC use measurements of inert gas concentrations in known and unknown volumes. For example, the helium (He) dilution technique involves attachment of the subject to a closed respiratory circuit of known volume (V_0), containing a known concentration of He (He_0), at the end of expiration (FRC). The subject breathes from the closed circuit for a period sufficient to distribute the He throughout the entire volume of the circuit and the lungs ($V_0 + FRC$), and then the resulting He concentration (He_{end}) is measured. Hence,

$$He_0 \times V_0 = He_{end} \times (V_0 + FRC). \tag{19.5}$$

Thus,

$$FRC = \frac{V_0 \times (He_0 - He_{end})}{He_{end}}. \tag{19.6}$$

Open circuit multiple breath washout techniques are other examples of *gas dilution* methods used to measure FRC. By measuring the concentration of an inert indicator gas at the airway opening and the flow of respired gas, FRC can be calculated from:

$$FRC = \frac{V_{tracer}}{\left[Tracer_{initial} \right]}, \tag{19.7}$$

where V_{tracer} is the volume of tracer gas washed out and [$Tracer_{initial}$] initial tracer gas concentration prior to washout, where the volume of tracer gas washed out is calculated by integrating the product of flow and tracer gas concentration over time.

The most common of the *multiple breath washout* techniques is the nitrogen (N_2) washout technique.[25] Measurement of lung volume using *gas dilution* allows indirect assessment of ventilation inhomogeneity; the longer the time for equilibration of gas concentrations, the greater the inhomogeneity.

Measurement of lung volumes using modern imaging modalities such as magnetic resonance imaging (MRI)[26] and phase contrast x-ray imaging,[27] although currently restricted to research applications, shows promise for providing much more detailed, real-time measurements of lung volumes.

A comprehensive review of lung volumes, techniques for their measurement and their variation in disease can be found in the American Thoracic Society's *Consensus statement on measurements of lung volumes in humans*.[28]

19.2.10 Measurements of the Respiratory System Function

The most useful index of airway function is measurement of the *forced expiratory volume in 1 s (FEV₁) divided by the forced vital capacity (FVC)* as obtained by spirometry.

$$\text{Respiratory\_Sytem\_Function} = \frac{\text{FEV}_1}{\text{FVC}}. \qquad (19.8)$$

Measurement involves having the subject inhale maximally (to total lung capacity) and then exhale forcefully, as quickly as possible, to residual volume. In healthy subjects, over 75% of the FVC can be exhaled within 1 s. Reductions in FEV_1/FVC reflect increased resistance to airflow, and are therefore useful for the diagnosis of asthma and chronic obstructive pulmonary disease. These forced respiratory maneuvers are dependent on subject cooperation, making them unsuitable for use in infants and some patient populations.

Airway function may also be assessed using the *forced oscillatory technique*, which allows determination of the mechanical properties of the lungs by examining the simultaneous pressure and airflow responses of the respiratory system to externally applied oscillations.[29] The pressure–flow relationship yields a measure of respiratory system impedance (Z_{rs}), which includes contributions from all tissue elements within the lungs. In practice, a range of oscillatory frequencies are applied to the respiratory system and Fourier analysis is applied to extract individual frequency components. Extraction of information about lung mechanics from such data is complex and requires inverse modeling.[30]

Numerous theoretical models have been applied to the analysis of measurements of impedance using the forced oscillatory technique, with model elements representing airway and tissue compartments within the lung. The most descriptive of these models is the *constant-phase model*[30] developed by Hantos et al.[31]:

$$Z_{rs} = R_N + i2\pi fI + \frac{G - iH}{(2\pi f)^{\alpha}}, \qquad (19.9)$$

where R_N is a Newtonian flow resistance approximating airway resistance, I is inertance due to the mass of gas in the airways, G reflects tissue energy dissipation (or tissue damping), and H reflects energy storage (or elastance). The constant-phase model allows partitioning of lung mechanics into airway (R_N and I) and tissue (G and H) components. The current clinical applications of the forced oscillatory technique, and its potential for aiding diagnosis and providing an understanding of the pathophysiology of lung disease, are addressed by LaPrad and Lutchen.[32]

A distinct benefit of the forced oscillatory technique is that it does not require patient cooperation, making it useful for assessment of respiratory function in all types of subjects.

Diffusing capacity of the lung refers to the ability of gas to diffuse across the blood–gas barrier, and is commonly assessed by measuring the diffusing capacity of the lung for carbon monoxide using a single breath technique to yield information about the blood–gas barrier and pulmonary capillaries.[33]

Relatively simple tests, based on adding load during various respiratory maneuvers, exist for the assessment of the strength and endurance of the respiratory muscles. In addition, respiratory muscle electromyogram measurement is used to examine respiratory muscle function in various musculoskeletal and neurological diseases. These tests are described in a review of the mechanics of the respiratory muscles by Ratnovsky et al.[34]

19.2.11 Gas Exchange

The behavior of the respiratory gases, namely oxygen and carbon dioxide, is dictated by standard properties of gases as described by the *ideal gas law* (after Boyle and Charles), *Dalton's law of partial pressures* and *Henry's law of gas solubility* in liquid.

The levels of oxygen and carbon dioxide in the blood are influenced by the partial pressures of these gases in respired air, mixing of inspired air with residual air in the lung, and diffusion of the gases across the alveoli into the blood in the capillaries.

Ambient air contains 21% oxygen and 79% nitrogen, and a negligible amount of carbon dioxide (<0.001%). Thus, at normal atmospheric pressure (760 mm Hg), the partial pressure of inspired oxygen is ~160 mm Hg (0.21 × 760) and that of carbon dioxide is ~0.3 mm Hg. As air is inspired, it is heated to body temperature and saturated with water vapor (at 37°C, water vapor pressure is 47 mm Hg). Thus, within the lungs, the partial pressure of oxygen in inspired air would be ~150 mm Hg. In reality the volume of air inhaled with each breath (i.e., the tidal volume; ~500 mL) is mixed with the air that remains in the lungs at the end of the preceding expiration (i.e., the FRC; ~3l), which itself contains oxygen. On average, in healthy subjects, the partial pressure of oxygen at the exchange surface of the lungs is ~100 mm Hg and varies little across the respiratory cycle, reflecting a balance in supply of oxygen by inspired air, its mixing with the airways, and the removal of oxygen by the pulmonary circulation.

Variability in the ventilation/perfusion ratio results in regional differences in alveolar partial pressures of oxygen in the upper and lower regions of the lungs of ~130 mm Hg and 90 mm Hg, respectively.[35] Abnormal ventilation/perfusion ratios in disease can similarly affect oxygen supply to the gas-exchange surface of the lungs. Although the carbon dioxide content of inspired air is negligible, air mixing within the airways and the continual addition of carbon dioxide form the pulmonary circulation result in an average alveolar partial pressure of carbon dioxide of ~40 mm Hg.

Gas transport across the alveolar wall occurs by diffusion. As blood flows through the pulmonary capillaries, oxygen and carbon dioxide partial pressures equilibrate with those of the air in the alveoli. *Henry's law* states that the volume, and therefore the concentration ([Gas]) of a gas dissolved in a liquid, is proportional to its partial pressure (*P*):

$$[Gas] = K \times P, \qquad (19.10)$$

where *K* is a solubility constant. The thin alveolar and pulmonary capillary walls provide a minimal barrier to diffusion of oxygen and carbon dioxide, so partial pressures rapidly equilibrate. This rapid equilibration is described by *Fick's law*, which states that the rate of diffusion is directly proportional to the concentration gradient and surface area for exchange, and is inversely related to thickness of the exchange surface.

In recent years the concept of *diffusional screening* in the lung has been proposed, whereby interactions between gas molecules and the alveolar surface limit the ability of other molecules to diffuse across the alveolar surface. The concept has consequences for understanding mechanisms of respiratory reserve and the regulation of pulmonary perfusion.[36]

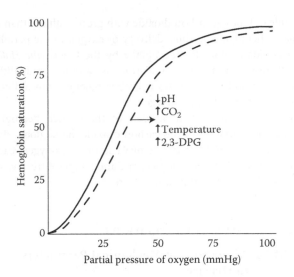

FIGURE 19.3 The oxygen–hemoglobin dissociation curve. Right shift of the curve occurs due to reductions in pH and increases in the partial pressure of carbon dioxide (CO_2; the Bohr effect), and increases in temperature and 2,3-diphosphoglycerate.

19.2.12 Gas Transport in the Blood

The total content of oxygen and carbon dioxide in the blood is not determined simply by the amount of the gases dissolved in the blood, which would be insufficient for transport of sufficient amounts to sustain cellular respiration. The majority of oxygen transported in the blood is bound to hemoglobin on red blood cells and only a small proportion (<2% at a partial pressure of 100 mm Hg) is dissolved. The dissolved gas concentration provides an osmotic pressure (Π), using the Morse equation[37,38]:

$$\Pi = iMRT, \qquad (19.11)$$

where $R = 0.08206$l atm/mol K = 8.314472 J/mol K is the gas constant, *M* the molarity of the dissolved gas, and *T* the temperature.

The majority of carbon dioxide is carried in the blood in the form of bicarbonate, with small amounts of the total carbon dioxide content dissolved in the blood or bound to proteins in the blood (forming carbamino compounds), the most important of which, in this respect, is hemoglobin (thus, carbaminohemoglobin is formed).

19.2.12.1 Oxygen Transport and Tissue Delivery

The ability of oxygen to bind to hemoglobin on red blood cells is dependent on the partial pressure of oxygen and pH of the blood, which can be illustrated using the *oxygen–hemoglobin dissociation curve* (Figure 19.3). At a partial pressure of 100 mm Hg (i.e., that in the arterial circulation after alveolar exchange), hemoglobin is almost completely (~98%) saturated with oxygen. In the pulmonary capillaries, as oxygen dissolves into the blood, it rapidly binds to hemoglobin, thus moving out of solution to facilitate further diffusion.

As blood circulates through the body, oxygen diffuses into the tissues. At arterial partial pressures of oxygen (>80%) there is little variation in the oxygen saturation of hemoglobin so the total oxygen content of the blood remains high despite a reduction in oxygen partial pressure in the arteries. In the tissues, as oxygen diffuses out of the blood to support cellular respiration, the partial pressure of oxygen in the blood falls to a point at which the relationship between oxygen and hemoglobin results in the liberation of large amounts of oxygen from the red blood cell reserve for only small reductions in partial pressure. Diffusion of oxygen to the peripheral tissues is aided by the affinity of myoglobin in the tissues, which maintains a high affinity for oxygen at low partial pressures, and effects of relatively low pH and higher carbon dioxide levels (*the Bohr effect*) and temperature in the peripheral tissues, which favors a right shift in the oxygen–hemoglobin dissociation curve. The organophosphate, 2,3-diphosphoglycerate, which increases in red blood cells in states of chronic hypoxia, also causes a right shift in the oxygen–hemoglobin dissociation curve to aid tissue oxygen delivery in such states.

19.2.12.2 Carbon Dioxide Transport and Delivery

In the peripheral tissue, the relatively low pH, high partial pressure of carbon dioxide, and high temperature that aids dissociation of oxygen from hemoglobin facilitates uptake of carbon dioxide by the blood. Formation of bicarbonate in the blood as carbon dioxide levels increase in the periphery liberates hydrogen ions, most of which are buffered in the blood. Some hydrogen ions, however, bind to hemoglobin and a conformational change in the molecule reduces oxygen binding (thus, the Bohr effect is mediated largely by changes in pH). Deoxygenated hemoglobin is capable of binding carbon dioxide with greater affinity than oxygenated hemoglobin. Thus, delivery of oxygen in the peripheral tissues aids carbon dioxide uptake by the blood (*the Haldane effect*). In the pulmonary circulation, as carbon dioxide diffuses from the blood it dissociates from hemoglobin, thus aiding the binding of hemoglobin to oxygen.

Detailed mathematical models of the oxygen–hemoglobin dissociation curve (and the carbon dioxide–hemoglobin dissociations curve) and the effects of variations in oxygen, carbon dioxide, pH, 2,3-diphosphoglycerate and temperature have been described by Dash and Basingthwaite.[39]

19.2.13 Clinical Correlations

19.2.13.1 Alterations of Mechanical Parameters in Disease

The measurement of respiratory mechanics may provide an index of clinical stability, the ability to detect a change in clinical condition, insight into the most appropriate treatment strategy, a means of assessing response to treatment and information for prognostication and follow-up.

In broad terms, diseases that impair airflow are categorized as obstructive, whereas those that limit lung expansion are described as restrictive. A patient with lung disease can be assigned to one or other of these categories by measuring the ratio between the FVC and the forced expired volume in 1 s (FEV_1): a low value is typical of patients with obstructive disease as their flow is reduced but they typically have normal lung volumes. In contrast, restrictive diseases limit both airflow and lung volume; hence the FVC/FEV_1 ratio usually remains in the normal range.

19.2.13.2 Alveolar Expansion

A long-standing simplification of the alveolar structure represents these distal airspaces as independent, balloon-like structures that rely on positive pressure for inflation. As such, the law of Laplace (as it relates to spheres) has been applied to describe the forces dictating alveolar inflation and the need for surfactant in the alveoli.

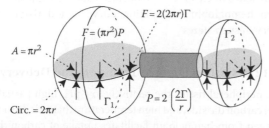

$$A = \pi r^2$$
$$F = (\pi r^2)P$$
$$F = 2(2\pi r)\Gamma$$
$$P = 2\left(\frac{2\Gamma}{r}\right)$$
$$Circ. = 2\pi r$$

FIGURE 19.4 Unstable quasi-steady-state balance for two spheres with different size radii (r) connected by a tube. The tension (Γ) in the circumference of the wall of the sphere provides a force which is counter acted by the force applied on the flat surface area of a hemispherical section by the pressure. The sphere on the left has a radius r_1 and the sphere on the right is defined by r_2. The smaller bubble will try to empty out into the larger bubble.

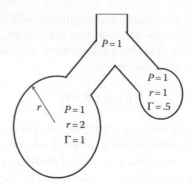

FIGURE 19.5 Schematic representation of alveoli of different sizes connected while coated with lung-surfactant. The use of lung-surfactant creates a steady-state and equalizes the pressure (P) by conforming the surface tension (Γ) using Laplace's law and avoiding smaller alveoli to collapse.

19.2.13.3 Law of Laplace

The law of Laplace states that the pressure in an alveolus is directly proportional to the surface tension in the curved wall of the alveoli and inversely proportional to the radius of the alveoli. This law is derived from the force balance as illustrated in Figure 19.4 and outlined in Equation 19.12.

Laplace's law states that

$$P = \frac{2\Gamma}{r}, \tag{19.12}$$

where P is the pressure required to keep the alveolus in an open (inflated) state, Γ is the surface tension, and r is the radius of the alveoli. If this were true, smaller alveoli would tend to empty into larger alveoli.

However, in real life, alveoli are interconnected polygonal structures that share common walls that are flat in shape.[40] The resulting connective tissue matrix, supported by elastin and collagen fibers at the septa, creates an omnidirectional tension that increases as the alveolus becomes smaller, resisting further collapse. This radial traction is often referred to as interdependence and implies that it is the tension of the adjacent tissue, rather than the ability of surfactant to counteract Laplace's law, that prevents different-sized alveoli from collapsing into each other and maintains the homogeneity of the lung parenchyma (Figure 19.5).

During spontaneous breathing, the lungs are inflated by negative pressure transmitted to the pleural surface by the expansion of the chest wall. The three-dimensional transmission of force throughout the lung tissue causes volume expansion, resisted by delicate connective tissue matrix of the lung parenchyma, and to a lesser extent also the elastic properties of the pleura and airways. At low lung volumes, lung expansion is also resisted by the surface tension of the thin layer of fluid lining the alveoli and the airways. The surfactant has a variable surface tension that is directly proportional to the surface area.

19.2.13.4 Lung Volume

Accurate bedside measurement of lung volume has attracted considerable interest over the last decade, with the appreciation of the critical role that avoidance of atelectasis (lung collapse) and over-distension play in averting ventilation-induced lung injury (VILI). FRC is the volume of greatest interest to intensivists. During mechanical ventilation, FRC is determined by the lung volume at end expiration, which is influenced by the level of positive end-expiratory pressure (PEEP). It is therefore more appropriate to talk about the end expiratory lung volume (EELV) during mechanical ventilation. Chest radiographs provide a crude index of EELV but are insensitive to small but clinically relevant changes in volume. While computed tomographic (CT) images inform the clinician about alveolar collapse and effect of ventilator strategies on atelectasis, it is not a bedside technique, and exposes the ventilated patient to risk of transport and significant exposure to radiation. Noninvasive techniques have attracted research attention over the last two decades. Multiple breath inert gas washout may provide good estimates of lung volume but has limited repeatability and reproducibility in young infants, and will underestimate the volume of a lung with significant gas trapping.[41] Whole body plethysmography includes the volume of gas trapped in the lung but is impractical for the ventilated patient. Other methods that have been used in a clinical research setting to determine relative changes in lung volume during initial stabilization on a ventilator include respiratory inductive plethysmography (RIP) and electrical impedance tomography (EIT).[42] The technology and ease of use of these techniques at the bedside remain elusive for the average clinician.

19.2.14 Transit Time

Transit time refers to the length of time that blood stays in the capillaries. At rest, the stroke volume normally replaces the entire pulmonary capillary blood volume. For an adult, the transit time of blood in the pulmonary capillaries is the pulmonary capillary volume (~75 mL) divided by the flow (cardiac output; ~6 L/min) or 0.75 s. As cardiac output increases more than the pulmonary capillary volume during exercise, transit time decreases.

19.2.15 Diffusing Capacity

Any impairment of the air–blood barrier (thickness or surface area) can create a diffusion limitation. This would include atelectasis, pulmonary consolidation, or airway closure restricting airflow. The pulmonary interstitium must also be thin for efficient gas exchange: diseases such as interstitial fibrosis can create stiff lungs with thickened membranes that limit airflow and gas diffusion between the blood and alveoli. In some patients, this limitation may only be unmasked during exercise, during which their transit time is decreased; hence, the exercise test is an important component of assessing lung function. Additionally, the ability of the lungs to exchange gas is dependent on the available surface area that forms the pulmonary membrane. Measurement of diffusion capacity is achieved using diffusion-limited carbon monoxide. Reduced pulmonary diffusion capacity is impaired by any disorder that destroys the membrane or that limits the diffusion of oxygen or carbon dioxide across it such as emphysema.

19.2.16 Impaired Alveolar Oxygenation

Fick's equation shows that the driving pressure $(P_1–P_2)$ is an important determinant of gas transfer. A decrease in alveolar PO_2 decreases the driving pressure causing diffusion, resulting in a decrease in diffusion rate. At high altitude, even normal lungs will have some diffusion limitation during exercise due to decreased atmospheric PO_2. As the alveolar PO_2 is also influenced by the alveolar PCO_2, patients who hypoventilate and who have a high alveolar PCO_2 may also have low alveolar PO_2.

19.2.17 Perfusion Limitation

Normally, O_2 uptake is complete within ~0.25 s and is not diffusion limited. Once O_2 transfer is complete, the only way to transfer more O_2 is to increase pulmonary blood flow. Hence under normal situations, oxygen transfer is perfusion limited. This contrasts with carbon monoxide, which is diffusion limited (hence making it the ideal gas for measuring diffusing capacity). However, in the presence of disease, or high altitude, diffusion limitation can cause O_2 transfer to become diffusion limited.

Extrapulmonary factors including body position, body size, oxygen consumption, ventilation, cardiac output, acid/base state, the oxygen–hemoglobin dissociation curve, body temperature, and atmospheric pressure can also alter diffusing capacity and arterial oxygen levels, even when the lungs themselves are unchanged. These factors may also influence the extent to which arterial oxygen is altered by changes in inspired oxygen content.

19.3 Ventilation

Ventilation is the process by which a sufficient volume of air is moved in and out of the lungs in order to sustain aerobic metabolism and clear waste CO_2. To maintain arterial concentrations of O_2 and CO_2 within physiological limits, the volume of gas transported each minute (i.e., minute volume or minute ventilation) is tightly regulated by a nonlinear multiinput, multioutput feedback system operating between the controller (which includes the neural, muscular and mechanical processes and receptors involved in generating airflow) and the plant (represented by the gas-exchanging processes of the lungs and body tissues). A system delay (the sum of all the latencies in the forward and feedback information flows) is a third feature of this feedback system, and essentially represents the time for blood to flow from the lungs to the chemoreceptors. The physical and temporal isolation of the feedback sensors from the lung contributes to respiratory control instability. This instability may be modulated by the gain with which the controller responds to the feedback signals (blood gas levels) by correcting the ventilator pattern. High controller gain may result in overcorrection, whereas insufficient gain will result in a system that is easily perturbed by external influences and which may decompensate. The plant gain refers to the responsiveness of P_ACO_2 or P_AO_2 to a unit change in ventilation and essentially describes the damping inherent in the system. Large volume lungs will absorb a set volume change more readily than a small volume lung: a given volume change is more damped in a large volume lung, implying low plant gain and greater system stability.

The respiratory neural control network is highly adaptive, and rather than behavior typical of a hard-wired neural controller, it demonstrates neural plasticity similar to the mechanisms underlying learning and memory in the higher brain.[43] This plasticity includes integral, differential, statistical, and memory-type plasticity.[44] The coexistence of integral and differential types of plasticity appears unique to the respiratory control system within the brain, and has been used to explore integral–differential neural computation, sometimes referred to as "brain calculus."[43] The dynamic integral–differential signal processing and long-term memory capabilities effectively filter the afferent and efferent signals that modulate the respiratory control system in response to the physiological and pathophysiological changes frequently encountered in health and disease.[43]

19.3.1 Brainstem Respiratory Center

The mammalian respiratory control system has several groups of cells located within the medullary and pontine brainstem.[45]

Neurons responsible for afferent data processing, rhythm generation, and shaping of the output waveform are located in three major groups within the medulla oblongata (see Figure 19.6).

19.3.2 The Medullary Neurons

19.3.2.1 Dorsal Respiratory Group

Located in the nucleus Tractus Solitarius, this neuronal group receives afferent signals arising from central and peripheral chemoreceptors, higher cortical inputs and pulmonary mechano- and stretch-receptors[45] to stimulate diaphragmatic inspiratory muscle activity.

19.3.2.2 Ventral Respiratory Group

Premotor processing for inspiratory[46] and expiratory[47] muscle activities appears to be coordinated by neuronal groups located within cranial and caudal parts of the ventral respiratory group of cells lying bilaterally and just laterally to the nucleus Ambiguus through much of the medulla's length.

19.3.2.3 PreBötzinger Complex

The pre-Bötzinger complex (pre-BötC) also resides within the medulla and is the principal site of respiratory rhythm generation.

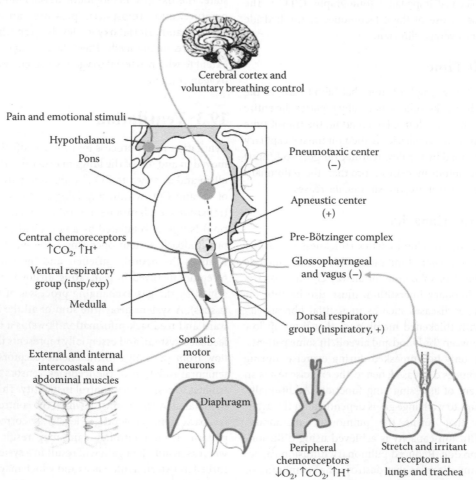

FIGURE 19.6 Influences on the brain respiratory center.

The retro-trapezoid nucleus (RTN) may be functionally and anatomically indistinct from the more recently identified parafacial respiratory group (pFRG). Recent evidence suggests that these sites may act as a phase-locked coupled oscillator for the generation of the respiratory rhythm, with inspiratory and expiratory activities governed by the pre-BötC and RTN/pFRG loci, respectively.[48]

19.3.2.4 The Pontine Neurons

In addition to the medulla, electrophysiological and anatomic studies indicate that pontine respiratory neurons contribute to respiratory control. Signals to this area are primarily received from communicating neurons in the medulla.

19.3.3 Respiratory Rhythm Generation

Traditionally, the inspiratory phase of the respiratory cycle is considered the "active" phase, responsible for driving the respiratory pump musculature. Whereas expiration was considered previously as a passive process (especially in the adult), it exhibits active characteristics, with laryngeal and airway smooth muscular playing an important role in regulating postinspiratory airflow. This "active expiratory phase" is supported by the identification of brainstem neuronal burst activity during both inspiratory and expiratory segments of the breath cycle.

The electrical output pattern generated by the brainstem controller is characterized by the timing of action potential burst activity identified in the Bötzinger and pre-Bötzinger regions. Neuronal groups forming the pre-BötC located within the rostral ventro-lateral medulla are intrinsically rhythmogenic[49,50] and play an important role in generating cyclical inspiratory muscle activity.[51] Action potential bursts occurring during the inspiratory phase comprise frequencies that decrement, augment, or remain constant over time.

During expiration there is an "early" postinspiratory phase, characterized by decrementing frequency of neuronal firing and a "late" preinspiratory phase during which bursts of activity are augmented and increase in frequency over time.[45] Feedback control of muscle tone during early postinspiration may regulate expiratory resistance and airflow, influencing gas mixing.[52] Preinspiratory neurons exist in the rostral Bötzinger complex and likely determine in part, the biphasic nature of expiratory activity.[50]

Pre-Bötzinger inspiratory neurons may communicate with preinspiratory neurons within the rostral Bötzinger complex in an inhibitory fashion. Bötzinger complex neurons located within the rostral ventro-lateral medulla mediate the expiratory phase, under the influence of inhibitory cells from the pre-Bötzinger pre-Inspiratory group.[45] Neurons of the RTN or pFRG may represent a second anatomical location for respiratory rhythm generation. Cells within this group are primarily preinspiratory in function and discharge prior to the onset of the inspiratory burst. These cells are intrinsically rhythmogenic and are actively inhibited during the inspiratory phase.[48] These separate but interactive neuron groups thus represent an interactive two-rhythm generator with feedback regulation. It has been postulated that the pre-BötC represents the major site for respiratory

rhythm generation in the adult mammal.[48] The coupling between these generators may be influenced by development,[53] hypoxia,[54] or other alterations to the internal milieu. Pontine noradrenergic neurons may also play a role in regulating rhythmogenesis in response to sleep, behavioral changes, cardiovascular alterations, and environmental influences.[55]

19.3.3.1 Cortical Influences

The output of the respiratory center can be modulated by cortical influences arising from voluntary control of the rate and depth of breathing, as might occur during speaking, singing, coughing, and so on. Emotional influences and pain receptors may act to alter respiratory activity via the hypothalamus.

19.3.3.2 Chemical Influences

The feedback to the brainstem of information about arterial PCO_2 and PO_2 is vagally mediated primarily by central chemoreceptors located in the floor of the fourth ventricle and peripheral chemoreceptors within the wall of the aortic and carotid bodies.[56] The response to arterial PCO_2 is likely indirect through the response of the central and peripheral chemoreceptors to acid–base balance and H^+ concentration. Importantly, central chemoreceptors are responsive to alterations in CO_2 within the cerebro-spinal fluid only, since the blood–brain barrier prevents diffusion of H^+ from the blood to the brain.

The peripheral chemoreceptors also respond to arterial PO_2. The carotid bodies provide tonic input to the respiratory controller, which increases in the presence of hypoxia. Carotid body stimulation causes increased activity of inspiratory $R\alpha$ neurons, inhibition of late inspiratory activity and of $R\beta$ neurons, and excitation of postinspiratory neurons.

19.3.3.3 Mechanosensitive and Thermal Influences

Thermal and tactile stimuli also generate vagal feedback responses to the respiratory controller. Nonchemically mediated stimuli to the respiratory feedback system induce inhibitory reflex activity with consequent reduction in ventilation. Stimulation of pulmonary stretch receptors has a powerful inhibitory effect on inspiration. In piglets, stimulation of rapid and slowly adapting pulmonary stretch receptors activates terminal connections located in the various nuclei of the tractus solitarius, including the ventrolateral nucleus of the dorsal respiratory group.[48] The resulting increased firing rate may be responsible for the elicitation of the Hering–Breuer reflex, as it appears that this response is stronger in infants than in adults. Further tactile reception is also received from receptors located in the upper airway (larynx, hypopharynx) and many of these inputs have an inhibitory effect on the respiratory control center.

19.4 Artificial Ventilation

Artificial ventilation refers to the movement of gas into and out of the lungs by an external source in direct communication with the patient. Whereas the external source may be a resuscitation bag or other forms of continuously inflating device, more commonly, it is provided by a mechanical ventilator. The goal of mechanical ventilation is to produce a minute volume that meets the physiological requirements of gas exchange while avoiding ventilator-associated lung injury and other physiological compromises including impairment of cardiac function and systemic circulation, adverse extrapulmonary consequences of positive pressure (e.g., cerebral damage) and increased patient discomfort and anxiety.

19.4.1 The Operation of a Mechanical Ventilator

The basic function of a mechanical ventilator is to transmit applied energy in a predetermined manner to perform useful work for the patient. They have four main mechanical characteristics including input power, power conversion and transmission, a control system, and a power output (pressure, volume, and flow waveforms).[57]

19.4.2 The Input Power

Electric and pneumatic (compressed gas) are the only two sources of power input used in commercially available mechanical ventilators.

19.4.2.1 Electric Power Conversion and Power Transmission

The input power is converted to a gas pressure that is delivered in a predetermined, controlled manner (by the control circuit) via the ventilator's power transmission drive mechanism. The goal is to either assist or to replace the patient's muscular effort required to perform the work of breathing (output) in order to generate a gas flow through the airways and to inflate the lungs.

Most often, a positive pressure (relative to atmospheric pressure) is delivered to the airway opening by a positive pressure ventilator. However, the application of negative pressure to the body surface (usually rib cage and abdomen) has been used in negative pressure ventilators with clinical effect.

19.4.3 Electric Control Scheme

Ventilators can be described as pressure controlled, volume controlled, or flow controlled. The control variables of these include pressure, volume, flow, and time. The control circuit includes mechanical, pneumatic, fluidic, electric, and electronic components. A logical classification of ventilator systems and the general principles of ventilator operation were defined in detail, recently, by Chatburn, and are outlined below.[57,58]

Understanding of the interaction between the ventilator and the patient and of the nature of the control scheme that best describes the ventilator's operation can be gained from simple physical models that have electrical analogs. In its simplest form, the equation of motion describes the respiratory system as a single compartment model with a rigid flow-conducting tube (resistor) connected in series with an elastic compartment (capacitor). The pressure (voltage) required to generate a gas flow (current) that will enter the airway and increase lung volume (charge) can be expressed as

$$P_{rs} = P_E + P_R + P_I, \tag{19.13}$$

where P_{rs} is the transrespiratory pressure, P_E is the pressure required to overcome the elastic load (recoil), P_R represents the pressure required to overcome the flow resistive impedance load, and P_I corresponds to the pressure required to overcome load arising from the inertance. This model can be expanded to include the mechanical properties of the patient circuit, tracheal tube, large airways, lung tissue, and chest wall.

In the intubated respiratory system, 90% of the inertance and 50% of the Newtonian resistance arise from the tracheal tube.[59]

For the purpose of ventilation at conventional breathing frequencies, the contribution of inertance is negligible, and can be discounted, so that the equation of motion is most often written as

$$P_{rs} = P_E + P_R. \tag{19.14}$$

During mechanical ventilation, the transrespiratory pressure represents the sum of the applied ventilator pressure (P_{vent}) and that generated by the muscles of the chest wall (P_{mus}). The elastic pressure component is the product of the elastance, El, defined as

$$El = \frac{\Delta P}{\Delta V} \tag{19.15}$$

and volume (EV), while the resistive pressure component is the product of the resistance, R, expressed as

$$R = \frac{\Delta P}{\Delta \dot{V}}, \tag{19.16}$$

and the flow.

Considering these relationships we can rewrite Equation 19.14 as follows:

$$P_{vent} + P_{mus} = EV + R\dot{V}. \tag{19.17}$$

Clinicians more commonly think in terms of respiratory system compliance (C_{rs}) than elastance; hence, an alternative form of Equation 19.17 can be rewritten as

$$P_{vent} + P_{mus} = \frac{V}{C} + R\dot{V}. \tag{19.18}$$

Given that the transrespiratory pressure determines the volume and flow to be delivered to the patient, it is clear from Equation 19.18 that the pressure generated by the muscles of the chest wall can subtract from or add to the ventilator pressure. The extent of support required from the ventilator (total, partial, or none) depends on the amplitude and nature of the patient's respiratory effort. Where significant patient–ventilator asynchrony exists, the respiratory muscles create a pressure that opposes that of the ventilator, reducing the effectiveness of the flow and volume that would otherwise have been delivered to the patient.

A second important point arising from Equation 19.18 is that the control scheme for ventilation can be defined in terms of either pressure, volume, or flow—but only one of these three variables at a time.[58] The control variable remains independent of changes in condition. For a pressure-controlled system, the pressure waveform will be maintained despite changes in resistance and elastance, but the resultant flow and volume will depend on the pressure waveform and the mechanical properties of the respiratory system. The reverse holds true for volume and flow-controlled ventilation, when airway pressure becomes the dependent variable. Despite this limitation, it is possible for the control mode to change not only breath-to-breath, but also intrabreath.[57]

In addition to defining the control scheme for inspiration, the equation of motion also provides information on the basis of expiratory flow. Expiration is normally passive, with no contribution from the ventilator or the respiratory muscles. Equation 19.17 then becomes:

$$-R\dot{V} = EV, \tag{19.19}$$

highlighting the importance of the stored elastic energy for determining expiratory flow.

19.4.4 Phase Variables

Phase variables can be used to describe how a ventilator starts, sustains, and ends both inspiration and expiration.

19.4.4.1 Trigger Variable

The ventilation commences breath delivery when the trigger variable reaches a preset value. The variable that is preset determines whether the ventilator mode is described as pressure, volume, flow, or time-triggered. As baseline pressure need not change with flow triggering, this approach reduces the work of breathing by the patient to initiate a breath.[60] The sensitivity of the trigger variable set on the ventilator determines how much work the patient has to perform to initiate the breath. The response time reflects the time required for signal processing and overcoming the mechanical inertia of the drive mechanisms. A short response time is critical to achieve optimal synchrony with the spontaneous patient inspiration.

19.4.4.2 Limit Variable

Phase limit variables will reach and maintain a preset level prior to the end of inspiration. Within a given breath, the limit variable may be any one or a combination of pressure, volume, or flow. Such limits effectively set the maximum value for the variable of interest and need to be differentiated from backup cycling mechanisms that are activated when a pressure or time threshold is met.

19.4.4.3 Cycle Variable

Cycle variables are used to end the inspiratory phase and can include pressure, volume, flow, or time, in addition to manual cycling. The simplest example of cycle variables is time-cycling, whereby expiratory flow commences as soon as a preset inspiratory time has passed. Pressure cycling implies the delivery of flow until the preset pressure is achieved, at which point flow direction reverses and expiratory flow commences. Likewise, volume cycling implies the delivery of flow to the patient until the preset volume passes through the control valve. If expiratory flow does not commence immediately after achievement of either the preset volume or flow, then the ventilator is cycled by the preset inspiratory time, and an inspiratory hold will have been set. Flow cycling is most often utilized in pressure support ventilation (in which pressure is the control variable), whereby the patient respiratory effort determines the duration of the inspiratory flow. During flow cycling, the initial flow is high and decreases exponentially (in the absence of a contribution from active respiratory muscles) until the preset flow limit is achieved. The flow-cycle threshold may be adjusted by the operator, and in some ventilators, is also automatically compensated for by the presence of a tracheal tube leak to avoid excessively long inspiratory times.

19.4.4.4 Baseline Variable

Most commonly, pressure is the parameter controlled during expiration and is measured and set relative to atmospheric pressure. A positive baseline pressure is referred to as PEEP.

19.4.5 Power Outputs

The outputs of a ventilator are pressure, volume, and flow. The patterns of each of these waveforms depend on the control scheme of the ventilator. Idealized waveforms for different ventilator patterns were described by Chatburn.[57] Pressure waveforms include rectangular, exponential, sinusoidal, and oscillating. Volume waveforms are usually described as ascending ramp or sinusoidal. Flow waveforms include rectangular, ascending ramp, descending ramp, and sinusoidal. The impact of specific inspiratory flow waveforms on shear stress and lung injury is not fully understood.

In practice, the ventilator output waveforms are often distorted by extraneous noise such as flow turbulence and vibration. Instrument error and inaccurate calibrations can influence the pressure, volume, and flow that are received by the patient. The compliance and resistance of the patient circuit may contribute to a discrepancy between set parameters and delivered outputs. Pressure will be lower at the patient airway opening than that measured within the ventilator, while expiratory manifold flow and volume will be greater than what is delivered to the patient.

The circuit and patient resistance can be modeled in series as they share the same flow, but have different pressure drops arising from their different resistances. While the circuit resistance contributes to the overall resistive load seen by the ventilator, it is small when compared to the patient resistance, and can normally be disregarded.

In contrast, the patient circuit compliance is in parallel with the compliance of the respiratory system as both compliances experience the same pressure change, but will fill with different volumes over the same inspiratory time, resulting in different flows. The effect of this parallel connection is that the total compliance (circuit + respiratory system) is greater than the sum of the two components on their own. When circuit compliance exceeds the respiratory system compliance, this can markedly reduce the proportion of volume delivered to the patient—a scenario that needs to be considered carefully when ventilating with volume-controlled mode (particularly those with low compliance such as the preterm infant). In the absence of a flow sensor positioned at the patient "wye," the circuit compliance needs to be known, and the ventilator adjusted to account for the decrease in delivered volume due to gas compression in the patient circuit. When the tidal volume is set and monitored within the ventilator, the actual delivered volume (V_{del}) can be calculated using the following equation to correct for the compliance of the patient circuit (C_{PC}):

$$V_{del} = \frac{V_{set}}{1 + C_{PC}/C_{rs}}, \qquad (19.20)$$

where C_{PC} is measured by dividing the pressure resulting when a known volume is delivered into the circuit with the patient "wye" occluded. From Equation 19.20, it is clear that when the patient circuit is totally noncompliant, the $V_{del} = V_{set}$. However, when patient compliance is large and not compensated for by the ventilator, then the V_{del} is significantly less than V_{set}. This is particularly important in the neonate, where C_{PC} may be three times larger than C_{rs}, even with circuit optimization using small bore stiff tubing and small volume humidifiers.[57] In that scenario, only a quarter of the V_{set} would be delivered to the patient, with the remaining 75% distending the patient circuit. This example highlights the importance of monitoring the V_{del} at the patient airway opening in subjects with low compliance.

19.4.6 Modes of Mechanical Ventilation

Ventilation modality primarily informs the user about the nature of delivered breath and variables relevant to the inspiratory phase.[61] The mode can be identified and classified by specifying the breathing pattern, control type, and control strategy (phase variables and operational logic).

19.4.7 Breathing Pattern

Breath delivery is controlled in some measure by pressure, volume, flow, or time, or a combination of these variables. Volume control results in a more stable minute ventilation than pressure control if the lung mechanics are unstable. However, while a stable breath–breath minute ventilation may be the goal, there is some evidence that varying tidal volume and respiratory rate on a breath to breath basis may be beneficial.[62] In contrast, pressure

control may enhance patient–ventilator synchronization because inspiratory flow is not restricted by a preset value.

Breath type may be *mandatory* or *spontaneous*. In a spontaneous breath, the start, and duration of the breath is determined by the patient, whereas a mandatory breath is any breath whereby the ventilator triggers and/or cycles the breath. Mandatory breaths that receive a contribution from patient effort are called *assisted*. During high-frequency oscillatory ventilation (HFOV), short mandatory breaths may be superimposed on a longer spontaneous breath.

Three different breath sequences can be defined. These include continuous mandatory ventilation (CMV) where all breaths are mandatory, and spontaneous breaths are not allowed between the mandatory breaths. Continuous spontaneous ventilation is the opposite of CMV—whereby all breaths are spontaneous. Intermittent mandatory ventilation (IMV), on the other hand, allows spontaneous breaths between the mandatory breaths. If the mandatory breath is triggered by the patient, then it is referred to as synchronized IMV (SIMV). Newer ventilators use an active exhalation valve, which makes it possible for spontaneous breaths to occur during a mandatory pressure-controlled breath.

19.4.8 Control Type

Control circuits are required to measure and regulate the ventilator outputs of pressure, flow, and volume. The advent of microprocessors has increased the complexity of such control circuits to facilitate more precise and flexible management of breathing cycle variables, with the resultant proliferation of ventilation modalities. Engineering control systems are based on three different types of subsystems: a controller (which directs changes in output), a plant (the subsystem being controlled) whose output is called the controlled variable (what is being regulated), and an effector, which is the power element that modifies the plant to some desired set point under instruction from the controller. Whereas early jet ventilators were based on open-loop control circuits (set input driving pressure, variable delivered pressure and flow), modern ventilators generally use closed-loop control to sustain the driving pressure and flow under constant changing external conditions (e.g., respiratory impedance). Closed-loop control circuits utilize a sensor-measured output variable as a feedback signal that is compared to the controller-set parameter. Where there is a difference between the two (determined as an error by a comparator), the controller drives the system toward reducing the magnitude of the error and achieving the target output value (e.g., the use of pressure as a feedback signal to drive ventilator gas flow). This closed-loop approach facilitates continuous and automatic adjustment of the ventilator performance to minimize the impact of external disturbances. The feedback system may be electrical (e.g., electronic pressure transducer) or mechanical (regulator/valve).[63]

19.4.9 Control Strategy

At a further level, the phase and pattern of each breath are determined by the control strategy, which is influenced by the trigger mechanism (machine/patient), the set limits (pressure/volume/flow/time—depending on the control variable), and the cycle.[64] Understanding how a ventilator is controlled is perhaps the most useful approach to understanding different ventilator modalities.

The evolution of closed-loop control circuits used in ventilator technology was reviewed extensively by Chatburn in 2004[63] and is summarized below. There are now at least seven different control systems that serve as the foundation for the multitude of different ventilator modalities available commercially. Each additional control system adds complexity and distances the operator from the decision-making process and direct control of breath-to-breath ventilation, essentially moving from the traditional tactical control toward strategic—and ultimately—intelligent control systems.

19.4.9.1 Set-Point Control

Set-point control is the minimum level of control and essentially constrains the output to match a set input. For example, if a fixed pressure or flow limit is set by the operator, the ventilator responds by maintaining a consistent pressure or flow waveform.

19.4.9.2 Auto-Set-Point Control

Taking set-point control a step further, the ventilator can decide whether flow- or pressure-controlled breath should be delivered, according to operator set priorities. For example, the ventilator may set out to deliver a certain volume, but be switched to pressure-controlled if a maximum pressure setting is reached before the desired volume is achieved (or start with a pressure-controlled breath and switch to flow-controlled).

19.4.9.3 Servo Control

Servo control introduces flexibility by tracking a changing (i.e., nonconstant) input. This approach is useful in modes such as proportional assist, where the magnitude of the patient's own flow generation guides the amplification of the synchronized ventilator output. The rationale for this approach is to reduce patient effort to overcome normal breathing load, while the ventilator assistance is proportionally adjusted to account for disease-imposed increased workload.

19.4.9.4 Adaptive Control

Adaptive control facilitates the automatic adjustment between breaths of one set point by the ventilator to maintain a different set point determined by the operator. The most common example of this control mode is the automatic adjustment of the pressure limit to achieve a set tidal volume over several breaths.

19.4.9.5 Optimal Control

Optimal control involves automatic determination of the most favorable volume and pressure set points according to the best value of some performance function (determined via a mathematical model) such as minute volume or even exhaled carbon dioxide.[65] In this experimental form of optimal control, the ventilator estimates the patient's minute volume requirements to the extent that operator input may only be required for setting inspired fractional oxygen, PEEP, and alarm settings.

19.4.9.6 Knowledge-Based Control

This approach seeks to couple mathematical models with human expertise, to increase the capacity of the ventilator to control mode variables. Such systems may incorporate "expert algorithms" in parallel with fuzzy logic to drive ventilator function. Examples include systems that are controlled to deliver the lowest level of patient support ventilation (PSV) to maintain a group of physiological variables (e.g., respiratory rate, tidal volume, heart rate, oxygen saturation, and exhaled CO_2) such as respiratory rate within expert predetermined (knowledge-based) acceptable ranges.[66-69] The drawback of this knowledge-based control is that the expert rule-based algorithms are static.

19.4.9.7 Intelligent Control

In the experimental setting, artificial neural networks have been incorporated into ventilator control systems to support decision making. Their fascinating promise is to further advance the knowledge-based control to support a dynamic system that is capable of learning through experience and storing this knowledge in a database that constantly evolves, and which could potentially be pooled with that created from other patients within and between units. This approach assigns a weight to different input variables, effectively adjusting the strength to which they influence the ventilator output. Depending on the response to the ventilator output, each ventilator input may vary on a breath-to-breath basis. The neural network "learns" from these responses, and adjusts the strength of the input variable in order to achieve a desired outcome.

19.4.10 Novel Modes of Ventilatory Support

Whereas mechanical ventilation is normally considered as assisted delivery of normal breathing patterns (breath size and respiratory rate), the last 25 years have seen the emergence of high-frequency ventilation (HFV)—a novel mode of ventilation that has theoretical advantages for gas exchange and lung volume recruitment.

19.4.10.1 High-Frequency Ventilation

While HFV has been in clinical use in neonatal ventilation for over 25 years, it has been adapted for adult ventilation only over the last decade. Unlike ventilation at conventional breathing rates, distending lung volume (which influences oxygenation) and gas transport/mixing (which influences ventilation and CO_2 removal) are largely independent of each other during HFV. Lung volume is controlled by changing mean airway pressure, whereas ventilation is controlled by the frequency and amplitude of the ventilator waveform. Although this independence is not absolute in the collapsed or overdistended lung, or when the duration of the inspiratory and expiratory cycles are not equal, this independence of oxygenation and ventilation confers a significant practical advantage for HFV over conventional ventilation.

HFV utilizes frequencies between 3 and 15 Hz to deliver tidal volumes equal to or less than anatomical dead space volume to ventilate the lung. These small tidal volumes imply reduced cyclic stretch of the lung and theoretically reduced lung injury. Although less lung injury has been shown in animal studies, this has been difficult to translate to improved long-term outcomes when HFV has been applied to human infants in randomized controlled trials.[70] It is possible that this may be in part due to barotrauma. Moderately high mean airway pressures are routinely used during HFV, to recruit and optimize distending lung volume. The high-frequency oscillations are superimposed on top of this mean pressure. Whereas in the highly compliant lung, the pressure waveform is damped markedly between the airway opening and the lung parenchyma, a much higher proportion of the waveform amplitude is transmitted to the low-volume, noncompliant lung. In contrast, in the presence of high peripheral airway resistance, very small pressure oscillations reach the distal lung, but proximal ventilator units may be exposed to rapidly changing pressure waveforms with peak pressures that match or exceed those used during conventional ventilation.[71]

Whereas ventilation is determined by minute volume:

$$V_{\text{minute}} = V_{\text{T}} \times R, \tag{19.21}$$

at conventional frequencies, ventilation is highly efficient in HFV, where ventilation efficiency CO_2 elimination is proportional to

$$V_{\text{T}}^b \times v^a, \tag{19.22}$$

where *a* and *b* approximate 1 and 2, respectively.[72–75] The efficiency is in part due to the complex mechanisms of gas transport during HFV. Unlike conventional ventilation, bulk convection plays a relatively small role in gas transport during HFV. Other gas transport mechanisms at play include turbulent flow and radial mixing, asymmetric inspiratory and expiratory velocity profiles, branch angle asymmetry, pendelluft (between neighboring units with differing time constants), laminar flow with Taylor dispersion (beyond approximately airway generation 8), cardiogenic mixing, collateral ventilation between alveolar units via the pores of Kohn, and molecular diffusion in the acinar zone. These gas transport mechanisms are reviewed in detail elsewhere.[71] Studying the damping of the pressure waveform and the tidal volume delivery during HFV has provided some valuable insights into optimal ventilator frequencies and levels of PEEP that would minimize overdistension of the alveolar unit while also subjecting the lung to minimal pressure oscillations.[71,76] In the overdamped lung, optimal frequency is the corner frequency (f_c), where

$$f_c = \frac{1}{2\pi RC}. \tag{19.23}$$

Hence diseases characterized by high resistance are more optimally ventilated at lower frequencies than those with low compliance.

There are two main types of HFV: high-frequency oscilatory ventilation (HFOV) and high-frequency jet ventilation (HFJV). True HFOV utilizes either a piston or diaphragm/loudspeaker system to create an active inspiratory and expiratory phase. The active expiratory phase means that very short expiratory times do not result in gas trapping, providing the mean airway pressure is high enough to stent the airways open and avoid the formation of choke points. One of the major differences between HFJV and HFOV is that only the inspiratory phase is active and expiration (like conventional ventilation) relies on passive relaxation of the chest wall, when there are no external forces acting on the respiratory system. HFJV thus utilizes slightly lower frequencies than HFOV. It reportedly has some advantages for inhomogeneous lung disease and air leak; however, this has not been demonstrated in a randomized controlled trial.

19.4.11 Clinical Application of Mechanical Ventilation in Adults, Children, and Neonates

Ultimately, mechanical ventilators need to be capable of delivering flow to a wide range of subjects with varying age, disease etiologies, and mechanical constraints. Most often, mechanical ventilation in the intensive care unit is instituted due to impending respiratory failure; however, some patients with normal lungs require mechanical ventilation as a supportive therapy during anesthesia or treatment of a serious illness including acute brain injury.

Age and body size are important considerations in the formation of a ventilation strategy. Lung development commences during fetal life and peak lung function is achieved during early adulthood. Extremely preterm infants, born less than 26 weeks gestation (term is ~40 weeks), are in the late cannalicular stage of lung development. They have an established airway tree, but have yet to develop the gas-exchange unit. From the cannalicular stage, they will progress into the saccular stage of development, when primitive air sacs develop and Type II epithelial alveolar cells begin to secrete surfactant into the airspaces. By about 36 weeks gestation, the alveolarization stage commences, with the emergence of secondary septa that continues through until ~2–3 years of age. Alveolarization occurs in parallel with microvascular maturation and thinning of the alveolar walls that are necessary for efficient gas exchange.[77] While there is a 20-fold increase in gas-exchanging surface area during alveolarization, the human lung at birth contains only a fraction of the total number of alveoli of the adult lung. The adult complement of alveoli is attained by the age of about 2–3 years. After this time, the alveoli principally increase in volume, until aging phenomena develop, leading to progressive destruction of lung tissue from the age of ~35 years onwards.

It is clear from the above brief overview of lung development that the mechanical properties of the respiratory system may differ markedly at different ages, and that an understanding of these changes is critical to effective and safe ventilation. These developmental differences have implications for the fragility of the lung to VILI. The preterm airway/alveolar attachments are extremely fragile, and susceptible to stretch injury associated with barotrauma and volutrauma, which may result in loss of the mechanical interdependence that is crucial to the maintenance of lung volume. In the absence of an expansive gas-exchange ventilation compartment (absence of alveolar structures), they have a high dead space to tidal volume ratio:

$$\frac{V_d}{V_T}. \tag{19.24}$$

This results in highly inefficient ventilation patterns, using rapid shallow respiration that does not overdistend the small lung capacity, while generating sufficient minute ventilation to clear carbon dioxide from the airways.

Because the interaction between the ventilator and the lungs is highly dependent on the mechanical properties of the respiratory system, a thorough understanding of the pathophysiology of the disease being treated is essential to selection and optimization of the ventilator strategy. The passive expiratory time constant τ of the lung is an important consideration, particularly with respect to setting respiratory rate.

$$\tau = R_{rs} + C_{rs}. \tag{19.25}$$

Subjects with obstructive and predominantly airway occlusive disorders have long time constants and need to be ventilated at low ventilator rates to ensure sufficient time for full and complete expiration. The failure to achieve this promotes the development of intrinsic PEEP and hyperinflation, often referred to as gas trapping. In contrast, the presence of a low respiratory system compliance (C_{rs}) leads to short time constants, and these patients may be safely ventilated at higher respiratory rates. It is beyond the scope of this chapter to discuss an exhaustive list of different disease pathologies. Instead, we highlight five common pathologies that cover many of the different ventilator approaches (the pressure volume relationships in these diseases are illustrated in Figure 19.7).

19.5 Respiratory Diseases

19.5.1 Asthma

Asthma is defined by the presence of airway hyperreactivity normally resulting from a persistent inflammatory process in response to one or more different stimuli in a genetically susceptible individual. It is characterized by mucosal edema, mucus secretion, epithelial damage, and bronchoconstriction. Bronchoconstriction results in increased airway resistance and impaired expiration, leading to hyperinflation (raised lung volume) at rest; however, the increased lung volume tends to enlarge the airways, aiding expiration by increasing the elastic recoil of the chest. Increased airway resistance is reversible in asthma, although remodeling of the airways secondary to the inflammatory process may result in long-term irreversible changes in airway mechanics. Lung compliance is not altered significantly, unless there is substantial

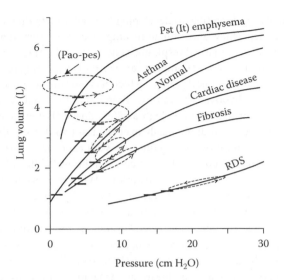

FIGURE 19.7 Static and dynamic lung volume–pressure curves illustrating the effects of certain diseases on lung compliance, lung volumes, and airway resistance. The upper horizontal bar on each volume–pressure curve represents a typical adult value for the FRC, and the lower bar the residual volume. Increased airway resistance is the predominant feature in asthma and emphysema (obstructive diseases), while decreased compliance is most obvious in pulmonary fibrosis and RDS (restrictive diseases). Cardiac patients show some changes in both characteristics.

collapse of portions of the lung. Respiratory failure from severe asthma is a potentially reversible, albeit life-threatening condition. These patients are predisposed to the development of gas-trapping, which is associated with poor outcome. Mechanical ventilation strategies must therefore avoid the development of this complication in asthmatic patients.

19.5.2 Emphysema

Emphysema is characterized by overinflation of the airspaces distal to the terminal bronchiole either from dilatation or from destruction of their walls accompanied by loss of elasticity. The destructive changes in the bronchiolar walls and loss of the lung parenchyma supporting the bronchioles promote airway collapse and increased lung compliance. The accompanying collapse of the large airways may dominate the clinical picture.

19.5.3 Fibrosis

Fibrosis leads to a thickening of the pulmonary membranes, resulting in decreased compliance and reduced diffusing capacity without associated increases in airway resistance. Exertional dyspnea (shortness of breath during exercise) is a characteristic feature of this disorder.

19.5.4 Respiratory Distress Syndrome

Respiratory distress syndrome (RDS) is the major cause of respiratory disease in the newborn infant. Preterm babies are particularly susceptible to developing RDS after birth due to surfactant deficiency and immature lung structure. They develop clinical signs of marked inspiratory effort, sternal retraction (due to increased chest wall compliance), expiratory grunting (breathing out against a closed glottis—which has the effect of providing PEEP), and hypoxia or cyanosis. Blood shunting may increase from the normal 3% to as high as 80%. Survival after neonatal RDS has increased dramatically over the last two decades due to antenatal maturation of lung function with exposure of the fetus to maternal glucocorticoid treatment, and the availability of postnatal exogenous surfactant preparations. Without exogenous surfactant, and appropriate treatment, the baby's lungs develop thickened glassy (hyaline) membranes in the alveoli, which in the past led to RDS often being called "hyaline membrane disease."

Acute RDS (ARDS) also develops in older children and adults. ARDS is characterized by bilateral pulmonary infiltrates, stiff lungs, and hypoxemia that is not due to cardiac failure. Some of the common conditions or factors that can injure the lungs and lead to ARDS either directly or indirectly include sepsis, pneumonia, severe bleeding (usually secondary to body trauma), breathing in harmful fumes or smoke, and aspiration of vomited stomach contents. ARDS can be exacerbated by ventilator-induced injury.

19.5.5 Cardiac Disease

Breathing difficulties are often the first sign of cardiac disease, likely due to back-pressure from the left side of the heart, and subsequent overfilling of the lungs with blood. This results in loss of lung volume and reduces lung compliance and may also impair diffusing capacity due to thickening of the alveolo-capillary membrane.

References

1. Ochs M, Nyengaard JR, Jung A, Knudsen L, Voigt M, Wahlers T, et al. The number of alveoli in the human lung. *Am J Respir Crit Care Med* 2004; 169(1): 120–124.

2. Gehr P, Bachofen M, and Weibel ER. The normal human lung: Ultrastructure and morphometric estimation of diffusion capacity. *Respir Physiol* 1978; 32(2): 121–140.

3. Metzger RJ, Klein OD, Martin GR, and Krasnow MA. The branching programme of mouse lung development. *Nature* 2008; 453: 745–750.

4. Weibel ER and Cloth MD. *Morphometry of the Human Lung*. Berlin-Gottingen-Heidelberg: Springer-Verlag; 1963.

5. Tsuda A, Henry FS, and Butler JP. Gas and aerosol mixing in the acinus. *Respir Physiol Neurobiol* 2008; 163(1–3): 139–149.

6. Tsuda A, Rogers RA, Hydon PE, and Butler JP. Chaotic mixing deep in the lung. *Proc Natl Acad Sci USA* 2002; 99(15): 10173–10178.

7. van Ertbruggen C, Hirsch C, and Paiva M. Anatomically based three-dimensional model of airways to simulate flow and particle transport using computational fluid dynamics. *J Appl Physiol* 2005; 98(3): 970–980.

8. Lumb A. *Nunn's Applied Respiratory Physiology*. 6th Edition. Edinburgh: Elsevier; 2005.

9. Stocks J, Godfrey S, Beardsmore C, Bar-Yishay E, and Castile R. Society EATFoSfIRFTERSAT. Plethysmographic measurements of lung volume and airway resistance. ERS/ATS task force on standards for infant respiratory function testing. European Respiratory Society/American Thoracic Society. *Eur Respir J* 2001; 17(2): 302–312.

10. Suki B and Bates JH. Extracellular matrix mechanics in lung parenchymal diseases. *Respir Physio Neurobiol* 2008; 163(1–3): 33–43.

11. Suki B, Ito S, Stamenovic D, Lutchen KR, and Ingenito EP. Biomechanics of the lung parenchyma: Critical roles of collagen and mechanical forces. *J Appl Physiol* 2005; 98(5): 1892–1899.

12. Weibel ER, Federspiel WJ, Fryder-Doffey F, Hsia CC, König M, Stalder-Navarro V, et al. Morphometric model for pulmonary diffusing capacity. I. Membrane diffusing capacity. *Respir Physiol* 1993; 93(2): 125–149.

13. Maina JN and West JB. Thin and strong! The bioengineering dilemma in the structural and functional design of the blood-gas barrier. *Physiol Rev* 2005; 85(3): 811–844.

14. Weibel ER and Knight BW. A morphometric study on the thickness of the pulmonary air-blood barrier. *J Cell Biol* 1964; 21: 367–384.

15. Crapo JD, Young SL, Fram EK, Pinkerton KE, Barry BE, and Crapo RO. Morphometric characteristics of cells in the alveolar region of mammalian lungs. *Am Rev Respir Dis* 1983; 128(2): S42–S46.

16. Creuwels LA, van Golde LM, and Haagsman HP. The pulmonary surfactant system: Biochemical and clinical aspects. *Lung* 1997; 175(1): 1–39.

17. Fuhrman BP, Smith-Wright DL, Kulik TJ, and Lock JE. Effects of static and fluctuating airway pressure on intact pulmonary circulation. *J Appl Physiol* 1986; 60(1): 114–122.

18. Evans AM. Hypoxic pulmonary vasoconstriction. *Essays Biochem* 2007; 43: 61–76.

19. Stocks J and Quanjer PH. Reference values for residual volume, functional residual capacity and total lung capacity. *Eur Respir J* 1995; 8(3): 492–506.

20. Frazer DG, Lindsley WG, Rosenberry K, McKinney W, Goldsmith WT, Reynolds JS, et al. Model predictions of the recruitment of lung units and the lung surface area-volume relationship during inflation. *Ann Biomed Eng* 2004; 32(5):756–763.

21. Suki B. Fluctuations and power laws in pulmonary physiology. *Am J Respir Crit Care Med* 2002; 166(2): 133–137.

22. Namati E, Thiesse J, de Ryk J, and McLennan G. Alveolar dynamics during respiration: Are the pores of Kohn a pathway to recruitment? *Am J Respir Cell Mol Biol* 2008; 38(5): 572–578.

23. Hülskamp G, Pillow JJ, Dinger J, and Stocks J. Lung function tests in neonates and infants with chronic lung disease of infancy: Functional residual capacity. *Pediatr Pulmonol* 2006; 41(1): 1–22.

24. DuBois AB, Botelho SY, Bedell GN, Marshall R, and Comroe JH. A rapid plethysmographic method for measuring thoracic gas volume: A comparison with a nitrogen washout method for measuring functional residual capacity in normal subjects. *J Clin Invest* 1956; 35(3): 322–326.

25. Newth CJ, Enright P, and Johnson RL. Multiple-breath nitrogen washout techniques: Including measurements with patients on ventilators. *Eur Respir J* 1997; 10(9): 2174–2185.

26. Eichinger M, Tetzlaff R, Puderbach M, Woodhouse N, and Kauczor HU. Proton magnetic resonance imaging for assessment of lung function and respiratory dynamics. *Eur J Radiol* 2007; 64(3): 329–334.

27. Kitchen MJ, Lewis RA, Morgan MJ, Wallace MJ, Siew ML, Siu KK, et al. Dynamic measures of regional lung air volume using phase contrast x-ray imaging. *Phys Med Biol* 2008; 53(21): 6065–6077.

28. Clausen J and Wanger S. Consensus statement on measurements of lung volumes in humans: American Thoracic Society; 2003.

29. Pride NB. Forced oscillation techniques for measuring mechanical properties of the respiratory system. *Thorax* 1992; 47(4): 317–320.

30. Bates JH and Suki B. Assessment of peripheral lung mechanics. *Respir Physio Neurobiol* 2008; 163(1–3): 54–63.

31. Hantos Z, Adamicza A, Govaerts E, and Daróczy B. Mechanical impedances of lungs and chest wall in the cat. *J Appl Physiol* 1992; 73(2): 427–433.

32. LaPrad AS and Lutchen KR. Respiratory impedance measurements for assessment of lung mechanics: Focus on asthma. *Respir Physiol Neurobiol* 2008; 163(1–3): 64–73.

33. Hsia CC. Recruitment of lung diffusing capacity: Update of concept and application. *Chest* 2002; 122(5): 1774–1783.

34. Ratnovsky A, Elad D, and Halpern P. Mechanics of respiratory muscles. *Respir Physiol Neurobiol* 2008; 163(1–3): 82–89.

35. West JB. *Respiratory physiology: The essentials.* 8th Edition. Philadelphia: Lippincott Williams & Wilkins; 2008.

36. Felici M, Filoche M, Straus C, Similowski T, and Sapoval B. Diffusional screening in real 3D human acini—a theoretical study. *Respir Physiol Neurobiol* 2005; 145(2–3): 279–293.

37. Van't Hoff J. Études de Dynamique chimique, 1884.

38. Van't Hoff J. L'Équilibre chimique dans les Systèmes gazeux ou dissous à l'État dilué, 1885.

39. Dash RK and Bassingthwaighte JB. Blood HbO_2 and $HbCO_2$ dissociation curves at varied O_2, CO_2, pH, 2,3-DPG and temperature levels. *Ann Biomed Eng* 2004; 32(12): 1676–1693.

40. Prange HD. Laplace's law and the alveolus: A misconception of anatomy and a misapplication of physics. *Adv Physiol Educ* 2003; 27(1–4): 34–40.

41. Hulskamp G, Lum S, Stocks J, Wade A, Hoo AF, Costeloe K, et al. Association of prematurity, lung disease and body size with lung volume and ventilation inhomogeneity in unsedated neonates: A multicentre study. *Thorax* 2009; 64(3): 240–245.

42. Wolf GK and Arnold JH. Noninvasive assessment of lung volume: Respiratory inductance plethysmography and electrical impedance tomography. *Crit Care Med* 2005; 33(3 Suppl): S163–S169.

43. Poon CS. Neural plasticity of respiratory control system: Modeling perspectives. *Conf Proc IEEE Eng Med Biol Soc* 2005; 6: 5847–5849.

44. Poon CS and Siniaia MS. Plasticity of cardiorespiratory neural processing: Classification and computational functions. *Respir Physiol* 2000; 122(2–3): 83–109.

45. Duffin J. Functional organization of respiratory neurons: A brief review of current questions and speculations. *Exp Physiol* 2004; 89(5): 517–529.

46. Stornetta RL, Sevigny CP, and Guyenet PG. Inspiratory augmenting bulbospinal neurons express both glutamatergic and enkephalinergic phenotypes. *J Comp Neurol* 2003; 455(1): 113–124.

47. Shen L, Peever JH, and Duffin J. Bilateral coordination of inspiratory neurons in the rat. *Pflugers Arch* 2002; 443(5–6): 829–835.

48. Feldman JL and Del Negro CA. Looking for inspiration: New perspectives on respiratory rhythm. *Nat Rev Neurosci* 2006; 7(3): 232–242.

49. Onimaru H, Arata A, and Homma I. Firing properties of respiratory rhythm generating neurons in the absence of synaptic transmission in rat medulla *in vitro. Exp Brain Res* 1989; 76(3): 530–536.

50. Onimaru H and Homma I. A novel functional neuron group for respiratory rhythm generation in the ventral medulla. *J Neurosci* 2003; 23(4): 1478–1486.

51. Del Negro CA, Morgado-Valle C, and Feldman JL. Respiratory rhythm: An emergent network property? *Neuron* 2002; 34(5): 821–830.

52. Paton JF and Dutschmann M. Central control of upper airway resistance regulating respiratory airflow in mammals. *J Anat* 2002; 201(4): 319–323.

53. Paton JF and Richter DW. Maturational changes in the respiratory rhythm generator of the mouse. *Pflugers Arch* 1995; 430(1): 115–124.

54. St-Jacques R, Filiano JJ, Darnall RA, and St-John WM. Characterization of hypoxia-induced ventilatory depression in newborn piglets. *Exp Physiol* 2003; 88(4): 509–515.

55. Hilaire G, Viemari JC, Coulon P, Simonneau M, and Bevengut M. Modulation of the respiratory rhythm generator by the pontine noradrenergic A5 and A6 groups in rodents. *Respir Physiol Neurobiol* 2004; 143(2–3): 187–197.

56. Feldman JL, Mitchell GS, and Nattie EE. Breathing: Rhythmicity, plasticity, chemosensitivity. *Annu Rev Neurosci* 2003; 26: 239–266.

57. Chatburn RL. Physical basis of mechanical ventilation. In: MJ Tobin (Ed.), *Principles and Practice of Mechanical Ventilation*, 2nd edition, pp. 37–128. New York: McGraw-Hill; 2006.

58. Chatburn RL. Classification of ventilator modes: Update and proposal for implementation. *Respir Care* 2007; 52(3): 301–323.

59. Dorkin HL, Stark AR, Werthammer JW, Strieder DJ, Fredberg JJ, and Frantz ID, III. Respiratory system impedance from 4 to 40 Hz in paralyzed intubated infants with respiratory disease. *J Clin Invest* 1983; 72(3): 903–910.

60. Jansson L and Jonson B. A theoretical study on flow patterns of ventilators. *Scand J Respir Dis* 1972; 53(4): 237–246.

61. AARC. Consensus statement on the essentials of mechanical ventilators—1992. American Association for Respiratory Care. *Respir Care* 1992; 37(9): 1000–1008.

62. Suki B, Alencar AM, Sujeer MK, Lutchen KR, Collins JJ, and Andrade JS, Jr, et al. Life-support system benefits from noise. *Nature* 1998; 393(6681): 127–128.

63. Chatburn RL. Computer control of mechanical ventilation. *Respir Care* 2004; 49(5): 507–517.

64. Branson RD and Chatburn RL. Technical description and classification of modes of ventilator operation. *Respir Care* 1992; 37(9): 1026–1044.

65. Laubscher TP, Frutiger A, Fanconi S, and Brunner JX. The automatic selection of ventilation parameters during the initial phase of mechanical ventilation. *Intensive Care Med* 1996; 22(3): 199–207.

66. Dojat M, Harf A, Touchard D, Lemaire F, and Brochard L. Clinical evaluation of a computer-controlled pressure support mode. *Am J Respir Crit Care Med* 2000; 161(4 Pt 1): 1161–1166.

67. Nemoto T, Hatzakis GE, Thorpe CW, Olivenstein R, Dial S, and Bates JH. Automatic control of pressure support mechanical ventilation using fuzzy logic. *Am J Respir Crit Care Med* 1999; 160(2): 550–556.

68. Bates JH, Hatzakis GE, and Olivenstein R. Fuzzy logic and mechanical ventilation. *Respir Care Clin N Am* 2001; 7(3): 363–377.

69. East TD, Heermann LK, Bradshaw RL, Lugo A, Sailors RM, Ershler L, et al. Efficacy of computerized decision support for mechanical ventilation: Results of a prospective multicenter randomized trial. *Proc AMIA Symp* 1999: 251–255.

70. Henderson-Smart DJ, Cools F, Bhuta T, and Offringa M. Elective high frequency oscillatory ventilation versus conventional ventilation for acute pulmonary dysfunction in preterm infants. *Cochrane Database Syst Rev* 2007; (3): CD000104.

71. Pillow JJ. High-frequency oscillatory ventilation: Mechanisms of gas exchange and lung mechanics. *Crit Care Med* 2005; 33(3 Suppl): S135–S141.

72. Fredberg JJ. Augmented diffusion in the airways can support pulmonary gas exchange. *J Appl Physiol* 1980; 49(2): 232–238.

73. Jaeger MJ, Kurzweg UH, and Banner MJ. Transport of gases in high-frequency ventilation. *Crit Care Med* 1984; 12(9): 708–710.

74. Rossing TH, Slutsky AS, Lehr JL, Drinker PA, Kamm R, and Drazen JM. Tidal volume and frequency dependence of carbon dioxide elimination by high-frequency ventilation. *N Engl J Med* 1981; 305(23): 1375–1379.

75. Slutsky AS, Kamm RD, Rossing TH, Loring SH, Lehr J, Shapiro AH, et al. Effects of frequency, tidal volume, and lung volume on CO_2 elimination in dogs by high frequency (2–30 Hz), low tidal volume ventilation. *J Clin Invest* 1981; 68(6): 1475–1484.

76. Venegas JG and Fredberg JJ. Understanding the pressure cost of ventilation: Why does high-frequency ventilation work? *Crit Care Med* 1994; 22(9 Suppl): S49–S57.

77. Burri PH. Structural aspects of postnatal lung development—alveolar formation and growth. *Biol Neonate* 2006; 89(4): 313–322.

78. Mansoor M. Amiji and Beverly J. Sandmann. *Applied Physical Pharmacy*, pp. 54–57. McGraw-Hill Professional.; 2002; Van't Hoff.

79. Salmon RB, Primiano FP Jr, Saidel GM, and Niewoeher DE, Human lung pressure-volume relationships: Alveolar collapse and airway closure, *J Appl Physiol* 1981; 51(2): 353–362.

IV

Bioelectrical Physics

20

Electrodes

Rossana E. Madrid,
Ernesto F. Treo,
Myriam C. Herrera, and
Carmen C. Mayorga
Martinez

20.1 Measurement of Electrical Potential with Electrodes

Physiological systems generate signals of different types: electrical, mechanical, chemical, magnetic, and thermal.[1]

The cell physiological activity in different biological structures is associated with electrical processes, such as the ionic interchange at the cellular membrane or active mechanisms such as metabolism.[2] Although most cells present bioelectric events, only a few present electrical potential changes that reveal its physiological operation.[3] These signals are of great interest in biomedical applications, since they can provide information about the operation of the biological system.

In all cases, the first contact between the electronic and biological worlds is the *transducer*. This device "transforms" the energy commanding the biological phenomenon to be measured into another type of energy, which can be registered, processed, and displayed.

The way these signals can be restored depends on its origin and characteristics. In some cases, electrodes can do it, but in other cases, more specific transducers or sensors are necessary. The development of biosensors had a great impact due to the level and variety of signals that can be detected.[4] Biosensors are devices consisting of two parts, the biological transducer, in contact with the sample, and the physical transducer, which can be an electrode or more complex transducers. In biomedical applications, biosensors are well known for the detection of glucose and other analytes. Since 1962, when Clark and Lyons proposed for the first time the concept of electrodes with a deposited enzyme, enzymatic sensors have had great development, and even today, research continues into this fascinating area.[4,5]

Electrodes are the simplest transducers, which transform ionic currents, usually present in biological media, into electronic currents, which are then registered, amplified, modified, and displayed in the electronic device. The contact zone between the electrode and the biological sample is the known electrode–electrolyte interface (EEI). This EEI generates an impedance and a direct current potential (the half-cell potential). This EEI impedance markedly influences the instrumentation to be employed in the measurement of bioelectric potentials and hence is widely studied in an effort to diminish it.[6-12]

However, this EEI impedance is not always an "interference." Such is the case, for example, of impedance microbiology, where the interface impedance is very sensitive to microorganisms growth.[13] It is useful, therefore, to quantify microorganisms that grow in a liquid sample.[14] Another example is the glass pH electrode, which measures the half-cell potential of the glass–sample interface.[15]

20.2 Recording Biopotentials on the Skin

In most cases, biopotential-recording electrodes are placed over the skin.[16] Human skin is an organ including three main layers that form a protective barrier and keep the body isolated from its environment (Figure 20.1). The skin also helps to maintain a constant body temperature and has only about 0.07 inches (2 mm) thickness. The outermost layer—called *epidermis*—is a tough protective layer and plays the most important role in the electrode–skin interface. The second layer (located under the epidermis) is called *dermis*; it contains sweat glands, sebaceous glands, blood vessels, nerve endings, and hair follicles. Under these two skin layers is a third fatty layer of *subcutaneous tissue*.

Epidermis is constantly renewing itself and it has three sublayers: the outer is the *stratum corneum* (SC), which consists of dead cells (flat keratinous material) with high impedance (a barrier). The characteristics of the electrode–skin system will be treated later in this chapter.

20.3 Electrode Types

Electrodes can be of different types depending on their application. They can be classified as *noninvasive* and *invasive*. The first group includes surface electrodes, which are placed over the

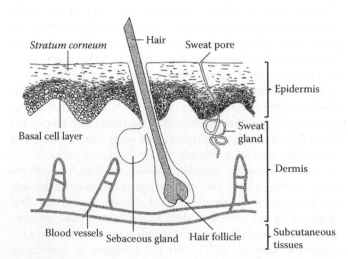

FIGURE 20.1 Cross-sectional diagram of the human skin, showing its layers and components.

patient's skin. In the second group, needle and implantable electrodes are included. We will not present traditional electrodes, which can be found very well explained in other books, but novel electrodes will be presented.

Noninvasive electrodes: They can be metallic plates, discs, disposable foam-pad electrodes, often used with electrocardiograph monitoring apparatus, or electrodes with special shapes depending on its applications. They are placed over the skin, and rely on the presence of an electrolytic solution/gel to conduct the bioelectrical signal from the surface of the skin to and through the electrode. Success and failure of these gel-based electrodes are largely dependent on the hydration level of the electrode. A dried-out electrode will yield a signal dominated by artifact. Further, gel-based electrodes are impaired by the high electrical impedance of the outermost layer of the skin, the SC. From 1935, many electrode jellies and pastes have been commercially available for different applications.[17] To overcome this high impedance, aggressive skin preparation techniques such as abrasion with emery paper are standard clinical practice when using gel-based electrodes. But for long-time monitoring or long determinations, this electrolyte can also produce skin irritations and infection problems. A good alternative is dry electrodes, which are designed to avoid the use of electrolytic paste. There are three types of dry electrodes: (a) contact electrodes, (b) noncontact electro-optic sensors, and (c) contact electro-optic sensors.[18]

Type (a) electrodes employ a special contact surface (either machined or nano-machined) to improve direct, dry contact to the skin. They are manufactured with a surface micro-structure that functions as the sensing element of the electrode. These micro-features augment the electrode/skin interface by mechanically connecting the skin and the electrode, thus facilitating the transmission of the bioelectric signals from the body through the electrode and reducing motion artifact. Innovative manufacturing processes enable both the monolithic design and the fabrication of the micro-features.[19] Another interesting example of a dry contact electrode is the one made of conductive rubber, which can be integrated in clothing for monitoring purposes. These dry electrodes use sweat produced by the glands of the SC for a conductive bridge from the skin to the electrode. The conversion from ionic to electronic current takes place at the rubber electrode surface.[20] There is another innovative dry electrode fabricated polymerizing a conductive hydrogel within a piece of corn-shaped foam.[18] By adding a cetyl trimethyl ammonium chloride molecule, which has a positively charged hydrophilic head and a long hydrophobic tail with 18 carbons to the hydrogel, the skin resistance can be reduced significantly without the need of skin preparation.[18]

Noncontact dry electrodes consist of the measurement of a capacity formed between an insulating surface of the electrode and the skin. As the electrode functions as an insulator, no net charge flow occurs, and therefore there is a capacitive coupling between the electrode and the skin. These electrodes are connected to ultra-high impedance amplifiers. They have the disadvantage that movements produce interferences due to the variation of the capacitive coupling. It is also very important to use an effective insulator material, looking for nontoxicity, porosity, or avoiding possible problems that could arise from sweat, which can degrade insulation.[2]

Finally, the contact electro-optic sensor uses the biopotential from the body to modulate laser light supplied by a fiber optic input. It converts the modulated light to an electrical signal that provides traditional EEG- and ECG-type output.[21] It uses a light interferometer to divide light into two paths, and then it uses the biopotential from the body to change the optical lengths of one or both paths, so that the interferences generated produce a change in the intensity of the original incoming light. These are dry electrodes that do not need the use of gel and have very high (on the order of 10^{14} Ω) device impedance. This feature is due to the use of the interferometer, which takes advantage of the high frequencies of light. These dry electrodes were initially developed for the U.S. Army Aeromedical Research Laboratory for physiological status monitoring of Army pilots.[21]

Invasive electrodes: The first and simpler electrodes of this type were the needle electrodes for percutaneous measurement of biopotentials. With the emergence of new technologies, this invasiveness was reduced today. New electrodes are micromachined devices designed to pierce only the outer skin layer, or to inflict the least possible damage to the tissue.[22–26]

With the development of new technologies and the rise of miniaturization, microelectrodes have had great development in biomedical applications. One example is the micro-multi-probe electrode array, which can be used to measure neural signals and can be applied inside the body.[22] These small electrodes present important impedance, which influences the capability of recording neural signals. The authors report an impedance of 1.41 MΩ at 1 kHz. But at higher frequencies the impedance of the electrode decreased gradually. This impedance was comparable with those measured using a traditional glass needle electrode.

Other types of microelectrodes are surface electrodes, which can penetrate the SC of the skin. An interesting example is a spiked (microneedles) electrode array for the measurement of surface biopotential.[23,26] These spikes can penetrate the skin without pain sensation, minimizing the variations of skin impedance as well as body motion. The goal of the spiked electrode is to pierce the SC and to penetrate the electrically conducting stratum germinativum in order to circumvent the high impedance characteristics of the SC. Skin preparation and electrolytic gel are therefore not needed. This type of electrode presents lower electrode impedance since, despite the smaller area, it "eliminates" the isolating SC layer.

20.3.1 Electrode–Electrolyte Interface

Electrodes are transducers of ionic into electronic current. A pure electrolyte is a solution with ions. A living tissue has an intracellular and an extracellular conductor electrolyte containing free ions that are free to migrate. The contact region

TABLE 20.1 Half-Cell Potentials for Common Electrode Materials at 25°C

Metal and Reaction	Potential E^0 (V)
$Al \rightarrow Al^{3+} + 3e^-$	-1.706
$Zn \rightarrow Zn^{2+} + 2e^-$	-0.763
$Cr \rightarrow Cr^{3+} + 3e^-$	-0.744
$Fe \rightarrow Fe^{2+} + 2e^-$	-0.409
$Cd \rightarrow Cd^{2+} + 2e^-$	-0.401
$Ni \rightarrow Ni^{2+} + 2e^-$	-0.230
$Pb \rightarrow Pb^{2+} + 2e^-$	-0.126
$H_2 \rightarrow 2H^{2+} + 2e^-$	+0.000 by definition
$Ag + Cl^- \rightarrow AgCl + e^-$	+0.223
$2Hg + 2Cl^- \rightarrow Hg_2Cl_2 + 2e^-$	+0.268
$Cu \rightarrow Cu^{2+} + 2e^-$	+0.340
$Cu \rightarrow Cu^+ + e^-$	+0.522
$Ag \rightarrow Ag^+ + e^-$	+0.799
$Au \rightarrow Au^{3+} + 3e^-$	+1.420
$Au \rightarrow Au^+ + e^-$	+1.680

between the electrode and the electrolyte (solution or tissue) is known as EEI.

When a metal electrode is placed in an electrolyte solution, a charge distribution is created next to the EEI. This localized charge distribution causes an electrical potential, called the half-cell potential. There is a double charge layer, where the charges are distributed according to different models proposed by different authors.[27]

The half-cell potential is basically due to two factors: the development of free or induced surface charges in excess or defect with respect to both phases (metal and solution); and the development of a layer of dipoles oriented through the electrode surface. It depends on the metal, the concentration of ions in solution, and the temperature. This half-cell potential cannot be measured, because the reference electrode against which the potential is measured also introduces another half-cell potential. Therefore, a hydrogen standard electrode is used whose half-cell potential is assigned by convention to be zero at all temperatures.[15] In this way, there are tabulated the standard potential versus the hydrogen standard electrode of a great variety of electrode materials, as can be seen in Table 20.1.

20.3.2 Electrical Double Layer

There are many electrochemical phenomena occurring in the EEI in equilibrium conditions, that is, without an external potential applied. Concepts of the electrochemical thermodynamics and kinetics are applied to study the reactions that take place in this region, including the electrical double-layer theory. The model for this electrical double layer was proposed for the first time by Helmholtz in 1853.[27] This model considered the electrical double layer as a parallel plate capacitor. In this case, charges in the electrode and in the solution have the same value but opposite sign. In the solution, charges are confined to a layer at a distance d from the electrode (Figure 20.2a). This model considers C as a constant

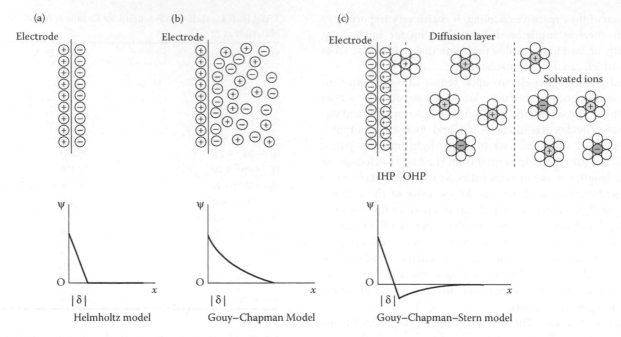

FIGURE 20.2 Double-layer models. (a) The Helmholtz model; (b) the Gouy–Chapman model; (c) the Gouy–Chapman–Stern model. ψ: potential.

and do not explain its dependence on voltage. It also disregards some phenomena such as diffusion and adsorption.

Gouy and Chapman present independently in 1910 and 1913 the first detailed model for the double layer (Figure 20.2b). This model takes into account the diffusion of ions in the vicinity of the electrode. Diffusion is the process resulting from random motion of molecules, which produces a net flow of matter due to a concentration difference. According to this model, the excess ions are nonuniformly distributed as in the Helmholtz model. The "thickness" of this so-called "diffuse layer" is variable, but it is typically around the order of magnitude of a millionth of a centimeter. In this model, the ions are considered as punctual charges, so it assumes that the ions are infinitely small and can get infinitely close to the surface of the electrode. The model is valid for diluted solutions and low potentials. Stern proposed in 1924 modifications to the Gouy–Chapman model and postulated that the ions could not reach the surface of the electrode beyond its ionic radius. The Stern modification is essentially a combination of the Helmholtz and Gouy–Chapman models, and it is called the de Gouy–Chapman–Stern model (Figure 20.2c). It assumes a plane of the closest approach, named the internal Helmholtz plane, which is the distance between the metallic electrode and the nucleus of the specifically adsorbed species. This most internal layer is also defined as the *compact layer*. The ions of opposite charge as that of the electrode, which are solvated, can approach the electrode up to a distance defined as the outer Helmholtz plane (OHP). The electrostatic interaction between the solvated ions and the electrode is independent of the chemical nature of the ions; so they are said to be nonspecifically adsorbed. Attached to this zone is a Gouy–Chapman-type diffuse layer, which is the zone between the OHP and the bulk of

the solution.[28–30] The transport of charged molecules/ions in this region may be due to two phenomena: concentration gradients and electric field. The latter produces a migration of charged molecules/ion due to the external electric field applied to the electrodes. The general term for both these transport processes is *electrodiffusion*.[31]

Reactions that take place in the EEI are heterogeneous processes, whose kinetics depends on mass transfer, charge transfer velocity, as well as the oxide–reduction reactions occurring there.

20.3.3 Physical Models of the EEI

The first and simplest model, frequently used to model the EEI, is a parallel resistance and capacitance model (the double-layer capacitor and the charge transfer resistance).

The *charge transfer resistance* (R_{ct}) considers the physical hindrance faced by the electrons when they move from and to the electrodes. A more complex and recent model, like the Randles one, also considers the difficulty of ions to diffuse from the bulk of the solution to the electrode surface, modeled by the diffusion impedance Z_w, also called the Warburg impedance.[32] Z_w is a complex quantity having real and imaginary parts that are equal. This impedance is proportional to the reciprocal of the square root of the frequency ($1/\sqrt{\omega}$). At high frequencies, therefore, this term is small, since diffusing reactants do not have to move very far. Consequently, the diffusion process is only observed at low frequencies, where the reactants have to diffuse farther, thereby increasing the Warburg impedance. The low-frequency region can provide information concerning charge transfer and diffusion phenomena and the high-frequency part defines the solution resistance. If the electrodes are in contact with the skin instead of being immersed in an electrolytic solution, the Warburg impedance turns negligible since there is no problem with the diffusion of ions.

Solution resistance is often a significant factor in the impedance of an electrochemical system. When modeling the electrochemical system, any solution resistance (R_m) between the reference electrode and the working electrode must be considered. The resistance of an ionic solution depends on the ionic concentration, type of ions, temperature, and the geometry of the area in which current is carried. The Randles model for the EEI impedance is depicted in Figure 20.3.

Figure 20.4a shows the complete equivalent circuit for the impedance measured between two electrodes submerged in an electrolyte. Considering a bioelectric event measured with electrodes, the interface impedance is modeled in the same manner, but the tissue can be modeled in a different way. The extracellular, intracellular, and membrane resistances are taken into account, as well as the membrane capacitance. The bioelectric event also generates a potential that can be modeled as a potential source called the *biological electric potential* (BEP) (Figure 20.4b).

The Randles model holds only for ideally smooth interfaces and does not take into account electrode rugosities. On the other hand, the interface behaves in a nonlinear way for high current densities. The bibliography does not explain the genesis of this nonlinearity, and a constant phase angle (CPA) element is often used instead of the C_{dl} to take care of this nonlinearity.[8] Ruiz et al. proposed a model to predict the nonlinear response of the interface without a CPA element (large amplitude signals

>10 mV). This model is based on the geometrical structure of a fractal net. Liu[33] models the interface using Cantor's bar, and only includes the electrochemical parameter C_{dl}. The Ruiz model incorporates also the R_{ct} in parallel with C_{dl} and draws the model in a tridimensional way, as can be seen in Figure 20.5. This is a novel model that considers both electrochemical and geometrical aspects of the EEI without the use of any CPA element. In this model, geometry affects R_{ct} behavior, because the alternate current produces a voltage drop that depends on the fractal level (i.e., roughness or rugosity factor).[8] The model fits very well to experimental data.

20.3.4 Polarizable and Nonpolarizable Electrodes

The electrode must be an electrical conductor and must act as a source or sink of electrons by the interchange with molecules in the EEI. It must also be electrochemically inert. A great variety of materials can achieve these conditions. For example, platinum or gold, but there are other metals, not so noble, that can be chemically or electrochemically modified to produce the desired surface over them. This modification can be made to produce specific sensors or to reduce the electrode–electrolyte impedance. The latter effect is more important as the size of the electrode decreases. It has been reported that the iridium oxide (IrOx) can reduce the EEI impedance in two orders of magnitude.[34,35] The important decrease in EEIZ should be caused by the presence of the thin electrodeposited iridium oxide film (EIROF), which have a high electronic conductivity (very low R_{ct}) and an increased surface area (big C).[34] It is possible to obtain this kind of electrodeposit over surfaces of different metallic substrates like gold, platinum, titanium, and stainless steel. The use of low-cost substrates like stainless steel to obtain EIROF electrodes with very low EEI impedance makes this a promising material to be used as an electrochemical transducer in microelectronics applications.[34]

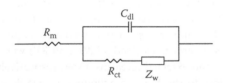

FIGURE 20.3 The Randles model for the EEI impedance. R_m: medium resistance; R_{ct}: charge transfer resistance; and C_{dl}: double-layer capacitance.

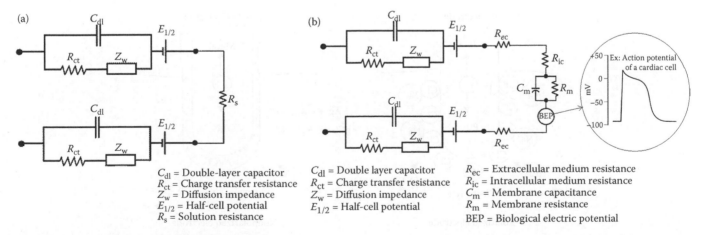

C_{dl} = Double-layer capacitor
R_{ct} = Charge transfer resistance
Z_w = Diffusion impedance
$E_{1/2}$ = Half-cell potential
R_s = Solution resistance

C_{dl} = Double layer capacitor
R_{ct} = Charge transfer resistance
Z_w = Diffusion impedance
$E_{1/2}$ = Half-cell potential

R_{ec} = Extracellular medium resistance
R_{ic} = Intracellular medium resistance
C_m = Membrane capacitance
R_m = Membrane resistance
BEP = Biological electric potential

FIGURE 20.4 (a) Equivalent circuit for the impedance measured between two electrodes submerged in an electrolyte. (b) Equivalent circuit for the impedance measured between two electrodes on the skin (biopotential measurement).

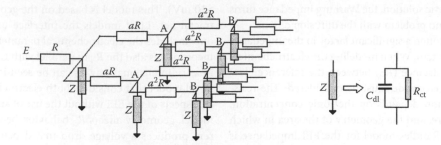

FIGURE 20.5 Tridimensional model of EEI fractal net. R: medium resistance of the first level; a: scale factor; R_{ct}: charge transfer resistance; C_{dl}: double-layer capacitance. A and B are equipotential points.

There are many factors to take into account in the selection of materials for electrodes. Some of these factors are the potential range to be used in the measurement and the speed of charge transfer. Electronic transfer can considerably change from one material to another.

In this sense, electrodes can be perfectly polarizable and perfectly nonpolarizable. The first are those in which the R_{ct} is so big that they can be considered as an open circuit; therefore no electronic current flows through the EEI. The electrode behaves as a capacitor. In these circumstances, the phenomena that take place in the interface are related only to the rearrangement of ions and/or dipoles, changing the double layer composition and structure with the applied voltage.[29] An example of this kind of electrodes is the platinum metal electrode, where no net transfer of charge across the metal–electrolyte interface occurs (Figure 20.6a). There is a range of applied potential where any electrochemical reaction occurs. As this material is relatively inert, it is very difficult to be oxidized or dissolved. There are no overpotentials for this type of electrodes.

On the other hand, there are systems where it is impossible to establish a range of potentials where the electrode behaves as perfectly polarizable electrode. They are the perfectly nonpolarizable ones, in which R_{ct} is a short circuit. The current freely passes through the EEI as can be seen in Figure 20.6b. Reference electrodes, as the calomel or Ag/AgCl electrodes, are nonpolarizable electrodes. Reference electrodes will be discussed later in this chapter.

Nonpolarizable electrodes are used in recording, while polarizable electrodes are used in stimulation. In practice, neither of these ideal configurations can be produced, but the examples showed can approach the ideal very well. Other types of electrodes have an intermediate value of R_{ct}.

20.3.5 Frequency Response of the EEI

The frequency response of the EEI will be presented with an example of the measurement of the most common electrodes used in ECG and EEG measurements versus frequency. The impedance was measured with a tripolar system connected to a Solartron S1258W frequency response and electrochemical analyzer. The tripolar system is depicted in Figure 20.7. The working and reference electrodes (WE and RE) were placed on the internal and external forearms' faces, respectively. Beneath the skin close to the WE, an acupuncture needle served as a reference electrode (RE; diameter 0.22 mm, length 25 mm). The

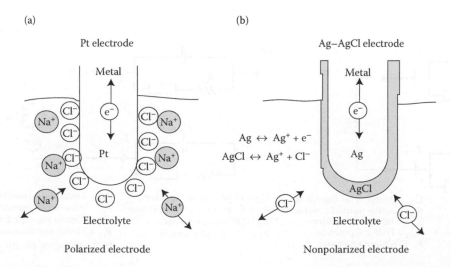

FIGURE 20.6 Polarizable and nonpolarizable electrodes. (a) Platinum electrode: no net transfer of charge across the interface—capacitive; (b) Ag–AgCl electrode: unhindered transfer of charge—resistive.

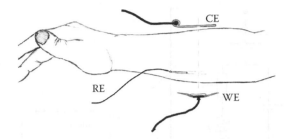

FIGURE 20.7 Positioning of the electrodes during EEI measurements.

counter electrode (CE) was made of a high area (15 cm²) stainless steel with conductive gel and was secured with a rubber band. The impedance was measured between the RE and the WE, with 15 points per frequency decade. The skin was gently abraded with smooth sandpaper and thereafter cleaned firmly with alcohol. The WE was changed to one of five different types: a typical 3 M ECG monitoring electrode, the same electrode with some conductive gel added, a secondary needle electrode, an EEG Ag/AgCl cup electrode, and a stainless-steel plate electrode. The impedance modulus was normalized against the electrode area and the results are presented in Figure 20.8.

Only the needle interface resembles a typical EEI electrochemical measurement, with a linear plot in the log–log scale. Furthermore, it achieved the lowest average value. The monitoring electrode showed a 50 kΩ stable impedance in the low-frequency range (0.1–100 Hz), which was reduced to its half by adding some extra electrolytic gel. The bentonite + EEG electrode system presented an impedance 10 times greater than the monitoring electrode, quite stable in the 0.1–50 Hz range. The highest impedance value was measured with the stainless-steel electrode, despite having used electrolytic gel.

From these typical responses, it can be seen that the frequency response of an EEI presents high- and low-frequency impedance, given by the capacitive characteristic of the interface (see Figure 20.4a). The Warburg impedance is negligible since the electrodes are in contact with the skin. At high frequencies the impedance is purely resistive. Therefore, the total impedance measured takes into account only the R_{tc} and the resistance of the body. The frequency dependence of the EEI impedance is very useful in many applications, like impedance microbiology, which make use of these characteristics to measure the interface impedance to quantify microorganisms.[14,36] Other applications may be impedance pneumography or impedance cardiography, where a measurement with a high-frequency carrier signal (on the order of 10 KHz) is very useful to rescue the signal of interest and, in contrast to the previous case, to avoid or diminish the EEI impedance.

20.4 Skin Impedance

The characterization of skin impedance is useful to model the human skin. Rosell et al. found that gel-coated skin show impedances from approximately 100 Ω to 1 MΩ, depending on the frequency of input pulses. At lower frequencies, that is, at 1 Hz, the impedance varied from 10 kΩ to 1 MΩ and at higher frequencies (100 kHz to 1 MHz), the impedances were about 220 Ω and 120 Ω, respectively. If the SC is abraded, the skin impedance is drastically reduced.[16,37] The same was found true of the skin's capacitive characteristics.

In addition to electrical parameters changing the impedance of the skin, mechanical deformations also change the skin impedance. A stretch in the surface of the skin results in a decrease of skin potential between the inside and outside of the barrier layer from about 30 to 25 mV. For 1 cm² of skin, the barrier layer forms approximately 50 kΩ impedance of skin.[38] The change in skin potential is a result of skin impedance changing with respect to surface stretches.

The skin's capacity lies within the range of 0.02–0.06 μF/cm². Much of this capacity lies within the SC. If the SC is gradually removed, the skin capacitance is reduced with every layer that is removed. When the SC is completely removed, the skin capacitance is reduced to nearly zero.[39] The individual cell membranes

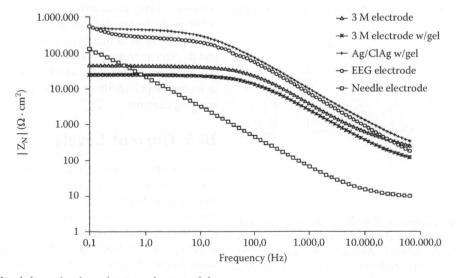

FIGURE 20.8 Normalized electrode–electrolyte impedance modulus.

within the skin's layers are one explanation for the skin's high capacity.

20.4.1 Skin Circuit Model

An equivalent circuit model for the human skin impedance cannot be expressed as a simple passive circuit because the skin shows nonlinear and time-variant characteristics. The simplest model that can be used to represent skin impedance is a parallel network consisting of a capacitor C_p and a resistor R_p representing the epidermis layer, followed by a series resistor R_s representing the impedance of the subdermal tissue.[39,40] Figure 20.9a shows this equivalent circuit model. Figures 2.9b and 2.9c show the current response of skin to square-wave constant-voltage pulses: (b) with an intact SC and (c) with a removed SC. Without the SC, the current is approximately limited by R_s. The skin capacitive component produces a clear derivative response and it is a possible cause of distortion in biosignal recorders. A more complete model would consider the skin as composed of numerous layers of cells, each having a capacitance and a conductance in parallel with the previous model.

The determination of parameters of the equivalent circuit (Figure 20.9a) is difficult because their numerical values depend greatly on skin preparation and the site of measurement.[41] For 0.8 cm² of dry, glabrous (hairless) skin cleaned with ethanol and water, the parameters are $R_s = 2–200$ kΩ, $R_p = 100–500$ kΩ, and $C_p = 50–1500$ pF.[42,43]

20.4.2 Electrode–Electrolyte–Skin Model

In order to understand the behavior of one biopotential record electrode over the skin, it is necessary to analyze the electrode-electrolyte–skin interface. Generally, a good electrolytic contact (ohmic) between skin and electrodes is established by a conductive gel, which contains Cl⁻ as the main anion. The equivalent circuit for the resulting interface is presented in Figure 20.10. The oversimplified model includes four parts: (1) the EEI represented by E_e, R_e, Z_{we}, and C_e (previously described in detail); (2)

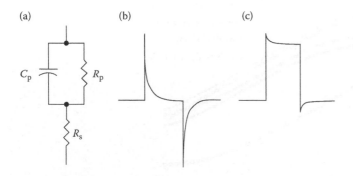

FIGURE 20.9 Current response of skin to square-wave constant-voltage pulses. (a) Equivalent circuit, (b) response of skin with SC intact, (c) response of skin with SC removed. (Adapted from Reilly JP. 1992. *Electrical Stimulation and Electropathology.* Cambridge: Cambridge University Press.)

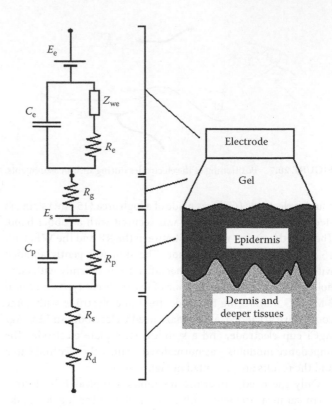

FIGURE 20.10 The electrode–electrolyte–skin model. The electrical circuit on the right represents: (a) electrode impedance, (b) electrolyte impedance, and (c) skin impedance. See text for details.

the gel or electrolyte resistance R_g; (3) the skin interface (between gel and epidermis) represented by E_s and R_s and the skin tissue itself represented by R_p in parallel with C_p; and finally, (4) the deeper layer of skin, which have, in general, a resistive behavior, represented by R_d.

The skin is an active organ that responds to different stimuli. Therefore, when biological events are needed to be accurately recorded using surface electrodes, the complex electrode-electrolyte–skin interface previously described imposes its characteristics. Many parameters can modify it, and the components of the model will reveal changes. The electrode type (i.e., shape, area, dry, and wet), material, number of electrodes to be used, and its placement are possible sources of changes.[44] In this way, it is important to consider whether the electrodes are used only to record biopotentials or are also used to stimulate tissues by applying currents.

20.5 Current Levels

When measuring biological impedances, it is necessary to inject electric current into the biological sample to be studied. From a physiological point of view, an excitable tissue should not be stimulated. An excitable tissue is sensitive to the passage of an electric current through it, and if the current is adequate, an action potential will be triggered. It is of great importance which value, frequency, and type of current must be used. In terms of hazard, the most significant effect of current on the body has to

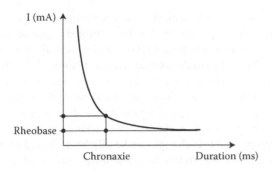

FIGURE 20.11   Strength–duration threshold curve.

do with the cardiac and nervous systems. Further details about biological stimulation can be found in Geddes and Baker.[27]

To understand the characteristics that an adequate current must have, Lapicque's curves must be understood.[45] These curves are the strength–duration threshold ones (Figure 20.11), which relate the amplitude of a rectangular stimulus (in mA) with its duration, applied to an excitable tissue in order to obtain response. The threshold intensity is called *rheobase*; currents lower than rheobase will never trigger a response no matter what its duration is. On the other hand, if the duration of the pulse is too short, no matter how big the amplitude is, no response will be evoked. *Chronaxie* is defined as the pulse duration corresponding to a current amplitude twice the rheobase, and offers a simple and practical form to compare excitable tissues.[45] All amplitudes or pulse durations falling above the curve will trigger a response. The type of response depends on the tissues involved. The same stimulus can produce sensation, pain, contractions, or damage according to which tissue is applied. Nerve fibers and cardiac muscles, for example, are very sensitive. Vagal stimulation and cardiac fibrillation are nondesirable effects, because they can produce cardiac arrest or eventually death.

On the other hand, perception threshold changes with frequency. The lowest threshold values are obtained at stimulating frequencies below 100 Hz. Taking into account the current amplitude, perception is present between 0.3 and 10 mA rms. Between 10 and 100 mA, nerves and muscles are strongly stimulated, producing contractions, pain, and fatigue. Below 15 mA, the person can still voluntarily release his hands. It is called the "let-go current." Let-go currents for women are about two-thirds the values for men. Higher currents will produce respiratory arrest, marked fatigue, and pain, depending on the current path. Sustained contraction of muscles will force the hand to grasp the wire firmly, thus worsening the situation by securing excellent contact with the wire. The person will be completely unable to let go of the wire. Fibrillation is produced with currents higher than 50 mA until 5 or 6 A rms. Figure 20.12 shows the "let-go current" versus frequency. Fibrillation threshold is still considered by many authors.[46] Higher values are capable of producing deep and severe burns in the body due to power dissipation across the body's electrical resistance. It is very important to ensure that the current amplitudes and frequencies employed remain within safety values.

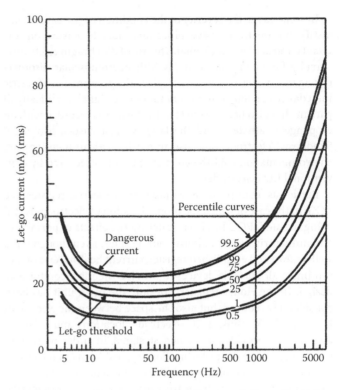

FIGURE 20.12   Let-go current versus frequency.[98] (Reproduced from Webster JG. 1998. *Medical Instrumentation: Application and Design.* New York: John Wiley & Sons Inc. With permission.)

20.6 Electrodes Placement

Electrodes placement is a matter of concern in almost all electrophysiological analyses. There are many considerations for electrode locations that are intimately related to the bioelectrical source and the acquisition system, but there are still two conditions that are maintained for all register: The closer to the source the electrodes are located, the better is the the signal-to-noise ratio (SNR) achieved (considering electrical signals from other tissues as noise). The closest the electrodes are placed to each other, the smallest the signal strength registered. Electrooculogram, electroretinogram, electrodermogram, and superficial muscles EMG are examples where the electrodes can be placed close enough to the source.

Locating electrodes close to the bioelectric source also provides an additional advantage, for it can precisely examine the properties of a single part of an entire muscle or tissue. For example, ECG provides meaningful information about the conduction system of the heart. Given a ventricular arrhythmia, the ECG specialist can infer the approximate location of the ectopic source, but cannot provide us further details about it. Cardiac mapping, as performed with a multielectrode intracavitary catheter, can be useful to locate the ectopic source. Unfortunately, taking the electrodes close to the electric source implies invasive maneuvers, as epicardial or endocardial mapping, cortical and subdermal EEGs, intramuscular EMG, or intracellular measurements.

In some cases, the severity of the condition of the patient may justify minimally invasive procedures just to ensure on-line quality and artifact-free signal. The use of diaphragm electromyography for monitoring patients with neuromuscular disorders during their recovery is being accepted. The performance of gastric impedance as a mucosal injury monitoring signal in critically ill patients has also been evaluated. The former is recorded with an esophageal catheter[47] and the latter with a nasogastric one.[48] Subdermal electrodes are also tested in critical/intensive care environments to avoid the constant attention required by conventional EEG electrodes.[49]

But let us analyze the most popular bioelectric register, the ECG, in the case of which the proper location of electrodes is still being debated. ECG has been suffering the effects of portability and miniaturization for almost four decades. Even though it is a well-known and established register, current health trends are to minimize all portable devices to provide the most comfortable design. In the Holter register, not only the main unit has been considerably reduced, but also the space between the electrodes is reduced.[50,51] Thus, ECG electrode locations must be re-analyzed to provide the strongest signal in order to sustain good-quality signal registering. For a typical 5 cm interelectrode distance, the intersubject variability has been found to be significant. However, judging from average values, the best orientation for a closely located bipolar electrode pair is diagonally on the chest. The best locations for QRS-complex and P-wave detection are around the chest electrodes of the standard precordial leads V2, V3, and V4 and above the chest electrodes of leads V1 and V2, respectively.[50] Figure 20.13 depicts the location of the precordial leads on the axial and frontal planes of the chest.

EMG, another popular surface bioelectric signal, is not fully standardized yet despite its acceptance. Commonly, misplacement during EMG records can yield cross-talk electric activity between muscles located close to the one being analyzed. Nonconventional electrode arrays are being developed with higher spatial selectivity with respect to the traditional detection systems. These are intended to noninvasively detect the activity of a single motor unit,[52] thus reducing the cross-talk problem. However, a lack of standardized placement also produces low reproducibility and reduces the intrasubject and intersubject comparison. Bony landmarks have been proposed for a standard EMG electrode placement, but for a limited group of muscles. The best location has been found to be between the innervation zone (IZ) and distal/proximal muscle tendons.[53] In the case of dynamic contractions, the muscle shifts with respect to the skin and the recording electrodes. As a consequence, the IZ can move by a 1–2 cm depending on muscle type and joint angle variation. The register obtained will change reflecting the IZ movement, and this change can be falsely attributed to muscle activity. Thus, electrodes location must also be observed in dynamic contractions.

20.7 Bipolar versus Unipolar

The instrumentation amplifier (IA) records the difference in electrical fields between two electrodes. In a unipolar configuration, one electrode is placed over the region of interest and the other electrode is located at some distance away from the tissue. These electrodes are usually referred to as the exploratory and indifferent electrodes, respectively. By convention, the signal from the recording exploratory electrode is fed into the positive input of a differential amplifier and the signal from the indifferent electrode is fed into the negative input.

Unipolar signals represent electrical activity from an entire region and only the electrical events at the exploratory electrode are described. The remote anode contributes negligible voltage, since its location can be considered to be external or is located at some distance. As an advantage, unipolar records are not altered by the orientation of the activation wave, because of their virtually indifferent anode, and are considered to yield larger and morphologically consistent electrograms. In standard ECG registering, the indifferent electrode is replaced by Wilson's terminal (see Figure 20.13). The major disadvantage of unipolar recordings is that they contain substantial far-field signal generated by depolarization of tissue remote from the recording electrode. This far-field signal is of high importance in cardiac analysis and will be briefly discussed later.

The bipolar signal is recorded as the potential difference between two closely spaced electrodes in direct contact with the tissue; it is essentially the sum of the signal from one unipolar

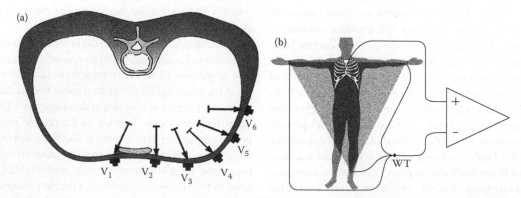

FIGURE 20.13 Location of the precordial leads: (a) cross-section of the chest and (b) frontal plane with the scheme of unipolar register. WT stands for Wilson's terminal.

electrode and the inverted signal measured from the other electrode. Typically, a bipolar record exhibits a large voltage deflection each time the dipole wave passes close to each electrode. As a result, bipolar and unipolar recordings have different characteristics. In the bipolar arrangement, distant muscle activity affects each electrode of the bipole equally, but its amplitude is inversely proportional to the third power of the distance between the recording site and the dipole.[54] Consequently, the differential amplifier usually cancels out these far-field signals.

The smaller far-field effect of the bipolar electrode renders it superior in rejecting electromagnetic interference and distant physiological activity. By the attenuation or enhancement of selected events the bipolar electrode can provide a superior SNR to that of a unipolar one. It attenuates not only electrical noise but also the physiological signals, which should be treated as noise. However, bipolar recordings are very sensitive to the orientation of the two electrodes and the activation wavefront, as it determines the amplitude and slew rate of the acquired signal. The orientation can cause either a full subtraction or an addition of the two unipolar amplitudes at each pole. When the electrodes are aligned perpendicular to the depolarization wavefront, the two monopolar signals are registered with a small delay in between. The subtraction of both signals (as performed by the IA) produces a huge negative deflection as a result, as shown in Figure 20.14. When the electrodes are parallel (or normal) to the wavefront, each electrode will be exposed to similar nonshifted electrical fields, monopolar signals will be subtracted, and the resulting bipolar signal will be significantly attenuated. Thus,

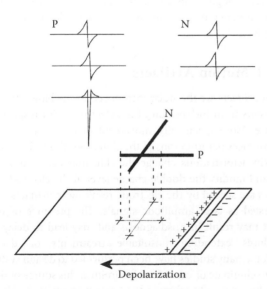

FIGURE 20.14 Bipolar acquisition of a depolarization with electrodes placed perpendicular (P direction) and normal (N direction) to the propagation wavefront. P and N stand for perpendicular and normal, respectively. Each plot shows both unipolar electrogram (upper and medium trace) and their difference (lower trace). (Redrawn from Stevenson WG and Soejima K. 2005. *Journal of Cardiovascular Electrophysiology* 16: 1017–1022.)

there is a greater probability of obtaining a signal much larger or much smaller than average.

It has been demonstrated that unfavorable electrode orientation can produce low-voltage bipolar signals, even in the presence of high voltages across the tissue. In contrast to unipolar signals, the direction of wavefront propagation cannot be reliably inferred from the morphology of the bipolar signal.[55] With bipolar recordings the signal of interest may be beneath either the distal or proximal electrode (or both) of the recording pair. This is a relevant consideration for catheter ablation that is performed by applying radiofrequency (RF) energy only at the distal electrode, because the tissue generating the signal at the proximal electrode is less likely to be damaged.

EMG exploits the bipolar configuration to avoid cross-talking and far-field potentials of other muscles. In long muscles, the electrodes are placed parallel to the muscle direction. Bipolar configuration is very useful when the electrodes are close to each other. Classical ECG does not take advantage of the bipolar configuration and, in essence, the signals it provides are very similar to the unipolar derivations. The main difference is the angle where the cardiac vector is projected.

There is no clear criterion whether the unipolar or bipolar method is the most appropriate to study the cardiac electrophysiology. There are many publications with no strong conclusions about it, and there are still defenders and detractors on both sides.

The most controversial application is cardiac mapping. By convention, an electrogram is the electrical signal picked up at the heart surface. During a cardiac mapping procedure, many (easily more than 200) electrograms are acquired simultaneously. Each electrogram is processed to determine the activation time of the tissue located below each electrode. The activation time of all electrograms is combined to provide information about the whole electric activity of the region being examined. Thus, researchers mostly focused on detecting the activation time.

Simultaneous acquisition of unipolar and bipolar electrograms has already been performed and analyzed. The main conclusion is that activation could not be unambiguously detected by any of the two types of electrograms. The results demonstrate that complementary information regarding local activations and diastolic potentials can be derived from unipolar and bipolar recordings and suggest that both electrode configurations should be used in multichannel cardiac mapping systems.[56] Similar results are obtained in other procedures. The maneuver can be achieved more efficiently and with less RF energy when it is guided by both unipolar and bipolar recordings than by bipolar recordings alone.[57]

Unipolar and bipolar arrays are early compared for cardiac activity sensing in the first pacemakers. DeCaprio et al.[57] reported an increased SNR of ventricular depolarization, by using bipolar instead of unipolar. However, the price of low noise was a clinical sensing failure (2% of the cases), which was unable to trigger the pacemaker.

Most modern pacemakers use the stimulation electrode to also sense the auricular activity. Thus, the electrode placement and lead design are being tested simultaneously to obtain the best achievable signal. The bipolar sensing signal's quality has been fairly improved reducing far-field R waves by means of changing the lead type and interelectrode distance, without deterioration of the P wave.[58-62]

20.8 Noise Sources

Measurements with electrodes have several sources of noise: the most outstanding are the galvanic skin response (GSR), the fluctuations of half-cell potentials, electrodes contamination, and movement artifacts. Another possible source of noise is related to the devices used to measure the bioelectric events, like leads and amplifiers connected to the electrodes. A single analysis will be presented including SNR for bioelectric events.

20.8.1 Galvanic Skin Response

The first item to analyze is a term incorporated in the early 1900s by many scientists, the GSR, also known as the electro-dermal response, or the sympathetic skin response. From a physical point of view, GSR is a change in the electrical properties of the skin in response to different kinds of stimuli (emotional and physical). Commonly, GSR appears when two electrodes are attached over the skin, mainly on the palm or fingertips. In this case, changes in the voltage measured from the surface of the skin are recorded.

20.8.2 Skin Resistance Response

There has been a long history of electro-dermal activity research; already in 1888, Charles Feré, a French neurologist, reported two phenomena: a decrease in skin resistance—called the skin resistance response—and a change in skin potential—called the skin potential response—both in response to sympathetic nervous stimulation (increased sympathetic activation).[63,64]

The most accepted explanation for these electro-dermal phenomena has been suggested to be the activation of the sweat glands due to the stimulus. In the clinical neurophysiological literature, most investigators accept the phenomenon without understanding exactly what it means, but knowing that GSR is highly sensitive to emotions in some people, GSR determinations are considered a simple and reproducible test of the function of the sympathetic system.[65-68]

At this time, the problem is to analyze whether this phenomenon is a source of noise in recording biopotentials using electrodes over the skin. During sympathetic stimulation, sweat glands become more active and sweat secretion rises thereby hydrating the SC (epidermis), reducing temporally the magnitude of R_p in Figure 20.10. The range of these changes is reported to be 5 to 25 kΩ at different electrode placements.[69]

On the other hand, the skin electrical potential E_s in Figure 20.10 has negative values on the order of 50 or 60 mV at rest. A transient change is observed during sympathetic activation, usually with biphasic or triphasic waves that vary substantially depending on the experimental conditions.[70]

Finally, beyond what has been said on electro-dermal phenomena and their extensive use in areas like neuro-psycho-physiological research and the clinic, and in the light of the present knowledge on the electrode–electrolyte–skin interface, both subjects are complete and univocally related. A new attempt is mandatory to improve and update the terminology between them.

20.8.3 Electrode–Skin Conductivity

Another parameter that affects the electrode–skin equivalent circuit is the type and integrity of skin and electrolyte characteristics.[71] About *electrode–skin conductivity*, most surface metallic electrodes require an abrasion on the skin, and an electrolyte–gel between them, as has been previously stated. Another source of noise is the *electrode potential fluctuations*. A half-cell potential similar to that generated on the EEI is generated when an electrode is placed over the skin.

Aronson and Geddes[72] demonstrated in 1985 that even a small contamination on a single metal electrode produces large fluctuating potentials. Controlled contaminants were placed on single metal electrodes—Cu, Zn, and Ag/AgCl—and recordings were made versus a clean electrode of the same metal. In all cases, the fluctuating potentials recorded with the *contaminated electrode* were much larger than those when no contaminant was present. They explained this phenomenon from the effect of mechanical disturbance of ions in the double layer when contaminants are added.

In the same way, any process that modifies the electrical double layer can cause lack of stability. So is the case with movements or shifts of a recording electrode over the subject/patient skin, even though contact jellies/pastes are used or when attempts have been made to measure potentials with electrodes on moving subjects.[73]

20.8.4 Motion Artifacts

Motion artifacts are the noise introduced to the bioelectric signal that results from the motion of the electrode with respect to the interface. More specifically, movement of the electrode or lead wire produces deformations of the skin around the electrode site. The skin deformations change the skin impedance and capacitance and modify the double-charge layer at the electrode. These changes are sensed by the electrode resulting in artifacts that are manifested as large-amplitude signals. The presence of motion artifact may result in misdiagnosis and may lead to delay of the procedures extending unsuitable treatments or decisions. Therefore, many studies have been performed to design systems in order to eliminate or considerably diminish this source of noise.

The first step is to improve the contact between the electrode and the skin using conductive jellies[71] and to cause an abrasion on the skin.[74] Another way is to locate, if possible, electrodes in body cavities that contain fluids/electrolytes (such as vagina, urethra, rectum, esophagus, mouth, nose, ear, and intracardiac). The second step is trying specially designed electrodes in order to reduce this artifact. Different electrode designs arise, some of which are as follows: *Recessed electrodes*: A metal electrode is inserted in a different material cup filled with jellies that holds the metal electrode to a short distance from the skin. In this case, the electrolyte–skin interface must be stable. Recessed electrodes have been demonstrated to have a remarkably high electrical stability.[75] *Adhesive electrodes*: They are similar to previous ones but, in addition, they include a pregelled or solid gel conducting adhesive material between the metallic electrode and the skin in order to reduce skin potentials due to skin stretch.[76] Alternatives to conventional wet electrodes are dry electrodes and capacitive electrodes, both of which have already been described in this chapter.

For many decades, several authors have worked to design special electrodes to diminish motion artifact in many applications, such as ECG[73,77] or EEG.[78] In other cases, some of them have studied and designed artifact detection strategies proposing intelligent systems for artifact suppression.[76,79] Electrode technology and movement detection devices are being used to monitor biological parameters in citizen medicine at home.[80] Smart clothes and textiles with noninvasive sensors/electrodes were designed for home healthcare or for illness prevention.[81,82] Considering that motion artifacts cause unbalance between electrode–skin impedance, major problems can arise in relation to the biological events recorders.[11]

20.8.5 Noise Related to Recording Devices

When two electrodes are placed on the subject skin in order to faithfully record biological signals, the impedance appearing between these electrodes will depend mainly on the electrode material and area, the EEI and the current through this interface. Bioelectrode materials define half-cell potentials that represent each electrode, while the bioelectrode area is inversely related to the impedance between the electrodes and the skin (the smaller the area, the greater the electrode impedance).

Figure 20.15 depicts the EEI model for two electrodes over the skin (similar to that of Figure 20.4b) and considers the recording amplifier (Biol Amp) with an input impedance R_{in}. The system behavior will show a typical impedance–frequency characteristic of Figure 20.8. At low frequencies, capacitive impedance is high and the system shows a resistive behavior (Figure 20.17) where

$$BEP = [(R_s + R_p) + (R_1 + Z_{w1}) + (R_2 + Z_{w2}) + R_{in}]I. \quad (20.1)$$

or

$$BEP = R_{low} I, \quad (20.2)$$

where I is the current derived from the biological tissue, $R_s + R_p$ is the component due to the subject, $(R_1 + Z_{w1})$ and $(R_2 + Z_{w2})$ are the components due to electrodes 1 and 2, respectively, R_{in} is the component due to the recording amplifier input impedance, and R_{low} is the low-frequency impedance.

Moreover, it should be recognized that a successful measurement of the biological signal implies accurately measuring the

biological voltage without drawing current from the tissue.[27] This condition is obtained when

$$R_{in} \gg [(R_s + R_p) + (R_1 + Z_{w1}) + (R_2 + Z_{w2})]. \quad (20.3)$$

Currently, amplifying devices present high input impedance, that is, $R_{in} = 10^{12}\ \Omega$ for INA121, an FET-input low-power IA, Burr-Brown; therefore, it is not a problem. We must consider that the electrode impedance grows inversely with the electrodes area forcing us to use extremely high impedance devices for very small electrodes. The last considerations must be strongly attempted for low-level signals.

On the other hand, if $R_{in} \approx 10[(R_s + R_p) + (R_1 + Z_{w1}) + (R_2 + Z_{w2})]$ or less, the equivalent circuit is a resistive divider causing loss of amplitude ("load effect"), which is unacceptable. Considering the model presented in Figure 20.15, at high frequency, Z_{w1} and Z_{w2} are small and the capacities ($C1$, $C2$, and C_p) behave like short circuit; then, the total impedance is equal to $R_s = R_{high}$ or the series subject resistance (R_{high} is the high-frequency impedance), which is less than at low frequency R_{low} (Figure 20.8). In middle frequencies, the behavior is complex.

20.9 Electrode–Skin Impedance Mismatch

As stated previously, a biopotential differential amplifier is used to record biological signals, using two electrodes over the skin, as shown in Figure 20.15. In order to analyze the effect of electrode-skin impedance mismatch, there are two elements to consider:

Electrode impedance differences: The electrode impedance should vary by differences in the effective contact area or by the use of electrodes of different materials or by the electrode location. Note that in the electrode impedance can be included the gel impedance.

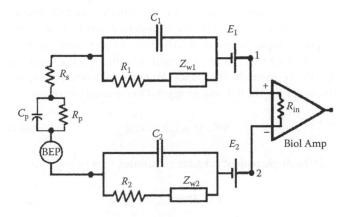

C = Double layer capacitor   1-Electrode 1
R = Charge transfer resistance   2-Electrode 2
Z_w = Diffusion impedance
E = Half-cell potential

FIGURE 20.15 Equivalent circuit of two electrodes applied on the subject skin.

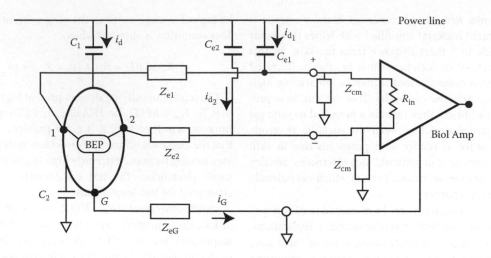

FIGURE 20.16 Esquematic representation of the network interference. A subject is connected via electrodes to a biological amplifier within a line power interference environment. See definitions in the text.[56,93]

Half-cell potential differences: Mainly caused by the use of different electrode materials or by the contamination of one of them as it has been mentioned previously.

Figure 20.16 schematizes a usual arrangement to amplify biopotentials via surface electrodes. A biological generator (BEP) is included in a volume conductor where electrodes are placed. A differential input amplifier is connected to those electrodes. The amplifier input impedance R_{in} and the common mode input impedances Z_{cm} are shown. Z_{e1} and Z_{e2} represent the electrode impedance from electrodes 1 and 2.

In addition, the biological element is capacitively coupled to the power line (220 V, 50 Hz) (C_1) and to the ground (C_2). Other possible capacitive coupling pathways are between the power line and the lead wires represented by C_{e2} and C_{e1}. Displacement currents will circulate through these capacitive elements; i_d flows via C_1 from the biological tissue to ground, probably through the return electrode G because it has the least impedance pathway. Therefore, i_{d1} and i_{d2} will circulate through the electrode impedances Z_{e1} and Z_{e2}, respectively, considering that there is a high amplifier input impedance. All these displacement currents ($i_{d1} + i_{d2} + i_d$) will return to ground also through the tissue via Z_{eG}, the G electrode impedance. Moreover, as a result of capacitive coupling, the net voltage applied to the amplifier results as

$$V^+ - V^- = i_{d1}Z_{e1} - i_{d2}Z_{e2}. \qquad (20.4)$$

If the displacement currents are similar, $i_{d1} = i_{d2} = i_d$

$$V^+ - V^- = i(Z_{e1} - Z_{e2}). \qquad (20.5)$$

Clearly, there is a differential input voltage due to the capacitive coupling and it is dependent on the electrode impedance mistmach ($Z_{e1} - Z_{e2}$).[11]

As stated previously, i_d flows via C_1, generating a voltage between the tissue and the ground, called "common mode

potential" V_{cm}, where

$$V_{cm} = i_d Z_{eG}. \qquad (20.6)$$

Then, the voltage at the amplifier inputs (points + and -) can be written as

$$V^+ = [Z_{cm}/(Z_{e1} + Z_{cm})]V_{cm}, \qquad (20.7)$$

$$V^- = [Z_{cm}/(Z_{e2} + Z_{cm})] \cdot V_{cm}. \qquad (20.8)$$

Finally, the total equivalent difference result,

$$V^+ - V^- = V_{cm}\{[Z_{cm}/(Z_{cm} + Z_{e1})] - [Z_{cm}/(Z_{cm} + Z_{e2})]\}. \qquad (20.9)$$

Since both electrode impedances are much smaller than Z_{cm}, Equations 20.8 and 20.9 reduce to

$$V^+ - V^- = V_{cm}[(Z_{e2} + Z_{e1})]/Z_{cm}]. \qquad (20.10)$$

Again, the electrode impedance imbalance ($Z_{e2} - Z_{e1}$) produces an output due to common mode interference.[83] Two methods are possible in order to reduce this interference level: to increase the amplifier common mode rejection and/or to diminish the electrode impedance mismatch using, for example, low-impedance electrodes. In conclusion, capacitive coupling and electrode impedance imbalance are able to induce interferences at biopotential recorders.

20.10 Electrochemical Electrodes

In recent years, research and development of electrochemical sensors for biomedical applications, environmental science, and industry has gained great importance. This development has focused primarily on finding new materials for the design of these electrodes. It is very important that the manufacture material must respond selectively, rapidly and reproducibly to changes of activity of the analyte ion. Although no electrode is absolutely specific in response, nowadays it is possible to find strong selective electrodes.

20.10.1 Electrode Classification

According to the material used for manufacturing them, electrodes can be classified into metal and membrane electrodes.

20.10.1.1 Metal Electrodes

There are four types of metal electrodes:

a. *First class electrodes:* These electrodes are sensitive to their cations. When the metal comes in contact with the dissolved cation, an electrochemical equilibrium with a given potential is generated in the EEI. This involves a single electrochemical reaction, for example, such as with copper:

$$Cu^{2+} + 2e^- \leftrightarrow Cu(s). \qquad (20.11)$$

These electrodes have low selectivity, poor stability, and low reproducibility and are limited by the pH of the solution and the presence of oxygen. Some materials (Fe, Cr, W, Co, and Ni) do not provide reproducible potential.[15] These electrodes are not used in potentiometric analysis, except for the following systems: Ag/Ag^+ and Hg/Hg^{2+} in neutral solution, and Cu/Cu^{2+}, Zn/Zn^{2+}, Cd/Cd^{2+}, Bi/Bi^{3+}, Tl/Tl^+, and Pb/Pb^{2+} in solutions without oxygen.[84]

b. *Second class electrodes:* These electrodes are able to sense the activity of an anion. A well-known example is silver (Ag), which recognizes the halide activity, such as Cl^- or pseudohalogens anions (compounds that resemble the halogen elements). The formation of a deposit of silver halide in the metal surface is required; this layer will be sensitive to the activity of the halides (Cl^-).[15] Another example is the use of mercury (Hg) electrodes to measure the activity of Y4-anion of ethylenediaminetetraacetic acid (EDTA).[15]

c. *Third class electrodes:* These metal electrodes determine the activity of a different cation. An example of these electrodes is the use of mercury to determine the activity of Ca^{2+}. In this case, a small amount of Hg (II) is introduced into the solution containing Ca^{2+} and EDTA. Since Ca^{2+} is in excess in the solution and the formation of metal–EDTA complex is reasonably large, the concentration of $[CaY^{2-}]$ and $[HgY^{2-}]$ will be approximately constant at equilibrium. The generated potential will depend only on the Ca^{2+} present in the solution. In this way, mercury electrode functions as a third class electrode for the Ca^{2+}. This electrode is important in the EDTA potentiometric titrations.[15]

d. *Metallic electrodes for redox systems:* This type of electrodes are constructed using inert metals such as Pt, Au, Pd, C, or other inert metals and are used to sense any oxidation–reduction system. These electrodes act as source or sink of electrons transferred from the redox system present in the solution.[15] It is possible to measure, for example, the potential of a platinum electrode immersed in a solution containing cerium(III) and cerium(IV). The platinum electrode is an indicator of the activity of Ce^{4+}.[15] But in the process of electron transfer, inert electrodes are not often reversible. Thus, these electrodes do not respond in a predictable way to many of the potential semireactions of known standards.[84] Nonreproducible potential is obtained with a platinum electrode immersed in a solution of thiosulfate ions and tetrationato, because the electron transfer process is slow and not reversible at the electrode surface.[15]

These inert metals are commonly used as substrates for the development of biosensors for the determination of redox reactions that occur as a result of the interaction between the enzyme and the analyte.[85] One example among many is the development of a biosensor for the detection of hemoglobin. This is based on a self-assembled monolayer electrode containing redox dye, and Nile Blue as an electron-transfer mediator, immobilized on a gold substrate.[86]

20.10.1.2 Membrane Electrodes

The membrane electrodes are known as specific electrodes, or ion-selective or pION electrodes. The potential generated in this type of electrode is known as binding potential. It is developed on the membrane, which separates the analyte dissolution from the reference dissolution.[87] These electrodes must be minimally soluble, so they are generally constructed from materials of large molecules or molecular aggregates such as silica glass or polymer resins. The low solubility of inorganic compounds, such as the case of silver halides, is also used for this purpose. The membrane must be slightly conductive. Finally, it must be selective to the analyte; in this case either the membrane or some species of the matrix of the membrane must be selective to the analyte.[15]

There are five types of membranes divided into two groups (see Figure 20.17). The literature reports many optimizations of membrane electrodes in order to improve response time, sensitivity, and reproducibility upon sensing the activity of the analyte. One of the most promising developments is the modification of polymeric membranes with nano-particles and particles of different nature, such as the use of particles of gold, platinum, zeolite, and so on. The latter is used for manufacturing ISFET selective to ammonia. These particles increase the conductivity of the electrodes to improve their response time.[88]

Nowadays, the use of metal oxides such as SbO_3, PbO, PtO_2, IrO_2, RuO_2, OsO_2, TaO_2, RhO_2, TiO_2, and SnO_2 to sense pH changes is widely prevalent.[89] IrO_2 is the most promising, since it is more stable, has fast response, high selectivity, and very low EEI impedance, properties that make them highly reactive.[89–92] The use of this oxide as a substrate for the immobilization of glucose oxidase in the development of biosensors is also known.[93] TiO_2, PbO, NiO are also reported in the literature for the development

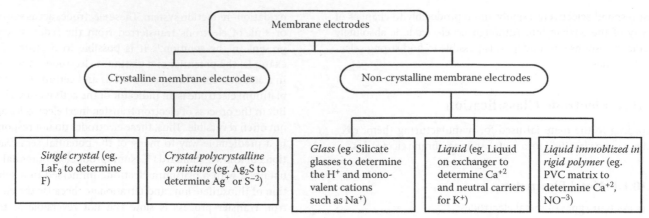

FIGURE 20.17 Classification of membrane electrodes.

of biosensors.[94] These materials are characterized by their high reactivity, which optimizes the response time compared with other materials.

The selection of materials and manufacturing techniques is critical for the development of sensors. It depends on functionality; consequently, future developments in sensor and biosensor design will focus on the technology of new materials.

20.11 Reference Electrodes

20.11.1 Standard Hydrogen Electrode

Many years ago, electrodes of hydrogen gas were used, not only as reference electrodes but also as pH electrodes.[15,27,28] The surface potential of platinum depends on the activity of the hydrogen ion in solution and on the partial pressure of hydrogen used to saturate the dissolution.[95] Figure 20.18 shows the standard hydrogen electrode (SHE).

In the SHE the reactive surface is a layer of platinum black. This layer is obtained through a rapid process of electroreduction of H_2PtCl_6.[28] The platinum black reflects light and provides a large surface area, ensuring a quick occurrence of the reaction on the surface of the electrode. This phenomenon is due to the decrease of the EEI impedance. The flow of hydrogen is used to maintain the EEI saturated with this gas.[15]

$$2H^+ + 2e^- \leftrightarrow H_2(g). \qquad (20.12)$$

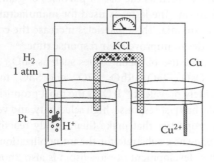

FIGURE 20.18 Standard hydrogen electrode.

The SHE is electrochemically reversible under certain conditions. Therefore, it can function as an anode or a cathode depending on the semicell with which it is connected. The hydrogen is oxidized to hydrogen ions when the electrode is an anode. The reverse reaction takes place when it is a cathode.[15,95] The SHE potential depends on the temperature, the activity of hydrogen ion in the dissolution (equal to unity), and the hydrogen pressure in the electrode surface (1 atm). These parameters should be defined in order to function as a reference electrode. By convention, the potential of this electrode is set to zero at all temperatures.[15,95]

20.11.2 Silver/Silver Chloride Electrode

For electrochemical purposes, the SHE is very important but is hardly practical. The preparation of the electrode surface and the control of reactants activities make it impractical for routine measurements. Instead, the Ag/AgCl reference electrode is simple to build, more robust and easier to use. This electrode can be manufactured by applying an oxidizing potential to a silver wire immersed in a solution of diluted hydrochloric acid. Consequently, a thin layer of AgCl adheres strongly to the wire. The wire is then introduced into a saturated solution of potassium chloride (KCl). A salt bridge connects the KCl dissolution and the analyte solutions.[15,27,28] In the current commercial design, the internal solution of many reference electrodes is an aqueous gel. Like the electrode inside the glass electrode, the potential of the external reference is controlled by the concentration of chloride in the filling solution. Because the chloride level is constant, the potential of the reference electrode is fixed. The potential does change if temperature changes. The half-cell potential with respect to SHE is +0.223 V and is determined by the following half-reaction:

$$AgCl(S) + e^- \leftrightarrow Ag(S) + Cl^-. \qquad (20.13)$$

Compared with the calomel reference electrode or SHE, these electrodes can be used at temperatures above 60°C; they are easy to manufacture and are practical. They can be implemented in microelectronic design. They are known as pseudo-reference electrodes integrated into amperometric microsensor designs and are built using standard photolithographic technology.[96]

20.11.3 Calomel Electrode

The calomel reference electrode, also named as the mercury/mercurous chloride electrode, is widely used in electrochemistry. It is very stable and simple to manufacture and use. The metal used is mercury, which has a high resistance to corrosion and is fluid at ambient temperature.[97] This electrode has two compartments (Figure 20.19). The internal tube contains the calomel paste composed of mercurous chloride (Hg_2Cl_2), liquid mercury (Hg), and saturated potassium chloride (KCl). A platinum wire is submerged in this paste making the electric contact. The second external compartment contains a solution of KCl with a concentration of 0.1 or 3.5 M or saturated. Both compartments are connected by a salt bridge. A second salt bridge makes the electric contact of the CE with the sample to be analyzed.[15]

The one that contains a saturated solution is known as the saturated calomel electrode (SCE), which is the most used since it is easy to construct. The disadvantages compared to those electrodes with known KCl concentration are its high temperature coefficient and the slow stabilization of the potential with changes in temperature. The SCE has a potential relative to the hydrogen electrode of +0.268 V.

The reduction reaction occurring at the CE corresponds to the reduction of mercury (I).

$$Hg_2Cl_2(s) + 2e^- \leftrightarrow 2Cl^- + 2Hg(l). \qquad (20.14)$$

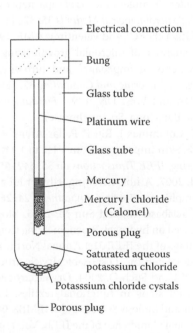

FIGURE 20.19 Saturated calomel reference electrode (SCE).

20.12 Specific Applications of Electrochemical Electrodes

20.12.1 Glass Electrode for pH Measurements

Since the early 1930s, pH was determined by measuring the potential difference produced in a glass membrane between the solution of unknown pH and a reference solution with constant acidity. Two decades later, the measurements were optimized with the invention of the vacuum tube. Nowadays, glass membrane electrodes for various ions such as Li^{2+}, Na^{1+}, K^{1+}, and Ca^{2+} can be obtained. Selectivity and sensitivity of the glass membrane electrode are based on the composition of the membranes (see Table 20.2). Corning glass 015 is used for pH measurements. This membrane is selective to hydrogen ions, and has a good response up to pH 11. At higher pH values, it is sensitive to Na^+ and other monovalent cations. Nowadays, formulations including barium and lithium instead of sodium and calcium are used. These changes in the formulation will respond selectively to the H^+ ions at higher pH values.[15]

The standard design of pH glass electrodes consists of two semicells (Figure 20.20): one is the glass electrode and the other is a reference Ag/AgCl or CE immersed in a solution of unknown pH. The glass electrode consists of a thin membrane of pH-sensitive glass located at the lower end of a tube of thick-walled glass or plastic.[15]

The inner reference electrode is an Ag/AgCl reference electrode containing a small amount of a saturated KCl solution.

TABLE 20.2 Compositions of some Glass Ion-Selective Membranes

Analyte	Composition (%)						
	SiO_2	Al_2O_3	Na_2O	Li_2O	CaO	Fe_2O_3	P_2O_5
pH	72		22		6		
Li^+	60	25		15			
Na^+	71	18	11				
K^+	69	4	27				
Ca^{2+}	3	6	6		<1	16	67

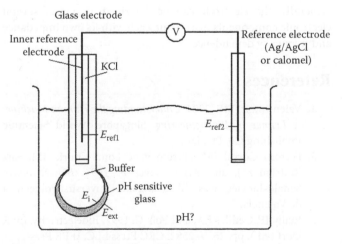

FIGURE 20.20 pH glass electrode.

This electrode is immersed in a buffer solution, usually of pH 1. The other side of the glass bulb is exposed to the solution of unknown pH. The internal reference is connected to one of the terminals of a device for measuring the potential. The external reference electrode is connected to the other terminals.[15,27,95]

In the pH electrode, four potentials are generated (Equation 20.15). Two of these, E_{ref1} and E_{ref2}, are the inner and the external reference electrodes potentials. The third potential E_i corresponds to the internal half-cell potential between the glass-sensitive membrane and the internal buffer. Finally, E_{ext}, the most important, is the unknown potential between the glass-sensitive membrane and the sample solution. This is not constant and is the only one that changes with the pH of the sample. All other potentials are constant. The liquid–liquid junction potentials are minimized by the use of the salt bridges. These potentials are negligible because the saturated solutions used contain ions of the same mobility, thus minimizing the formation of concentration gradients, and hence of half-cell potentials.[27]

$$V = E_{ref1} + E_{ref2} + E_i + E_{ext}. \qquad (20.15)$$

It is very important for the hydration of the surface of a glass membrane. The amount of water involved is about 50 mg per cubic centimeter of glass.[15] Nonhygroscopic glass does not respond to pH changes, so if the membrane dries, it loses its pH sensitivity. However, it is a reversible effect and recovers the response of the glass electrode after soaking in water.[15,87,95]

In the glass electrodes, the electrical conductivity within the hydrated gel layer is due to the movement of hydrogen ions. However, sodium ions are the charge carriers in the dry layer of the membranes.[15]

20.13 Summary

In electrical mapping there are many factors that influence the accuracy and signal strength. The choice of electrode material as well as the surface treatment play a role, and are interdependent. Additionally, the choice of electrode is important for the anticipated frequency range acquisition of the signal in question. Generally, the electrode can be described as one of several electronic components in a circuit, each with its own impedance and frequency dependence.

References

1. Valentinuzzi ME. 2004. *Understanding the Human Machine: A Primer of Bioengineering*. Singapore: World Scientific Publishing Co. Pte. Ltd.
2. Ferrero Corral JM, Ferrero y de Loma-Osorio JM, Saiz Rodríguez J, and Arnau Vives A. 1994. *Bioelectrónica. Señales bioeléctricas.* Valencia, Spain: Universitat Politécnica de Valencia.
3. Reilly JP, Geddes LA, and Polk C. 1997. Bioelectricity. In: R Dorf (ed.), pp. 2539–2582. CRC Press LLC, IEEE Press.
4. Zhang X, Ju H, and Wang J. 2008. *Electrochemical Sensors. Biosensors and their Biomedical Applications*. Amsterdam: Elsevier Inc.
5. Wang J. 2002. Glucose biosensors: 40 years of advances and challenges. *Electroanalysis* 13(12): 983–988.
6. Feldman Y, Polygalov E, Ermolina I, Polevaya Y, and Tsentsiper B. 2001. Electrode polarization correction in time domain dielectric spectroscopy. *Measurement Science and Technology* 12: 1355–1364.
7. Felice C, Madrid R, and Valentinuzzi M. 2005. Amplifier spurious input current components in electrode–electrolyte interface impedance measurements. *BioMedical Engineering OnLine* 4: 22.
8. Ruiz GA, Felice CJ, and Valentinuzzi ME. 2005. Non-linear response of electrode-electrolyte interface at high current density. *Chaos, Solitons & Fractals* 25: 649–654.
9. Silva ISS, Naviner JF, and Freire RCS. 2006. Compensation of mismatch electrodes impedances in biopotential measurement. Medical measurement and applications, 2006. MeMea 2006. *IEEE International Workshop on*, pp. 33–36.
10. Spinelli EM, Pallas-Areny R, and Mayosky MA. 2003. AC-coupled front-end for biopotential measurements. *Biomedical Engineering, IEEE Transactions on* 50: 391–395.
11. Spinelli EM, Mayosky MA, and Pallas-Areny R. 2006. A practical approach to electrode–skin impedance unbalance measurement. *Biomedical Engineering, IEEE Transactions on* 53: 1451–1453.
12. Valverde ER, Arini PD, Bertran GC, Biagetti MO, and Quinteiro RA. 2004. Effect of electrode impedance in improved buffer amplifier for bioelectric recordings. *Journal of Medical Engineering & Technology* 28: 217–222.
13. Felice CJ, Madrid RE, Olivera JM, Rotger VI, and Valentinuzzi ME. 1999. Impedance microbiology: Quantification of bacterial content in milk by means of capacitance growth curves. *Journal of Microbiological Methods* 35: 37–42.
14. Madrid RE, Felice CJ, and Valentinuzzi ME. 1999. Automatic on-line analyser of microbial growth using simultaneous measurements of impedance and turbidity. *Medical and Biological Engineering and Computing* 37: 789–793.
15. Skoog DA and West DN. 1997. *Fundamentos de química analítica*. Barcelona, Spain: Thomson.
16. Rosell J, Colominas J, Riu P, Pallas-Areny R, and Webster JG. 1988. Skin impedance from 1 Hz to 1 MHz. *Biomedical Engineering, IEEE Transactions on* 35: 649–651.
17. Cox JM. 2007. A non-irritating jelly for long-term electrocardiographic monitoring. *Anaesthesia* 24: 267–268.
18. Zhao J, Sclabassi RJ, and Sun M. 2005. Biopotential electrodes based on hydrogel. Bioengineering Conference, 2005. Proceedings of the IEEE 31st Annual Northeast, pp. 69–70.
19. Orbital Research Inc. Disposable dry electrodes. 2006.
20. Mühlsteff J and Such O. 2004. Dry electrodes for monitoring of vital signs in functional textiles. Engineering in Medicine and Biology Society, 2004. IEMBS '04. 26th Annual International Conference of the IEEE, Vol. 1, pp. 2212–2215.

21. Kingsley SA, Sriram SS, Boiarski AA, and Gantz N. High-impedance optical electrode. Srico, Inc. Columbus, OH, 09/898,402(6,871,084 B1). 2005.

22. Chen CH, Yao DJ, Tseng SH, et al. 2009. Micro-multi-probe electrode array to measure neural signals. *Biosensors and Bioelectronics* 24: 1911–1917.

23. Griss P, Enoksson P, Tolvanen-Laakso HK, et al. 2001. Micromachined electrodes for biopotential measurements. *Microelectromechanical Systems, Journal of* 10: 10–16.

24. Griss P, Tolvanen-Laakso HK, Merilainen P, and Stemme G. 2002. Characterization of micromachined spiked biopotential electrodes. *Biomedical Engineering, IEEE Transactions on* 49: 597–604.

25. Kordas N, Manoli Y, Mokwa W, and Rospert M. 1994. The properties of integrated micro-electrodes for CMOS-compatible medical sensors. Engineering in Medicine and Biology Society, 1994. Engineering Advances: New Opportunities for Biomedical Engineers. Proceedings of the 16th Annual International Conference of the IEEE, pp. 828–829.

26. Li-Chern P, Wei L, Fan-Gang T, and Chun L. 2002. Surface biopotential monitoring by needle type micro electrode array. Sensors, 2002. Proceedings of IEEE, Vol. 1, pp. 221–224.

27. Geddes LA and Baker LE. 1989. *Principles of Applied Biomedical Instrumentation*. New York: John Wiley.

28. Menolasina Monrreal S. 2004. *Fundamentos y aplicaciones de electroquímica*. Mérida: Universidad de Los Andes, Consejo de Publicaciones.

29. Villullas HM, Ticianelli EA, Macagno VA, and Gonzalez ER. 2000. *Electroquímica: Fundamentos y aplicaciones en un enfoque interdisciplinario*. Córdoba: Universidad de Córdoba.

30. Zanello O. 2003. Fundamentals of electrode reactions. In: *Inorganic Electrochemistry: Theory, Practice and Applications*. Cambridge, UK: Royal Society of Chemistry.

31. Grimnes S and Martinsen OG. 2000. *Bioimpedance and Bioelectricity Basics*. New York: Academic Press.

32. Warburg von E. 1899. Ueber das verhalten sogenannter unpolarisirbarer elektroden gegen wechselstorm. *Annalen der Physik und Chemie* 67: 493–499.

33. Liu SH. 1985. Fractal model for the ac response of a rough interface. *Phys. Rev. Lett.* 55: 529.

34. Mayorga Martinez CC, Madrid RE, and Felice CJ. 2008. Electrochemical and geometrical characterization of iridium oxide electrodes in stainless steel substrate. *Sensors and Actuators B: Chemical* 133: 682–686.

35. Meyer RD, Cogan SF, Nguyen TH, and Rauh RD. 2001. Electrodeposited iridium oxide for neural stimulation and recording electrodes. *Neural Systems and Rehabilitation Engineering, IEEE Transactions on* 9: 2–11.

36. Madrid RE and Felice CJ. 2005. Microbial biomass estimation. *Critical Reviews in Biotechnology* 25: 97–112.

37. van Boxtel A. 1977. Skin resistance during square-wave electrical pulses of 1 to 10 mA. *Medical and Biological Engineering and Computing* 15: 679–687.

38. Webster JG. 1984. Reducing motion artifacts and interference in biopotential recording. *Biomedical Engineering, IEEE Transactions on* BME-31: 823–826.

39. Reilly JP. 1992. *Electrical Stimulation and Electropathology*. Cambridge: Cambridge University Press.

40. Kaczmarek KA and Webster JG. 1989. Voltage–current characteristics of the electrotactile skin–electrode interface. Engineering in Medicine and Biology Society, 1989. Images of the Twenty-First Century, Proceedings of the Annual International Conference of the IEEE Engineering in, pp. 1526–1527.

41. Tam H and Webster JG. 1977. Minimizing electrode motion artifact by skin abrasion. *Biomedical Engineering, IEEE Transactions on* BME-24: 134–139.

42. Szeto AYJ and Saunders FA. 1982. Electrocutaneous stimulation for sensory communication in rehabilitation engineering. *Biomedical Engineering, IEEE Transactions on* BME-29: 300–308.

43. Szeto AYJ and Riso RR. 1990. Sensory feedback using electrical stimulation of the tactile sense. In: RV Smith and JHJ Leslie (eds), *Rehabilitation Engineering*, pp. 29–78. Boca Raton: CRC Press.

44. Searle A and Kirkup L. 2000. A direct comparison of wet, dry and insulating bioelectric recording electrodes. *Physiological Measurement* 21: 271–283.

45. Valentinuzzi ME. 1996. Bioelectrical impedance techniques in medicine. Part I: Bioimpedance measurement. In: J Bourne (ed.), *Critical Reviews in Biomedical Engineering*, Vol. 24(4–6). New York: Begell House, Inc.

46. Lawo T, Deneke T, Schrader J, et al. 2009. A Comparison of Chronaxies for Ventricular Fibrillation Induction, Defibrillation, and Cardiac Stimulation: Unexpected Findings and Their Implications. *Journal of Cardiovascular Electrophysiology*.

47. Luo YM, Moxham J, and Polkey MI. 2008. Diaphragm electromyography using an oesophageal catheter: Current concepts. *Clinical Science* 115: 233–244.

48. Beltran NE. 2006. Gastric impedance spectroscopy in elective cardiovascular surgery patients [abstract]. *Physiological Measurement* 27: 265.

49. Ives, John R. 2005. New chronic EEG electrode for critical/intensive care unit monitoring. *Journal of Clinical Neurophysiology* 22(2): 119–123.

50. Puurtinen M, Viik J, and Hyttinen J. 2009. Best electrode locations for a small bipolar ECG device: Signal strength analysis of clinical data. *Annals of Biomedical Engineering* 37: 331–336.

51. Vaisanen J, Puurtinen M, and Hyttinen J. 2008. Effect of Lead Orientation on Bipolar ECG Measurement. *Proceedings of the 14th Nordic-Baltic Conference on Biomedical Engineering and Medical Physics*, Vol. 20, pp. 343–346.

52. Farina D and Cescon C. 2001. Concentric-ring electrode systems for noninvasive detection of single motor unit activity. *Biomedical Engineering, IEEE Transactions on* 48: 1326–1334.

53. Rainoldi A, Melchiorri G, and Caruso I. 2004. A method for positioning electrodes during surface EMG recordings in lower limb muscles. *Journal of Neuroscience Methods* 134: 37–43.

54. Biermann M, Shenasa M, Borggrefe M, Hindricks G, Haverkamp W, and Breithardt G. 2003. The interpretation of cardiac electrograms. In: M Shenasa, M Borggrefe, and G Breithardt (eds), *Cardiac Mapping*, pp. 15–40. Blackwell Publishing.

55. Stevenson WG and Soejima K. 2005. Recording techniques for clinical electrophysiology. *Journal of Cardiovascular Electrophysiology* 16: 1017–1022.

56. Tada H, Oral H, Knight BP, et al. 2002. Randomized comparison of bipolar versus unipolar plus bipolar recordings during segmental ostial ablation of pulmonary veins. *Journal of Cardiovascular Electrophysiology* 13: 851–856.

57. DeCaprio V, Hurzeler P, and Furman S. 1977. A comparison of unipolar and bipolar electrograms for cardiac pacemaker sensing. *Circulation* 56: 750–755.

58. de Groot JR, Schroeder-Tanka JM, Visser J, Willems AR, and de Voogt W. 2008. Clinical results of far-field R-wave reduction with a short tip-ring electrode. *Pacing and Clinical Electrophysiology* 31: 1554–1559.

59. de Voogt W, Van Hemel N, Willems A, et al. 2005. Far-field R-wave reduction with a novel lead design: Experimental and human results. *Pacing and Clinical Electrophysiology* 28: 782–788.

60. Inama G, Massimo S, Padeletti L, et al. 2004. Far-field R wave oversensing in dual chamber pacemakers designed for atrial arrhythmia management. *Pacing and Clinical Electrophysiology* 27: 1221–1230.

61. Lewicka-Nowak E, Kutarski A, browska-Kugacka A et al. 2008. Atrial lead location at the Bachmann's bundle region results in a low incidence of far field R-wave sensing. *Europace* 10: 138–146.

62. Nash A, Fröhlig G, Taborsky M et al. 2005. Rejection of atrial sensing artifacts by a pacing lead with short tip-to-ring spacing. *Europace* 7: 67–72.

63. Feré C. 1888. Note sur les modifications de la resistance electrique sous l'influence des excitations sensorielles et des emotions. *Comptes Rendus Société de Biologie* 40: 217–218.

64. Tarchanoff J. 1890. Ueber die galvanischen Erscheinungen in der Haut des Menschen bei Reizungen der Sinnesorgane und bei verschiedenen Formen der psychischen Th+ñtigkeit. *Pfl++gers Archiv European Journal of Physiology* 46: 46–55.

65. Arunodaya GR and Taly AB. 1995. Sympathetic skin response: A decade later. *Journal of the Neurological Sciences* 129: 81–89.

66. Gutrecht JA. 1995. Sympathetic skin response. *Jouranl of Clinical Neurophysiology* 12: 396.

67. Reitz A, Schmid DM, Curt A, Knapp PA, and Schurch B. 2002. Sympathetic sudomotor skin activity in human after complete spinal cord injury [abstract]. *Autonomic Neuroscience*.102: 78–84.

68. Vetrugno R, Liguori R, Cortelli P, and Montagna P. 2003. Sympathetic skin response. *Clinical Autonomic Research* 13: 256–270.

69. Schwartz MS and Andrasik F. 2005. *Biofeedback. A Practitioner's Guide.* New York: Guilford Press.

70. Tarvainen MP, Karjalainen PA, Koistinen AS, and Valkonen-Korhonen MV. 2000. Principal component analysis of galvanic skin responses. Engineering in Medicine and Biology Society, 2000. Proceedings of the 22nd Annual International Conference of the IEEE, Vol. 4, pp. 3011–3014.

71. Tatarenko L. 1975. The problem of skin–electrode processes during medical electrography. *Biotelemetry* 2: 324–328.

72. Aronson S and Geddes LA. 1985. Electrode potential stability. *Biomedical Engineering, IEEE Transactions on* BME-32: 987–988.

73. Wiese SR, Anheier P, Connemara RD, et al. 2005. Electrocardiographic motion artifact versus electrode impedance. *Biomedical Engineering, IEEE Transactions on* 52: 136–139.

74. Burbank D and Webster J. 1978. Reducing skin potential motion artefact by skin abrasion. *Medical and Biological Engineering and Computing* 16: 31–38.

75. de Talhouet H and Webster JG. 1996. The origin of skin-stretch-caused motion artifacts under electrodes. *Physiological Measurement* 17: 81–93.

76. Ottenbacher J, Jatoba L, Grossmann U, Stork W, and Müller-Glaser K. 2007. ECG electrodes for a context-aware cardiac permanent monitoring system. *Proceedings of the World Congress on Medical Physics and Biomedical Engineering 2006*, Vol. 14, pp. 672–675.

77. Tong, DA. Electrode systems and methods for reducing motion artifact. Southwest Research Institute. 10/352,204 (6,912,414). 2009.

78. Taheri BA, Knight RT, and Smith RL. 1994. A dry electrode for EEG recording. *Encephalography and Clinical Neurophysiology* 90: 376–383.

79. Kearney K, Thomas C, and McAdams E. 2007. Quantification of motion artifact in ecg electrode design. Engineering in Medicine and Biology Society, 2007. EMBS 2007. 29th Annual International Conference of the IEEE, pp. 1533–1536.

80. Such O. 2007. Motion tolerance in wearable sensors—The challenge of motion artifact. Engineering in Medicine and Biology Society, 2007. EMBS 2007. 29th Annual International Conference of the IEEE, pp. 1542–1545.

81. Axisa F, Schmitt PM, Gehin C, et al. 2005. Flexible technologies and smart clothing for citizen medicine, home healthcare, and disease prevention. *Information Technology in Biomedicine., IEEE Transactions. on* 9: 325–336.

82. Dittmar A, Axisa F, Delhomme G, and Gehin C. 2004. New concepts and technologies in home care and ambulatory monitoring. *Studies in Health Technology and Informatics* 108: 9–35.

83. Degen T and Loeliger T. 2007. An improved method to continuously monitor the electrode–skin impedance during bioelectric measurements. Engineering in Medicine and

Biology Society, 2007. EMBS 2007. 29th Annual International Conference of the IEEE, pp. 6294–6297.

84. Biblioteca digital de la Universidad de Chile. Sistemas de servicios de información y bibliotecas SISIB. 2009.

85. Zhang S, Wright G, and Yang Y. 2000. Materials and techniques for electrochemical biosensor design and construction. *Biosensors and Bioelectronics* 15: 273–282.

86. Kuramitz H, Sugawara K, Kawasaki M, et al. 1999. Electrocatalytic reduction of hemoglobin at a self-assembled monolayer electrode containing redox dye, Nile Blue as an electron-transfer mediator. *Analytical Sciences* 15: 589–592.

87. Alegret S., del Valle M., and Merkoci A. 2004. *Sensores Electroquímicos*. Barcelona, Spain: Servei de Publicaciones.

88. Bakker E. 2004. Electrochemical sensors. *Analytical Chemistry* 76: 3285–3298.

89. Yao S, Wang M, and Madou M. 2001. A pH electrode based on melt-oxidized iridium oxide. *Journal of the Electrochemical Society* 148: H29–H36.

90. Ges IA, Ivanov BL, Schaffer DK, et al. 2005. Thin-film IrOx pH microelectrode for microfluidic-based microsystems. *Biosensors and Bioelectronics* 21: 248–256.

91. Marzouk SAM, Ufer S, and Buck RP, et al. 1998. Electrodeposited iridium oxide pH electrode for measurement of extracellular myocardial acidosis during acute ischemia. *Analytical Chemistry* 70: 5054–5061.

92. Mayorga Martinez CC, Madrid RE, and Felice CJ. 2009. A pH sensor based on a stainless steel electrode electrodeposited with iridium oxide. *Education, IEEE Transactions on* 52: 133–136.

93. bu Irhayem E, Elzanowska H, Jhas AS, Skrzynecka B, and Birss V. 2002. Glucose detection based on electrochemically formed Ir oxide films. *Journal of Electroanalytical Chemistry* 538–539: 153–164.

94. Salimi A, Sharifi E, Noorbakhsh A, and Soltanian S. 2007. Immobilization of glucose oxidase on electrodeposited nickel oxide nanoparticles: Direct electron transfer and electrocatalytic activity. *Biosensors and Bioelectronics* 22: 3146–3153.

95. Togawa T, Tamura T, and Ake Öberg P. 1997. *Biomedical Transducers and Instruments*. New York: CRC Press.

96. de Almeida FL, Burdallo I, Andrade Fontes MB, and Jiménez Jonquera C. 2008. Desarrollo de un pseudo-electrodo de referencia de Ag/AgCl integrado en un microsensor de silicio. Sixth Ibero-American Congress on Sensors IBERSENSOR 2008.

97. Boyes W. 2002. *Instrumentation Reference Book*. Burlington, UK: Elsevier Butterworth-Heinemann.

98. Webster JG. 1998. *Medical Instrumentation: Application and Design*. New York: John Wiley & Sons Inc.

Recording of Bioelectrical Signals: Theory and Practice

Juha Voipio

21.1 Background

Bioelectricity has been studied since the days of Luigi Galvani (1737–1798) and Alessandro Volta (1745–1827).[1] Before the development of modern electronics, the sensitivity and limited speed of electromechanical recording devices hindered accurate monitoring of bioelectrical signals. However, the basis for some of the nowadays widely used applications were laid over a century ago, when the first human electrocardiogram was published,[2] and when spontaneous electrical activity of the brain was reported in some animal species[3] and in humans.[4] Hermann von Helmholtz measured the conduction velocity of a nerve impulse already in the nineteenth century,[1] and the use of glass micropipettes as intracellular microelectrodes was demonstrated in 1946.[5] A major breakthrough in research at the cellular level was the development of the voltage clamp technique.[6,7] This method allowed quantitative studies on time and voltage-dependent ionic currents and conductances of excitable cell membranes.[8] More recently, the patch clamp method[9,10] provided a means for real-time monitoring of currents carried by single protein channels on cell membranes. Today, genuine bioelectrical signals can be faithfully acquired using different kinds of electrodes, which provide the interface between recording devices and the live object. In what follows, the discussion will start with bioelectrical signals, how they arise at the cellular level, how larger scale signals are generated, what the properties of signals are, and how they can be measured. Finally, this chapter will review in more detail some technical aspects of

recording and the properties of some electrode materials commonly used in commercial electrodes.

21.2 Origin of Bioelectrical Signals

A driving force in a closed circuit makes current flow. If we consider bioelectrical signals, most but strictly speaking not all of them, are consequences of ionic currents flowing through specific pathways of excitable cell membranes. Other mechanisms play a minor role, and include, for example, changes in the three-dimensional structure of charged protein molecules or mechanical movements at tissue level, but such mechanisms will not be discussed here.

The functional barrier that separates a cell's cytoplasm from the extracellular fluid is called the plasma membrane. Its basic structure is a self-organizing phospholipid bilayer that forms a few nanometers thick nonpolar layer, thereby preventing the passage of hydrophilic ions and creating a high specific capacitance of about $1 \mu F/cm^2$ across the membrane. Membrane-bound proteins mediate a large number of functions related to transport of substances, catalysis of reactions, detection of signals, and transfer of information across and along the membrane. With regard to bioelectrical signals, two classes of membrane proteins are of crucial importance: ion transporters that generate and maintain electrochemical gradients of ions across membranes, and ion channels that mediate transmembrane currents by providing conductive pathways for ions[8,11] (Figure 21.1).

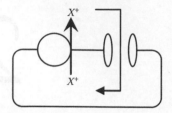

FIGURE 21.1 A schematic illustration of a cell with an active transport mechanism that extrudes ion X^+ from the cell and thereby generates a nonequilibrium distribution that tends to drive the ion back to the cell through conductive pathways (ion channels).

Different types of ion channels mediate currents that do seldom reverse at zero membrane potential (the membrane potential of a cell, V_m, is defined as the intracellular potential when extracellular potential is taken as zero). This is because channel-mediated net ion fluxes are driven not only by the membrane potential but also by the concentration gradient of the permeating ion species (see the appendix for the equilibrium condition). It is important to note that this has (at least) two consequences. First, a cell membrane can generate inward and outward currents depending on what ion channel populations are active at different times. Second, if the spatial distribution and/or activation level of ion channel types differs at different locations along the cell, current will flow in a loop and generate intracellular and extracellular voltage gradients. In general, intracellular membrane potential changes are shaped by the resistance (R) and capacitance (C) of the plasma membrane that are in parallel (Figure 21.2), whereas the extracellular space is a resistive volume conductor in which the evoked voltage gradients are directly proportional to current. While there are several mechanisms that account for nonlinearities,[8] Ohm's law (modified with an offset to take into account the reversal of current at nonzero membrane potential; see the appendix) is often used to approximate channel-mediated currents. For instance, in excitable animal cells such as neurons or muscle cells, open sodium (Na^+) and potassium (K^+) channels mediate inward and outward currents, respectively. On the other hand, neurotransmitter activation of anion channels that mediate synaptic inhibition in the brain may result in either an inward or an outward current at a

cell's resting membrane potential if the cell is actively maintaining a high or a low intracellular chloride concentration.[11]

21.2.1 Action Potentials and Associated Extracellular Signals

Regenerative spikes or action potentials are generated by many kinds of excitable cells. An action potential is triggered when depolarization (a positive shift in the cell's membrane potential) causes Na^+ (in some cases Ca^{2+}) channels to open to such an extent that the resulting inward current gives rise to further depolarization and opening of more Na^+ channels. Repolarization follows because of Na^+ channel inactivation and opening of K^+ channels, which mediate an outward current. Such voltage-sensitive gating of ion channels is due to forces induced by changes in the high transmembrane electric field (~10 MV/m) on fixed charges of membrane proteins.[12] A local action potential in an elongated cell gives rise to a significant intracellular voltage gradient. A longitudinal current will flow to adjacent regions and cause there a local depolarization by charging the membrane capacitance. This process results in a current loop where local transmembrane currents consist of ionic and capacitive components, and it accounts for the propagation of action potentials in neurons and muscle fibers. As illustrated in Figure 21.3, an extracellular microelectrode placed in the vicinity of an axon observes a traveling action potential as a current source–sink–source sequence or a traveling double dipole (quadrupole), and consequently, as a rapid triphasic voltage deflection (also see Chapters 3 and 4). The consequences of individual current components underlying signals in elongated cellular structures may appear puzzling unless an illustrative equivalent circuit is drawn, as a local positive shift in membrane potential results from an outward capacitive current but from an inward ionic current. An excellent review and quantitative treatment of electric current flow in excitable cells is provided by Ref. [13].

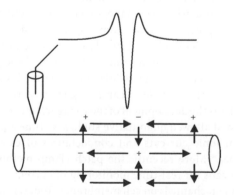

FIGURE 21.3 Schematic illustration of the signal generated in the extracellular space by an action potential that travels to the left along the axon. Action potential is marked by + within the axon, arrows indicate local current loops, and + and – signs indicate polarity of extracellular potential shifts. The trace shows a triphasic waveform that would be detected with an extracellular microelectrode in the close vicinity of the axon.

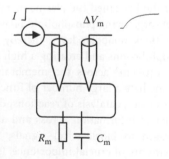

FIGURE 21.2 Passive electrical properties of a cell membrane consist of a parallel RC circuit. A step current I injected into the cell causes a shift $\Delta V_m = R_m I$ in membrane potential V_m that develops exponentially with a time constant $\tau_m = R_m C_m$.

Unit activity is a term that refers to extracellular spikes recorded with a microelectrode that is inserted in nervous tissue *in vivo* or *in vitro*. Unit activity or unit spikes reflect action potentials in individual neurons and they have amplitudes up to a few hundred microvolts. A few cells in the vicinity of the electrode may be distinguished on the basis of discrete amplitudes of their unit spikes. Unit spikes may appear in recordings as monophasic or biphasic spikes because of a number of reasons including insufficient signal-to-noise ratio, filtering-induced distortion, as well as location and distance of the electrode with respect to the cell's structure. Synchronous firing of closely packed neurons results locally in much larger amplitude extracellular spikes called population spikes (pop spikes), which can have amplitudes up to many millivolts in brain areas such as the hippocampus where cell bodies of pyramidal neurons are aligned together (for typical extracellular population responses observed within brain slices, see Reference [14]). Unit activity recording with microelectrodes or electrode arrays is used widely in basic research when studying activity in neuronal networks (for review, see Ref. [15]). Dedicated multichannel instruments, electrode arrays, and software tools for storage and analysis of unit activity are nowadays commercially available.

Action potentials in skeletal or cardiac muscles generate signals that can be recorded on the body surface (electromyogram and electrocardiogram, respectively). However, brain-borne signals on the scalp (electroencephalogram) mainly arise from postsynaptic currents, which will be discussed in what follows next.

21.2.2 Postsynaptic Currents and Associated Extracellular Signals

Chemical synaptic transmission is the main form of neuron-to-neuron communication. Synaptic activation of excitatory postsynaptic ion channels in dendrites (branched projections of a nerve cell that receive and integrate input signals) results in an inward current that spreads within the cell mainly towards its soma (cell body; Figure 21.4), leaks out via different kinds of ion channels including inhibitory postsynaptic channels (and to a much lesser extent as a capacitive current during the rising phase of the intracellular response) and returns along the extracellular space. This loop of volume current generates a potential gradient in the resistive extracellular space, that is, it appears as a dipole to an external observer. In current source-density analysis (CSD)[16,17] the spatiotemporal patterns of extracellular potentials are recorded within tissue using microelectrodes or microelectrode arrays and analyzed in order to identify and localize individual transmembrane current pathways that contribute to the signal. In fact, active maintenance of the electrochemical gradients of ions that provide the driving force of inwardly and outwardly directed postsynaptic currents plays a major role in the energy consumption of the brain,[18] which indirectly indicates the crucial role of postsynaptic currents in information processing in the brain.

The laminar organization of cortical neurons[19] favors the generation of extracellular signals that can be detected on the scalp. Loops of synaptic currents along dendrites organized in a laminar array generate an open field source that can be detected as a dipole even from a distance, whereas a closed field source would be hard to detect with a distant electrode (Figure 21.5). Therefore, synaptic currents via dendrites with a perpendicular orientation with the brain surface[19] form the main source of brain-borne signals that can be detected on the scalp.[20] Another example of an open field dipole source is the heart, which can be modeled as a single dipole (heart dipole or heart vector) that goes through periodic changes in its magnitude and three-dimensional direction (vectorcardiography[21]).

21.2.3 Electroencephalography

Electroencephalography (EEG) refers to the measurement of the brain's electrical activity using electrodes on the scalp. Since the amplitudes of EEG signals are in the microvolt range, EEG is carried out as a differential measurement in order to eliminate mains frequency interference and other common mode noise sources. It is customary to use many recording electrodes against one common reference electrode positioned, for example, at the left mastoid (behind the ear lobe), while the ground of the floating preamplifier circuitry can be coupled to any site in the body

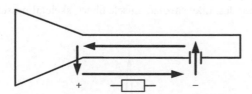

FIGURE 21.4 Generation of extracellular potentials by synaptic currents. Opening of excitatory cation channels by a neurotransmitter in the dendrite of a neuron gives rise to an inward current that spreads mainly toward soma (soma and dendrite are schematically shown with a triangle and a bar, respectively) and returns via the extracellular space generating in its distributed resistance a potential deflection. The extracellular potential will shift to negative and positive directions in the vicinity of the current sink and source, respectively.

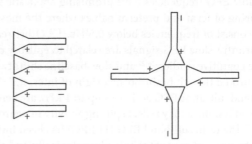

FIGURE 21.5 A schematic drawing of the orientation of a group of dipole sources in an open field source (left) and a closed field source (right). For details, see text.

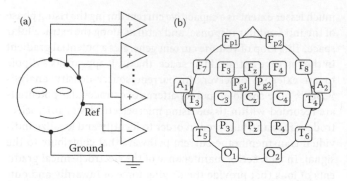

FIGURE 21.6 (a) Coupling of a floating multichannel differential amplifier with a common reference to record EEG from the scalp. (b) Electrode locations in the international 10–20 system for EEG.

(Figure 21.6a). Off-line re-referencing with software tools allows representation of EEG signals between any electrode pairs. A defined selection of pairs of electrodes for displaying EEG data is called a montage. For instance, signals between adjacent electrodes are displayed in bipolar montage.[22]

EEG electrodes have standard locations on scalp. The widely used international 10–20 system defines positions of 21 electrodes[22] (Figure 21.6b) and is sufficient for many clinical applications. A much higher number of electrodes (e.g., 128 or 256) is used for high-density EEG. Electrodes are often integrated to flexible head caps, which assure quick and correct positioning of the electrodes.

A typical human EEG signal from a healthy adult subject is up to 100 μV in amplitude and its clinically most relevant frequency content is within 0.1–100 Hz.[23] Therefore, the EEG bandwidth is usually limited using band-pass filtering with high-pass and low-pass cutoff frequencies at 0.1–1 and 35–100 Hz, respectively. Within these limits, characteristic EEG rhythms are classified using Greek letters ($\delta < 3.5$ Hz, θ 4–7.5 Hz, α 8–13 Hz, β 14–30 Hz, and $\gamma > 30$ Hz). It is worth noting that the commonly used narrow EEG bandwidth was fixed many decades ago. AC-coupled amplifier inputs solved technical problems associated with DC-unstable electrodes and drifting baselines, and low-pass filtering reduced noise. More recently, the recording bandwidth has been extended down to 0 Hz, which requires DC-stable electrodes and DC-coupled amplifiers with a wide dynamic range to avoid saturation by electrode offset potentials. Infra-slow EEG frequencies have promising applications in the monitoring of term and preterm babies where the most salient events consist of frequencies below 0.5 Hz.[24,25] Other reasons to monitor infra-slow EEG signals are related to epilepsy diagnostics and cognitive studies,[26,27] and slow baseline shifts can have a non-neuronal though brain-borne origin of interest.[24,28] On the other hand, ultrafast EEG activities up to 1 kHz are being studied nowadays, since they reflect spiking activity in neuronal networks.[29] The term full-band EEG (FbEEG) has been introduced to denote measurement of the full, physiologically, and clinically relevant EEG bandwidth.[24,30]

The electrical activity of the human brain can be recorded not only noninvasively but also invasively. A typical application

where invasive recording can be necessary is preparative investigation prior to epilepsy surgery. Depth electroencephalography is performed using metal needle electrodes inserted deep into brain parenchyma.[31] Electrocorticography (ECoG) measures brain activity from the surface of the brain.[32] ECoG can be performed during operation using, for example, ball-shaped gold electrodes brought gently in contact with the exposed brain surface. Subdural grid and strip electrodes usually have stainless steel or platinum–iridium contacts with 5–20 mm distances and embedded in a flexible Silastic plate.[33] Invasive techniques provide a better spatial resolution and record much larger amplitudes, but are of course demanding and used only when necessary.

21.2.4 Electrocardiography

From a technical point of view, electrocardiography (ECG), that is, measurement of the heart's electrical activity with electrodes on the skin, is less demanding than EEG. This is because electrodes can be mounted on hairless skin and the amplitude of the signal is tens of times higher, while the frequency range is comparable to that in EEG. A typical ECG signal has a peak-to-peak amplitude of up to 5 mV and its recording bandwidth can be limited with band-pass filtering to 0.5–40 Hz in monitoring applications and to 0.05 to >100 Hz in diagnostic applications. Like in EEG, electrodes are mounted to pre-defined positions, such as in the standard 12 lead ECG. Detailed descriptions of electrode placements and interpretation of normal and pathophysiological ECG signals can be found in any standard textbook on electrocardiography.

Each heart beat generates a characteristic sequence of waves which are named with the letters P, Q, R, S, and T (Figure 21.7). The rapid high-amplitude triphasic wave is called the QRS complex and it corresponds to the rapidly spreading depolarization of the ventricles. The QRS complex consists of frequencies that go up to and above the mains frequency. Therefore, one should be aware of possible distortion of the QRS waveform if low-pass or notch filtering is used in order to remove noise at mains frequency.

21.2.5 Electromyography

Electromyography (EMG) is the measurement of electrical activity in muscles. Like cardiac muscle fibers, skeletal muscle fibers

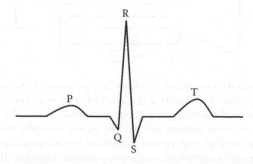

FIGURE 21.7 Naming the sequence of waves in a typical ECG signal with letters P–T.

fire action potentials. The rate of action potentials in muscle fibers depends on activity in the motor neurons that innervate them. One motor neuron innervates several muscle fibers, and this ensemble forms what is called a motor unit. The contraction force of a muscle is controlled by the number of activated motor units and their firing rate, which can go up to several tens of Hz depending on the muscle. It is worth pointing out that the frequency content of an EMG signal depends on the recording method and goes up to much higher frequencies than the maximum firing rate, because currents underlying muscle action potentials are very brief and have a wide frequency content.[34]

Electromyograms can be recorded using electrodes on the skin (surface EMG) or with invasive needle electrodes (intramuscular EMG). From a technical point of view, intramuscular EMG is an easy method due to the much higher (mV level) signal amplitudes. Intramuscular EMG also has much better spatial resolution than surface EMG and is therefore well suited to detection of motor unit activity.

21.2.6 Bioelectrical Signals Generated by the Eye

The eyeball is a dipole. In addition, the retina at the back of the eye contains photoreceptor cells in a neuronal network. Thus, bioelectrical signals generated by the eye can be classified into two main types. Electro-oculography (EOG)[35] monitors changes in the amplitude and orientation of the dipole of the eye, whereas electroretinography (ERG) is used to detect signals that mainly reflect activity in retinal cells.

The eye is a dipole with the front of the eye being electrically positive with respect to the back. The standing potential across the eyeball arises mainly from the pigment epithelium, which is a layer of epithelial cells coupled together with tight junctions (preventing passive leak and conductance via the paracellular route). It forms the nonreflecting black inner surface of the transparent eyeball. Since the basolateral and apical membranes (the two faces) of the epithelial cells have different membrane potentials, a transepithelial potential difference prevails across the epithelium.

EOG is used widely to record eye movements with electrodes mounted horizontally or vertically adjacent to the eye. Eye movements are also a common artifact in EEG recordings. The amplitude of the standing potential difference across the eye (the corneo-fundal potential) shows light-dependent changes in a time scale of minutes, providing a means to detect pathophysiological changes in the pigment epithelium in clinical applications. It is evident that slow EOG signals are prone to signal distortion unless the measurement is performed using DC-coupled equipment and DC-stable electrodes (see below).

Electroretinograms are recorded to test retinal function. Various types of electrodes including contact lens electrodes and moist cotton wick electrodes can be used. The choice of the electrode type does have an effect on the signal, especially on its amplitude. For positioning of the reference electrode and other standard procedures, the reader is advised to see Ref. [36]. ERG

is recorded usually in AC-coupled mode (i.e., the electrodes do not need to be DC stable) with a bandwidth of 0.3–300 Hz or wider. The ERG response of a dark adapted eye to a flash of light has a characteristic waveform with two very prominent phases called the *a*-wave and the *b*-wave. Since the cellular origins of different components of ERG responses are known, ERG is used for diagnostic purposes. In addition to single flashes, other stimulation paradigms such as flicker and pattern stimulations are often used.

21.3 Electrodes for Recording Bioelectrical Signals

Electrodes for recording bioelectrical signals can be classified in several ways. There are electrodes for clinical applications and for basic research, invasive and noninvasive electrodes, microelectrodes and macroelectrodes, recording electrodes and stimulation electrodes, ion-sensitive electrodes, and so on. While the variety of electrode designs is ever increasing, understanding electrodes is largely based on only a few basic concepts.

Most bioelectrical signals are voltage signals, and thus most electrodes should faithfully couple the potential signal in the biological signal source to a metal conductor. In case the aim is to record only fluctuations in potential, recording can be performed in AC-coupled mode. However, if also the steady-state baseline potential (standing potential) is of interest, the electrode must be DC stable, or in other words, a reversible electrode is needed. Problems associated with inadequate DC stability of electrodes, resulting in drifting baselines and amplifier saturation during past decades, seem to account for the prevalence of recording in AC-coupled mode in clinical practice. Therefore, electrodes that are good for conventional recording of EEG or ECG may fail completely when used to record in DC-coupled mode. As far as voltage recording is concerned, this issue is crucial compared to, for example, differences in noise level. Therefore, what follows will first focus on the operation of reversible and nonreversible electrodes.

21.3.1 Reversible and Nonreversible Electrodes

Modeling the electrode–electrolyte interface (EEI) is discussed in detail in Chapter 20 of this book. In brief, reversible electrodes maintain a constant potential difference across the EEI even when a continuous current is passed through. Irreversible electrodes often show an unstable potential difference across the EEI, and even minor current amplitudes cause large changes in potential. Although this classification uses two extreme and idealized categories, many electrodes seem to behave in real recording situations in either way. This is largely because of the very limited bandwidths used in biomedical recordings. Therefore, albeit an oversimplification, the EEI in polarizable electrodes can in most cases be modeled as a capacitor, whereas the EEI in a reversible electrode is best described by an electromotive force in series with a resistor, that is, as a simple battery (Figure 21.8).

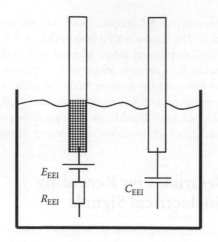

FIGURE 21.8 Simplified model of a reversible Ag/AgCl (left) and a nonreversible metal (right) electrode.

Metal electrodes made of stainless steel, tungsten, gold, platinum and so on. are typical polarizable electrodes. In practice, the silver–silver chloride (Ag/AgCl) electrode is the only reversible electrode used in biomedical applications. There are promising new electrode technologies based on conducting polymers or silicon arrays, but at least in my hands the DC stability of such novel electrodes has proven poor. When an Ag/AgCl electrode is immersed in a solution that contains chloride ions (Cl^-), a steady electrode potential $E_{Ag/AgCl}$ develops immediately. This is based on equilibration across the EEI of the reaction shown in Equation 21.1:

$$Ag(s) + Cl^- \rightleftarrows AgCl(s) + e^-, \qquad (21.1)$$

where Ag(s) and AgCl(s) are solid silver and silver chloride, respectively, and the negative charge is carried in the electrolyte by Cl^- and in the solid silver by e^-. This reaction proceeds in both directions very easily, giving the electrode the capacity to maintain a fixed electrode potential even at moderate current densities across the EEI. It is worth noting that the EEI in an Ag/AgCl electrode may be viewed as a hypothetical membrane that lets only chloride ions to pass through. Therefore, the Ag/AgCl electrode provides a potential signal that is a measure of the electrochemical potential of chloride ions in the electrolyte solution (the free energy of Cl^- in the fluid phase, see the appendix). In other words, a signal obtained from an Ag/AgCl electrode may be caused by a shift in the electrical potential of the fluid phase or by a change in the Cl^- concentration, or both. The chloride-sensitive $E_{Ag/AgCl}$ can be approximated at common temperatures of 25–40°C by Equation 21.2:

$$E_{Ag/AgCl} = 60 \text{ mV } \log_{10}[Cl^-] + E_0, \qquad (21.2)$$

where $[Cl^-]$ is the concentration of Cl^- and E_0 is constant. From this equation, it is evident that Ag/AgCl electrodes are optimal for measurement of bioelectrical signals only if the electrolyte they are exposed to has a constant Cl^- concentration. This

accounts for the use of chloride-containing pastes or gels between skin and Ag/AgCl electrodes in many clinical applications, as well as the use of Cl^--containing electrolyte solutions in glass electrodes, such as intracellular microelectrodes (see below) or reference barrels of standard laboratory pH electrodes. It should be noted that drying of electrode gel between skin and an electrode causes an increase in Cl^- concentration and will introduce drift in baseline in case the recording mode is DC-coupled.[37]

Ag/AgCl electrodes are available as sintered electrodes that have a block or pellet of solid AgCl giving the electrode a long lifetime (even up to years). On the other hand, an electrode can be made by immersing two pure silver wires in a Cl^--containing aqueous electrolyte solution (100 mM HCl or even NaCl) and passing a low DC current through for some tens of seconds (e.g., 0.1–1 mA for chloriding 1 cm of 0.25 mm wire).

If we consider the use of an Ag/AgCl electrode in combination with a DC-coupled amplifier that has a high input impedance, it is evident from the equivalent circuit shown in Figure 21.9a that changes in the potential of the fluid phase are coupled to the amplifier input without any distortion at any reasonable frequencies. On the other hand, the capacitance across the EEI in a polarizable electrode (and its large value due to the close vicinity of the two conducting phases) couples voltage signals appearing in the fluid phase effectively to the input of a high-input-impedance amplifier. Therefore, in biomedical applications where electrodes are used to record signals at a typical clinical bandwidth of, for example, 1–100 Hz, one does not necessarily notice any difference in the quality of the signal obtained with Ag/AgCl electrodes or with polarizable metal electrodes.

Electrodes are often mounted on skin. In addition to a series resistance, skin can generate (slow) signals under certain conditions. This issue has been discussed elsewhere in this book (Chapter 20 by Madrid et al.), and is therefore omitted here. However, skin will be considered as a simple series resistance in

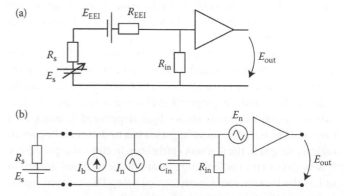

FIGURE 21.9 (a) Equivalent circuit of a signal source (E_s and R_s) coupled via an Ag/AgCl electrode to an ideal amplifier with a nearly infinite input resistance. The reference/ground electrode has been omitted from the equivalent circuit. (b) A model of an amplifier input with five nonideal features shown. The triangle represents an amplifier. A resistance R_s of the signal source shown on the left includes tissue resistances as well as the resistance of the electrode interface. For details, see text.

the discussion below, since its value is significant and can be even orders of magnitude higher than that of the EEI unless the skin is prepared prior to the recording.

21.3.2 Polarization and Noise in Electrodes Caused by Amplifier Input

The observed behavior of electrodes cannot be understood without considering the circuit that is formed by the signal source, the electrode, and the input of the amplifier. For this discussion, we may simply view the bioelectrical signal source as a voltage source in series with a resistance that represents tissue resistances (including skin resistance in the case of skin electrodes). A reversible electrode will be modeled as a resistor (the electrode potential is omitted here since it would only add a constant offset to the recorded signal) and an irreversible electrode as a capacitor. An ideal amplifier would reproduce the signal it sees at its input without any current flow through the input and without distorting the signal by any means. This is, however, not the case with real amplifiers.

Figure 21.9b illustrates a typical way of modeling five nonideal features that are present in real amplifiers. Four of these are in interaction with the biological signal source, whereas the fifth (the voltage noise) represents noise contaminating the measured signal at later stages within the amplifier. In brief, the triangle in the figure is an ideal amplifier, which together with the five components at the input represent a real amplifier. Real amplifiers show resistive leak from the input to ground (R_{in}), stray capacitances to ground are unavoidable (C_{in}), and inputs leak current in a manner that is best modeled as a current generator (bias current). The often used term input impedance is here modeled as a parallel RC circuit from the input to ground. The input resistance is seldom a problem in present-day amplifiers, but in case if it is not orders of magnitude higher than the total resistance of the signal source and the electrode, a voltage divider is formed and the signal amplitudes will be attenuated. The same mechanism may attenuate signals at higher frequencies in case the input capacitance is high enough to decrease input impedance. Bias currents via amplifier inputs are typically modeled with two current generators. The mean value of the bias current (I_b) is shown as the bias current generator in the figure, whereas the temporal fluctuations of the current are represented as the noise current (I_n) that has a zero mean value. These currents are important in at least two ways. First, the noise current will flow through the EEI, the electrode gel, and the biological tissues generating in all series resistances or impedances a voltage signal that is detected by the amplifier. If external noise sources are properly eliminated, the noise current of the amplifier is in many cases (e.g., in EEG) the source of noise that dominates. Amplifier stages other than the input add, of course, some noise to the signal, and these noise sources are lumped together in the figure as V_n. The second important consequence of the bias current is polarization of the electrodes. In case the electrode behaves as a pure capacitor, a steady current I_b will charge up the EEI until the local high electric field across the EEI will activate some forced mechanism(s) of charge transfer.

Values of I_b vary even between individual amplifiers (and operational or instrumentation amplifiers) of the same type. Therefore, the same polarizable metal electrode may show different kinds of polarization when tested with different amplifiers, or even with different channels of the same amplifier. In addition, slow components of the noise current (slow fluctuations of I_b) will generate fluctuations in the polarization of the electrode, which can be seen as low frequency noise even in AC-coupled recordings where the bandwidth is extended to frequencies much below 1 Hz.

In addition to the amplifier-dependent mechanism discussed above, electrode polarization in metal electrodes is affected by the affinity of the electrode surface to ions in the fluid phase and the composition of the fluid (see chapter by Madrid et al. in this book).

21.3.3 Commercial Skin Electrodes for Biomedical Recordings

Both reversible Ag/AgCl electrodes and nonreversible (polarizable) metal electrodes are commercially available in a large number of configurations. The simple and straightforward principles discussed above hold to all these electrodes, but general awareness of these issues is weak, especially among clinicians. To give some examples, precious metal electrodes (gold) as well as a combination of Ag/AgCl electrodes and a Cl⁻-free gel have been recommended for DC-coupled recordings from skin. An evaluation of a reasonable number of widely used commercial electrodes[37] clearly indicated that metal electrodes (stainless steel, gold, etc.) show drifting baselines with low-frequency noise and various degrees of polarization in DC recordings. On the other hand, Ag/AgCl electrodes had superior DC stability and low-frequency noise, but only if the electrode gel applied under the electrode contained a significant concentration of chloride. The initially surprising observation that gold-plated silver electrodes can upon time attain some of the features of Ag/AgCl electrodes was fully accounted for by the thinness of the gold plating that along time allowed small patches of silver to be exposed to the gel and to get spontaneously chlorided.[37]

Polarizable electrodes in combination with AC-coupled amplifiers with a bandwidth starting from much below 0.1 Hz have been used to record EEG signals lasting seconds or longer. While this approach may be successful, care should be taken to verify that the capacitance across the EEI in combination with the input of the amplifier does not produce a high-pass filter. In other words, extending the use of DC-noncompatible electrodes to near-DC recordings requires verification of the true low frequency response of the system. One way of doing this is to make a block of NaCl-agar and couple a square test signal to it from a function generator.[37] The electrode–amplifier combination can be tested easily by recording known signals from such a volume conductor.

21.3.4 Stimulation Electrodes

One application of metal electrodes is stimulation of excitable cells and tissue. To give only two examples of the wide variety of applications, stimulation can be done using very brief (≪1 ms) voltage or current pulses to stimulate bundles of axons in brain slice preparations in basic research, or in nerve conduction studies with invasive electrodes in clinical neurophysiology. The high specific capacitance of EEIs in metal electrodes is an advantage in stimulation, but with increasing current densities other mechanisms of charge transfer become evident in the form of bubbling or significant deterioration of the electrode. Compatibility with biological tissues and safety in invasive human applications are important questions.[38] Widely used materials include platinum–iridium and tungsten. The stimulation waveform can be optimized to reduce long-term net charge transfer and the consequent electrochemical decomposition of the stimulation electrode.[38]

21.3.5 Microelectrodes

Glass capillary microelectrodes are manufactured using glass tubing (usually of borosilicate glass) by pulling two parts of the tubing apart when the glass is softened by local heating.[39] Commercial micropipette pullers produce tips with various shapes and dimensions. Sharp micropipettes can have outer tip diameters below 100 nm, making visual inspection of the structure of the tip with a light microscope impossible. Depending on the application, tip diameters below and above 1 μm are common. Sharp microelectrodes are sharp micropipettes filled with an electrolyte solution, which is in contact with an Ag/AgCl electrode that can be a wire electrode inserted into the capillary or a sintered electrode in a pipette holder. Sharp microelectrodes have been used as a basic tool for recording voltages and passing currents in intracellular electrophysiology for decades. Since the small tip diameter limits (but does not fully prevent) solution leak through the pipette tip, a concentrated solution of KCl (e.g., 3 M) is often used as the filling electrolyte. The advantage of using KCl is that K^+ and Cl^- ions have almost equal mobilities, which is a prerequisite together with the high concentration when avoiding the generation of a liquid junction potential at the interface where the filling fluid and biological fluids meet.[10] For most applications, the passive electrical properties of a sharp microelectrode can be modeled as an RC low-pass filter. The resistance of sharp microelectrodes ranges from approximately 1 MΩ to hundreds of MΩ and it concentrates to the outermost tip. On the other hand, a capacitance (~pF) is formed across the glass wall of the pipette. If only the very tip of a microelectrode is inserted into a cell, this electrode capacitance is between the signal within the electrode and the signal ground level present in the extracellular fluid. The RC time constant of microelectrodes would suggest that intracellularly recorded fast action potentials are seriously distorted. However, microelectrode amplifiers are equipped with an adjustable positive feedback circuit (called capacitance neutralization or compensation, or negative capacitance), which solves the problem.

Patch pipettes[10,39] are glass capillary micropipettes with a blunt tip that can be gently pressed against a cell membrane. Sealing of the glass tip with a cell's membrane is facilitated by applying gentle suction. Patch electrodes are used to study current–voltage relationships across a tiny patch of membrane. In this way, currents via a single ion channel (membrane protein) can be monitored in real time from a living cell or from an excised membrane patch.[9,10] Another common application of patch electrodes is called recording in whole-cell mode. In this method, the pipette is filled with a solution that mimics the natural intracellular fluid (cytosol), and the membrane patch at the pipette tip is ruptured by a pulse of suction. This results in intracellular perfusion of the cell, which has advantages and disadvantages. From a technical point of view, whole cell recording provides access into a cell through a series resistance that is roughly an order of magnitude lower than with sharp microelectrodes (~10 MΩ versus ~100 MΩ) and thereby allows much easier implementation of techniques in which current is being passed into a cell while recording its membrane potential (voltage and current clamp techniques). Since patch pipettes are filled with a more dilute solution than sharp micropipettes, liquid junction potentials at the pipette tip can easily exceed 10 mV, and must therefore be taken into account as a source of error. However, liquid junction potentials can be easily estimated and cancelled from the results.[10]

21.3.6 Ion-Sensitive Microelectrodes

Glass micropipettes can be used as an electrode body with liquid membrane solutions to make ion-sensitive (micro)electrodes.[40,41] Their operation is based on sealing the tip region of the pipette with a nonpolar solvent (to prevent free ion movements) that contains a lipophilic ionophore with an ion-specific binding site on it. The ionophore is responsible for an ion species-specific permeability and the generation of a Nernst potential across the membrane. Ready-to-use liquid membrane solutions are commercially available for most inorganic ions present in biological systems (Na^+, K^+, H^+, Ca^{2+}, etc.) and also CO_2 microelectrodes can be made using liquid membranes.[42] Liquid membrane ion-sensitive microelectrodes have far better selectivity to the ion of interest than ion-sensitive glass membranes. An exception is pH (or H^+): pH glass (micro)electrodes have even better selectivity than their liquid membrane counterparts; on the other hand, pH microelectrodes with a glass membrane are tricky to manufacture and their response times are slow. Detailed instructions for making liquid membrane ion-sensitive microelectrodes have been published.[40] Ion-sensitive microelectrodes require special amplifiers,[40] since their resistances are typically from a few GΩ to above 100 GΩ.

Appendix

Electrochemical Potential, the Nernst Equation, and the Driving Force of Ions across Cell Membranes

The free energy change of ions moving across a membrane (in a cell or in an ion-sensitive electrode) from one concentration and electrical potential to another is quantified by the electrochemical potential. For an ion species j, the electrochemical potential $\tilde{\mu}_j$ is given by Equation 21.3:

$$\tilde{\mu}_j = \mu_0 + RT \ln a_j + z_j F \psi, \tag{21.3}$$

where a_j is the activity of the ion,[40] z_j is its charge, ψ is the electrical potential, μ_0 is constant (standard state electrochemical potential at unit activity and zero electrical potential), and R, T, and F have their usual meanings. The equilibrium distribution of an ion species across a membrane prevails when $\tilde{\mu}_j$ has no gradient. The usual way of presenting this condition is the Nernst equation, which can be easily derived by writing $\tilde{\mu}_j$ equal on two sides of the membrane ($\tilde{\mu}_{j,i} = \tilde{\mu}_{j,o}$ where subscripts i and o denote inside and outside of a cell) and solving the electrical potential difference E_j, as expressed by Equation 21.4:

$$E_j = \frac{RT}{z_j F} \ln \frac{a_{j,o}}{a_{j,i}}. \tag{21.4}$$

The chemical (concentration) and electrical components of the electrochemical gradient result in equilibrium when the membrane potential V_m of the cell is equal to E_j. A deviation of V_m from E_j tends to drive a conductive flux of ions across the membrane. Therefore, and in case the current–voltage relationship can be approximated using Ohm's law, the ionic current across the membrane is given by Equation 21.5:

$$I_j = G_j \left(V_m - E_j \right) = \frac{1}{R_j} \left(V_m - E_j \right), \tag{21.5}$$

where the term in parentheses is usually called the driving force of the current, and inward currents through the membrane have negative values.

References

1. Goldensohn ES. Animal electricity from Bologna to Boston. *Electroencephalogr. Clin. Neurophysiol.* 1998; 106: 94–100.
2. Waller AD. A demonstration on man of electromotive changes accompanying the heart's beat. *J. Physiol.* 1887; 8: 229–234.
3. Caton R. The electric currents of the brain. *Br. Med. J.* 1875; 2: 278.
4. Berger H. Über das Elektroenkephalogram des Menschen. *Arch. Psychiat. Nervenkr.* 1929; 87: 527–570.
5. Graham J and Gerard RW. Membrane potentials and excitation of impaled single muscle fibers. *J. Cell. Comp. Physiol.* 1946; 28: 99–117.
6. Hodgkin AL, Huxley AF, and Katz B. Ionic currents underlying activity in the giant axon of the squid. *Arch. Sci. Physiol.* 1949; 3: 129–150.
7. Hodgkin AL, Huxley AF, and Katz B. Measurements of current-voltage relations in the membrane of the giant axon of *Loligo*. *J. Physiol. (Lond.).* 1952; 116: 424–448.
8. Hille B. *Ion Channels of Excitable Membranes*, Third edition. Sunderland, MA: Sinauer Associates; 2001.
9. Neher E and Sakmann B. Single-channel currents recorded from membrane of denervated frog muscle fibers. *Nature* 1976; 260: 799–802.
10. Sakmann B and Neher E. *Single-Channel Recording*, Second edition. New York: Plenum Press; 1995.
11. Payne JA, Rivera C, Voipio J, and Kaila K. Cation-chloride co-transporters in neuronal communication, development and trauma. *Trends Neurosci.* 2003; 26: 199–206.
12. MacKinnon R. Potassium channels. *FEBS Lett.* 2003; 555: 62–65.
13. Jack JJB, Noble D, and Tsien RW. *Electric Current Flow in Excitable Cells.* Hong Kong: Oxford University Press; 1983.
14. Johnston D and Amaral DG. Hippocampus. In: GM Shepherd (Ed.), *The Synaptic Organization of the Brain*, Fifth edition, pp. 455–498. New York: Oxford University Press; 2004.
15. Buzsáki G. Large-scale recording of neuronal ensembles. *Nat. Neurosci.* 2004; 5: 446–451.
16. Nakagawa H and Matsumoto N. Current source density analysis of ON/OFF channels in the frog optic tectum. *Prog. Neurobiol.* 2000; 61: 1–44.
17. Johnston D and Wu SMS. *Foundations of Cellular Neurophysiology*, Fifth edition. Cambridge, MA: The MIT Press; 2001.
18. Buzsáki G, Kaila K, and Raichle M. Inhibition and brain work. *Neuron* 2007; 56: 771–783.
19. Douglas R, Markram H, and Martin K. Neocortex. In: GM Shepherd (Ed.), *The Synaptic Organization of the Brain*, Fifth edition, pp. 499–558. New York: Oxford University Press; 2004.
20. Spekmann EJ and Elger CE. Introduction to the neurophysiological basis of the EEG and DC potentials. In: E Niedermeyer and FL Da Silva (Eds), *Electroencephalography: Basic Principles, Clinical Applications and Related Fields*, Fifth edition, pp. 17–30. Philadelphia, PA: Lippincott Williams & Wilkins; 2005.
21. Plonsey R and Barr RC. *Bioelectricity: A Quantitative Approach*, Second edition. New York: Kluwer Academic/ Plenum Publishers; 2000.
22. Reilly ER. EEG recording and operation of the apparatus. In: E Niedermeyer and FL Da Silva (Eds), *Electroencephalography: Basic Principles, Clinical Applications and Related Fields*, Fifth edition, pp. 139–160. Philadelphia, PA: Lippincott Williams & Wilkins; 2005.
23. Niedermeyer E. The normal EEG of the waking adult. In: E Niedermeyer and FL Da Silva (Eds), *Electroencephalography: Basic Principles, Clinical Applications and Related Fields*, Fifth edition, pp. 167–192. Philadelphia, PA: Lippincott Williams & Wilkins; 2005.
24. Vanhatalo S, Voipio J, and Kaila K. Infraslow EEG activity. In: E Niedermeyer and FL Da Silva (Eds), *Electroencephalography: Basic Principles, Clinical Applications and*

Related Fields, Fifth edition, pp. 489–494. Philadelphia, PA: Lippincott Williams & Wilkins; 2005.

25. Vanhatalo S, Palva JM, Andersson S, Rivera C, Voipio J, and Kaila K. Slow endogenous activity transients and developmental expression of K^+–Cl^- cotransporter 2 in the immature human cortex. *Eur. J. Neurosci.* 2005; 22: 2799–2804.

26. Vanhatalo S, Holmes MD, Tallgren P, Voipio J, Kaila K, and Miller JW. Very slow EEG responses lateralize temporal lobe seizures: An evaluation of non-invasive DC-EEG. *Neurology* 2003; 60: 1098–1104.

27. Monto S, Palva S, Voipio J, and Palva JM. Very slow EEG fluctuations predict the dynamics of stimulus detection and oscillation amplitudes in humans. *J. Neurosci.* 2008; 28: 8268–8272.

28. Voipio J, Tallgren P, Heinonen E, Vanhatalo S, and Kaila K. Millivolt-scale DC shifts in the human scalp EEG: Evidence for a nonneuronal generator. *J. Neurophysiol.* 2003; 89: 2208–2214.

29. Curio G. Ultrafast EEG activities. In: E Niedermeyer and FL Da Silva (Eds), *Electroencephalography: Basic Principles, Clinical Applications and Related Fields*, Fifth edition, pp. 495–504. Philadelphia, PA: Lippincott Williams & Wilkins; 2005.

30. Vanhatalo S, Voipio J, and Kaila K. Full-band EEG (fbEEG): A new standard for clinical electroencephalography. *Clin. EEG Neurosci.* 2005; 4: 311–317.

31. Niedermeyer E. Depth electroencephalography. In: E Niedermeyer and FL Da Silva (Eds), *Electroencephalography: Basic Principles, Clinical Applications and Related Fields*, Fifth edition, pp. 733–748. Philadelphia, PA: Lippincott Williams & Wilkins; 2005.

32. Quesney LF and Niedermeyer E. Electrocorticography. In: E Niedermeyer and FL Da Silva (Eds), *Electroencephalography: Basic Principles, Clinical Applications and Related Fields*, Fifth edition, pp. 769–776. Philadelphia, PA: Lippincott Williams & Wilkins; 2005.

33. Lesser RP and Arroyo S. Subdural electrodes. In: E Niedermeyer and FL Da Silva (Eds), *Electroencephalography: Basic Principles, Clinical Applications and Related Fields*, Fifth edition, pp. 777–790. Philadelphia, PA: Lippincott Williams & Wilkins; 2005.

34. Beck TW, Housh TJ, Johnson GO, Cramer JT, Weir JP, Coburn JW, and Malek MH. Does the frequency content of the surface mechanomyographic signal reflect motor unit firing rates? A brief review. *J. Electromyogr. Kinesiol.* 2007; 17: 1–13.

35. Brown M, Marmor M, Vaegan, Zrenner E, Brigell M, and Bach M. ISCEV standard for clinical electro-oculography (EOG) 2006. *Doc. Ophthalmol.* 2006; 113: 205–212.

36. Marmor MF, Fulton AB, Holder GE, Miyake Y, Brigell M, and Bach M. ISCEV standard for full-field clinical electroretinography (2008 update). *Doc. Ophthalmol.* 2009; 118: 69–77.

37. Tallgren P, Vanhatalo S, Kaila K, and Voipio J. Evaluation of commercially available electrodes and gels for recording of slow EEG potentials. *Clin. Neurophysiol.* 2005; 116: 799–806.

38. Geddes LA and Roeder R. Criteria for the selection for materials for implanted electrodes. *Ann. Biomed. Eng.* 2003; 31: 879–890.

39. Brown KT and Flaming DG. *Advanced Micropipette Techniques for Cell Physiology*, Ninth edition. Chichester: Wiley; 1992.

40. Voipio J, Pasternack M, and MacLeod K. Ion-sensitive microelectrodes. In: D Ogden (Ed.), *Microelectrode Techniques—The Plymouth Workshop Handbook*, Second edition, pp. 275–316. Cambridge: The Company of Biologists Limited; 1994.

41. Voipio J. Methods of studying pH: Ion-sensitive microelectrodes. In: K Kaila and B Ransom (Eds.), *pH and Brain Function*, pp. 95–108. New York: Wiley-Liss; 1998.

42. Voipio J and Kaila K. Interstitial Pco_2 and pH in rat hippocampal slices measured by means of a novel fast CO_2/H^+-sensitive microelectrode based on a PVC-gelled sensor. *Pflugers Arch.* 1993; 423: 193–201.

Diagnostic Physics

V

22

Medical Sensing and Imaging

Robert Splinter

22.1 Introduction

One of the fundamental aspects of medical diagnostics is acquiring the information to base a clinical decision on. The information is obtained from either a one-dimensional signal or a two-dimensional signal: an image. Signals can be chemical, electrical, acoustic, or electromagnetic in nature. This chapter will provide an introduction to the physics and engineering of various sensing and imaging concepts and devices and the limitations associated with the signal acquisition and signal processing.

Successful treatment of many diseases depends on diagnosis in the early stages of the disease. Both sensing and imaging can be performed by means of various mechanisms.[1-4] There are two basic methods for sampling internal tissue in order to determine the presence and progression of tissue degeneration. A biopsy can be performed, in which tissue is sampled at random and surgically removed for testing. Alternatively, an imaging technique, either invasive or noninvasive, can be used to obtain information on the tissue microstructure. If the images provide sufficient detail and can be obtained for a large enough volume of tissue, the imaging approach is advantageous because it does not require removal of tissue and does not introduce a significant time delay for the testing of biological samples.

There are several imaging techniques that can provide near real-time and three-dimensional imaging of biological samples. Some techniques used for imaging are electromagnetic (optical, x-ray, magnetic resonance imaging [MRI], thermography); other techniques are acoustic (ultrasound), chemical, and electrical.

Among the most prolific are MRI, x-ray computed tomography, and high-frequency ultrasound. Each finds applicability for specific imaging problems. However, none of these techniques are capable of achieving resolution at the cellular level, which requires spatial resolution of less than 10 μm. In order to diagnose certain diseases in their early stages, this level of resolution is required.

The techniques used to convey the obtained information in a useable image or signal will rely on signal processing, and conversion into a display. The use of biosensors can provide *in vivo* monitoring of specific biological activities such as blood–gas analysis, blood biochemistry composition: for example, blood–sugar monitoring next to electrical activity under normal operating conditions (e.g., halter monitor) or pacing feedback for cardiac pacing/defibrillator devices. One specific example of *in vivo* sensing is described in Chapter 41, illustrating the concept of the "lab-on-a-chip."

22.2 The Nature of a Signal

The definition of a signal is a succession of numbers.[1,3,5] The numbers usually follow a timed sequence and can be either real or complex. A signal can be analog (continuous in both amplitude and time), discrete (continuous in amplitude but split into individual time segments), or digital (both the time axis and the amplitude are separated in segmental steps: discrete values). Examples of analog signals are: temperature recorded by a mercury thermometer, speed of a car expressed by a dial, and airflow flow measured by a cup anemometer. Examples of discrete signals are: heart rate and blood flow measured by a pulsed laser Doppler flow sensor. Digital signals are primarily electronic conversions of analog and discrete signals; however, some biological signals are digital by nature, such as the action potential. However, on closer examination, the action potential does have an analog structure, as shown in Chapter 4.

A two-dimensional signal, such as an image, has gray-scale or color values that are place and time dependent; the optical image acquired by our eyes is in fact a digital signal with a sampling frequency on the order of 10 frames per second (acquisition rate of 10 Hz) that is spatially distributed over the retina of the eye.[6] The fact that the eye is using digital data acquisition becomes evident when observing a spoked wheel, which at certain speeds appears to revolve backward. The frame rate of the eye may capture a random spoke at a location at a negative angle with respect to a previous random spoke, making it seem that the wheel is revolving backward while the car is moving forward.

22.3 Physical and Physiological Principles of Sensing and Imaging

The measured signals are limited by various factors, such as noise, resolution, and device limitations.[3,4,7,8] The device used to acquire and process the biological or anatomical signal has a conversion mechanism that transforms the input signal into a signal we understand or can display. The system obtaining the input signal and displaying the output signal can be described by a series of processes.

In most cases it is assumed and validated within specific boundary conditions that the system operates in a linear mechanism. This specification allows the following mechanisms to be applied: superposition, magnification, convolution, impulse response, and time and spatial spread.[1,3,9,10]

In current imaging technology the majority of signals are digitized, requiring signal processing under specific guidelines such as sampling rate, bandwidth limitations in addition to filtering techniques relying on spatial and frequency transforms.

The final image generation may require reconstruction algorithms, projections, and compression. This mathematical process describing the elementary processes involved in data acquisition and signal processing that will create a reliable output with known distortion limits and controlled noise is generally referred to as "System Theory."[1-5]

In order for an imaging system to be reliable and accurate, the following two criteria need to be satisfied:

1. The system needs to be linear; the sum of two objects needs to equal the sum of these two objects imaged individually.
2. The system needs to be place invariant; the image cannot be a function of where the object was placed in the field of view of the detectors. This entails that no distortions are allowed.

22.3.1 System Theory

The formation of an image is the transformation of a spatial or temporal distribution O to a spatial or temporal distribution P. The distribution "O" is the object or source and the distribution "P" is the projection or image. The object of imaging is the transformation of the coordinate system of the object to the coordinate system of the image. In general the object has a three-dimensional coordinate system matching the three-dimensional structure of the item of interest. The image will initially only have either a one- or a two-dimensional coordinate system, since the imaging modality will need to reconstruct the three-dimensional structure from multiple one- or two-dimensional data arrays. The image will generally have at least one dimension less than the object.

The data array depends on the type of imaging modality used, and is technologically limited by the current state-of-the-art available detectors for the specific imaging signal in question. In the case of ultrasonic imaging, the object is the organ to be examined, the transfer function uses the amount of reflected pressure wave intensity and the image is the electrical signal generated by the piezoelectric detector upon return. The electrical signal is processed and displayed on a screen as the ultimate image. The ultimate image is a two-dimensional distribution of gray scales representing the acoustic densities in a projected two-dimensional plane of the source. The perceived two-dimensional distribution, however, may not accurately represent a true two-dimensional slice within the source. Distortions and signal processing will create the best interpretation of the signals generated and obtained by the imaging device.

Even our vision is not truly three-dimensional, since we cannot look inside an opaque object; however, we can detect the phase difference in path length traveled by light from a curved surface; this way our brain interprets the detection of the relative distances of the observed object to each other. This in fact also uses a reconstruction algorithm.

In imaging, the main concern is when is an image good enough, which in most cases translates to "as good as is attainable with the current technological means." In most cases the medical image may not be an exact physical likeness image but an image portraying the anatomical geometry or biological functionality of the object under investigation.

System theory describes the mechanism leading from the source to the projection. The stages of the image formation use data acquisition and signal processing algorithms that are described by transfer functions.

In case the input is an impulse or binary situation, the source will have an event that is either present or not present at one location in space or time, depending on the phenomenon. A spatial distribution can be associated with our vision, whereas a temporal distribution will be associated with our hearing for instance.

The impulse response of the signal acquisition system is represented as a time or point-spread function. Any object can be collected by acquiring an infinite number of spatial or temporal events by the imaging mechanism.

22.3.1.1 Time and Point-Spread Function

The image formation relies on a transfer mechanism that probes the object and relates this through the imaging means into a response.

If the input to the imaging system is an all-or-nothing point source (impulse) in the origin of the coordinate system, this is referred to as a Dirac pulse $\delta(0)$. The coordinate system in this case is time. The response by the imaging mechanism creates an impulse response, which has both a temporal and a spatial spread, $h(t)$ or $h(x', y')$, respectively. Any arbitrary input will not be an infinitesimally small source, but a collection of an infinitesimally large number of Dirac-pulse sources spread out over three-dimensional space. The reconstructed image is now a convolution of the input signal with the impulse response $f(t)$ in the time domain expressed as

$$g(t) = f(t) \otimes h(t), \qquad (22.1)$$

where $g(t)$ is the output of the device, or an intermediate output step within the signal processing algorithm. In case multiple steps of signal and image processing are involved, the output will be the result of all the individual steps.

22.3.1.2 Convolution

The process of combining various aspects of the image formation by having the operator applied to the source is performed mathematically by convolution. Convolution is similar to cross-correlation, only applied to the signal by means of detection. Convolution combines two phenomena to produce a third, that is, the base signal, and the perception function (the system performing the observation) provides the output signal. Convolution can be seen as a mathematical means of scanning over the base signal with a finite window (impulse response) to perform the output signal transform at each point in time or space; this is necessary due to spatial and temporal influences in the system providing the output. The concept of convolution is described in detail by signal processing textbooks, and the general principles will be discussed briefly. Convolution operation between two signals for instance at different time slots $g_1(t)$ and $g_2(t)$ is defined as

$$g_1(t) \otimes g_2(t) = \int_{-\infty}^{+\infty} g_1(t)g_2(t-\tau)\,d\tau = \int_{-\infty}^{+\infty} g_1(t-\tau)g_2(t)\,d\tau. \quad (22.2)$$

Calculation of convolution operations is a time-consuming process requiring many complicated mathematical computations. However, the mathematical process can generally be simplified by performing the Fourier transform from time to rate of occurrence, the frequency domain.

22.3.1.3 Fourier Transform

In order to proceed with the signal processing leading to the image formation, the Fourier principle is briefly described. The Fourier transform is a mathematical algorithm that converts a spatial or temporal sequence of values into a series of repetitious phenomena that can be analyzed mathematically with simpler arithmetic functions. The Fourier transform highlights the periodicity of a signal event rather than the time distribution. The Fourier transform thus becomes an arithmetic superposition of symmetric functions. On the other hand, a single occurrence of an event can be viewed as the superposition of an infinitesimal number of periodic events around one location or time frame. The mathematical analysis of a single event is rather complex; however, the personal interpretation by a physician can be rather swift. The personal interpretation as a matter of fact uses the brain to perform a Fourier transform in the deductive reasoning process as well.

Symmetric functions used in Fourier transform are for instance trigonometric functions (sin and cosine). The superposition principle again relates to addition of events rather than complex algebraic operations.

The Fourier transform of the output signal $g(t)$ is defined as follows:

$$G(f) = \int_{-\infty}^{+\infty} g(t)\ \mathrm{EXP}[-j2\pi\nu t]\,dt, \quad (22.3)$$

where ν is the frequency of the phenomenon and j is the imaginary number. Note that the integration is taken with respect to time. The Fourier transform of the time-dependent output signal is no longer a function of time. The summation or integration can be over one single period if the process is repetitive in time.

Consider two processes that are mutually dependent and result in a sequential process. This is expressed by multiplication in the time domain, which transforms into a convolution of the two sequential stages in the frequency domain:

$$g_1(t)g_2(t) \Rightarrow \int_{-\infty}^{\infty} G_1(\lambda)G_2(f-\lambda)\,d\lambda = G_1(f) \otimes G_2(f), \quad (22.4)$$

where the Fourier transforms are, respectively, $g_1(t) \Leftrightarrow G_1(f)$ and $g_2(t) \Leftrightarrow G_2(f)$.

When two events are taking place in independent conditions, the events $f_i(t)$ will have a signal described by the superposition principle as

$$g_1(t) + g_2(t) = [f_1(t) + f_2(t)] \otimes h(t) = f_1(t) \otimes h(t) + f_2(t) \otimes h(t). \quad (22.5)$$

In case the input is a two-dimensional signal from a detector with two degrees of freedom, the generalized input becomes $\delta(0,0)$, with a response function: $h(x',y')$, which is called the point-spread function, since the detector has physical limitations imposed minimally by the Heisenberg uncertainty principle, in addition to further limitations of the detection device due to the construction (e.g., size, response time, defraction limits, and quantum response, among others). Again an arbitrary object is a collection of an infinitesimal number of Dirac pulses (Figure 22.1), which makes the images formation the convolution of the signal collected from the object convolved with the point-spread function as represented by Equation 22.6 and illustrated in Figure 22.2

$$i(x', y') = o(x, y) \otimes h(x, y) \quad (22.6)$$

or in the frequency or Fourier domain:

$$I(\xi', \eta') = O(\xi, \eta) \times H(\xi, \eta), \quad (22.7)$$

where $H(\xi, \eta)$ is called the modulation transfer function, which is the Fourier transform of the point-spread function.[11,12]

Another more tangible example of the use of Fourier transforms in mental signal processing is the number of dots per unit

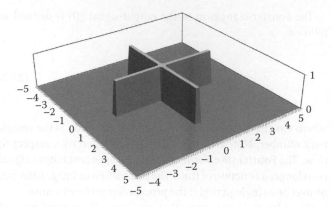

FIGURE 22.1 Hypothetical sharp delta-line distribution in two dimensions representing an infinitely precisely defined object, either in the temporal or spatial domain.

area in newspaper gray-scale print. The eyes and brain will perform an inverse Fourier transform of the dot matrix to present a gray-scale image, see Figure 22.3, converting the spatial periodicity ranges at representative gray-scale levels (Figure 22.4).

Mathematically, imaging is most convenient in the Fourier domain, since the convolution is then only a multiplication. There is only one drawback to operating in the time domain with respect to operating in the place domain: namely, time functions are causal and they exist only from a certain point in time onward:

$$\text{if } h(t) = 0 \quad \text{if } t < 0. \tag{22.8}$$

Place functions, however, are universal.

Adding a Dirac-pulse response to the origin $\delta(x, y) = (0, 0)$ may not be the easiest mechanism to derive a solution. It is usually not true that

$$h(x', y') = 0 \quad \text{if } x' < 0 \text{ or } y' < 0. \tag{22.9}$$

The point-spread function will spread in both positive and negative directions of the coordinate system by equal amounts, while the impulse response function only acts in positive time (resulting from the initiation onward).

This means that in order to provide a proper imaging algorithm, the transformation of a place function to the frequency

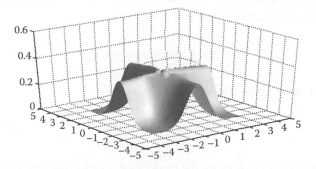

FIGURE 22.2 Graphic representation of a two-dimensional line spread image of the object in Figure 22.1 resulting from the transfer function of the imaging system.

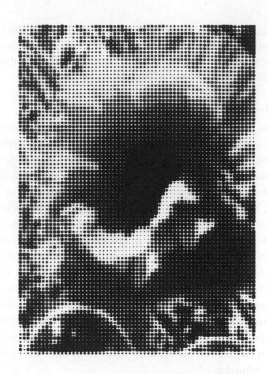

FIGURE 22.3 Exaggerated newspaper print half-tone gray-scale example of a Corn poppy, *Papaver rhoeas*. The standard newspaper print generally consists of 85 lines of dots per inch (Note: linear scale) using only black ink.

FIGURE 22.4 Smooth, continuous gray-scale image of the poppy from Figure 22.3. The eye, in combination with the brain, performs a Fourier transform to provide a visual gray perception of the newspaper halftone image.

domain needs to result from a Fourier transform. This is in contrast with the time function transform through a Laplace transform to the momentum or energy domain, which is not allowed in place function transforms.

Since the majority of imaging systems are two or more dimensional, the point-spread function needs to be identical in all dimensions. If this is the case the system is isotropic. In case the system is not providing the same response in all directions it is anisotropic. For this reason the line-spread function is introduced for inherently anisotropic imaging systems.

An example of an inherently anisotropic imaging system is the cathode ray television display, where there are different ways the image is projected in horizontal and vertical directions, and there is a different spatial frequency associated with both directions.

22.3.1.4 Temporal Transfer Function

$$i(x', y', t) = o(x, y, t) \otimes \otimes \otimes h(x, y, t). \quad (22.10)$$

22.3.1.5 Static Transfer Properties

The essential components of image formation that describe the official relationship between the object and the image need to be elaborated on. The focus of this section is the image transfer of still-standing objects alone. The reconstruction of the image relies on the device used to acquire the information carrier: the signal of interest. Additionally, contrast and areal confinement of the signal in a two-dimensional distribution determine the final outcome.

One of the most important issues in image formation is noise. Other aspects of image formation are resolution, sampling rate, artifacts, and distortion. All the parameters and boundary conditions influencing the projected output of the signal are described next. The concept of noise can be described with the introduction of the following three concepts:

1. Information carrier
2. Contrast
3. Pixel

22.3.1.6 Information Carrier Concept

Every imaging system conveys information. The information is conveyed from the object to the image. The mechanism of conveying the information relies on energy transfer. The energy transfer is the means of imaging: for example, light, x-ray, acoustics, or electricity. In most cases the energy is quantified in discreet packages of information. The magnitude of the energy quanta is directly correlated to the quality of the signal and hence the image quality. The greater the information carrier energy quanta, the fewer the quanta in a stream of information.[13] The statistical variation in the quanta is noise. Noise is a direct threat to the image quality. The lower the intensity of the energy stream of information, the greater the impact of the noise. The low intensity carries the same statistical error as a larger intensity; however, there is only a small opportunity for calculation by two opposing

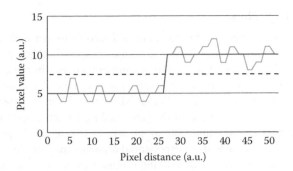

FIGURE 22.5 Noisy step function, two discrete levels of information (e.g. blue/green or soft/loud). The average value at each level and for entire signal stream is also illustrated.

noise segments. A large energy content per quanta also results in a lower stream of quanta, even though the signal intensity will be large.

22.3.1.7 Concept of Contrast

"Contrast" means the level of discrimination between individual data points. Two data points next to each other can be distinguished from each other when a device-specific level of contrast is exceeded. For instance the human nose can distinguish smell intensities that are 1.3 times each other's value. The eye can discriminate brightness (intensity) levels within the same view that are approximately 1% apart. Generally the concept of contrast for any detection mechanism can be described by Weber's Law:

$$\frac{\Delta I}{I} = c, \quad (22.11)$$

where ΔI is the incremental intensity of the phenomenon, I is the base line intensity, and c is a constant that is representative of the detection mechanism. This in turn converts into a logarithmic detection function representing the perception of contrast:

$$\text{Contrast} = k_1 \ln\left[\frac{I}{I_0}\right], \quad (22.12)$$

which is often referred to as the Weber–Fechner law and k_1 is a function of the detection/imaging system.

Figure 22.5 illustrates the noise around two discrete levels of information, as well as the average value of the signal.

22.3.1.8 Pixel Concept

The spatial distribution of information in the display, or the raster distribution, is referred to as pixel. The pixel is the smallest section that has a uniform intensity distribution and cannot be subdivided. The word "pixel" is a contraction of the words: "picture" and "element," denoting the discrete distribution of information elements in the composition of the projection of the output picture. The file that contains the axial distribution of the pixel values in both horizontal (x axis) and vertical (y axis) directions is called a bitmap. Each array data point can hold gray-scale values (intensity) or red, green, and blue information combined.

22.3.1.9 Noise

Generally, noise can be distinguished into Gaussian and non-Gaussian noise. Gaussian noise is an even spread of data points around an average value due to random influences. The noise distribution around the average value obeys a normal probability function and is also referred to as "white noise." Some influences affecting the magnitude of the generated signal are temperature and external radiation. The amplitude of the signal drops off from the average with an exponential decaying magnitude. Non-Gaussian noise on the other hand is noise that has specific external influences that contaminate the Gaussian probability distribution to form a skewed distribution of amplitudes. The non-Gaussian noise is more difficult to filter and requires complex signal processing techniques, such as wavelet transform.

Noise is defined as the random fluctuations of a signal due to specific uncertainty principles. Electrical noise can be generated by temperature fluctuations and electrical interference from nearby electronic components and devices. Temperature is defined as the average kinetic energy of all molecules in motion. Since molecules carry electrical charges the electrical fields and associated external voltage fluctuations will provide a source for noise, especially when measuring extremely small signals, of the order of magnitude of the thermal motion.[14–16] The temperature is derived from the Boyle–Gay-Lusac law, as shown in Equation 22.13:

$$PV = nRT, \tag{22.13}$$

where P is the equivalent pressure of a group of molecules in a solid, V is the volume containing the molecules, n is the number of moles of "gas," $R = 8.314472$ J/mol K the universal gas constant, and T is the temperature of the molecular gas.

This translates into the following expression for the average kinetic energy (KE_{avg}):

$$\frac{3}{2} kT = \overline{\frac{1}{2} mv^2} = KE_{avg} \tag{22.14}$$

where $k = 1.3806503 \times 10^{-23}$ m² kg/s² K is the Boltzmann constant, m is the mass of the molecule(s), and v is the velocity of the moving molecules. Note that due to the diverse mix of molecules and velocities the average makes perfect sense. As such the temperature can be derived. The electrical influence of the moving charges is explained in Chapter 11 under Gauss's law.

Optical noise can be the result of deformations in the instrumentation, birefringence, imperfections in the lens, and so on. It is frequently difficult to discriminate between noise and distortion. Distortion is primarily introduced by the device, while noise can originate in the sample as well as the device. One major source of noise is the mismatch in signal and detection device. When the signal is too small to be properly detected by the imaging device, the output will be primarily noise. One common noise factor is the "snow" on a television screen resulting from climatologic influences and for instance passing conduction objects such as a plane, or scratches on a CD or album. Noise is often relative, for instance

listening to one person speak in a crowded room may be hampered by the conversations of the crowd, but the information may still be useful, but not appropriate for the goal of the listener.

22.3.1.10 Resolution

Resolution describes the ability to distinguish two adjacent points with confidence. For optical imaging this is defined by the Abbey principle (also the Rayleigh criterion). The Rayleigh criterion describes the angular separation between two points that can be seen as an individual source. The optical resolution is a direct result of the point-spread function of the optical device (diffraction, for instance) as well as the imaging device, the eye. The eye has finite detectors: rods and cones, which will be one of the limiting factors on resolution. The Rayleigh criterion is given without derivation as

$$\theta_{min} = 1.22 \frac{\lambda}{D}, \tag{22.15}$$

where θ_{min} is the minimum angle of separation that can be observed, λ is the wavelength of the light source, and D is the diameter of the opening of the final exit window of the device.

On the other hand, the frequency resolution is limited by the Nyquist theorem, which states that the highest frequency resolution is directly dependent on the sampling frequency of the signal acquisition device.

22.3.1.11 Sampling Rate

In order to correctly identify a periodic phenomenon, such as a wave, the single wave needs to be sampled in at least two locations per period, thus positively identifying the crest and the trough. This translates in a sampling frequency f_s of minimum twice the highest frequency f_o in the signal. This principle can also be applied to spatial frequency

$$f_o < \frac{f_s}{2} = f_n = \text{Nyquist frequency.} \tag{22.16}$$

In case a signal has bandwidth and this needs to be digitized, the following conditions will apply.

The bandwidth for speech is approximately 100–3300 Hz, yielding a bandwidth of 3200 Hz. The bite rate of signal processing is determined by the number of events (#events) per second: the baud rate multiplied with the data acquisition-dependent bits per event (B_e). The number of bits per event is expressed in the binary mode as

$$B_e = {}^2\text{log}(\text{#events}). \tag{22.17}$$

The ²log is indicative of the digital conversion, illustrating that the 8 bit color coding in digital format translates into $2^8 = 256$ levels of color density, while 24 bits account for three colors (red, green, and blue) all at 8 bit density.

When considering a signal with bandwidth H and a finite number of discrete levels V, the maximum data rate D_r is given by the Nyquist theorem as

$$D_r = 2H \, {}^2\text{log V bits/s.} \tag{22.18}$$

Note: the signal will be deteriorated by noise, which needs to be included in the determination of the bit rate evaluation as described by Shannon's theorem (also the Shannon–Hartley theorem).

22.3.1.12 Artifacts

Artifacts influencing the diagnostic values are widespread and will need to be evaluated on an individual basis. Artifacts will introduce a systematic error that can be filtered or removed after the fact when properly identified. For instance during the electro-encephalogram (EEG) the brain signal will be mixed with muscle signals such as eye movement. The eye movement can be removed by deliberate placement of reference electrodes in the vicinity of the eyes. Electronic devices can also produce interference with medical devices, such as mobile phones in a hospital setting. Optical artifacts can be light sources in the periphery of an optical device, such as an examination room light during an eye exam on pupil constriction (pupil reflex). Heart rate and blood pressure measurements may come out differently when administered by a male or female nurse.

22.3.1.13 Distortion

Distortions are a primary function of the device used to obtain the data and are due to malfunctions or out-of-calibration conditions.

22.3.1.14 Compromises

The final item in the discussion of signal analysis is the fact that the diagnostic conditions often require compromises to be made due to the perceived danger to the patient. The main issue is acquiring the correct and complete information to derive at a clinical decision. Often the data acquisition is a compromise between accuracy (available state-of-the-art equipment), available time, and available equipment (limiting the number of vital signs to be measured). The medical device design requirements should focus on minimizing the equipment compromises and provide the most accurate and complete signal under limiting circumstances.

22.4 Summary

Image formation is a function of the source and the imaging mechanism; every system introduces a response curve, the point-spread function. Additional factors impacting the image are noise and distortions. Both noise and distortion create an error that can have far reaching consequences.

References

1. Enderle J, Blanchard S, and Bronzino J. *Introduction to Biomedical Engineering.* Burlington, MA: Elsevier Academic Press; 2000.
2. Fleckenstein P and Tranum-Jensen J. *Anatomy in Diagnostic Imaging.* Philadelphia, PA: W.B. Saunders Company; 2001.
3. Najarian K and Splinter R. *Biomedical Signal and Image Processing.* Boca Raton: CRC Press; 2005.
4. Wells PNT. *Scientific Basis of Medical Imaging.* New York: Churchill Livingstone; 1982.
5. Kak AC and Slaney M. *Principles of Computerized Tomographic Imaging.* Philadelphia, PA: Society of Industrial and Applied Mathematics; 2001.
6. Wandell BA. *Foundations of Vision.* Sunderland, MA: Sinauer Associates; 1995.
7. Chow P-L, Khasminskii R, and Liptser R. On estimation of time dependent spatial signal in Gaussian white noise. *Stochast. Process. Appl.* 2001; 96: 161–175.
8. Parzen E. On estimation of a probability density function and mode. *Ann. Math. Statist.* 1962; 33(3): 1065–1076.
9. Barrett HH and Swindell W. *Radiological Imaging: The Theory of Image Formation, Detection, and Processing.* New York: Academic Press; 1981.
10. Shannon CE and Weaver W. *The Mathematical Theory of Communication.* Urbana: University of Illinois Press; 1959.
11. Ibragimov I and Khasminskii R. *Statistical Estimation: Asymptotic Theory.* New York: Springer; 1981.
12. Weir J and Abrahams PH. *Imaging Atlas of Human Anatomy.* Dordrecht: Mosby Inc; 2005.
13. Sorenson J and Phelps M. *Physics in Nuclear Medicine.* New York: Grune and Stratton; 1980.
14. Ashcroft NW and Mermin ND. *Solid State Physics.* Philadelphia: Saunders College Publishing; 1976.
15. Feynman RP, Leighton RB, and Sands M. *Lectures on Physics, Vol III: Quantum Mechanics.* Reading, MA: Addison-Wesley; 1965.
16. Bronzino JD. *Biomedical Engineering Handbook*, Second edition. Boca Raton: CRC Press; 2000.

23

Electrocardiogram: Electrical Information Retrieval and Diagnostics from the Beating Heart

Orsolya Kiss

23.1 Structure, Electrical Characteristics, and Function of the Heart

23.1.1 Cardiac Muscle

23.1.1.1 Characteristics, Specialties: Excitability, Contractility

Cardiac muscle is a tissue specialized for complex periodic electrical activity and repetitive contraction. Certain parts of the myocardium—such as pacemaker cells—perform automatic electrical impulse generation. Other parts—the conductive tissues—are responsible for aligned impulse conduction towards the working muscle, which is the main determinate of cardiac contractility.

23.1.1.2 Cardiomyocytes: Types and Subtypes

23.1.1.2.1 Pacemaker Cells

The dominant pacemaker cells of the heart are small cardiac cells located in the sino-atrial node (SA-node). The main characteristic of these pacemaker cells is automatic electrical activity. It should be noted that lower levels of the conductive tissues and generally every type of myocytes can act as pacemaker cells and take over the role of the main pacemaker in special pathophysiologic circumstances.

23.1.1.2.2 Conductive Tissues

Specialized conductive tissues form anatomical structures in the heart called the atrio-ventricular (AV)-node, the bundle of His, the left and right bundle branches, and the Purkinje fibers. These cells are responsible for the transmission of the depolarization wave throughout the whole cardiac muscle. Certain parts of this system have their own specific role in the aligned operation of the atria and the ventricles. While the conduction in the AV-node is slow, about 0.05 m/s, the bundle of His and the bundle branches conduct the impulses at a high velocity of about 2 m/s. The conduction velocity is the highest, about 4 m/s in the reticulation of large myocytes having tight cell-to-cell connections in the ventricles called the Purkinje fibers. Certainly, every myocyte has the ability of slow, cell-to-cell impulse conduction, which is, for example, the basis for depolarization wave propagation in the atria.

23.1.1.2.3 Working Muscle

Working muscle cells, the archetypal myocardial cells, constitute the main mass of the cardiac muscle. Atrial and ventricular working cells are similar in structure, although atrial myocytes are longer and thinner. At least three electrophysiologically and functionally different cell types compose the ventricular working muscle: subepicardial, mid-myocardial, and subendocardial cells. These histologically similar cells mainly differ in their electrophysiological, especially repolarization properties, with the M-cells having the longest action potentials. Basically, the main function of these working muscle cells is to establish myocardial contraction throughout a process called excitation–contraction coupling as detailed below.

23.1.2 Electrical Activities of the Heart

23.1.2.1 Electrical Impulse Generation

The electrical impulses controlling the heart are physiologically generated in the SA-node high up in the posterior wall of the right atrium near the superior vena cava. The basis of automatic electrical activity is that the pacemaker cell has negative electrical membrane potential in the resting state, spontaneously losing its negativity. This process—called spontaneous depolarization—starts action potential formation in the myocyte and a cell-to-cell wave of depolarization—a flow of electricity towards the neighboring cells. Afterwards, the cell recovers its negative potential by pumping negative ions into and positive ions out of the cells, with a process requiring energy and called repolarization. The frequency of these automatic depolarization–repolarization cycles is determined by the inner characteristics of the pacemaker cells and by neural, hormonal, and metabolic regulation.

23.1.2.2 Electrical Impulse Conduction

The depolarization wave proceeding from the SA-node basically spreads over the atria by cell-to-cell conduction with a conduction velocity of about 0.5 m/s. Following the activation of the whole atrial muscle, action potentials enter the AV-node localized in the infero-posterior region of the inter-atrial septum, above the tricuspid valve. This specialized conductive tissue is normally the only way for the depolarization wave to enter the ventricles. The main function of the AV-node is to delay the impulse propagation

into the ventricles until the atrial contraction is completed. In the ventricles, the excitation wave proceeds towards the bundle of His running along the posterior border of the interventricular septum and propagates over the bundle branches. Right, left anterior, and left posterior bundle branches are localized in the interventricular septum and divide into the Purkinje fibers branching in the inner ventricular walls of the heart beneath the endocardium. The rapid impulse conduction in these fibers, mentioned above, is the main principle for the synchronous contraction of the ventricles. Normally, the activation of the ventricles starts in the septum, spreads through the endocardial wall of the myocardium, and ends in the epicardial region.

23.1.2.3 Cardiac Excitation Process, Electrical Origin of the Cardiac Cycle

Once a cardiac muscle cell is electrically activated, it results in the contraction of the cell occurring slightly later. This process is known as excitation–contraction coupling. During the action potential, intracellular Ca^{2+} concentration increases due to the opening of voltage-sensitive L-type Ca^{2+}-channels, Na^+–Ca^{2+} exchange, and Ca^{2+}-induced Ca^{2+} release from the sarcoplasmic reticulum (SR). Free Ca^{2+} is the main substrate of the contractile system, the thin and thick filaments in the myocyte. Free Ca^{2+} binds to troponin C, a protein localized in the thin filament, and initiates structural changes in the myofilaments. This results in the interconnection and shift of the thin and thick filaments causing the contraction of the cell. The contraction of the myocyte comes to an end with a process called relaxation. Cardiac relaxation mainly starts through the activation of Ca^{2+}-ATPase pumps and the Na^+–Ca^{2+} exchanger resulting in the reintroduction of Ca^{2+} into the SR and the decrease of cytosolic Ca^{2+} concentration.

23.2 Electrocardiogram: Electrical Signal of the Cardiovascular System

23.2.1 Brief History of the Development of Electrocardiogram Recording

The first observations of the electric current accompanying each heart beat were made in 1842 by Carlo Matteucci. Marey used a capillary electrometer to record the electrical activity of the frog heart in 1876. The first human electrocardiogram (ECG) was recorded with Lippmann's capillary electrometer in 1887 by Augustus D. Waller. This simple ECG revealed only two deflections indicating the ventricular events. However, the term "electrocardiogram" was established later in 1893 by Willem Einthoven as well as the terms P, Q, R, S, and T denoting different deflections of the ECG. Einthoven invented a new galvanometer for the accurate recording of the ECG and published the first ECG recorded with it in 1902. He received the Nobel prize in 1924 for his life's work in developing the ECG.

23.2.2 Origin of the ECG

The ECG is a graphic image of the sum of the electric currents generated in the heart. These currents spread over the tissues surrounding the heart and reach the surface of the body. Therefore they can be recorded using surface electrodes and visualized after appropriate amplifying. As the mass of the pacemaker cells and the specialized conductive tissues is small, the ECG mainly represents the electrical activity of the working muscle.

23.2.3 ECG Electrode Placement

The standard 12-lead ECG measures the heart from 12 different directions throughout the recording of the potential difference between electrode pairs (Figure 23.1).

Bipolar leads are placed on the limbs: *Lead I*: positive electrode set on the left arm, negative electrode set on the right arm, *Lead II*: positive electrode set on the left leg, negative electrode set on the right arm, and *Lead III*: positive electrode set on the left leg, negative electrode set on the left arm. Einthoven's triangle is an approximate presentation of the three limb lead vectors with the heart in the center as shown in Figure 23.1. Unipolar leads provide information about the difference in potential between positive limb or chest electrodes and an indifferent electrode having very low ("zero") potential as the negative electrode. In case of the conventional unipolar chest leads, the positive electrodes are placed on the chest wall (Figure 23.1): *aVR*: right arm lead, *aVL*: left arm lead, *aVF*: left leg lead, *V1*: fourth intercostal space, next to the right margin of the sternum, *V2*: fourth intercostal space, next to the left margin of the sternum, *V3*: midway between V2 and V4, *V4*: fifth intercostal space, in the midclavicular line, *V5*. left anterior axillary line, at the horizontal level of V4 and *V6*: left midaxillary line, at the horizontal level of V4.

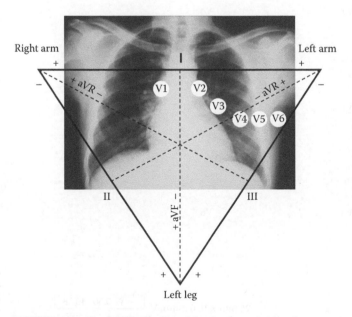

FIGURE 23.1 The 12 ECG leads and the Einthoven's triangle.

23.2.4 Special Leads Used in Specific Diagnostic and Therapeutic Interventions

Holter ECG leads are suitable for long-term—most often 24–48 h—noninvasive monitoring of transitional ECG changes. Right thoracic ECG leads (Vr3–Vr6) placed on the right chest as a symmetric image of their left ventricular matches can unequivocally reveal right ventricular myocardial damage. Because of the close anatomical location of the heart, as well as low impedance and high conductivity, leads placed into the esophagus can provide ECG signals more sensitive of low-voltage signs such as early atrial activation or local myocardial ischemia. During electrophysiological testing or therapeutic interventions, special electrodes are inserted into the heart. Right and left atrial and ventricular intracardiac ECG signs can be recorded, as well as His bundle electrograms or coronary sinus electrograms. Implanted intracardiac electrodes of pacemaker and implantable cardioverter-defibrillator devices also provide intracardiac ECG signs for arrhythmia analysis and treatment. Recent technological advancements led to the development of systems combining three-dimensional anatomic imaging and electrical activation or voltage mapping (Figure 23.2). These systems have the ability to visualize the heart chambers and anatomic landmarks as well as diagnostic and mapping catheters without using fluoroscopy, while providing three-dimensional snapshots or continuous images of electrical events in the heart.

23.3 Twelve-Lead ECG

The 12-lead ECG presents temporospatial details about the heart's electrical activity (Figure 23.3). Limb leads (Lead I, Lead II, Lead III, aVR, aVL, aVF) provide information about cardiac electrical activity in the frontal plane, while chest leads (V1–V6) show views of the horizontal plane of the heart. Arising from the polarity arrangement of the electrodes, depolarization waves proceeding towards an electrode result in positive deflections, waves moving away from an electrode provide negative deflections on the ECG. Certainly, as the electrical events are reciprocal, a repolarization wave heading away from an electrode will cause positive signs on the ECG lead.

23.3.1 QRS Axis

In general, the QRS axis means the average direction of cardiac electrical activity. The most commonly applied frontal plane QRS axis indicates the average direction of ventricular activation wave fronts in the frontal plane (Figure 23.4). Conventionally, an axis pointing horizontally to the left is 0° and an axis pointing vertically down is +90°. The normal QRS axis is between −30° and +90°. If it is more negative than −30°, left axis deviation is present, while right axis deviation means a QRS axis more positive than +90° (Figure 23.5). Certain leads indicate concrete

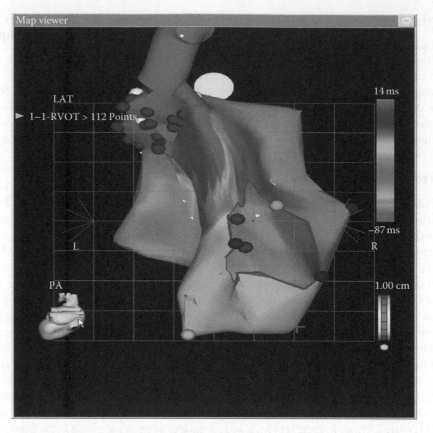

FIGURE 23.2 A 3D mapping of a right ventricular outflow tract tachycardia.

directions in the frontal plane as seen in Figure 23.1. Generally, the QRS axis is near to the lead with the largest positive deflection, 180° away from the lead with the largest negative deflection and perpendicular to the isoelectric lead.

23.3.2 Coherent Leads and Represented Cardiac Areas

Certain leads together represent well-defined anatomical areas of the heart. The inferior leads (II, III, aVF) provide the most

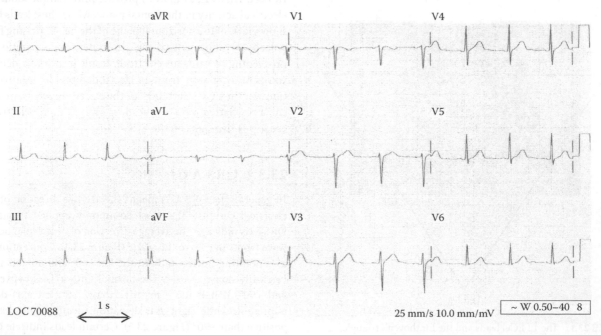

FIGURE 23.3 Normal ECG.

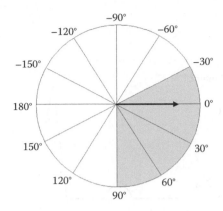

FIGURE 23.4 Determination of the frontal plane QRS axis. The gray area indicates the possible directions of the normal QRS axis. The arrow shows a QRS axis of 0°.

information about the inferior cardiac areas. The lateral leads (I, aVL, V5, V6) mainly look at the cardiac electrical activity of the lateral wall of the left ventricle. The anterior leads (V1, V2) face the anterior surface of the heart, while the septal leads (V3, V4) map electrical wave fronts from the direction of the interventricular wall. Lead aVR views the endocardial surface and the right atrium.

23.4 Development of the Specific Parts of the ECG

23.4.1 Origin of the P-Waves

The P-wave represents the sequential depolarization of the right and left atria (Figure 23.6).

23.4.2 Origin of the QRS Complex

The QRS complex is generated by the depolarization of the right and left ventricles (Figure 23.6). The Q-wave (if present) is the initial negative deflection of the QRS complex, the R-wave is the first upward deflection of the QRS complex, while the S-wave is the first negative deflection after the R-wave. If there is an additional positive deflection in the QRS complex, it is called R′-wave. Normally, the depolarization process of the ventricles goes on simultaneously due to the fast impulse conduction through the bundle branches and the Purkinje fibers, resulting in narrow QRS complexes.

23.4.3 The ST-Period, the T-Waves, and the U-Waves

The ST-segment of the ECG represents a physiologically uneventful period of the cardiac electrical activity between the depolarization and repolarization processes of the ventricles. The T-waves map the repolarization of the ventricles (Figure 23.6). The origin of the U-waves sometimes following the T-waves on the ECG is not yet clear. Presumably, these small waves indicate the development of after depolarizations in the heart.

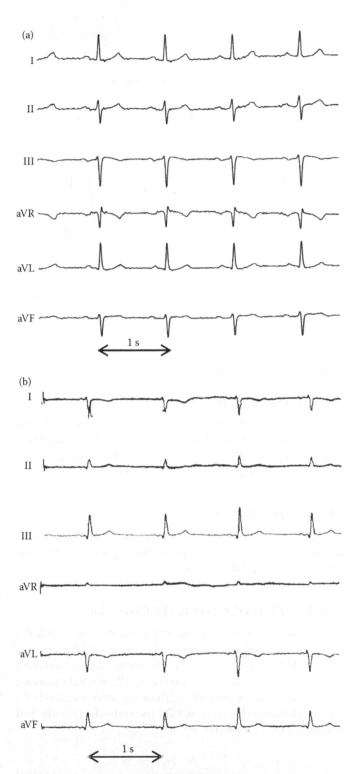

FIGURE 23.5 Fascicular blocks. (a) Left axis deviation in left anterior hemiblock. (b) Right axis deviation in left posterior hemiblock.

23.4.4 Relevant Intervals on the ECG

During the assessment of the ECG, there are some time intervals that should be analyzed (Figure 23.6). The shortening or more often the extension of these ECG parts may refer to the origin of electrical or structural heart diseases.

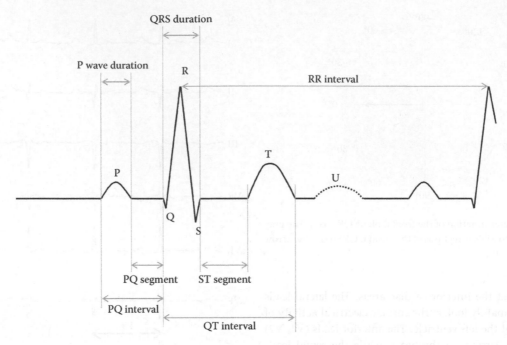

FIGURE 23.6 Specific parts and relevant intervals of the ECG.

23.4.4.1 PQ Interval

The PQ interval shows the time interval from the onset of atrial depolarization (beginning of the P-waves) to the onset of ventricular depolarization (beginning of the QRS complex). The PQ interval is physiologically between 0.12 and 0.2 s.

23.4.4.2 QRS Duration

The QRS duration is measured between the start of the Q-wave and the end of the S-wave. Normally, the length of the QRS complex is between 0.07 and 0.10 s.

23.4.4.3 QT (QTc) Interval, QT Dispersion

The QT interval represents the entire duration of ventricular depolarization and repolarization. It is measured between the onset of the Q-wave and the end of the T-wave. Because its length is strongly influenced by the heart rate (HR), normal values are determined by correcting for it. There are several methods for the calculation of this corrected QT interval—the QTc, the best known is Bazett's formula:

$$QTc = \frac{Qt}{\sqrt{RR}} \, [s].$$ (23.1)

The QTc interval is normal if <0.44 s. The QT dispersion is the difference between the maximum and minimum values of QT time in any two leads of the standard 12-lead ECG. Measurement of QT dispersion is a noninvasive method to quantify cardiac repolarization inhomogenity and arrhythmia risk. Normal QT dispersion is less than 0.05 s.

23.4.4.4 PP Interval

The PP interval means the distance between two consecutive P-waves on the ECG, representing the duration of atrial cycle, and indicating atrial rate.

23.4.4.5 RR Interval

The RR interval is the time duration between two consecutive R-waves, representing the duration of the cardiac cycle, and indicating ventricular rate.

23.5 Periodicity of the ECG

23.5.1 Heart Rate

HR means the number of cardiac cycles taking place in 1 min. The HR can be calculated from the RR interval if it is constant: when the ECG is recorded at a speed of 25 mm/s, HR = 60/RR (in s). Normal resting HR of an adult is between 60 and 100 min⁻¹. HR under 60 min⁻¹ is called bradycardia, while HR more than 100 min⁻¹ is called tachycardia. A clinically significant bradycardia is stated when bradycardia causes symptoms such as fainting or dizziness. HR increases during physical or psychological stress, the rate of increase mainly depends on the age, sex, and fitness of the individual. In case when the HR increase is reduced or fails during exercise, chronotropic incompetence is established. On the other hand, when the increase in sinus rate is causeless or excessive, inappropriate sinus tachycardia is the diagnosis.

23.5.2 Heart Rate Variability

Heart rate variability (HRV) refers to the beat-to-beat alterations in ventricular rate. The simplest measures of HRV are the time domain measures: either the momentary HR values or the time intervals between consecutive QRS complexes ("normal-to-normal," NN intervals) are determined in a chosen time interval. From a series of measurements carried out over longer time periods, complex statistical time domain measures can be calculated. For example, the mean NN interval, the mean HR, or the difference between the longest and shortest NN intervals can be determined. The standard deviation of the NN intervals (SDNN), the standard deviation of the average NN intervals (SDANN), and the square root of the mean squared differences of successive NN intervals (RMSSD) are commonly used statistical parameters. HRV represents one of the quantitative markers of autonomic influence on the heart: a decrease in HRV indicates increased sympathetic and/or reduced vagal effect on the heart. The reduction of HRV has been documented in several cardiac and noncardiac diseases. Decreased HRV has been observed in acute myocardial infarction (AMI) and in chronic heart failure as well as in diabetic neuropathy. The assessment of HRV can be used for risk stratification after AMI.

23.5.3 Changing ECG Morphology, Electrical Alternans

Electrical alternans means a repetitive variation in the morphology, amplitude, duration, and/or direction of any component of the ECG. Alternant ECG signs can be caused by repolarization alternans (affecting the ST-segment, the T- and/or the U-waves) or conduction and refractoriness alternans (affecting the P-wave, the PR interval, the QRS complex, and/or the RR interval). As an artifact of ECG recording, "electrical alternans" can result from the mechanical movement and alteration in the position of the heart. Alternant T-waves often accompany QT lengthening, rapid changes in HR, electrolyte imbalance, or diseases such as cardiomyopathies or acute pulmonary embolism. An ST-segment alternans often occurs in acute myocardial ischemia or infarction, and usually means alternating measures of ST-elevation. The significance of T-wave and ST-alternans lies in the fact that both phenomena can be predictors of life-threatening ventricular arrhythmias. Conduction alternans on the ECG is a result of the alternation of impulse propagation through one or more specialized conductive tissues in the heart. It may appear in cases of myocardial ischemia, chronic heart diseases, pulmonary embolism, atrial fibrillation (AF), or Wolff–Parkinson–White syndrome.

23.6 Normal ECG

Normal ECG is considerably variable due to age, habit, weight, and several other known or not defined individual features of the person (Figure 23.3). All uncommon ECG patterns should be denoted on the documentation and should be evaluated considering the previous ECG recordings and documentation as well as the current signs and symptoms of the patient.

23.6.1 Rhythm

Since the physiological pacemaker of the heart is the sinus node, sinus rhythm is a crucial component of the normal ECG. Sinus rhythm means that the wave front depolarizing the whole heart originated from the sinus node.

23.6.2 Frequency

Normal sinus rate of an adult is between 60 and 100 min^{-1} at rest. A few healthy people, especially sportsmen, often show a resting HR less than 60 min^{-1} due to the predominance of the parasympathetic control of sinus node impulse generation. On the other hand, sinus tachycardia is often observed in the normal ECG due to mental stress or physical exercise. The variability of sinus rate with respiration may exceed 15% and sometimes can reach 50%. In young healthy individuals, this phenomenon called sinus arrhythmia is usually physiological.

23.6.3 The Frontal Plane QRS Axis

The average direction of ventricular activation—the frontal plane vector of myocardial depolarization—is normally between -30° and +90° (Figure 23.4).

23.6.4 The P-Waves

In case of sinus rhythm and normal activation sequence of the atria, P-waves are always positive in lead I, and always negative in lead aVR. Furthermore, P-waves are often positive but sometimes are biphasic (positive-negative) in leads II, III, and aVF. In lead aVL, P-waves can be positive, negative, or biphasic (negative–positive). The P-wave is usually positive in the horizontal leads V5–V6, and positive or biphasic (positive–negative) in lead V1.

23.6.5 The QRS Complex

The QRS amplitude shows a remarkable individual variability: not only the size of the heart chambers and the muscle mass, but also the distance and tissues between the electrodes and the heart are determinates. The normal QRS axis is between -30° and +90°, which means that the QRS is mostly positive in leads I and II. Most often, small r- and big S-waves (rS complexes) are present in leads V1 and V2. In V3, the r-wave increases, but it is usually still smaller than the S-wave. The lead where the amplitude of the R-waves becomes bigger than the S-waves (usually V3 or V4) is called the transition zone. In leads V5 and V6, small q-waves and large R-waves (qR complexes) can be observed, often without s-waves.

23.6.6 The Q-Waves

For the great significance of pathologic Q-waves on the ECG, the knowledge of normal Q-waves is important. The normal Q-wave is shorter than 0.04 s, and lesser than 2 mm in amplitude.

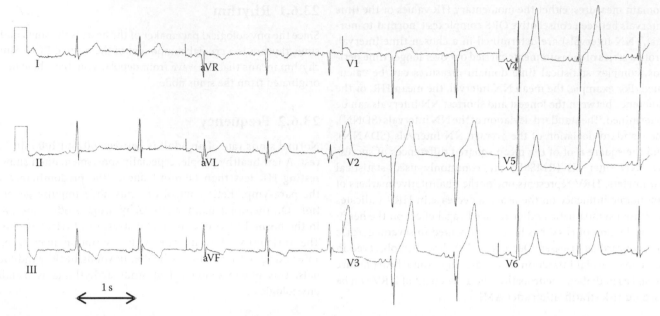

FIGURE 23.7 Left ventricular hypertrophy. R-wave in V5: 3 mV (30 mm), S-wave in V1: 1.8 mV (18 mm). The tall, symmetric T-waves in the precordial leads refer to increased vagal tone.

Normally, QS-waves can be present in V1, V2 and sometimes in aVR leads. The presence of isolated Q-waves in leads II, III, or aVF is sometimes a normal variant. These Q-waves are usually shorter than 0.03 s, are not accompanied by Q-waves in the other inferior leads, and they often disappear on deep inspiration. Moreover, small, narrow Q-waves often appear in some of the lateral leads (I, aVL, V4, V5, or V6) accompanying septal depolarization. Considering the depolarization followed by repolarization described in Chapters 4 and 5, the repolarization follows next.

23.6.7 Repolarization

23.6.7.1 The ST-Segment

As the elevation or depression of the ST-segment can be the main signal of myocardial infarction (MI) or ischemia—as detailed below—the recognition of harmless ST changes is important. The most common normal ST-variant is the elevation of the J-point (the point marking the end of the QRS complex and the beginning of the ST-segment). The 0.5–1.5 mm elevation of the J-point is frequently observed in V2 and V3 leads resulting in ST-segment elevation of up to 0.3 mV, especially in sinus bradycardia. The phenomenon called early repolarization—consisting of J-point and ST-segment elevation of up to 0.4 mV, starting up from the downstroke of the R-wave and observable in inferior and anterior leads—is usually a difficult differential diagnostic problem. In case of early repolarization, the ST-segment usually shows an upward *concave* configuration and notching of the J-point in one or more leads.

23.6.7.2 The T-Wave

Normal T-waves are generally not taller than 10 mm in chest and 5 mm in limb leads. The T-waves are normally positive in

lead I and are commonly positive in lead aVL, in the inferior leads and in leads V2–V6. Otherwise, normal T-waves are often negative in lead III and are highly variable (negative, positive, or isoelectric) in lead V1. Isolated T-wave inversion with preceding negative QRS complexes, especially in lead aVL, III, or aVF, usually does not refer to myocardial ischemia. In respect to the morphology, T-waves are normally asymmetric as they show a slow upstroke and a faster downstroke. However, as a normal variant, tall, positive, and often symmetric T-waves may appear in the precordial leads due to sinus bradycardia and increased vagal tone (Figure 23.7).

23.6.7.3 The U-Wave

As mentioned above, the U-waves are not a regular component of the ECG. However, they can appear on the normal ECG as small flat positive waves following the T-waves. Although the absence of the U-waves has no significance, their appearance sometimes indicates pathologic processes such as low potassium levels.

23.6.7.4 Intervals

As mentioned before, the most important intervals on the normal ECG are as follows: normal PQ interval is between 0.12 and 0.2 s, normal QRS duration is between 0.07 and 0.10 s, and the normal QTc interval is less than 0.44 s.

23.7 Abnormalities of ECG in Cardiac Diseases

23.7.1 Deviating ECG

In general, all changes on the ECG compared to the previous recordings should be considered carefully. However, many times

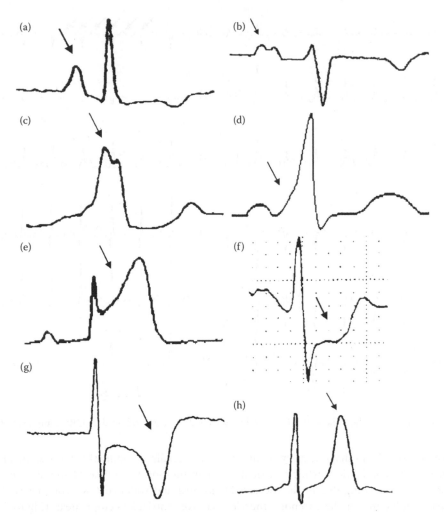

FIGURE 23.8 Some common changes of the ECG. (a) P pulmonale, (b) P mitrale, (c) QRS widening and distortion, (d) delta wave on the QRS complex, (e) ST elevation, (f) ST depression, (g) T inversion, and (h) peaked T wave.

the preceding ECGs are not available or had never been done. Some of the changes are unequivocally pathologic, some of them need detailed attention.

23.7.1.1 Specific Changes of the P-Wave

The inversion of the P-waves (which usually means the turn of generally positive deflections to negative ones) indicates that the atrial depolarization has an unusual direction. This is the situation when atrial activation starts from an atrial site distinct from the SA-node or when the anatomic position of the heart is irregular (e.g., dextrocardia). Moreover, an increase in the duration and/or amplitude of the P-waves together with changes in shape can be observed in case of atrial hypertrophy (Figure 23.8a and b) Sometimes the P-waves cannot be revealed on the ECG due to sino-atrial block or in case of junctional rhythm. Much more often, P-waves are replaced by F-waves (atrial flutter) or f-waves (AF) as detailed below (Figures 23.9 and 23.10).

23.7.1.2 Specific Changes of the PQ Interval

The PQ interval is reduced in cases of Wolff–Parkinson–White syndrome (Figure 23.11) or junctional rhythm. On the other hand, a prolongation of the PQ interval can be observed in some types of heart block (Figures 23.12 and 23.13).

23.7.1.3 Specific Alterations of the QRS Complex

Except for small Q-waves that can be present on the normal ECG, the appearance of wide and deep Q-waves (duration ≥0.04 s, amplitude: >2 mm or >25% of R-wave amplitude) on the ECG generally refer to MI (see below). The pathological Q-wave does not disappear during deep inspiration—in contrast to the normal one. The QRS complex may be broad (more than 0.10 s in duration) and bizarre in shape either in sinus rhythm or during ventricular rhythm. In case of sinus rhythm, right or left ventricular bundle branch block (Figures 23.8c, 23.13, and 23.14) and intraventricular conduction disorders

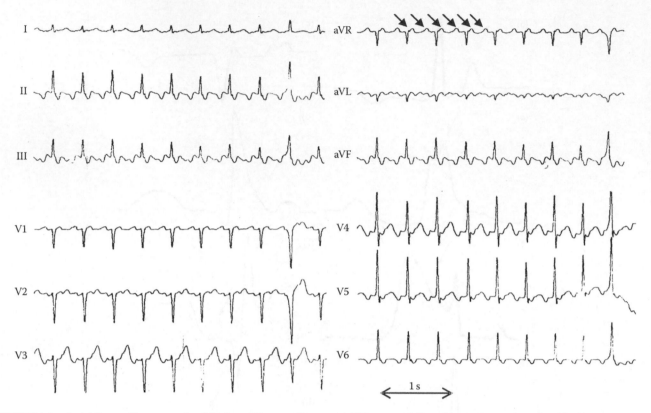

FIGURE 23.9 Atrial flutter. The arrows show the flutter F waves. Every second F-wave is followed by QRS complexes (2:1 block).

may result in the abnormality of QRS in shape and length due to the changes in normal impulse conduction, while ventricular hypertrophy may cause QRS broadening due to the increased time needed for impulse conduction in the extended muscle mass. On the other hand, when pathological electrical impulse generation starts from the ventricles, the depolarization wave front propagates in the ventricular working muscle instead of the specialized conductive tissues. This process is much slower than the normal course of impulse conduction, causing malformed and widened QRS complexes. This is the situation in case of ventricular ectopic beats (Figure 23.15a), idiopathic ventricular rhythm, different types of ventricular arrhythmias (Figures 23.15 through 23.18), or during ventricular pacemaker stimulation.

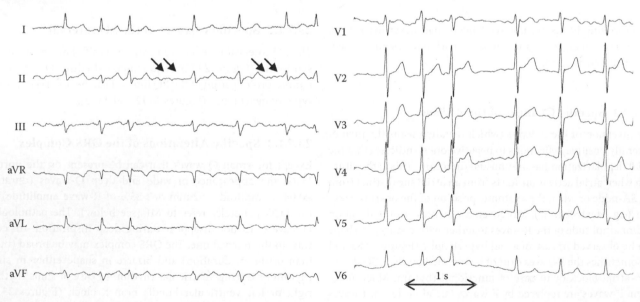

FIGURE 23.10 Atrial fibrillation. The arrows show some of the low-amplitude F-waves.

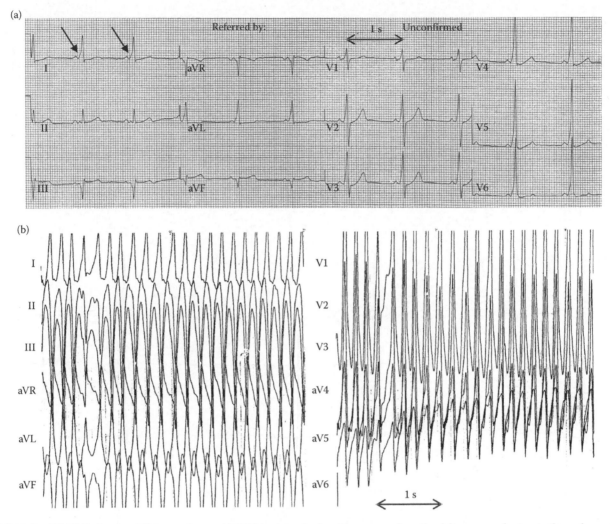

FIGURE 23.11 Wolff–Parkinson–White syndrome. (a) ECG in sinus rhythm. The arrows show the delta waves referring to the early ventricular activation. (b) FBI tachycardia developing on the basis of atrial fibrillation in the WPW syndrome.

23.7.1.4 Specific Changes of the ST-Interval

The changes of the ST-interval are measured 0.04 s (1 mm in case of 25 m/s paper speed) after the J-point, and are compared to the baseline (the line from the end of the T-wave to the start of the P-wave). The elevation of the ST-segment is considered significant if it exceeds 1 mm in the limb leads or 2 mm in the precordial leads. Acute ST-elevation myocardial infarction (STEMI) causes ST-elevation in coherent ECG leads facing the infarct (Figures 23.8e, 23.19, and 23.20). Typically, this kind of ST-elevation shows upward convexity. Acute vasospastic (Prinzmetal) angina leads to similar ECG changes. However, ST-elevation of any kind from a patient having chest pain must be considered as a sign of MI until proven otherwise. It is important to emphasize that there are multiple other diseases that cause ST-segment elevation. Pericarditis causes diffuse moderate ST-elevation (in several or all ECG leads) usually showing an upward concave shape. Closer inspection may reveal depressed pQ intervals and that the ST is not elevated compared to the T-P true isoelectrical line. Myocarditis also can result in extended nonspecific ST-elevation (or even depression) usually small in amplitude (Figure 23.21). Moreover, the causes of QRS widening as detailed above also may cause bizarre ST-elevation or depression in several leads. Usually, an ST-depression is abnormal when it is higher in amplitude than 0.5 mm, horizontal or descending, and appears in coherent leads (Figures 23.22 and 23.23). Left or right ventricular "strain" results in descending ST-depression followed by inversed T-waves in leads that look at the affected hypertrophied ventricle. Ischemic ST-depression is usually horizontal or slightly descending and is often followed by symmetric T-inversion (Figures 23.22 and 23.23). Digitalis therapy may cause a typical "sagged" shape ST-depression without T-inversion in several leads that are not definitely coherent. The ascending ST-depression often accompanying sinus tachycardia is a harmless phenomenon.

23.7.1.5 Specific Changes of the T-Waves

The evaluation of T-wave deviations is difficult as there can be several reasons for nonspecific T-wave changes on the ECG. The

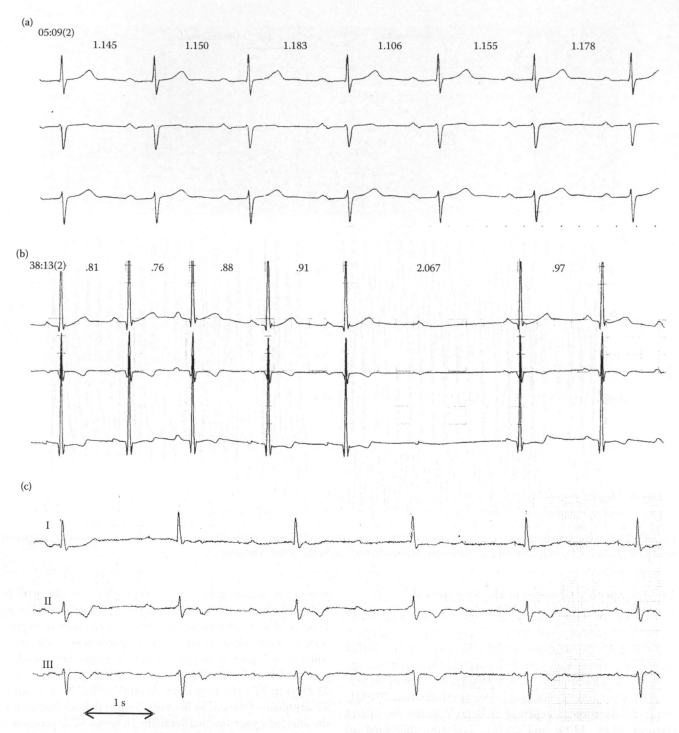

FIGURE 23.12 Atrioventricular blocks. (a) A first-degree AV-block (Holter leads I, II, and III); (b) a Wenkebach-type (Mobitz I) second-degree AV-block (Holter leads I, II, and III); and (c) a third-degree AV-block (AV-dissociation), junctional rhythm with a rate of 36 min⁻¹.

alterations of other parts of the ECG, as well as anamnestic data and other clinical findings help in finding the appropriate diagnosis. Flattened T-waves or slight T-inversion is a common and nonspecific finding. Hyperventilation, smoking, or other individual features may cause this variation. However, flattened T-waves together with the low voltage of all other ECG complexes may also appear in hypokalemia or myxoedema. Ventricular hypertrophy and bundle branch blocks are often accompanied by T-wave inversion, as well as myocardial ischemia (Figures 23.8g, 23.13, 23.14, and 23.23). In early MI, tall and peaked T-waves appear on the ECG (Figure 23.19) followed by T-inversion as the time progresses. Hyperkalemia also results in tall and peaked T-waves but without specific ST-changes that are usually observable in MI (Figure 23.8h).

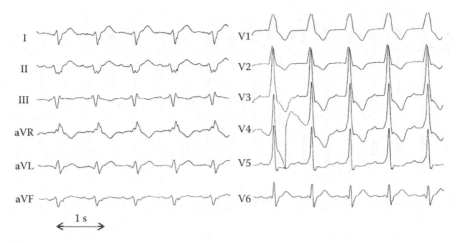

FIGURE 23.13   Right bundle branch block with a first-degree AV-block.

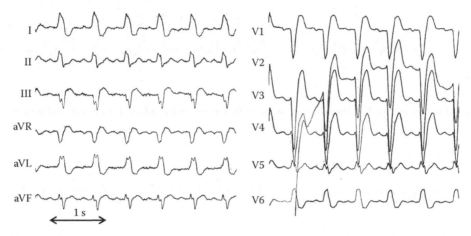

FIGURE 23.14   Left bundle branch block.

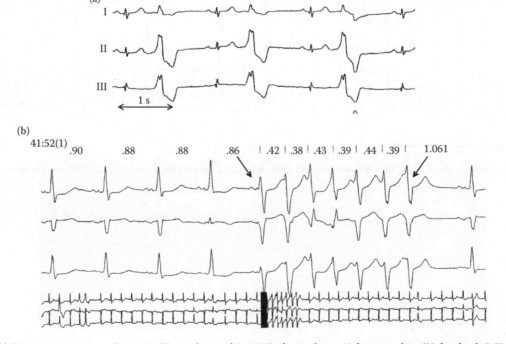

FIGURE 23.15   (a) Bigeminic monomorphic VES. (b) A polymorphic nsVT of seven beats. Holter recording (Holter leads I, II, and III).

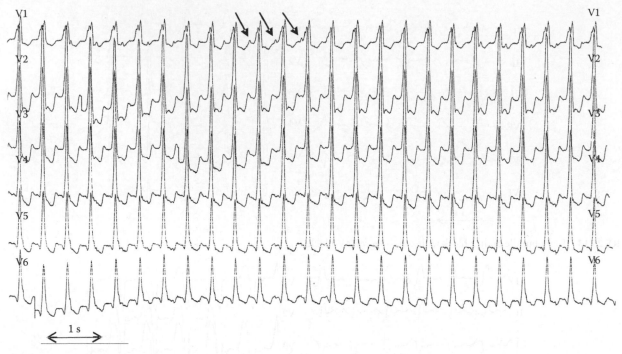

FIGURE 23.16 Sustained monomorphic VT. The arrows show the atrial P waves independent of the broad- and bizarre-shaped QRS complexes (AV-dissociation).

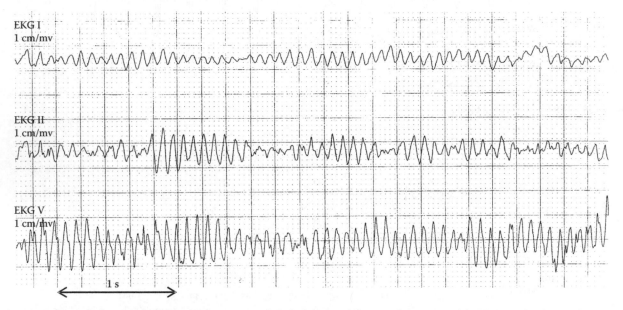

FIGURE 23.17 Torsade de pointes VT. The QRS twist around the baseline and the periodic varying of the QRS amplitude is present.

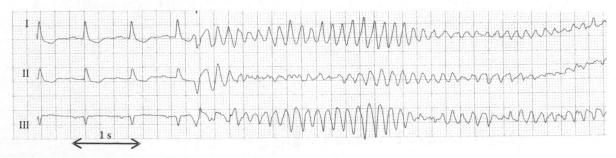

FIGURE 23.18 A fast polymorphic VT progrediating into ventricular fibrillation.

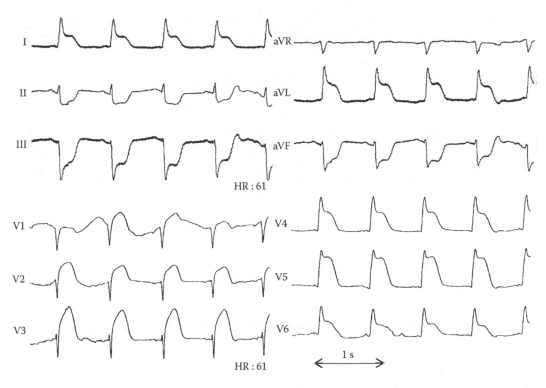

FIGURE 23.19 Acute extensive anterior myocardial infarction. ST-elevation in leads I, aVL, and leads V1–V6.

23.7.1.6 Specific Changes of the U-Waves

The U-waves may become prominent in hypokalemia or during digitalis therapy. The significance of solitary U-wave abnormalities is not yet ascertained, although may refer to abnormalities of repolarization.

23.7.1.7 Specific Changes of the QT (QTU) Interval

A long QT interval (>0.44 s), as a congenital or acquired, transient phenomenon is a sign of impaired, prolonged repolarization. Its main significance is that long QT syndrome is associated with an increased risk of ventricular arrhythmia formation (particularly torsade de pointes, see below) and sudden death. Short QT syndrome, a recently described and rare entity is characterized by a QTc interval of <0.30 s. Similarly to long QT syndrome, short QT syndrome also has a predisposition for ventricular tachyarrhythmias and sudden death.

23.7.2 Arrhythmias

23.7.2.1 Sinus Node Diseases, SA-Blocks

Sinus arrest (Figure 23.24), the transient loss of the SA-node electrical activity, means a pause without a P-wave and a QRS complex. SA-block is present when the excitation of the atria from the SA-node is delayed or blocked. First-degree SA-block is not reflected on the ECG. Second-degree SA-block means the temporary, third-degree SA-block means the complete dropout of the SA-node action potential propagation through the atria

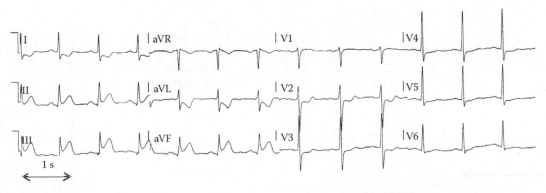

FIGURE 23.20 Acute inferoposterior myocardial infarction. ST-elevation in leads II, III, and aVF, ST-depression in leads V2 and V3.

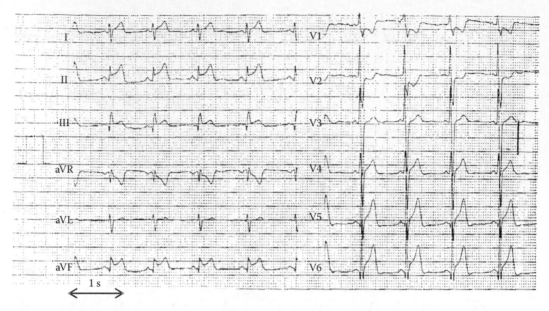

FIGURE 23.21 A 12-lead ECG of a patient suffering from myocarditis. Generalized ST-T abnormalities.

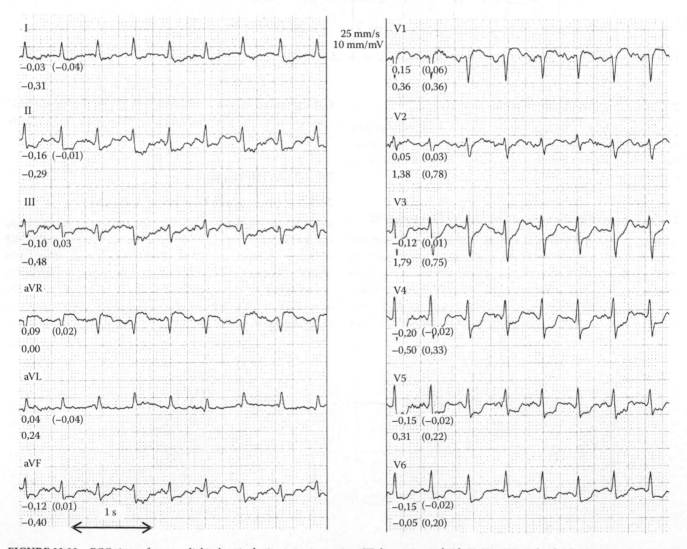

FIGURE 23.22 ECG signs of myocardial ischemia during exercise testing. ST-depression in leads II, III, aVF, and in leads V3–V6.

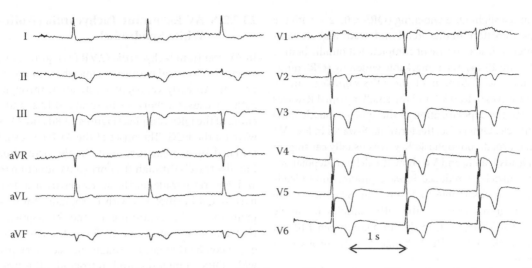

FIGURE 23.23 Non-ST-elevation myocardial infarction. Severe ST-T changes: ST-depression and T-wave inversion in almost all ECG leads.

and appear as the occasional or persistent absence of the P-waves, similar to the signs of sinus arrest.

23.7.2.2 AV-Blocks

In case of AV-blocks, the impulse propagation from the atria to the ventricles slows down or fails. First-degree AV-block means a slower but uninterrupted AV conduction (Figures 23.12a and 23.13), the PQ-interval lengthens and exceeds 0.2 s. Second-degree AV-block is the case when a transient block of the AV conduction occurs. Type I (Wenckebach or Mobitz I type) second-degree AV-block is the diagnosis when the continuous prolongation of the PR interval precedes the nonconducted P-wave (Figure 23.12b). In case of type II (Mobitz II) second-degree AV-block, the PR intervals and the RR intervals before and after the pause are constant. Second-degree AV-block often occurs as 2:1, 3:1, and so on. block, which means every second or third, and so on. P-wave is blocked (Figure 23.9). In case of third-degree AV-block, the AV conduction is completely blocked, QRS complexes appear independently of the atrial P-waves (Figure 23.12c).

23.7.2.3 Right Bundle Branch Block

When the conduction in the right bundle branch is blocked, the widening and distortion of the QRS complex appears on the ECG (Figure 23.13). Terminal R′-waves, usually rSR′ complexes in lead V1, terminal R-waves in lead aVR and terminal S-waves in leads I, aVL, and V6 are typical ECG signs. The ST-T-waves are usually opposite to the terminal part of the QRS. The QRS duration is ≥ 0.12 s in complete right bundle branch block (RBBB), and is between 0.10 and 0.12 s in incomplete RBBB.

23.7.2.4 Fascicular Blocks, Left Bundle Branch Block

Left anterior fascicular block (left anterior hemiblock) causes marked left axis deviation and slight QRS widening (QRS < 0.12 s) (Figure 23.5a) with ECG signs as follows: rS complexes in the inferior leads, small preserved septal q-waves, usually qR complexes in the lateral limb leads, poor R progression in leads V1–V3, and deep S-waves in leads V5–V6. Left posterior fascicular block (left posterior hemiblock) (Figure 23.5b) results in right

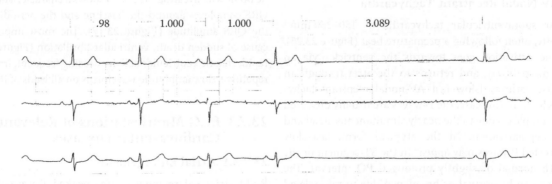

FIGURE 23.24 Sinus arrest with a significant pause of 3089 ms. Holter recording (Holter leads I, II, and III).

axis deviation and a slight QRS widening (QRS < 0.12 s) with the next ECG signs: rS pattern in the lateral limb leads and qR complexes in the inferior leads. In case of complete left bundle branch block (LBBB, Figure 23.14), the remarkable widening (QRS duration > 0.12 s) and the distortion of the QRS complex can be observed. The Q-waves in lead V1 and the smaller primal R-waves in lead V6 refer to the depolarization of the smaller muscle mass of the right ventricle. Otherwise, the terminal S-waves in lead V1 and the terminal broad, monophasic R-waves usually causing an "M" pattern in leads I, aVL, and V6 show the delayed depolarization of the left ventricle. A reduced R-wave progression in leads V1–V3 may also be observable in LBBB. The ST-T-waves are opposite to the terminal waves of the QRS complex in simple LBBB. In incomplete LBBB, the QRS duration is between 0.10 and 0.12 s with similar QRS but less ST-T changes than those observed in complete LBBB.

23.7.2.5 Supraventricular Tachycardias

These arrhythmias arise from the atria or the AV junction, usually appearing with narrow QRS complexes on the ECG. A single supraventricular extrasystole means one premature beat, a couplet or a triplet means two or three premature beats following each other. A nonsustained supraventricular tachycardia (SVT) is a supraventricular rhythm disturbance of more than three consecutive beats and less then 30 s duration, while a sustained SVT is longer than 30 s. Moreover, the premature beats may occur in bigeminic, trigeminic, and so on forms, which means that every second, third, and so on beat is premature.

23.7.2.6 Atrial Flutter, AF

Atrial flutter (Figure 23.9) appears on the ECG with the so-called saw-tooth atrial flutter waves (F-waves) of 250–300 min⁻¹, which are best seen in leads II, III, and aVF. A 2:1 AV-block usually appears in atrial flutter causing a regular ventricular rate of about 150 min⁻¹. AF (Figure 23.10) means rapid, chaotic atrial activity of >300 min⁻¹, resulting in the absence of the normal P-waves and the appearance of small f-waves on the ECG. The ventricles beat irregularly in AF.

23.7.2.7 AV-Nodal Reentrant Tachycardia

This regular supraventricular tachycardia of 150–250 min⁻¹ starts abruptly, often following a premature beat (Figure 23.25). Typically, the activation goes towards the ventricles via an AV-nodal slow pathway, and returns to the atria through an AV-nodal fast pathway [slow–fast AV-nodal reentrant tachycardia (AVNRT)]. In this case, the P-waves are usually obscured by the QRS complexes due to the nearly simultaneous atrial and ventricular depolarization. In the atypical form, fast–slow AVNRT, inverted P-waves may appear in the ST-segment or the T-waves with normal or slightly prolonged PQ interval. The QRS complex can be normal or broadened due to rate-related aberrant conduction.

23.7.2.8 AV Reentrant Tachycardia (Wolff–Parkinson–White Syndrome)

In AV reentrant tachycardia (AVRT) (Figure 23.11), one or more accessory electrical pathways somewhere in the AV-junction are present. An early ventricular activation through the accessory pathway causes a short PR interval (<0.12 s) and an initial widening of the QRS complex (Figures 23.8d and 23.11a) called delta wave on the ECG. The onset of the AVRT is usually abrupt and initiated by a premature beat. In the orthodromic form, the impulse travels through the normal AV junction to the ventricles and goes retrograde on the accessory tract to the atria, causing narrow QRS complexes with retrograde P-waves at the end of them or at the beginning of the ST-segment (RP < 0.10 s). Conversely, an antidrome AVRT is the case when the anterograde conduction goes through the accessory path resulting in wide QRS complexes and retrograde P-waves with a long RP-time (RP > 0.10 s).The fast, wide complex tachycardia turning out due to the fast conduction of electrical impulses in AF through the ventricles by the aberrant pathway is called FBI (fast, broad and irregular) tachycardia (Figure 23.11b).

23.7.2.9 Tachyarrhythmias of Ventricular Origin

These rhythm disturbances originate from the ventricles, typically appearing with broad QRS complexes on the ECG. The terms single ventricular premature beats (VES, ventricular extrasystole), couplets (coupled VES), triplets, nonsustained (nsVT), and sustained VT, bigeminic and trigeminic VES are used with the same naming conventions as mentioned at supraventricular tachycardias (Figure 23.15). A ventricular rhythm of 100–120 min⁻¹ is called accelerated idioventricular rhythm. A ventricular tachycardia (VT) may have a rate of 120–240 min⁻¹. The VT is termed monomorphic when it consists of identical or very similar QRS complexes (Figure 23.16), and typically has a regular rhythm. Idiopathic ventricular outflow tract tachycardia (Figure 23.2), a monomorphic VT, usually arises from the right ventricular outflow tract and shows LBBB pattern and right axis deviation. Bundle branch reentrant ventricular tachycardia usually originates from the left bundle branch causing an RBBB pattern monomorphic ventricular tachycardia and relatively short QRS complexes during the VT. A polymorphic VT shows beat-to-beat variation in QRS complexes and a usually irregular rhythm (Figure 23.15b). Torsade de pointes tachycardia (TdP VT) shows a characteristic twist of the QRS complexes around the baseline and the periodic varying of the QRS amplitude (Figure 23.17). The most important direct cause of sudden death, ventricular fibrillation (Figure 23.18), is a rapid, chaotic ventricular activity causing frequent, irregular, nonrepetitive, slow amplitude waveforms on all leads of the ECG.

23.7.3 ECG Manifestations of Relevant Cardiovascular Diseases

23.7.3.1 Atrial Enlargement

Right atrial enlargement causes peaked P-waves (also called "P pulmonale") in leads II, III, and aVF (Figure 23.8a). In case of

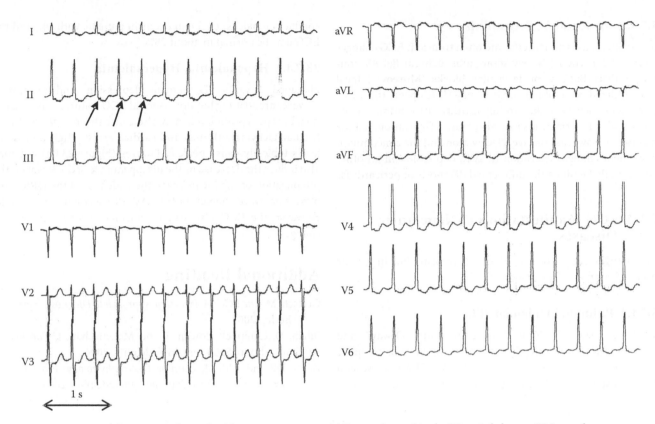

FIGURE 23.25 AV nodal re-entry tachycardia. The arrows point at atrial P waves located in the ST period close to QRS complexes.

left atrial enlargement, a notched, m-shaped P-wave (also called "P mitrale") appears in leads I and II (Figure 23.8b).

23.7.3.2 Ventricular Hypertrophy

Right ventricular hypertrophy (RVH) causes dominant R-waves instead of the physiologically dominant S-waves in leads facing the right ventricle. On the contrary, deep S-waves develop in the left ventricular leads. A slight increase in QRS duration, right axis deviation as well as ST-T changes (directed opposite to QRS duration) may also occur in RVH. Left ventricular hypertrophy (Figure 23.7) causes R-wave amplitude increase, usually ST-depression, and T-inversion in the left ventricular leads with S-wave amplitude increase in the right ventricular leads.

23.7.3.3 Myocardial Ischemia, Prinzmetal Angina

The most common ischemic ECG sign is significant ST-depression (horizontal or descending), often accompanied by T-wave abnormalities (flattening, inversion) (Figures 23.8f, g and 23.22). Coronary vasospasm causes transitory ST-elevation on the ECG.

23.7.3.4 Acute Myocardial Infarction

Generally, the first ECG signs of MI are increased T-wave amplitude and width. In case of transmural injury, ST-elevation evolves associated with the T-wave changes mentioned above.

Later, the myocardial necrosis can be indicated by the development of pathologic Q-waves—mainly in cases of STEMI—together with the decrease of the ST-elevation and the inversion of the T-waves. In the chronic stage of MI, ST-T changes usually disappear and pathologic Q-waves (if any) may be the only ECG sign of previous MI. Newly developed LBBB, although masking its typical ECG signs, may also refer to AMI. In STEMI, the significant ST-elevation appears in coherent leads facing the infarct and typically shows upward convexity (Figures 23.19 and 23.20). The specific ECG changes appear as follows: anterior MI: leads V2–V4, anteroseptal MI: leads V1–V4, septal MI: leads V1–V2, extensive anterior MI: leads V1–V6 (Figure 23.19), anterolateral MI: leads V2–V6, I, aVL, lateral MI: I, aVL, V5–V6, inferior MI: II, III, aVF (Figure 23.20), right ventricular MI: right chest lead V4R, posterior MI: reciprocal changes in leads V1–V3 (Figure 23.20). Moreover, reciprocal ST-depression in leads opposite to the site of the infarction accompanies the above-mentioned ST-elevation in most of the cases. In case of non-ST-elevation myocardial infarction (nSTEMI) (Figure 23.23), partial damage of the myocardial wall is not accompanied by pathologic ST-elevation. The usual ST-T changes—ST-depression (commonly downward convex or straight), T-inversion (typically symmetrical); and other ECG signs (arrhythmias, conduction disturbances, etc.)—are not specific to MI. In nSTEMI, the appearance of ST-T changes usually does not allow the exact localization of the affected myocardial region.

23.7.3.5 Myocarditis, Pericarditis

Myocarditis causes nonspecific and often transient ECG changes (Figure 23.21). The ECG may show sinus tachycardia, AV conduction disturbances, or fascicular blocks. Moreover, focal or generalized ST-T abnormalities may develop on the ECG. ST-segment elevation due to myocarditis can mimic acute transmural MI. Pericarditis may also cause ST-elevation, but the generalized, upwards concave ST-elevation and the concomitant PR-segment changes (elevation in aVR, depression in all other leads) usually facilitate the differential diagnosis of pericarditis.

23.7.4 ECG Manifestations of Noncardiac Diseases

ECG changes can also provide useful information in a wide variety of noncardiac diseases.

23.7.4.1 Pulmonal Embolism (PE)

The classical signs of PE are S-waves in lead I, Q-waves and T-inversion in lead III ($S_1Q_3T_3$ pattern), incomplete or complete RBBB and T-wave inversion in leads V1–V4. However, several other nonspecific ECG changes (such as sinus tachycardia, conduction blocks, etc.) can occur in patients with PE, and the ECG may be normal in about 25%.

23.7.4.2 Hypokalemia, Hyperkalemia

Flattened T-waves, ST-segment depression, and prominent U-waves are striking but not essential ECG features of hypokalemia. Low potassium levels may also cause the formation of ventricular tachyarrhythmias and cardiac arrest. High potassium levels typically cause peaked T-waves (Figure 23.8h) but may also induce the decrease or the disappearance of the P-waves, the prolongation of the PR interval, the widening of the QRS complex, sine wave-shaped QRST, AV dissociation, or asystole. However, the ECG also can be normal in severe potassium changes.

Additional Reading

Gertsch, M, *The ECG: A Two-step Approach to Diagnosis*, Springer, Berlin, 2004.

Julian, D, Campbell-Cowan, J, and McLenachan, J, *Cardiology*, 8th edition, Saunders, Philadelphia, 2004.

Zipes, DP and Jalife, J, *Cardiac Electrophysiology: From Cell to Bedside*, 4th edition, Saunders, Philadelphia, 2004.

24

Electroencephalography: Basic Concepts and Brain Applications

Jodie R. Gawryluk and
Ryan C.N. D'Arcy

24.1 Overview

The objective of this chapter is to provide an overview of electroencephalography (EEG) in terms of basic concepts and brain applications. This represents a general survey of the field, which identifies major areas of interest from a biomedical engineering point of view. For more details on either the technical or neuroscience perspectives, the interested reader should consult specific reviews on these topics.[1,2]

24.2 Basic Concepts

24.2.1 Brief History of Electroencephalography

The brain monitors and controls nearly every human function, including sensory, perceptual, and cognitive functions. In 1929, Hans Berger recorded the first electrophysiological correlates of human brain activity by placing electrodes on the scalp and measuring voltage changes over time.[3] The device was called an "elektrenkephalogram," which translates to electric brain recorder or electroencephalogram (EEG). At the time, these findings were controversial, but the technique gained acceptance as the findings were replicated in subsequent studies.[4–6] Soon after the advent of EEG, evoked potentials (EPs) were reported when the brain's response to a stimulus or event was further isolated as small electrical potentials embedded within the EEG

signals.[7] The popularity of EEG increased significantly in the early 1960s, with the ability to average large amounts of data using computers. Closely following this technological advance, the first derivatives of EEG could also be used to study higher level cognitive processing (event-related brain potentials or ERPs).[8] Since then, the use of EEG as a powerful brain imaging tool has become increasingly widespread in both the basic and clinical neurosciences.

24.2.2 Overview of Human Brain Function

The human brain has an average of 85 billion nerve cells (or neurons), each with between 1000 and 10,000 synaptic connections.[9] Neurons have three main component parts: dendrites (input), cell bodies, and axons (output). While the brain can be partitioned in many ways, the main structural divisions include the brainstem, subcortex, and cortex. The cortex has gyral and sulcal folds (to allow for increased surface area), and spans over two hemispheres (Figure 24.1).

Within the cortex, the primary projection neurons are pyramidal cells.[10] Pyramidal neurons have an asymmetric shape (reflecting that of a pyramid). It is this asymmetry that forms the basis for the open field geometric configuration necessary to generate the volume conducted electrical potentials recorded at the scalp (Figure 24.2).

(a)   (b)

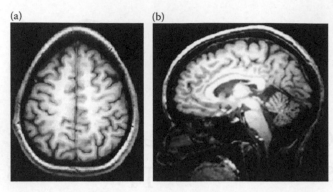

FIGURE 24.1 Magnetic resonance images of the human brain in axial (a) and sagittal (b) orientations. Note the two hemispheres and the highly convoluted surface.

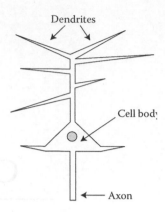

FIGURE 24.2 Pyramidal neuron.

In order to understand how electrical activity is measured at the scalp, it is important to first review how neurons communicate through electrical signals.[2] The resting potential of a neuron, −70 mV, is created by an ionic gradient between the inner membrane and the external cellular environment (largely through voltage-gated ion channels and Na^+/K^- pumps). The main mode of neural communication occurs through action potentials, which represent rapid and supra-threshold depolarization (above -55 mV). Action potentials are "all-or-none" phenomena, whereby spiking waves (peaking at approximately +40 mV) are propagated along axonal membranes to presynaptic terminals. When action potentials reach the terminals, a release of chemical neurotransmitters is triggered into the synapse (the space or gap between neurons). The binding of neurotransmitters to postsynaptic membranes leads, in turn, to graded postsynaptic potentials, which can be either excitatory or inhibitory. Postsynaptic potentials may or may not result in the transmission of an action potential, depending entirely on the nature of the preceding input.[11]

When neurotransmitters depolarize a neuron (i.e., excitatory input), positive charges flow in at the distal sites of the dendrites (creating a current "sink") and negative charges flow out at sites near the cell body (creating a current "source").[12] This creates a bioelectric situation in which there exists a dipole with negative and positive charges. As pyramidal cells are organized together with the same orientation, the open field circuit that arises from each cell sums and is volume conducted towards sensors placed on the scalp (Figure 24.3).

EEG sensors at the scalp measure changes in voltage over time. While the sources of measured signals can theoretically arise from both action potentials and postsynaptic potentials, there are a number of factors that limit the recording of electrical activity from action potentials.[2] First, action potentials occur very quickly (1 ms) and therefore are often not synchronized in time. Postsynaptic potentials, on the other hand, occur over 10–100 ms and are synchronized. Second, the orientations of axons also vary and therefore the signals are often not summed spatially. In contrast, the dendrites and cell bodies that give rise

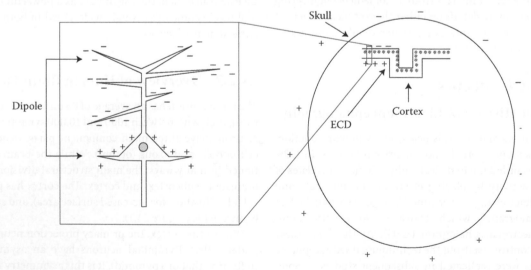

FIGURE 24.3 During depolarization, negative charges at distal dendrites and positive charges near the cell body create a dipole. When a group of cells in the cortex are active, the dipoles summate to form an equivalent current dipole or ECD. The orientation of the ECD determines the distribution of voltages on the scalp.

to postsynaptic potentials tend to be uniformly organized, which allows for the spatial summation of electric fields. And, third, action potentials create quadrapolar fields (rather than dipolar fields), which give rise to an additional cancellation component.

If EEG signals arise primarily from the synchronous activation of postsynaptic potentials in pyramidal neurons, the next question is how many cells are involved? There are approximately 1,000,000 pyramidal neurons in 1 cubic millimeter of cortical tissue.[13] When many cells are active, the considerable number of individual dipoles with the same orientation sum together to form a single large dipole. This dipole is commonly referred to as an equivalent current dipole (ECD).

When ECDs are present in a conductive medium such as the brain, volume conduction occurs causing the signal to spread out towards the surface following the path of least resistance. In this case, a dipole is akin to one individual doing "the wave" at a stadium, while the ECD signal spread through volume conduction is similar to a crowd doing "the wave" around a stadium. Consequently, the voltage measured at the scalp is a reflection of the number of dipoles as well as their locations, orientations, strengths over time, interactions, and the resistance characteristics of the head. Currently the resolution of localizing individual ECD activity is on the level of cms.[14]

Typically, the voltage being recorded in EEG is quite small (μV–mV range) and changes in time occur within milliseconds. The main frequency range varies depending on the nature of the study, with a typical bandpass in EEG being 0.01–100 Hz and a typical bandpass in EP/ERPs being 0.1–30 Hz.[2] However, many applications exist in which other frequency ranges are explored (e.g., >30 Hz). The traditional strength of EEG resides in its temporal resolution, which is limited only by the sampling rate of the analog-to-digital converter and data management capabilities (neither of which pose significant barriers at the current time). However, a number of significant advances have been made in estimating the spatial localization of intracranial generators, which stem largely from increased number of channels (up to 256) and improving the source localization methods (e.g., dipole, current source density, and realistic head models). Two fundamental challenges that limit the ability to resolve spatial ECDs are the inverse problem and the spatial blurring of electrical potentials that pass through tissue and bone.[14]

24.3 Nuts and Bolts of Measuring Brain Potentials

Measurement of EEG involves basic biomedical and electrical engineering concepts. The core concepts behind the technology are fairly well established and are therefore reviewed briefly.[15]

Electrodes in contact with tissue create a standing potential difference (measured in mV) as well as impedance (measured in kΩ). If the standing potential differences and impedances are constant and approximately equal between electrodes, then there is little inherent current flow to the input circuits of the

amplifiers. Electrodes are typically made out of stainless steel, tin, gold, or silver coated with silver chloride.[16] Different types of electrodes have different standing potential characteristics, and therefore different recording characteristics. As a general rule, this necessitates that the type of electrodes should be kept constant. Similarly, impedances should be kept constant and as low as possible (<5–10 kΩ). This will help to avoid a potential-divider circuit with the input impedance of the amplifier. Prior to electrode placement, the site should be cleaned and prepared to ensure a good connection. A conductive gel is applied to each electrode site and the impedance is verified prior to recording. Depending on the purpose of EEG measurement, the electrodes may be held in place with a glue like substance (collodion is commonly used for long duration clinical recordings) or an elastic cap with embedded electrodes (Figure 24.4).[16] Electrodes are often held in place using adhesive collars and tape (e.g., electrodes around the eyes for an electro-oculogram or EOG). A variety of different electrodes exist, the selection of which depends on the specific application.[17]

To collect usable data, at least three electrodes are needed: one for active recording; one as a reference; and one ground electrode.[2] However, most manufactured systems include between eight and 256 channels. The names and locations of electrodes are specified by the International 10/20 system (except for high-density systems).[18] The International 10/20 system provides a consistent convention for 19 electrodes (plus the ground and the reference) in which the nasion (landmark at the top of the nose), inion (landmark extrusion at the back of the head), and preauricular (in front of the ears) points are used as anatomical landmarks. The distances between electrodes are either 10% or 20% of the distances between these anatomical landmarks, which accommodates for relative differences in head size. Electrode locations are specified by the particular combinations of letters and numbers. Letters indicate regions (F: frontal, C: central, T: temporal, P: parietal, O: occipital), which are then specified in terms of hemisphere (z: midline, even number: right, odd number: left) and distance from the midline (Fz: frontal midline, C3: central left medial; T8, temporal right distal; Figure 24.5). All electrodes are connected via standard leads and an

FIGURE 24.4 A single electrode.

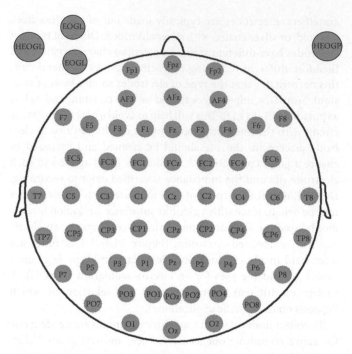

FIGURE 24.5   Diagram of high-density EEG electrode recording placement array with International 10-20 labels (where appropriate).

FIGURE 24.6   A subject set up for EEG recording.

electrode interface device (often called an electrode board adaptor or headbox) to either analog or digital amplifiers.[19]

The EEG signal is typically amplified by a factor of 1000–100,000.[2] This is done using differential amplifiers, in which signals from an active electrode go to one input and signals from a reference electrode go to the other input. Differential input to one is amplified, while common input to both is attenuated (common mode rejection).[20] Amplifiers are typically evaluated in terms of gain, which is a measurement of the output/input magnitude ratio and a useful description of the amplifiers capabilities.[16]

The method (or approach) for connecting active and reference electrodes to the differential amplifiers is often referred to as a recording montage.[2] There are three montage types that are commonly used: (1) common reference—the reference electrode is placed in a relatively neutral area (e.g., the nose) and is referenced to all other EEG channels; (2) average reference—activity from all EEG electrodes is averaged and used as a reference; and (3) bipolar—electrodes are paired and common noise is attenuated (e.g., this is commonly used in clinical recordings and EOG). In contemporary recordings, even though it is possible to set different montages, data are typically collected with a common reference so that montages can be selected or changed off-line. A person set up for EEG recording is depicted in Figure 24.6.

EEG signals are also bandpass filtered, either at the amplification stage and/or during off-line data processing. High-pass filters (low-frequency cutoff) are used to attenuate low-frequency drifts, which can, for instance, arise from changes in the standing potential differences at the skin (e.g., sweat). Common high-pass filter settings range from 0.01 to 0.1 Hz. Low-pass filters (high-frequency cutoff) are also used to attenuate frequencies outside of the range of interest. Common low-pass filter settings range from 30 to 100 Hz.[16] Often, a large source of noise comes from AC line current associated with nearby electrical supply (60 Hz in North America).[2] Accordingly, most EEG systems use a selective (or notch) filter to attenuate the AC line current artifact. The filters do not attenuate all frequencies outside the specified range completely. Rather, filter characteristics are specified in terms of a roll-off and expressed in terms of dB of attenuation per octave (e.g., 24 dB/octave). The design of a filter attempts to achieve a tradeoff between a narrow roll-off while avoiding passband or stop band ripple.

EEG signal is measured continuously and exists in analog form after the amplification process.[20] In order to store, display, and manipulate data on a computer, signals must be converted from an analog to a digital format by defining the waveforms with numerical values. Data are converted at a fixed rate with a constant inter-sample interval (ISI) or dwell time (the shorter the ISI, the faster the sampling rate).[16] The sampling rate is a critical factor, which is dictated by the requirements of Nyquist theorem. The Nyquist–Shannon sampling theorem states that an analog signal can be reconstructed if the sampling rate exceeds two times the highest frequency in the original sample.[20] Often the minimum sampling rate used is much greater than twice the highest frequency but is limited mainly by hardware and data storage capabilities.

If the sampling rate is set too low, information will be lost and artificial low frequencies may be introduced to the data (i.e., aliasing). Sampling skew can occur when there is a time lag between the sampling of each channel; to reduce sampling skew many systems sample all channels simultaneously.[2]

Analysis of EEG data typically focuses on the identification of specific events (e.g., abnormal spikes) or frequencies. A normal EEG recording for an adult in a relaxed state can be viewed in Figure 24.7. Although EEG data are most commonly presented in the time domain, it is possible to Fourier transform the data into the frequency domain. Data can be represented in the frequency domain by plotting the amplitude and phase at each

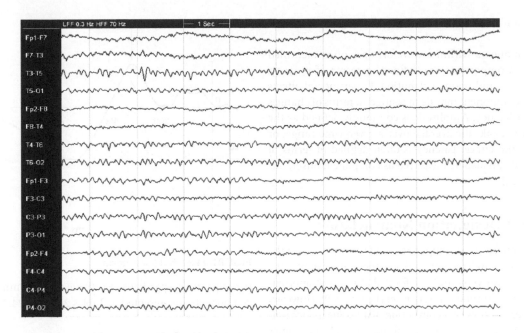

FIGURE 24.7 Normal EEG for an adult in a relaxed state.

frequency. Five well-known frequency bands of interest are presented in Table 24.1 and Figure 24.8.

24.4 Evoked Potentials and Event-Related Brain Potentials

The difference between collecting EEG data and EPs/ERPs relates primarily to signal averaging. EPs represent the brain's electrical reception of and response to an external stimulus (e.g., auditory, visual, and somatosensory). ERPs are acquired in much the same way as EPs, but can be differentiated conceptually in terms of measuring internal mental processes rather than external sensory processes (i.e., they are "related" to instead of "evoked" by an event). In time, EPs occur early on in basic information processing, whereas ERPs refer to later and more complex processing. Both EPs/ERPs are small electrical potentials (μV) embedded within background EEG activity. When stimuli are presented while EEG data are recorded, it is possible to extract the segments (or epochs) surrounding each stimulus presentation. Averaging the EEG activity in these epochs removes

random noise while maintaining stimulus-locked activity (i.e., increases the signal-to-noise ratio). As a result, signal averaging improves the detection of EPs/ERPs as a function of the number of trials averaged together. EPs are typically obtained by averaging 100s–1000s of trials because the responses are derived from deeper within the brain (i.e., far-field potentials). ERPs tend to originate from cortical sources (i.e., near-field potentials), so they typically require less averaged trials (approximately 30–100). EP/ERP waveform data may also be submitted to a number of postprocessing methods to improve data quality (e.g., artifact rejection/correction, baseline correction, additional filtering, et cetera). The resultant waveforms are typically analyzed in terms of components of interest, with polarity and amplitude on the

TABLE 24.1 Common EEG Frequency Bands

Name	Frequency (Hz)	Normal Presence
α	8–13	Present when eyes are closed
β	>13	Present when eyes are open or closed, sensitive to drugs/medications
θ	4–7	Present when drowsy
δ	<4	Present when in deep sleep
γ	20–100	Present when engaged in certain types of information processing

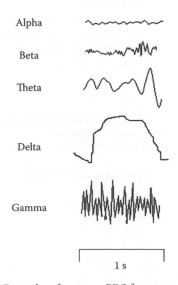

FIGURE 24.8 Examples of common EEG frequency patterns.

y-axis (in μV) and time on the *x*-axis (ms) at a given electrode location.[21,22]

Specific components within a waveform are often identified and examined. Components are defined in terms of polarity, amplitude, latency, and scalp distribution. Experimental effects on a particular component may be measured in terms of changes in amplitude or latency. Amplitude is commonly defined as the maximum at the peak, the average over a given time range, or the area under the curve. Latency is often defined in terms of peak latency, but can also be measured as component onset, offset, or time to peak.

24.4.1 Signal Designation Nomenclature

In terms of nomenclature, a component is typically named on the basis of its polarity and prototypical latency. For example, the N300 denotes a negative-going waveform that peaks at approximately 300 ms after stimulus presentation (Figure 24.9). However, not all components follow this convention, with some names reflecting form (ELAN or early left anterior negativity) and other names reflecting function (MMN or mismatch negativity). Experimental component effects are typically evaluated using statistical analyses such as a within-subject repeated-measures analysis of variance and data are often presented as the grand average across

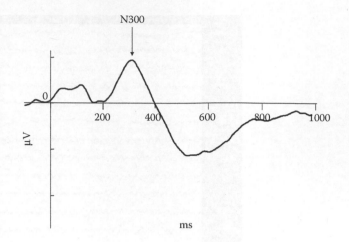

FIGURE 24.9 N300 waveform.

participants (at a particular electrode site).[2] Common EP and ERP components are presented in Tables 24.2 and 24.3.

24.4.2 High-Density Electrode Arrays and Spatial Source Analysis

Peaks or components in the waveform can also be considered in terms of the scalp distribution (i.e., scalp topography) or

TABLE 24.2 EP: Sensory Functions

Component	Latency	Type of Processing	Localization
C100 (visual; polarity varies)[23]	Peak = 80–100 ms	Visual processing sensitive to contrast and spatial frequency	Posterior midline (V1)
P100 (visual)[24]	Peak = 100–130 ms	Visual processing sensitive to stimulus contrast and direction of spatial attention	Lateral occipital
N100 (visual)[25]	Subcomponent 1: peak = 100–150 ms Subcomponent 2: peak = 150–200 ms	Present in visual discrimination tasks, sensitive to attention	Subcomponent 1: anterior Subcomponent 2: posterior (parietal/lateral occipital)
VEP[26,27]	Varies as a function of age	Processing of visual stimuli	Occipital
P100 (auditory)[28]	Peak = 50 ms	Auditory processing, sensitive to attention	Medial frontal
N1 (auditory)[29]	Subcomponent 1: peak = 75 ms Subcomponent 2: peak = 100 ms Subcomponent 3: peak = 150 ms	Auditory processing, sensitive to attention	Temporal
Mismatch negativity[30]	Peak = 160–220 ms	Auditory processing, sensitive to presentation of identical stimuli with occasional mismatching stimuli (present when subject is passive)	Central midline sites
BAEPs[27,31]	Peak = 10 ms (five peaks present after each stimulus presentation)	Sensitive to auditory click stimuli presented 10–70 times per second over thousands of trials (amplitude low)	Wave I: 8th cranial nerve to brain stem Wave II: not always present Wave III: lower pons IV and V: upper pons and lower midbrain
SEPs[32]	Varies by type of stimulation	Sensitive to peripheral nerve stimulation at a frequency of 3–5 Hz	Upper extremity SEPs N13: activity in cervical cord P14: activity in caudal medial lemniscus N18: activity in brainstem or thalamus N20: activity in primary somatosensory cortex Lower extremity SEPs LP: activity in lumbar cord N34: activity in brainstem and thalamus P37: activity of primary somatosensory cortex

TABLE 24.3 ERPs: Perceptual and Cognitive Functions

Component	Latency	Type of Processing	Localization
P200[33]	Peak = 130–250 ms	Visual processing of simple target features	Anterior and central
N170[34,35]	Peak = 170	Sensitive to face stimuli and inverted faces	Lateral occipital (greater on the right)
Vertex positive potential (VPP)	VPP Peak = 150–200 ms	VPP sensitive to highly familiar stimuli, including words	VPP: central midline
P300[36,37] Subcomponent a	Peak = 300 ms	Subcomponent a: sensitive to unpredictable and infrequent changes in stimulus presentation; involved in context updating	Subcomponent a: frontal
Subcomponent b		Subcomponent b: present during infrequent task relevant stimuli	Subcomponent b: parietal
N400[38,39]		Processing of language, sensitive to volations of semantic expectancies	Central and parietal sites (slightly larger over right hemisphere)—Appears to be generated by left temporal lobe (points medially to create right focus)
P600[39]	Peak = 600 ms	Processing of language, sensitive to syntactic violations	Frontal
Error-related negativity[40] (negative going wave sometimes followed by positivity)	Peak = 50–100 ms after motor response	Processing which occurs when mistakes are recognized in a task or when negative feedback is given	Frontal and central sites

estimated spatial location of the intracranial generators (i.e., source analysis). Scalp topography is inferred through interpolation of voltage iso-contour lines between electrode sites (with more electrodes providing better spatial maps). Most scalp distribution maps use spline interpolation, but it is also possible to estimate the current source density maps. Ohm's law ($V = IR$) dictates that current flowing between two points is directly related to differences in potential and inversely related to the resistance between the two points. In reference to measuring voltages on the head, this means that current flowing across the skin (the path of lesser resistance relative to the skull) can be described as a current density. Assuming that resistance is constant, current densities can be computed using two-dimensional spatial derivatives of the voltage map. This highlights local details of the spatial maps, serving in essence to de-blur the topographic data. Although surface (and transformed) maps can be useful in highlighting differences between components, they should not be interpreted as reconstructions of intracranial source activity.

To estimate intracranial sources, a model is needed that describes how signals are generated within the brain and subsequently conducted to the outside of the head. This is commonly referred to as source localization.[41] To elaborate, if (1) the location and orientation of a dipole was known and (2) the manner in which the electrical activity was conducted to the scalp was known, then it would be possible to determine the voltage recorded at a fixed point on the scalp. This is referred to as the forward solution.[42] However, the reverse situation occurs in EEG. The voltage distributions at fixed points on the scalp are known, but location and orientation of dipoles must be determined. This is referred to as the inverse problem.[43] The problem is under-defined, as there are an infinite number of possible solutions and a finite number of electrodes. As a result, a unique

solution cannot be defined and constraints must be applied to the model.[2]

Despite the challenges inherent in the inverse problem, a number of advances in technical and mathematical approaches have greatly improved the spatial resolution of source analysis. Technical advances in localization have resulted largely from higher-density EEG systems (128 and 256 channel arrays). Mathematical approaches to source analysis are generally divided into two methods (both rely on similar assumptions related conductivities and head models). Dipole modeling approaches estimate the number, location, and orientation of ECDs and evaluate the model in terms of goodness-of-fit with the measured data (e.g., Brain Electromagnetic Source Analysis or BESA).[44,45] This approach typically requires constraints based on anatomy and/or prior knowledge. Distributed source modeling divides the brain into volume elements (or voxels) and estimates the current densities across all voxels (e.g., LOw-Resolution Electromagnetic Tomographic Analysis or LORETA; e.g., Figure 24.8).[46–48] This approach typically requires mathematical and/or anatomical constraints. There are a number of advantages and disadvantages to either type of source analysis approach,[49] with the choice of a particular approach depending on the specifics of the application.

24.5 Clinical Applications

This section describes various clinical applications where EEG monitoring can provide assistance in medical decision making.

24.5.1 Electroencephalogram

The following are examples of EEG monitoring conditions.

24.5.1.1 Epilepsy

EEG is an invaluable tool in the diagnosis epilepsy, detecting seizure activity, classifying seizure types, and assessing seizure frequency.[50] In particular, an important feature of EEG is the ability to conduct long-term, noninvasive monitoring for seizure activity.[51] The EEG data are often used to help localize the focus of seizure activity prior to surgical intervention. It is also routinely used in the operating room to assess whether the tissue being resected is epileptogenic. Typically, seizure activity can be detected visually or by software programs, and presents as spikes or sharp waves (either during a seizure or between seizures; Figure 24.10). The standard epilepsy EEG application uses the International 10/20 system, with a bipolar montage (longitudinal and transverse chains of linked electrodes).[50] Such set ups should not include the use of muscle or notch filters as these may filter out low-frequency spikes of interest to the diagnostician. For recordings over long durations (sometimes weeks), EEG is often collected on a telemetry unit that allows for continuous video recording of patient's behavior.[50,52] When long-term monitoring is performed, it is common to use a portable system that allows for mobility.

24.5.1.2 Sleep Assessment

Our knowledge of sleep has expanded greatly as a direct result of EEG recordings.[53] EEG studies have revealed the different stages of sleep. Wakefulness can be recognized by low-voltage waves (10–30 μV) with mixed frequency bands. As shown in Table 24.1, α, β, θ represent distinct levels of alertness/arousal during wakefulness in which the EEG changes from high-frequency low-amplitude activity (beta) to activity with increasingly lower frequencies and larger amplitudes (α and θ; Figure 24.11). Theta

is also referred to as Stage 1 sleep, and is characterized by slow eye movements with mixed EEG frequencies and amplitudes in the 2–7 Hz range. Stage 2 sleep is characterized by sleep spindles (12–14 Hz) and K complexes (sharp negative wave followed by a slow positive going wave). Stage 3 sleep is characterized by low-frequency activity (<2 Hz) with amplitudes above 65 μV. Stage 4 sleep is similar to Stage 3, but has waves with slightly higher amplitudes. Perhaps the most well-known stage is rapid eye movement (REM) sleep, which is commonly associated with dreaming. Similar to awake EEG, EEG during REM sleep is characterized by low voltages and mixed frequencies with a sawtooth pattern (which is why it is often referred to as paradoxical sleep). These sleep stages are now known to cycle through the night.

Sleep assessments are often referred to as polysomnography. Polysomnography typically uses an EEG setup similar to that used in epilepsy in combination with a variety of other techniques including electromyography, electrocardiography, and assessments of oxygen saturation and air flow.[54] EEG can be used to identify dysomnias (disturbances in the regular pattern of sleep) or parasomnias (sleep disorders). A single night of polysomnography is often enough to determine whether an individual may be experiencing narcolepsy or sleep apnea among other disorders.

There are several mechanisms available to provide maximum value assessment of brain activity. The following sections will describe some of the available techniques.

24.5.2 Evoked Potentials

The main EP methods are reviewed below. For more details on any or all of these methods and the related clinical

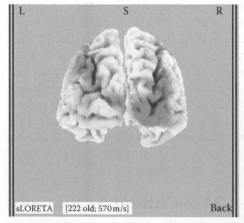
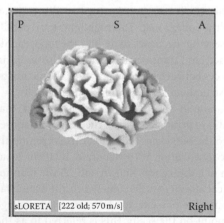
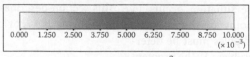

FIGURE 24.10 Source maps derived from LORETA.

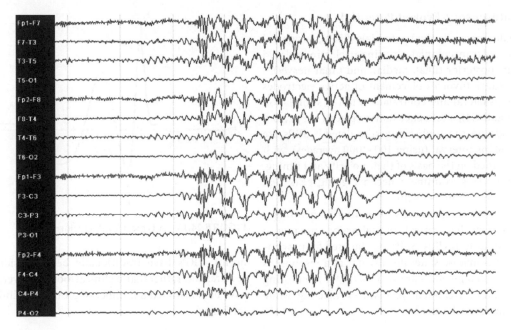

FIGURE 24.11 Generalized spike and wave activity seen in a patient with epilepsy.

applications, the reader is directed towards more comprehensive reviews.[16,20]

24.5.2.1 Brainstem Auditory-Evoked Potentials

Brainstem auditory-evoked potentials (BAEPs) are used commonly in the evaluation of both hearing and brain stem function (especially useful in infants).[27] BAEPs typically involve only one active electrode on the vertex of the head (often at Cz) to record the EP from a series of auditory tones.[55] In healthy individuals, the resultant waveform has a series of well-characterized components that occur within 10 ms from stimulus onset (using 1000–2000 tones). The first three waves (waves I, II, and III) originate from the auditory nerve, cochlear nuclei, and superior olive, respectively. Waves IV and V are thought to originate further along the auditory pathway, in the lateral lemniscus and inferior colliculus (respectively).

Extensive normative data have been collected for each of these waves, which makes it possible to identify changes in either the latencies or amplitudes of BAEP peaks classified as "abnormal."[27] Given that they reflect both peripheral (auditory nerve) and central (brainstem) function, BAEPs have a wide variety of clinical applications. For instance, they are used in the detection of hearing loss, tumors, and central nervous system inflammation/insult. They are also used for the assessment of comatose patients, with the ability to provide prognostic information about recovery of consciousness and outcome.

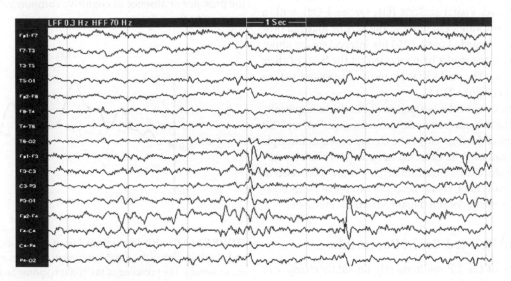

FIGURE 24.12 EEG data from an adult during sleep.

24.5.2.2 Somatosensory-Evoked Potentials

During the somatosensory-evoked potentials (SEPs), peripheral nerves are stimulated transcutaneously using an electrode placed over a nerve (e.g., posterior tibial, medial, and ulnar nerves).[56] The typical procedure involves placement of active electrodes on sites that are ipsilateral and contralateral to the stimulation (usually near C3 and C4), a reference may be placed at Fpz, and a ground electrode on the limb being stimulated.[56] SEPs are recorded over a 50 ms interval (using 100–1000 averages), with the peaks being named on the basis of their polarity and timing. For example, many labs examine a negative-going peak that occurs 20 ms after the stimulus (N20).

SEPs are often used intraoperatively to monitor for neurological dysfunction and localize critical cortical regions in order to avoid damage. Similar to BAEPs, SEPs may be used for prognosis of patients in a comatose state; the absence of a response is indicative of poor outcome. SEPs are also useful for monitoring the progression and/or treatment of diseases affecting white matter, such as multiple sclerosis.[32]

24.5.2.3 Motor-Evoked Potentials (MEPs)

In contrast to SEPs, motor evoked potentials (MEPs) involve recording from muscles during stimulation of motor cortex. Stimulation of motor cortex can be done by either directly stimulating exposed cortex or through transcranial stimulation (magnetically or electrically).[57,58] MEPs are done intraoperatively to monitor the functional integrity of key motor systems (e.g., the pyramidal tract).[59,60] They are also used in transcranial magnetic stimulation (TMS) to establish magnetic stimulation thresholds.

24.5.2.4 Visual Evoked Potentials

There are a number of different types of visual-evoked potentials (VEPs), which typically depend on the specific form of visual stimulation. Visual stimuli such as light flashes or pattern reversal checkerboards are commonly used. Active electrodes are typically placed over visual regions (O1, Oz, and O2), with a ground at the vertex and a reference at the ears or mastoid processes.[61] VEPs are recorded over a 250 ms interval (using about 100 averages), with the peaks being named on the basis of their polarity and timing. The peaks that are most often examined are the N75 and P100 (Figure 24.13).

VEPs are often used in the evaluation of blindness (especially in infants) and the detection of optic nerve damage.[27] Also, VEP latency delays are characteristic of the changes seen in multiple sclerosis, which can even be detected in patients without symptoms of visual dysfunction. Additionally, VEPs are often used to study the basic aspects of visual perception. Figure 24.13 presents a typical waveform obtained using visual EP methods.

24.5.3 Event-Related Brain Potentials

Given that many of the EP methods rely on subtle changes in latency or amplitude characteristics (compared against the

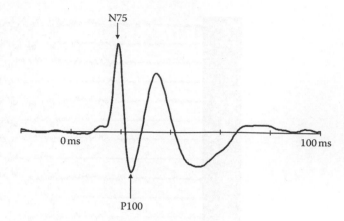

FIGURE 24.13 A typical VEP waveform.

norm), ERPs have traditionally been considered more of a research tool rather than a clinical tool. However, recent advances have demonstrated a number of promising clinical applications for ERPs.

24.5.3.1 Cognitive Assessment

Cognitive ERP assessment has emerged as an important method for evaluating patients who cannot be assessed with traditional neuropsychological tests.[62-64] Neuropsychological tests rely on the ability for the patient to respond behaviorally (either an oral or motor response). However, brain damage is often associated with concomitant impairments in the ability to respond behaviorally (e.g., stroke and traumatic brain injury).[65]

In these situations, intact cognitive functioning can be detected through the presence of well-known ERP components. For example, the N400 component, which is sensitive to semantic comprehension, may be used to evaluate language function following left hemisphere stroke. Additionally, ERPs can be exceptionally useful in the assessment of consciousness.[66] Specifically, in cases where the patient is unable to communicate, ERPs can be used to evaluate level of awareness based on the presence or absence of cognitive components (e.g., the N400; Figure 24.14).

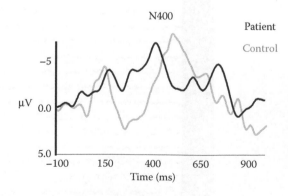

FIGURE 24.14 ERP data collected from a noncommunicative patient and a healthy control during the presentation of semantically incongruent sentences. The presence of the N400 response indicates that stimulus processing/comprehension has occurred.

24.6 Emerging Applications

Despite the fact that EEG has been around for almost a century, there is still much to be discovered from passive recordings of the brain's electrical activity. Highlighted below are promising applications that demonstrate the richness and complexity inherent in the EEG signal as well as the essential requirement to integrate EEG data with data from other imaging modalities (i.e., multimodal imaging).

24.6.1 Resting State EEG

The idea that the brain is never fully at rest is not new. Indeed, the previous review of EEG rhythms demonstrates this fact elegantly. However, there has been a recent resurgence of studies in this area, particularly with respect to investigating functional connectivity. Functional imaging is increasingly playing a more important role in determining the relationships between different brain areas.[67,68]

Functional connectivity can be evaluated either through attempts to map connectivity in the whole brain or through the use of prior knowledge to analyze a restricted set of regions.[69] These assessments typically rely on correlational or covariance analyses. In EEG, the most common method involves coherence analysis of oscillations recorded from different scalp sensors.[68] However, electrophysiological methods have also given rise to cortico-cortical EP methods,[70] which can be used to demonstrate bidirectional connections between different brain regions.

24.6.2 High-Frequency EEG

Until recently, high-frequency EEG oscillations were not well studied, generally because the amplitudes of these waves are relatively small.[71] Nevertheless, findings from emerging studies indicate that synchronized activity around 40 Hz, known as Gamma activity, can provide valuable insight into brain function. Gamma band activity (20–100 Hz) is associated connectivity between the thalamus and the cortex. It has been linked to the process of binding features of an object together—a process that is regulated by attention[72] and memory.[73] Furthermore, gamma activity has been observed to vary in a number of psychiatric and neurological disorders.[71] For instance, individuals with positive symptoms of schizophrenia show increased gamma activity, and those with negative symptoms show decreased gamma activity. An increase in gamma band frequency has also been observed in individuals with epilepsy and is thought to be associated with the experience of an aura. In addition to gamma activity, investigations are also beginning to examine ultra-high-frequency oscillations (>100 Hz).

24.6.3 Event-Related Desynchronization/Synchronization

Alpha band EEG activity is considered to be representative of cortical inactivation. However, studies have revealed that short-lasting event-related desynchronization (ERD) of frequencies in the alpha band are indicative of cortical activation. In contrast, event-related synchronization (ERS) of frequencies in the alpha range have been thought to signify a deactivated area or inhibited network.[74] Interestingly, ERS can be elicited when subjects withhold the execution of a response, leading to the idea that ERS reflects top-down inhibitory control.[75] To date, the majority of studies on ERD/ERS have focused on the measurement of cortical activation and deactivation using a motor control model. As this area expands to include other aspects of cognition, measuring ERD/ERS may allow for a better understanding of the complex cortical networks that underlie processes such as memory and attention.

24.6.4 Brain Computer Interface

EEG is a major component in the development of brain computer interface devices. The central idea is to create neural prosthetics to replace lost functions in a damaged brain. For example, individuals with locked-in-syndrome are cognitively intact and conscious but are unable to communicate due to damage causing severe paralysis. Research has shown that the presence of a P300 ERP component can be used to select letters on a monitor in order to relay a message.[76] Work in this area has also examined slow cortical potentials, rhythms (e.g., beta), and other ERPs, with the objective of training patients to manipulate these EEG features in order to control their environments.

24.6.5 EEG and Advances in Neuroimaging

EEG was among the first physiological measures of brain function. Since then, a number of other noninvasive methods have been developed to study brain function. These can be divided into two broad categories: (1) those that take advantage of electrophysiological properties of the brain, such as EEG, magnetoencephalography (MEG), and TMS; and (2) those that utilize hemodynamic/metabolic properties of the brain, such as functional magnetic resonance imaging (fMRI), positron emission tomography (PET), single photon emission computed tomography (SPECT), and near infrared spectroscopy (NIRS). Hemodynamic techniques like fMRI are indirect measures of neural activity, which fundamentally limits their temporal resolution.[77] However, they are capable of providing higher spatial resolution, and therefore complement the high temporal resolution of EEG. This approach reflects a general principle in the field, namely, that no one technology is sufficient but rather investigations of brain function through the integration of information from multiple modalities (i.e., multimodal imaging).

In this respect, one of the closest technologies to EEG is MEG. EEG and MEG share many, but not all, fundamental characteristics. Unlike EEG, MEG measures the extremely weak magnetic fields that are generated from current dipoles (approximately $10^{-13/-14}$ T).[13] In order to measure these small magnetic fluctuations (i.e., increase the signal-to-noise ratio), MEG uses a combination of super-conducting quantum interference devices to improve signal detection along with passive and active noise cancellation.[13] Because magnetic fields can pass through the skull and scalp relatively unaffected, MEG is able to provide a higher spatial and similar temporal resolution relative to EEG. Given the close relationship between EEG and MEG, it is likely the two technologies will continue to become increasingly integrated as instrumentation and applications develop.

24.7 Summary

EEG represents an established, straightforward method of examining brain function. The fundamentals of EEG are relatively well understood from the biomedical engineering and neuroscience perspectives. The technology has given rise to a number of derivative methods (e.g., EPs), analyses (e.g., intracranial source analysis), and applications (e.g., cognitive ERP assessment). It is relatively inexpensive, practical, and robust, which makes it ideal for many applications that would not be otherwise possible (e.g., brain computer interface).

References

1. Niedermeyer, F. and Lopes da Silva, F.H. *Electroencephalography: Basic Principles, Clinical Applications and Related Fields.* Baltimore, MD: Lippincott Williams & Wilkins, 2004.
2. Luck, S.J. *An Introduction to the Event-Related Potential Technique.* Michigan: MIT Press, 2005.
3. Berger, H. Ueber das electrenkephalogramm des menschen. *Archives fur Psychiatrie Nervenkrankheiten*, 87: 527–570, 1929.
4. Adrian, E.D. and Matthews, B.H.C. The Berger rhythm: Potential changes from the occipital lobes in man. *Brain*, 57: 355–385, 1934.
5. Jasper, H.H. and Carmichael, L. Electrical potentials from the intact human brain. *Science*, 81: 51–53, 1935.
6. Gibbs, F.A., Davis, H., and Lennox, W.G. The electroencephalogram in epilepsy and in conditions of impaired consciousness. *Archives of Neurology and Psychiatry*, 34: 1133–1148, 1935.
7. Davis, H., Davis, P.A., Loomis, A.L., Harvey, E.N., and Hobart, G. Electrical reactions of the human brain to auditory stimulation during sleep. *Journal of Neurophysiology*, 2: 494–499, 1939.
8. Walter, W.G., Cooper, R., Aldridge, V.J., McCallum, W.C., and Winter, A.L. Contingent negative variation: An electric sign of sesorimotor association and expectancy in the human brain. *Nature*, 203: 380–384, 1964.
9. Williams, R.W. and Herrup, K. The control of neuron number. *The Annual Review of Neuroscience*, 11: 423–453, 1988.
10. Nolte, J. *The Human Brain: An Introduction to its Functional Anatomy.* Missouri: Mosby Year Book Inc., 1993.
11. Kandel, E., Schwartz, J., and Jessell, T. *Principles of Neural Science.* New York: McGraw-Hill Medical, 2000.
12. Rugg, M.D. *Cognitive Neuroscience.* Cambridge, MA: MIT Press, 1997.
13. Hamalainen, M., Hari, R., Ilmoniemi, R.J., Knuutila, J., and Lounasmaa, O.V. Magnetoencephalography—theory, instrumentation, and applications to noninvasive studies of the working human brain. *Reviews of Modern Physics*, 65: 413–497, 2003.
14. Gevins, A., Leong, H., Smith, M.E., Le, J., and Du, R. Mapping cognitive brain function with modern high-resolution electroencephalography. *Trends in Neuroscience*, 18: 429–436, 1995.
15. Cooper, R., Osselton, J.W., and Shaw, J.C. *EEG Technology.* London: Butterworths, 1974.
16. Chiappa, K.H. *Evoked Potentials in Clinical Medicine.* Massachusetts: Lippincott-Raven, 1997.
17. Binnie, C.D., Cooper, R., and Mauguiere, F. *Clinical Neurophysiology: EEG Paediatric Neurophysiology, Special Techniques and Applications.* Amsterdam: Elsevier Health Sciences, 2003.
18. American Clinical Neurophysiology Society Guideline 5: Guidelines for Standard Electrode Position Nomenclature, 2006.
19. Najarian, K. and Splinter, R. *Biomedical Signal and Image Processing.* Boca Raton, FL: The CRC Press, 2006.
20. Regan, D. *Evoked Potentials in Psychology, Sensory Physiology and Clinical Medicine.* London, UK: Taylor & Francis, 1972.
21. Handy, T.C. *Event-Related Potentials: A Methods Handbook.* Cambridge, MA: MIT Press, 2004.
22. Cacioppo, J.T., Tassinary, L.G., and Berntson, G.G. *Handbook of Psychophysiology.* Cambridge: Cambridge University Press, 2007.
23. Clark, V.P., Fan, S., and Hillyard, S.A. Identification of early visually evoked potential generators by retinotopic and topographic analyses. *Human Brain Mapping*, 2: 170–187, 1995.
24. Di Russo, F., Martinex, A., Sereno, M.I., Pitzalis, S., and Hillyard, S.A. Cortical sources of the early components of the visual evoked potential. *Human Brain Mapping*, 15: 95–111, 2002.
25. Herrmann, C.S. and Knight, R.T. Mechanisms of human attention: Event-related potentials and oscillations. *Neuroscience and Biobehavioural Reviews*, 25: 465–476, 2001.
26. Sokol, S. Visually evoked potentials: Theory, techniques and clinical applications. *Survey of Ophthalmology*, 21: 18–44, 1976.
27. Nuwer, M.R. Fundamentals of evoked potentials and common clinical applications today. *Electroencephalography and Clinical Neurophysiology*, 16: 2719–2733, 1998.
28. Boop, F.A., Garcia-Rill, E., Dykman, R., and Skinner, R.D. The P1: Insights into attention and arousal. *Pediatric Neurosurgery*, 20: 57–62, 1994.
29. Woods, D.L. The component structure of the N1 wave of the human auditory evoked potential. *Electroencephalography and Clinical Neurophysiology Supplement*, 44: 102–109, 1995.
30. Näätänen R, Paavilainen P, Rinne T, and Alho K. The mismatch negativity (MMN) in basic research of central auditory processing: A review. *Clinical Neurophysiology*, 18: 2544–2590, 2007.

31. Legatt, A.D., Arezzo, J.C., and Vaughan, H.G. The anatomic and physiological bases of brain stem evoked potentials. *Neurologic Clinics*, 6: 681–704, 1988.

32. Cruccua, G., Aminoffb, G., Curioc, J.M., Gueritd, R., Kakigie, F., Mauguiere, F.J., Rossing, P.M., Treedei, R.D., and Garcia-Larreaf, L. Recommendationf for the clinical use of somatosensory-evoked potentials. *Clinical Neurophysiology*, 119: 1705–1719, 2008.

33. Luck, S.J. and Hillyard, S.A., Electrophysiological correlates of feature analysis during visual search. *Psychogphysiology*, 31: 291–308, 1994.

34. Jeffreys, D.A. A face-responsive potential recorded from the human scalp. *Experimental Brain Research*, 78: 193–202, 1989.

35. Herrmann, M.J., Ehlis, A.C., Muehlberger, A., and Fallgatter, J. Source localization of early stages of face processing. *Brain Topography*, 18: 77–85, 2005.

36. Squires, N.K., Squires, K.C., and Hillyard, S.A. Two varieties of long-latency positive waves evoked by unpredictable auditory stimuli. *Electroencephalography and Clinical Neurophysiology*, 38: 387–401, 1975.

37. Polich, J. Updating P300: An integrative theory of P3a and P3b. *Clinical Neurophysiology*, 118: 2128–2148, 2007.

38. Kutas, M. and Hillyard, S.A. Reading senseless sentences: Brain potentials reflect semantic incongruity. *Science* 207: 203–205, 1980.

39. Friederici, A.D. Event-related brain potential studies in language. *Current Neurology and Neuroscience Reports*, 4: 466–470, 2004.

40. Gehring, W.J., Goss, B., Coles, M.G.H., Meyer, D.E., and Donchin, E. A neural system for error-detection and compensation. *Psychological Science*, 4: 385–390, 1993.

41. Koles, Z.J. Trends in EEG source localization. *Electroencephalography and Clinical Neurophysiology*, 106: 127–137, 1998.

42. Hallez, H., Vanrumste, B., Grech, R., Muscat, J., De Clercq, W., Vergult, A., D'Asseler, Y., et al. Review on solving the forward problem in EEG source analysis. *Journal of Neuroengineering and Rehabilitation*, 4: 46, 2007.

43. van Oosterom, A. History and evolution of methods for solving the inverse problem. *Journal of Clinical Neurophysiology*, 8: 371–380, 1991.

44. Scherg, M., Vajsar, J., and Picton, T. A source analysis of the human auditory evoked potentials. *Journal of Cognitive Neuroscience*, 1: 336–355, 1989.

45. Scherg, M. and von Cramon, D. A new interpretation of the generators of BAEP waves I-V: Results of a spatio-temporal dipole model. *Electroencephalography and Clinical Neurophysiology*, 62: 290–299, 1985.

46. Pascual-Marqui, R.D., Esslen, M., Kochi, K., and Lehmann, D. Functional imaging with low-resolution brain electromagnetic tomography (LORETA): A review. *Methods & Findings in Experimental & Clinical Pharmacology*, 24 supp C: 91–95, 2002.

47. Pascual-Marqui, R.D., Michel, C.M., and Lehmann, D. Low resolution electromagnetic tomography: A new method for localizing electrical activity in the brain. *International Journal of Psychophysiology*, 18: 49–65, 1994.

48. Pascual-Marqui, R.D. Standardized low-resolution brain electromagnetic tomography (sLORETA): Technical details. *Methods & Findings in Experimental & Clinical Pharmacology*, 24 supp D: 5–12, 2002.

49. Michel, C.M., Murray, M.M., Lantz, G., Gonzalez, S., Spinelli, L., and Grave de Peralta, R. EEG source imaging. *Epilepsia*, 43: 80–93, 2002.

50. Cascino, G.D. Video-EEG monitoring in adults. *Epilepsia*, 43: 80–93, 2002.

51. Harkness, W. Temporal lobe resections. *Childs Nervous System*, 22: 936–944, 2006.

52. Tatum, W.O. Long-term EEG monitoring: A clinical approach to electrophysiology. *Journal of Clinical Neurophysiology*, 18: 442–455, 2001.

53. Susmakova, K. Human sleep and sleep EEG. *Measurement Science Review*, 4: 59, 2004.

54. Binnie, C.D., Prior, P.F. Electroencephalography. *Journal of Neurology, Neurosurgery and Psychiatry*, 57: 1308–1319, 1994.

55. American Clinical Neurophysiology Society Guideline 9C: Guidelines on Short-Latency Auditory Evoked Potentials, 2006.

56. American Society Clinical Neurophysiology Guideline 9A: Guidelines on Evoked Potentials, 2006.

57. Penfield, W.G. and Boldrey, E. Somatic motor and sensory representation in the cerebral cortex of man as studied by electrical stimulation. *Brain*, 60: 389–443, 1937.

58. Merton PA, Morton HB. Stimulation of the cerebral cortex in the intact human subject. *Nature*, 285: 227, 1980.

59. Nagle, K.J., Emerson, R.G., Adams, D.C., Heyer, E.J., Roye, D.P., Schwab, F.J., Weidenbaum, M., et al. Intraoperative monitoring of motor evoked potentials: A review of 116 cases. *Neurology*, 47: 999–1004, 1996.

60. Curra, A., Modugno, N., Inghilleri, M., Manfredi, M., Hallet, M., and Berardelli, A. Transcranial magnetic stimulation techniques in clinical investigation. *Neurology*, 59: 1851–1859, 2002.

61. American Clinical Neurophysiology Society Guideline 9B: Guidelines on Visual Evoked Potentials, 2006.

62. Connolly, J.F., Major, A., Allen, S., and D'Arcy, R.C. Performance on WISC-III and WAIS-R NI vocabulary subtests assessed with event-related brain potentials: An innovative method of assessment. *Journal of Clinical and Experimental Neuropsychology*, 21: 444–464, 1999.

63. Marchand, Y., D'Arcy, R.C., and Connolly, J.F. Linking neurophysiological and neuropsychological measures for aphasia assessment. *Clinical Neurophysiology*, 113: 1715–1722, 2002.

64. Connolly, J.F., Marchand, Y., Major, A., and D'Arcy, R.C. Event-related brain potentials as a measure of performance on WISC-III and WAIS-R NI similarities sub-tests. *Journal of Clinical and Experimental Neuropsychology*, 28: 1327–1345, 2006.

65. Connolly, J.F., Mate-Kole, C.C., and Joyce, B.M. Global aphasia: An innovative assessment approach. *Archives of Physical Medicine and Rehabilitation*, 80: 1309–1315, 1999.

66. Gawryluk, J.R., D'Arcy, R.D., Connolly, J.F., and Weaver, D.F. Improving the clinical assessment of consciousness in the electric brain, *BMC Neurology*, provisionally accepted, 2009.

67. Buchel, C. Perspectives on the estimation of effective connectivity from neuroimaging data. *Neuroinformatics*, 2: 169–173, 2004.

68. Shibasaki, H. Human brain mapping: Hemodynamic response and electrophysiology. *Clinical Neurophysiology*, 119: 731–743, 2008.

69. Rogers, B.P., Morgan, V.L., Newton, A.T., and Gore, J.C. Assessing functional connectivity in the human brain by fMRI. *Magnetic Resonance Imaging*, 25: 1347–1357, 2007.

70. Matsumoto, R., Nair, D.R., and LaPresto, E. Functional connectivity in the human language system: a cortico-cortical evoked potential study. *Brain*, 127: 2316–2330, 2004.

71. Herrmann, C.S. and Demiralp, T. Human EEG gamma oscillations in neuropsychiatric disorders. *Clinical Neurophysiology*, 116: 2719–2733, 2005.

72. Fell, J., Fernandex, G., Klaver, P., Elger, C.E., and Fries, P. Is synchronized neuronal gamma activity relevant for selective attention? *Brain Research Reviews*, 42: 265–272, 2003.

73. Bertrand, O. and Tallon-Baudry, C. Oscillatory gamma activity in humans: A possible role for object representation. *International Journal of Psychophysiology*, 38: 211–223, 2000.

74. Neuper, C., Wortz, M., and Pfurtscheller, G. ERP/ERS patterns reflecting sensorimotor activation and deactivation. *Progress in Brain Research*, 159: 211–222, 2006.

75. Klimesch, W., Sauseng, P., and Hanslmayr, S. EEG alpha oscillations: The inhibition timing hypothesis. *Brain Research Reviews*, 53: 63–88, 2007.

76. Furdea, A., Halder, S., Krusienski, D.J., Bross, D., Nijboer, F., Birbaumer, N., and Kubler, A. An auditory oddball (P300) spelling system for brain-computer interfaces. *Psychophysiology*, 2009.

77. Jezzard, P., Matthews, P.M., and Smith, S.M. *Functional MRI: An Introduction to Methods*. New York: Oxford University Press, 2001.

25

Bioelectric Impedance Analysis

Alexander V. Smirnov,
Dmitriv V. Nikolaev,
and Sergey G. Rudnev

25.1 Introduction

Bioelectric impedance analysis (BIA) is a noninvasive, portable, and relatively inexpensive method widely used in health-related studies and clinical medicine. When applied to body composition analysis, BIA provides estimates of morphological and physiological parameters such as fat- and fat-free mass (FFM), body cell mass, tissue hydration, basal metabolic rate, and others. The use of bioelectrical impedance diagnostics relies on the influence of certain diseases and homeostatic conditions on the local cellular electrical circuitry and the respective changes in behavior with respect to the applied current frequency pattern. Typically, BIA includes electrodes positioning on the borders of body segment under study and measurement of impedance parameters at single or multiple frequencies using a specialized measurement device—the bioelectric impedance analyzer. The most popular areas of BIA application include objective control of obesity treatment procedures as well as estimates of patients' hydration status in intensive care units, including hemodialysis. This chapter will describe the electronic foundation of BIA and will provide several clinical applications. The technical basis of BIA measurements is described followed by examples of specific diagnostic applications.

25.2 Resistance and Impedance

The term "electrical resistance" was suggested by George Ohm. In 1825–1827, he had found experimentally the relation between electric current I (amps), potential difference U (volts), and resistance (ohms), known in its modern form as Ohm's law:

$$I = \frac{U}{R}. \tag{25.1}$$

The opposite, or inverse, of resistance, $G = 1/R$, is called conductance and is measured in siemens (S).

Figure 25.1a shows an electric circuit that consists of a resistor and a capacitor in series. In the case of alternative current, it is characterized by the complex impedance Z:

$$Z = R - jX_C, \tag{25.2}$$

where X_C is the reactance—the imaginary part of impedance that reflects capacitive properties of the circuits, and j is the imaginary unit defined as $j^2 = -1$.

The absolute value of reactance X_C equals

$$X_C = \frac{1}{\omega C}, \tag{25.3}$$

where ω is the angular frequency (rad/s) that is related to the phenomenological frequency f (Hz) as $\omega = 2\pi f$, and C is the capacitance (farad, F).

The phase of alternating voltage on the resistance U_R matches the phase of alternating current I. The minus sign in Equation 25.2 indicates that the voltage to current lag on the capacitance U_C is 90°. The vector diagram of the voltage U in this complex plane is shown in Figure 25.1b.

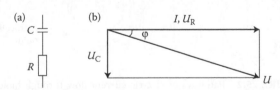

FIGURE 25.1 (a) Resistor and capacitor in series and (b) the vector diagram of current and voltage on the circuit elements.

The impedance on a complex plane can also be characterized in terms of magnitude, or absolute value, Z, and phase angle (PA) φ_Z:

$$Z = \sqrt{R^2 + X_C^2},$$
$$\varphi_Z = -\mathrm{arctg}\frac{X_C}{R}. \tag{25.4}$$

In the case of alternating current, Ohm's law (Equation 25.1) has more complicated form: the magnitude of impedance Z is substituted for R, whereas the values of I and U are replaced by active, or root-mean-square, values of current and voltage, respectively.

In electrical engineering, along with capacitive reactance X_C, an inductive reactance X_L is formed. Voltage across a coil preceeds current by 90°. The role of the inductive component in biological impedance is questionable and usually neglected.

The resistance reflects an ability of the conductive medium to allow heat dissipation of electric current energy as expressed by the electrical energy dissipation. In biological tissues the medium is represented by extracellular and intracellular liquid volumes (Figure 25.2).

The reactance characterizes dielectric properties of biological objects exerted by cell membranes and tissue interfaces and, hence, the ability to store energy. Under the influence of the electric field the accumulation of charges occurs on the interface of cells and tissues that balance an applied field.

The reciprocal of complex impedance, $Y = 1/Z$, is called admittance. Admittance is also expressed in a real and imaginary part (both measured in siemens):

$$Y = G + jB_C. \tag{25.5}$$

Tissue as such can be characterized by a representative electronic circuit. In an equivalent circuit diagram the components of complex admittance are combined in parallel. The phase of alternating current through the capacitor preceeds the voltage by 90° since an imaginary part of the admittance is positive.

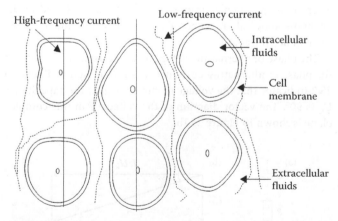

High-frequency current

Low-frequency current

Intracellular fluids

Cell membrane

Extracellular fluids

FIGURE 25.2 Pathways of electric current flow through biological tissue: the low-frequency current passes mainly through the ECW space; the high-frequency current penetrates the cell membrane and is thus conducted through both the ECW and ICW spaces.[1]

In order to derive expressions relating the real and imaginary parts of **Z** and **Y**, the following expression can be formulated:

$$Y = \frac{1}{R - jX_C} = \frac{(R + jX_C)}{(R - jX_C)(R + jX_C)} = \frac{R + jX_C}{R^2 + X_C^2}. \tag{25.6}$$

Hence,

$$G = \frac{R}{R^2 + X_C^2}, \quad B_C = \frac{X_C}{R^2 + X_C^2}. \tag{25.7}$$

The inverse transform gives the resistance (R) and imaginary admittance reactance (X_C)

$$R = \frac{G}{G^2 + B_C^2}, \quad X_C = \frac{B_C}{G^2 + B_C^2}. \tag{25.8}$$

From Equations 25.4, 25.5, and 25.7 the magnitude Y and PA φ_Y of the admittance can be written as

$$Y = \frac{1}{Z}, \quad \varphi_Y = -\varphi_Z. \tag{25.9}$$

The impedance depends on the frequency of alternating current. The simplified 2R–1C parallel model of biological objects is shown in Figure 25.3a, where R_e, R_i are the resistances of extracellular water (ECW) and intracellular water (ICW) compartments, and C_m is the total capacitance of the cell membrane. If the frequency of electric current grows, then, according to Equation 25.3, the capacitive reactance X_C decreases, and the current passes preferentially through the intracellular liquid space. One can show in this case that X_C and R are frequency dependent:

$$R = \frac{R_e + \omega^2 C_m^2 R_e R_i (R_e + R_i)}{1 + \omega^2 C_m^2 (R_e + R_i)^2},$$
$$X_C = \frac{\omega C_m R_e^2}{1 + \omega^2 C_m^2 (R_e + R_i)^2}. \tag{25.10}$$

Figure 25.3b shows frequency dependencies of R, X_C, Z, and φ according to Equation 25.10 with the parameters $R_e = R_i = 400$ ohm, and $C_m = 4$ nF in the frequency range 10^0–10^8 Hz. Only the absolute values of X_C and φ are shown. At high and low frequencies, the values X_C and φ approach zero. The frequency at which X_C is maximal is called the characteristic frequency (f_c). As is seen in Figure 25.3b, essential changes in impedance take place at frequencies 10^3–10^6 Hz. This range forms an impedance dispersion domain. The pair $Z(f)$ and $\varphi(f)$ is called a Bode diagram.

Figure 25.3c shows an impedance locus—the frequency dependence of impedance in the impedance plain, which is also called a hodograph, Nyquist plot, or Wessel diagram.

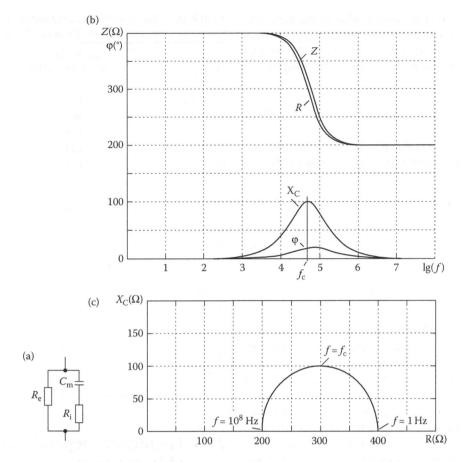

FIGURE 25.3 (a) The $2R$–$1C$ parallel model; (b) frequency dependencies of impedance, resistance, reactance, and PA; and (c) impedance locus in the impedance plane.

If the frequency changes from zero to infinity, then, for the circuit shown in Figure 25.3a, the impedance locus forms a semicircle (Figure 25.3c).

25.3 Impedance of Biological Tissues

To describe electrical properties of the biological objects, specific values are used that are independent of the object's size and shape. One value in particular is the specific resistance ρ measured in ohm m (Ωm). For direct electric current, the specific resistance of a cylinder with length L and cross-sectional area S is determined from the relation

$$R = \rho \frac{L}{S}. \tag{25.11}$$

The inverse value,

$$\sigma = \frac{1}{\rho}, \tag{25.12}$$

is called specific conductance and is measured in Ω^{-1} m^{-1} or S/m.

Other critical parameters in tissue impedance imaging are the capacitive properties of tissues, characterized by permittivity, or dielectric constant, ε, which is measured in F/m. The capacity of a plane capacitor consisting of two plates with area S and a dielectric layer between them with thickness d is equal to

$$C = \varepsilon \frac{S}{d}. \tag{25.13}$$

Here, $\varepsilon = \varepsilon_r \varepsilon_0$, where $\varepsilon_0 = 8.85 \cdot 10^{-12}$ F/m is the permittivity of free space, and ε_r is a nondimensional relative permittivity of the given dielectric—a measure of molecular polarization under the action of an electric field.

To characterize the conductive and capacitive properties of materials simultaneously, the complex values of resistivity, conductivity, and permittivity need to be considered[2]:

$$\rho = \rho' - j\rho'', \tag{25.14}$$

$$\sigma = \sigma' + j\sigma''. \tag{25.15}$$

$$\varepsilon = \varepsilon' - j\varepsilon''. \tag{25.16}$$

The parameter ε' determines polarization of the dielectric, accumulation of energy in it, and ε'' characterizes energy loss in the dielectric placed in an alternating electric field. Complex resistivity and complex permittivity are linked together as follows:

$$\sigma = j\omega\varepsilon. \qquad (25.17)$$

It can be shown that the real and imaginary values for conductivity and resistivity as well as permittivity, respectively, are interrelated.

$$\sigma' = \omega\varepsilon'', \quad \sigma'' = \omega\varepsilon',$$

$$\rho' = \frac{\sigma'}{\sigma'^2 + \sigma''^2}, \quad \rho'' = \frac{\sigma''}{\sigma'^2 + \sigma''^2}, \qquad (25.18)$$

$$\sigma' = \frac{\rho'}{\rho'^2 + \rho''^2}, \quad \sigma'' = \frac{\rho''}{\rho'^2 + \rho''^2}.$$

Thus, having frequency dependencies for ε' and ε'', one can obtain real and imaginary parts for complex resistivity and conductivity. The respective magnitudes and PA can be calculated similarly to Equation 25.4. Equation 25.12 holds true for the magnitudes of complex resistivity and conductivity.

There are a lot of reports describing specific impedance parameters of biological tissues. The differences between published values for the same tissue can be explained by the different frequencies used, different measurement techniques as well as a variety of other factors.

Mean values and confidence intervals of specific resistance for various tissues are shown in Table 25.1. The specific resistance is proportional to tissue hydration[3] since homeostatic mechanisms exist that maintain ion concentration in electrolytes.

TABLE 25.1 The Mean Values and Confidence Intervals (95%) of Specific Resistance for Various Tissues of the Human Body[3]

Tissue	Specific Resistance (Mean Value) (Ωm)	95% Confidence Interval (Ωm)
Cortical	4.64	3.6–5.97
Cancellous	176	123–252
Fat	38.5	30.5–48.7
Skin	3.29	2.55–4.24
Bone	1.24×10^6	$(0.91–1.69) \times 10^6$
Blood	1.51	1.20–1.91
Lung	1.57	1.22–2.02
Uterus	2.19	1.70–2.82
Breast	3.39	2.49–4.63
Bladder	4.47	2.88–6.93
Muscle (along fibers)	2.4	1.55–3.72
Muscle (across fibers)	6.75	4.35–10.5
Liver	3.42	2.96–3.96
Kidney	2.11	1.6–2.78
Spleen	4.05	3.07–5.35
Heart	1.75	1.33–2.31
Thyroid	1.83	1.18–2.83
Tongue	3.33	2.15–5.17
Testis	1.45	0.93–2.24
Ovary	2.24	1.44–3.47

25.4 Frequency Dependence of Impedance

Resistivity, conductivity, and permittivity of biological tissues strongly depend on the frequency of the alternating current. An example of such dependence for muscle tissue[4] is shown in Figure 25.4.

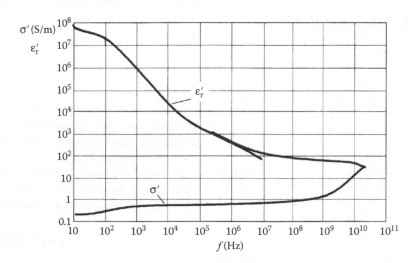

FIGURE 25.4 Frequency dependence of the specific conductance σ' and the relative permittivity ε_r' for muscle tissue. (Reproduced from Gabriel, S., Lau, R.W., and Gabriel, C. *Phys. Med. Biol.*, 1996, *1*: 2251–2269. With permission.)

The specific conductance rises with increasing frequency. At relatively low frequencies this pattern can be explained by a decrease in reactance and progressive involvement of ICW space. At frequencies higher than 1 GHz, an increase in conductance is explained by other mechanisms.[5]

The relative permittivity decreases with increasing frequency. Herman Schwan distinguished three frequency domains with different relaxation mechanisms.[5] At frequencies from Hz to some kHz the decrease in ε_r' occurs due to various effects at cell surfaces, as well as cell membranes and intracellular organelles. This frequency range is called α-dispersion.

Another dispersion mechanism, called β-dispersion, takes place in the frequency range between 10 kHz and 100 MHz. It is explained by the Maxwell–Wagner effect—a decrease, with frequency, of effective permittivity of a laminated (heterogeneous) dielectric. A gradual decrease in polarizability of large protein molecules also occurs.

The frequency range 100 MHz–100 GHz (γ-dispersion) is characterized by the progressive decrease in polarization of all protein molecules and, at the upper range limit, of water molecules. At present, the δ-dispersion subdomain is also distinguished (frequency range between 100 MHz and 1 GHz), where the decrease in permittivity is associated only with the polarization of protein molecules.[6]

The observed frequency dependencies of biological tissues impedance differ from those shown at Figure 25.3a for a simple 2R–1C parallel model. The studies carried out from 1920 through 1930 showed the need to match the data and the model by introducing the special circuit element that produces a phase shift that is not depending on frequency—the so-called constant phase element (CPE). The hodograph of an electrical circuit with the CPE represents a circular arc with a center depressed below the resistance axis.[7]

In 1940 Kenneth S. Cole suggested equations that describe the frequency dependence of biological impedance according to two electrical models.[8] One of them is shown in Figure 25.5a, where R_∞ is the resistance at infinite frequency (corresponds to parallel connection of R_e and R_i in Figure 25.3a), $\Delta R = R_0 - R_\infty$, R_0 is the resistance at zero frequency (corresponds to R_e). The Cole equation is written as

$$Z = R_\infty + \frac{\Delta R}{1 + (j\omega\tau_Z)^\alpha},$$

(25.19)

where α is a nondimensional parameter, τ_Z is the time constant that determines the characteristic frequency of the circuit: $f_c = 1/2\pi\tau_Z$. The constant phase shift of CPE equals $\alpha\pi/2$. The impedance of CPE is written as

$$Z_{CPE} = \Delta R(j\omega\tau_Z)^{-\alpha}.$$

(25.20)

The hodograph for $R_\infty = \Delta R = 200$ Ohm and $\alpha = 0.8$ is shown in Figure 25.5b. The value τ_Z does not influence the form and

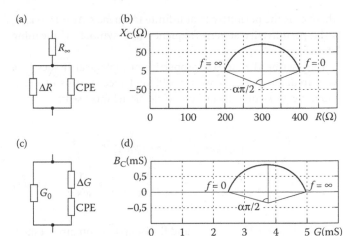

FIGURE 25.5 (a) Series model with CPE and (b) the related impedance locus in the impedance plane; (c) parallel model with CPE and (d) the related impedance locus in the conductance plane.

size of the hodograph, but only the location of frequencies on the hodograph. If $\alpha = 1$, then CPE converts to the "ideal" capacitor and the hodograph takes the form of a semicircle as in Figure 25.3c. If α decreases, then the center of the hodograph shifts down. Finally, if $\alpha = 0$, then CPE converts to the "ideal" resistance, and the hodograph collapses into a point on the resistance axis.

Another circuit is shown in Figure 25.5c. Here $G_0 = 1/R_0$ is the conductance at zero frequency, and ΔG, which is the difference between conductances at infinite and zero frequencies, is expressed as $\Delta G = G_\infty - G_0 = 1/R_\infty - 1/R_0$.

The Cole equation is written for complex conductivity:

$$Y = G_0 + \frac{\Delta G}{1 + (j\omega\tau_Y)^{-\alpha}}.$$

(25.21)

In this case, the impedance of CPE is calculated as

$$Z_{CPE} = \frac{1}{\Delta G}(j\omega\tau_Y)^{-\alpha}.$$

(25.22)

The hodograph is shown in Figure 25.5d with the same parameter values as before.

When describing the same object by means of these three respective circuits, the parameter α in Equations 25.19 and 25.21 will be the same, but the time constants τ_Y and τ_Z are not equal.

A similar equation to impedance, called the Cole–Cole equation, exists for complex permittivity[9]:

$$\varepsilon = \varepsilon_\infty + \frac{\Delta\varepsilon}{1 + (j\omega\tau_c)^{1-\alpha}},$$

(25.23)

where ε_∞ is the permittivity at infinite frequency, $\Delta\varepsilon = \varepsilon_\infty - \varepsilon_s$, ε_s is the permittivity at zero frequency (the symbol "s" meaning "static").

To describe the electric properties of biological tissues in a wide frequency range covering different mechanisms of relaxation, the four-component Cole–Cole model was used[4]:

$$\varepsilon = \varepsilon_\infty + \frac{\Delta\varepsilon_1}{1 + (j\omega\tau_1)^{1-\alpha_1}} + \frac{\Delta\varepsilon_2}{1 + (j\omega\tau_2)^{1-\alpha_2}}$$

$$+ \frac{\Delta\varepsilon_3}{1 + (j\omega\tau_3)^{1-\alpha_3}} + \frac{\Delta\varepsilon_4}{1 + (j\omega\tau_4)^{1-\alpha_4}} + \frac{\sigma_i}{j\omega\varepsilon_0}. \qquad (25.24)$$

Indices 1 and 2 relate to γ- and β-dispersion domains, respectively, 3 and 4 to α-dispersion domain, and finally, and σ_i is the frequency-independent ionic conductivity.[10]

The necessity of using CPE in electrical models of biological objects could be explained by a number of physical effects and properties, such as the behavior of a double electric layer at cell and tissue interfaces. The value of α relates to the distribution width of time constants in a heterogeneous system consisting of many frequency-dependent components.[2] Computer analysis of 3D cell models showed that the distribution of relaxation times can be explained by the dispersion of the molecules' orientation.[11]

The link between the parameter α and extracellular space morphology was studied.[12] Estimation of α using computer modeling of hierarchical organization of cells into clusters approximated well the experimental data.

25.5 Bioimpedance Measurement Mechanisms

The conventional descriptions of bioimpedance measurements are based on Ohm's law, from which the magnitude of total impedance can be written as

$$Z_m = \frac{U_m}{I_g}, \qquad (25.25)$$

where U_m is the potential difference. The current generator (CG) (Figure 25.6) maintains a constant current I_g regardless of Z_m.

Voltmeter Vm is calibrated in terms of resistance to exclude the necessity of using formula 25.25.

For complex impedance measurements an alternating current is used, and the voltmeter has to measure not only potential difference, but also phase shift relative to current. The magnitude of impedance can be assessed from the potential difference, and the phase shift determines the value of PA.

This method is widely used in *in vivo* measurements in the frequency range from 1 kHz to 1 MHz. For higher frequencies, other measurement methods exist. The studies on propagation or reflection of high-frequency electromagnetic waves in biological objects are carried out in the frequency range up to about 100 GHz. Also, a time domain analysis of transient processes is used when the system is affected by voltage jump. But these methods are used mainly in *in vitro* measurements.[2]

The bioimpedance measurement device interacts with biological objects through electrodes. The nature of the electric current in metal conductors and biological tissues is different. In wires and electrodes, the current is carried by electrons, whereas in biological tissues the current is carried by ions moving in electrolyte solutions. So, at the electrode–tissue interface various physical and chemical processes take place influencing the results of measurement. The equilibrium potential difference is established at the electrode–electrolyte contact area with the formation of a double electric layer on the boundary surface (Figure 25.7). At the same time, atoms of electrodes can move into the electrolyte, and, in turn, electrolyte ions can deposit on the electrode's surface.

In bioimpedance measurements, the dissolving of the electrode is inadmissible, so electrodes made of silver, platinum, and other inert metals are used.

When the current is propagated through the electrode–tissue interface, redox reactions occur there that provide a replacement of electric current carriers. As a result, the potential difference between the electrode and the electrolyte changes. This is called electrode polarization—an undesirable factor because of requiring compensation by means of an additional voltage.

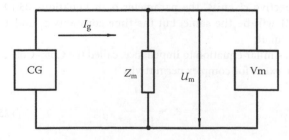

FIGURE 25.6   Electrical scheme of bioimpedance measurement.

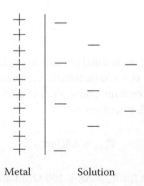

Metal   Solution

FIGURE 25.7   The double electric layer at the electrode–tissue interface.

The electrodes of pure inert metals have significant polarization voltage since their atoms are not involved in reactions. To reduce polarization, the presence of electrode atoms in the electrolyte is necessary. The widely used Ag/AgCl electrodes comply with this requirement. The AgCl layer is provided on the surface of the silver electrode, and the ions Cl⁻ are present in tissue. The AgCl layer increases or decreases depending on the direction of the current, so the polarization voltage is insignificant.

The magnitude of the electrode's contact area impedance decreases with frequency, so the influence of the electrode on the impedance of biological objects at frequencies of 5 kHz or higher is insignificant. But at lower frequencies the polarization and impedance of the electrode can substantially contribute to measurement error.

Special problems arise when the electrodes are imposed on the skin surface. The dry and very thin (10–20 μm) horny layer (stratum corneum) has very high specific resistance to direct current (10^4-10^5 Ω m). The alternating current-specific resistance gradually decreases with frequency growth up to about 10^2 Ω m at 1 MHz. The resistivity of moistened skin is decreased. The presence of sweat ducts also reduces resistivity, but their influence is irregular.

High impedance of the horny layer makes serious complications for bioimpedance measurements. To overcome them, it is necessary to clean the skin before the measurement and to apply a special gel or salt solution in order to impregnate the horny layer and to reduce its resistivity. Special electrodes with micropoints are also used, which penetrate the horny layer and contact directly with the highly conductive epidermis.

The impedance of the electrode–tissue contact area consists of electrode-gel connection impedance, gel impedance, and gel–horny layer impedance taken in series. These components show complex frequency dependencies. In general, the magnitude of impedance decreases with frequency.

In summary, the impedance of the electrode–tissue contact area can substantially influence the results of bioimpedance measurement. In a bipolar measurement scheme (Figure 25.8a), the same pair of electrodes 1 and 2 is used for current and voltage. The magnitude of total impedance is determined as

$$Z_{tot} = Z_m + Z_{c1} + Z_{c2}, \qquad (25.26)$$

where Z_{c1} and Z_{c2} are impedances of the respective electrode–tissue interfaces. Because of the uncertainties related to their accurate assessment, the area of bipolar measurement scheme application is limited.

In a widely used tetrapolar measurement scheme (Figure 25.8b), the current is propagated through the current electrodes 1 and 2, and the voltage is measured between the electrodes 3 and 4. If the input resistance of voltmeter is significantly higher than the magnitude of bioimpedance, then there is no current through electrodes 3 and 4, and the voltage drop on impedances Z_{c1} and Z_{c2} is negligible.

25.6 Whole-Body and Segmental Bioimpedance Measurements

One example of diagnostic imaging is the use of segmental bioimpedance measurement. Segmental bioimpedance measurement can be used for the estimation of body composition in specific body regions, for the assessment of blood and other fluids redistribution between various body compartments, as well as for the development of polysegmental measurement schemes.

For segmental measurements of biological impedance, it is necessary to select an appropriate scheme of current and potential electrodes' placement. A simplified equivalent electrical schematic of the human body is shown in Figure 25.9. The following notation is used: H—head, R—right arm, L—left arm, F—left leg, N—right leg, and T—trunk. For example, the measurement scheme with current electrodes connected to the left and right arms and the voltage measured between the left arm and the left leg is denoted as LR/LF. In this case, the measured voltage is proportional to the magnitude of impedance Z_L since there is no current across the circuit elements Z_T and Z_F. The impedance value of the same body segment can be assessed using HL/LR, LR/LH, LF/LR, and other measurement schemes. The trunk impedance can be measured, for instance, using LF/RN, HF/RN, and LN/RF schemes. Similarly, the measurement schemes for other body segments can be selected.

Polysegmental BIA is based on a successive connection of current and voltage electrodes to different pairs of electrodes for measuring the impedances of body segments. In practice,

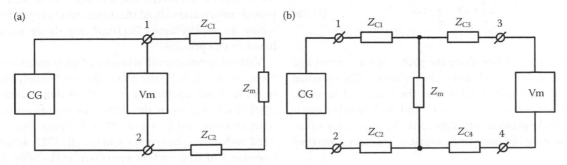

FIGURE 25.8 Electrical schemes of bipolar (a) and tetrapolar (b) bioimpedance measurements.

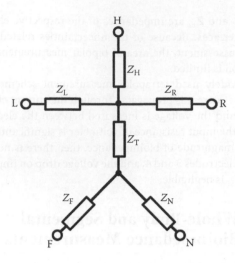

FIGURE 25.9 Simplified equivalent electrical scheme of the human body.

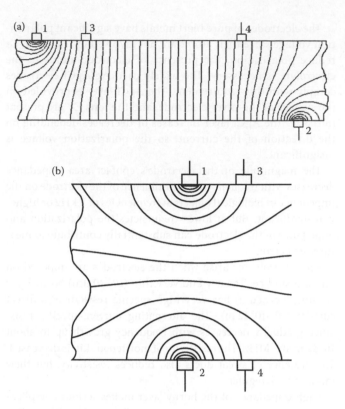

FIGURE 25.10 Equipotential lines: the current flows along (a) and across (b) the homogeneous biomaterial.

different methods of bioimpedance measurements of the same body segment produce different results. For example, the magnitude of trunk impedance is significantly higher when using the HF/RN measurement scheme compared to the LF/RN scheme.

The results of modeling electric current flow through hypothetical 2D objects are shown in Figure 25.10. The current flows between the electrodes 1 and 2. If the measured object is isotropic, then the vector of current density is determined as

$$J = E/\rho, \tag{25.27}$$

where ρ is the specific resistance (generally, a function of coordinates), and E is the vector of the electric field intensity. The spatial 3D pattern of the electric field can be represented by equipotential surfaces, or contours of constant potentials. In the 2D case, they degenerate into equipotential lines. The current flows along the electric field lines (not shown), which are perpendicular to equipotential lines.

The measured voltage equals the potential difference between the equipotential surfaces close to the electrodes 3 and 4, respectively. It is determined as

$$U = \int_L E \, dl = \int_L \rho J \, dl, \tag{25.28}$$

where the integral is taken along the path L of the current and dl is an element of the path along the trajectory. The observed impedance is an averaged value over the measured segment. Areas with greatest current density and high specific resistance provide the greatest contribution to the impedance value, primarily because the impedance measure provides a weighed average.

Figure 25.10a clearly shows that the measured impedance value will agree closely with the impedance of body segment between the voltage electrodes only if they are positioned sufficiently far from the current electrodes. Otherwise, if the voltage is measured between highly curved equipotential lines, they will approximate the boundaries of each segment with a significant error. A more thorough analysis must take into account that the voltage electrodes distort the field pattern as the potential on their surface is constant, meaning that the measurement technique itself influences the outcome.

Figure 25.10b illustrates the case of electrode placement on the opposite sides of the object. Moving the voltage electrodes 3 and 4 away from the current electrodes 1 and 2 enables the determination of the resistance of deep tissue areas. A sufficiently large number of electrodes on the body surface and alternating use of them as a current and a voltage electrode, respectively, provide information about the distribution of specific resistance within an area/volume. Electrical impedance tomography is based on this principle.

Various conventional schemes of bioimpedance measurement are shown in Figure 25.11. The most widely used scheme in professional equipment for bioimpedance analysis is the RN/RN scheme, when the electrodes are placed on the right wrist and right ankle (Figure 25.11a). Figure 25.11b shows the hand-to-hand measurement scheme (RL/RL), which gives an impedance of arms and the upper part of the body. The devices

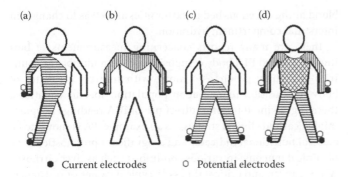

● Current electrodes   ○ Potential electrodes

FIGURE 25.11 The conventional schemes of BIA measurements: (a) wrist-to-ankle (RN/RN); (b) hand-to-hand (RL/RL); (c) leg-to-leg (FN/FN); and (d) polysegmental eight-point scheme.

that realize the leg-to-leg measurement scheme FN/FN (Figure 25.11c) give estimates of lower-body impedance and have the form of a platform scale with four stainless steel foot pads (electrodes) on the top. The polysegmental eight-point measurement scheme is shown in Figure 25.11d. Electrodes are placed on all the extremities, in which case various sets of segmental impedance measurements are available: for example, the combination of RF/RN, LN/LF, RF/LF, LN/RN, LN/RF, and RF/LN measurement schemes. Each determination of the first four parameters provides the sum of impedances of one extremity and trunk by itself, and the last two provide estimates of trunk impedance. The impedance of each respective extremity is then determined by subtraction. Other combinations of measurements are also used, which provide an impedance of various body segments.

25.7 Bioelectric Impedance Analysis and Body Composition

BIA is the most widely used diagnostic method for the *in vivo* human body composition assessment.[13,14] The classic body composition model subdivides body mass (BM) into the sum of fat mass (FM) and FFM. BIA of body composition is based on the principle that lean tissue conducts an electric current better than fat tissue. According to Equation 25.11, the resistance R of a cylindrical homogeneous conducting body of uniform cross-sectional area S, length L, and resistivity ρ is equal to

$$R = \frac{\rho L}{S} = \frac{\rho L^2}{SL} = \frac{\rho L^2}{V}, \quad (25.29)$$

where V is the volume of the conducting space, a volume of water containing electrolytes that conduct the electric current in the body. Hence, this volume can be written as

$$V = \frac{\rho L^2}{R}. \quad (25.30)$$

The factor L^2/R in the right-hand side of Equation 25.30 is called impedance index. Thomasset[15] and Hoffer[16] showed that, although human body is not a uniform cylinder, the volume of total body water (TBW) is indeed proportional to the impedance index, and the formula 25.30 can be used as a reliable predictor for the assessment of TBW. It has also been shown that the accuracy of TBW estimates can be improved by the inclusion of additional predictor variables including BM and age. So, the typical predictive equation for TBW has the form

$$\mathrm{TBW}_{\mathrm{BIA}} = \frac{\rho' \mathrm{Ht}^2}{R} + k_{\mathrm{BM}}\mathrm{BM} + k_{\mathrm{Age}}\mathrm{Age} + \cdots, \quad (25.31)$$

where Ht is the body height. For most studies, the term Ht^2/R in the right-hand side of Equation 25.31 is the best single predictor of TBW and FFM, accounting for 59–98% of total variance.[17] Parameters of such prediction equations are obtained using multiple regression analysis against TBW estimates in the same subjects by an appropriate reference method (isotope dilution). Many software packages such as MS Excel®, Statistica, and Matlab® offer standard tools that can be used for statistical analysis. The accuracy of the predictive equation is characterized by the value of standard error of estimate (SEE) compared to estimates obtained on the same subjects using a reference method.

In human subjects estimates of fat-free BM from TBW are based on the relatively constant ratio between these values: TBW/FFM ≈ 0.737 ± 0.036.[4,5] Other diagnostic applications of hydration values are observed for most mammal species. FM can then be assessed by subtraction:

$$\mathrm{FM} = \mathrm{BM} - \mathrm{FFM}. \quad (25.32)$$

As the impedance of biological objects depends on the frequency of alternating current, the parameters of Equation 25.31 relate to some constant frequency. In *single-frequency* BIA, the frequency of 50 kHz is generally used.

TBW consists of two distinct compartments: ECW and ICW. On the equivalent 2R–1C parallel model shown in Figure 25.3a, the "extracellular" path of the electric current is represented by the resistor R_e. For ECW assessment, it is necessary to measure the resistance to a direct current, because in this case cell membranes are impermeable, and the presence of ICW compartment does not influence the results of impedance measurement. So, the resistance to a direct current R_0 (at zero frequency) equals the ECW resistance:

$$R_0 = R_e. \quad (25.33)$$

On the other hand, if the frequency grows infinitely, the reactance tends to zero; thus the resistance R_∞ is determined by the electrical scheme with the resistors R_e and R_i only in parallel:

$$R_\infty = \frac{R_e R_i}{R_e + R_i},\tag{25.34}$$

thus providing a predictor for TBW. But the measurements at zero and infinite frequencies are impossible, so, in *dual-frequency* BIA, sufficiently low (5 kHz) and high (500 kHz) frequencies are used for estimation of ECW and TBW, respectively. Another approach is utilized in *multiple-frequency* BIA, with approximation of theoretical values of R_0 and R_∞ from the reactance-to-resistance plot on the impedance plane.

Predictive equations for ECW similar to Equation 25.31 are also used, in particular, for the estimation of adequate body hydration in pathological conditions.

The volume of ICW compartment is then determined as

$$ICW = TBW - ECW.\tag{25.35}$$

In a similar way, prediction equations for other body composition parameters, such as FFM, body cell mass, and skeletal muscle mass, are suggested. Published SEE values for estimation of FFM range from 1.8 to 3.6 kg and fall within the "excellent" to "good" interval on the suggested accuracy scale for body composition analysis.[18]

An equivalent electrical model of the human body was considered[19] based on the Hanai mixture theory[20] in which biological tissue is considered as a suspension of spherical cells in a conductive medium. Here, tissue resistivity and body geometry alter relationships in basic models shown in Figures 25.3 and 25.5. But the resulted equations for ECW and ICW showed no clear advantage over those obtained using other models.[20]

The most widely used for body composition assessment are single-frequency BIA analyzers, which utilize the conventional tetrapolar wrist-to-ankle measurement scheme (Figure 25.11a), as well as hand-to-hand and leg-to-leg measurement schemes (Figures 25.11b,c). Dual-frequency and multifrequency analyzers enable additional estimation of ECW and ICW. Segmental BIA is another development of bioimpedance analysis. For this, BIA analyzers utilizing the eight-point measurement scheme are manufactured (Figure 25.11d).

25.8 Bioelectric Impedance Analysis Applications

Well before the development of BIA body composition analysis, the measurements of impedance were also used in impedance cardiography for the assessment of central blood volume and degree of lungs hydration. In occlusion impedance plethysmography, changes in basic impedance were related to changes in blood filling of venous bed in extremities as well as to changes in interstitial compartment hydration.

There are many reports concerning measurements of bioimpedance and BIA body composition from a number of countries and study groups.[21-23] Along with indirect estimates of body composition relying upon population-specific equations, there is growing interest in direct use of BIA readings in epidemiological and clinical studies. As an example, PA is shown to be a useful nutritional indicator and a significant prognostic factor of clinical outcome in various conditions such as liver cirrhosis, AIDS, cancer, and critical illness.[24] Typical values of resistance, reactance, and PA for various body segments in a healthy man are shown in Figure 25.12.

If the conventional wrist-to-ankle measurement scheme of single-frequency BIA is used, then the values R and X_C in a healthy population can vary significantly (250–800 and 20–80 Ω, respectively), but the ratio R/X_C is less varying (of the order 7–11). Figure 25.13 shows typical age dependencies of population reference values for PA in healthy men and women.

Lower values are not infrequent in chronic catabolic diseases, which reflect cell death or increased permeability of cell membrane. Clinical norms for PA are described.[24] Higher values (compared to normal) are usually observed in athletes.

Typical values of PA for local body parts are represented in Table 25.2.

In sports medicine, BIA estimates of body composition can have significant diagnostic value reflecting the effectiveness of the training process (Figure 25.14).

The specific diagnostic value of monitoring changes in electric parameters of body regions is to identify different physiological and pathophysiological processes, such as changes in tissue hydration, edema, and so on. The rate of changes can be described by the characteristic time τ. Most rapid changes occur in the redistribution of venous blood between body regions, in response to orthostatic exposure ($\tau = 0.5$-1.5 s) (Figure 25.16b). The characteristic time for volume changes of interstitial fluids spans normally from tens of minutes to hours. The degree of intracellular hydration changes more slowly, except the cases of tissue chemical burn; τ values for body fat have the order of days. The patterns of temporal changes of impedance and water volume in the torso and shank during a tilt test are shown in Figures 25.15 and 25.16.

BIA was used for the evaluation of the severity of illness in pediatric patients after cardiac surgery.[25] Postoperative changes in bioelectrical impedance relative to preoperative data (Z/Z_0) were assessed (Figure 25.17). The initial decrease immediately after surgery (D0) caused by surgical stress itself was not directly related to the prognosis. In those patients who survived, the impedance index showed significant increase toward preoperative values 16 (D1) and 40 h (D2) after surgery. A sustained decrease of the impedance index was associated with high mortality risk.

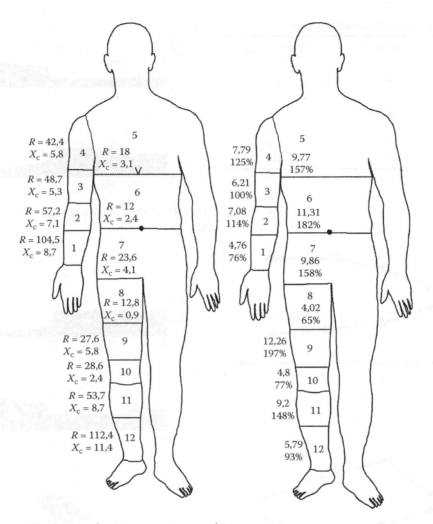

FIGURE 25.12 Body segment distribution of resistance, reactance, and PA.

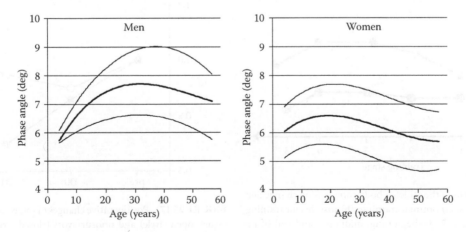

FIGURE 25.13 Population reference values of PA by age and sex (mean ± s.d.).

TABLE 25.2 Typical Values of PA for Local Body Parts

Body Parts	PA (grad)	PA/PA$_{wrist-to-ankle}$
Thumb	2.2	0.3
Index finger	2.5	0.4
Middle finger	2.1	0.3
Ring finger	2.8	0.4
Little finger	2.4	0.4
Palm	3.8	0.6
Foot arch	5.3	0.8

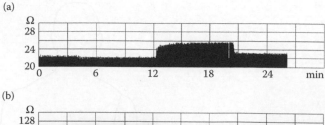

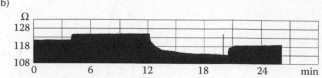

FIGURE 25.15 Temporal changes of impedance during a tilt test: (a) torso and (b) shank.

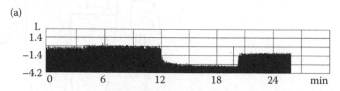

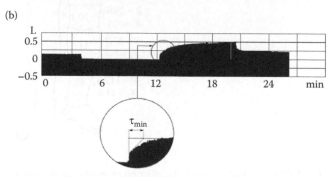

FIGURE 25.16 Temporal changes of water volumes during a tilt test: (a) torso and (b) shank.

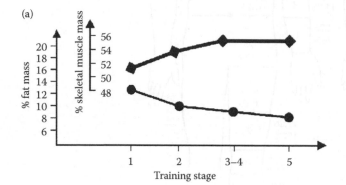

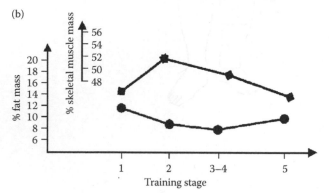

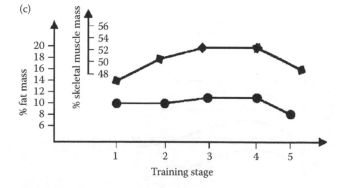

FIGURE 25.14 Patterns of the athlete's body composition change during the training process: (a) optimal training; (b) insufficient training; and (c) acute overload. 1, 2, 3—beginning, midpoint, and end of the preparatory stage; 4, 5—the beginning and end of sports competition.[25]

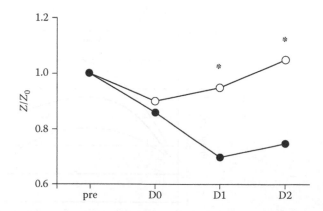

FIGURE 25.17 Postoperative changes of impedance index (Z/Z_0) in survivors (open circle) and nonsurvivors (closed circle). (Reproduced from Shime N. et al. *Crit. Care Med.*, 2002, *30*: 518–520. With permission.)

References

1. Cornish, B.H., Jacobs, A., Thomas, B.J., and Ward, L.C. Optimizing electrode sites for segmental bioimpedance measurements. *Physiol. Meas.*, 1999, *20*: 241–250.
2. Grimnes, S. and Martinsen, O.G. *Bioimpedance and Bioelectricity Basics* (2nd edition). London: Academic Press, 2008.
3. Faes, T.J.C., van der Meij, H.A., de Munck, J.C., and Heethaar, R.M. The electric resistivity of human tissues (100 Hz–10 MHz): A meta-analysis of review studies. *Physiol. Meas.*, 1999, *20*: R1–R10.
4. Gabriel, S., Lau, R.W., and Gabriel, C. The dielectric properties of biological tissues: II. Measurements in the frequency range 10 Hz to 20 GHz. *Phys. Med. Biol.*, 1996, *41*: 2251–2269.
5. Schwan, H.P., in: J.H. Lawrence and C.A Tobias (Eds.), *Advances in Biological and Medical Physics*. NY: Academic Press, 1957, *5*: pp. 147–209.
6. Feldman, Y. and Hayashi, Y. *Proc. XII Int. Conf. on Electrical Bioimpedance & V Int. Conf. on Electrical Impedance Tomography*. Gdansk, Poland, 20–24 June 2004, *1*: pp. 13–16.
7. Cole, K.S. Electrical phase angle of cell membranes. *J. Gen. Physiol.*, 1932, *15*: 641–649.
8. Cole, K.S. *Cold Spring Harbor Symp. Quant. Biol.*, 1940, *8*: 110–122.
9. Cole, K.S. and Cole, R.H. Dispersion and absorption in dielectrics. I. Alternating current characteristics. *J. Chem. Phys.*, 1941, *9*: 341–351.
10. Gabriel, C. and Gabriel, S. Compilation of the dielectric properties of body tissues at RF and microwave frequencies. Available at http://www.emfdosimetry.org/dielectric/Title/Title.html
11. Furuya, N., Kawamura, T., Sakamoto, K. et al. *Proc. XII Int. Conf. on Electrical Bioimpedance & V Int. Conf. on Electrical Impedance Tomography*. Gdansk, Poland, 20–24 June 2004, *1*: pp. 145–148.
12. Ivorra, A., Genesca, M., Hotter, G., and Aguilo, J. *Proc. XII Int. Conf. on Electrical Bioimpedance & V Int. Conf. on Electrical Impedance Tomography*. Gdansk, Poland, 20–24 June 2004, *1*: pp. 87–90.
13. Heymsfield, S.B., Lohman, T.G., Wang, Z.M., and Going, S.B. *Human Body Composition* (2nd edition). Champaigh, IL: Human Kinetics, 2005.
14. Heyward, V.H. and Wagner, D. *Applied Body Composition Assessment* (2nd edition). Champaigh, IL: Human Kinetics, 2004.
15. Thomasset, A. Bio-electrical properties of tissue impedance measurements. *Lyon Med.*, 1962, *207*: 107–118.
16. Hoffer, E.D. Correlation of whole-body impedance with total body water volume. *J. Appl. Physiol.*, 1969, *27*: 531–534.
17. Kushner, R.F. Bioelectrical impedance analysis: A review of principles and applications. *J. Am. Coll. Nutr.*, 1992, *11*: 199–209.
18. Lohman, T.G. *Advances in Body Composition Assessment*. Champaign, IL: Human Kinetics, 1992.
19. De Lorenzo, A., Andreoli, A., Matthie, J., and Withers, P. *J. Appl. Physiol.*, 1997, *82*: 1542–1558.
20. Hanai, T. in: P.H. Sherman (Ed.), *Emulsion science*. London: Academic Press, 1968, pp. 354–477.
21. Bosy-Westphal, A., Danielzik, S., Dorhofer, R.-P. et al. Phase angle from bioelectrical impedance analysis: Population reference values by age, sex, and body mass index. *J. Parenter. Enteral Nutr.*, 2006, *30*: 309–316.
22. Barbosa-Silva, M.C.G., Barros, A.J.D., Wang, J. et al. Bioelectrical impedance analysis: Population reference values for phase angle by age and sex. *Am. J. Clin. Nutr.*, 2005, *82*: 49–52.
23. Nikolaev, D.V., Smirnov, A.V., Bobrinskaya, I.G., and Rudnev, S.G. *Bioelectric Impedance Analysis of Human Body Composition*. Moscow: Nauka, 2009 (in Russian).
24. Selberg, O. and Selberg, D. Norms and correlates of bioimpedance phase angle in healthy human subjects, hospitalized patients, and patients with liver cirrhosis. *Eur. J. Appl. Physiol.*, 2002, *86*: 509–516.
25. Martirosov, E.G., Nikolaev, D.V., and Rudnev, S.G., *Technologies and Methods of Human Body Composition Assessment*. Moscow: Nauka, 2006 (in Russian).
26. Shime N., Ashida H., Chihara E. et al. Bioelectrical impedance analysis for assessment of severity of illness in pediatric patients after heart surgery. *Crit. Care Med.*, 2002, *30*: 518–520.

26

X-ray and Computed Tomography

Ravishankar Chityala

26.1 Introduction

X-rays are a form of electromagnetic radiations like light, radio waves, and so on. They have a wavelength of 10–0.01 nm. X-rays were discovered accidentally by Wilhelm Conrad Roentgen, a German physicist, while studying cathode ray tubes.

He found out that most of the materials allowed the new ray to pass through and also left a shadow on a photographic plate. His work on the new ray was published as "On a New Kind of Rays" and was subsequently awarded the first Nobel Prize in Physics, in 1901.

Since Roentgen's days, x-rays have found very widespread uses and are used across different fields such as radiology, geology, crystallography, astronomy, and so on. In the field of radiology, x-rays are used in fluoroscopy, angiography, computed tomography (CT), and so on. Today, many of the noninvasive surgeries are performed under x-ray guidance providing a new "eye" to the surgeons.

In this chapter, we will look at methods of generating x-rays, x-ray detection, different radiology procedures such as fluoroscopy and angiography. We will work on the details of CT including methods for reconstruction and CT artifacts.

26.2 Radiation

An x-ray machine requires an x-ray generator or tube, a material through which the x-ray traverses and an x-ray detector to measure the intensity of the incoming rays. We will begin with the discussion of radiation by studying the x-ray generation process using an x-ray tube.

26.3 X-ray Tube Construction

An x-ray tube consists of four major parts. They are an anode, a cathode, a tungsten target, and an evacuated tube to hold the three parts together as shown in Figure 26.1.

The cathode is the negative terminal that produces electrons that will be accelerated toward the anode. The filament is heated by passing current, which generates electrons by a process of thermionic emission, defined as emission of electrons

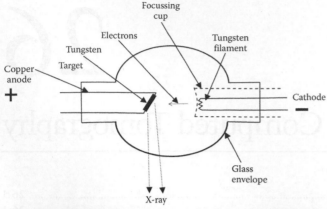

FIGURE 26.1 (Top) The components of an x-ray tube. (Bottom) X-ray tubes. (Reproduced from Siemens AG, Healthcare Sector. With permission.)

by absorption of thermal energy. The number of electrons produced is proportional to the current impressed upon it. The filament is generally made from tungsten to withstand high temperatures and also for its malleability. The electron produced is focused by the focusing cup, which is maintained at the same negative potential as the cathode. The glass enclosure in which the x-ray is generated is evacuated, so that the electrons do not interact with other molecules and can also be controlled independently and precisely. The focusing cup is maintained at a very high potential in order to accelerate the electrons produced by the filament. The anode is the positive electrode and is bombarded by the fast-moving electron. It is generally made from copper, so that the heat produced by the bombardment of the electron can be properly dissipated. A tungsten target is fixed on the anode. The fast-moving electrons knock out the electrons from the inner shells of the tungsten target. This process results in generation of x-rays.

There are two constructions for the anode, a stationary and a rotating type. The rotating anode has a longer life than the

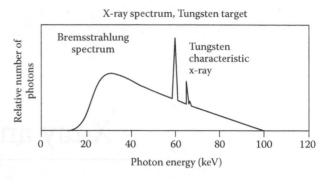

FIGURE 26.2 The x-ray spectrum illustrating characteristic radiation and the Bremsstrahlung spectrum.

stationary anode, as the area exposed to the electrons varies continuously.

26.3.1 X-ray Generation Process

There are two different types of spectra generated when x-rays are produced.[1] The general radiation or Bremsstrahlung "Braking" spectrum is a continuous radiation and the characteristic radiation is a discrete spectrum as shown in Figure 26.2.

When the fast-moving electron produced by the cathode moves very close to the nucleus of the tungsten atom (Figure 26.3), the electron decelerates and the loss of energy is emitted as radiation. Most of the radiation is at higher wavelengths and hence dissipated as heat. The electron is not decelerated completely by one tungsten nucleus and, hence, at every stage of deceleration, radiation of lower wavelength or higher energy is emitted. Since the electrons are decelerated or "braked" in the process, this spectrum is referred as the Bremsstrahlung spectrum.

From the energy equation, we know that Equation 26.1 holds true:

$$E = \frac{hc}{\lambda} \tag{26.1}$$

FIGURE 26.3 Production of the Bremsstrahlung or braking spectrum.

where h is the Planck constant, c the speed of light, and λ the wavelength of the x-ray. The product of h and c is equal to 12.4 if E is measured in keV and λ is measured in armstrong (Å). Thus Equation 26.1 becomes Equation 26.2:

$$E = \frac{12.4}{\lambda} \qquad (26.2)$$

The inverse relationship between E and λ implies that a shorter wavelength produces a higher energy x-ray and vice versa. For an x-ray tube powered at 112 kVp, the maximum energy that can be produced is 112 keV and, hence, the corresponding wavelength is 0.11 Å. This is the shortest wavelength and also the highest energy that can be achieved during the production of the Bremsstrahlung spectrum. Most of the x-rays, however, will be produced at much higher wavelength and consequently lower energy.

The second type of radiation spectrum (Figure 26.4) results from an electron orbit interaction within the tungsten structure, called the characteristic radiation. The fast-moving electrons could also eject the electron from the k-shell (inner shell) of the tungsten atom. Since this shell is unstable due to the ejection of the electron, the vacancy is filled by an electron from the outer shell. This is accompanied by the release of x-ray energy. The energy and wavelength of the electron are dependent on the binding energy of the electron whose position is filled. Depending on the shell, these characteristic radiations are referred to as K, L, M, and N characteristic radiation and thus the characteristic K and L radiation is shown in Figure 26.2. Because of the short wavelength of x-ray radiation, x-ray will interact on an atomic and a molecular level, thus ionizing any molecules in its path, hence requiring a vacuum environment as well.

26.3.2 X-ray Attenuation

Attenuation can be defined as the reduction in the intensity of the x-ray beam as it traverses matter by either the absorption or deflection of photons in the beam. The attenuation is quantified by using the linear attenuation coefficient (μ), defined as the attenuation per centimeter of the object. The attenuation is directly proportional to the distance traveled and the incident intensity. The intensity of the x-ray beam after attenuation is given by the Lambert–Bear law (Figure 26.5) expressed as

$$I = I_0 e^{-\mu \Delta x} \qquad (26.3)$$

where I_0 is the initial x-ray intensity, I is the exiting x-ray intensity, μ is the linear attenuation coefficient of the material, and Δx is the thickness of the material. The law also assumes that the input x-ray intensity is mono-energetic or monochromatic.

Monochromatic radiation is characterized by photons of single intensity, but in reality all radiations have photons of varying intensity and, hence, are polychromatic and therefore have spectra similar to Figure 26.2. Polychromatic radiation is characterized by photons of varying energy (quality and quantity), with the peak energy being determined by the peak kilovoltage (kVp). The mean energy of the polychromatic radiation is between 2/3 and 1/2 of its peak energy. When a polychromatic radiation passes through a matter, the longer wavelengths with lower energy is preferentially absorbed, resulting in the fact that the remaining transmitted photons are high-energy photons and hence the beam hardens during its passage and is referred to as beam hardening.

In addition to the attenuation coefficient, a material can also be defined using the half-value layer. It is defined as the thickness of material needed to reduce the intensity of the x-ray beam by half. Hence, from Equation 26.3 for a thickness $\Delta x = \text{HVL}$ (half value layer), Equation 26.4 can be derived:

$$I = \frac{I_0}{2} \qquad (26.4)$$

Hence,

$$I_0 e^{-\mu \text{HVL}} = \frac{I_0}{2} \qquad (26.5)$$

$$\mu \times \text{HVL} = 0.693 \qquad (26.6)$$

$$\text{HVL} = \frac{0.693}{\mu} \qquad (26.7)$$

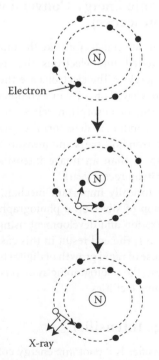

FIGURE 26.4 Production of characteristic radiation.

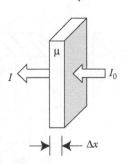

FIGURE 26.5 Lambert–Beer law for monochromatic radiation and for a single material.

For a material with a linear attenuation coefficient of 0.1/cm, the HVL is 6.93 cm. This implies that when a monochromatic beam of x-rays passes through the material, its intensity drops by half after passing through 6.93 cm of that material.

26.3.3 The Lambert–Beer Law for Multiple Materials

For an object with n elements (Figure 26.6), the Lambert–Beer law is applied in cascade,

$$I = I_0 e^{-\mu_1 \Delta x} e^{-\mu_2 \Delta x} e^{-\mu_3 \Delta x} \cdots e^{\mu_n \Delta x} = I_0 e^{-\sum_{i=1} \mu_i \Delta x_i} \qquad (26.8)$$

When the logarithm of the intensities is taken, we obtain

$$p = -\ln\left(\frac{I}{I_0}\right) = \sum_{i=1}^{n} \mu_i \Delta x_i = \int \mu(x) \mathrm{d}x \qquad (26.9)$$

in the continuous domain. Using this equation, we see that the value p, the projection image expressed in energy intensity, corresponding to the digital value at a specific location in that image, is simply the sum of the product of attenuation coefficients and thicknesses of the individual components. This is the basis of x-ray and CT, both of which will be discussed shortly.

26.3.4 Factors Determining X-ray Attenuation

The energy of the beam is one of the factors that determine the amount of attenuation. As we have seen earlier, lower energy beams get preferentially absorbed compared to higher energy beams.

The density of a substance through which x-rays pass makes a significant contribution to the attenuation. The higher density substance like bone attenuates x-rays more than a lower density substance like tissue. Also different types of tissues have different densities and hence different attenuation, resulting in different contrast on the x-ray image.

The number of electrons per gram in a material determines its x-ray stopping power. A material with higher number has higher

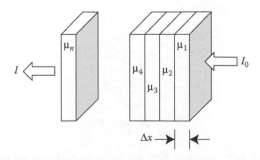

FIGURE 26.6 Lambert–Beer law for multiple materials.

probability of interacting with the x-rays. The number of electrons per gram is given by Equation 26.10

$$N = \frac{N_o Z}{A} \qquad (26.10)$$

where N is the number of electrons per gram, N_o is the Avogadro number, Z is the atomic number, and A is the atomic weight of the substance. Since the Avogadro number is a constant, the number of electrons per gram is dependent only on Z and A.

26.4 X-ray Detection

So far, we have discussed the x-ray generation using an x-ray tube, the shape of the x-ray spectrum and also studied the change in x-ray intensity as it traverses a material due to attenuation. These attenuated x-rays have to be converted into a human-viewable form. This conversion process can be achieved either by exposing them on a photographic plate to obtain an x-ray image or viewed using a TV screen or converted into a digital image, all using the process of x-ray detection, respectively. There are three different types of x-ray radiation detectors in practice, namely ionization, fluorescence, and absorption.

26.4.1 Ionization Detection

In the ionization detector, the x-rays ionize the gas molecules in the detector and, by measuring the ionization, the intensity of x-rays is measured. One example of such a detector is the Geiger–Muller counter.[2]

26.4.2 Photonic Energy Conversion Detection (Fluorescence)

In image intensification detectors like the image intensifier, the x-rays are converted into electrons that are accelerated to increase their energy.[1-3] The electrons are then converted back into light and are viewed on a TV or a computer screen.

One mechanism of absorption relies on the photoelectric effect. In a semiconductor detector, the x-rays are converted directly into electron hole pairs and measured[4]. These detectors occupy lesser space than an image intensifier and hence are gaining popularity in recent years.

The other historically important mechanism of absorption uses the oxidation of silver on a photographic plate, which is exposed after fixation and development using standard photographic techniques. The end result in this case is a hardcopy of the image. Because of rapid growth of digital technology and the lower overall cost, the photographic system is being replaced at most x-ray imaging facilities.

26.4.3 Image Intensifier

The image intensifier is a photonic energy conversion detector. The image intensifier (Figure 26.7) consists of (1) input phosphor

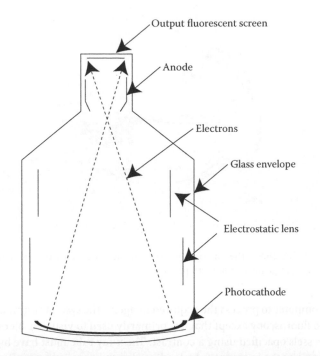

Output fluorescent screen

Anode

Electrons

Glass envelope

Electrostatic lens

Photocathode

FIGURE 26.7 (Top) The components of an image intensifier. (Bottom) Image intensifier tubes. (Reproduced from Siemens AG, Healthcare Sector. With permission.)

and a photocathode, (2) an electrostatic focusing lens, (3) an accelerating anode, and (4) an output fluorescent screen.

The x-ray beam passes through the patient and enters the image intensifier through the input phosphor. The phosphor generates light photons after absorbing the x-ray photons. The light photons are absorbed by the photocathode and electrons are emitted. The electrons are then accelerated by a potential difference toward the anode. The anode focuses the electrons on to an output fluorescence screen, which emits the light that will then be either displayed using a TV screen or recorded in an x-ray film.

The input phosphor is made of cesium iodide (CsI) and is vapor deposited to form a needle-like structure, which acts like a fiber optic cable and prevents the diffusion of light and thereby improves the resolution. Also, it has a greater packing density and hence a higher conversion efficiency even with smaller thickness (needed for a good spatial resolution). Also, it has a higher effective number and hence can absorb x-rays better in the diagnostic radiology range. The photocathode emits electrons when light photons are incident on it. The anode accelerates the electrons. The higher the acceleration, better the conversion of electrons to light photons at the output phosphor. The input phosphor is curved so that electrons travel the same length towards the output phosphor. The output phosphor is silver-activated zinc–cadmium sulfide. The output can be viewed using a series of lens on a TV or it can be recorded on a film.

26.5 Multiple-Field Image Intensifier

The field size is changed by changing the position of the focal point, the point of intersection for the left and right electron beams. This is achieved by increasing the potential in the electrostatic lens. Lower potential results in the focus being close to the anode and hence the full view of the anatomy is exposed to the output phosphor. At higher potential, the focus moves away from the anode and hence only a portion of the input phosphor is exposed to the output phosphor. In both cases, the sizes of the input and output phosphor remain the same, but in the smaller mode, a portion of the image from the input phosphor is removed from the view due to a farther focal point.

In a commercial x-ray unit, these sizes are specified in inches. A 12-inch mode will cover a larger anatomy while a 6-inch mode will cover a smaller anatomy. Exposure factors are automatically increased for smaller image intensifier modes to compensate for the decreased brightness as a result of magnification.

Since the electrons travel large distances during their journey from the photocathode to the anode, they are affected by the earth's magnetic field. The earth's magnetic field changes even for small motion of the image intensifier and, hence, the electron path gets distorted. The distorted electron path produces distorted images on the output fluorescent screen. The distortion is not uniform but increases as we move toward the edge of the image intensifier. Hence the distortion is more significant for a large image intensifier mode than for a smaller image intensifier mode. The distortions can be removed by careful design and material selection or, more preferably, by using image processing algorithms.

The image intensifier can be bundled with an x-ray tube, a patient table, and a structure to hold all these parts together, to create an imaging system. Such a system could also be designed to revolve around the patient table axis and provide images in multiple directions to aid diagnosis. Examples of such a system, namely fluoroscopy and angiography, are discussed below.

26.6 Fluoroscopy

The first generation fluoroscope consisted of a fluoroscopic screen (Figure 26.8) made of copper-activated cadmium sulfide that emitted light in the yellow–green spectrum of visible light.[1,2]

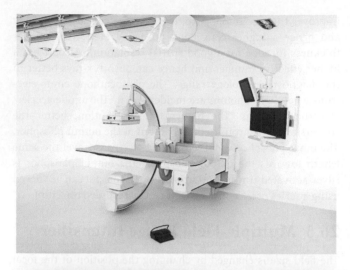

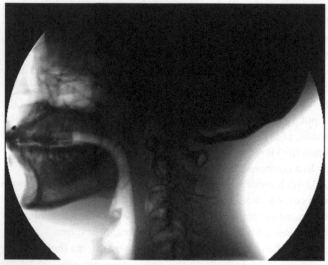

FIGURE 26.8 (Top) The fluoroscopy machine (Courtesy of Siemens Medical Solutions). (Bottom) Image of a head phantom acquired using an image intensifier system. (Reproduced from Siemens AG, Healthcare Sector. With permission.)

The examination was so faint that it was carried out in the dark room with the doctor dark-adapting their eyes prior to examination. Since the intensity of fluorescence was less, rod vision in the eye was used and, hence, the ability to differentiate shades of grey was also less. These problems were alleviated with the invention of the image intensifier discussed earlier. The image intensifier allowed intensification of the light emitted by the input phosphor; it could be safely and effectively used to produce a system that could generate and detect x-rays and also produce images fit enough for human consumption using televisions and computers.

26.7 Angiography

A digital angiographic system (Figure 26.9) consists of an x-ray tube, an image intensifier-based detector, a light diaphragm to control the light gain, a video camera to record the image, and a

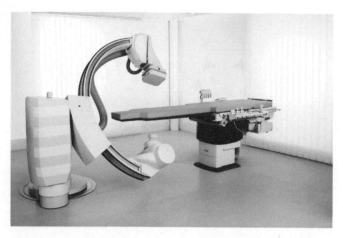

FIGURE 26.9 The angiography system. (Reproduced from Siemens AG, Healthcare Sector. With permission.)

computer to process the acquired image.[1,2] The system is similar to fluoroscopy except that it is primarily used to visualize blood vessels opacified using a contrast. The x-ray tube must have big focal spot to prevent tube loading due to constant generation of x-ray. It must also provide a constant output over time. The image intensifier must also provide a constant acceleration voltage, to prevent variation in gain during acquisition. In general, the voltage cannot be controlled accurately and hence a diaphragm is used to control the amount of light entering the TV camera.

The computer controls the whole imaging chain and also performs digital subtraction in the case of digital subtraction angiography (DSA) on the images obtained.[1] The computer controls the x-ray technique so that uniform exposure is obtained across all images. The computer obtains the first set of images without the injection of contrast and stores them as a mask image. Subsequent images obtained under the injection of contrast are stored and subtracted from the mask image to obtain the image with the vessel alone.

26.8 Computed Tomography

The fluoroscopy and angiography discussed so far produce a planar projection image, which is a shadow of the part of the body under x-rays. The image may also contain other organs/structures that impede the ability to make a clear diagnosis. In such cases, a slice through the patient would provide an unimpeded view of the organ of interest. The system that produces this virtual slice is called the CT.

Sir Godfrey N. Hounsfield and Dr. Allan McCormack developed CT independently and later shared the Nobel Prize for Physiology in 1979. It is also referred to as Computed Axial Tomography or Computer Assisted Tomography or a CAT scan although CT has become the most prevalent name in recent years.

The normal x-ray systems studied so far produce a shadow or projection of the object on a 2D plane and, hence, the 3D depth information is permanently lost. CT solves that problem by acquiring x-ray images all around the object. A computer then processes these images to produce a map of the original object using a process called reconstruction. The utility of this technique became so apparent that an industry quickly developed around it and still continues to be an important diagnostic tool for physicians and surgeons.

In honor of Sir Hounsfield's efforts, the unit of measure of the attenuation coefficient in CT is Hounsfield unit (HU).

26.8.1 Image Reconstruction

The basic principle of image reconstruction is that the internal structure of an object can be determined from multiple projections of that object. In the case of CT reconstruction, the internal structure being reconstructed is the spatial distribution of the linear attenuation coefficients (μ) of the imaged object. Mathematically, Equation 26.9 can be inverted by the reconstruction process to obtain the distribution of the attenuation coefficients. Ideally, the x-ray beam is mono-energetic, and the geometry of the imaging system is well characterized. In current clinical CT technology, however, the x-ray beam is not mono-energetic and the geometry is not well characterized. These result in error in the reconstructed image, commonly referred to as artifacts and will be discussed later.

In clinical CT, the raw projection data are often a series of 1D vectors obtained at various angles for which the 2D reconstruction yields a 2D attenuation coefficient matrix. There are other modalities where the raw projection data are acquired as a series of 2D matrices obtained at various angles for which the 3D reconstruction yields a 3D distribution of the attenuation coefficient volume. For the sake of simplicity, the reconstructions discussed in this chapter will focus on 2D reconstructions and, hence, the projection images are 1D vectors unless otherwise specified.

We will also describe the central slice theorem, which gives the relationship between the raw data after logarithmic transformation and the reconstructed image (Equation 26.9). The underlying assumption in all these discussions is that the data have been preprocessed for any nonidealities and hence has no error at the input.

26.8.2 Parallel Beam CT

The original method used for acquiring CT data used parallel-beam geometry such as shown in Figure 26.10. As shown in the figure, the paths that the x-rays take from the source to the detector are parallel to each other. An x-ray source was collimated to yield a single x-ray beam, and the source and detector were translated along the axis perpendicular to the beam to obtain the projection data (a single 1D vector for a 2D CT slice). After the acquisition of one projection, the source–detector assembly was rotated and subsequent projections were obtained.

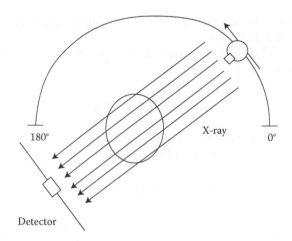

FIGURE 26.10 Parallel beam geometry.

An analytical solution was subsequently obtained and forms the basis for all the CT reconstruction techniques used. It is based on the central slice theorem or the Fourier slice theorem.[5]

26.9 Central Slice Theorem

Consider the object shown in Figure 26.11 to be reconstructed. The original coordinate system is x–y and when the detector and the x-ray source are rotated by an angle θ, their coordinate system is defined by x'–y'. In this figure, R is the distance

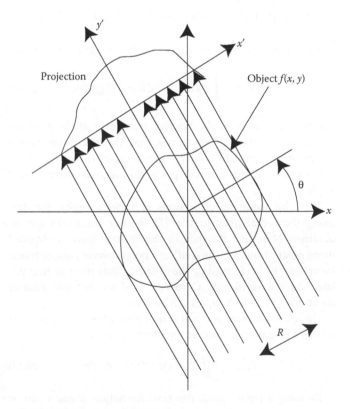

FIGURE 26.11 Central slice theorem.

between the iso-center (i.e., center of rotation) and any ray passing through the object.

After logarithmic conversion, the x-ray projection at an angle (θ) is given by

$$g_\theta(R) = \iint f(x, y)\delta(x\cos\theta + y\sin\theta - R)\,dx\,dy \qquad (26.11)$$

where δ( ) is the Dirac-Delta function.[6]

The Fourier transform of the distribution is given by

$$F(u, v) = \iint f(x, y)e^{-i2\pi(ux+vy)}\,dx\,dy \qquad (26.12)$$

where u and v are frequency components in perpendicular directions. Expressing u and v in the polar coordinate, we obtain $u = v\cos\theta$ and $v = v\sin\theta$, where v is the radius and θ is the angular position in the Fourier space.

Substituting for u and v and simplifying yield the following equation

$$F(v, \theta) = \iint f(x, y)e^{-i2\pi v(x\cos\theta + y\sin\theta)}\,dx\,dy \qquad (26.13)$$

The above equation can be rewritten as

$$F(v, \theta) = \iiint f(x, y)e^{-i2\pi vR}(x\cos\theta + y\sin\theta - R)\,dR\,dx\,dy \qquad (26.14)$$

Rearranging the integrals yields

$$F(v, \theta) = \int \left(\iint f(x, y)\delta(x\cos\theta + y\sin\theta - R)\,dx\,dy \right)e^{i2\pi vR}\,dR \qquad (26.15)$$

From Equation 26.11, we can simplify the above equation as

$$F(v, \theta) = \int g_\theta(R)e^{-i2\pi vR}\,dR = \text{FT}(g_\theta(R)) \qquad (26.16)$$

where FT( ) refers to the Fourier transform of the enclosed function. Equation 26.16 shows that the radial line along an angle θ in the 2D Fourier transform of the object is the 1D Fourier transform of the projection data acquired at that angle θ.

Thus, by acquiring projections at various angles, the data along the radial lines in the 2D Fourier transform can be obtained. Note that the data in the Fourier space is obtained using polar sampling. Thus, either a polar inverse Fourier transform must be performed or the obtained data must be interpolated onto a rectilinear Cartesian grid so that fast Fourier transform (FFT) techniques can be used.

However, another approach can be also taken. Again, $f(x, y)$ is related to the inverse Fourier transform, that is,

$$F(x, y) = \iint F(u, v)e^{-i2\pi(ux+vy)R}\,du\,dv \qquad (26.17)$$

By using a polar coordinate transformation, u and v can be written as $u = v\cos\theta$ and $v = v\sin\theta$.

To effect a coordinate transformation, the Jacobian is used and is given by Equation 26.18:

$$J = \begin{vmatrix} \partial u/\partial v & \partial u/\partial\theta \\ \partial u/\partial v & \partial u/\partial\theta \end{vmatrix} = \begin{vmatrix} \cos\theta & -v\sin\theta \\ \sin\theta & v\cos\theta \end{vmatrix} = v \qquad (26.18)$$

Hence

$$du\,dv = |v|\,dv\,d\theta \qquad (26.19)$$

Thus,

$$f(x, y) = \iint f(v, \theta)e^{i2\pi v(x\cos\theta + y\sin\theta)}|v|\,dv\,d\theta \qquad (26.20)$$

Using Equation 26.16, we can obtain

$$f(x, y) = \iint \text{FT}(g_\theta(R))e^{i2\pi v(x\cos\theta + y\sin\theta)}|v|\,dv\,d\theta \qquad (26.21)$$

$$f(x, y) = \iiint \text{FT}(g_\theta(R))e^{i2\pi vR}\delta(x\cos\theta + y\sin\theta - R)|v|\,dv\,d\theta\,dR \qquad (26.22)$$

$$f(x, y) = \iint \left(\int \text{FT}(g_\theta(R))|v|e^{i2\pi vR}\,dv \right)\delta(x\cos\theta + y\sin\theta - R)\,dR \qquad (26.23)$$

The term in the braces is the filtered projection, which can be obtained by multiplying the Fourier transform of the projection data with v in the Fourier space or equivalently by performing a convolution of the real space projections and the inverse Fourier transform of the function v. Because the function v looks like a ramp, the filter generated is commonly called the "ramp filter."

Thus,

$$f(x, y) = \iint \text{FP}(R, \theta) \cdot \delta(x\cos\theta + y\sin\theta - R)\,dR \qquad (26.24)$$

where FP(R, θ) is the filtered projection data at location R acquired at an angle θ and is given by

$$\int \text{FP}(R, \theta) = \int \text{FT}(g_\theta(R))|v|^{i2\pi vR}\,dv \qquad (26.25)$$

Once the convolution or filtering is performed, the resulting data are reconstructed using Equation 26.25. This process is referred to as the filtered backprojection (FBP) technique and is the most commonly used technique in practice.

26.10 Fan-Beam CT

The fan-beam CT scanners (Figure 26.12) have a bank of detectors, with all detectors being illuminated by x-rays simultaneously for every projection angle. Since the detector acquired images in one x-ray exposure, it eliminated the translation at each angle. Since translation was eliminated, the system was mechanically stable, and the exposure time of the scanner was

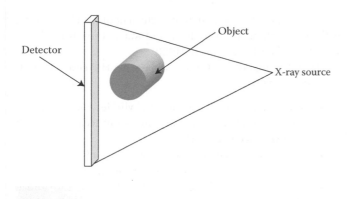

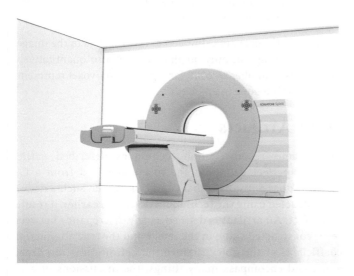

FIGURE 26.12 (Top) Fan-beam geometry. (Bottom) The fan-beam CT machine. (Reproduced from Siemens AG, Healthcare Sector. With permission.)

much reduced. However, x-ray scatter in the object reduced the contrast in the reconstructed images compared with parallel beam reconstruction. But these machines are still popular due to faster acquisition time, which allows the reconstruction of moving objects like slices of the heart in one breath hold. The images acquired using fan-beam scanners can be reconstructed using a rebinning method that converts fan-beam data into parallel beam data and then uses the central slice theorem for reconstruction. Currently, this approach is not used and is replaced by a direct fan-beam reconstruction method based on FBP.

A fan-beam detector with one row of detecting element produces one CT slice. The current generations of fan-beam CT machines have multiple detector rows and can acquire 8, 16, 32 slices, and so on in one rotation of the object and are referred as multislice CT machines. The benefits are faster acquisition time compared with a single slice and covering a larger area in one exposure.

With the advent of multislice CT machines, a whole body scan of the patient can also be obtained. Figure 26.13 is the axial slice

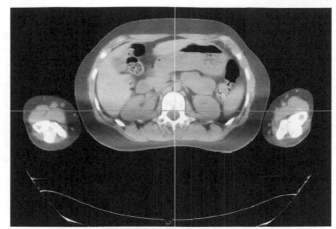

FIGURE 26.13 Axial CT slice.

of a patient in the region around the kidney. It is one of the many slices of the whole body scan shown in the montage in Figure 26.14. These slices were converted into a 3D object (Figure 26.15) using Mimics™.

26.11 Cone Beam CT

Cone beam acquisition or CBCT (Figure 26.16) consists of a 2D detector instead of 1D detectors used in the parallel and fan-beam acquisitions. As with the fan beam, the source and the detector rotate relative to the object, and projection images are acquired. The 2D projection images are then reconstructed to obtain 3D volume. Since a 2D region is imaged, cone-beam-based volume acquisition makes use of x-rays that otherwise would have been blocked. The other advantages are potentially faster acquisition time, better pixel resolution, and isotropic (the same voxel size in the x, y, and z directions) voxel resolution. The most commonly used algorithm for cone-beam reconstruction

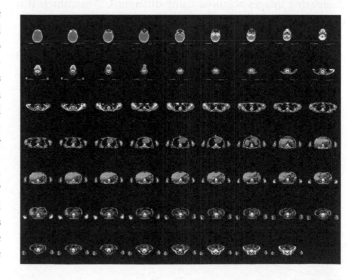

FIGURE 26.14 Montage of all the slices of the patient. One of the axial slices in the montage is shown in Figure 26.13.

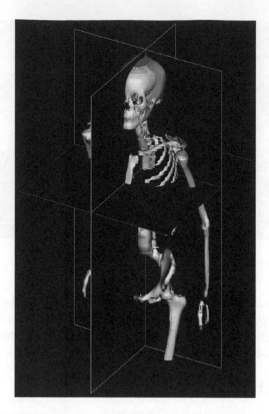

FIGURE 26.15 3D object created using the axial slices shown in the montage. The 3D object in gray is superimposed on the slice information for clarity.

is the Feldkamp algorithm,[7] which assumes a circular trajectory for the source and the flat detector and is based on FBP.

26.11.1 Hounsfield Unit

HU is the system of unit used in CT that represents the linear attenuation coefficient of an object. It provides a standard way of comparing images acquired using different CT machines. It is defined as

$$HU = \left(\frac{\mu - \mu_w}{\mu_w}\right) \times 1000 \qquad (26.26)$$

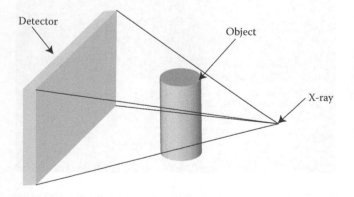

FIGURE 26.16 Cone beam geometry.

where μ is the linear attenuation coefficient of the object and μ_w is the linear attenuation coefficient of water. Thus, water has an HU of 0 and air has an HU of -1000.

The following are the steps to obtain the HU equivalent of a reconstructed image.

- A water phantom consisting of a cylinder filled with water is reconstructed using the same x-ray technique as the reconstructed patient slices.
- The attenuation coefficients of water and air (present outside the cylinder) are measured from the reconstructed slice.
- A linear fit is established, with the HU of water (0) and air (-1000) being the ordinate and the corresponding linear attenuation coefficient measured from the reconstructed image being the abscissa.
- Any patient data reconstructed are then mapped to HU using the determined linear fit.

Since the CT data are calibrated to HU, the data in the images acquire meaning not only qualitatively but also quantitatively. Thus, an HU number of 1000 for a given pixel/voxel represents quantitatively a bone in an object.

26.12 Artifacts

The various reconstruction techniques described earlier assumed perfect data. But the image acquired from a CT machine does not satisfy that condition. When such a data is reconstructed using the techniques described earlier, it results in artifact. An artifact is defined as any discrepancy between the reconstructed value in the image and the true attenuation coefficients of the object.[8] Since the definition is too broad and could encompass many things, the discussions of artifacts are generally limited to clinically significant errors. CT is more prone to artifacts than conventional radiography, as multiple projection images are used. Hence errors in different projection images cumulate to produce artifacts in the reconstructed image. These artifacts could annoy the radiologist or in some severe cases hide important details that could lead to misdiagnosis.

These artifacts could be eliminated to some extent during acquisition. They can also be removed by preprocessing projection images or postprocessing the reconstructed images. There are no generalized techniques for removing artifacts and hence new techniques are devised depending on the application, anatomy, and so on. The artifacts cannot be completely eliminated but can be reduced by using correct techniques, proper patient positioning, and improved design of the CT scanner or by software provided with the CT scanners.

There are many sources of error in the imaging chain, that results in artifacts. They can generally be classified into artifacts due to the imaging system or artifacts due to the patient. In the following discussion, the geometric alignment, offset, and gain correction are caused by the imaging system, while the scatter and beam hardening artifacts are caused by the nature of the object or patient being imaged.

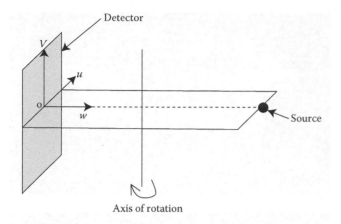

FIGURE 26.17 Parameters defining a cone beam system.

26.12.1 Geometric Misalignment Artifacts

The geometry of a CBCT system is specified using six parameters, namely the three rotation angles (angle corresponding to the *u*, *v*, and *w* axes in Figure 26.17) and three translations along the principal axis (*u*, *v*, and *w* in Figure 26.17). Error in these parameters results in ring artifact,[3,9] double wall artifact, and so on, which are visual and hence cannot be misdiagnosed as a pathology. However, very small errors in these parameters can result in blurring of edges and hence misdiagnosis of the size of pathology or shading artifacts that could shift the HU number. Hence, these parameters have to be determined accurately and corrected prior to reconstruction.

26.12.2 Scatter

An incident x-ray photon ejects an electron from the orbit of the atom and, consequently, a low-energy x-ray photon is scattered from the atom. The scatter photon travels at an angle from its incident direction (Figure 26.18). These scattered radiations are detected but are considered as primary radiation. They reduce the contrast of the image and produce blurring.

The effect of scatter in the final image is different for conventional radiography and CT. In the case of radiography, the images have poor contrast but in the case of CT, the logarithmic transformation results in a nonlinear effect.

Scatter also depends on the type of the image acquisition technique. For example, fan-beam CT has less scatter compared with a cone-beam CT due to smaller height of the beam.

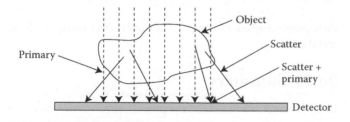

FIGURE 26.18 Scatter radiation.

One of the methods to reduce scatter is the air-gap technique. In this technique, a large air gap is maintained between the patient and the detector. Since the scattered radiation at a large angle from the incident direction cannot reach the detector they will not be used in the formation of the image. In general, it is not always possible to provide an air gap between the patient and the detector. So grids or postcollimators made of lead strips are used to reduce scatter.[1,8] The grids contain space that corresponds to the photodetector being detected. The scattered radiation arriving at a large angle will be absorbed by lead and only primary radiations arriving at a small angle from the incident direction are detected. The third approach is software correction.[10,11] Since scatter is a low-frequency structure causing blurring, it can be approximated by a number estimated using the beam-stop technique.[8] This, however, does not remove the noise associated with the scatter.

26.12.3 Offset and Gain Correction

Ideally, the response of a detector must remain constant for a constant x-ray input at any time. But due to temperature fluctuations during acquisition, nonidealities in the production of detectors and variations in the electronic readouts result in a nonlinear response in some detectors. These nonlinear responses result in the output of that detector cell being inconsistent with reference to all the neighboring detector pixels. During reconstruction, they produce ring artifacts with their center being located at the iso-center.[8] These circles may not be confused with a human anatomy, as there are no parts that form a perfect circle, but they degrade the quality of the image and hide details and hence need to be corrected. Moreover, the detector produces some electronic readout, even when the x-ray source is turned off. This readout is referred to as "Dark Current" and needs to be removed prior to reconstruction. Mathematically, the flat field and zero offset corrected image (IC) is given by Equation 26.27:

$$IC(x, y) = \frac{IA - ID}{IF - ID}(x, y) \times Average(IF - ID) \quad (26.27)$$

where IA is the acquired image, ID is the dark current image, and IF is the flat field image, which is acquired by the same technique as the acquired image with no object in the beam. The ratio of the differences is multiplied by the average value of (IF – ID) for gain normalization. This process is repeated for every pixel. The dark field images are to be acquired before each run, as they are sensitive to temperature variations.

Other software-based correction techniques based on image processing are also used to remove the ring artifacts. They can be classified as pre- and postprocessing techniques. The preprocessing techniques are based on the fact that the rings in the reconstructed images appear as vertical lines in the sinogram space. Since no feature in an object except those at the iso-center can appear as vertical lines, the pixels corresponding to vertical lines can be replaced using estimated pixel values. Even though the process is simple, the noise and complexity of human anatomy present a big challenge in the detection of vertical lines.

The other correction scheme is the postprocessing technique.[8] The rings in the reconstructed images are identified and removed. Since ring detection is primarily an arc detection technique, it could result in overcorrecting the reconstructed image for features that look like an arc. So the supervised ring removal technique, where there is consistent presence of inconsistencies across all views, is used for ring artifact removal. To determine the position of pixel corresponding to a given ring radius, a mapping that depends on the locations of the source, the object, and the image is used.

26.12.4 Beam Hardening

The spectrum (Figure 26.2) does not have a unique energy but has a wide range of energies. When such an energy spectrum is incident on a material, the lower energy gets attenuated faster as they are preferentially absorbed than the higher energy. Hence a polychromatic beam becomes harder or richer in higher energy photons as it passes through the material. Since the reconstruction process assumes an "ideal" monochromatic beam, the images acquired using a polychromatic beam produce cupping artifacts.[12]

The cupping artifact is characterized by a radial increase in intensity from the center of the reconstructed image to its periphery. Unlike ring artifacts, this artifact presents a difficulty, as it can mimic some pathology and hence can lead to misdiagnosis. They also shift the intensity values and hence present difficulty in quantification of the reconstructed image data. They can be reduced by hardening the beam prior to reaching the patient, using a filter made of aluminum, copper, and so on. Algorithmic approaches for reducing these artifacts have also been proposed.

26.12.5 Metal Artifacts

Metal artifacts are caused by the presence of materials that have a high attenuation coefficient when compared with pathology in the human body. These include surgical clips, biopsy needles, teeth fillings, implants, and so on. Because of their high attenuation coefficient, they produce beam-hardening artifacts (Figure 26.19) and are characterized by streaks emanating from the metal structures. Hence techniques used for removing beam hardening can be used to reduce these artifacts.

In Figure 26.19, the top image is a slice taken at a location without any metal in the beam. The bottom image contains an applicator. The beam hardening causes streaking artifact that not only renders the metal to be poorly reconstructed but also adds streaks to nearby pixels and hence makes diagnosis difficult.

Algorithmic approaches to reducing these artifacts have been proposed.[8,13,14] A set of initial reconstructions is performed without any metal artifact correction. From the reconstructed image, the location of metal objects is then determined. These objects are then removed from the projection image to obtain synthesized projection (Figure 26.20). The synthesized projection is

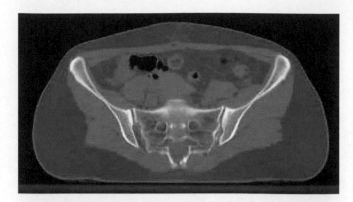

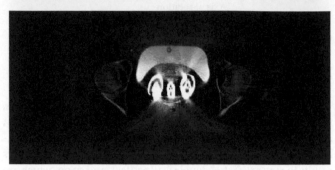

FIGURE 26.19 (Top) Slice with no metal in the beam. (Bottom) Beam hardening artifact with strong streaks emanating from a metal applicator.

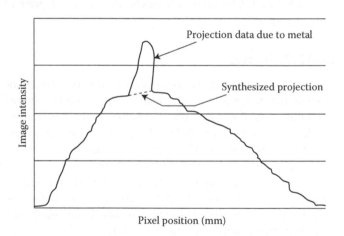

FIGURE 26.20 Illustration of metal artifact removal by using projection synthesis. The figure is a plot of the profile across the projection images including pixels containing the projection of the metal.

then reconstructed to obtain a reconstructed image without metal artifacts.

26.13 Summary

In this chapter, we discussed the process of generation of x-rays and the detection of x-rays using various methods such as the image intensifier. These 2D detection techniques have provided a means to see the inside of objects and are more commonly used to see the inside of the human body.

CT provided the first real 3D view of the human body. We delved into the details of CT, discussing the various schemes such as the parallel-, fan-, and cone-beam methods of image acquisition. The central slice theorem that forms the basis for CT reconstruction was also discussed. The various artifacts that result from the image acquisition and reconstruction in CT were also discussed. Although the artifacts like scatter were discussed under CT, they affect other modalities such as fluoroscopy, angiography, and so on. The scatter correction still produces blurring in those images, although the effect is more pronounced in CT due to subsequent processing.

This chapter is not an encyclopedic view of the world of x-rays and CT but it does provide an overview of the methods and materials. The author recommends that interested readers should read the various books and papers referenced all over this chapter.

References

1. T. S. Curry III, J. F. Dowdy, and R. C. Murry Jr, *Christensen's Introduction to Physics of Diagnostic Radiology*, Lea & Febiger, Philadelphia, PA, 1984.
2. A. Macovski, *Medical Imaging*, Prentice Hall Inc., Upper Saddle River, NJ, 1983.
3. R. Fahrig and D. W. Holdsworth, Three-dimensional computed tomographic reconstruction using a C-arm mounted XRII: Image-based correction of gantry motion nonidealities, *Med. Phys.* 27, 30–38, 2000.
4. J. H. Siewerdsen, D. J. Moseley, S. Burch, S. K. Bisland, A. Bogaards, B. C. Wilson, and D. A. Jaffray, Volume CT with a flat-panel detector on a mobile, isocentric C-arm: Pre-clinical investigation in guidance of minimally invasive surgery, *Med. Phys.* 32, 241–254, 2005.
5. A.C. Kak and M. Slaney, *Principles of Computerized Tomographic Imaging*, IEEE Press, Hoboken, NJ, 1988.
6. Bracewell, R. The impulse symbol. In *The Fourier Transform and Its Applications*, 3rd edition, ch. 5. NY: McGraw-Hill, pp. 69–97, 1999.
7. L. Feldkamp, L. Davis, and J. Kress, Practical cone beam algorithm. *J. Opt. Soc. Am.*, A6, 612–619, 1984.
8. J. Hsieh, *Computed Tomography: Principles, Design, Artifacts, and Recent Advances*, SPIE, Bellingham, WA, 2003.
9. Y. Cho, D. J. Moseley, J. H. Siewerdsen, and D. A. Jaffray, Accurate technique for complete geometric calibration of cone-beam computed tomography systems. *Med. Phys.* 32, 968–983, 2005.
10. L. A. Love and R. A. Kruger, Scatter estimation for a digital radiographic system using convolution filtering. *Med. Phys.*, 14(2), 178–185, 1987.
11. B. Ohnesorge, T. Flohr, and K. Klingenbeck-Regn, Efficient object scatter correction algorithm for third and fourth generation CT scanners. *Eur. Radiol.*, 9, 563–569, 1999. Search Council of Canada, Tech. Rep. NRCC Rep. PIRS-701, 2003.
12. J. Barrett and N. Keat, *Artifacts in CT: Recognition and Avoidance*, MHRA, 2003.
13. G. Wang, D. L. Snyder, J. A. O'Sullivan, and M. W. Vannier, Iterative deblurring for CT metal artifact reduction. *IEEE Trans. Med. Imaging*, 15, 657–664, 1996.
14. P. M. Joseph and R. D. Spital, A method for correcting bone induced artifacts in computed tomography scanners. *J. Comput. Assist. Tomogr.*, 2, 100–108, 1978.

Confocal Microscopy

Robert Splinter

27.1 Introduction

Confocal microscopy is an ingenious method for detection of fluorescence. This microscope allows a specimen to be optically sectioned into many focal planes and this improves the resolution of the resulting micrograph and aids the researcher in determining the spatial localization of a given protein within a tissue or cell. The confocal microscope utilizes laser technology for these scanning and fluorescence excitation functions. In this chapter, the laser technology will be described first followed by the role of lasers in the confocal microscope. Next the mechanism of action of this microscope will be outlined, followed by a discussion of biological techniques that utilize fluorescent microscopy.

27.2 Laser Technology

The acronym laser stands for light amplification by the stimulated emission of radiation. There are six types of lasers, classified by the lasing medium utilized in each. The six types of laser are: chemical, gas, liquid or dye, semiconductor, solid state, and free-electron. Dye lasers use complex organic dyes such as rhodamine 6G, which can also be used for their fluorescent properties, in a liquid solution or suspension. These lasers are capable of a wide variety of wavelengths. Semiconductive lasers, or diode lasers, are very small and use a low amount of power. They can be found in CD burners and laser printers.

There are three general components to a simple laser setup: two mirrors (one partially reflective and one fully reflective), an energy source for excitation, and the lasing medium. The lasing material determines the wavelength(s) that will be produced and the properties of that specific laser type and is generally a material of high energy and low density. The setup of a simple laser can be seen in Figure 27.1.

In order to produce the laser beam, the lasing material must be excited by an energy source, generally electricity, which is called pumping. Electrons within the atom of the lasing medium will move from a low energy level, commonly called the ground state, to a higher energy level, which is the excited state. The amount of electricity applied to the system as well as the properties of the atoms themselves determines how high of an energy level the electron will "jump" to. Because electrons in the higher energy levels are not stable, spontaneous decay will cause the electrons to drop down to their original ground state, releasing a photon. Photons running parallel within the lasing tube will be reflected off the fully reflective mirror and go on to stimulate electrons in their higher energy states to decay and lose another photon, thus amplifying the signal within the tube. At the partially reflective mirror, only a certain amount of photons will be reflected while the rest will pass through. The photons that pass through this mirror make up the visible laser beam.

Photons are little "packets of energy," which make up the laser light. The wavelength of the emitted photons is determined by the lasing material as well as the amount of energy applied to the

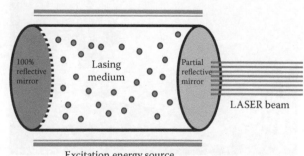

FIGURE 27.1 Diagram of a simple laser.

atoms through the energy supply. All photons released have several advantageous properties. They are monochromatic, meaning that they all have the same wavelength and therefore will have the same color in the visible spectrum. Because of their monochromatic nature, photons from one lasing material are also described as coherent and collimated. "Coherent" refers to the synchronized phase of the light waves, while "collimated" refers to the parallel nature of the laser beam. In comparison, white light is made up of many different wavelengths of light and will spread out from a given source, such as the light seen coming from a flashlight or overhead projector. Because many different wavelengths make up white light, these wavelengths will be continuously interfering with each other in both positive and negative ways.

27.2.1 Confocal Microscopy's Mechanism of Action

Confocal microscopy utilizes laser technology because of the specific properties of laser light. As described above, ordinary light microscopes only utilize white light. Because of this, all the focal planes of a specimen are seen at the same time, including planes, which are out of focus. This causes a loss of detail in the imaging process. The confocal microscope can be seen in Figure 27.2. A typical laser scanning confocal microscopy system consists of the conventional microscope and a confocal unit above it, and they are both connected to a detector system that collects the signals sent from the unit and ultimately reconstructs those images in three dimensions. A laser beam is produced from above the specimen being examined. Light coming from the laser source, will move through an X-Y scanner. This scanner will focus the laser light at specific targets on the viewed specimen. The laser light will then move from the X-Y scanner onto a dichroic mirror. Dichroic mirrors are designed so that some wavelengths of light can pass, while other wavelengths will be reflected. Each laser source within the confocal part will have its own dichroic mirror designed to reflect its specific wavelength to the specimen on the stage. The laser beam passing through the scanning mirrors is turned into a

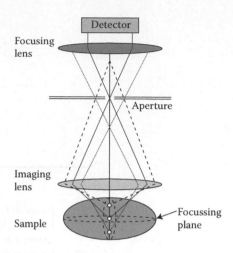

FIGURE 27.3 Cartoon illustrating the 3D slicing mechanism of the confocal microscope. The focal plane provides the spatial resolution in depth, and the scanning of the laser beam yields the 2D resolution within the focal plane. Note: only the solid drawn lines are providing imaging capability.

scanning beam. The scanning beam passes through the objectives, which diffract light, focusing it onto the specimen. Because this is a laser light source, the light will not spread and will be focused intensely on a single spot on the specimen. The real key is in the pinhole or confocal aperture. An illustration of the confocal mechanism is shown in Figure 27.3. The aperture in the confocal microscope is a small hole, which only allows light from a single focal plain to be seen by the detector and ultimately taken to the computer for processing and visualization. Both the condenser and objective lenses are focused to a common point. All light from other unfocused plains are eliminated. This process is repeated a number of times along the specimen, reading only small fields within the plane with the laser point in the $X–Y$ direction.

Another advantage of this type of microscopy is that one can also cut optical sections in the Z plane from the top of the 3D specimen to the bottom. In this way, multiple sections can be compiled together in order to reconstruct the original 3D image. A representation of this phenomenon can be seen in Figure 27.4.

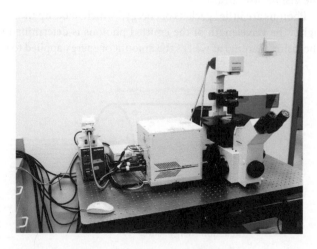

FIGURE 27.2 Diagram of a confocal microscope.

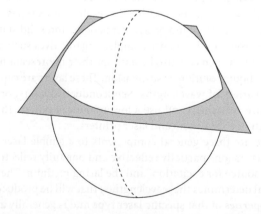

FIGURE 27.4 Confocal imaging is scanned horizontally in one imaging plane at-a-time, thus dissecting the biological entity layer-by-layer.

All optical sections in the *Z* plane are high resolution and high focus. All of these attributes make the confocal microscope a very powerful tool to be used in scientific research.

27.3 Immunofluorescence

One of the many bioassays that can be visualized on the confocal microscope is immunofluorescence. Immunofluorescence is the use of antibodies to label a specific protein within a sectioned tissue or within a population of cells. Before delving into the description of how this process, the immunology that led scientists to develop this type of assay needs to be explained.

This type of microscopy is also named fluorescent microscopy due to the use of fluorochromes to label the protein to be localized. Fluorochromes are specialized molecules that absorb certain wavelengths of light, called excitation, while emitting a different wavelength, called emission. This process is very similar to the excitation involved in producing laser light. The atoms of the fluorochrome are excited by a certain wavelength, which causes the electrons to move from their ground state orbital rings into a higher energy state. These electrons will then decay spontaneously, causing the release of a photon of a different wavelength. This wavelength is generally longer and is less intense than the excitation wavelength. So laser light hits the specimen, causing that particular fluorochrome to fluoresce. This fluorescent wavelength will be focused onto the dichroic mirror by the objective lens and will be able to pass through to the detector and ultimately to the computer for processing. Some examples of commonly used fluorochromes can be seen in Table 27.1.

Fluorescein, commonly called FITC, is generally used to tag antibodies, which are used to find specific localization of proteins in tissues and cells. This process will be explained in more detail in the following section. As shown in the chart, FITC's atoms will become excited when light in the 490 nm wavelength, or blue light, is reflected onto it. It will then emit light of a wavelength of 525 nm, which will be seen as green on the computer screen. Propidium iodide (PI) is a DNA stain that is most commonly used to visualize the nucleus in normal cells, or the fragmented nucleus in apoptotic cells. Calcein is a water-soluble dye, which is taken into cells and cleaved by a certain enzyme. This enzyme ensures that the molecule cannot leave the cytoplasm and thus can be used as an indicator of live versus dead cells in a population. This dye can also be used for cell swelling assays in live cells where the calcein, and the signal, will be more dilute the more water is introduced into the cell.

27.3.1 Mechanism of Action for Immunofluorescent Imaging

Inside the body is the immune system. The immune system is a series of cell types, duct systems (lymphatic duct system), and primary and secondary organs, which lead the defense of the body against foreign invaders and disease. This system can be divided into two specific systems, which have different jobs in defense. The first system is cell-mediated immunity. This type of immunity is generally used for intracellular pathogens, or bacteria and viruses that are found within the cells of any biological organism. It is led by the T cell and usually destroys the invaded cells to stop the spread of infection. The second arm of immunity is called humoral immunity. A cell type called the B cell leads this type of immunity. The B cell is most famous for producing soluble proteins that are called antibodies. Antibodies are used to tag foreign particles in the body, so the rest of the immune system can recognize them and get rid of them in an organized fashion. The immune system is formed through a complex process known as positive and negative selection. During this process, immune cells are "taught" to recognize all of the proteins in the body as "self," meaning resident and naturally occurring. This means that self-proteins are safe. The antibodies will tag anything not recognized as self. For this job, the antibodies must be very specific for the protein they are made to recognize. Antibodies have a specific protein motif, which can be seen in Figure 27.5. It is made of two light chains at the top and two heavy chains at the bottom, named so after the different centrifugation techniques that founded them. The F_c region or F constant region in the antibody is highly conserved among species. This means that all rabbits have an immunologically similar constant region in all of their antibodies. The second region is the F_{ab} region or fragment antigen binding region. This is really the business end of the antibody, used for binding to the protein that the antibody is against.

Scientists have learned to exploit the specificity of these antibodies to find proteins of interest. An overview of the process can be seen in Figure 27.6. To begin with, the specimen is fixed. The fixative is used to preserve the architecture of the tissue so that native localization can be determined. The section is then placed into paraffin to preserve it and sectioned to be put onto slides. Once the specimen is affixed to the slide a solution of the diluted antibody is added on top and allowed to incubate. The

TABLE 27.1 Selected Fluorochromes with Excitation and Emission Wavelengths

Fluorochrome	Absorbed λ	Emitted λ
Fluorescein (FITC)	490 nm	525 nm
Propidium iodide	536 nm	617 nm
Calcein	370 nm	435 nm

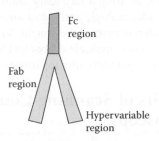

FIGURE 27.5 Diagram of fluorescent tag labeling for protein identification by means of specific receptor antibody used for attachment.

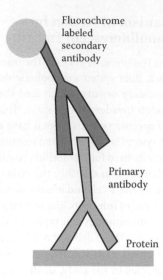

Fluorochrome labeled secondary antibody

Primary antibody

Protein

FIGURE 27.6 Diagram of fluorescent tag labeling for protein identification by means of a specific fluorescent emission based on the fluorochrome.

first antibody added is called the primary antibody because it is specific for the protein. These antibodies are made in a simple but ingenious fashion. If one is looking for a mouse protein, such as AQP-1, one must simply purify this protein and inject into an animal such as a rabbit. Because the rabbit will not recognize the mouse protein as part of itself, it will make antibodies against this particle. The researcher will then bleed the rabbit, and purify the antibodies specific for AQP-1. These antibodies will now be known as rabbit anti-mouse AQP-1 antibodies. Because antibodies are proteins, we cannot necessarily see them, so the next step is to add a secondary antibody, which is conjugated with a specific fluorochrome (which has been discussed earlier). The primary antibody will be washed off the specimen to eliminate nonspecific binding. The secondary antibody is added and again allowed to incubate. The secondary is made in much the same way as the primary antibody. Take the rabbit's primary antibody and inject it into another animal such as a goat. The goat will recognize the rabbit antibody as foreign and make antibodies against that antibody. These antibodies will be goat anti-rabbit F_c antibodies. The process of localizing proteins in this way is referred to generally as indirect immunodetection because the fluorochrome is indirectly attached to the protein one is looking for. Indirect immunodetection's biggest advantage is the amplification that occurs by using a secondary antibody. When these antibodies are made, multiple antibodies are generated, which are specific for different parts of the protein. So when the groups of antibodies are applied, multiple antibodies bind to the protein, making the signal bigger and easier to visualize.

27.4 Benefits of Scanning Confocal Microscopy

Laser scanning confocal microscopy offers several advantages: (1) it provides higher resolution of images, (2) it allows the use of

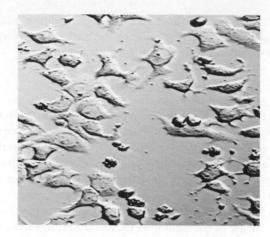

FIGURE 27.7 Confocal imaging example of tagged staphylococcus aureus cell grouping. (Courtesy of Elizabeth Jablonsky.)

higher magnifications, (3) it produces three-dimensional images, and (4) it minimizes stray light because of the small dimension of the illuminating light spot in the focal plane. Although this system provides higher resolution, it is still limited in resolution because this is a limitation in the properties of light itself. An example of the level of detail achieved with confocal microscopy imaging is provided in Figure 27.7.

To place confocal microscopy in perspective, the following is an illustration of the biological aspects involved in the field of imaging. Typically, to make a cell culture, osteoblasts are isolated from 2-day-old mice. Three million cells are cultured in 75 cm² tissue culture flasks at 37°C. Cells reach confluence in approximately 3 days at which point they are split into two new 75 cm² tissue culture flasks. Once cells reach approximately 80% confluence, they are ready for the next step in the experimental procedure. In order to monitor the progress of this mechanism, frequent observations are needed. Confocal microscopy is one of the tools that can be used to provide the feedback needed.

27.5 Summary

The confocal microscope uses a scanning laser to probe the 3D composition of a sample. The use of fluorescent tags is commonly used to selectively probe for specific proteins in the sample composition. An aperture provides the means of filtering for depth by eliminating out-of-focus rays to be detected.

Confocal microscopy has attracted considerable interest because it achieves up to approximately two fold better lateral resolution and better sectioning in the longitudinal axis as compared with the conventional light microscope. In principle, confocal microscopy should permit the characterization of ion concentrations in situations where wide field microscopy cannot, such as deep within tissues.

Immunofluorescence can also be applied to confocal microscopy. Confocal microscopy is an ingenious method for the detection of fluorescence. This microscope allows a specimen to be optically sectioned into many focal planes, which improves the resolution of the resulting micrograph and aids the researcher in determining the localization of a given protein within a tissue or cell.

Additional Reading

Chen, Q. and Zhang, X.-C., Electro-optic THz imaging, In: *Ultrafast Lasers: Technology and Applications*, Fermann, Galvanauskas, Sucha, Eds. Chapter 11, New York, NY: Marcel Dekker, Inc., 2002.

Choma, M.A., Optical Coherence Tomography: Applications and Next-Generation System Design, Ph.D. Dissertation, Raleigh, NC: Duke University, 2004.

Dunn, R.C., Near-field scanning optical microscopy, *Chem. Rev.* 99, 2391–2927, 1999.

Fercher, A.F., Hitzenberger, C.K., Kamp, G., and El-Zaiat, S.Y., Measurement of intraocular distances by backscattering spectral interferometry, *Optics Communications*, 117, 43–48, 1995.

Giancoli Physics, 5th edition. Upper Saddle River, NJ: Prentice Hall, 1998.

Gibson, A.P., Austin, T., Everdell, N.L., Schweiger, M., Arridge, S.R., Meek, J.H., Wyatt, J.S., Delpy, D.T. and Hebden, J.C., Three-dimensional whole-head optical tomography of passive motor evoked responses in the neonate, *Neuroimage*, 30(2): 521–528, 2005.

Huang, D., Swanson, E.A., Lin, C.P., Schuman, J.S., Stinson, W.G., Chang, W., Hee, M.R., Flotte, T., Gregory, K., and Puliafito, C.A., Optical coherence tomography, *Science*, 254, 1178–1181, 1991.

Enderle, J., Blanchard, S., Bronzino, J., and Burlington, M.A., *Introduction to Biomedical Engineering*, New York, NY: Academic Press, 2000.

Jiang, H., Frequency-domain fluorescent diffusion tomography: A finite element algorithm and simulations, *Appl. Opt.*, 37, 5337–5343, 1998.

Jiang, H., Paulsen, K.D., Osterberg, U.L., and Patterson, M.S., Frequency-domain optical image reconstruction for breast imaging: Initial evaluation in multi-target tissue-like phantoms, *Medical Physics*, 25, 183–193, 1998.

Jiang, H., Paulsen, K., and Osterberg U., Enhanced optical image reconstruction using DC data: Initial study of detectability and evaluation in multi-target tissue-like phantoms, *Phys. Med. Biol.*, 43, 675–693, 1998.

Kuby Immunology, 4th edition. New York, NY: W.H. Freeman and Company, 2000.

Luo, M.S.C., Chuang, S.L., Planken, P.C.M., Brener, I., Roskos, H.G., and Nuss, M.C., Generation of terahertz electromagnetic pulses from quantum-well structures, *IEEE J Quantum Electr.*, 30, 1478–1488, 1994.

Mirabella, F., Internal reflection spectroscopy, In: *Practical Spectroscopy Series*, vol. 15, Marcel Dekker, New York, 1992.

Najarian, K. and Splinter, R., *Biomedical Signal and Image Processing*, CRC-Press, Boca Raton, 2005.

Rost, F.W.D., *Fluorescence Microscopy*. Cambridge University Press, Cambridge, 1992.

Simhi, R., Gotshal, Y., Bunimovich, D., Sela, B-A., and Katzir, A., Fiber-optic evanescent-wave spectroscopy for fast multicomponent analysis of human blood, *Appl. Opt.*, 39, 3421–3425, 1996.

Splinter, R., Optical methods in medical treatment and diagnostics, *Res. Adv. Photochem. and Photobiol.*, 1, 43–75, 2000.

28

Magnetic Resonance Imaging

El-Sayed H. Ibrahim and
Nael F. Osman

28.1 Introduction

Magnetic resonance imaging (MRI) is a tomographic imaging technique that produces images of internal physical and chemical properties of the body by measuring the emitted nuclear magnetic resonance (NMR) signals. In a typical MRI exam, the patient lies inside the machine and a radiofrequency (RF) signal is emitted into the patient body, which responds by emitting another RF signal.[1] The received signal is recorded and processed to yield the MRI image.

The NMR phenomenon was reported by Bloch and Purcell in 1946. In 1973, MR imaging was made possible by Lauterbur and Mansfield. First clinical scanners appeared in the early 1980s. MRI uses electromagnetic waves with frequencies in the RF range, as shown in Figure 28.1, which have much lower energies than those used in ionizing radiation, like x-rays and computed tomography.[2]

MRI has many advantages and flexibilities over other medical imaging modalities: no ionizing radiation, no radioactive materials, high level of tissue contrast, arbitrary image orientation, high spatial resolution, and various physical parameters to be imaged.

28.2 MRI Scanner

Three types of magnetic fields are involved in an MRI experiment (Figure 28.2): main static magnetic field, B_0, responsible for tissue magnetization; RF magnetic field, B_1, responsible for signal excitation; and gradient magnetic field, G, responsible for signal localization.[3] Computer units are used for data entry, signal processing, and image reconstruction. The MRI scanning system involves no moving parts. An MRI scanner costs about one to two million dollars.

28.2.1 Main Magnet

The main magnet generates a strong homogeneous magnetic field around the object to be scanned. Higher field strengths enable a

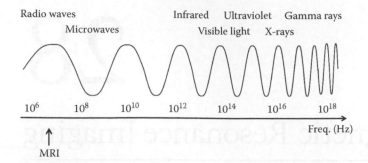

FIGURE 28.1 The electromagnetic spectrum. The waves used in MRI are in the RF range, with frequencies (and energy) much less than that used in x-rays or nuclear medicine.

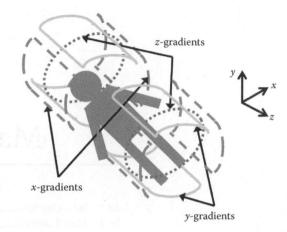

FIGURE 28.3 The gradient coils. Inside the magnet, the patient is surrounded by three orthogonal sets of gradient coils that change the magnetic filed in the *x*- (dashed), *y*- (solid), and *z*- (dotted) directions.

high signal-to-noise ratio (SNR), high resolution, and short scan time. However, higher field strengths are concomitant with inhomogeneity problems and high cost. MRI scanners in clinical use have magnetic field strength ranging from 0.5 Tesla (T) to 3 T. It should be noted that 1 T = 10,000 Gauss, where the earth's magnetic field is only half a Gauss. There are three types of main magnets: permanent, resistive, and superconducting. Permanent magnets are always on. They have low initial and maintenance costs. In resistive magnets, electric currents in the wires produce the magnetic field. This type of magnets can be turned on and off. Most scanners in use today have superconducting magnets, which operate near absolute zero temperature (-270°C). At this temperature, there is no resistance in thesuperconducting wire. Cryogens (liquid nitrogen and helium) are required for cooling. Superconducting magnets, despite being expensive and having large fringe fields, have higher field strengths and are more homogeneous than other magnet types.

28.2.2 Gradient Coils

The gradient system is composed of three orthogonal gradient coils, as shown in Figure 28.3. A gradient coil produces a spatially varying linear weak magnetic field in the *z*-direction, superimposed on the main magnetic field. Gradients are used for signal localization, which is necessary for image formation.

28.2.3 RF Coils

The RF system is composed of transmitter and receiver RF coils. A coil is generally composed of multiple loops of conducting wire that is used to generate or detect a magnetic field. The transmitter coil generates an RF magnetic field required for signal excitation. The receiver coil detects the NMR signal emitted from the body. Some coils are both transmitter and receiver (transceiver). There are RF coils designed for almost every part of the body (Figure 28.4).

The body coil is a fixed part of the scanner, which is usually used for signal transmission. The head coil is a helmet-like device (bird cage), which generates a homogeneous field around the head. There are also breast and extremity coils. Surface coils

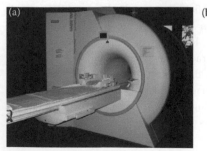
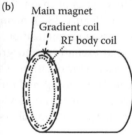

FIGURE 28.2 The MRI scanner. (a) A picture of an MRI scanner. (b) Inside the scanner, the patient is surrounded by the RF body coil (dotted), the gradient coils (dashed), and the main magnet (solid), which all lie under the scanner cover.

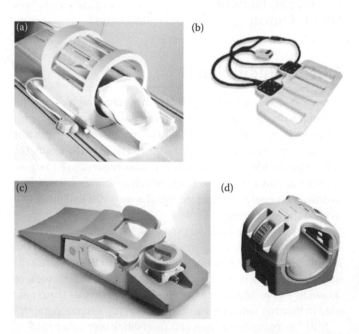

FIGURE 28.4 RF coils: (a) head coil; (b) phased-array surface coil; (c) breast coil; and (d) knee coil.

are used for imaging superficial parts of the body. They improve SNR because the coil picks up noise only from its region of sensitivity, not from the whole object as with the body coil. Phased-array coils are composed of multielements, which allows for parallel imaging with reduced scan time.

28.2.4 MRI Safety

MRI safety is an important issue. All persons in the scanner's surroundings should be aware of the magnetic field effects. Accidents could happen due to inappropriate actions or misuse of the scanner. The important thing to remember is that loose metallic objects act as projectiles if they are close enough to the magnet. They fly toward the magnet isocenter damaging anything on their path. Magnetic materials, like credit cards, get demagnetized. Sensitive devices, like watches, may be messed up. Extra precautions should be taken by the scanner operator to avoid harmful effects on the patient. The coil wires should be straightened such that no loops are formed. Wire loops may cause local burns to the patient. The protocol parameters are monitored by the scanner software so that the patient is not hurt. Specific absorption rate is observed to avoid high patient energy deposition (heating) from RF pulses with large flip angles. Furthermore, slew rate is observed to avoid patient nerve stimulation from steep gradient slopes.

28.3 Applications

MRI has successfully been applied for imaging almost all body parts. Examples of MRI images are shown in Figure 28.5. Brain imaging includes detecting multiple sclerosis, tumor, stroke, fiber tractography, and brain functions. Cardiac imaging includes measuring the heart structure, function, and viability. Breast imaging includes detecting cancer, cysts, and silicone implants. Musculoskeletal applications include imaging bones, ligaments, tendons, and cartilage. MR angiography is used for imaging blood vessels in all body parts. MR spectroscopy (MRS) is used for measuring metabolic components in the prostate, heart, brain, and liver.

28.4 NMR Phenomenon

Comprehensive understanding of the NMR phenomenon requires one to have a background in quantum physics. Fortunately, classical physics can be used to explain NMR, which will be adopted in this chapter. Nuclei that possess angular momentum (rotation around their axes) are called spins.[4] These are nuclei with odd number of protons, neutrons, or both. Hydrogen is an example of such elements, which is usually used in MRI imaging due to its abundance in the human body. A spin acts like a tiny magnet because a spinning charged particle creates an electromagnetic field, as explained in Figure 28.6.

Without external magnetic field, spins inside the body are randomly oriented, as demonstrated in Figure 28.7. Thus, they cancel each other out, resulting in zero net magnetization. However, in the presence of an external magnetic field (B_0), the spins align themselves in its direction (Figure 28.7). Approximately half the spins point up in the direction of B_0 (z-direction), and half point down in the opposite direction. The spins pointing up (parallel) are more than those pointing down (anti-parallel), with resulting net magnetization (M) pointing up in the B_0 direction (longitudinal direction).

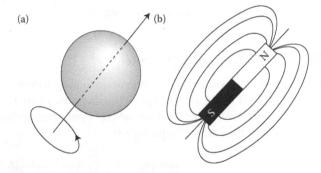

FIGURE 28.6 The spin concept. (a) The proton (spin) rotating around its axis (spinning) acts as a tiny magnet (b).

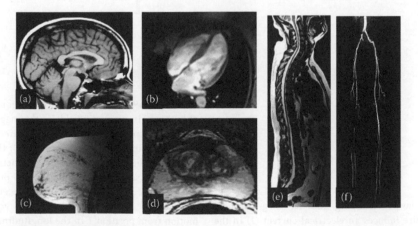

FIGURE 28.5 Example of MRI images: (a) head; (b) heart; (c) breast; (d) prostate; (e) spine; and (f) vascular.

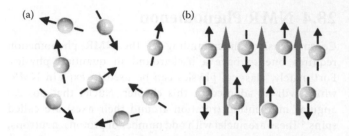

FIGURE 28.7 The effect of main magnetic field. (a) In the absence of the main magnetic field, the spins are randomly oriented and no net magnetization is created. (b) In the presence of the main magnetic field (long arrows), the spins align themselves either parallel or anti-parallel to it, with net magnetization in the main field direction.

28.5 Signal Generation: Mode of Operation

28.5.1 Faraday's Law

Faraday's law of induction[4] states that a changing magnetic field induces an electric current in a receiver coil oriented perpendicular to the magnetic field (Figure 28.8). (Note that the opposite is also true: an electrical current flowing in a coil produces a magnetic field perpendicular to the coil, which is the concept used in superconducting and resistive magnets.)

28.5.2 The Larmor Equation

The magnetization vector (M) needs to be oscillating to intercept the receiver coil and generate signal. This happens when M is tipped into the transverse xy plane, where it precesses around the z-axis with the Larmor frequency ω[1]:

$$\omega = \gamma B. \tag{28.1}$$

where γ is the gyromagnetic ratio and B is the magnetic field strength. Thus, the transverse magnetization component M_{xy} is the part that generates the signal in the receiver coil.

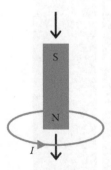

FIGURE 28.8 Faraday's law. A changing (moving) magnetic field perpendicular to a loop wire induces an electrical current (I) in the wire.

28.5.3 RF Excitation

The magnetization M is tipped into the transverse plane by applying an RF excitation pulse in the coil, which generates an external weak magnetic field (B_1) perpendicular to M, for example in the x-direction, as shown in Figure 28.9.

It should be noted that the B_1 field is not stationary; it is rotating in the xy plane around the z-axis with the Larmor frequency ω_0 in the same direction as spin precession, which creates the resonance condition. During the period of time when B_1 is on, M lies under the influence of two magnetic fields: a strong stationary field (B_0) in the z-direction and a weak rotating field (B_1) in the xy plane orthogonal to B_0. This results in that M simultaneously precesses quickly around the z-direction with the Larmor frequency ω_0 and slowly around the x-direction with frequency ω_1, where

$$\omega_0 = \gamma B_0 \tag{28.2}$$

and

$$\omega_1 = \gamma B_1. \tag{28.3}$$

This results in a spiral motion of M from the z-direction toward the xy plane, as explained in Figure 28.9. Because $\omega_1 \ll \omega_0$, M ends up tipped toward the xy plane with flip angle depending on the RF pulse strength and duration. Excitation RF pulses are referred to by the angle by which the magnetization vector is tipped: for example a 45° RF pulse tips the magnetization vector half way into the transverse plane, while a 90° RF pulse tips the magnetization vector all the way into the transverse plane, as shown in Figure 28.10.

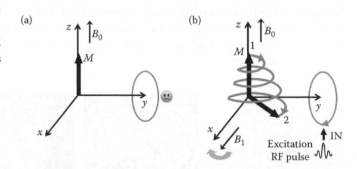

FIGURE 28.9 The effect of B_1 excitation field. (a) In the presence of only the main magnetic field B_0, the magnetization vector M is (stationary) oriented in the longitudinal direction, and does not intersect the conducting wire loop (gray circle); thus no electrical current is induced in the wire. (b) A weak B_1 magnetic field is generated orthogonal to B_0 by sending an (excitation) RF pulse in the coil, such that B_1 rotates in the xy plane with frequency ω_0. Under the combined (B_0 and B_1) magnetic field, the magnetization vector M undergoes a spiral motion from position 1 in the longitudinal direction to position 2 in the transverse plane.

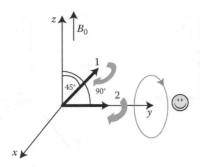

FIGURE 28.10 Magnetization position after the RF pulse application. The magnitude and duration of the B_1 excitation RF pulse determines the flip angle by which the magnetization M (thick arrow) is tipped into the transverse plane. Two example positions are shown: (1) 45° and (2) 90°. After the magnetization vector M is tipped into the transverse plane, it rotates around the B_0 direction with the Larmor frequency ω_0, intersecting the loop wire (gray circle) and inducing an RF electrical signal in the wire (coil). Note that the magnetization now has an oscillating transverse component, which intersects the coil and generates a signal.

28.5.4 The Bloch Equation

The Bloch equation describes the magnetization vector behavior under the NMR phenomenon[1]:

$$\mathrm{d}M/\mathrm{d}t = M \times \gamma B, \tag{28.4}$$

where M is the magnetization vector, B the magnetic field, t the time, and γ the gyromagnetic ratio. The operator $\times$ represents vector multiplication.

28.5.5 Rotating Frame of Reference

To simplify explaining the effect of the RF pulse on M, the rotating frame of reference is introduced[4] to avoid representing M precessing around the z-direction in the figure. Let $x'y'z'$ be the coordinates of the rotating frame: z' coincides with z; however, the $x'y'$ plane rotates clockwise around the z-axis with the Larmor frequency ω_0. From the $x'y'z'$ point of view, M does not precess around the z'-axis when the B_1 field is on. The only observed motion is the rotation of M around the x'-axis from the longitudinal z-direction toward the $x'y'$ plane, as shown in Figure 28.11.

28.5.6 Magnetization Relaxation

When the B_1 field is switched off, the net magnetic field is again B_0, and M tries to re-align itself back in the B_0 direction through relaxation, which has two independent components: longitudinal and transverse,[4] as shown in Figure 28.12. It is during the relaxation period that the transverse magnetization intersects the coil and generates an RF signal.

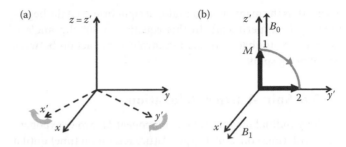

FIGURE 28.11 Rotating frame of reference. (a) The laboratory (stationary) xyz and rotating $x'y'z'$ frames of reference coincide in the longitudinal direction ($z = z'$). While the xy plane is stationary, the $x'y'$ plane rotates around the z-axis with ω_0 frequency. (b) The effect of the B_1 field, as shown on the $x'y'z'$ frame. The spiral motion of the magnetization vector M is not shown on the rotating frame of reference because both M and the $x'y'z'$ frame experience the same rotational motion around z-axis. In the $x'y'z'$ frame, the B_1 effect is to tip M directly into the $x'y'$ transverse plane.

28.5.7 Spin–Spin Relaxation

The transverse magnetization component M_{xy} decays exponentially with time constant T_2 (spin–spin relaxation time) until it vanishes after about five T_2's:

$$M_{xy} = M_0\, e^{-t/T_2}, \tag{28.5}$$

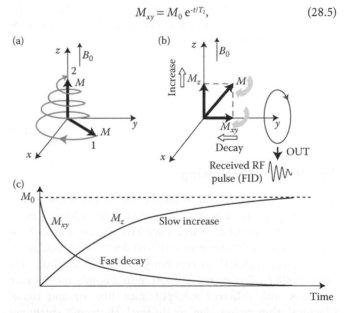

FIGURE 28.12 Magnetization relaxation. (a) After the B_1 field is switched off, the magnetization vector experiences a spiral motion (in the xyz laboratory frame of reference) going from position 1 in the transverse plane to position 2 in the longitudinal direction. (b) The magnetization vector is composed of two components: longitudinal M_z and transverse M_{xy}. During this relaxation period, M_z and M_{xy} experience different and independent relaxation behaviors. M_z grows exponentially with time constant T_1, whereas M_{xy} decays exponentially with time constant T_2. The rotating transverse component M_{xy} intersects the receiver coil (loop wire) and induces an electrical current signal, which is recorded and used to construct the MRI image. (c) The relaxation curves. T_1 is always greater than T_2; thus the transverse decaying rate of M_{xy} is always faster than the longitudinal growth rate of M_z.

where M_0 is the magnetization value at equilibrium (at the beginning of the experiment). In this equation, a 90° flip angle is assumed. Note that T_2 relaxation involves interaction between neighboring spins.

28.5.8 Spin–Lattice Relaxation

The longitudinal magnetization component M_z grows exponentially with time constant T_1 (spin–lattice relaxation time) until it reaches the equilibrium value M_0 after about five T_1's:

$$M_z = M_0 (1 - e^{-t/T_1}), \qquad (28.6)$$

where a 90° flip angle is assumed. T_1 involves energy exchange between spins and the surrounding lattice. T_1 is always greater than T_2; thus the transverse decay occurs faster than the longitudinal growth. It should be noted that the length of the magnetization vector does not remain the same during relaxation because longitudinal and transverse relaxations are independent of each other and occur with different rates. During relaxation, M rotates around the B_0 direction (in the xyz coordinates) as long as the transverse component does not vanish. Actually, this is the time period when the NMR signal is recorded as the oscillating magnetization induces an electrical signal in the receiver coil. The signal is called free induction decay (FID) (Figure 28.12). Note that the relaxation or decaying rate, R, is simply the inverse of the time constant, that is,

$$R_1 = \frac{1}{T_1} \qquad (28.7)$$

and

$$R_2 = \frac{1}{T_2} \qquad (28.8)$$

28.5.9 Spin Dephasing

Beside T_2 relaxation, there is another factor that causes additional M_{xy} decay, which is magnetic field inhomogeneity. Generally, neighboring spins do not have the same exact resonance frequency (off-resonance effects) due to main field inhomogeneities, material susceptibility, or chemical shift. The susceptibility effect is most severe near boundaries between materials with different susceptibilities, like air and tissue. Chemical shift arises due to different electronic shieldings among molecules with different molecular structures. As neighboring spins feel slightly different effective magnetic fields, they precess with slightly different frequencies (around ω_0) when tipped into the transverse plane. With time, phase shift builds up between the spins. They start to dephase and cancel each other out, resulting in an additional decay of M_{xy}, as shown in Figure 28.13. This decay can be expressed as an exponential decay with time constant T_2'. Thus, in total, M_{xy} experiences an exponential decay with time constant T_2^*, such that

$$\frac{1}{T_2^*} = \frac{1}{T_2} + \frac{1}{T_2'} \qquad (28.9)$$

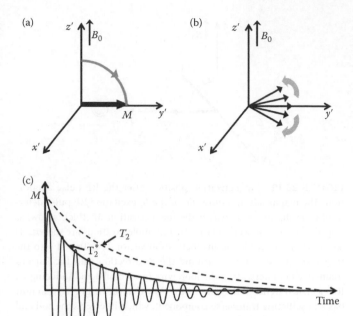

FIGURE 28.13 Spin dephasing. (a) Immediately after the RF pulse is switched off, the magnetization vector M now lies in the transverse plane. All the spins are in phase and point in the same direction (y' here). (b) With time (microseconds), different spins precess with slightly different frequencies and phase difference starts to build up between them, leading to additional decay of the transverse magnetization component. (c) The received signal (oscillating with RF frequency) experiences the combined T_2^* decay. T_2^* relaxation is the combination of T_2 relaxation and spin dephasing. T_2^* decaying rate is faster than that of T_2.

In other words, each spin experiences decay with time constant T_2. However, the whole signal (all spins inside the voxel) decays with time constant T_2^*. The smallest rate of M_{xy} decay occurs with time constant T_2, when the external magnetic field inhomogeneity effect is corrected for, as in spin echo (SE) imaging.

28.5.10 Signal Contrast

T_1, T_2, and proton density (PD) are intrinsic properties of tissue.[3] PD represents how many protons exist per unit volume of the tissue. Different tissues produce different signal intensities based on these properties. It should be mentioned that T_1 increases with field strength, while T_2 is almost invariant under the range of field strengths in clinical use.

Several control parameters can be adjusted in the imaging protocol to change the resulting image contrast. Examples of these parameters are: repetition time (TR), echo time (TE), and RF pulse flip angle (alpha). The imaging pulse sequence consists typically of repeated application of RF pulses, each followed by a period of data acquisition [signal readout (RO)], as shown in Figure 28.14. TR is the time interval between consecutive RF pulses. TE is the time interval from the RF pulse to the center of data acquisition. The imaging flip angle determines the degree by which M is tipped

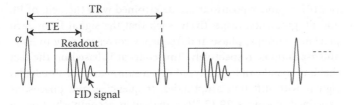

FIGURE 28.14 A simple pulse sequence. Multiple signals are acquired to construct the MRI image. The excitation RF pulse (with flip angle alpha) is applied repetitively with the appropriate gradients. An RF signal is received after every RF pulse. The time between consecutive RF pulses is called TR. The time from RF pulse to the center of data acquisition is called TE.

into the transverse plane. To obtain a T_1-weighted image, TR and TE have to be short (Table 28.1). To obtain a T_2-weighted image, TR and TE have to be long. Finally, to obtain a PD-weighted image, TR has to be long and TE has to be short.

28.6 Image Formation

The formation of an image relies on a wide range of parameters and boundary conditions. All the factors influencing the image formation in addition to image quality are described next.

28.6.1 Spatial Encoding

Without magnetic gradients, the received signal contains information from the whole body. To be able to reconstruct the image, the x, y, and z coordinates of the signal have to be specified,

TABLE 28.1 TR and TE settings for different image contrasts

Contrast	TR	TE
T_1	Short	Short
T_2	Long	Long
PD	Long	Short

which is the role of the magnetic field gradients. In an MRI scan, the RF pulse is applied multiple times while changing the gradient values to obtain multiple signals necessary to create the image. When a gradient is applied in a certain direction, it produces a weak linear change in the main magnetic field in that direction (Figure 28.15), which allows for deciphering the spatial information from the received signal.[1] One gradient field is applied in each direction: slice selection (SS), phase encoding (PE), and frequency encoding (FE) (or RO) in the z-, y-, and x-directions of the prescribed slice, respectively. The logical slice coordinates SS, PE, and RO are related to the magnet physical coordinates x, y, and z by a 3×3 rotation matrix, where rotation in three dimensions (3D) is determined by the axis of rotation and the rotation flip angle.

28.6.2 Slice Selection

The SS gradient (G_z) is applied in the z-direction of the prescribed slice. It causes a linear change in the main magnetic field in this direction. Thus, spins with different z-positions have different Larmor frequencies. An excitation RF pulse is applied at the same time when G_z is on, which is called slice-selective RF pulse. This RF pulse excites a certain slice of the body, whose location and thickness depend on the range of frequencies the RF pulse contains (bandwidth, BW), as illustrated in Figure 28.16. The slice thickness can be changed by adjusting the RF BW or the gradient value. Slice position can be changed by adjusting the RF center (carrier) frequency or the gradient value. In contrast to slice-selective pulses, if the RF pulse is applied without an accompanying G_z, it excites the spins in the whole body and is called nonselective RF pulse.

28.6.3 In-Plane Localization

With slice-selective excitation, the received signal comes only from the prescribed slice. However, the exact location of the signal inside the slice (row and column positions) is not known.

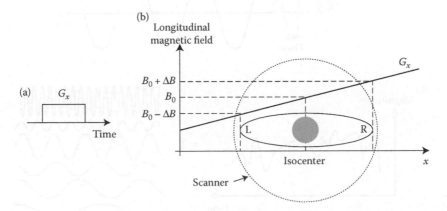

FIGURE 28.15 Gradient magnetic field. (a) A gradient pulse is applied in the x-direction. (b) As long as the gradient pulse is on, it produces a linear change in the longitudinal magnetic field in that direction. The scanner (large dotted circle) is shown with a patient lying inside (assume the gray circle is the patient head 😊). The right (R) and left (L) sides of the patient feel different magnetic fields due to gradient G_x. At the isocenter of the scanner, the magnetic field is the original value B_0 due to the main field only.

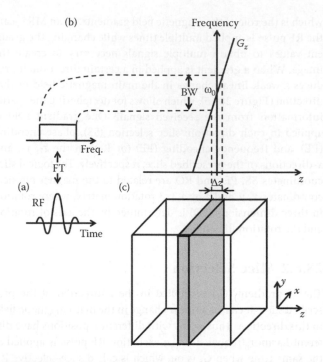

FIGURE 28.16 SS gradient. (a) An RF excitation pulse is applied at the same time when the SS gradient (G_z) is on. (b) The RF pulse contains a range of frequencies (BW), around the Larmor frequency ω_0, which corresponds to the desired slice thickness Δz. (c) The gray slice is selected to be imaged. The steepness of G_z determines slice thickness.

In MRI, y- and x-positions are determined with the help of PE and FE gradients, respectively, such that the signal from each pixel has a unique phase and frequency corresponding to its y- and x-positions, respectively. This concept is based on the fact that any signal can be analyzed into a summation of sinusoidal signals with different amplitudes, frequencies, and phases, as explained in Figure 28.17. This analysis is accomplished using Fourier transform (FT).

28.6.4 Phase Encoding

After spins in the selected slice are excited (tipped into the transverse plane) by the slice-selective RF pulse, a PE gradient, G_y, is applied in the y-direction for a short period of time before signal readout. G_y causes spins to precess with different frequencies based on their location in the y-direction. After G_y is switched off, spins with different y-positions have different phases, as illustrated in Figure 28.18. Later, when the received signal is decomposed into its sinusoidal components using FT, each component will have a specific phase corresponding to its y-position in the image. It should be noted that the area under the PE gradient pulse determines the degree of modulation, which is different for each data RO (in each TR). Typically, the gradient amplitude is changed with fixed pulse duration. Alternatively, the gradient amplitude could be kept constant while changing the pulse duration.

28.6.5 Frequency Encoding

In order to encode information in the x-direction, a FE gradient, G_x, is applied in the x-direction during signal RO. G_x changes the spin

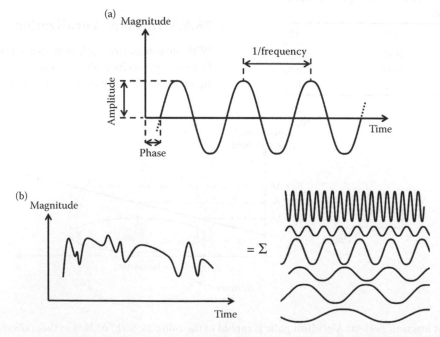

FIGURE 28.17 Fourier analysis. (a) A sinusoidal signal is determined by three parameters: amplitude, frequency, and phase. (b) Any arbitrary signal can be reconstructed as the sum of sinusoidal waves with different amplitudes, frequencies, and phases. FT is used to analyze any signal into its sinusoidal components.

(a)

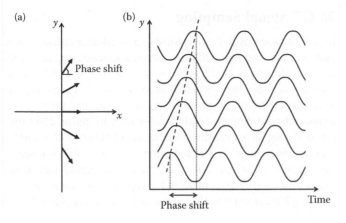

FIGURE 28.18 Phase encoding. (a) The phase encoding gradient introduces phase shift between spins with different y-positions. (b) Note how the position of the signal peak is shifted in the y-direction (dashed line). This idea is used to resolve spatial information of the received signal in the y-direction.

frequency, such that spins with different x-positions have different frequencies, as shown in Figure 28.19. Similar to phase encoding, when the received signal is analyzed with FT, each component will have a specific frequency, relating it to the correct x-position in the image. Usually, the fast Fourier transform (FFT) algorithm is used for signal analysis instead of FT to reduce processing time.

28.6.6 k-Space

The signals, $s(t)$, received from the prescribed slice are used to fill what is called the k-space,[4] from which the image $m(x, y)$ is reconstructed by applying FT (Figure 28.20):

$$m(x, y) \overset{\text{FT}}{\longleftrightarrow} s(t). \tag{28.10}$$

The k-space variables k_x and k_y in the x- and y-directions, respectively, are given by

$$k_x(t) = \frac{\gamma}{2\pi} \int_0^t G_x(\tau)d\tau, \tag{28.11}$$

$$k_y(t) = \frac{\gamma}{2\pi} \int_0^t G_y(\tau)d\tau, \tag{28.12}$$

where G_x and G_y are the FE and PE gradients, respectively.

k-space has certain features and properties. The important point to remember is that each point in k-space gives information about the whole image. This means that there is no direct relationship between the k-space center and the image center, or between the k-space periphery and the image periphery. The value at the k-space center has the highest signal intensity and determines the image contrast. It is acquired with zero phase accumulation and equals the area under the signal in the object domain. The peripheries of k-space have the lowest signal intensity and determine the image details.

It should be noted that both k-space and the image space are complex, that is, each signal has two components: real and

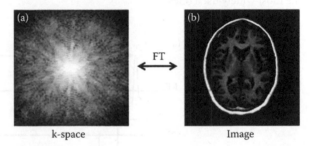

FIGURE 28.20 FT. (a) The acquired signals are used to fill the k-space. (b) FT is applied to k-space to generate the MRI image.

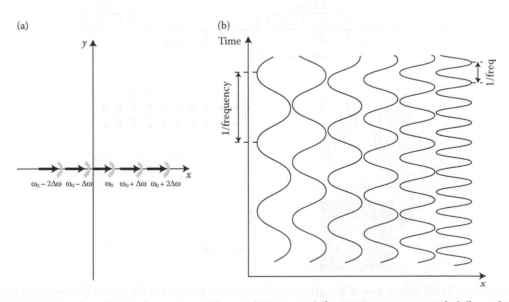

FIGURE 28.19 Frequency encoding. (a) The frequency encoding gradient causes different spins to precess with different frequencies based on their x-position. (b) This idea is used to resolve spatial information of the received signal in the x-direction.

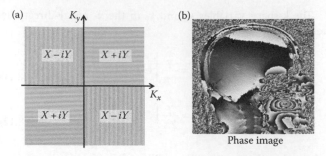

(a)

(b)

Phase image

FIGURE 28.21 Properties of k-space. (a) Conjugate symmetry of k-space. Theoretically, only a fraction of k-space can be acquired, and the rest of k-space can be determined using the conjugate property. (b) Because of the complex nature of k-space, four types of images can be reconstructed: magnitude, phase, real, and imaginary. Shown is a phase image of the head.

imaginary. k-space has the conjugate (Hermitian) symmetry property (Figure 28.21), that is, $(x + iy)^* = x - iy$, where * means complex conjugate. Four types of images can be reconstructed from the k-space data: magnitude, phase, real, and imaginary images, as shown in Figure 28.21. Magnitude (modulus) images are the most widely used ones.

28.6.7 Signal Sampling

In the simplest way, k-space is filled in a rectilinear fashion, such that each received signal fills one line of k-space during each TR, as shown in Figure 28.22. If the image size is 256 × 256 pixels, then k-space has 256 lines (rows), each composed of 256 samples. Thus, the excitation RF pulse has to be applied 256 times, each followed by a different G_y value. After each RF pulse, 256 samples of the signal are captured during data RO (while G_x is on) to fill one line in k-space. The sampling time (TS) of each k-space line = $256 \times \Delta T$, where ΔT is the sampling period (the time interval between consecutive RO samples). The sampling rate (reading BW) is equal to the inverse of the sampling period:

$$BW = \frac{1}{\Delta T}. \tag{28.13}$$

The Nyquist law states that the reading BW must be at least twice the maximum frequency in the signal, which depends on the object size. Otherwise, aliasing artifact appears in the image.

28.7 Pulse Sequences

An MRI pulse sequence is a sequence of RF pulses and gradients applied repeatedly during an MRI scan. Different pulse

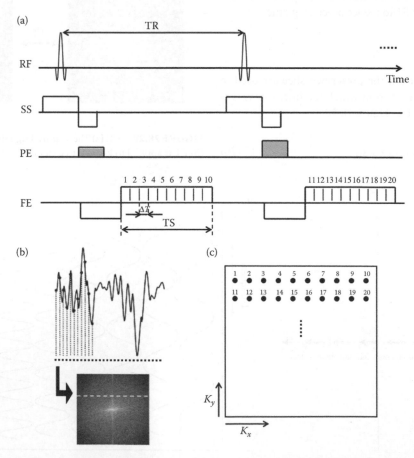

FIGURE 28.22 k-Space. (a) Simple pulse sequence. Signal is sampled during data RO for each TR. (b) The received signal is sampled to fill one line of k-space. (c) k-Space is filled in a rectilinear fashion (one line is filled for each TR). The time lapse between consecutive samples in the FE (K_x) and PE (K_y) directions is ΔT and TR (in milliseconds), respectively.

sequences have different designs and features.[5] Examples of pulse sequences are: gradient echo (GE), spin echo (SE), and inversion recovery (IR).

28.7.1 Gradient Echo

The saturation recovery GE pulse sequence (Figure 28.23) consists of a series of excitation RF pulses, each followed by data acquisition. Three gradient pulses appear in the sequence for spatial encoding. One gradient pulse is applied at the same time of the RF pulse for slice selection. Another gradient pulse (PE) is applied before each data acquisition to resolve spatial information in the y-direction. The third gradient pulse (RO or FE) is applied during data acquisition to resolve spatial information in the x-direction. The purpose of the negative half lobe gradient (RO prephaser) before the RO gradient is to dephase the spins in advance, such that at time TE, there is zero accumulated phase from the RO gradient, and thus maximum signal is obtained. The same concept applies to the negative half lobe gradient (SS rephaser) after the SS gradient, which cancels the phase dispersion caused by the SS gradient pulse. The recorded signal is maximum at the center of data acquisition, and it determines the image contrast. The imaging sequence is called full saturation when the flip angle equals 90° and partial saturation when it is less than 90°.

28.7.2 Spin Echo

The SE pulse sequence (Figure 28.24) is similar to GE, except that each data acquisition is preceded by a 180° refocusing RF pulse,

located in the middle between the excitation RF pulse (usually 90°) and the center of data acquisition. The purpose of the 180° RF pulse is to cancel the effect of field inhomogeneity by reversing the phase shift accumulated between spins midway before data acquisition. Thus, M_{xy} experiences T_2 (not T_2^*) decay, which results in stronger signal. This signal is called "echo" in contrast to FID. Acquiring the MRI signal from an echo has advantages over acquiring the FID because it allows sufficient time for gradient switching and separates signal excitation from the receiving part. In SE, TE can be changed without suffering from T_2^* decay, which is useful for acquiring T_2-weighted images. However, the added 180° pulse increases scan time and adds to patient power deposition.

28.7.3 Inversion Recovery

The IR pulse sequence (Figure 28.25) is similar to GE, but each excitation RF pulse is preceded by a 180° inversion RF pulse. The time interval between the inversion pulse and the excitation pulse is called the inversion time (TI). IR sequences are T_1-weighted, and TI determines how heavily the image is T_1-weighted. The effect of the inversion pulse is to flip the magnetization vector in the negative longitudinal direction ($-z$), from where it recovers to reach equilibrium back in the positive z-direction. When recovering from negative to positive, M crosses zero at a certain time point. Tissue suppression can be implemented by setting TI as (magnetization recovery is assumed within TR)

$$\mathrm{TI} = 0.693 T_1. \tag{28.14}$$

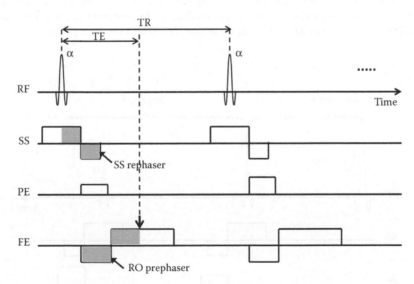

FIGURE 28.23 GE pulse sequence. The RF pulse is applied in the presence of a SS gradient in the z-direction to excite a certain slice of the body. After slice excitation, the PE gradient is applied in the y-direction to resolve spatial information in this direction. The frequency encoding (FE) gradient is then applied in the x-direction during signal RO to resolve spatial information in this direction. The SS gradient is followed by a negative rephaser gradient lobe with half the area of the SS gradient (shaded parts have equal areas) to cancel spin dephasing caused by the SS gradient. The RO gradient is preceded by a negative prephaser gradient lobe with half the area of the RO gradient (shaded parts have equal areas) to cancel the phase dephasing induced by the RO gradient at the center of data RO (vertical arrow). The set of RF pulses and accompanying gradients are repeated every TR with increasing PE gradient to acquire the signals needed to reconstruct the MRI image.

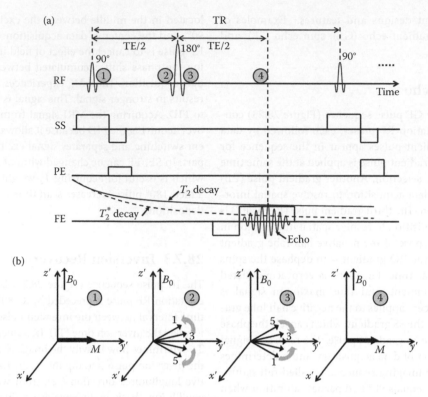

FIGURE 28.24 SE pulse sequence. (a) Pulse sequence diagram. A 90° RF excitation pulse is applied in the presence of an SS gradient in the z-direction to excite a certain slice of the body. PE gradient is then applied in the y-direction to resolve spatial information in this direction. A 180° refocusing RF pulse is applied midway (splitting TE in half) between the 90° RF pulse and the center of data acquisition. FE gradient is applied in the x-direction during data RO to resolve spatial information in this direction. The SS gradient is followed by a rephaser gradient lobe to cancel spin dephasing introduced by the SS gradient. The RO gradient is preceded by a prephaser gradient lobe to cancel spin dephasing caused by the FE gradient at the center of data acquisition. Because of the refocusing RF pulse, the recorded signal (echo) experiences T_2 (not T_2^*) decaying rate resulting in higher signal. (b) Immediately after the 90° pulse (point 1), the magnetization now lies in the transverse plane with all spins pointing in the same direction. Immediately before the 180° pulse (point 2), the spins experienced phase difference during TE/2 period of time. Immediately after the 180° pulse, the spins are now flipped 180° with respect to the y' axis, when they will rephase in the opposite direction for another TE/2 time period. At the center of data acquisition (point 4), the spins are again aligned in the same direction. The only decay that M experiences between points 1 and 4 is due to T_2 relaxation.

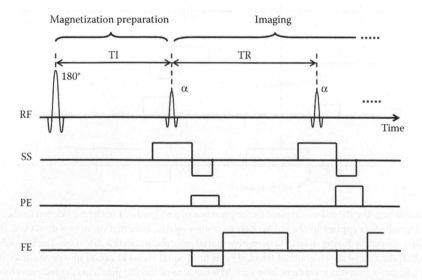

FIGURE 28.25 IR sequence. The imaging part is the same as in the GE sequence, but it is preceded by a 180° inversion RF pulse. The time between the 180° pulse and the first imaging RF pulse is called TI.

28.7.4 Volumetric Imaging

3D imaging is the direct way to achieve volumetric imaging.[4] There is no slice selection in 3D imaging. Rather, a thick chunk (or slab) is excited, as shown in Figure 28.26. A second PE loop in the z-direction is added outside the conventional y-direction PE loop, as shown in Figure 28.26. A 3D matrix in k-space is filled, and FT is applied to transform the matrix into a set of parallel 2D thin images. Slice thickness in the z-direction is determined based on the slab thickness and number of z phase-encodes. In 3D imaging, thin slices are achievable without SNR sacrifice because signals in k-space arise from the whole excited slab. 3D imaging needs longer scan time and results in higher SNR than 2D imaging.

28.8 Fast Scans

28.8.1 Segmented Acquisition

There are several techniques to reduce scan time,[2] which is necessary for dynamic imaging and breath-hold scans. One approach is to acquire more than one k-space line each TR. This technique is known as segmented k-space acquisition, as shown in Figure 28.27. Usually, TR is not long enough for full magnetization recovery. Instead, short TRs are used with small flip angles. After several TRs, a steady-state condition is established, such that the same magnetization value is reached at the beginning of each TR.

28.8.2 Echo Planar Imaging

The extreme case of segmented k-space acquisition is to acquire all k-space lines after only one excitation RF pulse (one shot acquisition), which is the idea of the echo planar imaging (EPI) pulse sequence, shown in Figure 28.28. However, this requires significant hardware capabilities besides the pronounced effect of T_2-weighting on k-space data. Alternatively, interleaved acquisition could be implemented with slight increase in scan time, but less hardware demands. In EPI, phase encoding is increased by G_y blips in between alternating G_x gradients. Thus, k-space is swept horizontally from line to line with alternating directions. EPI advantages include fast scan time, no motion artifacts, and dynamic imaging capabilities. EPI disadvantages include the need for fat suppression techniques (to avoid chemical shift artifacts), increased patient power deposition, and potential phase errors. EPI has been used in cardiac, abdominal, dynamic, diffusion, and perfusion applications.

28.8.3 k-Space Trajectories

Another approach to reduce scan time is to use faster paths, rather than rectilinear acquisition, to fill k-space; for example radial or spiral trajectories (Figure 28.29). In this case, k-space data have to be regridded to a rectangular grid before FT can be applied. Each k-space trajectory has its own advantages as well as

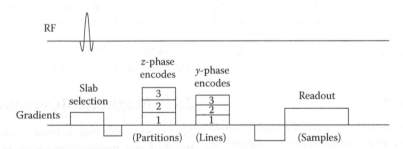

FIGURE 28.26 3D volumetric imaging. A thick slab is firstly excited. Then, two loops of phase encodes are implemented. The outer loop encodes consecutive partitions in the z-direction. The inner loop is the regular one, which encodes information in the y-direction. The RO gradient is implemented to acquire the samples in k-space and encode information in the x-direction.

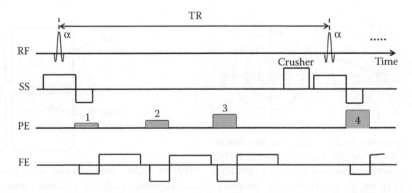

FIGURE 28.27 Segmented k-space acquisition. Multiple lines (three here) are acquired every TR. Note that the PE gradient (gray area) changes before each data acquisition to read a new line in k-space. A crusher gradient is added at the end of each TR to destroy the remaining transverse magnetization before the next RF pulse, where the next three k-space lines will be acquired.

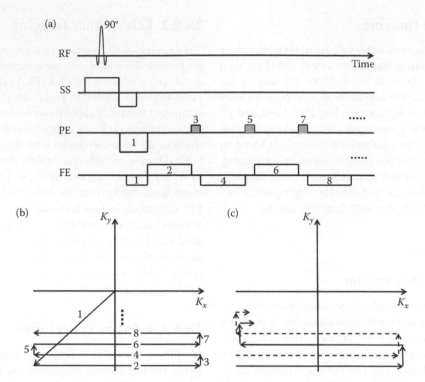

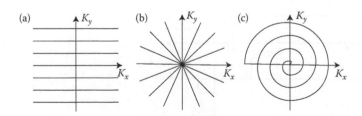

FIGURE 28.28 EPI pulse sequence. (a) Pulse sequence. Only one RF pulse is required to acquire the whole k-space. The FE gradients have alternating polarities to sweep k-space in alternating horizontal directions (from left to right and right to left), as shown in (b). PE gradient blips (dashed areas) are introduced between the FE gradients to move in k-space upward from line to line, as shown in (b). The numbers show the sequence of applied gradients and their effect on traversing the k-space in (b). (c) Interleaved EPI (two interleaves here). Two RF pulses are required in this case. The odd (solid trajectory) k-space lines are acquired after the first RF pulse. Then, the even (dashed trajectory) lines are acquired after the second RF pulses. In this case, the PE gradient blips will have double the areas as those in (b) because they skip every other line.

artifacts. Different k-space trajectories could also be applied in 3D imaging.

28.8.4 Parallel Imaging

Scan time could be reduced by applying parallel imaging. In this case, a phased-array receiver coil is used, which is composed of adjacent small coil elements that cover the body area of interest (Figure 28.30). Data are acquired simultaneously from all coil elements, and the image is constructed by combining the data together. Although it reduces scan time, parallel imaging reduces SNR as

well. SENSE and GRAPPA are famous parallel imaging techniques, which are based on the image space and k-space, respectively.

28.9 Magnetic Resonance Spectroscopy

MRS is a valuable tool that can be used to investigate molecular structure and metabolic processes of tissues.[1] Due to different molecular structures, protons in different molecules experience slightly different magnetic fields, and thus precess with slightly different frequencies (around the Larmor frequency). The role of MRS is to measure signal intensity at different frequencies

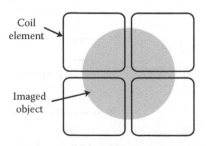

FIGURE 28.29 Different trajectories to fill k-space. (a) Rectilinear trajectory. The k-space is filled line by line. (b) Radial trajectory: the PE and FE gradients are applied simultaneously with different amplitudes to change the angle of the radial spoke. (c) Spiral trajectory: sinusoidal PE and FE gradients are applied at the same time with varying magnitudes. Interleaved spirals could be implemented. Radial and spiral acquisitions are faster than rectilinear.

FIGURE 28.30 Parallel imaging. A phased-array coil composed of four elements. Each coil element captures (sees) part of the object (gray circle). The four coil elements acquire data simultaneously, and the final image is reconstructed by combining the parts acquired by all elements.

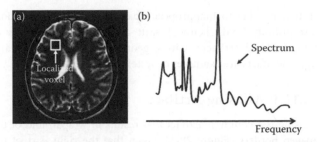

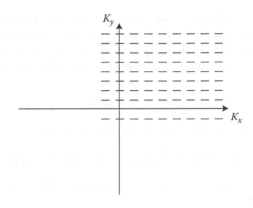

FIGURE 28.31 MRS. (a) A certain voxel is localized. (b) Frequency spectrum inside the specified voxel. Signal peaks are assigned to different metabolites based on frequency. The relative height of different signal peaks could determine the pathological condition of the tissue, for example cancer.

(spectrum) and thus determine the existence and relative magnitude of different metabolites inside the tissue (voxel-wise) (Figure 28.31).

High field strength results in large separation between different spectral peaks inside the voxel. MRS could be implemented on different nuclei: hydrogen (H), phosphorus (P), and sodium (Na). MRS of hydrogen is called proton MRS. Hydrogen is an ingredient of many molecules and is available in abundance in the body. MRS has many applications in the heart, brain, liver, breast, and prostate. It helps in staging, screening, and follow-up of patients with cancer.

28.10 Image Characteristics

28.10.1 Number of Signal Averages

The number of signal averages is the number of times the whole k-space is acquired (for the same image). When k-space is acquired n times, which increases scan time by a factor of n, SNR increases by square root of n. On the other hand, theoretically, only one-fourth of k-space is sufficient to reconstruct the image (Figure 28.32) due to the conjugate symmetry of k-space (secondary to the image space being real-valued), which reduces scan time at the cost of SNR loss. This is called partial k-space acquisition. However, practically, the image is usually not real-valued because of phase shifts from a variety of sources, and thus a little more of k-space coverage is needed to resolve the phase data.

28.10.2 Field of View and Resolution

The imaging field of view (FOV_x and FOV_y) and spatial resolution (Δx and Δy) depend on how k-space is filled.[3] FOV is equal to the inverse of the sampling period in k-space (Δk) and image resolution is equal to the inverse of k-space maximum extent (k_{max}), as shown in Figure 28.33:

$$FOV_x = \frac{1}{\Delta k_x},$$ (28.15)

$$FOV_y = \frac{1}{\Delta k_y},$$ (28.16)

$$\Delta x = \frac{1}{k_{x,max}},$$ (28.17)

FIGURE 28.32 Partial k-space acquisition. Due to the conjugate symmetry of k-space, only a part of k-space can be acquired and the rest of the k-space can be determined based on its conjugate symmetry. However, because the image is not usually pure real (phase shift is acquired), a little more of k-space than the theoretical limit is needed for correct image reconstruction.

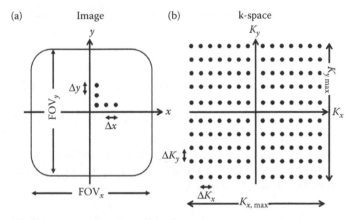

FIGURE 28.33 Relationship between image space and k-space. (a) Image with fields-of-view of FOV_x and FOV_y in the x- and y-directions, respectively. The image resolution is Δx and Δy in the x- and y-directions, respectively. (b) k-space extent is $K_{x,max}$ and $K_{y,max}$ in the K_x and K_y directions, respectively. k-space resolution is ΔK_x and ΔK_y in the K_x and K_y directions, respectively. There is an inverse relationship between the parameters in the image space and k-space: $\Delta x = 1/K_{x,max}$; $FOV_x = 1/\Delta K_x$. Similar equations exist for the y-direction.

$$\Delta y = \frac{1}{k_{y,max}}.$$ (28.18)

Thus, for a large FOV, k-space has to be densely filled, and for high spatial resolution, the outer peripheries in k-space have to be acquired.

28.10.3 Scan Time

The total scan time is

$$\text{Scan time} = \text{NSA}\ N_z\ N_y\ \text{TR}.$$ (28.19)

where NSA is the number of signal averages, N_z the number of phase encodes (partitions) in the z-direction (in 3D imaging), N_y

the number of phase encodes in the y-direction, and TR the repetition time. It should be noted that scan time depends on FOV and spatial resolution. Thus, scan time could be controlled by adjusting the imaging parameters that affect FOV and spatial resolution.

28.10.4 Signal-to-Noise Ratio

SNR determines the image quality, and is defined as

$$\text{SNR} \equiv \text{Signal mean/Noise standard deviation.} \quad (28.20)$$

SNR is proportional to the main magnetic field strength (B_0):

$$\text{SNR} \propto B_0. \quad (28.21)$$

Thus the images from a 3 T scanner have double the SNR of those from 1.5 T. SNR is also proportional to voxel size and total RO time:

$$\text{SNR} \propto \text{Voxel size} \sqrt{\text{Total readout time}}, \quad (28.22)$$

where

$$\text{Voxel size} = \Delta x \, \Delta y \, \Delta z, \quad (28.23)$$

$$\text{Total readout time} = \text{NSA} \, N_z \, N_y \, N_x \, \Delta T, \quad (28.24)$$

where Δx, Δy, and Δz are the voxel dimensions in the x-, y-, and z-directions, respectively, and ΔT is the sampling period. Note that the sampling time (TS) is

$$\text{TS} = N_x \, \Delta T. \quad (28.25)$$

By manipulating the parameters in the SNR equation, it can be shown that:

$$\text{SNR} \propto \text{FOV}, \quad (28.26)$$

$$\text{SNR} \propto 1/\text{Acceleration factor}, \quad (28.27)$$

where the acceleration factor determines the amount of time saving in parallel imaging. There are two sources of noise in MRI: body noise and coil noise. At low field strength, coil noise dominates, whereas at high field strength, body noise dominates. Coils are designed to contribute minimum noise, such that body noise dominates. It should be noted that noise is expressed as a random variable in image analysis.

28.11 Imaging Artifacts

Image artifacts occur due to the wrong selection of scan parameters, patient motion, or malfunctioning scanner hardware.[1] Aliasing, partial volume effect, and slice cross-talk are examples

of artifacts due to inappropriate parameter settings. Motion, susceptibility, and chemical shift artifacts are examples of patient-related artifacts. Gibbs, geometric, eddy current, and zipper artifacts are examples of system-related artifacts.

28.11.1 Aliasing Artifact

Aliasing is a common artifact that results in wraparound of the imaged borders (Figure 28.34), such that the right part of the object appears on the left side and vice versa. Aliasing occurs if the reading BW is not high enough (below the Nyquist frequency). It can be avoided by increasing FOV to cover the whole object being imaged; oversampling to avoid overlap of the image replicas in the object domain; localized excitation around the region of interest; or saturating the area outside the region of interest.

28.11.2 Partial Volume Effect

The partial volume effect occurs when the imaging voxel contains signals from more than one tissue (Figure 28.35). It can be

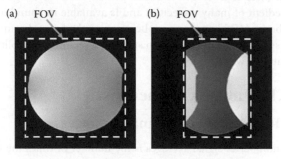

FIGURE 28.34 Aliasing artifact. (a) FOV (dashed line) is large enough, such that there is no aliasing. (b) The FOV is reduced in the FE direction, which results in image overlap and aliasing.

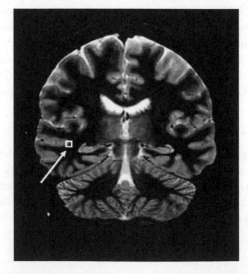

FIGURE 28.35 Partial volume effect. If the voxel size is not small enough (small white square), then it could contain more than one tissue (white and gray matters in this case), which produces mixed signal intensity of the voxel.

avoided by increasing the matrix size or reducing FOV to improve spatial resolution.

28.11.3 Slice Cross-Talk

Slice cross-talk means that the image of one slice contains signals from the adjacent slice in multi-slice imaging. This artifact occurs because the excited slice profile is not perfect, and thus it leaks and excites part of the adjacent slice. Solutions to cross-talk artifact include leaving gaps between adjacent slices or acquiring the odd-indexed slices first before the even-indexed slices.

28.11.4 Motion Artifacts

Patient motion results in ghosting artifacts (Figure 28.36), mainly in the PE direction, and it can be divided into random and repetitive motion. Motion artifacts are pronounced in the PE direction because it takes a whole TR to move one step in the PE direction, whereas it takes only ΔT to move one step in the FE direction, as shown in Figure 28.22. Random patient motion can be minimized by using fast imaging techniques or sedating the patient. Furthermore, swapping the PE and FE directions will switch the artifact location, which could move the artifacts away from the area of interest in the image. Respiratory artifacts can be suppressed by breath-holding. Artifacts from heart motion can be avoided by cardiac gating.

28.11.5 Susceptibility Artifacts

Susceptibility artifacts (Figure 28.37) result in signal void at certain areas in the image, such as at the tissue–air interface. It could also result from metallic implants inside the patient. One solution is to minimize the FOV to avoid including the problematic area in the image. Using frequency shimming and SE sequences also reduces susceptibility artifacts.

28.11.6 Chemical Shift Artifacts

Chemical shift artifact occurs because of the slight difference in the resonance frequency between molecules with different

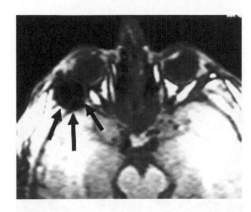

FIGURE 28.37 Susceptibility artifact. Note the signal void next to the eye ball (arrows).

chemical structures, like fat and water. The effect of chemical shift is pronounced at higher magnetic fields. Chemical shift artifact results in location shift (spatial displacement) between water and fat tissues (Figure 28.38). It can also result in signal void at the boundary between fat and water at certain TE values. One solution is to use fat suppression techniques. Tissue displacement resulting from chemical shift effects could be reduced by increasing the RO gradient. The TE could also be adjusted to avoid signal void at the tissue boundary.

28.11.7 Gibbs Artifact

Gibbs artifact results in ringing at the image boundaries, as shown in Figure 28.39. It results from abrupt truncating the high-frequency content of the signal. Solutions include using smooth windows for gradual data cut-off or reducing pixel size.

28.11.8 Geometric Artifacts

Geometric artifacts distort the shape of the imaged object. They result from gradient nonlinearity or gradient power drop-off. This type of artifacts is usually taken care of by the scanner manufacturer.

28.11.9 Eddy Currents

Eddy currents distort the image. They result from small electric currents generated when gradients are rapidly switched on and

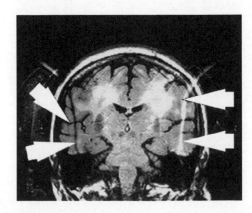

FIGURE 28.36 Ghosting artifact. Due to patient head motion, ghosts appear in the phase encoding direction (arrows).

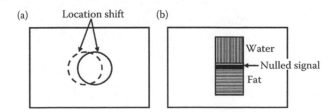

FIGURE 28.38 Chemical shift artifacts. (a) Location shift of the fat tissue (circle) from the solid to the dashed position due to chemical shift artifact. (b) The boundary between water and fat could be nulled (appears black) for certain values of TE when the water and fat spins are out of phase inside the same voxel due to chemical shift.

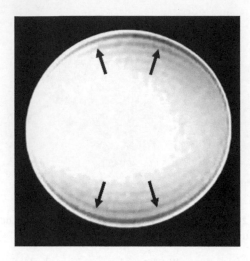

FIGURE 28.39 Gibbs artifact. It appears as ringing around the object boundary.

off. One solution is to reduce the gradient slew rate (the steepness of the ramp-up and ramp-down parts of the gradient pulse).

28.11.10 Zipper Artifact

Zipper artifact produces zipper lines along the FE direction. It results from RF noise in the area around the scanner. The first act is to ensure that the scanner door is firmly closed. Electronic devices inside the scanner room could also be the reason for this artifact.

References

1. Nishimura DG, *Principles of Magnetic Resonance Imaging*, Palo Alto, CA: Stanford University, 1996.
2. Hashemi RH and Bradley WG, *MRI the Basics*, Philadelphia, PA: Lippincott Williams & Wilkins, 1997.
3. Liang ZP and Lauterbur PC, *Principles of Magnetic Resonance Imaging*, Washington, DC: IEEE Press, 2000.
4. Haacke EH, Brown RW, Thompson MR, and Venkatesan R, *Magnetic Resonance Imaging Physical Principles and Sequence Design*, Hoboken, NJ: Wiley-Liss, 1999.
5. Bernstein MA, King KF, and Zhou XJ, *Handbook of MRI Pulse Sequences*, Amsterdam, the Netherlands: Elsevier, 2004.

Positron Emission Tomography

Robert Splinter

29.1 Introduction and Overview

Positron emission tomography (PET) is a nuclear imaging technique that produces images of the metabolic activity of living organisms by means of radioactive labeling. Introduction of a short-lived positron-emitting radiopharmaceutical is used to tag a specific metabolic activity that can subsequently be imaged noninvasively by specialized equipment. The radio-nucleotide that is administered contains a specified quantity of short-lived radioactively labeled chemical substances that are identical, or closely analogous, to naturally occurring substances in the body. Radioactive labeling means that one atom in a molecule is replaced by a radioactive Trojan horse, a radio-nucleotide. A detailed description of the radioisotope phenomenon is presented in Chapter 41. Specific examples of the radio-nucleotides used in PET imaging are the following: ^{11}C, ^{18}F, ^{13}N, and ^{15}O.

All the short-lived radioisotopes used in PET scans decay by positron emission. Positrons (β+) are positively charged electrons with the same mass as an electron and hence the same rest energy as an electron. Positrons are emitted from the nucleus of radioisotopes that are unstable because they have an excessive number of protons and a positive charge. A positron is the atomic equivalent of a positive electron, both in mass and in charge, except that the charge is opposite to that of an electron.

After introduction the isotope circulates through the bloodstream to reach the organ or tumor of interest.

Once the radioactively labeled substance has been absorbed, the location and quantity of uptake can be imaged, and a functional analysis of the organ in question can be made. The imaging is made possible by the positron annihilation with an electron, emitting two γ rays. The tomograph detects the γ rays and derives the isotope's location and concentration from the magnitude and detection sequence of the γ photons.

The PET images produced help physicians identify normal and abnormal activities in living tissue. Unlike computed tomography (CT), which primarily provides anatomical images, PET measures chemical changes in the tissue that occur before abnormalities are visible on a magnetic resonance imaging (MRI) or CT scan. The basic principles of CT imaging are outlined in Chapter 26. MRI does measure certain specific metabolic activities; however, the imaging capability is limited mostly to blood oxygenation consumption. PET recognizes these changes based on the cellular metabolism of the tissue. PET scans measure the amount of tracers distributed throughout the body. This information is subsequently used to create a three-dimensional image of tissue function from the acquired decay matrix.

29.2 Operational Mechanism

The radio-nucleotide isotopes will emit a positron after a relatively short half-life. The positron will virtually immediately annihilate with any of the many free electrons in the biological medium. The annihilation process releases a substantial amount of energy: >1 MeV. This energy is released as two γ quanta with 511 keV, emitted in an exactly perpendicular direction to each other. The energy balance between the positron annihilation process and the γ pair production is described in Equation 29.1:

$$E_{pos} + E^{kinetic} + E_e \geq 2E_\gamma \qquad (29.1)$$

The rest energy is determined by the difference in energy of the isotope and the product. In addition to the rest energy of the positron (E_{pos}) and the electron (E_e), respectively, there may be additional kinetic energy involved from positron or electron motion: $E^{kinetic}$. The nuclear energy follows from the quantum physics theorem that mass and energy are equivalent, providing the rest mass energy equivalent as illustrated by Equation 29.2:

$$E = mc^2, \qquad (29.2)$$

where E represents the rest energy of the element, m the atomic mass, and c the speed of light. Substituting Equation 29.2 into Equation 29.1 and using the photon energy

$$E = h\nu \qquad (29.3)$$

yield the following equation:

$$m_{pos}c^2 + m_e c^2 = 2h\nu_\gamma. \qquad (29.4)$$

Filling in the mass-equivalent energy for each component yields the energies in Equation 29.5:

$$511 \text{ keV} + E_{pos}^{Kinetic} + 511 \text{ keV} \geq 1.022 \text{ MeV}. \qquad (29.5)$$

Only concurrent detection of two 511 keV γ quanta is considered a recording of an emission, and as such an indication of a positron release. The positron will, by definition, have originated on the connecting line between the two detectors that registered the γ quanta. The concentration of isotopes will result in a cumulative recording in various directions of γ pairs from different locations within the organ over the half-life of the radio-nucleotide.

29.3 Noise and Contrast Features

The fact that two γ quanta need to be detected simultaneously and the γ quanta have distinct energy content makes this detection method very precise. There is no significant noise in this energy spectrum, nor will there be any background radiation that fits the criteria for detection. Emission from outside the slice of interest will also be negligible since it will never get registered by two detectors in the plane of observation.

The positron emission imaging technique has a very high contrast and very low noise without the requirement of specific engineering and design requirements.

29.4 Physical and Physiological Principles of PET

Because of the availability of various types of radioactive isotopes, the specific metabolistic changes resulting from assorted diseases make PET imaging particularly useful in the detection of malignancy in tumors.

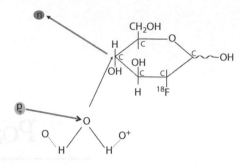

FIGURE 29.1 Production of fluorodeoxyglucose by proton impact on enriched water, which in turn is bound to 1,3,4,6-tetra-0-acetyl-2-0-trifluoromethanesulfonyl-β.

In order to understand the full use of the radioactive isotopes and the incorporation of each species into the cellular metabolism, the production of the isotope will be described first.

29.4.1 Production of Radio-Nucleotides

The radio-nucleotide can be produced in many different ways; the most prominent method is the use of an on-site cyclotron. Particle bombardment will produce the needed radionuclides. Fluorodeoxyglucose is the most commonly used radioisotope (Figure 29.1). The isotope degradation process is described in Chapter 41.

29.5 Isotope Detection

All the short-lived radioisotopes used in PET scans decay by positron emission. Positrons (β+) are positively charged electrons. Positrons are emitted from the nucleus of radioisotopes that are unstable because they have an excessive number of protons and a positive charge.

Positron emission stabilizes the fluoride nucleus by removing a positive charge through the conversion of a proton and a neutron, as shown in the chemical reaction

$$^{18}_{9}\text{F} \rightarrow {}^{0}_{+1}\text{e} + {}^{18}_{8}\text{O}. \qquad (29.6)$$

In the decay process, the radio-nucleotide is converted into an element that has an atomic number that is one less than the isotope. For radioisotopes used in PET scans, the element formed from positron decay is stable and would not have any remaining decay mechanisms. The distance a positron travels depends on the rest energy of the positron, as can be derived from Equation 29.5.[1]

Usually the distance a positron travels is limited to approximately 1 mm. The positron combines with an ordinary electron of a nearby atom in an annihilation reaction, forming positronium as an intermediate reaction product.

When a positron comes in contact with an electron, the annihilation process releases energy greater than 1 MeV. This reaction is governed by conservation of energy. The energy of the electron before interaction with a positron is foremost slightly more than the rest energy expressed by Equation 29.2. The momentum of the electron upon impact can be considered to be zero. After collision and annihilation, the produced γ radiation has an energy content given by Equation 29.7

$$E = h\nu'_\gamma \qquad (29.7)$$

and a momentum given by Equation 29.8

$$p' = \frac{h}{\lambda'}, \qquad (29.8)$$

where λ' is the γ radiation wavelength.

The scattered electron at this point has the energy given by Equation 29.9 with the accompanying momentum in Equation 29.10:

$$E'_e = E_{\text{kinetic}\_e}' = m'_e c^2, \qquad (29.9)$$

$$p'_e = c^{-1}\sqrt{E_e'^2 - (mc^2)^2}. \qquad (29.10)$$

PET scanning is hence an invasive technique since it requires injection of radioactive material into the patient. The radiation exposure is measured in Sievert (Sv) as explained in Chapter 41. The total radiation dose of a PET scan is generally small, approximately 7 mSv. In comparison, annual background radiation amounts to an average of 2.2 mSv; while a chest x-ray provides 0.02 mSv of exposure, it is up to 8 mSv for a CT scan of the chest. γ exposure is even worse at high altitudes; a frequent traveler or airplane crew is exposed to approximately 2–6 mSv per year.

29.6 PET Signal Acquisition

The radio-nucleotide isotopes will emit a positron after a relatively short half-life. The positron will virtually immediately annihilate with any of the many free electrons in the biological medium.

The mass of the positron and the electron combined equals the energy needed to produce a pair of γ photons. The annihilation energy is released as two γ quanta with 511 keV that are emitted in opposite directions to each other (180° ± 5°). The positron–electron annihilation process is presented in Figure 29.2. These γ photons have enough energy to allow them to travel out of the tissue and can be recorded by external detectors. The coincidence lines of γ emission provide a unique detection method for forming tomographic images with PET.

By time-resolved detection of two simultaneous γ photons, the connecting line of coincidence establishes one dimension of the location of the concentration of the radioisotopes. Any subsequent emissions from the same location will not necessarily follow the exact same path and lines of intersect can be established to yield the second dimension of the isotopes (Figure 29.3). Applying the lines of coincidence for all γ rays emitted can be used in standard tomographic techniques to resolve the image of the nucleotide concentrations.

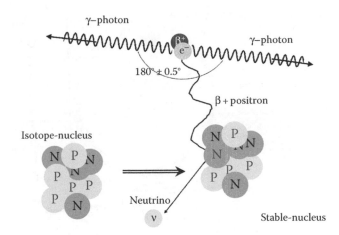

FIGURE 29.2 Annihilation of positron interaction with electron resulting in γ pair emission.

The PET imaging device has multiple rings with detector arrays specifically designed to capture simultaneous occurrences of γ photon emissions. A diagram of the configuration of a PET imager is shown in Figure 29.4.

29.6.1 γ Radiation Detectors

Certain crystals, such as bismuth germanate (BGO), convert the high energy 511 keV γ photons into lower energy light photons.[2] The visible light photons are collected by a semiconductor photosensor array, such as a CCD element (Figure 29.5). The charge coupled device (CCD) element subsequently converts the photon energy into electrical signals that are registered by the tomograph's electronics. This conversion and recording process occurs at the

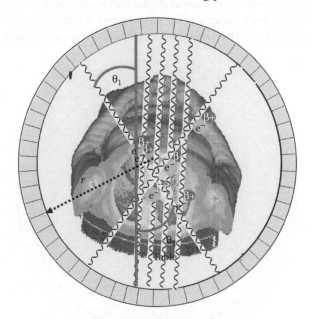

FIGURE 29.3 Cross-sectional data comparison at various angles used to form an image based on the cross-point array. (Reprinted from Najarian K and Splinter R, *Biomedical Signal and Image Processing*, Taylor & Francis Group, Boca Rotan, FL, 2006. With permission.)

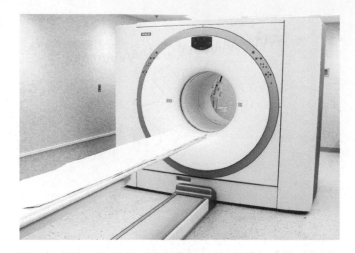

FIGURE 29.4 Siemens PET imaging equipment.

speed of light, allowing the scintillation events to be compared simultaneously among all opposing detectors. The detected spatial distribution of the positron-emitting radioisotope is a direct function of the local biochemical and physiological cellular activity.

A PET detector is divided into sections called gantry buckets. These buckets are subdivided into blocks. Each detector block consists of an array with an 8×8 detector configuration, called bins. This configuration provides each block with 64 detectors. Each gantry bucket in turn has four blocks, yielding 256 detectors per block. Over the circumference of the detector ring there are 16 buckets, which adds up to a total of 4096 detectors. Commercial PET systems have a minimum of two rings, providing a total of 8192 detectors. All these 8192+ detectors need to be operational at all times during a single scan. The detectors generally have a relatively wide acceptance angle and are closely spaced for maximum efficiency. Because of these two factors, it is not uncommon that the γ rays from one single annihilation process can be detected

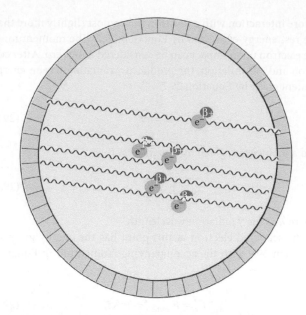

FIGURE 29.6 Schematic diagram of parallel beam detection configuration.

by two off-angle detectors, at less than perfect opposition. This crossover detection between neighboring detectors introduces a source of spatial error in the detection accuracy.

There are two distinct modes of detection: fan-beam and coincidence sampling. In fan-beam mode operation, one detector is paired to multiple detectors at various angles as illustrated in Figure 29.5. In the coincidence detection configuration, the detectors collect parallel to each other. The parallel mode of detection is illustrated in Figure 29.6.

The maximum error in location detection will occur under fan-beam detection, while the error in crossover detection differences under parallel ray detection is relatively limited.[3]

29.7 Applications of PET Scanning

PET is a valuable technique in the diagnostics of specific diseases and disorders by means of identifying metabolic deviations. For instance in oncology, the tracer (^{18}F) fluorodeoxyglucose is used because it mimics glucose in its metabolic stage and is avidly taken up and retained by most tumors. As a result, this technique can be used for the diagnosis and the planning and monitoring of the treatment of various tumors. Because of its specific sensitivity, PET is mostly used to diagnose brain tumors. Other PET applications in cancer diagnosis are for breast tumors, lung tumors, and colorectal tumors.

Since, generally, PET scans are more expensive than conventional imaging with CT and MRI, the use of PET for tumor diagnosis is not used too often.

Other applications of PET imaging are in neurology, based on an assumption that areas of high radioactivity are associated with brain activity. What is actually measured indirectly is the flow of blood to different parts of the brain, which is generally believed to be correlated and usually measured using the tracer (^{15}O) oxygen.

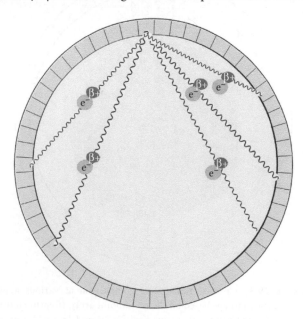

FIGURE 29.5 Schematic diagram of fan-beam detection.

For diagnostic applications in cardiology, the use of fluoro-deoxyglucose can be used to identify heart tissue after a heart attack to determine if there is any latent damage in the heart muscle. However, frequently, the relatively inexpensive single photon emission computed tomography (SPECT) is used for this purpose.

In psychology, PET is used to determine the stage of Alzheimer's disease and measure neurological activity in patients who suffer from amnesia. Other neurological diagnoses are in the study of epilepsy and stroke victims.

29.7.1 Anatomical Applications

PET provides no significant anatomical information primarily due to the fact that the uptake of radioisotopes in bone is too slow to get recorded. Based on the half-life of the isotopes and the assembly time for bone tissue, the radioactivity will have dropped below the detection level at the time of incorporation in the skeletal system.

Certain anatomical features that can be recognized are mainly due to the *a priori* metabolic activity that is well known for these organs such as in the brain and the heart. A PET image of the heart and vasculature is shown in Figure 29.7. Other anatomical features will need to be resolved by combining PET with a scanning technique that will provide irrefutable anatomic feedback. In this case, external markers will need to be used to register the two systems with respect to each other.

29.7.2 Functional Analysis

The functional imaging features of PET are most prominent in the diagnosis of cancer. Healthy tissue replenishes its cells by continuous regeneration, while old cells gradually die off. In

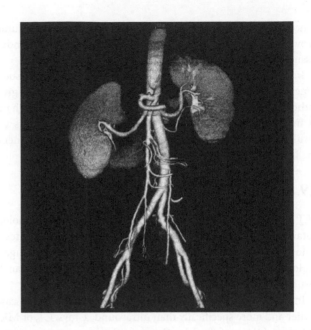

FIGURE 29.7   PET image of the kidneys and attached renal vasculature. (From General Electric Healthcare: GEHealthcare.com. With permission.)

both malignant and benign cancer, cells divide more rapidly than normal healthy cells. This process by itself will be identified under PET imaging due to the increased cellular metabolic rate. A functional image of the brain is illustrated in Figure 29.8.

The difference, however, between malignant and benign tumor growth is the fact that in malignant cancer cells the surrounding tissue is destroyed as well. In malignant tumors the

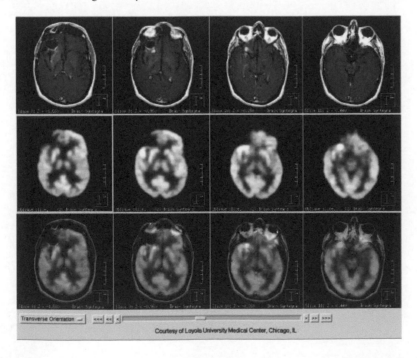

FIGURE 29.8   Functional PET image of the brain. (Reprinted from Najarian K and Splinter R, *Biomedical Signal and Image Processing*, Taylor & Francis Group, Boca Rotan, FL, 2006. With permission.)

cancer cell will migrate away from the organ in which the tumor growth is formed, and they will start infiltrating surrounding tissues and organs; this process is called metastasis.

In the absorption of radiopharmaceuticals, living tissue is generally identified by a homogeneous uptake of any given radioisotope, but fluorodeoxyglucose in particular. The high rate of metabolic activity in many types of aggressive tumors will result in an increased uptake of fluorodeoxyglucose, and hence the isotope ^{18}F will identify glucose metabolism.

29.8 Summary

PET scanners are capable of detecting areas of molecular biology detail via the use of radioisotopes that have different rates of uptake depending on the type of tissue involved. The change in regional blood flow in various anatomical structures can be visualized and relatively quantified with a PET scan.

In many cases the PET features will allow us to identify diseases earlier and more specifically than ultrasound, x-rays, CT, or MRI.

Some of the useful applications of PET scanning over scintillation imaging are examination of brain blood flow and local metabolism of specific organs.

PET can also help physicians monitor the treatment of disease. For example, chemotherapy leads to changes in cellular activity, and that is observable by PET long before structural changes can be measured by ultrasound, x-rays, CT, or MRI. A PET scan gives physicians another tool to evaluate treatments, perhaps even leading to a modification in treatment, before an evaluation could be made using other imaging technologies.

References

1. Levin CS and Hoffman EJ, Calculation of positron range and its effects on the fundamental limit of positron emission tomography system spatial resolution, *Physics in Medicine and Biology*, 1999, **44**: 781–799.
2. Dokhale PA, Silverman RW, Shah KS, Grazioso R, Farrell R, Glodo J, McClish MA, Entine G, Tran VH, and Cherry SR, Performance measurements of a depth-encoding PET detector module based on position-sensitive avalanche photodiode read-out, *Physics in Medicine and Biology*, 2004, **49**(18): 4293–4304.
3. Najarian K and Splinter R, *Biomedical Signal and Image Processing*, Taylor & Francis Group, Boca Rotan, FL, 2006.

In Vivo Fluorescence Imaging and Spectroscopy

Gregory M. Palmer and
Karthik Vishwanath

30.1 Introduction

Fluorescence imaging and spectroscopy is a highly active area of biomedical research operating at the interface between physics, chemistry, engineering, biology, and medicine. It fundamentally investigates the interaction between light and matter and offers several unique advantages, in particular:

- *High resolution*: approximately 200 nm for a conventional fluorescence microscope [1].
- *Sensitivity*: single molecule detection is possible [2].
- *Specificity*: probes exist that can specifically target a wide range of biological molecules, ions, and other compounds [1]. In addition, it is possible to probe the *interaction* between many compounds, using techniques such as fluorescence resonance energy transfer (FRET), and pump–probe spectroscopy [3].
- *In vivo*: many techniques are applicable to *in vivo* imaging, enabling an unparalleled view into the detailed interworking of complex biological phenomena.

The flexibility afforded by fluorescence techniques has made it a ubiquitous component of clinical and laboratory work. For example, fluorescence has become a standard technique in pathology, where immunohistochemistry and fluorescence *in situ* hybridization (FISH) techniques enable the pathologist to identify gene expression and customize treatment based on the patient's individual tumor type [4]. Fluorescence-based techniques have fueled a revolution in molecular biology, enabling scientists to probe the complex molecular pathways involved in many aspects of biological function and disease processes [2]. An overview of such techniques could easily fill their own volume (and have many times over), so this chapter will focus on a broad overview of the physical processes involved in fluorescence measurements, and a brief introduction into a few of the more common or interesting techniques used in the laboratory and clinic.

30.2 Nature of Light

The nature of light in classical theoretical physics was first elucidated by James Clerk Maxwell when he formulated the four partial differential equations governing electric and magnetic fields in space and time. Maxwell's equations predicted self-sustaining electromagnetic waves of given amplitude and frequency that moved in free space at the speed of light in vacuum [5]. These electromagnetic waves could explain well-known properties of light such as interference and diffraction but subsequently failed to explain experimentally observed phenomena of blackbody radiation and the photoelectric effect. Theoretical understanding of both these phenomena has shown that light can be thought to be composed of a "bunch" of discrete elementary particles—photons, which only exist at the speed of light and carry discrete amounts of energy and momentum [6]. The energy of a photon is given by the relation given by Equation 30.1:

$$E = h\nu, \tag{30.1}$$

where h is Planck's constant and ν the frequency of radiation. By assuming that electrons were localized around the nucleus of atoms in atomic orbitals with fixed energies, simple atomic spectra could be well described by considering that light carrying specific energies (i.e., at specified wavelengths) could be absorbed by electrons to move into higher energy excited states. They could then spontaneously decay back to ground state via the release of the extra energy in the form of photons of the same specific frequencies [7]. These insights have paved the way to understand the basis of light–matter interactions and gave rise to a quantitative theoretical framework to describe discrete atomic spectra. The same basic physical idea of absorption and remission of discrete frequencies of light by isolated atoms can be extended to explain the more complex phenomena of molecular fluorescence.

30.3 Measurement

Figure 30.1 shows a basic schematic of how fluorescence is measured. No matter what the technique is, there are a few basic principles common to all fluorescence-based measurements. The sample is illuminated with excitation light, which can be specified by its wavelength profile, polarization, temporal dynamics, intensity, spatial geometry, and focusing and steering optics. The common result is that light will reach the sample and be absorbed by a fluorophore, which will then emit fluorescent light across a characteristic wavelength spectrum. A detector will sense and quantify the remitted light, which can provide information about the concentration and distribution of fluorophores, as well as the presence of mediating agents that may modulate its fluorescence intensity or other characteristics (polarization, decay time profile, etc.).

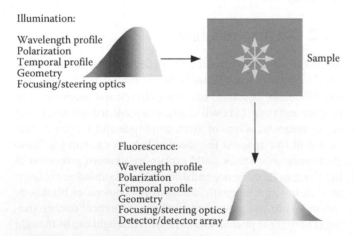

Illumination:

Wavelength profile
Polarization
Temporal profile
Geometry
Focusing/steering optics

Sample

Fluorescence:

Wavelength profile
Polarization
Temporal profile
Geometry
Focusing/steering optics
Detector/detector array

FIGURE 30.1 Schematic of a generalized measurement setup for fluorescence measurement showing the major instrument-dependent parameters. (Adapted from Hof, M., Hutterer, R., and Fidler, V., *Fluorescence Spectroscopy in Biology: Advanced Methods and Their Applications to Membranes, Proteins, DNA, and Cells*, Springer series on fluorescence 3, Vol. xix, 305pp. Berlin, NY: Springer, 2005.)

30.3.1 Absorption and Scattering

The utility of light-based sensing of biological material depends on the physical interactions between light and biological matter. Here, we will specifically focus on the phenomena of light absorption, scattering, and fluorescence. Fluorescence is the topic of this chapter, but the ability to measure this relies on our ability to couple light into and out of cells/tissues. Thus, both light scattering and nonradiative absorption play significant roles in determining the penetration depth, signal levels, and spatio-temporal resolutions that can be achieved in biological tissue.

Scattering in tissue is mostly elastic in nature (i.e., the incoming and scattered photons carry the same energies) and occurs due to microscopic inhomogeneities of the refractive index in tissues/cells. This is brought about by morphological features such as collagen fibrils, cellular organelles, and cell nuclei. Scattering by small particles can be described by the Mie theory, which provides an analytical solution to Maxwell's equations for an electromagnetic wave incident on a spherical, cylindrical, or ellipsoid particle [8]. For particles with scattering cross-sections less than ~λ/15 (such as macromolecules or striations in collagen fibrils) the Mie theory reduces to a simpler form of scattering, known as Rayleigh scattering. Rayleigh scattering in contrast to Mie scattering occurs isotropically and has a wavelength dependence proportional to $1/\lambda^4$; in other words, Rayleigh scattering increases greatly at shorter wavelengths. In general, it is the case that tissue scattering decreases toward longer wavelengths throughout the ultraviolet (UV)–visible–near-infrared (NIR) spectrum.

If the energy of an incident photon matches the difference between the excited and ground states of an electron, the photon can be absorbed by the molecule. Tissue has relatively high UV absorption, resulting in the low penetration depth of UV light in biological tissues. This, combined with the toxicity associated with UV radiation, makes UV wavelengths less than ideal for *in vivo* imaging. Tissue absorption in the visible wavelength range is dominated by hemoglobin for most tissue types. Hemoglobin absorption is dependent upon whether it is bound to oxygen. It peaks in the violet (~415–420 nm wavelength), and falls off in the NIR (>700 nm)—see Figure 30.2 for a plot. Water absorption is negligible in the visible, but becomes significant in the NIR above around 900 nm. This gap in the absorption bands of hemoglobin and water from approximately 700 to 900 nm leads to a spectral region where tissue is relatively transparent to light and has been referred to as the NIR window [9]. This "transparent" spectral region has been widely exploited for a variety of *in vivo* imaging techniques, as described later in the chapter.

30.3.2 Fluorescence/Phosphorescence

The final interaction that light can undergo with biological molecules that will be discussed here is fluorescence. The term fluorescence represents the phenomena where a particular molecular (or atomic) species (instantaneously) absorbs incident light at one wavelength and emits it after a short interval at a different, longer wavelength (or lower frequency). It can more formally be

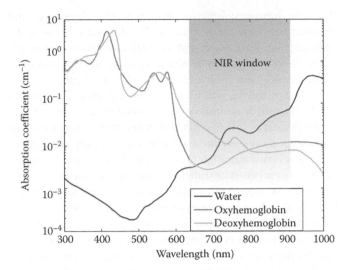

FIGURE 30.2 Absorption by hemoglobin (10 μM) [56] and water [57] is relatively low in the red to NIR range of around 700–900 nm, allowing for deeper penetration of light.

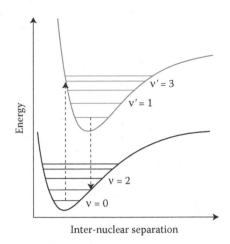

FIGURE 30.3 Representative sketch of the potential energy levels for the ground state and the first excited state for a typical fluorescent molecule. (Adapted from Redmond, R.W., in *Handbook of Biomedical Fluorescence*, M.-A. Mycek and B.W. Pogue, (Eds). New York: Marcel Dekker, 2003.)

defined as a radiative transition between two electronic states (typically, between the ground state and the first-excited state) possessing the same electronic spin multiplicity [10]. Most biological entities that exhibit fluorescence are organic molecules and have paired electrons in their ground state. They therefore have a total spin quantum number

$$S = \sum s = +\frac{1}{2} - \frac{1}{2} = 0, \tag{30.2a}$$

and consequently have a spin multiplicity, expressed by Equation 30.2:

$$M_s = 2S + 1 = 0. \tag{30.2b}$$

Molecules with $M_s = 1$ are considered to exist in singlet states; molecules with $M_s = 2$ are in doublet states (these molecules/radicals have $S = \frac{1}{2}$ because they typically have one unpaired electron in ground state), while molecules with $M_s = 3$ (these species have two unpaired electrons with same spin orientations) are considered to be in a triplet state [7,11]. The most common fluorescence phenomena represent electronic transitions between the singlet ground state (S_0) and the singlet first-excited (S_1) states in the fluorescing species, although it is possible to have fluorescence occurring as a result of electronic transitions between the second excited singlet state (S_2) and S_0 or as doublet–doublet transitions.

The principle of absorption of a photon and subsequent remission along with several important features of fluorescence can be illustrated using the Morse-potential energy diagram that denotes both the allowed electronic and vibrational levels. Figure 30.3 shows the allowed electronic potential energies of the ground state (black curve) and the first excited state (gray curve) of a typical molecular species capable of exhibiting fluorescence. The

horizontal lines represent allowed vibrational modes in each electronic state (labeled by the quantum numbers v/v') and the vertical lines show the radiative absorption/emission processes. Upon excitation by a photon of appropriate frequency, the electrons (initially occupying the lowest vibrational level of the ground state) absorb the incident radiative energy at a time scale of ~10^{-15} s. This time scale is much smaller than the time scale of molecular vibrations and thus the absorption is shown as a "vertical" line because the internuclear geometry before and after absorption will be almost identical as dictated by the Franck–Condon principle. In the excited state however, the equilibrium bond length is increased and the energy levels are therefore slightly shifted to the right (as shown by the blue curve). The vibrational level that the excited electron moves into corresponds to the vibrational mode that overlaps with the vertical transition from ground state ($v' = 3$ in Figure 30.3). As expected, in isolated molecules (when present as low-pressure gas, for instance) the emitted fluorescence would result from the symmetric, resonant transition from $v' = 3$ to $v = 0$. As a result of collisional interactions between the excited molecule, its neighbors, and/or solvent molecules in liquids and solids, however, the emission occurs from the lowest vibrational state ($v' = 0$) of the excited state back to one of the vibrational levels of the ground state. Thus, for an incident photon of frequency v_{ex}, and an emitted photon with frequency v_{em}, the energy balance is given by Equation 30.3:

$$h v_{ex} = h v_{em} + \Delta E_{IC}. \tag{30.3}$$

Here, ΔE_{IC} is the loss of electronic potential energy to vibrational modes of the neighboring atoms/molecules (which could be other fluorescing molecules or the molecules of the solvent).

This transition is again governed by the Franck–Condon principle—and thus the transition from the excited state will mostly occur where there is a vertical overlap between the $v' = 0$

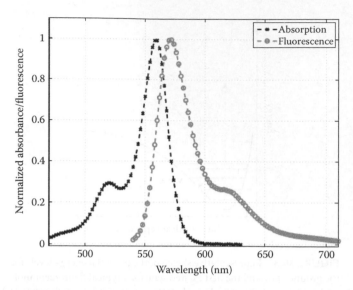

FIGURE 30.4 Normalized absorption and emission spectra of Rose Bengal dissolved in ethanol at basic pH. (Data obtained from Prahl, S.A., *Alphabetical Index of Photochem CAD Spectra*, 2007 [cited; Available from: http://omlc.ogi.edu/spectra/PhotochemCAD/html/alpha.html].)

vibrational state of the excited electronic state to the vibrational mode of the ground state (i.e., to v = 1 in Figure 30.3). As indicated above, the transitions for excitation and emission do not carry the same amount of energies and this difference between the wavelengths of the absorbed and the remitted photon is called the Stoke's shift. Thus, given ΔE_{IC}, and the wavelength of the excitation and emission photons, λ_{ex} and λ_{em}, respectively, the Stoke's shift can be calculated using Equation 30.4:

$$\lambda_{stroke} = \frac{\lambda_{ex}\lambda_{em}\Delta E_{IC}}{hc}. \tag{30.4}$$

Another important aspect of fluorescence to be noted here is that the vibrational energy level spacing in the ground state and the first excited state decreases with increasing energies, and these are identically spaced in both these states. This gives rise to the symmetric mirror image relationship observed between the absorption and the emission spectrum, shown in Figure 30.4 for Rose Bengal dissolved in ethanol and measured at a basic pH

[12]. The mirror symmetric shapes of the absorption and emission lines are clearly evident. The difference between the peak emission wavelength and the peak absorption wavelength is the Stoke's shift, and is about 12.5 nm for the molecule/solvent shown here (which results in ΔE_{IC} of 4.6 kJ/mol, where λ_{ex} = 559 nm and λ_{em} = 571.5 nm). Although most fluorophores exhibit the mirror symmetry rule between their absorption and emission spectra, this rule can be violated in some cases wherein the absorption and emission occur in different molecular species (arising at very low or high pH), or in molecules that are capable of forming excited-state dimers (excimers) with themselves [13]. Further, it is important to note that there is mirror symmetry between the absorption–emission spectra only when one considers the S_0–S_1 absorption, but not across the entire absorption spectrum. This occurs due to the fact that most organic (biological) fluorophores when excited to the S_2 states nonradiatively relax to the lowest vibrational level of the S_1 state, from where radiative emission takes place. This nonradiative relaxation from higher level states to S_1 is formally governed by Kasha's rule, which translates to a common observation that the emission spectra of most fluorophores are independent of the excitation wavelengths.

Besides radiative emission, there are other pathways available to excited-state molecules to relax back into the ground state. The classical figure used to denote these processes is called Jablonski's diagram and is shown in Figure 30.5 [13]. After absorption into the S_1, S_2, or higher excited states, the molecule can lose its energy in a variety of different ways. As mentioned before, excitation into energy levels higher than S_1 causes the molecule to quickly relax into the lowest vibrational state of S_1 by nonradiative transitions labeled internal conversion (as they occur between states with the same multiplicity). The probability of internal conversion is inversely proportional to the energy difference between the states and thus is least probable for the S_1–S_0 transition (as these states have the largest energy difference) [14,15].

As indicated in Figure 30.5, these nonradiative internal conversion processes occur on the time scale of a few picoseconds. From the lowest vibrational level of S_1, the excited state can decay back to the ground state via the radiative processes as discussed above. These radiative decay pathways contribute to the molecule's fluorescence and occur on the time scale of a few hundreds of picoseconds to nanoseconds. Besides these

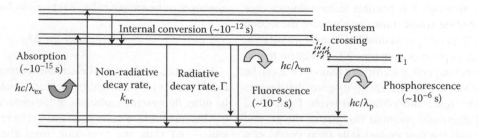

FIGURE 30.5 Jablonski's diagram showing available mechanisms for the relaxation of excited states of a fluorophore.

radiative pathways, the excited-state molecule can give up energy to the solvent or other molecules through molecular collisions and thereby relax to S_0 via nonradiative pathways. The other pathway shown in Figure 30.5 that serves to depopulate S_1 is the nonradiative intersystem crossing, which transfers the singlet S_1 state to a triplet T_1 state. Although selection rules in simple atomic models forbid transitions between states with different spin multiplicities, there exist mechanisms (such as spin–orbit coupling) in more complex molecular systems that allow these transitions [10]. The T_1 triplet state eventually relaxes back to the S_0 state through the radiative process of phosphorescence. Given that these transitions have much lower probabilities of occurring relative to fluorescence, the time scales of phosphorescence range from tens to thousands of microseconds.

30.3.3 Quantum Dots

Advances over the past few decades in manufacturing and engineering, specifically the widespread applications of integrated electronics to small-scale semiconductor crystals, have enabled the production of layered semiconductor materials whose spatial dimensions are controllable at nanometer scales. This has led to the development of quantum wells, quantum rods, and quantum dots. At these spatial scales, these materials have physical properties uniquely different from the bulk counterparts. The key property of quantum dots arises from the quantum confinement property—when the length of one dimension of the object becomes comparable with the radius of the Bohr exciton for that material, the nanostructure exhibits discrete energy levels which are excitable optically [16,17]. The energy band gap of the quantum dots is dictated by their geometry, thereby allowing "tuning" of the optical absorption and emission properties. The absorption profile of quantum dots is broad and composed of a series of overlapping bands that become weaker with increasing wavelengths. The emission spectrum is usually bell-shaped with an FWHM of ~30–40 nm and is independent of the excitation wavelength. Figure 30.6

shows representative absorption and emission spectra of commercially available quantum dots. Clearly, these fluorescent entities lack the mirror symmetry rules that govern molecular fluorophores since the exact electronic and vibrational level spacings are different in these cases.

30.3.4 Fluorescence Lifetimes and Quantum Yields

A typical pulse of light at 400 nm (assuming 20 µW of incident power) used to excite fluorescence in biological specimen has a mean photon flux of approximately 4×10^{13}/s incident on the sample. Assuming that there are more fluorophores than the average photon flux in the sample, we can postulate that this pulse excites a large number of fluorophores ($N(t)$) into the excited state. The excited state is depopulated stochastically and in the case of a simple two-level system can be written as Equation 30.5:

$$\frac{dN(t)}{dt} = -(\Gamma + k_{nr})N(t). \tag{30.5}$$

Here, Γ is called the radiative decay rate (governing the rate of radiative transitions from S_1 to S_0, see Figure 30.5) and k_{nr} is the nonradiative decay rate [13]. The excited state therefore depopulates exponentially with a characteristic lifetime τ_0 and quantum yield Q, where τ_0 and Q are given by Equation 30.6:

$$\tau_0 = \frac{1}{\Gamma + k_{nr}} \quad \text{and} \quad Q = \frac{\Gamma}{\Gamma + k_{nr}}. \tag{30.6}$$

We can therefore interpret τ_0 as the average time spent by the fluorophores in the excited state, while Q represents the efficiency of fluorescence for the molecule. As $k_{nr} \to 0$, Q approaches unity, indicating that there is one photon emitted for every photon absorbed; thus, the quantum yield can be defined as the number of photons emitted to those absorbed. Quantum dots

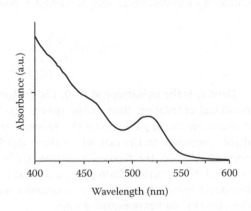
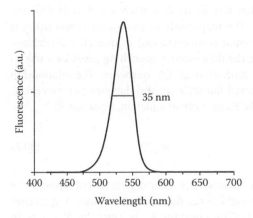

FIGURE 30.6 Sample absorption and fluorescence spectra from CdSe/ZnS quantum dots in H_2O in 50 mM borate buffer solution at pH 8.5. (Taken from www.oceannanotech.com for QD350.)

and a few brightly fluorescing molecules (such as rhodamine) have very high quantum yields.

Measurements of the quantum yield of unknown fluorophores can be quite a tedious task and are usually achieved by comparative absorption and emission measurements against reference fluorophores whose quantum yield is known [13]. Measurement of the fluorescence lifetime is achieved by measuring the fluorescence intensity [which is usually proportional to $N(t)$] over time. This can be done either directly in the time-domain using photon counting methods, a high-frequency oscilloscope, or time-gating techniques; or indirectly in the frequency domain by using intensity modulated excitation sources and measuring the phase difference between the incident source and remitted fluorescence [10,13]. Since both the fluorophore lifetime and quantum yield are closely related to the microscopic interactions taking place at the molecular scale, measurements of these parameters can provide a very sensitive means to detect changes in the fluorophores microenvironments. These parameters have been used to detect changes in fluorophore binding, pH, solvent viscosity, and resonant energy transfers [13].

Increases in the nonradiative rates have been studied both as fundamental fluorescent phenomenon and to extract information regarding the biochemical environment of fluorophores. The phenomenon of quenching causes a reduction in fluorescence intensity and can occur through a variety of reasons including excited-state reactions, ground-state complex formations, molecular transformations, and/or energy transfer. Fluorescence quenching has also widely been used as an indicator of the presence of molecular oxygen since oxygen is one of the most efficient quenchers of fluorescence [18]. There are two distinct classes of quenching processes that occur: collisional or dynamic quenching, and static quenching. In collisional quenching, there are molecular collisions between the excited-state fluorophore and the quencher, which causes the fluorophore to relax to the ground state in a nonradiative fashion. In static quenching however, a nonfluorescent complex is formed between the fluorophore and the quencher, which reduces the number of molecules capable of exhibiting fluorescence. Both cases of quenching require interactions at the molecular scale between the quencher and fluorophore and can therefore be used to study a variety of different phenomena [13]. The magnitude of fluorescence quenching is related to the amount (concentration) of quencher available—thus, quantifying the fluorescence quenching provides a way to measure the concentration of the quencher. The relationship between the observed fluorescence and the quencher concentration is given by the Stern–Volmer equation, Equation 30.7:

$$\frac{F_0}{F} = 1 + K_q[Q]. \tag{30.7}$$

Here, F_0 is the fluorescence intensity measured without the quencher present and F is the fluorescence intensity at quencher concentration $[Q]$. The constant K_q is given by $K_q = k_q\tau_o$ in the case of dynamic quenching. In this case, the quencher acts by increasing the nonradiative decay rate by $k_q[Q]$, thereby reducing both the quantum yield (hence fluorescence intensity) as well as the measured excited-state lifetime. In the case of static quenching, the quencher merely renders the ground-state fluorophores to become nonfluorescent entities causing a reduction in the fluorescence intensity while leaving the fluorescence lifetimes unchanged.

30.3.5 Anisotropy

Absorption of excitation energy involves an energy transfer between the vector electric field component of light and the electrical dipole moment of the molecule. The efficiency of this energy transfer depends on the alignment of these vectors; specifically the absorption is proportional to the cosine of the angle between them. Thus, excitation of a sample with linearly polarized light (light whose electric field component is fixed along a given arbitrary axis on a given plane) results in a preferential spatial orientation of the excited fluorescent molecules. Those fluorophores whose electrical dipole is aligned with the electric field of the incident light are most efficiently excited, while those in a perpendicular orientation are not; this is referred to as photoselection [2]. The emitted fluorescence can thus exhibit anisotropy. This anisotropy is typically characterized by making two measurements: one having an emission polarizer parallel to the excitation light, and one perpendicular. The anisotropy can then be quantified by Equation 30.8:

$$r = \frac{F_\parallel - F_\perp}{F_\parallel + 2F_\perp}, \tag{30.8}$$

where r is the anisotropy, $F_\parallel$ is the fluorescence intensity emitted parallel to the excitation polarization, and $F_\perp$ is that perpendicular to the excitation.

Once excited, the molecules will remain in the excited state for some period of time, during which they may tumble and change their orientation (and thus their anisotropy) prior to emitting a fluorescent photon. There is hence decay in the anisotropy over time depending on the rate kinetics with which the molecule tumbles. For a spherical particle in homogeneous medium, the anisotropy will decay according to a single exponential:

$$r = r_0 \exp\left(\frac{-t}{\tau_P}\right). \tag{30.9}$$

Here, r_0 is the anisotropy at $t = 0$, t is the time, and τ_p is the rotational correlation time of the sphere [2]. The value of r_0 depends on the angle between the ground- and excited-state dipole moments. In the case where these dipole moments are perfectly aligned, the theoretical maximum initial anisotropy is 0.4 [2]. Nonspherical fluorophores, such as proteins or other biological materials, exhibit more complex kinetics and are described by multiexponential decays.

The value of fluorescence anisotropy is that this is a measurement that is dependent on the ability of the molecule to tumble in space, so it depends on the size and shape of the fluorophore, its excited-state lifetime, and the viscosity of the surrounding medium. This enables probing of phenomena such as protein or antibody binding (which effectively increases the size of the fluorescence compound and thus decreases the anisotropy decay rate), membrane kinetics, and investigations into the spatial coordination between proteins or different regions on the same protein [2].

30.4 Fluorescence Resonance Energy Transfer

FRET involves the nonradiative transfer of energy between a donor and acceptor fluorophore (and may be considered as a specific case of quenching of the donor fluorescence by the acceptor). This is a near-field resonant energy transfer mechanism that is linked to the coupling between the excited-state dipole of a donor fluorophore and the ground-state dipole of an acceptor fluorophore. Effective coupling is dependent on the degree of overlap between the donor emission spectrum and the acceptor excitation spectrum, the quantum yield of the donor fluorophore, the relative orientation of the donor and acceptor dipoles, and their distance from one another. Based on this distance dependence, FRET efficiency can serve as an effective molecular ruler on a typical scale of approximately 1–10 nm [19–21].

FRET is a competing pathway from the donor excited state to the ground state. The rate constant can be characterized by Equation 30.10 [20]:

$$k_T = \frac{1}{\tau_D}\left(\frac{R_0}{R}\right)^6,$$ (30.10)

where k_T is the rate constant of FRET, τ_D is the donor fluorescence lifetime in the absence of an acceptor, R is the distance between the donor and acceptor, and R_0 is the distance at which 50% FRET efficiency is achieved, which typically ranges from 2 to 9 nm [2]. FRET efficiency is defined as the probability that a photon absorbed by a donor fluorophore will subsequently be transferred to the acceptor fluorophore. There are a variety of techniques for measuring this [21], but one relatively simple approach is to measure the donor fluorescence intensity of a sample containing the donor only (F_D), and again with both the donor and acceptor present (F_{DA}). Donor fluorescence and acceptor fluorescence can typically be separated by wavelength, with the acceptor emitting at longer wavelengths. In this case the efficiency is given by Equation 30.11 [20]:

$$E = 1 - \frac{F_{DA}}{F_D} = \frac{1}{1 + (R/R_0)^6}.$$ (30.11)

The extremely short separations at which this phenomenon occurs make FRET improbable for unconstrained fluorophores.

In typical applications, donor and acceptor compounds will be bound to two separate entities that undergo some interaction or binding that will bring the donor and acceptor into close proximity. Alternatively, the donor and acceptor fluorophores may be localized on the same compound that undergoes conformational changes in response to some event to alter the distance between the two fluorophores. To provide a few examples of applications, membrane fusion has been characterized by loading to different membranes with donor or acceptor molecules. When these membranes subsequently fuse, the fluorophores can intermix and FRET can occur [22]. FRET has also been used to elucidate the dynamics of protein folding [23–25]. Other examples include the chameleon calcium probe described in the molecular imaging section to follow.

30.5 Photobleaching

All fluorophores undergo a process of photo-inactivation upon exposure to high intensities of excitation light, when they lose the ability to exhibit fluorescence. This loss of fluorescence is termed photobleaching and is thought to occur due to a transition from the excited singlet state into the triplet state, wherein the molecule undergoes permanent covalent modifications that render it nonfluorescent. Each fluorophore has a number of characteristic excitation–emission cycles it can undergo before it photobleaches and this is dependent on the molecular structure, concentration, and the environment of the fluorophores. Occurrence of photobleaching in samples may be reduced by lowering the intensity of incident excitation, using a higher concentration of fluorophores and/or limiting the interval of exposure to the incident excitation.

The mechanism of photobleaching has been exploited in experiments that measure the fluorescence recovery after photobleaching (FRAP) to study the mobility and diffusion rates of fluorophores through membranes [26]. In these experiments, high-intensity excitation light (from either an arc lamp or a laser) is focused on a small diffraction limited spot (with radius of a few microns) on the sample to photobleach the fluorophores in the field of view. The light intensity is subsequently decreased and the spot is continually monitored in time to sense fluorescence emission. The fluorescence intensity increases with the diffusion of unbleached fluorophores into the field of view and eventually reaches steady-state equilibrium. By observing the time constants needed for the fluorescence to reach equilibrium values after photobleaching, it is possible to estimate the diffusion coefficients for the fluorophores in the measured environments [27].

30.6 Molecular Imaging/Biological Details Revealed

The principal theme of molecular imaging encompasses methods that allow the visualization and detection of biological processes as they occur within living cells at the molecular level.

Techniques implementing molecular imaging through appropriately labeled radiobiological, magnetically active, or ultrasound contrast agents for use with PET, SPECT, MRI, and ultrasound are being explored for clinical applications [28–30]. Optical techniques, however, offer several unique advantages, including (1) the use of nonionizing radiation and thus the ability to repeatedly scan tissues/cells without inflicting damage, (2) the typical costs of setting up, maintaining, and operating optical systems are far less in comparison with the more expensive MRI/PET/SPECT systems, (3) subcellular resolution is possible, (4) multichannel/multiplexing capability, (5) rapid acquisition times, and (6) a wide range of specific probes are available [30]. A number of excellent reviews and books expound on these techniques in detail [1,2,30–35] and only a few brief examples of which are provided here for the sake of brevity.

Early studies have investigated the use of nonspecific contrast agents such as fluorescein, indocyanine green (ICG), cresyl violet acetate, and toluidine blue and provided enhanced sensing of particular tissue structures such as vasculature, but suffer from issues of having high levels of nonspecific background signals that obscure signals from the site of interest. Recently, exogenous contrast agents have been engineered to provide highly targeted and specific labeling of molecular markers to reduce nonspecific background signals and act as signaling beacons marking disease at the cellular/subcellular levels.

The most direct and simple approach for improving the accumulation of contrast agents at a specific target site is to conjugate the fluorophore to a ligand (which can be peptides, proteins, or antibodies) that enables it to bind to a molecular target [36]. Such site-directed targeting is particularly effective in detecting small levels of tumorigenic cells as these cells often express specific surface receptors. Site-directed probes have been developed using both traditional organic fluorophores (such as ICG, Alexa Fluor 660, Cy5.5) and quantum dots [37–40]. The use of quantum dots is motivated by several desirable properties, including their characteristic broad excitation spectrum and narrow, tunable emission spectra and their relative resistance to photobleaching. Further, since these nanoparticles have large surface area, they can be modified to accommodate several probe molecules, thus enabling them to target more than one biomarker. However, the use of quantum dot-based molecular imaging techniques is still being actively researched given the potentially toxic nature of these entities.

Another significant development in the methods of molecular imaging involves the use of the green fluorescent protein (GFP) [41]. The use of this molecular marker has allowed labeling of specific proteins and ushered in a new era in fluorescence imaging of the function and structure of biological systems. Wild type GFP protein emits bright green fluorescence (centered at 509 nm) when excited in the blue at 495 nm. The fluorescence from wild type GFP has broad excitation and relatively weak fluorescence but it has subsequently been modified to have 35× brighter fluorescence in its enhanced GFP (EGFP) form. The discovery of GFP and engineering of EGFP and other mutants has allowed researchers to use it to label cells and proteins to specifically express fluorescence and therefore use it as a genetic beacon in a variety of different applications. Given its small molecular weight (~27 kDa) and relatively stable fluorescence properties under typical physiological conditions, GFP-based luminescence has found widespread application in our understanding of cellular structures and signaling pathways.

On a more fundamental level, perhaps the earliest and most widely utilized application of fluorescence detection of biochemistry is pH-sensitive fluorescence. Many of the most widely used fluorophores in biology, including fluorescein and GFP, are intrinsically pH sensitive over a physiologically relevant range. This is an artifact in many applications, but can be exploited to provide a sensitive indicator of pH. In the case of fluorescein, it is a weak acid, with protonated and deprotonated forms exhibiting different fluorescence properties [42]. GFP has been shown to exhibit similar properties due to simple protonation/deprotonation as well as conformational changes [43]. Other probes developed specifically for pH sensing include ratiometric sensors, where the ratio of fluorescence intensities at two different excitation or emission wavelengths can be calibrated to provide an absolute measurement of pH [44].

Another notable example of a chemical sensing probe that elegantly combines several of the techniques described above is that of the chameleon [Ca^{2+}] sensor [45]. This works using a FRET mechanism, where two GFP mutants (one that emits blue and one that emits green) are fused to either terminus of a calmodulin/M13 fusion protein. Calmodulin is a protein that acts as a calcium signal transducer. It binds calcium and in so doing changes its conformation, allowing it to interact with a wide range of targets. In the case of the chameleon probe, this causes a shift from an extended conformation to a compact globular form, which brings the two fluorescent proteins into close proximity. This is a reversible process, so determination of the dissociation constants of this process enables a quantitative means of assessing calcium concentration in a living cell. Because this probe is genetically encoded, it can be targeted to specific sites within the cell, enabling exquisite control over monitoring of calcium signaling in living cells and tissues [46].

30.7 *In Vivo* Imaging

The ultimate goal of all of the techniques described here is to deploy them in living cells and tissues to open a window into dynamic biochemical, morphological, and physiological processes. In so doing, one must recognize a number of potential pitfalls and limitations inherent to *in vivo* fluorescence imaging. First, tissue turbidity plays a significant role and is highly attenuating. This causes problems on two levels: (1) it affects the fluorescence intensity that is collected, making quantitative assessment of fluorophore concentration a challenge [47], and (2) it limits the penetration depth and resolution at which fluorescence can be probed in deep tissue. Second, fluorophores are typically distributed throughout the tissue in some manner; illuminating the tissue with light will excite fluorophores present anywhere within the tissue where the source light can reach, making localization of the source of fluorescence problematic. This second problem is compounded by tissue turbidity, since scattered light could have originated from distant locations. Third, a number of biological compounds are themselves fluorescent [48], which can often be a significant confounding factor. Fourth, phototoxicity must be considered, particularly for fluorophores excited in the UV. At longer wavelengths, heating can occur when using high-power illumination.

Confocal imaging is one widely employed technique that can overcome some of these limitations. Confocal imaging differs from standard fluorescence microscopy in that a laser point source is used to illuminate the sample. Passing through the objective, this is focused down to a point in the image plane. On the collection end, the light passes through a pinhole positioned at the conjugate image plane. Thus, any scattered light or fluorescence originating out of the focal plane is rejected (Figure 30.7). By controlling the size of the pinhole, the lateral resolution of a confocal microscope can somewhat exceed the typical Rayleigh resolution limit of $1.22\lambda/(2NA)$. Here, λ is the wavelength, and $NA = n\sin\theta$ denotes the numerical aperture and characterizes the acceptance angle of an objective, where n is the refractive index of the medium and θ is the maximum axial angle that can be accepted by the objective. The axial resolution is somewhat worse than this, typically $>0.4\ \mu m$ in practice [1].

A confocal imaging system thus allows for depth sectioning capability. A 3D image can then be constructed by raster-scanning in the lateral x–y plane using motorized mirrors, and stepping through the z plane, commonly using a motorized stage focus. This allows for detailed images of fluorescence to be acquired *in vivo*. As an example, Figure 30.8 shows a confocal image of the chemotherapy agent doxorubicin being taken up into muscle tissue. This image was acquired from an implantable window chamber in a mouse [49]. It can be seen that the drug is present in the blood vessels, and is accumulating in the cell nuclei.

30.8 Multiphoton Imaging

Another technique that is commonly employed for *in vivo* imaging is that of multiphoton fluorescence imaging [50,51]. In this technique, a nonlinear absorption process is utilized, where two or more photons are simultaneously absorbed to excite a fluorophore. So, for two-photon excitation, light of twice the single-photon wavelength (half the energy) is used to excite a given fluorophore, which allows for red or NIR wavelength excitation.

Multiphoton absorption is an extremely unlikely event (in comparison with the normal single-photon fluorescence described thus far) and therefore requires a high power density for excitation [51]. In order to achieve this, pulsed lasers with picosecond to femtosecond pulse widths are typically used to keep average power low and prevent tissue damage [51].

Multiphoton techniques offer several advantages: first, multiphoton excitation is efficient only for very high power densities, with a quadratic dependence on the intensity for two-photon excitation, so fluorescence is generated only within the focal point and thus retains spatial resolutions that are comparable

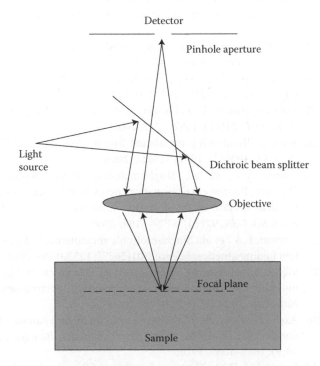

FIGURE 30.7 A simplified representation of a confocal imaging setup. The pinhole will allow unscattered fluorescent photons originating from the focal point to pass but will reject the majority of the out of plane or scattered photons since their conjugate image will not be formed at the pinhole aperture.

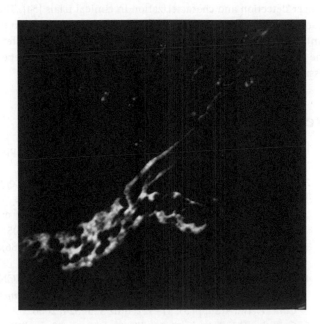

FIGURE 30.8 Confocal fluorescence image acquired from muscle tissue in an implantable dorsal skin fold window chamber after treatment with doxorubicin, a fluorescent chemotherapy agent. A large blood vessel contains the drug, which can also be seen accumulating in the cell nuclei outside the vessel. Also seen are the red blood cells, which appear dark. (Image courtesy of Ashley Manzoor and Mark Dewhirst.)

with that of laser scanning confocal imaging despite using longer excitation wavelengths [50]. This is in contrast with confocal imaging where fluorescence is generated throughout the sample and helps to minimize phototoxicity and photobleaching [51]. Second, longer wavelengths are typically less harmful to biological tissues [52], and can penetrate deeper (on the order of several hundreds of microns in tissue [53]).

30.9 Diffuse Fluorescence Imaging and Tomography

The high turbidity of tissue prevents deep penetration of ballistic photons, so probing deep tissue (>1 mm) requires illumination and collection using diffusely scattered photons. One application of growing interest is that of NIR fluorescence imaging [54]. The principles and advantages of NIR fluorescence imaging are the same as described above, with the main advantages being the low scattering and absorption properties of tissue in the NIR allow deeper light penetration and low background tissue autofluorescence [54]. The main difference in this technique is that there is no inherent depth sectioning capability built in. Tissue is illuminated at the excitation wavelength, and the emission light is collected using either a wide-field imaging technique, or using one or more point collectors such as optical fibers. These measurements can then be used as raw intensity data or further processed using inverse image reconstruction methods via the use of theoretical models of photon propagation in turbid media (such as the diffusion equation). Such techniques have been used in a wide variety of applications, ranging from small animal imaging to cancer detection and characterization in clinical trials [54]. The resolution of such techniques is necessarily much larger (mm to cm) relative to confocal or multiphoton microscopy, but offers the possibility of image reconstruction from significantly deeper tissue regions [55].

References

1. Prasad, P.N., *Introduction to Biophotonics*, Vol. xvii, 593pp., [8] p. of plates. Hoboken, NJ: Wiley-Interscience, 2003.
2. Hof, M., Hutterer, R., and Fidler, V., *Fluorescence Spectroscopy in Biology: Advanced Methods and Their Applications to Membranes, Proteins, DNA, and Cells*, Springer series on fluorescence 3, Vol. xix, 305pp. Berlin, NY: Springer, 2005.
3. Zewail, A.H., Laser femtochemistry. *Science*, **242**(4886): 1645–1653, 1988.
4. Carlson, R.W., et al., HER2 testing in breast cancer: NCCN task force report and recommendations. *J. Natl. Compr. Canc. Netw.*, **4**(Suppl 3): S1–S22; quiz S23–S24, 2006.
5. Griffiths, D.J., *Introduction to Electrodynamics*, 3rd edition. Upper Saddle River, NJ: Prentice-Hall, 1999.
6. Hecht, E., *Optics*, 4th edition, p. 698. San Francisco, CA: Addison-Wesley, 2002.
7. Griffiths, D.J., *Introduction to Quantum Mechanics*, 2nd edition. Upper Saddle River, NJ: Prentice-Hall, 2004.
8. Bohren, C.F. and Huffman, D.R., *Absorption and Scattering of Light by Small Particles*, Vol. xiv, 530pp. New York: Wiley, 1983.
9. Jobsis-vanderVliet, F.F., Discovery of the near-infrared window into the body and the early development of near-infrared spectroscopy. *J. Biomed. Opt.*, **4**(4): 392–396, 1999.
10. Redmond, R.W., Introduction to fluorescence and photophysics, in *Handbook of Biomedical Fluorescence*, M.-A. Mycek and B.W. Pogue, (Eds). New York: Marcel Dekker, 2003.
11. Shankar, R., *Principles of Quantum Mechanics*, New York: Plenum Press, 1994.
12. Prahl, S.A., *Alphabetical Index of Photochem CAD Spectra*, 2007 [cited; Available from: http://omlc.ogi.edu/spectra/PhotochemCAD/html/alpha.html].
13. Lakowicz, J.R., *Principles of Fluorescence Spectroscopy*, 2nd edition, p. 698. New York: Kluwer Academic/Plenum, 1999.
14. Freed, K.F., Irreversible electronic relaxation in polyatomic molecules. *J. Chem. Phys.*, **52**(3): 1345–1354, 1970.
15. Freed, K.F. and Tric, C., Collision dynamics of collision induced intersystem crossing processes. *Chem. Phys.*, **33**: 249–266, 1978.
16. Reed, M.A., Quantum dots. *Sci. Am.*, **268**(1): 118–123, 1992.
17. Zrenner, A., A close look on single quantum dots. *J. Chem. Phys.*, **112**(18): 7790–7798, 2000.
18. Rumsey, W., Vanderkooi, J., and Wilson, D., Imaging of phosphorescence: A novel method for measuring oxygen distribution in perfused tissue. *Science*, **241**(4873): 1649–1651, 1988.
19. Stryer, L., Fluorescence energy transfer as a spectroscopic ruler. *Ann. Rev. Biochem.*, **47**: 819–846, 1978.
20. Clegg, R.M., Fluorescence resonance energy transfer. *Curr. Opin. Biotechnol.*, **6**(1): 103–110, 1995.
21. Jares-Erijman, E.A. and Jovin, T.M., FRET imaging. *Nat. Biotechnol.*, **21**(11): 1387–1395, 2003.
22. Rizo, J., Illuminating membrane fusion. *Proc. Natl. Acad. Sci. USA*, **103**(52): 19611–19612, 2006.
23. Deniz, A.A., et al., Single-molecule protein folding: Diffusion fluorescence resonance energy transfer studies of the denaturation of chymotrypsin inhibitor 2. *Proc. Natl. Acad. Sci. USA*, **97**(10): 5179–5184, 2000.
24. Lipman, E.A., et al., Single-molecule measurement of protein folding kinetics. *Science*, **301**(5637): 1233–1235, 2003.
25. Weiss, S., Measuring conformational dynamics of biomolecules by single molecule fluorescence spectroscopy. *Nat. Struct. Biol.*, **7**(9): 724–729, 2000.
26. Axelrod, D., et al., Mobility measurement by analysis of fluorescence photobleaching recovery kinetics. *Biophys. J.*, **16**(9): 1055–1069, 1976.
27. Soumpasis, D.M., Theoretical analysis of fluorescence photobleaching recovery experiments. *Biophys. J.*, **41**(1): 95–97, 1983.
28. Weissleder, R., Molecular imaging: Exploring the next frontier. *Radiology* **212**(3): 609–614, 1999.

29. Massoud, T.F. and Gambhir, S.S., Molecular imaging in living subjects: Seeing fundamental biological processes in a new light. *Genes Dev.*, **17**(5): 545–580, 2003.

30. Weissleder, R. and Pittet, M.J., Imaging in the era of molecular oncology. *Nature*, **452**(7187): 580–589, 2008.

31. Lakowicz, J.R., *Principles of Fluorescence Spectroscopy*, 3rd edition, Vol. xxvi, 954pp. New York: Springer, 2006.

32. Thompson, R.B., *Fluorescence Sensors and Biosensors*, 395pp. Boca Raton: CRC/Taylor & Francis, 2006.

33. Pierce, M.C., Javier, D.J., and Richards-Kortum, R., Optical contrast agents and imaging systems for detection and diagnosis of cancer. *Int. J. Cancer*, **123**(9): 1979–1990, 2008.

34. Massoud, T.F. and Gambhir, S.S., Integrating noninvasive molecular imaging into molecular medicine: An evolving paradigm. *Trends Mol. Med.*, **13**(5): 183–91, 2007.

35. Ntziachristos, V., Bremer, C., and Weissleder, R., Fluorescence imaging with near-infrared light: New technological advances that enable *in vivo* molecular imaging. *Eur. Radiol.*, **13**(1): 195–208, 2003.

36. Rajendran, M. and Ellington, A.D., *In vitro* selection of molecular beacons. *Nucleic Acids Res.*, **31**(19): 5700–5713, 2003.

37. Lin, Y., Weissleder, R., and Tung, C.H., Novel near-infrared cyanine fluorochromes: Synthesis, properties, and bioconjugation. *Bioconjug. Chem.*, **13**(3): 605–610, 2002.

38. Hsu, E.R., et al., Real-time detection of epidermal growth factor receptor expression in fresh oral cavity biopsies using a molecular-specific contrast agent. *Int. J. Cancer*, **118**(12): 3062–3071, 2006.

39. Becker, A., et al., Receptor-targeted optical imaging of tumors with near-infrared fluorescent ligands. *Nat. Biotechnol.*, **19**(4): 327–331, 2001.

40. Gao, X., et al., *In vivo* cancer targeting and imaging with semiconductor quantum dots. *Nat. Biotechnol.*, **22**(8): 969–976, 2004.

41. Tsien, R.Y., The green fluorescent protein. *Annu. Rev. Biochem.*, **67**: 509–544, 1998.

42. Martin, M.M. and Lindqvist L., The pH dependence of fluorescein fluorescence. *J. Lumines.*, **10**(6): 381–390, 1975.

43. Kneen, M., et al., Green fluorescent protein as a noninvasive intracellular pH indicator. *Biophys. J.*, **74**(3): 1591–1599, 1998.

44. Whitaker, J.E., Haugland, R.P., and Prendergast, F.G., Spectral and photophysical studies of benzo[c]xanthene dyes: Dual emission pH sensors. *Anal. Biochem.*, **194**(2): 330–344, 1991.

45. Miyawaki, A., et al., Fluorescent indicators for Ca^{2+} based on green fluorescent proteins and calmodulin. *Nature*, **388**(6645): 882–887, 1997.

46. Truong, K., et al., Calcium indicators based on calmodulin-fluorescent protein fusions. *Methods Mol. Biol.*, **352**: 71–82, 2007.

47. Bradley, R.S. and Thorniley, M.S., A review of attenuation correction techniques for tissue fluorescence. *J. R. Soc. Interface*, **3**(6): 1–13, 2006.

48. Ramanujam, N., Fluorescence spectroscopy *in vivo*, in *Encyclopedia of Analytical Chemistry*, R. Meyers (Ed.), pp. 20–56. New York: Wiley, 2000.

49. Huang, Q., et al., Noninvasive visualization of tumors in rodent dorsal skin window chambers. *Nat. Biotechnol.*, **17**(10): 1033–1035, 1999.

50. Denk, W., Strickler, J.H., and Webb, W.W., Two-photon laser scanning fluorescence microscopy. *Science*, **248**(4951): 73–76, 1990.

51. Helmchen, F. and Denk, W., Deep tissue two-photon microscopy. *Nat. Meth.*, **2**(12): 932–940, 2005.

52. Svoboda, K. and Block, S.M., Biological applications of optical forces. *Annu. Rev. Biophys. Biomol. Struct.*, **23**(1): 247–285, 1994.

53. Oheim, M., et al., Two-photon microscopy in brain tissue: Parameters influencing the imaging depth. *J. Neurosci. Methods*, **111**(1): 29–37, 2001.

54. Frangioni, J.V., *In vivo* near-infrared fluorescence imaging. *Curr. Opin. Chem. Biol.*, **7**(5): 626–634, 2003.

55. Ntziachristos, V., et al., Looking and listening to light: The evolution of whole-body photonic imaging. *Nat. Biotechnol.*, **23**: 313–320, 2005.

56. Prahl, S., *Optical Absorption of Hemoglobin*, 2008 [cited 2008; Available from: http://omlc.ogi.edu/spectra/hemoglobin].

57. Segelstein, D., *The Complex Refractive Index of Water*. University of Missouri, Kansas City, 1981.

31

Optical Coherence Tomography

Kevin A. Croussore and
Robert Splinter

31.1 Introduction

Optical coherence tomography (OCT) has come as a promising optical technique for imaging live tissues with cellular and subcellular resolutions. The technique, based on low-coherence interferometry, is noninvasive and can operate in approximately real time while providing detailed information on tissue microstructure, and has proven successful in a variety of clinical applications. A detailed description of OCT systems and corresponding theory is presented. Applications are discussed, and derivatives of OCT that introduce polarization, phase, and spectral sensitivity are also described.

OCT has been developed initially at the Massachusetts Institute of Technology (MIT) that is capable of providing such resolution. OCT is based on the principle of low-coherence interferometry.[1] The technique will be described in detail next. Conventional OCT measures the amplitude of the light reflected from a biological sample and determines the longitudinal position of the reflection sites through low-coherence interferometry.[2] Light for performing OCT is typically delivered noninvasively by simple fiber optic and objective lens components, although invasive techniques using hypodermic needles and catheters have been developed. By scanning the sample or the delivery apparatus, two- and three-dimensional imaging is possible. This is analogous to ultrasound imaging. However, because OCT is an optical method, the resolution that can be obtained is orders of magnitude greater than that of even ultrahigh-frequency ultrasound. OCT thus represents an interesting and important advance in the science of clinical imaging.

Clinical applications of OCT include early diagnosis of macular edema, a disease that affects the human retina and can lead to blindness. OCT is particularly applicable in this case because the disease causes small holes to form in the retinal layer that cannot be detected by high-frequency ultrasound or other ocular diagnosis techniques.

In this chapter, the theoretical and practical considerations for performing optical coherence tomographic imaging will be presented and discussed. Section 31.2 introduces a conventional OCT system and provides information on coherence, interferometry, system operation (image acquisition process and detection methods), and general system properties and limitations. Section 31.3 describes the applications of four types of OCT imaging to physical problems: conventional (amplitude) OCT, phase- and polarization-sensitive OCT, and spectroscopic OCT.

31.2 The Conventional OCT System

In order to understand the operation of the optical coherence mechanism of action, the conventional OCT system will be discussed first to provide the theoretical basis. The conventional OCT uses the rudimentary principle of examining only the amplitude of the reflected signal.

OCT is based on the same principles as coherence domain reflectometry (CDR).[1] The operating principle relies on measuring the amplitude of the reflected light signal in a biological sample as a function of depth. Probing as a function of depth provides merely one-dimensional image formation. Combining this with a scanning probe, a three-dimensional image can be reconstructed. This mechanism is in principle similar to the way ultrasound imaging acquires an image (Chapter 32). The original OCT system, developed at MIT by Fujimoto and Company, is depicted graphically in Figure 31.1.

The basic system consists of a fiber optic Michelson interferometer with a low-coherence (broadband) light source. Optical

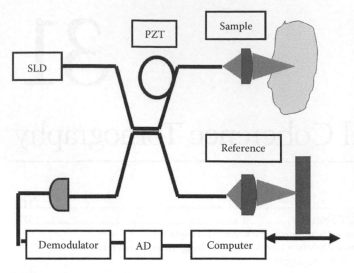

FIGURE 31.1 Original OCT design after Fujimoto. AD: analog-to-digital converter and SLD: superluminescent diode.

power from the light source, shown here as a superluminescent diode (SLD), is split approximately 50/50 between the two arms using a 2 × 2 fiber optic demultiplexer. In the sample arm, collimating and focusing optics direct light onto the biological sample to be imaged. Light is reflected from the sample at different depths and is allowed to pass unidirectionally through the fiber optic splitter, so that it reaches the detector but does not result in optical feedback to the light source, which could result in narrowing of the emission spectrum. Note that light reflected from varying depths within the sample will travel different distances, or equivalently, the light will have a variation in the optical path length. The reference arm contains a mirror, which reflects most of the light transmitted down the reference arm back to the detector. The position of the mirror is modulated in order to vary the optical path length along the reference arm. As will be shown below, the only time an interference signal is generated at the detector is when the optical path length of the reference arm matches that of the sample arm to within the coherence length of the source (i.e., resonance will occur). Scanning the reference mirror thus allows measurement of the amplitude of the signal generated from sample reflection as a function of depth within the sample. The remaining elements, which are the detection circuitry and the piezoelectric transducer (PZT), will be discussed later.

Because of the use of a broadband source, interference of all wavelengths is achieved at only one location. Under monochromatic single wavelength interferometry, the conditions for interference are satisfied every whole wavelength distance, providing no single location identification.

31.2.1 OCT Theory of Operation

The basic principles that should be explained are (1) coherence length and the generation of low-coherence light and (2) the operation of a low-coherence Michelson interferometer.

31.2.2 Light Sources and Coherence Length

In order to obtain the high resolution that is achieved using OCT, it is necessary to employ a light source that has a very short coherence length. The axial resolution in OCT is then limited to this length.

The coherence length of a source can be defined as the physical length in space over which one part of the electromagnetic (EM) wavetrain bears a constant phase relationship to another part. For practical measurements, the coherence length of a source is defined as the optical path difference (OPD) (between two EM waves) over which interference effects can be observed. The coherence length is thus the net delay that can be inserted between two identical waves that still allow observation of constructive and destructive interferences.

There are two important points to make in an elementary discussion of coherence length. The first concerns the relationship between the length of an EM wavetrain, which cannot be infinite, and the spectral content of the radiation. The second is the relationship between this spectral content and coherence length of the radiation.

All EM radiation must originate from an accelerating electric charge (Figure 31.2). The source depicted in Figure 31.2 is analogous to an electron that undergoes an atomic transition from an excited state to a lower state, as occurs in a laser or SLD.[3] Note that the duration of the wave packet is finite, as required by the finite time period of the electronic transition. This places a requirement on the spectral content of the wave packet that is best understood through an application of Fourier analysis. The Fourier series allows the construction of any periodic function through an (infinite) summation of sine and cosine functions with different amplitudes and frequencies. The number of component functions with nonzero amplitude depends on the length and structure of the original waveform. Using the Fourier integral,[4] the explicit relationship between the length of the pulse

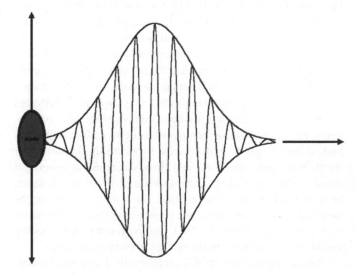

FIGURE 31.2 An electron that oscillates for a finite time period produces a wave packet that travels away from the source.

and its spectral content in frequency units can be obtained as described by Equation 31.1:

$$\Delta t \Delta \upsilon \geq \frac{1}{4\pi}. \qquad (31.1)$$

The product of the pulse duration, Δt, and its spectral width, $\Delta \upsilon$, has a minimum requirement, meaning that for a short pulse such as in Figure 31.2 there will be a large number of component frequencies present. The case above describes the minimum constant value that can be obtained as predicted by either Fourier analysis or quantum mechanics and occurs for a Gaussian distribution of the amplitudes of the component sine and cosine functions. It is important to note the consequences of this relationship for optical sources.

To obtain the relationship between spectral width and coherence length, we consider the effects of introducing a delay between two identical wave packets. Suppose that a single wave packet, such as in Figure 31.2, is partially reflected off two mirrors separated by a distance d. Each returning pulse will have traveled a different distance upon reaching a detector. Obviously, if the mirror separation d is greater than the length of the wave packet, no interference effects can occur as the two pulses will be detected at different times. Each component frequency of the two pulses produces an interference pattern at the detector, resulting in the superposition of a large number of such patterns. However, if the delay between the two pulses is long enough, one extreme frequency will exhibit an interference maximum at the same point that the other extreme frequency exhibits a minimum, resulting in constant intensity. For values of d greater than this, no interference can occur, either because the pulses will arrive separated by a time delay greater than the duration of the pulse or because the next, unrelated pulse will have arrived. This allows us to define the coherence length of the light as equal to the length of the wave packet.[5]

If a series of wave packets are generated, the coherence length depends on whether or not there is a constant phase relationship between the individual wave packets. In the case of a thermal source, atoms emit at random and with random frequency, so there is no relationship between pulses emitted from a single atom, and obviously not between different atoms. In the case of a continuous wave laser, however, the process of stimulated emission requires subsequent photons (or wave packets) to be emitted with the same frequency, phase, and polarization as the stimulating photon. The coherence length of a CW laser is then limited too by cavity instabilities that allow the laser center frequency or primary mode of oscillation to change.

Analysis of the location of interference maxima and minima as related to source spectral width in units of wavelength gives the approximate relationship in Equation 31.2[5]:

$$L_c = \frac{\lambda^2}{\Delta\lambda}. \qquad (31.2)$$

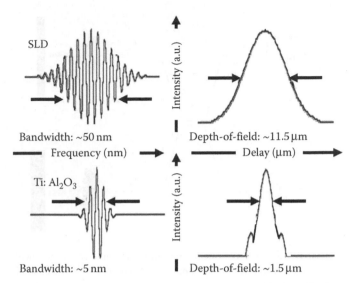

FIGURE 31.3 SLD and mode-locked laser interference patterns (left) and coherence envelopes. (After Drexler W, et al., *Opt. Lett.* 24, 1221–1223, 1999.)

The coherence length (L_c) is proportional to the square of the source center wavelength (λ) and inversely proportional to the spectral width ($\Delta\lambda$). The coherence length is the limit of the system's longitudinal resolution.

SLDs and mode-locked lasers have become the standard and high-resolution light sources of choice owing to their broad spectral bandwidth.[6] The two devices achieve broad spectral widths using different principles. A typical SLD can have a spectral width [full width half maximum (FWHM)] as large as 32 nm.[2] This leads to a coherence length in air of approximately 10–15 μm, with the resolution in tissue being the free-space resolution divided by the tissue's refractive index (n). This resolution is generally insufficient for imaging individual cells or intracellular structure. Mode-locked lasers, however, are capable of producing extremely short pulses. From Equation 31.1, it is obvious that this will result in a broad spectrum. Pulses as short as two optical cycles have been produced, with spectral widths of 350 nm corresponding to a free-space resolution of 1.5 μm, and less than 1 μ in tissue.[2] Figure 31.3 compares the interference patterns and coherence envelopes for each of the two sources, namely a SLD and a pulse titanium Sapphire laser.

31.2.3 Operation of the Fiber Optic Michelson Interferometer

The fiber optic system detection mechanism of action is analogous to a Michelson interferometer. We will make the initial theoretical approximations here that (1) the source is monochromatic and (2) dispersion effects are thus eliminated in analyzing the detected intensity as a function of reference mirror position. These approximations will be modified qualitatively in later sections to describe the operation of OCT. The additional approximation that the source emits plane waves, which is not truly a good approximation of the Gaussian single mode beam profile, will not be revised for the sake of simplicity.

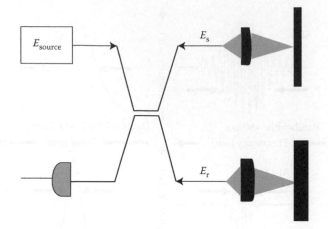

FIGURE 31.4 A fiber optic Michelson interferometer. Electric fields are defined in the text. The sample has been replaced with a single reflecting surface.

To begin, we consider the fiber optic interferometer shown in Figure 31.4.

The incident EM wave can be written for the electrical portion as

$$E_{source} = E_o \, e^{i(kz - \Omega t)}, \tag{31.3}$$

where Ω is the angular frequency of the electric field and E_o is the real amplitude of the wave traveling in the z-direction in a nondispersive, dielectric material. The quantity k is the (pure real) wave number, which is related to the wavelength of the source as provided by Equation 31.4:

$$k = \frac{2\pi}{\lambda_o} n. \tag{31.4}$$

The amplitude of the source electric field is split approximately 50/50 at the fiber optic coupler. To obtain the intensity transfer function of the interferometer, we will simply write expressions for the returning signals as

$$E_r = E_{ro} \, e^{i(kz_r - \Omega t)}, \tag{31.5}$$

$$E_S = E_{so} \, e^{i(kz_s - \Omega t)}, \tag{31.6}$$

where r-terms are the returning reference signal and s-terms are the returning sample signal. The terms z_r and z_s are the distances to and from the reference mirror and a specific reflection site within the sample, respectively. The amplitudes of the returning signals are no longer the same, as the total amount of light reflected from the sample will be very small, while nearly all of the light incident upon the reference mirror will be collected. The expression for the total electric field at the detector is given by the superposition of these two returning waves:

$$E_{total} = E_{ro} \, e^{i(kz_r - \Omega t)} + E_{so} \, e^{i(kz_s - \Omega t)}. \tag{31.7}$$

For the electric waves described above, the detected intensity can be calculated as

$$I = \frac{n}{2\mu_o c} \, E_{total} \, \overline{E_{total}}, \tag{31.8}$$

where n is the material refractive index ($n = 1.5$ for glass), μ_o is the magnetic permeability, c is the speed of light in vacuum, and the bar denotes complex conjugation of the electric field expression. The result after carrying out the complex conjugation and transferring the expression to trigonometric notation for clarity is given by Equation 31.9:

$$I = \frac{n}{2\mu_o c} \left(E_{so}^2 + E_{ro}^2 + 2\sqrt{E_{so} \, E_{ro}} \, \cos(kz_r - kz_s) \right). \tag{31.9}$$

The time dependence vanishes and the resulting expression contains an interference term, proportional to the cosine of the wave number k times the path length difference, $z_r - z_s$. By substituting for k and noting that

$$I_{so}, I_{ro} = \frac{n}{2\mu_o c} E_{so}^2, E_{ro}^2, \tag{31.10}$$

the detected intensity can be written as follows:

$$I = I_{so} + I_{ro} + 2\sqrt{I_{so} + I_{ro}} \, \cos\left[\frac{2\pi}{\lambda_o} n(z_r - z_s)\right]. \tag{31.11}$$

The detected intensity depends on the relative distance traveled by each of the electric waves, as captured by the interference (cosine) term. For an ideal case in which the returning electric field amplitudes are exactly equal, Equation 31.11 predicts the peak intensity to be four times the intensity of either wave, while the minimum is zero for perfectly destructive interference. In the case of OCT, the interference term is relatively small compared with the background due to the relative amplitudes of the reflected reference and sample signals.

The quantity in brackets under the cosine is the OPD, which denotes the phase delay between the returned respective signals. Interference effects are the direct result of adding waves that have a different phase at the detector. For values of the OPD less than the source coherence length, the two returning electric waves will still have a constant phase relationship, and interference effects can be observed. For larger values of OPD, the two waves will have a random phase relationship so that interference effects will be random, resulting in a constant intensity at the detector. In this case, Equation 31.11 does not apply. The period (in distance) over which the intensity varies from maximum to minimum, a variation known as a fringe shift, is of the order of half the source's center wavelength.[7,8]

Equation 31.11 was developed assuming a monochromatic wave. The effect of using a broadband source is to decrease the OPD over which Equation 31.11 applies. Interference effects are still observed within the coherence envelope, but they are blurred as the OPD exceeds the coherence length. Because interference effects will occur over a finite distance defined by the coherence length, the absolute axial position of the reflection site in the sample cannot be known with greater precision than the source coherence length. It can, however, be located repeatedly with much higher precision.[1]

As the thickness of the sample increases, if light is continuously reflected from different depths throughout the feature, the interference pattern will be broadened. This gives information about the thickness of the reflecting feature, but also requires that individual features be spaced further than one coherence length apart to be clearly resolved.

An important feature of OCT is the fact that the axial resolution is limited only by the source coherence length. This is in contrast to confocal microscopy, in which the axial resolution is determined by the numerical aperture of the lens. This allows the use of both large numerical aperture optics and fiber optic microlenses, so that highly adaptive imaging devices can be designed.

31.2.4 Image Acquisition Process

We have established that OCT measures the amplitude of the light signal reflected from the sample and determines the distance to the reflection point in order to construct two- or three-dimensional images. In this section, we will look in more detail at how the image is actually constructed and some of the issues that arise because of the nature of the measurement process.

The basic method for constructing an image can be described using Figure 31.1 as an example. Later systems introduce variations to the signal modulation and detection process that will be described throughout this chapter.[2,3]

When the OPD is greater than the coherence length of the source, no interference effects can occur. With both arms of the interferometer illuminated, the intensity at the detector is, in this situation, given by

$$I = I_{so} + I_{ro}, \qquad (31.12)$$

which is Equation 31.11 without the interference term. The two terms in the equation are the intensity of the light reflected from the sample (that which can be collected, at any rate) and the light reflected from the reference mirror.[9]

Light exits the fiber optic cable and is then focused onto the sample. For the fiber, the numerical aperture denotes the sine of the maximum angle at which light may enter the fiber and still meet the conditions for total internal reflection at the core–cladding interface. For the objective lens, it is the sine of the angle that the aperture subtends when viewed from an object at the focal point. In either case, the expression is given as Equation 31.13

$$NA = n \sin(\theta_i). \qquad (31.13)$$

Light that returns to the objective lens at an angle greater than the acceptance angle (θ_i) will not be collimated, and will thus fall outside the acceptance angle for the optical fiber.

Because of the law of reflection, all of the light that is reflected from the first layer, which is at the focal point, is reflected back to the objective lens and is collected by the fiber. Light rays backscattering from within the tissue that are not purely axial but fall within the acceptance angle of the imaging device at the distal tip of the fiber optic device are collected for imaging. At deeper layers, only the light incident normally will be collected by the fiber. Even in this simplified case, it can be seen that an explicit expression for the amount of light returned to the detector would be difficult to obtain. Variables for this case would include the depth of the reflecting layer, the reflection and transmission coefficients for each layer, the divergence angle of the beam, and so forth. In practical systems, effects such as absorption and the scattering coefficient of the sample material would have to be accounted for, further complicating the analysis. Nevertheless, it is clear that only a small amount of light reflected from each plane will be collected by the fiber and detected. It is also clear that the light reflected from each depth within the tissue travels a different distance. The fact that the matching optical path length of the reference arm will determine the collection depth allows the use of the reference signal to determine the longitudinal position of the reflected site.

Continuous scanning of the reference mirror modulates the length of the reference arm, performing a filtering mechanism (i.e., heterodyning). The positions of the backscattering sites are mapped by recording the interference envelope with respect to the position of the reference mirror. A single longitudinal scan thus generates a "one-dimensional" image of the reflection sites within the sample. The transverse dimension is equivalent to the spot size of the beam at the reflection site, which prescribes the lateral resolution.

The rate at which an OCT image can be obtained depends on the required resolution. OCT is capable of providing imaging on a time scale faster than the typical dynamics of the sample under observation.

The fiber optic compatibility of OCT makes the technique suitable for a wide variety of such implementations, including orthopedics, prostate, and cervical imaging.[10,11]

31.2.5 Detection Systems

Optical heterodyne detection and balanced detection are used to achieve the high sensitivity required for OCT imaging. The two are typically combined.[2,3] Consider the detected intensity as given by Equation 31.12. The equation shows two steady state (DC) components, due to reflection from the reference mirror and the total amount of light that is collected from the sample backscattering sites. For a given longitudinal scan of the reference mirror, corresponding to a single lateral and vertical position, we can consider at first approximation these two terms to be constant. As the beam is translated laterally, the magnitude of the reflected signal is subject to change in a random manner.

31.2.6 Optical Heterodyning

Both signals carry noise, primarily due to the amplitude noise of the source, which can be significant in the case of the mode-locked laser.[2]

Translation of the reference mirror causes a change in frequency of the reflected light, proportional to the mirror velocity. The modulated light returning from the mirror is combined with light returned from the sample, which does not undergo a frequency shift. The detection signal will display a photocurrent term proportional to the frequency difference between the two waves. This term is obviously periodic and is said to occur at the beat frequency between the two detected signals. The PZT in the sample arm is modulated at a high frequency, and stretches the fiber to change the path length of the sample arm. When the PZT is modulated at a frequency higher than the Doppler frequency generated by the mirror, this has the effect of amplitude modulating the interferometric signal. The system is thus analogous to any optical communication system, with the beat frequency serving as the carrier frequency and the PZT generating the transmission signal.[6,8]

The signal is demodulated at the summation of the frequency of the PZT and Doppler modulation. This isolates the amplitude modulated interferometric signal from noise outside a small bandwidth, which is determined by the detection electronics. The demodulator is followed by an envelope detector, which consists of an intermediate frequency (IF) filter, a squarer, a low-pass filter, and a threshold detector. The IF filter isolates the frequency difference term that is generated by the demodulator. Since this term is sinusoidal, a squarer is used to rectify the signal. The low-pass filter further reduces any high-frequency noise that might be on the signal, and the threshold detector measures the IF signal amplitude.

Using this technique, it is possible to detect a returning signal that is 5×10^{-10} of the incident signal, which enables an impressive 110 dB sensitivity or better.

Combining balanced detection with optical heterodyne detection is a powerful method for reducing the noise that is generated along with the interferometric signal.

Figure 31.5 illustrates the use of the heterodyning mechanism. The Doppler modulated interferometric signal, along with the dc components that are the reference signal and the normal reflection signal from the sample, is sent to two detectors D1 and D2. If we introduce amplitude noise associated with the SLD or laser source, the power spectrum can be represented by Equation 31.14:

$$P_t = P_s(\Omega, t) + P_r(\Omega') + f_n(t), \qquad (31.14)$$

where P_s and P_r are the optical power reflected from the sample and mirror, respectively, and $f_n(t)$ is the random noise source associated with the laser or SLD. The Doppler shifted reference signal occurs at optical frequency Ω'. This signal is incident on the two

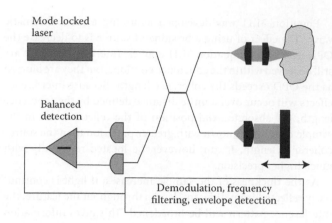

FIGURE 31.5 OCT system that combines optical heterodyne detection with balanced detection (similar to [3]).

identical detectors D1 and D2. The two photocurrents exhibit the beating term that is necessary for optical heterodyning, which, in the case of the system in Figure 31.5, occurs at 50 kHz due to the reference mirror velocity of 20 mm s^{-1}. The two respective photocurrents are sent to a subtracter. The subtracter introduces a phase shift of 180° for the beatnote on one of the signals, so that when the two are subtracted the resulting signal is approximated by Equation 31.15:

$$P_t = 2P_s(t)P_r \cos(\Omega - \Omega'). \qquad (31.15)$$

This effectively eliminates dc terms and the amplitude noise associated with the laser or SLD source. The remainder of the detection system provides frequency filtering and envelope detection. Such systems have been successful in producing a 93 dB dynamic range in an OCT device that was not optimized for the source spectral width.[3] A later system achieved 110 dB of sensitivity by combing optimization of the optics with polarization control and dispersion compensation.[2]

31.3 Other OCT Techniques

After the initial concept of reflectometry was applied to optical imaging, other optical properties such as polarization, spectral content, and phase shift under the influence of motion lead to the refinement of OCT for specific imaging applications. Every technique can highlight specific optical characteristics that can be associated with certain tissue properties, or can be used to enhance contrast and sensitivity.

31.3.1 Polarization-Sensitive OCT

Several classifications of biological tissues, such as muscles, tendons, and cartilage, have organized fibrous structures that lead to birefringence, or an anisotropy in refractive index that can be resolved using light of different polarizations to measure tissue

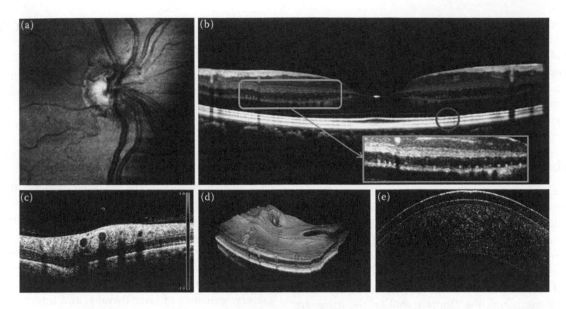

FIGURE 31.6 OCT images of the retina. (Courtesy of Bioptigen, Inc.) Images acquired using Bioptigen's SDOCT ophthalmic imaging system. (a) "OCT Fundus Image" constructed entirely from OCT volumetric data obtained in a single eye blink interval of over 5.7 s, consisting of 100 laterally displaced B-scans; the data were summed over an axial section of ~100 μm thickness; (b) averaged B-scan consisting of 1000 A-scans, 50 ms A-scan integration time, 17 images/s live display rate, broadband (170 nm) source, average of 80 frames, the inset showing capillaries in the inner nuclear layer; (c) Doppler image of the human retina; (d) volumetric rendering of three-dimensional retinal SDOCT dataset; and (e) cornea cross section of a contact lens wearer.

properties. Partial loss of birefringence in biological tissues is known to be an indicator of tissue thermal damage. One specific application of such an imaging system would include *in situ* measurement of burn depths to aid in treatment or removal of burned tissue. A variation of conventional OCT has been developed that shows promise in this capacity.[12]

When circularly polarized light is incident upon the sample, it will be returned to the detector in an elliptical polarization

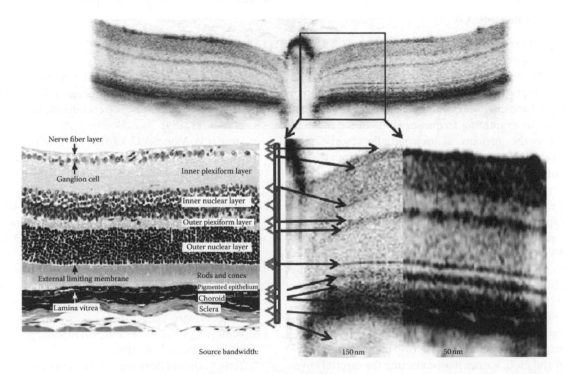

FIGURE 31.7 Ultra-high-resolution OCT image of the retina. (Courtesy of Bioptigen, Inc.) Images were acquired using Bioptigen's SDOCT ophthalmic imaging system. An ultra-high-resolution SDOCT image of wild-type C57BL/6 mouse retina is shown.

state affected by sample birefringence. When combined with linearly polarized light in the reference arm, the resultant projected polarization state can be resolved into horizontal and vertical components through phase-sensitive detection and OCT imaging can be performed on each state. The resulting interference patterns for each polarization depend on the product of imaging depth and refractive index.

31.3.2 Phase Resolved OCT

A technique called optical Doppler tomography or Doppler optical coherence tomography (DOCT) has been developed that allows simultaneous tomographic imaging and measurement of blood flow velocity.[13] Measurements of blood flow velocity are important for treating burns and determining burn depth, evaluating effectiveness of laser therapy and photodynamic therapy, and evaluation of brain injuries. Flow velocity can be calculated by measuring the Doppler shift in the fringe frequency of the interference pattern. The Fourier transform of the frequency shift gives information on the velocity. However, the minimum velocity sensitivity is inversely proportional to the Fourier transform window size, so that for high velocity sensitivity the scan rate must be slow or, equivalently, the spatial resolution must be poor. However, it is possible to decouple the position from the spectral analysis, hence maintaining high resolution on both aspects. It is currently possible to detect blood flow velocities as low as 10 μm/s with a spatial resolution of 10 μm in this manner.

31.3.3 Spectroscopic OCT

Perhaps the most valuable extension of conventional OCT is the development of spectroscopic OCT.[10,11,14] Generally, different tissues have different pigmentation or basically absorb different spectra. Optical spectroscopy can provide detailed information on the biological and chemical structures of tissues and cellular microstructure by examining frequency shifts in the reflected light as well as the intensity of preferentially absorbed and reflected wavelengths. Spectroscopic measurements can be used to obtain information about water absorption bands within a sample and, for example, hemoglobin oxygen saturation. Broadband emission generated by mode locking of solid-state lasers overlaps several important biological absorption bands, including water at 1450 nm, oxy- and deoxyhemoglobin between 650 and 1000 nm, and others at 1.3 and 1.5 μm. The availability of broadband sources centered at 800 nm permits imaging in highly vitreous samples, because of the low absorption of water in this wavelength range.

Combining spectroscopy with OCT imaging allows two- or three-dimensional spectroscopic imaging of live biological samples in a natural environment at extremely high resolution (see Figures 31.6 and 31.7). Rather than collecting the spectral envelope as in conventional OCT, collecting the full spectral interference pattern will be required.

31.4 Clinical Applications of OCT Imaging

The clinical applications of OCT are relatively limited but are very detailed and support noninvasive biopsy development.

31.5 Summary

In summary, we have analyzed OCT, an imaging technique that makes use of low-coherence interferometry. Conventional OCT, which measures the reflected signal amplitude, has already been demonstrated to be a viable diagnostic tool for several ocular diseases and has been shown to be effective in imaging both surface and deep tissues with subcellular resolution. The development of noninvasive, hand-held, and semiinvasive delivery devices has enabled entirely new regimes of imaging with this resolution, including in orthopedics, open surgery, guidance of laser therapy, and internal organs. The use of mode-locked lasers with ultra-broad bandwidths has enabled this resolution, and producing broader bandwidth sources using this method is a currently widely studied area of OCT research. Extensions of OCT to phase and polarization resolved imaging have been proven successful for measuring blood flow velocity and tissue birefringence, respectively, although the sensitivity of these systems can be somewhat less than that of conventional OCT. Spectroscopic OCT is a current area of active research in which the wealth of information generated through both interferometry and optical spectroscopy is used to obtain detailed biochemical and structural information from biological samples.

References

1. Huang D, Swanson EA, Lin CP, Schuman JS, Stinson WG, Chang W, Hee MR, Flotte T, Gregory K, Puliafito CA, and Fujimoto JG, Optical coherence tomography, *Science* **254**, 1178–1181, 1991.
2. Drexler W, Morgner U, Kärtner FX, Pitris C, Boppart SA, Li XD, Ippen EP, and Fujimoto JG, *In vivo* ultrahigh-resolution optical coherence tomography, *Opt. Lett.* **24**, 1221–1223, 1999.
3. Bouma B, Tearney GJ, Boppart SA, Hee MR, Brezinski ME, and Fujimoto JG, High-resolution optical coherence tomographic imaging using a mode-locked Ti:Al$_2$O$_3$ laser source, *Opt. Lett.* **24**, 1486–1488, 1995.
4. Eisberg R and Resnick R, *Quantum Physics*, 2nd Edition, pp. 75, D1–D3. New York, Wiley, 1985.
5. Young M, *Optics and Lasers*, 4th Edition, pp. 114–115. New York, Springer-Verlag, 1992.
6. Verdeyen J, *Laser Electronics*, 3rd Edition, pp. 722–725. New Jersey, Prentice-Hall, 1995.
7. Yariv A, *Optical Electronics in Modern Communications*, 5th Edition, pp. 421–423. New York, Oxford University Press, 1997.

8. Liu M, *Principles and Applications of Optical Communications*, pp. 754–763. Boston, MA: Irwin Professional Publishing, 1996.

9. Lexer F, Hitzenberger CK, Drexler W, Molebny S, Sattmann H, Sticker M, and Fercher AF, Dynamic coherent focus OCT with depth-independent transversal resolution, *J. Mod. Opt.* **46**, 541–553, 1999.

10. Fujimoto JG, et al., *Laser Medicine and Medical Imaging: OCT Technology*, Optics and Devices RLE Progress Report 143, 2001.

11. Fujimoto JG, et al., *Laser Medicine and Medical Imaging: Optical Coherence Tomography Technology*, Optics and Devices RLE Progress Report 142, 2000.

12. De Boer JF, Milner TE, Van Gemert MJC, and Nelson JS, Two-dimensional birefringence imaging in biological tissue by polarization-sensitive optical coherence tomography, *Opt. Lett.* **22**, 934, 1997.

13. Zhao Y, Chen Z, Saxer C, Xiang S, De Boer JF, and Nelson JS, Phase-resolved optical coherence tomography and optical Doppler tomography for imaging blood flow inhuman skin with fast scanning speed and high velocity sensitivity, *Opt. Lett.* **25**, 114, 2000.

14. Morgner U, Drexler W, Kärtner FX, Li XD, Pitris C, Ippen EP, and Fujimoto JG, Spectroscopic optical coherence tomography. *Opt. Lett.* **25**, 111, 2000.

Ultrasonic Imaging

Robert Splinter

32.1 Introduction and Overview

Ultrasound imaging uses sound waves to determine the mechanical characteristics of tissue. Sound waves are longitudinal waves that travel through matter using the medium as a carrier for the wave energy. This means that sound cannot travel in a vacuum. In longitudinal waves, the direction of wave propagation is parallel to the motion of the mechanism that forms the wave: particles, molecules, and so on. In contrast, electromagnetic waves (e.g., light, x-ray, and radiowaves) are transverse waves where the electric field and the magnetic field are perpendicular to each other, and the direction of propagation is perpendicular to the wave mechanism itself. Ultrasound waves are represented by medium compression and expansion, which form the crests and valleys, respectively, in the wave description, that is, pressure waves. Figure 32.1 illustrates the compression at the crest of the wave and the expansion at the valley.

Sound waves are generally classified based on the frequency of the waves. Infrasound waves are less than 20 Hz and cannot be heard by humans. Audible sound falls in the range between 20 Hz and 20,000 Hz. Any sound wave frequency above the limit of human hearing is technically considered ultrasound. However, diagnostic ultrasound generally uses frequencies ranging from 1 MHz up to 100 MHz.

32.2 Advantages of Ultrasonic Imaging

One of the advantages of ultrasound imaging is that it is relatively inexpensive, mainly due to the relatively basic technological basis of this imaging modality. Ultrasound imaging produces high-resolution images that rival another relatively common imaging modality: x-ray imaging, plus it can produce real-time images. The axial resolution is on the order of a fraction of millimeters, while the radial resolution depends on the beam

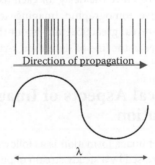

FIGURE 32.1 Wave form: the bottom represents the mathematical expression of a wave as a sine expression and the top represents the longitudinal compression and expansion illustrated by the separation of the vertical lines.

diameter. Additionally, ultrasound imaging can provide physiological data by applying the Doppler principle. A qualitative and quantitative description of blood flow can be derived to form physiological feedback. Because of its simple technology, portable units can easily be manufactured for universal applications.

Ultrasound is a very valuable diagnostic tool in medical disciplines such as cardiology, obstetrics, gynecology, surgery, pediatrics, radiology, and neurology.

32.2.1 Properties of Ultrasound Waves

As mentioned before, ultrasound imaging has reasonable similarities to x-ray imaging. Both rely on the assumption of rectilinear propagation for the purpose of image formation.

Even though ultrasonic waves undergo considerably more diffraction than x-ray, for imaging purposes the waves are assumed to travel along a straight line in first-order approximation.

On the other hand, also, there are considerable differences between ultrasound imaging and with x-ray imaging. Ultrasound has no reported side effects for biological imaging applications, whereas x-ray ionizes molecules and atoms since its energy approaches the atomic binding energies. The speed of an ultrasound wave is different in different tissues, whereas electromagnetic waves only have a relatively slight difference in speed of propagation between most biological tissues. Informally speaking, the wave spends more time passing through one type of tissue than another one. For instance, the speed of ultrasonic waves in soft tissue is much less than the speed of electromagnetic waves (including x-ray and light): for example, $v_{sound} = 1540$ m/s versus $c_{Electromagnetic} = 2.9979 \times 10^8$ m/s; however, a small but detectable variations in the speed of sound is capable of providing detailed structural information. The reason why the changes in speed of sound propagation are detected is the reflection from boundaries with different densities on either side. The density difference provides a means of reflection similar to a tennis ball bouncing off a wall. The material density will be described later as being related to the phenomenon of acoustic impedance.

In the ideal case, one can measure end-to-end delays and use tomographic techniques to calculate the delays in each point of the tissue structure. Once the delay for each tissue is computed, a very informative tomographic imaging technique can be applied that may reveal greater details about the physical properties of the tissues in the line of sight than x-ray can provide.

32.3 Physical Aspects of Image Formation

The mechanism of image formation is as follows. The ultrasound transducer is addressed by a steady stream of pulse trains that are originating from an electrical pulse generator. The expansion and contraction translate into a pressure wave train sequence that is coupled into an object that transmits, attenuates, reflects, and disperses these pressure waves. Any discrepancies in the acoustic impedance within the tissue will imply that there is a discontinuity or gradient in the structure (i.e., density) of the medium. This discontinuity or gradient will cause a mismatch in the wave equation and part of the wave will, as a result, be reflected, while the remaining part is transmitted and attenuated.

The reflected acoustic wave will return at the transducer site and will rely on conservation of energy to convert the mechanical energy back into electrical energy. The transducer thus records the electric signal belonging to a specularly reflected sound wave front. Reflections at an angle will propagate through the tissue to a location that will not correspond to a transducer site and thus will not get detected. The time delay between the generation and detection of the sound wave will indicate the distance traveled by the wave train, while attenuation of the magnitude of the pressure wave train will result from the acoustic gradients encountered and from conversion of mechanical energy into heat energy along the traversed path length.

32.3.1 Attenuation, Reflection, and Scattering

As waves travel through a medium that contains molecules packed in various densities, the molecules will be brought into oscillation and, hence, energy is depleted from the wave propagation. Attenuation of waves in biological tissues can occur by absorption, refraction, diffraction, scattering, interference, or reflection.

Since the ultrasound transducer can only detect sound waves that are returned to the crystal, the absorption of sound in the body tissue decreases the intensity of sound waves that can be detected.

Since most soft tissues transmit ultrasound at nearly the same velocities, refraction of ultrasound is usually a minor problem, although sometimes organs can appear to be displaced or have an incorrect shape due to the refraction of the ultrasound waves coming from or going to the transducer.

32.3.2 Generation of Ultrasound Waves

Ultrasound waves can be produced in two distinctly different modes of operation. The first and most often used mechanism is piezoelectrically. In piezoelectric ultrasound generation, a class of molecules with an unequal distribution of electric charges can be driven to oscillation by applying an external alternating electric field. As a result, the medium made up of these molecules changes shape in unison at the rhythm of the alternating current through the medium. The second technique used for ultrasound generation is magnetorestrictive based. A magnetic material can be made to change its shape under the influence of an external magnetic field. When this magnetic field is a periodically changing field with a constant period, the medium oscillates with the same identical frequency as the driving magnetic field.

Common transducer materials are, for instance, ceramic, barium titanate, lead–zirconate–titanate, quartz, and polyvinylidine difluoride (PVDF). The most commonly used ceramic material is a lead zirconate–titanate crystal. This crystal is sandwiched between two electrodes that provide a voltage across the thickness of the crystal. The crystal will hence change shape when a voltage is applied to it. Conversely, any ultrasonic transducer material will also produce a voltage when it changes shape. These crystals have this property because the dipole moments of all of the molecules are lined up such that any voltage applied in the direction of the dipole will either expand or compress the molecules. When the applied current is alternating at a fixed frequency, the crystal will expand and contract in a sinusoidal fashion with the same base frequency. When the electric pulse sent to the electrodes increases in magnitude, the intensity of sound output by the crystal increases.

The piezoelectric crystals produce sound waves with only milliwatts of power, which is still appropriate for diagnostic use. Since the emitted and received levels of acoustic power are very small, the intensity of the sound waves often provides a better mechanism to describe the magnitude of the sound. Intensity is defined as the power per unit area (mW/cm²).

32.3.3 Detection of Ultrasound Waves

Conversely, as an acoustic pressure wave reaches the same piezoelectric crystal, the change in shape of the crystal causes a voltage to be produced across that crystal, which is detected by the electrodes surrounding the crystal. At this point, the pressure signal is converted into a voltage spike of appropriate amplitude. This analog signal is then converted to a digital signal by a computer that analyzes the information by using algorithms to sort out the intensities, times, and directions of the echoes to estimate the location and characteristics of the interface that caused the echo. This digital signal is then sorted into the appropriate location within a matrix in the digital scan converter and eventually converted back to analog and shows up as a pixel of appropriate brightness on the monitor. Each beam, or pulse, of ultrasound generates pixels that correspond to the echoes and silences received by the transducer.

32.4 Definitions in Acoustic Wave Transfer

Acoustic waves share some of the descriptions that apply to all wave forms, only with the specific material properties for pressure, shear stress and strain, and kinetic energy associated with the respective media inserted.

32.4.1 Speed of Sound

The speed of sound propagation, v, is linked to the elastic modulus, K, of the material in question and the respective local density of the medium ρ as

$$v = \sqrt{\frac{K}{\rho}}. \tag{32.1}$$

In wave theory, the period of a wave, T, describes how much time there is between repetitions of an identical pattern in the time domain. The wavelength of the sound wave describes the spatial repetition pattern sequence. In general, the velocity of sound in any given tissue does not depend significantly on the frequency of that wave. Fascinatingly however, the higher the frequency, the greater the attenuation. Accordingly, lower frequencies of ultrasound penetrate deeper into the body. The wavelength is coupled to the period of the wave as

$$\lambda = cT. \tag{32.2}$$

The frequency describes how many times per second a pattern repeats itself, which is the reciprocal of the period:

$$f = \frac{1}{T}. \tag{32.3}$$

The frequency and wavelength of the wave are linked together by the speed of sound:

$$v = f\lambda, \tag{32.4}$$

where v is the velocity of the wave in the medium (m/s), f is the frequency of the wave (Hz or s^{-1}), and λ is the wavelength of the wave (m).

32.4.2 Wave Equation

Acoustic wave propagation in an infinite medium is governed by the wave equation. Assuming that the direction of wave propagation is in the z-direction, the wave equation of displacement as a function of time and place can be written as

$$\frac{\partial^2 u}{\partial z^2} = \frac{1}{v^2} \frac{\partial^2 u}{\partial t^2}, \tag{32.5}$$

where u describes the medium displacement in the z-direction and t denotes the time. A longitudinal or compressional wave propagating in the z-direction describes a displacement in the same direction as the direction of the wave propagation. The solution to Equation 32.5 can have many different forms, for instance,

$$u = u_0 e^{i(\omega t - kz)} \tag{32.6}$$

or

$$u = u_0 \sin(\omega t - kz) = u_0 \sin\left(\frac{2\pi t}{\lambda} - kz\right) = u_0 \sin(2\pi f t - kz), \tag{32.7}$$

where $\omega = 2\pi f$ is the angular frequency and $k = \omega/c$ is the wave number.

Another way of defining the speed of sound based on the compressibility of the medium in contrast to the elastic modulus is given by

$$v = \frac{1}{\sqrt{\rho G}}, \tag{32.8}$$

where ρ represents density in kg/cm³ and G is compressibility defined as the reciprocal of the change in pressure per unit change in volume with units cm²/dyn or m²/N.

$$G = \frac{\Delta V}{\Delta P}. \tag{32.9}$$

Equation 32.9 shows the importance of the speed of propagation in ultrasonic imaging, considering that the compressibility between fat and bone is significantly different.

For ultrasound wave propagation, the displacement itself does not carry the information that is needed to form a useful indicator of the tissue properties to generate an image that can be analyzed. Instead the pressure as a function of location is chosen as the primary attribute of interest. The wave equation can thus be rewritten as

$$\frac{\partial^2 P}{\partial z^2} = \frac{1}{v^2} \frac{\partial^2 P}{\partial t^2} \tag{32.10}$$

with solutions

$$P(z, t) = P_0 \sin(\omega t - kz) = P_0 \sin\left(\frac{2\pi t}{\lambda} - kz\right) = P_0 \sin(2\pi ft - kz). \tag{32.11}$$

In three-dimensional space, this transforms into

$$\nabla^2 P = \frac{\partial^2 P}{\partial x^2} + \frac{\partial^2 P}{\partial y^2} + \frac{\partial^2 P}{\partial z^2} = \frac{1}{v^2} \frac{\partial^2 P}{\partial t^2}. \tag{32.12}$$

with solutions

$$P(z, t) = P_0 \sin(\omega t - \boldsymbol{k} \cdot r) = P_0 \sin\left(\frac{2\pi t}{\lambda} - \boldsymbol{k} \cdot r\right)$$
$$= P_0 \sin(2\pi ft - \boldsymbol{k} \cdot r) = P_0 e^{i(\omega t - kz)} = P_0 e^{i(\omega t - k_x x - k_y y - k_z z)} \tag{32.13}$$

where the phase of the wave is given by

$$\phi = k \cdot r \tag{32.14}$$

The pressure is recorded by the transducer and translated into an electrical signal for signal processing and image formation.

32.4.3 Intensity

The intensity I, the power carried by the wave per unit area propagating in a medium, is the square of the complex amplitude of the wave. The amplitude of the wave is the pressure amplitude, with the maximum compression P_0 and time- and place-dependent

pressure $P = P_0 \sin[2\pi(\omega t - kz)$, yielding the intensity expressed by Equation 32.15:

$$I(z, t) = P(z,t)v(t) = \frac{P^2(z,t)}{Z}. \tag{32.15}$$

For a sinusoidal wave, the average intensity $\langle I(Z, t)\rangle$ is

$$\langle I(z, t)\rangle = \langle P_0 \sin(\omega t - kz)\rangle v_0(t) = \frac{P_0 v_0}{2}, \tag{32.16}$$

where P_0 and v_0 are the peak pressure and average velocity, respectively. The units for the intensity are W/m² or W/cm². The average intensity of ultrasound can also be expressed in terms of material properties as Equation 32.17:

$$I = \frac{1}{2} \frac{p_0^2}{Z} = \frac{1}{2} \frac{p_0^2}{\rho v}. \tag{32.17}$$

32.4.4 Frequency Content

The ultrasonic image is formed by launching an ultrasonic pulse into the body and then detecting and displaying either the echoes that are reflected or scattered from the tissues the wave encounters or how the pulse is transmitted and attenuated through the tissues. Basically, the instantaneous intensity and pressure when the pulse is on can be very high, while the time-average intensity is low; this is because the pulse duration τ is very short but the pulse repetition period (PRP) is relatively long (Figure 32.2). The frequency used for ultrasonic imaging is defined as the base frequency of each pulse train.

The energy carried by the ultrasound wave is

$$E = \frac{1}{v}\pi r^2(2\xi) = \frac{1}{2}\frac{Pv'}{v}, \tag{32.18}$$

where v' equals the displacement velocity of the acoustic vibrations, ξ is the particle displacement (which is a function of the

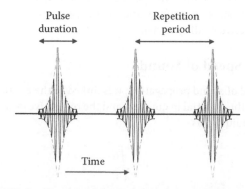

FIGURE 32.2 Ultrasound bursts emitted at intervals representing the pulse period, which is inversely proportional to the sound frequency. The pulses each have an identical frequency spectrum.

frequency), v is the speed of sound, p is the acoustic pressure amplitude, and r is the radius of the transducer.

The acoustic power can also be expressed as a function of the frequency of the expansions and contractions f, the changes in volume ΔV, and the pressure amplitude ΔP:

$$\text{Power} = f\Delta V\Delta P \qquad (32.19)$$

Because of the finite short lifetime of the acoustic wave pulse, the frequency content of those pulses will be relatively broad. In contrast, an infinitely long wave can easily be of single wavelength, but the shorter the pulse train, the greater the number of higher harmonics needed to be involved to generate this short wave pulse. Everywhere before and after the pulse, all the harmonics cancel each other out by definition; it is only this way that the pulse can be short with respect to the period of the wave itself (only a length of several periods is usually needed to justify the fact that the pulse is a stand-alone pulse).

32.4.5 Acoustic Impedance

Similar to the charge motion, in acoustic wave propagation the limiting factor is the general ability of motion. This ability of motion depends on the density and ease with which one secondary source of motion can transfer this motion to a neighboring secondary source. The characteristic acoustic impedance Z of a medium is the ratio of the pressure to the characteristic speed of propagation in the medium. The acoustic impedance is defined in Equation 32.20 as

$$Z = \frac{P}{v}. \qquad (32.20)$$

The units for the acoustic impedance are kg/(m²s) or Rayl (kg/m²s × 10⁻⁶). The compressibility can now also be expressed as

$$G = \frac{1}{Z}. \qquad (32.21)$$

The other primary factor in the amount of sound reflected is the difference in the sound-transmitting properties of the two tissues. The density of the tissue in kg/m³ (ρ) and the velocity of sound in that tissue in m/s (v) also determine the acoustic impedance of the tissue. The acoustic impedance of a tissue can be expressed by Equation 32.22:

$$Z = \rho v \qquad (32.22)$$

With Equation 32.1, this can also be written as

$$Z = \sqrt{\rho K}. \qquad (32.23)$$

32.4.5.1 Beer–Lambert–Bouguer Law of Attenuation

Attenuation or the decline in the amount of pressure by dP, as with any other streams of energy or particles, is directly proportional

TABLE 32.1 Selected Acoustic Tissue Parameters

Tissue	v at 20–25°C (m/s)	Z (kg/ m² × 10⁻⁶) (mRayl)	α at 1 MHz Attenuation Coefficient (dB/cm)
Water	1540	1.53	0.002
Blood	1570	1.61	0.2
Fat	1460	1.37	0.6
Muscle	1575	1.68	3.3
Liver	1550	1.65	0.7
Bone	4000	6	12
Air	343	0.0004	13.8

to the incident quantity I, the distance over which the absorption takes place dx, with a proportionality factor α.

$$dP = \alpha P\, dz. \qquad (32.24)$$

The proportionality factor α equals the absorption coefficient.

Solving this equation for a plane wave, the sound pressure decreases as a function of the propagation distance z, giving Beer's Law, the Beer–Lambert Law, or the Beer–Lambert–Bouguer Law of attenuation given by Equation 32.25:

$$P(z) = P_0 \exp[-\alpha z]. \qquad (32.25)$$

This is an empirical equation derived independently (in various forms) by Pierre Bouguer in 1729, Johann Heinrich Lambert in 1760, and August Beer in 1852.

Here, α is the attenuation coefficient (expressed in [dB/cm] or [neper/m] or [neper/cm]) and P_0 is the pressure at $z = 0$. The attenuation coefficients of different tissues and materials are listed in Table 32.1, as well as the speed of sound and the acoustic impedance. For n layers of tissues with thickness z_i and attenuation coefficient α_i, where i spans from 1 to n, the exponential factor in the above equation can be written as

$$-(\alpha_1 z_1 + \alpha_2 z_2 + \cdots + \alpha_n z_n). \qquad (32.26)$$

32.4.6 Reflection

The next issue in echo formation is the phenomenon of reflection itself. During its propagation, a sound wave will move through different media. At the interface between two acoustically different media the sound wave can be significantly affected. The acoustic density of the local medium or acoustic impedance will be discussed later in this chapter. Many secondary waves will be generated one of which is the reflected wave. The angle of reflection is equal to the angle of incidence (Figure 32.3). The portion of the sound wave that will go through the interface is called the refracted wave. When the interface is smaller than the wavelength, the molecules composing the interface act as point sources and will radiate spherical waves.

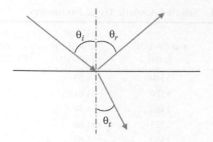

FIGURE 32.3 Reflection and refraction of a wave at an interface between two media with different densities. The reflected wave leaves the surface with the same angle to the normal as the incoming beam; the refracted beam exiting on the opposite side of the interface departs at an angle to the normal obeying Snell's law.

Two different kinds or reflection can be identified: specular reflection and diffuse reflection. The principles of specular reflection and diffuse reflection will be described in the following sections and their implications on image formation are addressed.

32.4.6.1 Specular Reflection

Specular reflection is caused by an instantaneous, stepwise change in acoustic impedance. In this case the acoustic wave will reflect according to the principle that the angle of incidence equals the angle of reflection. When sound travels from one medium (medium 1) into the next medium (medium 2) the speed of propagation of the sound waves will most likely be different. As a result the wave front will change its course and produce an angle with the normal (right angle locally to the interface of the two media) to the boundary that is different from the incident angle with the normal. This phenomenon was described by Willibrord Von Roijen Snell (1591–1626), who made his observations on the diffraction of light beams at larger angles and proposed his empirical Law of Refraction in 1621.

Refraction of sound waves occurs at the interface between two tissues, and this refraction obeys the optic principle Snell's Law shown in Equation 32.27.

$$\frac{\sin(\theta_1)}{\sin(\theta_2)} = \frac{v_1}{v_2},$$ (32.27)

where θ_1 is the incident angle with respect to Normal, θ_2 the transmitted angle with respect to Normal, v_1 the velocity of sound in the first medium, and v_2 the velocity of sound in the second medium. Since most soft tissues transmit ultrasound at nearly the same velocities, refraction of ultrasound is usually a minor problem, although sometimes organs can appear to be displaced or have an incorrect shape due to the refraction of the ultrasound waves coming from or going to the transducer. These are called refraction artifacts.

For an incident angle of θ_i with respect to the normal of the interface and a plane wave, the pressure reflection coefficient r and transmission coefficient t at a flat boundary between two media of different acoustic impedances Z_1 and Z_2 are given by

the ratio of the reflected pressure P_r to the incident pressure P_i or the transmitted pressure P_t versus the incident pressure, respectively:

$$r = \frac{P_r}{P_i} = \frac{Z_2 \cos \theta_1 - Z_1 \cos \theta_2}{Z_2 \cos \theta_1 + Z_1 \cos \theta_2},$$ (32.28)

$$t = \frac{P_t}{P_i} = \frac{2Z_2 \cos \theta_1}{Z_2 \cos \theta_1 + Z_1 \cos \theta_2},$$ (32.29)

where P_r, P_t, and P_i are the reflected, transmitted, and incident pressures, respectively

For normal incidence, Equation 32.28 reduces to

$$r = \frac{Z_2 - Z_1}{Z_2 + Z_1}.$$ (32.30)

So, the reflection at the interface of two media of very different acoustic impedances can be severe.

Since the difference in the acoustic impedances of the tissues is eventually squared, which tissue has the higher acoustic impedance does not make a difference in the amount of sound reflected. The ultrasound that is not reflected is transmitted through the interface and can penetrate deeper into the body.

However, considering a sharp discontinuity in acoustic impedance with finite dimensions, each point in the line connecting the edges of the opening is a secondary source that will re-emit the wave front that enters this region. Since each adjacent point is a source, interference will occur as the finite width beam emerges. In a broad beam plane wave, the interference will in fact maintain the plane wave itself. However, with the finite dimensions, the edge effects will play a significant role. This is different from a small discontinuity in speed of propagation at the interface of two media; in this case the light will refract while entering into the second medium.

32.4.6.2 Diffuse Reflection

Diffuse reflection results from a combination of changes in acoustic impedance, usually smaller in size or in the order of the size of the wavelength due to Equation 32.23. Often this will fall under the principle of scattering in ultrasound imaging. The ratio of reflected to refracted sound waves is dependent on the acoustic properties of both media. These properties are characterized by the acoustic impedance.

The intensity reflection coefficient is the square of the amplitude reflection coefficient. Amplitude can normally not be measured directly; however, intensity can be quantified instantaneously with most calibrated detectors. Since the difference in the acoustic impedances of the tissues is eventually squared, which tissue has the higher acoustic impedance does not make a difference in the amount of sound reflected. The ultrasound that is not reflected is transmitted through the interface and can penetrate deeper into the body.

A large difference in acoustic impedance between two body tissues will result in a relatively large echo. In fact, if the incident ultrasound is normal to the interface between medium 1 with impedance Z_1 and medium 2 with impedance Z_2, the percent of the ultrasound intensity reflected (%R) is determined by the following equation:

$$\%R = \left(\frac{Z_2 - Z_1}{Z_2 + Z_1}\right)^2 \times 100\%. \qquad (32.31)$$

Because most soft tissues have comparable acoustic impedance, only a small amount of the ultrasound is reflected, and most of the sound is transmitted. Therefore, structures that are deeper than the initial interface can also serve as a reflector and produce echoes. Because of this, ultrasound can penetrate deep into the body and provide information on interfaces beyond the first reflective surface. In fact, ultrasound can "bounce" between interfaces and the reflections from one surface can reach the transducer multiple times. Obviously, these multiple echoes can complicate the production of an image.

In order to sort out the series of echoes that a transducer receives, most diagnostic ultrasound devices operate in the pulse-echo mode. In this method, the piezoelectric crystal will emit an ultrasound beam subsequent to an electric impulse from a power source, called a pulser. The pulser will then wait until all of the echoes from that burst of ultrasound are collected before firing another beam. In this way, the device has a greater chance of sorting out the chain of echoes while reducing the buildup of echoes on top of each other.

Frequently, only the *ratio* of a reflected signal with respect to the source signal is of importance. Since practical signal ratios often cover a wide range, it is convenient to express ratios in decibel form. The intensity ratio of the detected signal I_2 measured with respect to the input signal I_1 yields an expression for the decibel formulated as equation 32.32 [decibels (dB)]:

$$dB = 10\log^{10}\left(\frac{I_2}{I_1}\right). \qquad (32.32)$$

Almost 99% of the energy incident upon the interface between the lung, which consists of air and soft tissue, is reflected. Thus ultrasound cannot be used for imaging anatomical structures that contain air, such as lung and bone. If the dimensions of a reflector are smaller than the wavelength of the sound wave, the reflector acts as a scatterer or as a secondary source. This secondary source is now a point source, radiating spherical waves, in contrast with the incident plane wave front.

Although all of these sources of attenuation need to be understood in order to estimate the location of structures inside the body, the creation of echoes in the body by reflection of the ultrasound beam is the basis for all ultrasound imaging. Sound waves are reflected at any interface between two tissues with different properties. Deflections from greater depths will have an increasingly higher probability of missing the transducer

partially or entirely when the acoustic wave is redirected during transmission or reflection at any angles deviating from the normal incidence. Therefore, the angle of incidence must be kept low ($\theta_i < 3°$) in order for the transducer to receive the information needed to form an image.

The time delay measured between the transmitted pulse and the received pulse represents the depth where the reflection came from and is based on the fact that distance traveled equals speed of propagation times time:

$$z = 2vt, \qquad (32.33)$$

keeping in mind that the acoustic wave travels the same distance in two directions: first in the inward direction and subsequently in the backward direction to the detector after reflection.

The probe–biological medium interface can also cause reflection. In order to minimize this, a gel layer with closely matching acoustic impedance is usually placed on the face of the probe between the crystal and the body. This matching layer reduces the difference in acoustic impedance between the probe and the skin. In addition, a copious amount of gel or oil must be used on the skin, not only to allow the probe to glide smoothly over the skin but also to exclude any air (which has a very high acoustic impedance) from the probe–skin interface.

32.4.6.3 Coupling Agents

To avoid loss of signal and maximum coupling efficiency of the ultrasound energy into the biological medium, acoustic impedance matching is implemented by applying a coupling agent. Several coupling agents are currently available for impedance matching: agar gel, water immersion, and a so-called bladder method, where a water-filled balloon is placed between the ultrasound probe and the surface of the biological medium. This balloon is pliable and conforms to the surface contour.

Additionally, a mismatch in acoustical impedance will cause heating at the interface due to vibrational absorption at the surface.

32.4.7 Dispersion

Diffraction of ultrasound is the spreading of the sound beam as it travels away from the transducer. Diffraction occurs at openings, ridges, or grooves and the effect is a function of both the dimensions of the irregularity and the wavelength of the sound. These openings, ridges, and grooves can be purely discontinuities in the acoustic impedance, while a physical opening has the visible attributes of the discontinuity in impedance (optical as well as mechanical).

The ultrasound beam formed by the transducer propagating into the tissue will have two zones: near field (the Fresnel region) and the far field (the Frauhofer region). While the near field is mostly composed of waves traveling in parallel, waves traveling in the far field are divergent. It is possible to use the diffraction properties of sound waves to focus the wave front into a smaller area. An acoustic lens is generally used for focusing, making the near field a little convergent. This will lead to an increase in the distance that the front wave will travel before becoming divergent and losing its resolution.

32.4.7.1 Scattering

As an ultrasonic wave penetrates a tissue, the ultrasonic energy is absorbed by the tissue and scattered from the primary direction whenever the wave encounters a discontinuity in acoustic properties.

Whereas diffraction is the gradual spreading of a beam of sound, scattering is a form of attenuation in which the sound is sent in all directions. Scattering occurs when the incident sound beam encounters small particles within the tissues. The sound waves interact with these particles and the waves are dispersed in every direction.

The ultrasonic scattering process in biological tissues is very complicated. Tissues have been treated both as dense distributions of discrete scatterers by some investigators and as continua with variations in density and compressibility by others.

32.4.8 Interference

Sound is also subject to the wave phenomenon of interference. This is where different waves are superimposed upon each other and the resulting wave is the sum of the individual waves. With respect to interference, the addition of waves that are in phase with each other produces a resultant wave with a greater intensity than either of the initial waves. Conversely, waves that are out of phase can "cancel" each other and diminish the intensity of the consequent wave.

The principles of interference are described by the mathematical addition of the wave equations of the wave phenomena merging in one single plane. The summation of acoustic waves is derived under the theory of superposition of waves.

32.5 Operational Modes of Image Formation

The pressure waves collected by the piezoelectric detector followed by conversion into an electronic signal can subsequently be recorded on an oscilloscope. The oscilloscope can represent the received signal in two main modes of operation: A-mode or B-mode. A-mode stands for amplitude mode, and B-mode represents brightness mode.

Additional operational modes are M-mode or TM-mode (memory-mode) and compound B-scan.

32.5.1 A-Mode Imaging

In A-mode the oscilloscope is triggered on the electronic excitation pulse that motivates the piezoelectric transducer to produce the mechanical action that produces the pressure wave; the received electronic signal in the delay period before the subsequent pulse represents the incoming amplitude of the conversion from mechanical to electronic signal.

The height of the pulse corresponds to the magnitude of the electric signal after passing through amplifiers.

32.5.1.1 Applications of A-Mode Imaging

A-mode is used when distances need to be calculated in relative accuracy.

Examples of A-mode imaging are found in neurology and in ophthalmology. In neurology the mid-line echo of the separation between the left and the right side of the brain often needs to be determined. The mid-line echo needs to be exactly in between both the sides of the scull; any deviation is an indication of brain hemorrhage (accident survivors). This method is fast and relatively inexpensive and accurate enough for a first diagnosis. In ophthalmology, the measurement of eye dimensions: length of the eye, position of the lens, and the location of foreign objects, can be measured in A-mode imaging. However, most of these diagnoses can also be done with greater precision and not much more expensively by laser interferometry measurement.

32.5.2 B-Mode Imaging

In B-mode imaging the amplitude of the collected electronic signal is represented by a relative brightness of the tracking dot on the screen. Initially this does not seem the most appealing manner of representing the data; however, this method of graphical presentation has been used to develop the way current ultrasound machines work by combining the single brightness display with additional time and place configurations of displaying.

32.5.3 M-Mode or TM-Mode Imaging

In M-mode the display keeps track of each scan by the transducer, while adding subsequent scans as a function of time to the same display. When an artifact that provides a reflection remains in one place, the brightness spot does not move in each subsequent scan line; however, when there is movement, the data created by the moving reflection will change position on the screen as well. An example of the bicuspid valves at the ventricular exit of the heart is shown in Figure 32.4.

32.5.4 Compound B-Scan Imaging

Similarly, if the transducer probe is moved and the interface at a given depth increases or decreases in depth with respect to the location of the probe on the outer surface, the inner topography can be revealed. The imaging system will record the position or direction of the probe and store this together with the brightness data in a memory bank. With this technique the adjacent brightness points can be seen as if they are connected and contour images can be revealed from various depths.

32.6 Real-Time Scanners

In modern ultrasound devices, transducers can "sweep" across an area of tissue without moving the probe at all. This allows

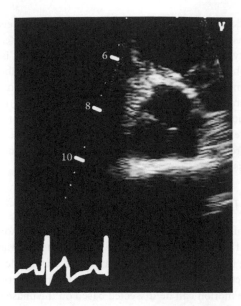

FIGURE 32.4 M-mode image of the heart, showing the bicuspid valve opening up.

technicians to get a real-time look at the body without moving the probe to every possible angle by hand. In order to accomplish this, either more than one transducing crystal is used in the probe, or the ultrasound produced from one crystal is "swept" across by mechanically moving the crystal itself by rocking it or by linear or rotational translation. Another method uses reflection of the ultrasound inside the transducer by means of a moveable deflector. By having an array of crystals (up to 256 separate piezoelectric transducers), ultrasound can be pulsed from one crystal at a time (sequential linear array) or from small groups of crystals at once (segmental linear array). Pivoting a single crystal inside the probe or using a pivoting acoustic mirror with a stationary crystal can produce a similar "sweeping" result.

Regardless of the type of probe used, more ultrasound beams per unit area (line density) translate into better lateral resolution. The information obtained from each beam is sent to a computer, which will smooth out the spaces between the lines by interpolation. Another factor that will influence the quality of an ultrasound image is how many scans of the tissue are made per second, or frame rate. Frame rates of greater than 20 frames per second allow the image to be displayed without a visible flicker on the monitor. As the line densities and frame rates increase, the images appear clearer.

Various different types of mechanical scanners have been developed, some with moving parts and a single ultrasound transducer, other with no moving parts and an array of transducers that can be triggered in a sequence. Both provide composite images that show the reflections in a single plane perpendicular to the surface of the compound probe. Additional rows of transducers or motions in two perpendicular directions provide the ability to reconstruct a three-dimensional image of reflections in brightness mode.

32.6.1 Resolution

Two distinct modes of resolution can be distinguished depending on the direction of the scan. The axial resolution is the level of distinction between subsequent layers in the direction of propagation of the sound wave. Lateral resolution is the discrimination between two points resulting from the motion of the transducer signal, perpendicular to the wave propagation.

32.6.1.1 Axial Resolution

The axial resolution is a direct function of the wavelength of the sound wave. A wave has a crest and a valley. The crest is maximal pressure and the valley represents minimal pressure. Thus the reflection can be maximal or minimal; everything in between will not be resolved. The axial resolution can thus never exceed the distance spanned by half a wavelength. The line-spread function in axial resolution is the half-wavelength.

The axial resolution can be maximized by mechanical filtering, thus reducing reverberations from the excitation by the incoming sound wave.

32.6.1.2 Lateral Resolution

One of the limiting factors in the sweeping scan is the physical dimension of the beam traveling into the biological medium. In general the transducer sends out a plane wave that starts out with the dimensions of the transducer itself. The recorder reflection is also acquired by the transducer with the same cross-sectional area. It will be virtually impossible to make any distinction between reflections that return with different intensities to a single transducer, which effectively takes the average intensity as the signal strength. The line-spread function in lateral resolution is the width of the sound beam.

It may not be possible to distinguish two adjacent transducers when they are separated by less than the beam diameter, since they will overlap sufficiently to avoid any place discrimination. The smaller the bundle of sound waves, the higher the lateral resolution.

An overview of the axial and lateral resolutions for various imaging frequency ranges is presented in Table 32.2.

32.6.2 Image Quality

The reconstruction of the image will rely on the details that can be distinguished between points at different depths or between points recorded adjacent to each other. Additional image quality

TABLE 32.2 Range of Penetration Depths and Resolutions for Bone Tissue

Frequency (MHz)	Range of Penetration Depths (cm)	Axial Resolution (mm)	Lateral Resolution
1	2–3	1.75	Depends on beam diameter
2	1–2	0.87	
3	0–1	0.58	

will depend on the discrimination of the magnitudes of reflection, since these provide material information and can add an additional dimension to the anatomical image.

As with any diagnostic tool, there are limitations to the accuracy of images that are produced. Many of the artifacts seen in ultrasound images are due to the reflective properties of the tissues being analyzed. One artifact arises when two reflective surfaces are near each other and the sound bounces, or reverberates, between them. In this case, each time the sound reverberates, some of the sound will transmit through the proximal interface and an echo will reach the transducer. Consequently, apparent structures are seen in the image at regular intervals descending down into the tissue although there may be no structures there at all, causing a "comet tail" effect in the image. Another artifact can be seen behind any highly reflective interface that only transmits a small amount of ultrasound. Oftentimes in this case, there is not enough sound energy to reach deeper tissues to produce echoes that are received by the transducer. Consequently, there is tissue beyond the highly reflective surface that the device cannot "see." The resulting areas appear dark on the image and are called reflective shadows. Echoes reaching the transducer that do not come from the opposite direction of the incident ultrasound beam cause another artifact, called displacement. This can occur when a beam of ultrasound is reflected off two or more reflective surfaces at angles that cause the echo to return to the transducer. In this case, an apparent structure will appear to be in a place where, perhaps, no structure exists.

32.6.2.1 Image Artifacts

There are several other image distorting phenomena that will affect the final image quality. Some of the image artifacts that will blur or add nonexistent features to the display are reverberation, shadowing, and displacement.

Reverberation describes the effects caused by two reflective surfaces near each other. As a result the sound bounces back and forth between them resulting in a "comet tail" effect in the displayed image. The sound appears to return from a sequence of layers, resulting from the multiple reflections occurring at delayed time intervals.

Shadowing is the effects caused by a highly reflective surface that reflects most of the sound, preventing observation of the volume behind this dense structure.

Last but not least, motion artifacts result in significant loss of information. Displacement results in objects appearing where they are not as a result of diffraction. Another displacement artifact is multipath displacement: (multiple-) reflection(s), diffraction, and refraction distort the image formation of a single article multiple times before being captured by the transducer.

32.7 Frequency Content

Several frequency aspects are used to retrieve information about the medium and its anatomical and physiological characteristics. The frequency content of ultrasound imaging has more than

one application. Primarily, the source frequency needs to be recognized by the detector for image formation by reflection; the secondary information embedded in the frequency is the frequency shift resulting from motion, which is described in the section on the Doppler effect. The third frequency component is the generation of second and third harmonics by resonance effects caused by particular tissue components. Harmonic frequency detection can reveal details on tissue structure and point out the presence of anatomical features. Harmonic imaging will be discussed in Section 32.7.2.

32.7.1 Doppler Effect

As part of reflection tomography, the reflection of moving particles in the plane of view will cause a change in wavelength resulting from the motion of the reflector during the time the wave is reflected. If the motion is slower than the period of the wave, the motion can be dissected based on the phase information received at each position of the object in motion.

When a wave is reflected/scattered from a moving object, its apparent frequency is changed: This is the *Doppler effect*. The Doppler effect is a phenomenon in which an observer perceives a change in the frequency of the sound emitted by a source when the source or the observer is moving or both are moving. The Doppler frequency shift, f_d, is related to the moving velocity v and the angle relative to the sound propagation direction.

An object (e.g., blood cell) that is capable of scattering the sound wave back in the direction it came from is moving at presumably constant velocity, or at least with an average velocity v, while the speed of sound through the surrounding tissue is v_s. The motion of the object is at an angle θ with the normal to the surface of the ultrasound probe, at a distance z (Figure 32.5).

Over a given time frame τ, the object moves towards the detector; thus each part of the wave reflected at any successive time interval will be released closer to the detector. The compound effect is that the wave is detected in a compressed form, shorter wavelength. The reflective object has become a secondary source. The distance traveled by the sound is now given by Equation 32.34:

$$z' = (v_s - v)\tau \qquad (32.34)$$

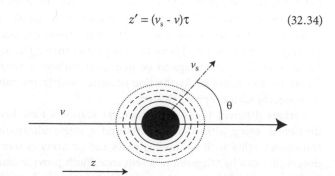

FIGURE 32.5 Illustration of a reverberating object emitting a mechanical (acoustic) wave as it travels from the left to right. The observer receives the emitted sound at an angle θ. This situation resembles standing at the edge of the railway while listening to a train approach.

instead of

$$z = v_s \tau. \qquad (32.35)$$

Comparing Equations 32.34 with 32.35 gives a time compression of the time required for the wave to travel the distance z' from the source to red blood cells as

$$\tau' = \left(1 - \frac{v}{v_s}\right)\tau. \qquad (32.36)$$

Note: The sound wave travels with respect to the moving object twice the distance, first from the probe to the blood cell, and then from the blood cell to the detector after reflection. Alternatively, the frequency observed by the red blood cell has decreased by the fractional time compression. While traveling towards the blood cell it encounters the cell at increasingly closer proximity as well, before the sounds are returned. The time compression will thus satisfy Equation 32.37:

$$\tau' = \left(1 - \frac{2v}{v_s}\right)\tau. \qquad (32.37)$$

Taking the component of the sound wave velocity only in the direction of motion of the red blood cell gives

$$\tau' = \left(1 - \frac{2v\cos\theta}{v_s}\right)\tau. \qquad (32.38)$$

The frequency detected by the red blood cell is now

$$f = f_0 \frac{1}{(1 - 2v\cos\theta/v_s)}. \qquad (32.39)$$

The wave reflected from the object has the following format when collected by the detector:

$$P = P_0 \cos\left[\omega_0\left(t + \frac{2z}{v}\right)\right]$$

$$= P_0 \cos\left[\omega_0\left(t \pm \frac{2tv_s\cos\theta}{v}\right)\right] \qquad (32.40)$$

$$\approx P_0 \cos\left[\omega_0\left(1 \pm \frac{2v_s\cos\theta}{v}\right)t\right] \quad \text{if } v_s \ll v. \qquad (32.41)$$

The frequency f of the reflected signal with reference to the original source frequency f_0 for both the situation where the red blood cell is moving towards the emitter/detector and,

respectively, moving away from the detector/receiver once it has passed it is now

$$f_0 - f = \left(\frac{2v_s\cos\theta}{v}\right), \qquad (32.42)$$

where the cosine turns negative once it has passed the probe.

So the *Doppler shift* from the transmitted frequency is

$$f - f_0 = \frac{2v_s\cos\theta}{v}. \qquad (32.43)$$

For scattering material velocities in the 0–1 ms^{-1} range, at ultrasound frequencies of 1 and 6 MHz the Doppler shift frequencies are in the 0–1.3 and 0–8 kHz ranges.

Imaging of Doppler frequency versus position can give useful indications of arterial blood flow in tissue regions and major organs. The Doppler principle is used for measuring blood flow velocity in blood vessels.

The Doppler effect states that changes in the beam–receptor distance will affect the frequency of the wave perceived by the receptor.

$$f = f_0\left(\frac{v_s}{v_s - v}\right). \qquad (32.44)$$

The frequency f is perceived by the receptor, f_0 is the frequency of the source, v_s is the velocity of sound, and v is the velocity of the moving beam or receptor. This effect can be used to image movements of acoustic scatters within the human body. Doppler imaging is widely used to visualize the movement of blood. It is used both for peripheral arteries and veins but also for cardiopathologies. Doppler imaging measures blood velocity, and detects blood flow direction. An example of a Doppler image illustrating ventricular flow is presented in Figure 32.6.

32.7.2 Harmonic Imaging

Harmonic imaging has been proven to increase resolution and to enable more enhanced diagnostic capabilities.

32.7.2.1 The Concept of the Use of Higher Harmonics

Superficial tissue structure, for example, skin, fat, and muscle, produces pulse distortion and aberration (Figure 32.7), similar to an ocean wave breaking near the beach. The breaking wave at the beach is a direct result of the fact that the speed of the mechanical wave is a function of the depth of the water and the frequency; the velocity of propagation increases more rapidly for higher frequencies as the water becomes shallower, causing the crest to outrun the base frequency and topple over. The sound wave emitted by the transducer is almost a perfect sine wave; however, the reflected wave that will be detected by the transducer is a much distorted wave. This wave includes not only the fundamental wave but also many other waves, one of which is

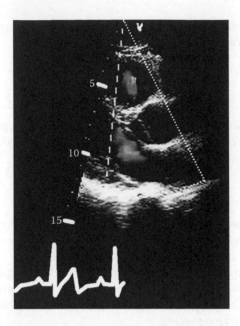

FIGURE 32.6 Doppler image of ventricular flow combined with a standard M-mode image of the heart.

the second harmonic wave, whose frequency is twice the frequency of the emitted wave. It has been known that the conventional imaging array produces significant nonlinear effects.

Acoustic nonlinearity is due to the pressure density (constitutive) behavior, meaning that high pressures are correlated with higher densities and lead to higher sound propagation velocities. In contrast low pressures are correlated with lower densities and lead to lower propagation velocities.

Most of the nonlinear characteristics of ultrasounds have been studied through computer simulations and based on mathematical modeling. One of the most recent mathematical models is the quasilinear model. A nonlinear equation that describes the propagation of sounds in soft tissues is

$$\nabla^2 P(r,t) - \frac{1}{v^2}\frac{\partial^2 P}{\partial t^2} + L(P) = \varepsilon\frac{\partial^2 (P^2)}{\partial t^2}, \qquad (32.45)$$

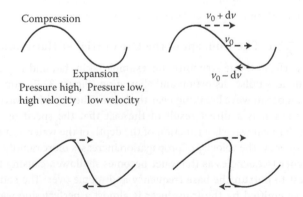

FIGURE 32.7 Illustration of the dispersion of a wave traveling through a medium with velocity of propagation dependent on the frequency of the wave.

where c is the speed of sound, L is a linear operator that accounts for loss, and ε is a small parameter and is very small in soft tissue (10^{-3}). The quasilinear solution is based on a perturbation of Equation 32.45, yielding Equation 32.46:

$$P(r,t) = P_1(r,t) + \varepsilon P_2(r,t) + O(\varepsilon^2). \qquad (32.46)$$

Thus

$$\nabla^2 P_1(r,t) - \frac{1}{v^2}\frac{\partial^2 P_1(r,t)}{\partial t^2} + L(P_1(r,t)) = 0, \qquad (32.47)$$

$$\nabla^2 P_2(r,t) - \frac{1}{v^2}\frac{\partial^2 P_2(r,t)}{\partial t^2} + L(P_2(r,t)) = \varepsilon\frac{\partial^2 (P_1(r,t)^2)}{\partial t^2}, \qquad (32.48)$$

where P_1 is the fundamental field and P_2 is the second harmonic wave. Assuming that

$$\varepsilon^2(P_1(r,t)) = O(\varepsilon^2). \qquad (32.49)$$

32.7.2.2 Generation of Harmonics

In addition to the natural by-product of excited oscillation, second harmonic waves can be generated by the use of contrast agents.

Contrast agents will lead to a much more homogeneous second harmonic wave with a higher intensity. The generation of higher harmonics is illustrated in Figure 32.8, resulting from reverberations.

Contrast agents are still under extensive research. The properties of an ideal contrast agent are as follows:

- Nontoxic, easily eliminated
- Administered intravenously
- Passes easily through microcirculation
- Physically stable
- Acoustically responsive: stable harmonics

32.7.2.3 Advantages of Second Harmonics

The frequency content of the harmonic signal is narrower and has lower intensity side lobes. This narrow bandwidth significantly

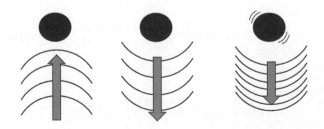

FIGURE 32.8 Cartoon mimicking the reverberation of an incident mechanical wave on an object embedded in a compliant medium. Every higher harmonic is a multiple of the base frequency of the incident wave.

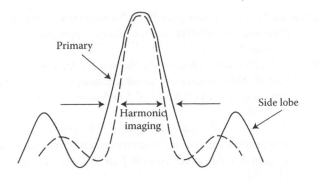

FIGURE 32.9 Frequency spectrum of the base wave and harmonic sound package, respectively.

increases the spatial resolution (Figure 32.9). In addition, the second harmonics are generated inside the tissue and thus suffer less from attenuation and dispersion. As a result the second harmonic signal is less deformed; yielding additional information about the tissues it passed through.

32.8 Bioeffects of Ultrasounds

Because of the mechanical nature and high-energy wave interaction, certain biological effects can be experienced: thermal effects, biochemical effects, and mechanical and cavitational effects. These effects take place on various levels in the organism: cellular and subcellular effects such as macromolecular, and membrane and organelles. The vibrational energy will by definition raise the tissue temperature locally as a result of the definition of temperature as the average kinetic energy. The temperature effect is frequently desired for destruction of kidney and gall stones. For imaging purposes any harmful temperature increase is avoided.

Two different types of mechanical and cavitational effects can be identified. The first order mechanical effect is accomplished by localized strain and shear stress, resulting from out-of-phase acceleration of components of the same biological structure. The local cellular acceleration can exceed 25,000 times the gravitational acceleration. When the focal point is relatively small compared to the local structural size, this may cause tear and twist effects. Examples are cellular membrane fatigue observed in red blood cells resulting in autolysis of the erythrocytes. Additionally ultrasound can also liquefy thixotropic structures, including mitotic and meiotic spindles. The secondary effect, cavitation, is the oscillatory motion of highly compressible volumes such as gas bubbles or "cavities." This feature is a direct function of the acoustic pulse duration and can disrupt red and white blood cells as well as epithelial cells.

On the cellular and subcellular levels, the cavitation process produces shear stress, which causes cellular disruption. For instance, the mitochondria have shown increased membrane permeability to water; however, no direct effect on the functionality of the mitochondria has been observed. The relatively long wavelength has not shown any influence on the atomic or molecular (DNA) level.

The biochemical impact is recognized in the influence of ultrasound to increase AST and ALT levels and decrease glutathione levels in blood. An increase in collagen synthesis has also been observed.

On the biological level there is no conclusive evidence on any effects on fetal growth. On another note, wound healing seems to be accelerated. In the eye, ultrasound can be used to treat mild cases of myopias, but has also shown to produce retinal damage and affect the cornea. Another biological effect from ultrasound exposure is damage to the inner ear.

Despite various reported potential effects of ultrasound under laboratory conditions, no epidemiological data can support any harmful side effects. Ultrasound imaging is still the most frequently used diagnostic technique.

32.9 Summarizing the Physical Aspects of Image Formation

The six main characteristics of ultrasonic image formation are the following: absorption, refraction, diffraction, scattering, interference, and most importantly, reflection. Absorption signifies the conversion of ultrasound energy into heat by friction. Refraction indicates the change of direction at the interface between media with different acoustic impedances according to Snell's law. Diffraction is an indication of the spreading of a beam of sound waves. Scattering describes the interaction with objects generally smaller than the wavelength of the sound. Interference explains the wave effect interaction due to superposition and results in constructive or destructive interaction on the wave level, which is correlated to amplification or destruction of intensity. The primary principle of ultrasonic imaging is caused by reflection, which sends echoes back to the transducer and supplies the signal for the derivation of an image based on discontinuities.

Additional Reading

Barnett S.B., Kossoff G., and Clark G.M. 1973. Histological changes in the inner ear of sheep following a round window ultrasonic irradiation. *J. Otolaryngol. Soc. Aust.* 3; 508–512.

Coakley W.T. and Dunn F. 1972. Interaction of megahertz ultrasound and biological polymers. In: *Interaction of Ultrasound and Biological Tissues*, J.M. Reid and M.R. Sikov (Eds), pp. 43–45, Ultrasound Biological Tissues Workshop Proc., Rockville.

Dunn F. and Pond J.B. 1978. Selected non-thermal mechanisms of interaction of ultrasound and biological media, In: *Ultrasound: Its Applications in Medicine and Biology*, F. J. Fry (Eds), Elsevier, Amsterdam.

Graham E., Hedges M., Leeman S., and Vaughan P. 1980. Cavitational bioeffects at 1.5 MHz. *Ultrasonics* 18; 224–228.

Hrazdira I., Raková A., Vacek A, and Horký D. 1974. Effect of ultrasounds on the formation of hematopoietic tissue colonies in spleen. *Folia Biol. (Praha)* 20; 430–432.

Jaffe B., Roth R.S., and Marzullo S. 1955. Properties of piezoelectric ceramics in the solid solution series lead titanate–lead zirconate-lead oxide: Tin oxide and lead titanate–lead hafnate, *J. Res. Natn. Bur. Stand. A Phys Chem 55*; 239–254.

Lizzi F.L., Coleman D.J., Driller J., Franzen L.A., and Jakobiec F.A. 1978. Experimental ultrasonically induced lesions in the retina, choroids and sclera. *Opthalmol. Visc. Sci. 17*; 350–360.

McCulloch M., Gresser C., Moos S., Odabashian J., Jasper S., Bednarz J., Burgess P., et al. 2000. Ultrasound contrast physics: A series on contrast echocardiography, Article 3. *J. Am. Soc. Echocardiogr. 13*(10); 959–967.

Mendez J., Franklin B., and Kollias J. 1976. Effects of ultrasounds on the permeability of D_2O through cellulose membrane. *Biomedicine 25*; 121–122.

Pospisilová J. and Rottová A. 1977. Ultrasonic effect on collagen synthesis and deposition in diff erently localized experimental granuloma. *Acta Chir. Plast. 19*; 148–157.

Pye D. 1979. Why ultrasounds? *Endeavour 3*; 57–62.

Reece E.A., Assimakopoulos E., Zheng X-Z., Hagay Z., and Hobbins J.C. 1990. The safety of obstetric ultrasonography: Concern for the fetus, *Ostet Gynecol 76*(3); 139–146.

Straburzynski G., Jendykiewicz Z., and Szulc S. 1964. Effect of ultrasonics on glutathione and ascorbic acid contents in blood and tissues. *Acta Physiol. Pol. 16*; 612–619.

Tsutsumi Y. 1965. A new portable echo-encephalograph, using ultrasonic transducers and its clinical applications. *Med. Electron. Biol. Eng. 2*; 21–29.

Varslot T., Dursun S., Johansen T., Hansen R., Angelsen B., and Torp H. 2003. Influence of acoustic intensity on the second-harmonic beam-profile, *IEEE Symposium on Ultrasonics, 2*(5); 1847–1850.

Ward B., Baker A.C., and Humphrey V.F. 1997. Nonlinear propagation applied to the improvement of resolution in diagnostic medical ultrasound. *J. Acoust. Soc. Am. 101*(1); 143–154.

Williams A.R. 1969. An electromagnetic modification of the Zimm–Crothers viscometer. *J. Sci. Instrum. 2*; 279–281.

Williams A.R. 1971. Hydrodynamic disruption of human erythrocytes near a transversely oscillating wire. *Rheol. Acta 10*; 67–70.

Williams A.R. 1983. Ultrasounds: Biological effects and potential hazards. London: Academic Press.

Wojcik G., Mould J., Ayter S., and Carcione L. 1998. A study of second harmonic generation by focused medical transducer pulses, *IEEE Ultrasonic Symposium Proc.*, Sendai, Japan, pp. 1583–1588.

Ziskin M.C., Bonakdapou A., Weinstein D.P., and Lynch P.R. 1972. Contrast agents for diagnosis ultrasounds. *Invest. Radiol. 7*; 500–505.

33

Near-Field Imaging

Kert Edward

33.1 Introduction

One of the most common imaging applications in medical diagnostics is the optical microscope. Optical imaging in general is one of the more accessible techniques compared to magnetic or electric imaging. Other optical imaging techniques besides the microscope include some specialized devices with near molecular resolution, whereas the light microscope is limited to cellular resolution. Before describing one of the high-resolution near-field imaging methods, namely, near-field optical microscopy, the operation and limitations of the optical microscope will be discussed below.

33.2 Optical Microscopy

The light microscope can be seen in most histology and scientific research labs.[1,2] The basic setup of a compound light microscope can be seen in Figure 33.1. The specimen is put onto a stage and white light ($\lambda = 400$–700 nm) is focused on the sample by a condenser. The compound microscope is so named because it utilizes two lenses to magnify the specimen as shown in Figure 33.2.

33.2.1 History

The now ubiquitous optical microscope has had an indelible impact on the field of biological imaging since its invention more than 400 years ago. This instrument allowed early researchers and scientists to examine objects that were otherwise invisible to the naked eye. However, the optical microscope like most instruments is not without its limitations. These limitations are due primarily to aberrations and diffraction. Optical aberrations refer to several types of image distortions that result from imperfections in lens design and the spectrum and wave nature of light.

Any instrument forming an image using some form of lens is bound by the laws of optics. The principles of lens image formation rely on the fact that a lens has a focal point where an infinitely far object will form an image. The distance from the

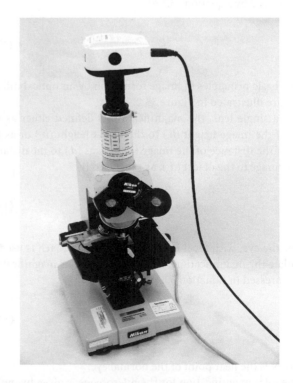

FIGURE 33.1 The optical compound microscope.

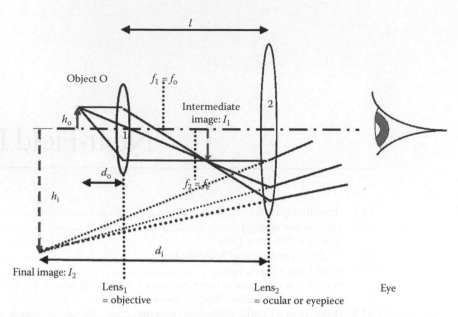

FIGURE 33.2 Diagram of the optical configuration for a compound microscope. The microscope uses a beam of visible light to illuminate the specimen. The object O is placed just beyond the focal length of the objective lens f_o. Light rays travel from the illumination source to the object and then to the objective lens. The rays are diffracted by the lens and an image I_1 is formed by the objective lens just past the focal length of the eyepiece f_e. The first image is real and inverted and is larger than the original object being examined. The image I_1 is magnified by the eyepiece into a very large virtual image I_2 and remains inverted. That second image is then seen with the observer's eye as if it were in the near point of the eye. Since the total length of the microscope is generally less than 25 cm, the ultimate image is perceived as if it were beyond the dimensions of the microscope.

central plane of the lens to this image is the focal length of the lens f. The power (P) of the lens is the inverse of the focal length as expressed by Equation 33.1.[2,3]

$$P = \frac{1}{f}. \tag{33.1}$$

The basic principles of image formation by an optical microscope are illustrated in Figure 33.2.

For a single lens, the magnification is defined either as the ratio of the image height (h_i) to the object height (h_o) or as the ratio of the distance of the image to the lens (d_i) to the distance of the image to the object (d_o), as expressed in

$$m_o \equiv \frac{h_i}{h_o} = \frac{d_i}{d_o}. \tag{33.2}$$

For the eye, with focal length f_e, another approach is used to calculate the magnification. The eye uses angular magnification M_e, expressed in Equation 33.3

$$M_e = \frac{N}{f_e}, \tag{33.3}$$

where N is the near point of the normal eye.

The total magnification for the microscope is given by multiplying the magnification of the objective with that of the ocular.

For example, if the objective magnifies 40 times and the ocular has a 10 times magnification, the total magnification will be 400 times.

In case the microscope is viewed by the eye, the magnification becomes

$$M = m_o M_e = \frac{h_i}{h_o} \frac{N}{f_e} = \frac{d_i}{d_o} \frac{N}{f_e}. \tag{33.4}$$

The compound microscope allows biologists to examine specimens that are microns in size. The specimen sizes in this range include cells and some of their organelles.

33.2.2 Diffraction Limit

The wave nature of light leads to a phenomenon known as diffraction. Diffraction can be simplistically described as the deviation of light from rectilinear propagation, which occurs when light is partially obstructed. This effect results in point sources appearing as airy disks rather than as distinct points.[2-4] Consider two such points in Figure 33.3a separated by a distance d_o. The airy disk patterns overlap but are clearly distinguishable. One can therefore logically ask: What is the minimum separation between the two points such that the airy disks are individually identifiable? This is defined as the resolution or diffraction limit of the system and a criterion was suggested by Lord Rayleigh. If the central maximum of one of the disks coincides with the first minima

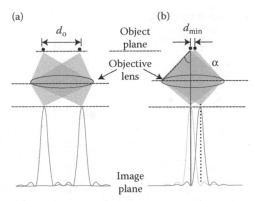

FIGURE 33.3 (a) The airy disk pattern for two clearly resolved points with separation do. If the points are gradually moved closer together, we arrive at the situation in (b) where the first minimum of the left airy disk coincides with the maximum of the disk on the right. This condition is referred to as the Rayleigh criterion.

of the other, as shown in Figure 33.3b, then the separation between the points corresponds to the minimum resolvable distance d_{min}. Diffraction effects reduce the theoretical maximum resolution of an optical system from 200 nm to about 500 nm.

33.3 Image Formation and Magnification

In any optical system, reflected or transmitted light from an object is altered by at least one optical element resulting in an image. To understand the process of image formation, it is instructive to consider the object as being composed of a large number of individual point sources. The path of light from each point source through the optical elements of the system onto an image plane determines the nature, size, and position of the final image. As an example, let us consider the simple case of the image formed by a convex lens. In Figure 33.4a, a beam of light parallel to the central axis converges to a focus at the focal plane.

If we apply this principle to the infinite points that make up an object, the result is an image of the object. This idea is illustrated in Figure 33.4b for two points on the object ox, at the tip and just below the midpoint. The image ox′ is inverted but, more importantly, is larger than ox. The object is said to have undergone a magnification. The magnitude of the magnification is given by the ratio of the heights of the image and the object.

Electromagnetic radiation has both a particle and a wave characteristic. The limiting factor of optical microscopy is derived from the wave phenomenon. When a plain wave is incident on an aperture the Huygens–Fresnel principle suggests that each portion of the aperture's surface area will provide a secondary point source that emits individual spherical waves. The spherical waves released from the edge of the aperture will travel with the same characteristics as the incoming wave.

The wave pattern emitted from all the points on the edge of the aperture creates an interference effect resulting in a radially periodic intensity function with maxima and minima. This interference pattern is called an airy disk (Figure 33.3). In case the aperture represents the diameter of a lens, the same holds true. For a lens with a diameter D, the angle in the radial direction with respect to the first-order minimum can be written as shown below[2,4]

$$\sin(\theta) = \frac{1.22\lambda}{D}, \tag{33.5}$$

which is often referred to as the Abbe condition. This condition describes the angular separation between the central first maximum and the adjacent first intensity minimum in the light transmitted from the aperture. In case two objects are imaged through this lens, they will be just resolvable when the intensity maximum of the first object coincides with the intensity minimum of the second image. This statement is also called the Rayleigh criterion.[2]

The resolving power of the lens now refers to the minimum lateral distance between two adjacent points that can be discriminated when observed through a lens. The definition of the resolving power is given by Equation 33.6

$$\text{Resolving power} = \frac{0.61\lambda}{n\sin(\alpha)},$$

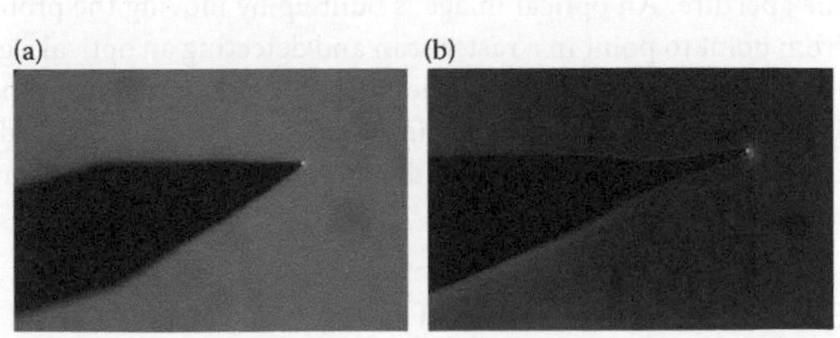

FIGURE 33.4 (a) A parallel ray converging to a focus by a convex lens. In (b), a convex lens is shown to form a magnified image ox′ of the object ox. The image is formed by tracing out the path of light from individual points on the object. The paths of two such points are shown in the figure.

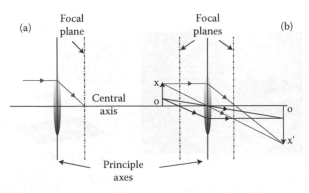

FIGURE 33.5 (a) An NSOM probe formed after etching. The tapered region is short, which allows for better throughput. (b) An NSOM probe formed after heating and pulling. In both images, light from an He–Ne laser launched into the single-mode fibers is seen to be emitted from the aperture.

where n is the index of refraction of the medium in which the images are observed and α is half the acceptance angle of the aperture. In theory the limit of the resolving power for the microscope is $\lambda/2$.[2]

In efforts driven by the desire to overcome this diffraction limit of optical imaging, one of the devices that was developed is the near-field scanning optical microscope (NSOM).

33.4 Near-Field Microscopy

The near-field scanning optical microscope (NSOM) is an optical imaging system that circumvents the previously mentioned diffraction limit.[3-7] The main idea involves imaging a sample probe with a subwavelength size in such close proximity, that the lateral extent of the emitted light is confined to the aperture's dimensions. The sample–probe separation for which this is true is referred to as the near-field. Diffraction is effectively eliminated from such a system over this region and the resolution is primarily limited by the aperture size.

33.4.1 History

Optical near-field imaging was realized through the pioneering research on the scanning tunneling microscope (STM) by the 1986 Nobel Prize winners Binning and Rohrer. In the NSOM, image acquisition is achieved point by point, while the sample is raster scanned in the near-field. A typical NSOM probe has an aperture size of about 50 nm and the probe–sample separation throughout a scan is typically 10 nm.[9]

33.4.2 NSOM Probe

Perhaps the most important component of the NSOM is the illumination aperture or probe. The resolution of the NSOM is directly related to the size and quality of the probe. It is fabricated from a single-mode optical fiber by etching in hydrofluoric acid (Figure 33.5a) or by heating the mid-region of a length of fiber with a carbon dioxide laser, followed by pulling until breakage (Figure 33.5b).

In both methods, the tip of the probe is coated with a thin layer of aluminum, which serves to further confine the size of the aperture. An optical image is built up by moving the probe from point to point in a raster scan and detecting an optical signal at each point while in the near-field. Maintaining the probe in the near-field is achieved using a feedback system. The feedback signal can be used to build up a simultaneous topography image.[6-8]

33.4.3 Instrumentation

In a typical near-field scanning optical microscope, the sample sits on an x, y, z precision stage while the probe is fixed. Alternatively, the sample can be fixed with the probe moving relative to the sample. The x and y movement is controlled by a computer. To avoid catastrophic tip–sample collision while

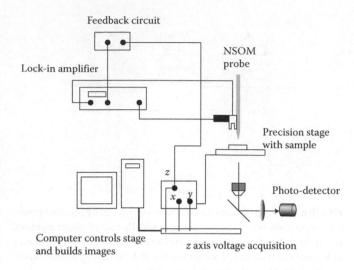

FIGURE 33.6 Schematic diagram showing the main components in a typical NSOM setup. The x and y axes of the stage are controlled by computer software. The z axis is controlled by an error shear-force feedback signal.

scanning in the near-field region, the tip–sample separation is kept constant via a feedback signal sent to the z-axis control. The schematic of such a system is shown in Figure 33.6. The lock-in amplifier aids in signal detection.

33.4.4 Feedback Systems

There are two main approaches for maintaining the probe in the near-field during a scan. The first is via shear-force feedback and the second is via optical feedback. In the former, the probe is attached to a quartz tuning fork as shown in Figure 33.7a.

This fork is electrically dithered at its resonant frequency by a piezoelectric transducer (PZT) element. A typical arrangement is shown in Figure 33.7b. As the probe is moved closer to the sample and into the near-field region, the oscillating probe experiences a damping effect. The nature of this effect is not well understood but the feedback system sends an error signal to the

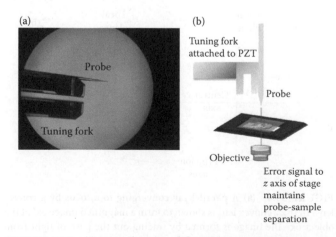

FIGURE 33.7 (a) Optical image of the probe-tuning fork assembly. (b) Schematic of the sensing element of the feedback circuit.

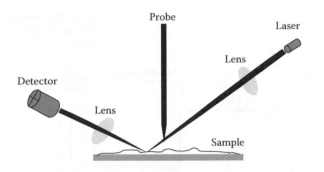

FIGURE 33.8 Schematic diagram showing all the main components in a near-field microscope setup.

z-axis control of the stage to maintain the damping at a constant value while in the near-field. Since the damping is a function of separation, this ensures that the separation is kept constant.

Another method for feedback relies on light reflected from the tip of a dithered probe. This is the feedback mechanism employed in an atomic force microscope (AFM) and a schematic is shown in Figure 33.8. This is further discussed in Chapter 34.

33.4.5 The Optical Image and the Topography Image

Two different images are simultaneously obtained during a single NSOM scan as shown in Figure 33.9. The first is an optical image and the second is a topography image. The optical image is obtained by monitoring the intensity of light transmitted through or reflected from the sample at each point in the scan. The final optical image is thus a two-dimensional array of intensity values.

The topography data are obtained from the feedback signal employed to maintain the tip in the near-field. For example, a large positive voltage to the stage will cause it to move upwards a large distance implying a low feature. A small voltage would cause a smaller shift upwards implying a tall feature. The feedback voltage at each point in the scan forms a two-dimensional array, which is used to reconstruct the surface morphology of the sample with very high resolution. During a scan, the sample moves in the *x* direction, followed by a small *y* increment and then back in the *x* direction (Figure 33.9). This continues until the scan is completed.

33.4.6 Illumination Modes

There are four main modes of operation of a conventional NSOM as illustrated in Figure 33.10. The most common operation is in the transmission mode. Opaque samples require operation in the reflection mode.

33.5 Imaging in a Liquid/Biological Imaging

The noninvasive imaging capability of the NSOM is a significant advantage over other high-resolution imaging devices such as the scanning electron microscope. The growing popularity of this instrument as an imaging tool in the biological science is due in part to this important utility. However, to truly understand the physiological processes of living cells, they have to be imaged *in vivo*. Live cells are typically found in an aqueous environment,

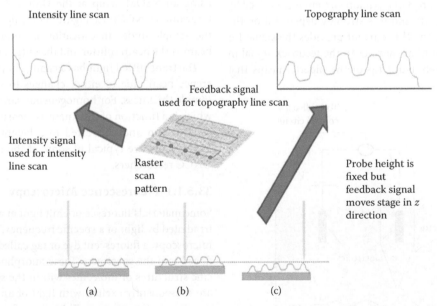

FIGURE 33.9 In an NSOM scan, intensity and voltage values are used to build up the optical and topography images. Each scan in *x* produces a line scan with the subsequent line scan produced for each step jump in *y*. The result is a two-dimensional array of intensity and feedback voltages. In images (a), (b), and (c), the position of the stage relative to the probe is indicated for three points along the *x*-axis. The horizontal line indicates the fixed position of the probe. Any voltage applied to the stage to maintain a constant tip-sample separation from the begining to the end of the scan is converted to a topography height.

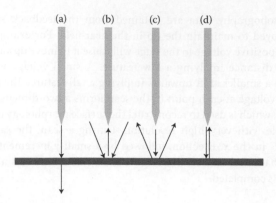

FIGURE 33.10 Illumination modes of NSOM (a) transmission, (b) collection and (c) illumination, and (d) illumination/collection modes.

which is problematic in near-field microscopy. A few of the concerns include damaging or movement of the sample by the probe and liquid-evaporation-induced artifacts. The latter occurs when probe damping becomes a function of both tip–sample separation and liquid level. A novel solution to the evaporation problem has been proposed by a group from Johns Hopkins University using a "liquid cell" replenished by siphon action.[11] Today, liquid cells can be obtained commercially from companies such as Nanonics Imaging Ltd. The presence of artifacts in the final images due to the aforementioned factors is still a problem but very promising results have been obtained in the last few years.[12,13]

A novel solution to near-field imaging in a liquid environment can be found in the scanning ion conductance microscope (SICM), which will be discussed in Chapter 35. This instrument employs a micropipette in an electrolytic solution as the sensing element as shown in Figure 33.11. An ion current is generated by the electrode in the pipette and it is strongly dependent on the pipette–sample separation. This current provides the signal for height regulation in a similar manner to the feedback signal in the NSOM. The relatively blunt tip of the pipette means that

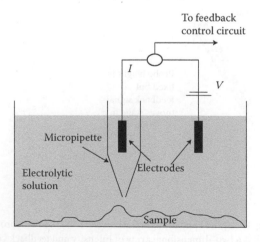

FIGURE 33.11 Schematic diagram of the sensing element and the basic setup of the SICM. A voltage (V) is applied to the electrodes; the ion current (I) is recorded as a function of tip–sample separation.

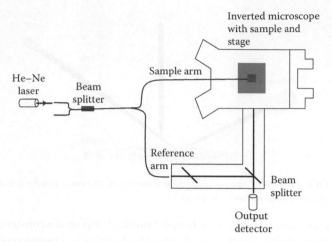

FIGURE 33.12 The diagram depicts a schematic of the phase NSOM.[10] The phase difference between light traversing the sample and reference arm is determined from the output signal.

there is little damage to the sample. In addition, liquid height regulation is not an issue.

33.5.1 Quantitative Phase NSOM

Biological cells are translucent, rendering them difficult to image without the use of an exogenous contrast agent. The contrast agent typically kills the cell and precludes the possibility of *in vivo* imaging. In 1993, two researchers at the Rochester Institute of Technology (Iravani and Crow) proposed a technique for incorporating phase imaging into an NSOM.[14] This technique was the inspiration for a novel procedure for quantitative phase imaging using an NSOM setup at the University of North Carolina at Charlotte (UNCC). In Figure 33.12, light transmitted through the sample under investigation is combined with a reference beam at the beam splitter and the output detected.

The transmitted light has a change in phase relative to the reference beam due to either changes in refractive index or the sample thickness. For homogeneous samples, the phase change, which is a function of thickness, is measured at each point in the NSOM scan and converted to a height value. The images in Figure 33.13 are typical of some of the results obtained by the UNCC researchers.

33.5.1.1 Fluorescence Microscopy

Some materials fluoresce or emit light at a lower frequency when irradiated by light of a specific frequency. In NSOM fluorescence microscopy, a fluorescent dye or tag called a fluorophore is introduced into the specimen. These fluorophores are targeted at specific structures or molecules within the specimen sample which are subsequently excited with light of an appropriate frequency resulting in flourescence. This allows for detection and localization of the targeted protein or molecule in question. The procedure provides excellent contrast while providing information on chemical composition and molecular structure. Fluorescence microscopy has found great utility in the field of biological

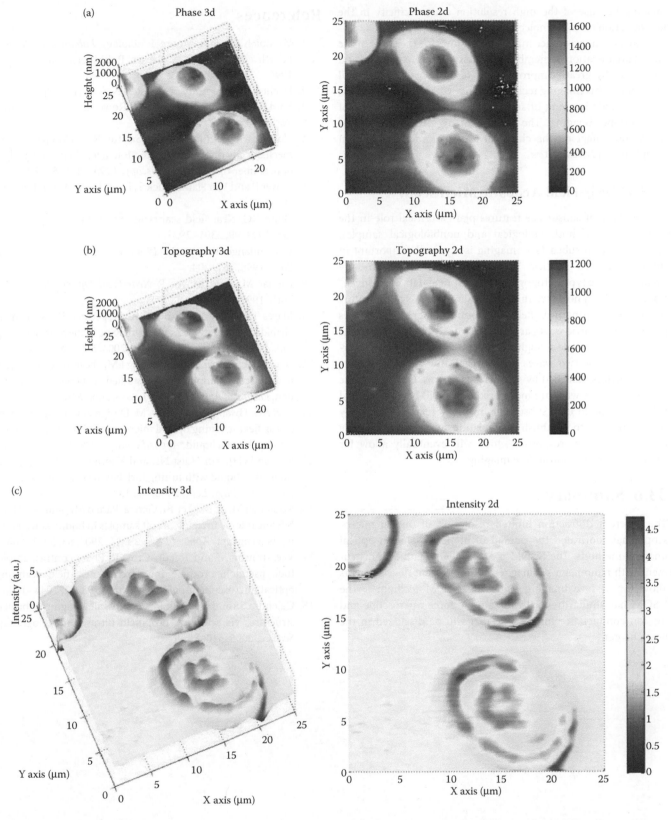

FIGURE 33.13   Phase NSOM images of fish's red blood cells. The two top images in (a) represent the 3d (left) and 2d (right) phase images of a collection of cells. The nucleus is clearly visible. The middle images in (b) represent the 3d (left) and 2d (right) topography images. The bottom images shown in (c) represents the 3d and 2d intensity images of the cells.

imaging because of the high resolution and specificity in the identification of single molecules.

Near-field fluorescence microscopy allows for the study of the lateral organization of membrane domains, protein clusters, and complexes by greatly improving the resolution of the original procedure. This imaging technique is not without its challenges, which include imaging of soft tissue and the interpretation of thick sample imaging. The fluorescent dyes may also have a deleterious effect on the chemical processes of the biological sample under investigation.

33.5.2 Near-Field Acoustic Microscopy

The imaging of subsurface features plays a critical role in the investigation of both biological and nonbiological samples. High-resolution subsurface imaging is especially important in intracellular cell studies. This information is substantially excluded in the typical topography and optical data of a conventional NSOM. A novel technique that allows access to this information is scanning near-field ultrasound holography.[15] In this procedure, acoustic waves are simultaneously launched onto the probe tip and sample at slightly different frequencies. The interference of these two waves forms a standing acoustic surface wave, which is modified by subsurface features. Changes in the frequency of the wave are monitored by an AFM cantilever and are used to build up a subsurface image. Research is currently being pursued to establish the relationship between feature depth and system response, which will eventually allow for three-dimensional subsurface imaging.

33.6 Summary

The inherent diffraction limited resolution of optical microscopes has motivated the pursuit of higher resolution optical imaging methods. The NSOM is an excellent candidate instrument with nanometer cellular resolution. Recent developments allow for *in vivo* imaging using the NSOM technique. The NSOM has similarities with the atomic force microscope and the ion-conductance microscope that will be discussed in the next chapter.

References

1. Vodopich DS and Moore R. *Biology Laboratory Manual*, Fourth edition, Wm. C. Brown Publishers, Dubuque, IA, 1996.
2. Hecht E, *Optics*, Addison-Wesley, New York, 1998.
3. Ford BJ, *Single Lens*, Harper and Row Publishers, Inc., New York, 1985.
4. Hoftman P, Dutoit B, and Salatte R-P, Comparison of mechanically drawn and protection layer chemically etched optical fiber tips, *Ultramicroscopy*, 1995, 61, 165–170.
5. Moyer P and Van Slambrouck T, *Laser Focus World*, 1993, 29, 105–109.
6. Dunn RC, Near-field scanning optical microscopy, *Chem. Rev.* 1999, 99, 2391–2927.
7. Wiesendanger R, *Scanning Probe Microscopy*, Springer, New York, 1998, 161–204.
8. Paeslar MA and Moyer P, *Near-Field Optics*, Wiley, New York, 1996.
9. Mayes TW, Riley MS, Edward K, Fesperman R, Williams S, Shahid U, and Suraktar A, Phase imaging in the near-field, *Cirp Annals Manufacturing Technology*, 2004, 53(1), 483–486.
10. Schmidt JU, Bergander H, and Eng LM, Shear force interaction in the viscous damping regime studied at 100pN force resolution, *Journal of Applied Physics Letters*, 2000, 87(6), 3108–3112.
11. Levi AG, Hwang J, and Edidn M, Design and optimization of a near-field scanning optical microscope for imaging biological samples in a liquid, *Applied Optics*, 1998, 37, 3574–3581.
12. Rensen WHJ, van Hulst NF, and Kammer SB, Imaging soft samples in liquid with tuning fork based shear force microscopy, *Appl. Phys. Lett.*, 2000, 77, 1557.
13. Koopman M, de Bakker BI, Garcia-Parajo MF, and van Hulst NF, Shear force imaging of soft samples in liquid using a diving bell concept, *Applied Physics Lett.*, 2003, 83, 5083–5085.
14. Vaez-Iravani M and Toledo-Crow R, Phase contrast amplitude pseudoheterodyne interference near-field scanning optical microscope, *Appl. Phys. Lett.*, 1993, 62, 1044–1046.
15. Gajendra SS and Vinayak PD, Nanoscale imaging of buried structures via scanning near-field ultrasound holography, *Science*, 2005, 310, 89–92.

34

Atomic Force Microscopy

Christian Parigger

34.1 Introduction and Overview

Microscopy usually comprises generation of images from objects using specialized instrumentation. In the field of atomic-molecular-optical (AMO) physics, the application of optical, electron, and scanning probe microscopy is common. The atomic force microscope (AFM) allows one to investigate objects at high resolution, typically 1000 times better than the optical diffraction limit. The AFM is a tool for measuring nano-scale materials by moving piezoelectrically controlled elements across a surface while mapping mechanical deflections with the aid of a laser beam. In this chapter, we present some historical background on scanning probe microscopy (SPM), address the basic principles of AFM, debate the driven harmonic oscillator, summarize operational details, mention examples of bioengineering applications, and conclude with a brief summary including suggestions for further reading.

34.2 Historical Background

The fundamental work that have led to the AFM is closely related to the scanning tunneling microscope (STM). The 1986 Nobel Prize in Physics honors the development of the STM at the IBM Research Laboratories in Zürich by both G. Binnig and H. Rohrer.

The STM uses a mechanical device to sense the structure of a surface. To this extent, the principle is the same as that of braille reading. In braille, it is the reader's fingers that detect the impressed characters, but a much more detailed picture of the topography of a surface can be obtained if the surface is traversed by a fine stylus, the vertical movement of which is recorded. What determines the amount of detail in the image—the resolution—is the sharpness of stylus and how well it can follow the structure of the surface. Horizontal resolution is approximately 2 Å and vertical resolution is approximately

0.1 Å. This makes it possible to depict individual atoms, that is, to study in the greatest possible detail the atomic structure of the surface being examined.

In 1986 it was evident that the STM is a technique of exceptional promise, and in 1986 it was evident that we have seen only the beginning of its development. In comparison, the AFM combines the use of a laser beam reflecting off a cantilever for the purpose of recording its mechanical motion during a scan of the surface. The initially reported AFM resolution[1] was 30 Å lateral and less than 1 Å vertical resolution. The lateral and vertical resolutions differ due to the use of piezodrives laterally and the use of springs vertically. The AFM resolution is typically one order of magnitude poorer than the one for the STM. Yet the AFM is much more flexible than the STM and can be used on both conducting and insulating samples.

34.3 Basic Principles

In the AFM, one measures the force between the tip and the sample, rather than the tunneling current in the STM. A sharp tip is mounted on a millimeter-sized cantilever. The displacement of the cantilever is measured typically by the reflection of radiation from a laser beam. A photodiode array records the deflection. With this configuration, picometer-scale displacements of the dip can easily be measured. Forces of typically 1 fN can be measured under special circumstances.

The simplest mode of operation is the contact mode. Here, the dip is dragged along the contact and the cantilever deflection is measured. The topology of the surface can be reconstructed from the recorded laser-beam deflections. However, the "tap-mode" or "drag-mode" can damage both the tip and the sample. In the non-contact or intermittent contact mode, the cantilever oscillates due to an applied driving force. The model of this AFM mode is a driven, damped harmonic oscillator discussed in the next section.

TEM image of the PointProbe® Plus tip apex.

FIGURE 34.1 The PointProbe® Plus (PPP) combines the well-known features of the proven PointProbe series such as high application versatility and compatibility with most commercial SPMs with a further reduced and more reproducible tip radius as well as a more defined tip shape. The typical tip radius of less than 7 nm and the minimized variation in tip shape provide more reproducible images and enhanced resolution. [After Nanonsenors: http://www.nanosensors.com/PointProbe_plus.pdf (2008).]

The change of resonance quality and shift of resonance are used to construct an image. In so-called tapping mode imaging, the tip "taps" the surface during the closest approach of the oscillation cycle, causing both a frequency shift and additional dissipation.

Traditionally, the AFM could not scan images as fast as a scanning electron microscope (SEM) that is capable of scanning at nearly real time. The relatively slow rate of scanning during AFM imaging often leads to thermal drift in the image, making the AFM microscope less suited for measuring accurate distances between artifacts on the image. AFM images can also be affected by hysteresis of the piezoelectric material and cross-talk between the (x, y, z) axes. However, newer AFM uses real-time correction software, for example, feature-oriented scanning or closed-loop scanners, which practically eliminate these problems. Some AFM also use separated orthogonal scanners (as opposed to a single tube), which also serve to eliminate cross-talk problems. Due to the nature of AFM probes, they cannot normally measure steep walls or overhangs (Figure 34.1). Specially made cantilevers can be modulated sideways as well as up and down (as with dynamic contact and noncontact modes) to measure the side walls, at the cost of more expensive cantilevers and additional artifacts. Resolution at the single-atom or single-molecule level becomes possible with AFM. One example is a biomedical application: Figure 34.2 presents images of the human cataract lens membrane in obtained samples.

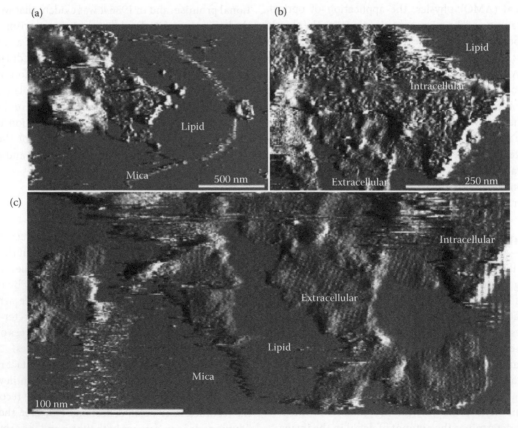

FIGURE 34.2 Overview deflection AFM images of cataract lens membranes: (a) large smooth lipid domains containing corrugated regions; (b) corrugated protein layers; (c) edge region of a membrane adsorbed to the mica showing the smooth lipid bilayer surface containing the square-patterned protein patches. Both intracellular and extracellular surfaces of the cell junction are present. (Reproduced from Buzhynskyy, N. et al. *J. Mol. Biol.* **374**, 162–169, 2007. With permission; Courtesy of Simon Scheuring.)

34.3.1 Driven Harmonic Oscillator

In a typical AFM mode of operation, an oscillatory force is applied to the cantilever. The oscillatory motion of the cantilever will be sensitive to forces that occur between the tip and the sample. For example, forces can be electrostatic, magnetic, Casimir type, or van der Waals type, to name a few. Critical in the response will be the location of the resonance and the quality factor of the resonance. Recorded (i) resonance frequency change and (ii) quality of oscillation allow one to generate a surface-plot-type representation of the surface.

A driven, damped harmonic oscillator can be represented by the ordinary differential equation for the angle of rotation, φ,

$$\ddot{\varphi} + 2\gamma\dot{\varphi} + \omega_0^2\varphi = f_0\cos\alpha t. \qquad (34.1)$$

Here, γ describes dissipation or damping of the oscillator, ω_0 is the natural frequency, f_0 is the driving amplitude at a frequency of α. The amplitude of oscillation is usually small, that is, $\varphi < \sin\varphi$. In other words, the cantilever can be described by the driven harmonic oscillator, see Equation 34.1.

The driven, damped harmonic oscillator involves both homogeneous and particular solutions; for nonvanishing damping, $\gamma > 0$, only the particular solution of the differential equation for $\varphi(t)$ survives. The homogeneous solution will depend on the initial conditions, for example, at time $t = 0$: $\dot{\varphi}(0) = v_0$ with initial velocity v_0, and $\varphi(0) = 0$ or initially at rest.

The transient phenomena will vanish for times $t \gg 1/\gamma$. We expect that the oscillator will move at an excitation frequency α, with amplitude $z(t) = z_\alpha \exp\{i(\alpha t - \theta)\}$, and with a phase shift θ. Consequently, as one uses the *ansatz* for the complex variable $z(t) = z_\alpha \exp\{i(\alpha t - \theta)\}$ with the oscillatory force $f_0 \exp\{i(\alpha t)\}$, one finds for the amplitude $|z_\alpha|$:

$$|z_-| = \frac{f_0}{\sqrt{(\alpha^2 - \omega^2)^2 + 4\gamma^2\alpha^2}}. \qquad (34.2)$$

It is customary to introduce scaled variables $\eta = \alpha/\omega_0$ and $d = \gamma/\omega_0$ and also the oscillator quality, $Q = 1/(2d)$. Furthermore, one defines the response function, $V(\eta, Q)$, as a function of η and the quality factor Q,

$$V(\eta, Q) = \frac{1}{\sqrt{(1 - \eta^2)^2 + (\eta/Q)^2}}. \qquad (34.3)$$

Figure 34.3 illustrates the driven, damped harmonic oscillator response. A maximum elongation of the oscillator is reached for the driving frequency $\eta = \sqrt{1 - (2/Q)^2}$, that is, near the eigenfrequency ω_0. The maximum amplitude equals the quality Q of the oscillator. Figure 34.3 also shows that higher quality oscillators show narrower resonances. For oscillators of quality $Q \approx 10^8$, the V-function is typically replaced by a Lorentzian curve shape, for example, for single atoms, and the quality factor Q is a measure of the natural lifetime. AFM oscillators are typically of quality $Q \approx 100$.[1] Recent developments indicate a quality of

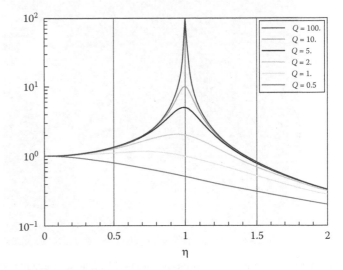

$V(\eta, Q)$

FIGURE 34.3 The driven, damped harmonic oscillator response.

$Q \approx 30{,}000$ for ultrahigh vacuum applications.[2] The Q-factor equals the full resonance width at half-maximum. For AFM applications both the location and width of the resonance is of interest for image generation.

34.3.2 Operational Details

The early results on AFM[1] showed an in-air lateral resolution of 30 Å and a vertical resolution of less than 1 Å. Significant progress in imaging allows one to obtain images at a rate of 20 ms,[3] or at nominal standard video rates. Resolving features today are on the order of a few nanometers with scan ranges of up to hundreds of microns.[4] The AFM's role in nanotechnology and biological imaging is to create highly magnified three-dimensional images of a surface. In this section, details of the forces and full mechanism of operation are summarized, as well as brief descriptions of typical resolutions.

The fundamental component in the AFM is a spring that is as soft as possible to obtain maximum deflection for a given force, but also the spring has to be stiff enough to minimize sensitivity from building near 100 Hz.[1] The resonant frequency, v_0, of the spring system is given by $v_0 = (1/2\pi)\sqrt{k/m_0}$, where k is the spring constant and m_0 is the effective mass that loads the spring. Springs with a mass less than 10^{-10} kg and a resonant frequency greater than 2 kHz can be fabricated. For displacements of 10^{-14} Å, a force of 2×10^{-16} N is required; this force can be reduced by two orders of magnitude when a cantilever with a Q of 100 (see the next section) is driven at its resonance frequency. AFM images are obtained by measuring the force on a sharp tip created by the proximity to the surface of the sample.

The AFM has been operated initially in four different modes relating to the specific feedback mechanism employed.[1] Feedback mechanisms serve to maintain constant force between the sample and the stylus. In the first mode, the sample is modulated; in the second and third modes the lever carrying the stylus

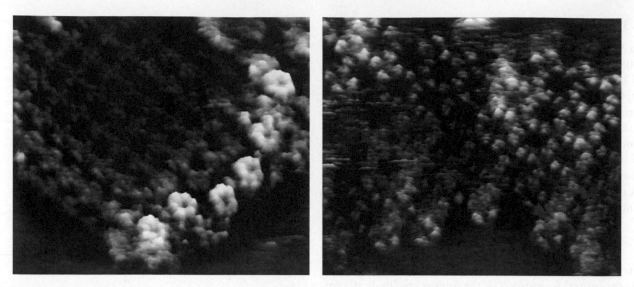

FIGURE 34.4 Images of membranes of a healthy (left, square AQP0 array with connexon channels at the edges) and cataract (right, square AQP0 array, connexons absent) lenses. (Reproduced from Buzhynskyy, N., et al., *J. Mol. Biol.* **374**, 162–169, 2007. With permission; Courtesy of Simon Scheuring.)

is driven at resonant frequency, and both the amplitude and phase of the modulation can be used to drive the feedback circuits. In the fourth mode, the AFM sample and tip are driven in the opposite direction. The force sensitivity at room temperature is limited to 10^{-15} N. Cooled systems, down to 300 mK, show a possible sensitivity of 10^{-18} N.

Measurement of AFM images can be accomplished at high speed using a passive mechanical feedback loop with a bandwidth of 2 MHz.[3] This would allow recording of images at a nominal video rate of 50 Hz. The sample is mounted on a crystal resonator, and an optical lever is used to measure deflections in response to piezoactuators. By tuning the magnitude of the "driver force" and the degree of damping, a high bandwidth feedback loop is created.

In standard AFM applications, however, it takes a few minutes to measure a sample, if the sample is properly prepared for AFM scanning.[4] The proper preparation includes that the sample adheres to the surface, the sample is clean, the sample dimensions are realistic in view of the range of the piezoactuators, and the sample must be rigidly mounted in the AFM stage. Basic elements of an AFM include (i) piezoelectric materials; (ii) force transducers that measure forces as low as 10^{-11} N between the probe and the surface; and (iii) feedback control to monitor a fixed relationship/force between the probe and surface. The most widely used force sensor for AFMs is a light lever sensor, where laser radiation is reflected off the cantilever to monitor the force variation. Practical application details are extensively elaborated in Reference [4]. Primary applications of AFMs include visualization, spatial metrology, and physical property maps. Also, AFM is used to measure a material's surface structure, including measurement of phase transitions, defects, nano-particles, coatings, carbon nanotubes, and of course cells and/or biomolecules, to name but a few examples.

34.4 Bioengineering Applications

In this section, an application of AFM is discussed in the study of the human cataract lens membrane at subnanometer scale resolution.[5] Human pathologies often originate from molecular disorders that can be investigated with views of single molecules. Membrane proteins aquaporin-0 (AQP0) and connexons form junctional microdomains between healthy lens core cells in which AQP0 form square arrays surrounded by connexons. Figure 34.4 illustrates high-resolution AFM images of healthy and cataract lenses.

The images illustrated in Figure 34.4 were collected by operating the AFM in contact mode at ambient temperature and pressure. Imaging was performed with a commercial Nanoscope-E AFM (Veeco, Santa Barbara, CA, USA) equipped with a 160 μm scanner (J-scanner) and oxide-sharpened Si_3N_4 cantilevers (length 100 μm; $k = 0.09$ N/m; Olympus Ltd, Tokyo, Japan). For imaging, minimal loading forces of ~100 pN were applied, at scan frequencies of 4–7 Hz using optimized feedback parameters and manually accounting for force drift.

In general, applications of the AFM in Life Sciences have great promise for measuring the surface structure of biological material.[4] The AFM is the only microscopy technique that allows one to collect nanometer scale images with the sample submerged in a liquid. With the AFM it is possible to measure the mechanical activity of a cell by simply placing the probe on the surface of the cell and monitoring the motion of the AFM cantilever. For AFM imaging the cell does not have to be coated, and in fact can still be alive when imaged. AFM imaging includes visualization of breast cancer cells, or as another example, imaging of biomolecules such as DNA as long as the bio-molecules are directly attached to a surface.

34.5 Summary

The AFM[1] and its precursor the STM[6] have become essential instruments for generating images by scanning a nanometer-sized tip across a surface, thereby obtaining nanometer resolution.[7,3] Important in the application of AFM is, however, sustenance of both the cantilever and the probe: basic principles of physics include the requirement of minimally, if any, destroying the instrument or the probe during the measurement. Use of the AFM tapping mode can certainly endanger and most likely destroy the cantilever; moreover, it can also scratch the surface. The choice of an appropriate cantilever is essential. The results in 1986 of the AFM capabilities showed 30 Å lateral and less than 1 Å vertical resolution without damage to the surface.[1] A minimally invasive or noncontact mode can be understood by using the driven, damped harmonic oscillator model. The speed of image acquisition for the AFM has been increased recently to 0.2 m/s while maintaining nanometer resolution, viz. 0.2×10^9 nm/s allows one to generate images of crystalline and molten polymer surfaces at rates of 50 Hz.[4] This will even further enhance the use of AFM as the most widely used form of SPM with applications on surface, materials, and biological sciences.[2,4,5]

References

1. Binnig, G., Quate, C.F., and Gerber, Ch., Atomic force microscope, *Phys. Rev. Lett.* **56**, 930–933, 1986.
2. Nanosensors, http://www.nanosensors.com product announcement in 2005 of Q30K-Plus®. Recent announcements reference the 2007 development of the PointProbe® Plus tip apex.
3. Humphris, A.D.L., Miles, M.J., and Hobbs, J.K., A mechanical microscope: High-speed atomic force microscopy, *Appl. Phys. Lett.* **86**, 034106, 1–3, 2005; and see references therein.
4. West, P., *Introduction to Atomic Force Microscopy: Theory, Practice, Applications*, http://www.afmuniversity.org (electronic book, 2007), and see continuing newsletters at http://www.pacificnanotech.com.
5. Buzhynskyy, N., Girmens, J.-F., Faigle, W., and Scheuring, S., Human cataract lens membrane at subnanometer resolution, *J. Mol. Biol.* **374**, 162–169, 2007.
6. The Royal Swedish Academy of Sciences, Press Release of 1986: *Nobel Prize in Physics*, 1986.
7. Kittel, C., *Introduction to Solid State Physics*, 8th edition, John Wiley & Sons, Hoboken, NJ, 2005, Chapter 18.

Scanning Ion Conductance Microscopy

Shelley J. Wilkins
and Martin F. Finlan

35.1 Introduction

Scanning ion conductance microscopy (SICM) is a form of scanning probe microscopy (SPM). SICM was developed by Hansma and coworkers[1] in 1989 and is a method for high-resolution noncontact imaging of living cells in physiological conditions.[2] Since its invention, SICM has been used to study a wide range of small living cells including contracting cardiac myocytes, kidney A6 epithelial cells, and neuronal synapses. With SICM, the scanning probe is a micropipette filled with electrolyte solution that is scanned over the surface of a sample immersed in electrolyte. The pipette–sample separation is maintained at a constant value by controlling the ion current that flows via the pipette aperture. This technique offers major advantages over other SPM techniques for investigating living cells. For example, SICM is a noncontact way of imaging with the probe maintained sufficiently far away from the surface so as not to make unintentional contact. The probe is hollow so it can be used to deliver molecules or ions to the surface. Furthermore, positive and negative pressure applied through the pipette can be exploited for noncontact mechanical stimulation of the sample.

35.2 Fundamental Principles of SICM

SICM is used to acquire topographical images of living cells immersed in electrolyte solution, often a physiological buffer. A schematic representation of SICM is shown in Figure 35.1. The scanning probe is a hollow glass or quartz nanopipette filled with electrolyte, which measures the ionic current passing through its tip aperture. The potential difference that drives the current through the pipette aperture is applied through reversible silver chloride electrodes, one inside the pipette and one in the sample's bathing solution. This flow of ions is sensitive to the distance between the tip and the sample, which decreases as the probe approaches the surface due to partial blockage of the pipette aperture. The sample, under the pipette, is raster scanned, while the probe–sample distance is controlled to maintain a constant ion current.

The nanopipette probes are easily produced from capillaries made either of fused silica or borosilicate glass. The glass capillary is symmetrically clamped into two pulling bars of a pipette puller and moderate force is applied on the pulling bars in parallel. Light from a CO_2 laser melts the glass as the pulling bars draw out the capillary. A hard pull leads to the formation of two pipettes whose inner tip diameter and taper length depend on the setting of the parameters of the pulling program. A typical nanopipette has an inner diameter of 100–150 nm and a small half cone angle of 3–6°. Fused silica offers the possibility of making pipettes with opening diameters below 15 nm.

The ion current is measured for DC voltages of up to 200 mV applied between the electrode in the nanopipette and that in the solution as a function of the sample–tip separation. It is amplified by means of a high-impedance operational amplifier and converted to a voltage signal over a resistance of approximately $10^8 \, \Omega$. This signal is then inputted into the control electronics where it is used for feedback control.

There can be some problems working in this constant current DC mode. Complex and unpredictable processes occurring at the electrodes induce drifts and fluctuations in the current circuits, leading to spurious changes in the control signal. Such fluctuations can lead to difficulties in maintaining the pipette–sample distance and may result in the tip crashing into the surface, damaging the specimen and breaking the probe. To avoid this occurrence, the distance of the probe from the surface has conventionally been kept purposefully large.[1,3] This, however,

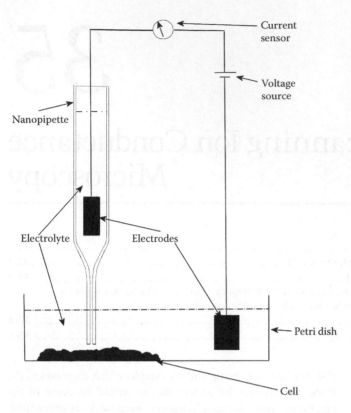

FIGURE 35.1 Schematic representation of the principles of scanning ion conductance microscopy (SICM).

results in reduced sensitivity and resolution since the efficiency of the size of the probe is related to the probe–sample distance.

To address this issue, a distance modulation control mechanism has been developed that uses a distance modulation to significantly increase the sensitivity and resolution of SICM.[4] With distance modulation feedback, it is possible to scan living cells continuously for more than 24 h without loss of control. During the pipette's approach to the sample, the pipette–sample distance is modulated, which introduces an AC component to the ionic current. In principle, this AC component can be detected and the amplitude used to control the pipette distance from the surface, but the SICM setup generates a substantial amount of electronic noise, which makes detection difficult. A lock-in amplifier solves this problem.

Lock-in amplifiers are used to measure the amplitude and phase of signals buried in noise down to a few nanovolts. The essential idea in signal recovery is that noise tends to be spread over a wider spectrum than the signal. By acting as a narrow

band-pass filter, lock-in amplifiers remove much of the unwanted noise while allowing through the signal that is to be measured. Lock-in measurements require a reference signal. A block diagram of a typical lock-in amplifier is shown in Figure 35.2.

Lock-in amplifiers consist of a phase-sensitive detector (PSD), which invariably consists of a mixer followed by a low-pass filter. In the mixer, a sinusoidal function of frequency f_{sig}, amplitude V_{sig}, and phase φ_{sig} is multiplied by a reference sinusoidal function of frequency f_L, amplitude V_L, and phase φ_{ref}. The value of the signals as a function of time t is given by Equations 35.1 and 35.2:

$$V_1 = V_{sig} \sin(2\pi f_{sig} t + \varphi_{sig}), \quad (35.1)$$

$$V_2 = V_L \sin(2\pi f_L t + \varphi_{ref}). \quad (35.2)$$

The output V_{PSD} of the mixer is simply the product of two sine waves:

$$V_{PSD} = V_1 V_2 = V_{sig} V_L \sin(2\pi f_{sig} t + \varphi_{sig}) \sin(2\pi f_L t + \varphi_{ref}), \quad (35.3)$$

$$= \frac{1}{2} V_{sig} V_L \cos[2\pi t(f_{sig} - f_L) + (\varphi_{sig} - \varphi_{ref})]$$

$$- \frac{1}{2} V_{sig} V_L \cos[2\pi t(f_{sig} + f_L) + (\varphi_{sig} + \varphi_{ref})]. \quad (35.4)$$

When integrated over a time much longer than the period of the two functions, the result is zero. If the PSD output is passed through a low-pass filter, the AC signals are removed. However, in the case where f_{sig} is equal to f_L, and the two functions are in phase, the average value is equal to half of the product of the amplitudes, which is a DC signal. In this case, the filtered PSD output will be represented as in Equation 35.5:

$$V_{PSD} = \frac{1}{2} V_{sig} V_L \cos(\varphi_{sig} - \varphi_{ref}). \quad (35.5)$$

In SICM, V_{sig} is amplified and used to drive the axial oscillation of the pipette. V_L is generated locally and is controllable in amplitude and phase. The output of the lock-in amplifier is essentially a DC signal, where the contribution from any signal that is not at the same frequency as the reference signal is

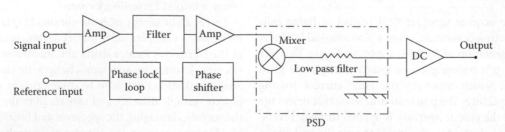

FIGURE 35.2 A typical lock-in amplifier.

attenuated to zero, as well as the 90° out of phase component of the signal that has the same frequency as the reference signal (because sine functions are orthogonal to the cosine functions of the same frequency). In practice there are other considerations, such as the input dynamic reserve and dynamic range of the input amplifier and mixer stage, which must be addressed by measures such as the use of a switching amplifier, heterodyning, various filter modes, and digital processing.

Sample preparation is a straightforward process for SICM. The cells to be imaged are simply grown in or transferred to a Petri dish filled with physiological buffer. The instrument cannot image floating cells or particles, so it is important that the cells adhere to or remain at the bottom of the dish through gravity or adhesion. The nanopipette is filled with electrolyte solution using a syringe and an electrode is inserted into the top of the pipette and lowered into the solution. A second electrode is placed in the Petri dish, as shown in Figure 35.1.

A three-dimensional piezotranslation stage produces highly precise movement of the probe in the *XYZ* directions (with *Z* being vertical) relative to the sample and vice versa. The *z*-separation of the sample–probe distance is kept at the inner pipette radius using the distance-modulated control. The modulation of the axial position of the nanopipette can be achieved either by modulation of the *z*-driver of the piezoelectric block to which the pipette is attached or by a separate piezodrive. This distance modulation provides a more stable distance control and offers a more sensitive method than measuring direct ionic current,[1,3] and allows the scanning probe to be operated only a few nanometers from the sample surface. A typical pipette approach curve showing modulated ion current with distance between the pipette tip and the sample surface is shown in Figure 35.3.

35.3 Applications of SICM

35.3.1 Ion Channel Studies in Live Cells

Technological advances in SICM and its compatibility with current methods for culturing and handling cells have rendered the technique ideal for noninvasive imaging of live cells.[3,5-8] Since 1997, the technique has been widely applied to a range of different cell types. Continuous high-resolution imaging of living cells can advance our understanding of the correlation between the outer cell membrane and its internal function. Furthermore, it facilitates the study of major physiological processes at a fundamental level that are still not understood.

More reliable and robust innovations in the technology[4] have enabled continuous high-resolution imaging of living cells left unattended for up to 24 h. Figure 35.4 shows an SICM image of live kidney epithelial cells where the movement of the microvilli can be monitored by continuous scanning. It can also respond to large motions of, for example, a contracting myocyte sufficiently fast so that the pipette never touches the surface of the moving cell. The typical resolution of SICM is 20 nm laterally and 5 nm vertically and is dependent on the internal diameter of the pipette. Pulling finer pipettes and improvements in the mechanical stability have enhanced the resolution down to 10 nm, enabling proteins to be imaged directly on both a flat surface and the surface of a live sperm cell.[9] The capability of resolving individual proteins directly embedded in a living cell membrane is a significant advancement. This level of resolution offers huge potential in studying dynamic molecular processes such as structural rearrangements, endocytosis, and spatial distributions of fixed or slowly diffusing proteins such as ion channels in the cell membrane. Figure 35.5 shows an image of live inner ear hair cell demonstrating the resolution capability of SICM.

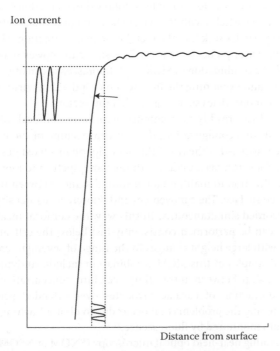

FIGURE 35.3 Modulated mode approach curve for SICM.

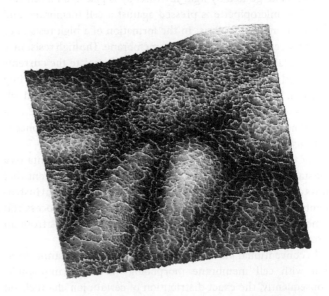

FIGURE 35.4 Modulated mode SICM image of live epithelial kidney cells showing microvilli. Scan area: 40 × 40 µm². (From Wilkins, S.J., et al. *Microscopy and Analysis.* 22 (2): 511–513. With permission.)

FIGURE 35.5 Modulated mode SICM image of live inner ear hair cell. Scan area: 8×8 µm. (From Wilkins, S.J., et al. *Microscopy and Analysis*. 22 (2): 511–513. With permission.)

The function and distribution of ion channel proteins inside a cell membrane are commonly studied by conventional patch clamp techniques. Patch clamp recording takes place with a glass micropipette that has an open tip diameter of about 1 µm, a size enclosing a membrane surface area or "patch" that often contains just one or a few ion channel molecules. This is sealed onto the surface of the cell membrane rather than inserted through it. The interior of the pipette is filled with a solution matching the ionic composition of the bath solution, as in the case of cell-attached recording, or the cytoplasm for whole-cell recording. A silver/silver chloride electrode is placed in contact with this solution and conducts electrical current to the amplifier. In this technology, a micropipette guided by light microscopy is placed on a cell surface. The micropipette is pressed against a cell membrane and suction is applied to assist in the formation of a high resistance seal between the glass and the cell membrane. The high resistance of this seal makes it possible to electrically isolate the currents measured across the membrane patch from the surrounding solution. It then becomes possible to "clamp" the voltage and follow the changes in the current through the micropipette as a function of the variation of parameters such as solution concentrations and ion content.

The term "clamping" refers to the circuitry, which maintains a constant voltage to an electrode and monitors the current that flows as the electrode environment is varied. Commercial instruments can be found that can computer-automate the process and remove effects such as instability and long-term drifts from an essentially simple computational process.

In conventional patch clamping, correlating ion channel location with cell membrane morphology is nearly impossible. Consequently, the exact distribution of certain ion channels on different functional areas of a cell is unknown. SICM can improve this correlation by using the combined scanning capabilities of SICM to locate and identify small cellular or subcellular

membrane structures with subsequent patch clamp recording.[6,10–13] In this technique, the SICM probe is also used for the patch clamp, thus ensuring precise positioning of the electrode relative to the cell topography. The functional location of active single ion channels can be compared with the spatial structure of the membrane, giving further insight into the working mechanisms of the cell. This permits recording of single ion channel activity in obscure regions conventionally difficult to patch clamp, such as the synapse of dendritic neurons, regions of a sperm, and the t-tubule of cardiac myocytes.

This so-called Smart patch clamp offers many advantages over conventional patch clamping. SICM pipettes are much smaller than those used for conventional patch clamping so they can be more precise. Structures do not need to be visible since the specific patch location can be selected from the topographical image using computer control. The pipette is always normal to the cell surface, leading to a high success rate in forming seals (>90%). Moreover, the nanopipette can be used to deliver defined chemical, electrical, and mechanical stimuli permitting investigations into mechano-gated ion channels[14] and the effects of localized drug application.[15]

35.3.2 SICM Plus Optical Microscopy Cell Surface Studies

Although SICM has the potential to resolve processes on a near molecular scale, chemical identification of individual molecules is impossible. Combinations of SICM with optical methods allow direct correlation of topographical and spatial information. An extension of SICM combined with scanning confocal microscopy enables simultaneous mapping of topography and quantitative fluorescence images of the cell surface.[10] This technique can be used, for example, to follow the interaction of single virus-like particles with the cell surface and to demonstrate that single particles sink into the membrane in invaginations similar to caveolae or pinocytic vesicles. This shows the powerful potential that a combination of SICM with fluorescence can offer as an insight into explaining the interaction of individual viruses and other nanoparticles such as gene therapy vectors.

SICM can readily be combined with laser confocal microscopy. In this configuration, the confocal volume of the microscope is focused at the tip of the nanopipette in a fixed position. The sample is moved underneath the nanopipette and moved in the z direction to maintain a constant distance between the tip and cell surface. The z movement and fluorescence intensity can be recorded simultaneously. In this way, the confocal measurement can be performed consistently just below the cell surface even with large height changes in the range of several micrometers. Examples of this SICM combination include studying the relationship between intracellular calcium concentration and the contraction of cardiac myocytes over extended periods, eliminating the problems that occur due to photobleaching with calcium monitored by fluorescence.[11]

Scanning near-field optical microscopy (SNOM or NSOM) has the capability to obtain simultaneous high-resolution topography and fluorescence images of the cell surface. However, few studies have evolved due to the difficulties in reliable regulation of the distance control. By using the SICM ion current to control the pipette–sample distance, simultaneous optical and SICM images can be used to correlate optical and spatial information.

35.3.3 Controlled Delivery of Molecules

Molecules in aqueous solution can easily be filled into the SICM probe. If the molecules are electrically charged, the electric field built up by the SICM bias electrodes makes appropriately charged molecules follow the electrophoretic forces. Since this force depends on the magnitude of the field and on the sign of the molecule's charge, controlled delivery of the molecule through the tip aperture is possible.

An example of this work demonstrated a combination of SICM with fluorescence detection to study the delivery of fluorophore-labeled DNA.[16] In this study, the application of a voltage pulse generated controlled pulses of DNA from the pipette down to just 20 molecules per pulse. The DNA was found to have a 100-fold increase in concentration at the tip of the pipette. In this experimental configuration, the electrode in the nanopipette was grounded with the voltage applied to the electrode in the bath. Negatively charged DNA flowed out of the pipette due to electrophoresis when a positive charge was applied. High electric fields were found to form at the tip, which were large enough to induce dipoles that acted to pull the molecule in the direction of the tip due to dielectrophoresis. When a negative voltage was applied, the electrophoretic force was opposed by the dielectric force at a certain distance from the pipette tip, resulting in a large increase in DNA concentration. Changing the potential resulted in a pulsed delivery of this high concentration of DNA from the pipette. The results suggest that this is a simple way to perform dielectrophoresis studies on biological molecules and to apply high electric fields without using metal electrodes that carry the complications of electrolysis.

35.3.4 Mechanical Sensitivity

Mechanosensitivity of cells is of huge importance in biomedicine and is essential for a number of physiological processes including touch, pain, hearing, growth, and osmoregulation. Mechanosensitive ion channels convert external mechanical force into electrical and chemical signals in the cell. By combining pneumatic and hydraulic pressure systems with SICM (Figure 35.6), highly localized pressure can be applied to the cell membrane. Atomic force microscopy (AFM) is another tool that has been used to map the local mechanical properties on a range of cells at the nanoscale. With AFM, however, the probe used to investigate the surface makes physical contact with the cell during scanning and force measurements. This contact often leads to contamination of the probe and damage to the soft cell surface. Moreover, it is difficult to establish the point at which contact between the probe and soft willowy cell surface has been made; hence, it is difficult to determine how much the cell has

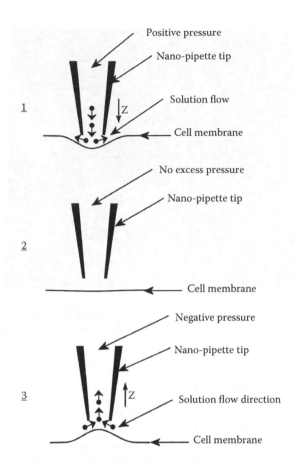

FIGURE 35.6 Noncontact mechanical stimulation of cells using SICM.

been deformed during the measurement and to measure cell topography without deformation.

There are several advantages of using SICM for mechanical stimulation over conventional methods, such as noncontact, minimal or no damage to the cell, and no production of artifacts.[17] SICM also permits application of pressure to a highly localized area, thus permitting the study of mechanoelectric transduction within small regions or structures (e.g., dendrites and synapses), which are conventionally difficult to stimulate independently. The mechanical stimulus can be applied repetitively with reliable results, producing accurate gradual responses in single cells, which permits before and after comparisons for studies such as drug dosing. Subsequent studies have quantified SICM mechanosensitive stimulation and recorded the mechanical response of a variety of cells including red blood cells, cardiac myocytes, neurons, and epithelial cells showing that the method can exert sufficiently small forces to resolve different mechanical elements of the cell. Since the cell can be imaged first at no pressure, measured mechanical properties can be related to topography. A typical SICM image of live neurons is shown in Figure 35.7. Utilizing hydraulic and pneumatic pressure in conjunction with patch clamping to measure the effect on ion channel function makes SICM the ideal vehicle for minimally invasive investigation of mechanosensitive ion channels.

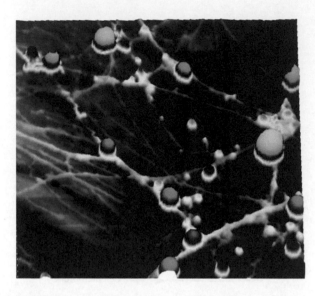

FIGURE 35.7 Modulated mode SICM image of live neurons. Scan area: 60 × 60 μm. (From Wilkins, S.J., et al. *Microscopy and Analysis*. 22 (2): 511–513. With permission.)

35.4 Summary

The examples presented here illustrate a method to address the need for nanoscale mapping of living cells in an aqueous environment. The unique feature that differentiates SICM from other nanoscale imaging modalities is its ability to produce high-resolution images by sensing ion currents, thus not requiring any physical contact with the surface during scanning. The nanopipette, with its capability for high-resolution imaging, can also be combined with optical and fluorescence imaging methods. It has the ability to allow the controlled deposition of molecules, which can be applied to enable drug and toxin delivery at target sites in living cells, but also DNA and protein delivery to any surface. All of these attributes make SICM a highly attractive method for functional imaging of cell dynamics.

References

1. Hansma, P. K., Drake, B. et al. The scanning ion-conductance microscope. *Science* 243: 641–643, 1989.
2. Prater, C. B., Hansma, P. K. et al. Improved scanning ion-conductance microscope using microfabricated probes. *Rev. Sci. Instrum.* 62 (11): 2634–2638, 1991.
3. Korchev, Y. E., Milovanovic, M. et al. Specialized scanning ion-conductance microscope for imaging of living cells. *J. Microsc.* 188 (Pt 1): 17–23, 1997.
4. Shevchuk, A. I., Gorelik, J. et al. Simultaneous measurement of Ca²⁺ and cellular dynamics: Combined scanning ion conductance and optical microscopy to study contracting cardiac myocytes. *Biophys. J.* 81 (3): 1759–1764, 2001.
5. Korchev, Y. E., Bashford, C. L. et al. Scanning ion conductance microscopy of living cells. *Biophys. J.* 73 (2): 653–658, 1997.
6. Korchev, Y. E., Negulyaev, Y. A. et al. Functional localization of single active ion channels on the surface of a living cell. *Nat. Cell. Biol.* 2 (9): 616–619, 2000.
7. Korchev, Y. E., Raval, M. et al. Hybrid scanning ion conductance and scanning near-field optical microscopy for the study of living cells. *Biophys. J.* 78 (5): 2675–2679, 2000.
8. Korchev, Y. E., Gorelik, J. et al. Cell volume measurement using scanning ion conductance microscopy. *Biophys. J.* 78 (1): 451–457, 2000.
9. Shevchuk, A. I., Frolenkov, G. I. et al. Imaging proteins in membranes of living cells by high-resolution scanning ion conductance microscopy. *Angew. Chem. Int. Ed.* 45 (14): 2212–2216, 2006.
10. Gorelik, J., Shevchuk, A.I. et al. Scanning surface confocal microscopy for simultaneous topographical and fluorescence imaging: Application to single virus-like particle entry into a cell. *Proc. Natl. Acad. Sci. USA* 99 (25): 16018–16023, 2002.
11. Gorelik, J., Harding, S. E. et al. Taurocholate induces changes in rat cardiomyocyte contraction and calcium dynamics. *Clin. Sci. (Lond)* 103 (2): 191–200, 2002.
12. Gorelik, J., Gu, Y. et al. Ion channels in small cells and subcellular structures can be studied with a smart patch-clamp system. *Biophys. J.* 83 (6): 3296–3303, 2002.
13. Gu, Y., Gorelik, J. et al. High-resolution scanning patch-clamp: New insights into cell function. *Faseb. J.* 16 (7): 748–750, 2002.
14. Sanchez, D., Anand, U. et al. Localized and non-contact mechanical stimulation of dorsal root ganglion sensory neurons using scanning ion conductance microscopy. *J. Neurosci. Methods.* 159 (1):26–34, 2007.
15. Gorelik, J., Zhang, Y. et al. Aldosterone acts via an ATP autocrine/paracrine system: The edelman ATP hypothesis revisited. *Proc. Natl. Acad. Sci.* 102 (42): 15000–15005, 2005.
16. Ying, L., Bruckbauer, A. et al. Programmable delivery of DNA through a nanopipette. *Anal. Chem.* 74 (6): 1380–1385, 2002.
17. Sanchez, D., Johnson, N. et al. Non-contact measurement of the local mechanical properties of living cells using pressure applied via a pipette. *Biophys. J.* 95: 3017–3027, 2008.
18. Wilkins, S.J., Finlan, M.F., and Finlan, C.S.G. Scanning ion conductance. *Microscopy and Analysis*. 22 (2): 511–513, 2008.

Quantitative Thermographic Imaging

John Pearce

The internal thermal energy of all objects at temperatures above 0 K results in emitted electromagnetic radiation from the surface. Tissues are very efficient radiators, nearly as efficient as an ideal, or "black body," radiator. We can estimate tissue temperatures from measurements of the thermal radiation; and, clinically, skin surface measurements can reveal anomalies in subsurface vasculature. This chapter summarizes the physical origins of thermal radiation and methods for estimating tissue temperatures from these measurements. The emphasis is placed on minimizing the error sources and calibration of the measurements.

36.1 Introduction to Thermal Radiation Physics

All electromagnetic radiation originates in the acceleration of charge. In antennas at radio frequencies, the charges are electrons pumped back and forth by the transmitter resulting in a radiated electromagnetic wave as a solution of the Helmholtz vector wave equations. In optical emission, the light photons are emitted as charge carriers relax (decelerate) from an excited energy state to a lower energy state (Figure 36.1). In thermal imaging, the emission comes from photons emitted by the relaxation of thermally excited atoms and molecules—that is, from internal thermal energy modes in the material under observation.

Electromagnetic radiation has both a wave nature (wave propagation and absorption) and a particle nature (photon emission and absorption). The two can be used nearly interchangeably, as

above, to explain the different aspects of thermal optics. In free space the propagation velocity of light is $c = 3 \times 10^8$ (m/s). The lowest energy photons (photon energy $U_p = hf$, where h is Planck's constant and f is the frequency) have the longest wavelength ($\lambda_0 = c/f$ where λ_0 is the free space wavelength), and the highest energy photons have the shortest wavelength.

36.1.1 Electromagnetic Radiation Spectrum

The electromagnetic (EM) spectrum describes radiated E-M energy in terms of frequency or wavelength, assuming propagation in free space. As can be seen in Table 36.1, the wavelengths associated with thermal imaging are approximately 10 times those of the

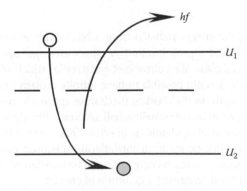

FIGURE 36.1 Energy state transition from higher energy U_1 to lower energy U_2 resulting in an emitted photon. Photon energy $hf = U_1 - U_2$, where h is Planck's constant and is equal to 6.627×10^{-34} (J s).

TABLE 36.1   Summary of the Electromagnetic Spectrum

Signal	Frequency Range	Wavelength in Air	Photon Energy
A-M radio	500–1600 kHz	188–600 m	3.3–10.6×10^{-28} J
F-M radio	88–108 MHz	2.8–3.4 m	0.58–0.72×10^{-25} J
Microwave oven	2.45 GHz	12.2 cm	16.2×10^{-25} J
HF radar	40 GHz	7.5 mm	265×10^{-25} J
Thermography	25–37.5×10^{12} Hz	8–12 µm	0.17–0.25×10^{-19} J
	60–100×10^{12} Hz	3–5 µm	0.40–0.66×10^{-19} J
Nd:YAG laser	283×10^{12} Hz	1.06 µm	1.9×10^{-19} J
Visible light	0.43–1×10^{15} Hz	300–700 nm	2.8–6.6×10^{-19} J

visible spectrum. The longer wavelengths make thermal image formation and collection much more difficult and expensive.

36.1.2 Planck–Einstein Radiation Law and Wien's Displacement Law

Near the end of the nineteenth century, there was some controversy on the relationship between the temperature of an ideal radiating body (a "black" body) and the measured radiant flux. The Stefan–Boltzmann relation had been derived from classical thermodynamics and successfully compared to experiment:

$$E_b(T) = \sigma_B T^4, \tag{36.1}$$

where E_b is the black body surface radiation (W/m²), σ_B is the Stefan–Boltzmann constant (5.67×10^{-8} W m⁻² K⁻⁴), and T is the absolute temperature (K). This relation was derived from the Carnot cycle consideration of a closed cavity with reflecting walls and moving piston.[1] Treating the enclosed photons as a gas, the work done in compression and expansion was related to the cavity wall temperature.[1,2]

Based on an analysis of the Doppler shift of photons reflected from a piston in motion, Wien proposed a wavelength relationship with an unspecified distribution function, f:

$$E_b(\lambda, T) = \frac{f(\lambda, T)}{\lambda^5}. \tag{36.2}$$

That is, the energy radiated from a black body at finite temperature should approach infinity as the wavelength approaches zero. This is the so-called ultraviolet catastrophe, which, of course, is aphysical: it is not possible to have infinite radiative energy at any wavelength. In the classical mechanics approach, energy can be exchanged in infinitesimally small amounts. Plainly, it cannot; therefore, classical mechanics is in error. This is one of the observations that Planck made in formulating quantum mechanics: there is a finite minimum energy that can be transferred from one particle to another, namely a quantum of energy.

Planck's Law describes the monochromatic (single wavelength) emissive power from an ideal black radiator, $W_b(\lambda, T)$, as a function of the body temperature as expressed by Equation 36.3:

$$W_b(\lambda, T) = \frac{2\pi h c^2 \, \lambda^{-5}}{\{e^{hc/\lambda kT} - 1\}} \ (\mathrm{W m^{-2} \, \mu m^{-1}}). \tag{36.3}$$

The denominator originates in the distribution function missing the classical mechanics formulation, and it limits the radiative energy to short wavelengths. In essence, Planck finalized the distribution function, $f(\lambda, T)$, by enforcing the constraint that energy could only be emitted in quanta. Figure 36.2 plots Equation 36.3 for an ideal radiator at representative temperatures. The radiation temperature of the sun (daylight) is around 5780 K (5500°C), while incandescent light is about 3200 K (2927°C).

As the temperature increases, the total area under the monochromatic emissive power curve increases, the maximum emission increases, and the wavelength at which the maximum emission occurs decreases. The product of the wavelength at which peak emission occurs, λ_{max}, and the surface temperature is a constant. This is Wien's Displacement Law:

$$\lambda_{max} T = 2898 \ \mu m \ K, \quad \lambda_{max} T = 5216 \ \mu m \ R. \tag{36.4}$$

Note that the peak wavelength of sunlight, 500 nm, corresponds to the center of the visible spectrum—certainly not coincidental—whereas incandescent lighting has its peak at 906 nm, in the near-infrared (IR) band. At body temperature (37°C), $\lambda_{max} = 9.34 \ \mu m$, whereas at 100°C, $\lambda_{max} = 7.77 \ \mu m$, both in the far IR band.

36.1.3 Wide-Band Radiation Power

For a black surface radiating into free space, all wavelengths are emitted from the surface. So, to calculate the total emitted surface flux, $E_b(T)$, one integrates $W_b(\lambda, T)$ over all wavelengths[1]:

$$E_b(T) = \int_0^\infty W_b(\lambda, T) \, d\lambda = \sigma_B T^4, \tag{36.5a}$$

where σ_B is the Stefan–Boltzmann constant:

$$\sigma_B = \frac{2\pi^5 k^4}{15 c^2 h^3} = 5.67 \times 10^{-8} \ (\mathrm{W m^{-2} \, K^{-4}}). \tag{36.5b}$$

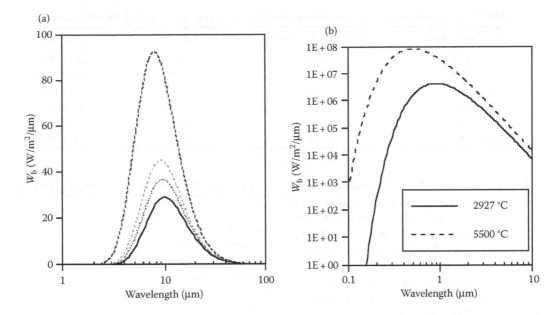

FIGURE 36.2 Monochromatic emissive power, $W_b(\lambda, T)$, as a function of wavelength for a black radiator at (a) 23, 37, 50, and 100°C and (b) 3200 and 5780 K.

36.1.4 Band-Limited Thermal Radiation Power

The two thermal imaging wavelength bands that are used at present are 3–5 μm and 8–12 μm. Thermographic imaging is limited to these two bands by photo-detector spectral response limitations and by atmospheric pass-band windows. The emissive power received is less than the wide-band power since only a portion of the spectrum is measured. To obtain the expected received power, the integral in Equation 36.5a is computed over the spectral band of the imaging device from λ_{min} to λ_{max}. In Figure 36.3 the two

imaging bands have been computed over the range of temperatures in Figure 36.2a. Note that the 8–12 μm band emissive power is much closer to a linear relationship than the 3–5 μm band. This is the reason that 3–5 μm band images appear more "contrasty" than 8–12 μm images for the same thermal scene. That is, in the 3–5 μm band, the lower temperature range occupies a smaller fraction of the total available video signal than the upper temperature range because (referring to Figure 36.2) the 3–5 μm band is on the steep part of the curve near room temperature and body temperature.

The monochromatic emissive power of a black body radiator, $W_b(\lambda, T)$, at typical room temperatures (296 K) has an emission

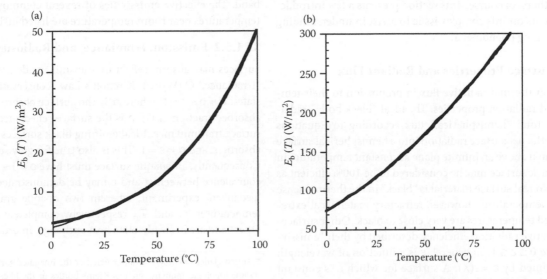

FIGURE 36.3 Band-limited emissive power (W/m²) for the two bands common in thermal imaging: (a) 3–5 μm and (b) 8–12 μm (curve fit = narrow line).

TABLE 36.2 Band-Limited Emissive Power versus Surface Temperature for a Black Body Radiator

Surface Temperature	Wide-Band Flux (W/m²)	3–5 mm	8–12 mm
280 K = 6.8°C	349	2.7 (0.8%)	85.3 (24%)
290 K = 16.8°C	401	4.03 (1%)	102 (25%)
300 K = 26.8°C	459	5.86 (1.3%)	121 (26%)
310 K = 36.8°C	523	8.34 (1.6%)	142 (27%)
320 K = 46.8°C	595	11.6 (2%)	164 (28%)
330 K = 56.8°C	672	15.9 (2.4%)	189 (28%)

peak at 9.8 μm, close to the center of the 8–12 μm imaging band. Consequently, there is much more power in the 8–12 μm band than in the 3–5 μm band, at ordinary temperatures. However, it turns out that the signal-to-noise ratio for the 3–5 μm band is about the same as that in the 8–12 μm band, for typical imaging devices, owing to stronger thermal noise components in the narrow band gap of 8–12 μm detectors. Table 36.2 compares the wide band, 3–5 μm band and 8–12 μm band emissive powers. For the 3–5 μm band (E_b in W/m² and T in °C):

$$E_b(T) = 5.675 \times 10^{-8} T^4 + 3.06 \times 10^{-5} T^3 - 5.702 \times 10^{-4} T^2$$
$$+ 0.1829T + 1.054; \quad r^2 = 1.000. \tag{36.6a}$$

For the 8–12 μm band (E_b in W/m² and T in °C):

$$E_b(T) = -9.466 \times 10^{-10} T^4 - 3.669 \times 10^{-6} T^3 - 1.10 \times 10^{-2} T^2$$
$$+ 1.373T + 75.1; \quad r^2 = 1.000. \tag{36.6b}$$

36.1.5 Thermal Radiative Heat Exchange between Bodies

Thermal radiation heat transfer is an important component of engineering practice. There are many excellent texts and reference works on the topic. Several that come to mind are those by Siegel and Howell,[3] Gebhart,[4] and Incropera and DeWitt.[5] There are many others, of course. This section presents a few introductory comments on this complex issue to assist in understanding thermal imaging phenomena.

36.1.5.1 Surface Properties and Radiant Flux

Surfaces emit thermal radiative flux in proportion to their temperature and radiation properties. The ideal "black body" radiator emits a total "hemispherical" flux according to Equations 36.6. This is the net surface radiation into a hemispherical surface (2π steradians) above an infinite plane at constant temperature in W/m². A black surface may be considered to be 100% efficient as a radiator. No real surface material is "black" in the thermal sense at ordinary temperatures; however, refractory materials at extremely elevated temperatures are very close to black. Other surfaces have an effective radiation efficiency described by their "emissivity," ε, where $0 \leq \varepsilon \leq 1$. In general, ε is a function of wavelength λ and is defined by $\varepsilon = \varepsilon(\lambda)$. A surface for which ε is constant is a so-called gray body. Most surfaces are adequately described by assuming gray body behavior: in the thermal imaging sense

TABLE 36.3 Emissivities of Common (Flat) Surfaces

Surface	Emissivity
Aluminum, polished	0.04
Aluminum, anodized	0.82
Stainless steels, polished	0.17–0.22
Platinum, polished	0.04
Cloth	0.75–0.90
Snow	0.82–0.90
Teflon	0.85
Pigment: cobalt blue (CO_2O_3)	0.87
Pigment: white acrylic paint	0.90
Pigment: zinc white	0.92
Pigment: green (Cr_2O_3)	0.95
Pigment: red (Fe_2O_3)	0.96
Pigment: white (Al_2O_3)	0.98
Glass, window	0.90–0.95
Skin	0.95
Water	0.96
Acetylene soot (or as a pigment)	0.99

Source: Incropera, F.P. and Dewitt, D.P., *Fundamentals of Heat and Mass Transfer*, 4th Edition, Wiley, New York, 1996; Fraas, A.P. and Ozisik, M.N., *Heat Exchanger Design*, Wiley, New York, 1965.

the surface need only be gray over the imaging wavelength band. The effective emissivities of several common surfaces for temperatures near room temperature are listed in Table 36.3.

36.1.5.2 Emission, Irradiance, and Radiosity

Surfaces may absorb, reflect or transmit incident radiation, the "irradiance," G (W/m²). Kirchoff's Law (conservation principle) states: $\alpha + \rho + \tau = 1$, where α is the surface absorbance (i.e., the absorbed fraction of G), ρ is the surface reflectance, and τ is the surface transmittance. Ideal emitting black surfaces are also ideal absorbers, and $\alpha = \varepsilon = 1$. This is also true for gray surfaces, $\alpha = \varepsilon$. Consequently, an opaque surface must have $\rho = 1 - \alpha = 1 - \varepsilon$. The equivalence between α and ε may be demonstrated by a simple "gedanken" experiment, wherein two opaque gray surfaces at temperatures T_1 and T_2, respectively, completely enclose each other* and are in steady-state heat exchange. In order to keep the

---

* At first glance one might wonder whether the imagined surfaces need to be some topologic anomaly, such as Klein's bottles or the like, to enclose each other. In fact, they can be two infinite planes separated by a small fixed distance to satisfy this constraint.

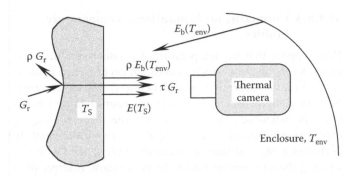

FIGURE 36.4 Surface radiosity, *J*, imaged by a thermal camera potentially consists of many components. Reflections from the enclosure at Tenv, transmission of the rear side irradiance, Gr, and surface emission, $E(T_s) = \varepsilon\sigma_B T_s^4$, all contribute to the thermal image.

net heat flux between the surfaces constant, $\alpha = \varepsilon$ must apply for each surface individually.

The "radiosity," J (W/m²), is the total radiation leaving a surface. It consists of emitted, reflected, and transmitted thermal radiation. A thermal imaging device absorbs its fraction of the total surface radiosity. One advantage of cooled detectors—many imaging detectors are cooled to liquid nitrogen temperatures, 77 K—is that the cooling makes them excellent absorbers with negligibly small radiosity of their own. Note that each component of surface radiosity has its origins in the temperatures of the source surface, as in Figure 36.4. In the figure, the enclosure is assumed to be complete and isothermal at T_{env}, so its irradiance is that of a black body (all complete enclosures have $\varepsilon = 1$). A fraction of the irradiance on the rear side of the surface, G_r, is transmitted, and the surface, at T_s, emits $E(T_s) = \varepsilon\sigma_B T_s^4$ if gray. In thermal imaging, the problem will consist of a complex radiation heat transfer calculation. For simplicity, we may assume that the rear side irradiance is from the enclosure; thus there are three surfaces interacting in this example.

36.1.5.3 Some Comments Regarding Heat Exchange by Thermal Radiation

Thermal imaging is essentially a thermal radiation heat exchange problem. A formal description of this three-body problem requires substantial analysis to include all of the embedded interrelationships, and is beyond the scope of this discussion. In practice, we make use of several simplifying assumptions to reduce the complexity of the problem. However, we must always be prepared to critically inspect the framework of the assumptions under which we are working. The possible number of interactions among the constituents of an imaging geometry is large and should not be underestimated. The simple example of Figure 36.4 illustrates the complexity of a typical imaging situation.

Typical thermal radiation heat exchange calculations account for all of the exchange mechanisms. Each of the surfaces only absorbs a fraction of the total energy leaving each of the other surfaces in the radiation exchange system. Thermal imaging is considerably simplified because the photo-detectors have a very narrow field of view and are cooled to low temperatures for noise management

reasons. This makes the imaging device an excellent absorber and very low power thermal radiation source. So, the complex formalism regarding the "absorption factors" between surfaces—B_{ij} = the fraction of radiation emitted by surface "i" and absorbed by surface "j," where $0 \le B_{ij} \le 1$—is not strictly necessary. Also, we can reasonably make the considerable simplifying assumption that all of the surfaces in the enclosure are opaque, so that $\alpha = (1 - \rho)$.

In Figure 36.4, define surface 1 to be the thermal camera photo-detector itself and surface 2 to be the surface under observation by the camera. The enclosure is then surface 3. For "good" thermal imaging we want to maximize B_{21} and minimize the other absorption factors. We can drive B_{31} to zero by using an instantaneous field of view (IFOV) that views surface 2 only. The surface under observation may reflect such things as people moving about in the enclosure and even the cold void created by the photo-detectors themselves.

36.1.6 Tissues as Gray Bodies and the Origin of Their Emitted Radiation

Many real bodies can be considered gray, at least over the imaging wavelengths. This is certainly true for tissues since their water content typically dominates their thermal radiation properties. The radiosity of an opaque gray body at temperature T_w will consist of the emitted radiation, $\varepsilon E_b(T_w)$, and reflected radiation, $\rho G(T_s) = (1 - \varepsilon)G(T_s)$, where T_w is the tissue temperature and T_s is the temperature of the surroundings.

36.1.6.1 Estimating Tissue Temperature from Surface Radiosity Measurements

We can make estimates of the tissue temperature from the measured radiosity if we are able to make a substantial number of simplifying assumptions, namely, that the enclosure is complete and isothermal and consists of only one surface at temperature T_s, then it is a relatively simple matter to correct the measured surface radiosity, J_m, to estimate the emitted thermal radiation for a black surface at T_w, $E_b(T_w)$:

$$J_m = \varepsilon E_b(T_w) + \rho G(T_s) \cong \varepsilon E_b(T_w) + (1 - \varepsilon)E_b(T_s), \quad (36.7a)$$

$$E_b(T_w) \cong \frac{J_m - (1 - \varepsilon)E_b(T_s)}{\varepsilon}. \quad (36.7b)$$

From the (band-limited) estimate of $E_b(T_w)$, it is a simple matter to calculate an estimate of the surface temperature. Calibration curve polynomials can be constructed from integration of Planck's Law, Equation 36.5a, as was done to obtain the curve fits in Equations 36.6. Note that the compensation mechanism in Equations 36.7 is not very effective for low emissivity surfaces since the subtraction of large numbers in the numerator and division by a small number enhances the uncertainty due to noise processes in the IR detector.

In practice, if reference black bodies are included in the scene at known temperatures, the video levels corresponding to the band-limited radiosity measurements are established. Including reference black bodies in the scene compensates for all the losses due to optical pathway attenuation, such as the effect of various lenses and mirrors used to collect the image. This is the best experimental practice.

36.1.6.2 The Origin of the Emitted Radiant Flux

Perfectly absorbing/emitting surfaces, $\varepsilon = 1$, where the absorption can truly be considered as a purely surface phenomenon, do not exist in a microscopic sense. In practical thermal imaging of tissues, the emitted signal may be considered to originate in the tissues closest to the surface, but over some finite depth. That is to say, tissues deep below the surface emit thermal radiation just as does the surface tissue. As the emitted flux propagates toward the surface it is absorbed by interaction with the surrounding tissues, the "lattice" as it were. The total emission from the surface consists of all the volumetrically emitted thermal flux that is not absorbed. To describe this, we may imagine a thermally black substance and define a uniform temperature-dependent volumetric emitted flux $P(T)$ (W/m^3) that is absorbed by an optical absorption coefficient, μ_a, as it propagates to the surface. The total surface-emitted flux is then the following:

$$E_b(T) = \int_0^\infty P(T)\, e^{-\mu_a z}\, dz = \frac{P(T)}{\mu_a}. \quad (36.8)$$

That is, the nonabsorbed portion finally gets emitted into the surroundings. Note that $P(T) = \mu_a E(T)$, whether it is a black substance or not.

As an example, the absorption coefficient for the CO_2 laser ($\lambda_0 = 10.6\ \mu m$) in water is approximately $\mu_a = 7.92 \times 10^4$ (m^{-1}). Consequently, we would expect that 63.2% of the observed emitted thermal radiation originates within the first 12.6 μm of the surface of the water. Tissue-emitted thermal flux is determined by its water content, and so is its effective "thermal emission depth" when viewed with a thermal camera. For a tissue water fraction of 55% by volume, we would expect that the penetration depth at that wavelength be approximately 23 μm. Consequently, an 8–12 μm thermal camera would get something like 86% of its signal from the 46 μm of tissue nearest the surface.

36.1.6.3 Correcting for Nonuniform Temperature Profiles

Now suppose that the temperature-dependent volume emission is a function of z—such a condition occurs when tissue is heated by high-absorption-coefficient lasers, such as at excimer laser wavelengths where the absorption coefficient may be an order of magnitude higher than over the thermal imaging band. As an example, suppose that the tissue temperature decreases along a strong exponential: $P(T, z) = P(T_w, 0)e^{-\gamma z}$, where γ, the temperature profile decay constant, is larger than μ_a. The total radiated surface thermal IR flux will be much lower than if the temperature were constant at the surface temperature T_w along the z-axis. The temperature indicated by a thermal camera measurement will be much lower than the true surface temperature, T_w.

$$E_{meas}(T) = \int_0^\infty P(T, z)\, e^{-\mu_a z}\, dz = \int_0^\infty P(T_w, 0)\, e^{-\gamma z} e^{-\mu_a z} dz = \frac{P(T_w, 0)}{\gamma + \mu_a}, \quad (36.9a)$$

$$P(T_w, 0) = (\gamma + \mu_a)E_{meas}(T). \quad (36.9b)$$

Consequently, an estimate of the emitted thermal flux that would have been measured if the temperature profile was a constant at T_w, $E_{meas}(T_w)$, can be obtained from

$$E_{meas}(T_w) = \frac{(\gamma + \mu_a)}{\mu_a} E_{meas}(T). \quad (36.10)$$

The actual emitted flux at T_w can be obtained by applying the correction factor $(\gamma + \mu_a)/\mu_a$ to the measured emitted flux. Note that for the example, where $\gamma = 10\ \mu_a$, the correction factor is quite significant, that is, a factor of 11. A gross underestimation of the actual surface temperature would be obtained from an uncorrected thermogram.

36.1.6.4 Specular and Lambertian Surfaces

There are two general types of reflecting surfaces: specular and diffuse. Specular surfaces have an angle of reflection equal to their angle of incidence (Figure 36.5a) and make good mirrors. Interestingly, surfaces that appear scratched and diffuse at visible wavelengths may be excellent specular mirror surfaces in an IR

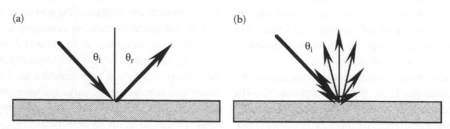

(a)   (b)

FIGURE 36.5 (a) Specular surface (mirror) for which the angle of incidence, θ_i, (with respect to the surface normal) equals the angle of reflection, θ_r. (b) A Lambertian surface has diffuse reflectance.

image because the longer wavelength makes the scratches invisible in the thermal image. Specular reflection is generated by electric field boundary conditions in plane wave propagation. For example, a plane wave normally incident on a surface has an electric field reflection coefficient, Γ_E, and transmission coefficient, T_E, of

$$\Gamma_E = \frac{\eta_1 - \eta_2}{\eta_1 + \eta_2} \quad T_E = \frac{2\eta_2}{\eta_1 + \eta_2}, \quad (36.11)$$

where η is the characteristic impedance of the medium (Ω), and the wave in medium 1 is incident on the surface of medium 2 (that is, $\rho = |\Gamma_E|^2$ and $\tau = |T_E|^2$ for the surface). The reflection of a plane wave at oblique incidence is a classic problem in electromagnetics.

Diffuse reflection is actually back-scattered incident radiant flux. A Lambertian surface (Figure 36.5b) has reflected intensity distributed in a cosine fashion: $I(\theta) = I_0 \cos(\theta)$ (W/radian). Most diffuse surfaces are Lambertian, or approximately so. Note, however, that the observed reflected radiation flux for a Lambertian surface is independent of viewing angle, θ, because the area of the surface in the IFOV of the IR detector, A, increases with increasing θ: $A = \text{IFOV}/\cos(\theta)$. As a consequence, diffuse surfaces will appear isothermal, essentially independent of the viewing angle, until θ approaches quite close to 90°.

36.2 Thermal Image Formation

Perhaps one of the most thorough treatments of the fundamentals of image acquisition and processing for biomedical engineers is that by Rosenfeld and Kak[6]—in particular, the chapter on Computed Tomography. The reader is referred to that reference for more complete discussions of these topics. A thermal image is a two-dimensional (2-D) measure of the surface radiosity, $J(x, y)$. The components of an imaging system include a photo-sensitive detector and some means of creating a 2-D plot of the received radiosity. The two dominant imaging strategies for image formation are: (1) a single photo-sensitive detector scanned over the scene to create a function of position—a flying spot scanner and (2) a focal plane array of staring IR sensors—a charge-coupled device (CCD) imaging device. The thermal camera output signal, $g(x, y)$, is a modified version of $J(x, y)$—modified by the spatial response characteristics of the camera, indicated by its point spread function, $h(x, y)$, as depicted in Figure 36.6. For a linear imaging device:

$$g(x, y) = \int_{-\infty}^{\infty}\int_{-\infty}^{\infty} f(x - \xi, y - \eta) h(\xi, \eta) d\xi d\eta, \quad (36.12a)$$

$$g(x, y) = \int_{-\infty}^{\infty}\int_{-\infty}^{\infty} \left\{ \int_{\lambda_1}^{\lambda_2} S(\lambda) J(x - \xi, y - \eta, \lambda)\, d\lambda \right\} h(\xi, \eta)\, d\xi\, d\eta,$$

$$(36.12b)$$

where $S(\lambda)$ is the spectral response of the photo-detector, $f(x, y)$ is the input image signal that results from $S(\lambda)$, $h(x, y)$ is the

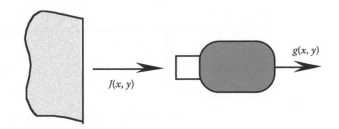

FIGURE 36.6 Surface radiosity, $J(x, y)$, imaged by a thermal camera with output $g(x, y)$. The output is determined by both the spectral response of the photo-detectors and their point spread function, $h(x, y)$.

point spread function of the detector(s) and image processors, and the dummy variables of integration, ξ and η, are used to compute the required convolution.

Each of the two image scanning strategies, the flying spot scanner and the 2-D staring CCD array, has its advantages and disadvantages; however, the image quality (spatial resolution) currently available from CCD cameras is unmatched by flying spot technology and has essentially displaced it. In both cases the critical element is the photo-detector, which must respond to the low-energy photons typical of thermal images. We begin with a survey of IR photo-detector technology and develop 2-D image formation and image processing strategies.

36.2.1 IR Photo-Detectors

The heart of the thermal imaging system is the IR photo-detector. The basic physical principles are not trivial, and are well treated in many texts on semiconductor physics, for example.[7,8] This chapter introduces a few of the important results. Typical photo-detectors convert incoming photons into electrical signals—either photo-voltaic, photo-diode, or photo-resistive detectors may be used. Photo-voltaic cells, for example, solar cells, generate a voltage and can put out limited power for charging batteries and similar energy storage applications. Typical conversion efficiencies are around 15% for very good solar cells. Photo-diodes convert incoming photons into reverse-bias current. That is, the diode current under reverse bias conditions is proportional to incoming radiant power. Photo-resistive elements convert the incoming photon power into a linear change in the resistance of the semiconductor by increasing the number density of charge carriers in the semiconductor. The operating principle of a charge-coupled device (CCD) staring sensor array is similar to the photo-diode, except that rather than generating a current, the photo-generated static (minority carrier) charge occurs in the depletion region of a MOSFET channel, is moved by a clock signal to the end of the line of detection MOSFET regions (i.e., "pixels"), and counted by a charge amplifier.

In all of the above devices, the incoming photon causes a charge carrier ("n" or "p" material—i.e., an electron or a hole) to jump from a nonconducting band energy level, for example, the valence band, U_v, to a (mobile) conducting energy level, U_c, as indicated in Figure 36.7a, in the photo-generation process. The inverse process, recombination, is the annihilation of the mobile carrier. Both

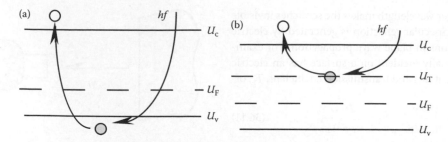

FIGURE 36.7 Photo-generation of charge carriers in an appropriate band gap detector. U_F is the "Fermi level." (a) The incoming photon, hf, generates a mobile carrier from the one in the valence band, U_v, intrinsic photo-generation. (b) Trap-level extrinsic photo-generation typical of a trimetal detector: the incoming photon (at much lower energy) interacts with a charge carrier in the trap level, U_T.

processes can either be direct (between U_v and U_c) or depend on "trap" levels between the conduction and valence bands, U_T (Figure 36.7b). If the charge jumps directly from the valence band, U_v, it is "intrinsic" photo-generation (Figure 36.7a); if the jump is from a trap level, U_T, it is "extrinsic" photo-generation (Figure 36.7b). In order for the energy transition to take place in a direct process, the energy of the photon, hf, must be more than the valence to conduction band gap energy, $U_c - U_v$. Consequently, the particular semiconductor materials used in a photo-detector are selected based on the wavelength (frequency, $f = c/\lambda$) of the desired photons. This is fairly easy to do for visible light detectors since the required band gap energies—around $3-7 \times 10^{-19}$ J or 1.9–4.4 eV—are typical of many semiconducting materials. At the long IR wavelengths of thermal imaging, the band gap energies cannot be made narrow enough, and intrinsic transition is not observed.

Traps (G–R centers, or Shockley–Read–Hall centers) are usually impurity atoms—gold (Au) is a typical example—but they can also be lattice defects, or impurity atoms caught in lattice defects. Thermal generation and recombination is dominated by the trap level mechanism. At IR wavelengths, the list of appropriate materials is quite short since for the longer wavelengths involved, the band gap must be narrow to respond to the low-energy photons—typically about $0.25-0.66 \times 10^{-19}$ J or 0.16–0.41 eV. The low photon energy of thermal IR wavelengths essentially precludes the use of photo-diode configurations.

Band gaps appropriate for the 3–5 μm band ($0.4-0.66 \times 10^{-19}$ J) may be obtained in either Indium–Antimonide (InSb) or Mercury–Cadmium–Telluride ($Hg_{1-x}Cd_xTe$, i.e., MCT) tri-metal detectors, where x is selected as appropriate for the band gap: x is typically 0.25–0.3 for the 3–5 μm band. Only MCT detectors are effective in the 8–12 μm band—x is approximately 0.2 or a little less. In the 8–12 μm band, the incoming photon generates a mobile charge carrier in the trap level (Figure 36.7b). IR photo-detectors may be constructed in photo-diode or photo-resistor configurations. Single photo-resistive elements are usually used for flying spot scanners, while 2-D CCD MOSFET photo-diode arrays make up staring (i.e., focal plane) sensors.

36.2.1.1 Photo-Resistive IR Detectors

In the 8–12 μm band, only MCT ($Hg_{0.8}Cd_{0.2}Te$) detectors are feasible since it is not possible to get narrow enough band gaps

in other materials. These detectors are usually used in the photo-resistive mode to get as fast a slew rate as is feasible for use in flying spot scanners. The photo-resistive detector slew rate is determined by the mobility of the charge carrier in the semiconductor lattice, μ_n or μ_p ($m^2V^{-1}s^{-1}$), and the width of the detector, w(m) (Figure 36.8), since the (vector) drift velocity of the charge carrier is $\mathbf{v}_d = \mu\mathbf{E}$ (m/s)—"p" charge carriers (holes) move parallel to $\mathbf{E}$ and "n" charge carriers (electrons) move anti-parallel to $\mathbf{E}$—and the carrier transit time to the Ohmic contacts is then determined by $\mathbf{E}$ and w.

In the photo-resistive mode, incoming photons photo-generate carriers and the electrical conductivity of the material is increased, according to Equation 36.13:

$$\sigma = \mu_n qn + \mu_p |q|p, \qquad (36.13)$$

where μ is the charge carrier mobility (m^2 V^{-1} s^{-1}), q is the electron charge (-1.602×10^{-19} C), n, p are the charge carrier number densities (number m^{-3}), and both contributions to conductivity are positive numbers since the mobility of an electron is negative. The increase in electrical conductivity due to mobile charge carrier generation results in a decrease in the electrical resistance of the semiconductor that is proportional to the incoming radiation power. The detectors are usually operated in a standard Wheatstone bridge circuit that is biased so that the detector is operated at an average current that gives the fastest response and greatest photo-sensitivity (Figure 36.9). The operating point bias current is determined experimentally by the manufacturer and must be individually set for each detector.

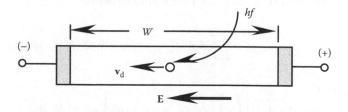

FIGURE 36.8 Photo-resistive detector. Vector electric field, $\mathbf{E}$, between Ohmic contacts, (+) and (−), sweeps the generated charge carrier at drift velocity, $\mathbf{v}_d$, across average distance $w/2$ to register as a change in electrical resistance. The Figure illustrates p-type charge carrier generation.

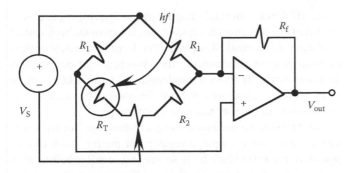

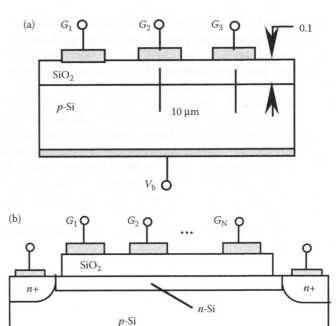

FIGURE 36.9 Wheatstone Bridge circuit for a photo-resistive detector typical in flying-spot thermal cameras. Photo-resistive detector, R_T, is in balanced bridge. A current-to-voltage amplifier configuration is used rather than a differential amplifier to preserve bandwidth and compensate thermocouple voltages at ohmic contacts.

36.2.1.2 Charge-Coupled Device (CCD) IR Detectors

CCDs are members of the larger class of charge transfer devices (CTDs). Charge-coupled imaging devices consist of a 2-D array of channels of MOSFET diodes, in the case of a surface channel device (SCCCD, Figure 36.10a), and transistors, in the case of the buried channel device (BCCCD, Figure 36.10b), held in deep depletion by gate bias voltages (i.e., bias with respect to the "backplane"). Discrete "potential wells" are depleted of majority carriers (electrons in n-type material and holes in p-type material) by the electric field under the gate (Figure 36.11). Impinging photons generate mobile minority carriers in the depletion region under the gates. The MOSFET gates are located in a line, and the respective gate bias voltages are "clocked" to move the photo-generated minority charge carriers from the generation region to a sensing amplifier located at the terminus of the line.

For the case of a surface channel CCD, as in Figure 36.11, photo-generated minority carrier charge (electrons in this example) reside in their deep-depleted potential well until a transient potential under adjacent gates moves them along. In the example, the "clock" sequence $V_3 > V_2 > V_1$ moves the electrons toward the end of the line of pixels where a charge amplifier is located. The recombination rate is indicated by the "minority carrier lifetime," τ_{n0} and τ_{p0}. Consequently, it is essential to achieve adequately long minority carrier lifetimes in the semiconductor to preserve the photo-generated charge until it arrives at the sense amplifier. The limit on the number of pixels in a CCD line is essentially determined by the G–R (generation–recombination) noise process, which is in turn established by the minority carrier lifetime. The typical "dynamic range" of a CCD imaging device—the ratio of the saturation signal to the noise floor—is about 10^2. This is in contrast to the human eye, which has a dynamic range of about 10^7.

36.2.1.3 Photo-Detector Response Characteristics

Both types of photo-detectors are cooled, typically to liquid nitrogen, LN_2, temperatures (77 K) either by immersion in LN_2 or by a thermo-electric cooler (i.e., a Peltier junction), or by a Joule–Thompson cooling device (i.e., by isentropic expansion of a high-pressure gas, similar to a standard refrigeration device). The

FIGURE 36.10 Structure of typical 2-D CCD imaging arrays. (a) A surface channel CCD is a line of closely spaced metal oxide semiconductor (MOS) capacitors (shown in cross-section, typical gate lengths are 30–150 μm)—charge transfer occurs in the p-region at the SiO_2 interface. (b) A buried channel CCD has a thin epi-layer of opposite doping—charge transfer takes place at the epi-layer p–n junction, away from the SiO_2 interface.

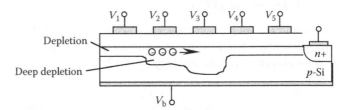

FIGURE 36.11 Charge transfer in a surface channel CCD. The deep depletion region boundary (solid line) is moved from left to right by successively biasing adjacent gates to shift the deep "potential wells": $V_4 = V_1$ and $V_3 > V_2 > V_1$ to move the photo-generated electrons to the end element, a charge amplifier that senses the magnitude of the charge.

purpose of the cooling is to reduce the thermal noise process in the detector substantially below the band gap energy—and it also makes the detector a more effective thermal radiation absorber by reducing the detector emittance. Noise processes are an important consideration in thermal imaging and will be discussed in some detail in a following section.

The most often used figure of merit for photo-detector response is the "Detectivity" D^* expressed as

$$D^* = \frac{A^{0.5} B^{0.5}}{\text{NEP}}, \tag{36.14}$$

where A is the detector area (m²), B is the bandwidth (Hz), and NEP is the noise equivalent power (i.e., the root mean square (rms) sinusoidal power of a modulated irradiance signal that is equivalent to the rms noise power in the detector). Whether the modulated source is a monochromatic or black body must be specified, and the modulation frequency as well. It is recommended that D^* be expressed as $D^*(\lambda, f, 1)$ or $D^*(T, f, 1)$, where λ is the wavelength (μm), f is the frequency of modulation (Hz), and T is the black body temperature (K), and the reference bandwidth is always 1 Hz. This figure of merit alone, however, is only part of the story. Equally important—and especially so for a flying spot imaging device—is the temporal response, because it determines the maximum slew rate possible. Consequently, a high D^* detector may not prove effective in practice, so some care with this figure of merit is advised.

Similarly, photo-conductive detectors are often characterized by their gain, given by Equation 36.15:

$$\text{Gain} = \frac{\tau}{t_r} = \frac{\tau \mu V}{w^2}, \tag{36.15}$$

where τ is the carrier lifetime, t_r is the transit time (s), μ is the mobility (m^2 V^{-1} s^{-1}), V is the applied voltage (V), and w is the distance between electrodes (m). Similarly to the case of D^*, a high gain alone does not always indicate an acceptable imaging device. For example, a long carrier lifetime gives a high gain; but the transit time in a detector may still be too long to give adequate bandwidth in a particular application.

36.2.2 Detector IFOV

The photo-detector has a characteristic "IFOV," β, which determines how much of the surface is in the acceptance angle of the detector at each instant of time (Figure 36.12a).

$$\beta = \tan^{-1}\left\{\frac{d}{D}\right\}. \tag{36.16}$$

Wide acceptance angles equate to a high numerical aperture (as in microscope objectives), and indicate a large IFOV, which is not a desirable feature in a thermal imaging device. In general, a narrow IFOV is much preferred for reasons previously outlined.

The IFOV for a thermal imaging photo-detector is purposely restricted to a very small acceptance angle to provide high spatial resolution. At thermal IR wavelengths the only practical optical material is solid germanium. The absorbance of germanium decreases markedly above a wavelength of 1.6 μm, whereas glass increases markedly above $\lambda = 1$ μm. Focusing optics are extremely expensive in the thermal wavelength range.

The IFOV can be measured using a rectangular "square wave" image source; for example, a regular slot pattern mask can be placed over a warm black body source (e.g., a water bath). For a uniform bar size and spacing, b, and uniform sensitivity over the IFOV, d, the output line scan signal will be a maximum amplitude triangular pattern when $b = d$ (Figure 36.12). If $2b = d$ a single trapezoid output line scan will result (no individual bars will appear). When measuring the IFOV it is important to ensure that the first trapezoidal maximum is determined by increasing D gradually until the trapezoidal pattern disappears into a triangle. This is because "pseudo-resolution" maxima (triangles) also occur when $3b = d$. The clues that this is happening are: (1) when $3b = d$ the output response is between 1/3 and 2/3 of the difference between the "light and dark" gray levels, as it were, and (2) the number of bars resolvable in the output image is reduced by 1; that is, for K bars in the original image plane, $K - 1$ bars will appear in the output line scan.

At least as important as the IFOV is the "point spread function," psf $= h(x, y)$, as in Equation 36.12. The two are related, of course, but in an indirect fashion. The psf can be used to calculate the output of a linear imaging system, $g(x, y)$, by convolution with the image plane signal, $f(x, y)$. The point spread function is defined as the output of the imaging device when the input is a Dirac delta function, $\delta(x, y)$:

$$h(x,y) = \int_{-\infty}^{\infty}\int_{-\infty}^{\infty} \delta(x - \xi, y - \eta)\, h(\xi, \eta)\, d\xi\, d\eta. \tag{36.17}$$

We cannot implement a Dirac delta experimentally, but we can create an edge. The analogous "edge spread function," esf $= e(x, y)$, can be measured with a sharp-edged mask over a thermally black source at elevated temperature. For example, if

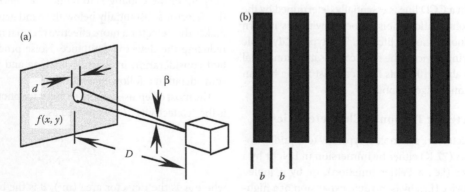

FIGURE 36.12 Photo-detector IFOV. (a) Definition. (b) Mask for its measurement.

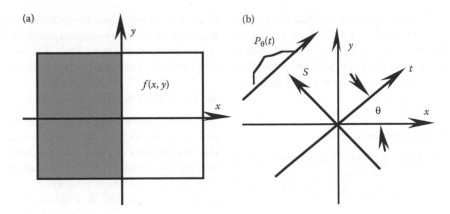

FIGURE 36.13 Edge spread function measurement. (a) Edge located on the *y*-axis, θ = 0. (b) The derivative of the edge spread function in the normal direction at angle θ is a "projection" of *h*(*x*, *y*) at that angle.

the edge is along the *y*-axis, Figure 36.13a, the input image is $f(x, y) = u(x)$ where "*u*" is a step function.

Note that the spatial derivative of the "step" output in the direction normal to the edge orientation, $\partial g/\partial n$ is the *x*-direction derivative, and $\partial g/\partial x$ is the "line spread function" $l(x, y)$ at an angle with respect to the *x*-axis of θ = 0. The integral along the line spread function (i.e., the "*s*" axis in Figure 36.13b) constitutes a "projection" of the 2-D point spread function at angle θ, $P_\theta(t)$, in the sense of a projection in computed tomography—that is, the Radon transform of $h(x,y)$[6]:

$$P_\theta(t) = \int_{-\infty}^{\infty} h(x, y)\,ds = \int_{-\infty}^{\infty}\int_{-\infty}^{\infty} \delta(x\cos(\theta) + y\sin(\theta) - t)h(x, y)\,dx\,dy,$$

(36.18)

where $P_\theta(t)$ is the projection integral at angle θ with respect to the *x* axis, and *t* is the coordinate normal to the integration direction, *s*. An ensemble of these projections at many different angles with respect to the *x*-axis (from 0 to 2π) constitutes an ensemble of parallel beam projections, $P_\theta(t)$, that can be inverse Radon transformed (say, for example, by a standard filtered back projection algorithm) to estimate $h(x, y)$. The example *x*-axis line spread function would be the projection for θ = 0. Integrating the line spread function along the *t*-axis reduces the noise in the estimate of the projection.

36.2.3 Photo-Detector Noise Processes

The noise processes that limit the sensitivity of photo-detectors are: (1) thermal (or Johnson) noise, (2) G–R noise, (3) shot noise, and (4) $1/f$ or flicker noise. The dominant noise process at high frequencies is thermal noise, at mid-frequencies it is G–R noise in the detector, while at very low frequencies $1/f$ noise dominates, but this process is only measurable in DC-coupled systems and not important in thermal imaging devices. Both thermal noise and G–R noise are temperature dependent, increasing with detector operating temperature, which is why most practical thermo-

graphic devices cool the detectors to 77 K using liquid nitrogen (LN_2). Noise sources are usually characterized by a description of their power spectral density, $S(f)$. Power spectral density is essentially the expected frequency distribution of the energy in a random signal. The power spectral density may be determined for either the voltage signal, $S_v(f)$ (V^2/Hz), or the current signal, $S_i(f)$ (A^2/Hz), and the noise signal square-voltages and square-currents are represented in the following equation

$$v_n^2 = \int_0^{\infty} S_v(f)\,df, \quad i_n^2 = \int_0^{\infty} S_i(f)\,df.$$

(36.19)

White noise has $S(f) = S_0$, a constant for all frequencies; which obviously cannot be true since the noise voltage or current are not infinite. In fact those processes are white only over the imaging bandwidth.

The noise process power spectral density can be measured for a particular imaging system, but some care is warranted. The definition of power spectral density is given by

$$S_v(\omega) = \lim_{T\to\infty}\left\{\frac{1}{T}E\{|\tilde{V}(\omega)|^2\}\right\}, \quad \text{where } \tilde{V}(\omega) = \int_{-\infty}^{\infty} \tilde{v}(t)e^{-j\omega t}\,dt,$$

(36.20)

where $E\{\}$ is the expectation operator, $V(\omega)$ is the Fourier transform of the time domain $v(t)$, and the "tilde" notation is used to denote a random variable. It looks from the definition that one should take the expectation over a very long data sequence to calculate $S(\omega)$. However, lengthening the data sequence only increases the frequency resolution of the transform; it does not improve the uncertainty in the estimate. Consequently, a much better estimate of $S(\omega)$ is obtained when the long noise data sequence is broken up into many shorter epochs, the discrete Fourier transform (DFT) of each data epoch is calculated individually, and those power spectra are averaged at each frequency to find the required expectation. The expectation is taken over the ensemble of Fourier transformed epochs (the square of the magnitude—i.e., the power spectrum).

The epochs need only be long enough to ensure that the desired frequency resolution is maintained. In the DFT of an N-long epoch, remember that the first component is the DC value (which must be zero for all noise processes due to physics), the second has the significance of "one cycle per sequence length," and the maximum frequency is the $N/2$ component.

36.2.3.1 Thermal or Johnson Noise

Thermal noise is inherent in all signal sources and arises because of random carrier excitation due to thermal energy modes in the semiconductor material. Thermal noise is white at low frequency (i.e., for $hf/kT << 1$):

$$S_v(f) = 4kTR, \quad S_i(f) = 4kTG, \quad (36.21)$$

where R is the photo-detector resistance (Ω) and $G = 1/R$ (S). Notice that for both formulations, the total noise power is $4\,kTB$ (W), where B is the bandwidth of the measurement. At room temperature, $hf = kT$ at approximately 10^{12} Hz—far above practical imaging bandwidths. For completeness, the actual broadband thermal noise spectrum is represented by

$$S_v(f) = \frac{4kTR}{\{e^{hf/kT} - 1\}}. \quad (36.22)$$

36.2.3.2 G–R Noise

G–R noise arises from spurious generation of carriers not due to thermal agitation or photo-generation. G–R noise is not white and has a power spectral density that depends on f^2 above the roll off corner frequency, $\omega_c = 1/\tau$:

$$S_i(f) = \frac{S_0\tau}{1 + \omega^2\tau^2}, \quad (36.23)$$

where τ is approximately the minority carrier lifetime. G–R noise usually dominates the measured power spectrum at lower frequencies in IR photo-detectors, while thermal noise dominates at high frequencies. G–R noise is often impulse-like in the time domain.

36.2.3.3 Shot Noise

Shot noise arises from current in semiconductor ohmic contacts and is of some importance in photo-resistive detectors. Because the bias currents of photo-resistive detectors are usually quite low, the shot noise process is typically less important than the other processes at high frequency; however, the overall noise process is impulse-like, so shot noise cannot be ignored. In any event, shot noise is approximately white with power spectral density given by Equation 36.24:

$$S_i = 2|q|I, \quad (36.24)$$

where q is the charge of an electron and I is the device current (A). It is important to note that shot noise is impulsive, or impulse-like, and also multiplicative, rather than additive, and so is a nonlinear noise process.

In summary, for frequencies above about 1 kHz G–R noise usually dominates the noise power spectrum. The overall noise is strongly influenced by the required imaging bandwidth—and this in turn is very different for flying spot and staring focal plane array sensors.

36.2.4 Flying Spot Scanners

Flying spot scanners move a single photo-sensor over a scene in two dimensions to form an image. This is typical of useful thermographic devices and warrants some discussion. The typical strategy is to scan a horizontal row, move the sensor back to its horizontal origin, shift down one row, and repeat the horizontal scan. This scan is a "raster" scan and is typical of video imaging devices and thermal cameras, as in Figure 36.14a. The flying spot

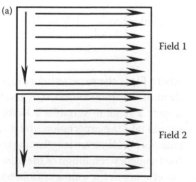
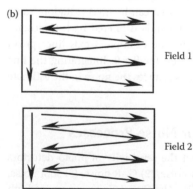

FIGURE 36.14 Typical raster scans. (a) Standard video raster: U.S. National Television System Committee (NTSC) standard line rate 15.75 kHz, field rate = 60 Hz, frame rate = 30 Hz, interlaced. (b) Inframetrics Model 525 raster scanned by oscillating mirrors: line rate = 7.875 kHz, field rate = 60 Hz, frame rate = 30 Hz, interlaced. The alternating line scan directions are compensated for by using a first-in first-out (FIFO) memory for the right-going scan line and a last-in first-out (LIFO) memory for the left-going scan. Each line is read out twice to achieve a standard video line rate for recording. The "fly-back" interval between fields is longer because the vertical scanning mirror must be physically moved back to the top of the image.

imager is scanned by oscillating first-surface mirrors, and its raster is shown in Figure 36.14b.

To get high-quality images and resolve small features in a scene using a flying spot scanner, the imager needs to have a large number of independently measurable samples in each raster line. The individually resolvable samples are called "pixels" or picture elements. The wider the acceptance angle (IFOV) of the detector, the fewer independent pixels in a line scan. Having a large number of samples in a line is equivalent to having a high-bandwidth, B, (i.e., high-resolution) imaging device:

$$B = (\text{lines/second}) \cdot (\text{pixels/line}). \qquad (36.25)$$

In the example Inframetrics Model 525 device of Figure 36.14b, the bandwidth of the detector was measured to be about 1.23 MHz. So, by Equation 36.25, there are about 156 independently resolvable thermal pixels in a line. A typical thermal video line is digitized to 256 samples, and approximately 40–50 of those are consumed by the horizontal sync pulse, so the 200+ remaining thermal pixels represent a slight overscanning of the line. Additionally, this is a very high bandwidth for an LN_2 cooled photo-resistive detector: the low temperatures significantly reduce the carrier mobility, μ, increasing the transit time of photo-generated carriers. The wide bandwidth required in these devices also increases their NEP, as in Equations 36.19.

36.2.4.1 Detector Slew Rate Limitations

Detector bandwidth and slew rate are two quite different specifications. The bandwidth characterizes the device's small signal response. The slew rate is the large signal rate at which the detector will respond to a step change in temperature. For the Inframetrics Model 525 8–12 μm example system, it takes about 3 pixels (744 ns) to slew 20°C—a slew rate of about 27×10^{6}°C/s. The thermal lag due to slew rate limitations may be significant when very small spots are imaged. The user should be sure that the measurement is not slew limited at the peak of such a spot. In the example flying spot system, the effect of slew rate limitations is moderated somewhat by adjusting the phase difference between the right- and left-going scans. Such an adjustment is always some sort of compromise, however.

36.2.4.2 Linear and Nonlinear Image Processing Strategies

The flying spot scanner is an interesting study in image processing. The standard approach in image processing is to use the "optimal" Wiener filter. This filter is optimal in the mean-square sense: that is, it minimizes the mean-square error (MMSE) in the image due to the noise process. The Wiener filter transfer function, $M(u, v)$, is[6]

$$M(u,v) = \frac{H^*(u,v)S_{ff}(u,v)}{S_{ff}(u,v)\left|H(u,v)\right|^2 + S_{nn}(u,v)} \approx \frac{H^*(u,v)}{\left|H(u,v)\right|^2 + 1/\text{SNR}},$$

$$(36.26)$$

where (u, v) are spatial frequencies associated with the image plane (x, y), H is the optical transfer function (OTF) of the imaging system—that is, the 2-D Fourier transform of the point spread function, $h(x, y)$—S_{ff} is the power spectral density of the (random) image ensemble, $f(x, y)$, and S_{nn} is the noise power spectral density, and finally, SNR is the signal-to-noise ratio, which can be used as a weak approximation to the required power spectra. Whether or not the MMSE filter turns out not to be very effective depends on the error criterion used. An MMSE criterion favors the MMSE filter, of course.

If the criterion is the maximum absolute error, however, another filter strategy is at least as effective, and in some ways superior. We first recognize that the noise process in a flying spot scanner is time-based, and only spatially distributed by the raster scan. The noise process is also impulse-like, and an order-statistic filter—specifically, the median filter—is more effective against impulse noise. The median filter replaces the pixel in the center of a sliding window with the median of the rank-ordered pixels in the window (i.e., the value in the center of the rank-ordered sequence). No calculations are performed; existing data are simply shifted around. The median filter has several distinguishing characteristics: (1) it eliminates impulses rather than preserving their energy, as all linear filters do, (2) a monotonic sequence of points is unchanged by a median filter. The first characteristic makes this filter very effective against the impulsive nature of the detector noise. The second characteristic is important in slew-limited data, since no additional distortion is introduced by the filter (because the noise process is inactive when the detector is slew limited, i.e., monotonic). The caution here is that the only parameter that can be set is the width of the rank order window, and the window width defines "impulse" for the filter. For a window $(2n + 1)$ pixels wide, an impulse is n or fewer pixels that are different from the surrounding windows, both of which are "constant neighborhoods" (i.e., either constant over at least $n + 1$ points, or monotonic).

Nonrecursive filters do not update the image data points as new calculations are made—recursive implementations update the image data points as they are calculated while the filter window is moving. The filtering strategy that has proven effective on flying spot thermal images is to use a 1-D nonrecursive median filter on the row data ($n = 2$ has proven useful) and a triangular-weighted nonrecursive window on column data ($n = 1$ has proven useful) to help smooth out phase errors between adjacent rows.

36.2.5 2-D Focal Plane Arrays: CCD Cameras

2-D focal plane CCD arrays comprise a very effective IR imaging strategy because many of the limitations of flying spot scanners are eliminated. The staring sensors do not need to have nearly as wide a bandwidth to provide the required spatial resolution. The effective "shutter time" may be relatively long compared to a flying spot system, giving a long time for both photon collection and signal averaging, reducing the overall noise envelope. Alternately, the image frame rate may be substantially higher with a CCD array, since it is limited only by the time required to clock out the rows of potential wells. In fact, the spatial resolution of the CCD array is limited only by the G–R noise process: the line of detectors may be as long as the SNR will permit. It is also reasonable to apply an MMSE filter on CCD array image data.

The downsides are two: (1) it is not feasible, at present, to construct a CCD array sensitive in the 8–12 μm band and (2) the individual pixel potential wells all have slightly different sensitivities, making calibration of the pixels more difficult. The first limitation means that the nonlinear nature of the signal is an important consideration. The second limitation is typically managed by periodically exposing the array to an internal constant temperature black source; and this also eliminates the effects of $1/f$ noise. Both limitations are very effectively addressed by modern embedded microprocessor systems and digital image processing methodology.

36.3 Clinical Applications

The most effective clinical application of thermographic imaging is for the visualization of vascular function, particularly in the extremeties. An excellent, thorough description of standard clinical methods may be found in Woodrough.[1] Resting skin temperature typically ranges between about 30 and 34°C depending on seasonal and room conditions. Women usually have a slightly higher skin surface temperature than men. The key element in clinical thermography is a carefully prepared patient in a well-controlled environment: no detectable air movement and an environmental temperature and humidity low enough to prevent sweating and other thermoregulatory responses. The patient must be completely relaxed and composed in a low-anxiety state. Patient education is central to obtaining reliable results: the general public does not always understand the completely passive nature of the measurement and the zero-health risk involved.

Vascular function in the extremities can be studied by the response to a warm challenge or a cold challenge, as reflected in the transient skin surface temperature distribution. Immersing the hands or feet in temperature-controlled liquids can be used to study peripheral vascular disorders. Figure 36.15 is a pseudo color version of the response to a warm challenge.[9] In a warm challenge the fingers rapidly warm the venous blood because the

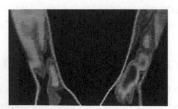

FIGURE 36.15 Venous thermogram 2 min after a 66°C warm challenge created by immersing the fingers in paraffin. Symmetric heat distribution by the veins is clearly identified. Lower arm and hand dark gray = 35–36°C, light gray surrounding fringe = 33–34°C, medium gray in hands = 31–32°C, light gray upper arm = 29–30°C. (From Chan, E.K.Y., Vessel extraction, visualization and tracing in thermographic venography, PhD Dissertation, Biomedical Engineering Program, The University of Texas at Austin, 1991. With permission.)

sphincters of the artero-venous shunts are not active and there is no peripheral vasoconstriction. Delayed or asymmetric warming is an indication of anatomical vascular damage. A cold stress test challenges the sympathetic nervous response and is used to detect peripheral artery disorders. The normal sympathetic response is to close the artero-venous shunts; a delayed bilateral, or asymmetric, rewarming response is indicative of sympathetic nerve pathology, such as Raynaud's phenomenon.[10] This test has also been used to confirm the carpal tunnel syndrome.

Growing tumors have a hyper-perfused outer shell leading to temperatures slightly elevated with respect to the surrounding normal tissue. Tumor detection with thermal imaging is somewhat hampered by the shallow effective depth of the measured signal. Tumors close enough to the surface to exhibit identifiable thermal signatures can generally be detected in other ways. However, asymmetric skin surface thermal patterns can identify a relatively shallow lesion in early stage development, and thermography has proven useful in identifying breast cancers, particularly in younger women where mammography is less effective. This modality does suffer from low specificity; however, thermal asymmetries also arise from vascular anomalies and additional differential diagnostic techniques are required to resolve vascular discrepancies.

36.4 Summary

IR thermal imaging is an effective method for estimating transient 2-D surface temperature fields. Careful planning of the calibration procedures is perhaps the most important exercise in designing these experiments. It must be properly appreciated that thermal imaging is, at heart, a radiation heat transfer problem. The nonlinear nature of the imaging band-limited emittance as a function of surface temperature must be included in the calculations. Almost all of the error sources result in underestimation of the actual surface temperature. Surface emissivity must be known and incorporated into the calculations. The optical pathway attenuation effects of mirrors and lenses can be compensated for by including black body reference sources in the image plane. Variations in the tissue temperature over the effective viewing depth of the imaging wavelengths used must be compensated for if significant. Additionally, the noise in the thermal images should be addressed with an appropriate post-acquisition filter strategy.

An example laser spot experiment is shown in Figure 36.16. The spot was generated in aorta, *in vitro*, with an argon laser ($\lambda = 514$ nm) at a power of 1 W with a beam diameter of 2 mm, Gaussian profile. The image shown was collected at 4.6 s into the 5 s activation. The image in Figure 36.16a is from an Inframetrics Model 525, 8–12 μm flying spot thermal camera. The laser spot thermal pattern is in the center of the image. Calibration black bodies flank the laser spot and are at 44.2 and 109.3°C as noted. The color scale of Figure 36.16b is in 5°C increments, calibrated with the nonlinear polynomial of Equation 35.6b. A 5-wide median filter has been applied to the row data, and a 3-wide triangular smoothing window has been applied to the columns. The maximum temperature is 100°C at the center of the spot.

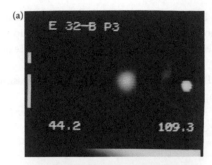

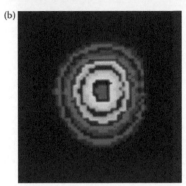

FIGURE 36.16 Example of thermal image processing. (a) Original black and white video screen image at $t = 4.6$ s. (b) Close up of the laser spot in the calibrated image. The maximum temperature in the center is 100°C, surrounding bands represent approximate 10–15°C intervals (original 5°C color data has been lost in this gray scale rendition).

As a final note, "colorization" of thermal images, as in Figure 36.16b, can conveniently be used to display desired features. Consequently, most thermal cameras have a colorization feature built-in. It must be borne in mind that colorization introduces additional uncertainty, however. Colorization is equivalent to a quantization process, just as when an analog image gray scale is digitized. In a constant step-size quantizer, it can be shown[9] that the mean square quantization error, MSQE, is given by

$$\text{MSQE} = \frac{\Delta^2}{12}, \qquad \text{RMSQE} = \frac{\Delta}{\sqrt{12}}, \qquad (36.27)$$

where Δ is the quantization step size. Colorized data have generally been quantized to a very few steps and thus have a high-root-mean-square quantization error, RMSQE. It is the best experimental practice to record gray level (black and white) thermal data and colorize on analysis, rather than to record colorized data in real time, because the original gray levels are lost on colorization and cannot be reconstructed for filtering or other post-processing. Notice, for example, that colorization in Figure 36.15 has obscured the venous track, and if only this image were available reconstruction of the vein would not be possible.

Nomenclature

c	Velocity of light propagation (3×10^8 m/s)
E	Surface emitted flux (W/m²)
$\mathbf{E}$	Vector electric field (V/m)
D^*	Detectivity of a photo-detector (m sqrt{Hz} W^{-1})
f	Frequency, or input image $f(x, y)$
G	Impinging surface irradiance (W/m²)
g	Output image, $g(x, y)$
h	Planck's constant (6.627×10^{-34} J s)
$h(x, y)$	The point spread function
H	Optical transfer function, $H(u, v)$
I	Light beam fluence density (W/m²)
J	Surface radiosity $= \rho G + E$ opaque surface (W/m²)
k	Boltzmann constant (1.381×10^{-23} J/K)
n, p	Charge carrier concentration (number m^{-3})
P	Volumetric radiation power density (W/m³)
q	Charge of an electron (-1.602×10^{-19} C) or net thermal radiation heat (W)
r	Radial coordinate (m)
$S(\lambda)$	Spectral sensitivity for a detector
$S(f)$	Power spectral density of a random process
T	Temperature (K or °C)
U	Energy (J)
$\mathbf{v}_d$	Vector drift velocity (m/s)
W	Surface monochromatic emissive power (W m^{-2} μm^{-1})
z	Coordinate in the direction of the beam (m)
α	Surface absorbance
β	Instantaneous field of view of a photo-detector (steradians)
ε	Surface emissivity
η	Characteristic impedance of a medium (complex) (Ω)
λ	Wavelength (m)
μ_a	Optical absorption coefficient (m^{-1})
μ_n, μ_p	Charge carrier mobility (m² V^{-1} s^{-1})
ρ	Surface reflectance
σ_B	Stefan–Boltzmann constant (5.6697×10^{-8} W m^{-2} K^{-4})
τ	Surface transmittance
ω	Angular frequency (radians/s) $\omega = 2\pi f$

Subscripts

b	Black body variable
c	Constant temperature conditions
m	Measured variable
0	Free space parameter
×	Ambient quantity

References

1. Woodrough, R.E., *Medical Infra-Red Thermography: Principles and Practice*, Vol. 7 of Techniques of Measurement in Medicine, Cambridge University Press, Cambridge, 1982.
2. Sears, F.W., *An Introducion to Thermodynamics, The Kinetic Theory of Gases and Statistical Mechanics*, Addison-Wesley, Reading, MA, 1953.
3. Siegel, R. and Howell, J.R., *Thermal Radiation Heat Transfer*, 2nd Edition, Hemisphere, Washington, 1981.
4. Gebhart, B., *Heat Transfer*, McGraw-Hill, New York, 1961.

5. Incropera, F.P. and DeWitt, D.P., *Fundamentals of Heat and Mass Transfer*, 4th Edition, Wiley, New York, 1996.

6. Rosenfeld, A. and Kak, A.C., *Digital Picture Processing*, Vol. 1, Academic Press, London, 1982.

7. Sze, S.M., *Physics of Semiconductor Devices*, 2nd Edition, Wiley-Interscience, New York, 1981.

8. Blakemore, J., *Semiconductor Physics*, Pergamon, New York, 1962.

9. Chan, E.K.Y., Vessel extraction, visualization and tracing in thermographic venography, PhD Dissertation, Biomedical Engineering Program, The University of Texas at Austin, 1991.

10. Ring, E.F.G., Raynaud's phenomenon: Assessment by thermography, *Thermology*, 3, 69–73, 1988.

11. Fraas, A.P. and Ozisik, M.N., *Heat Exchanger Design*, Wiley, New York, 1965.

37

Intracoronary Thermography

Frank Gijsen

37.1 Introduction

Atherosclerosis is the most frequent cause of heart disease,[1] and it can be subdivided into a number of developmental stages: one of the important ones involving the vulnerable plaque. Plaque rupture is the main cause of events such as the acute coronary syndromes of unstable angina, myocardial infarction, and sudden death.[1] The thin-cap fibroatheroma (TCFA) can be described as a vulnerable plaque, since its structure resembles that of ruptured plaques.[2] The major components of a TCFA are a lipid-rich, atheromatous core and a thin fibrous cap with macrophage infiltration.[2] It is suggested that the inflamed TCFA in coronary arteries show an increased temperature. The underlying heat production might be due to increased metabolic activity of the macrophages or due to enzymatic extracellular matrix breakdown. Detection of heat could possibly lead to detection of vulnerable plaques. Since the original studies by Cascells[3] on hot plaques, many clinical trials on intravascular coronary thermography were carried out to detect vulnerable plaque. Based on the reported temperature differences in these studies, it is important to study the origin and the clinical consequences of this thermal heterogeneity. Production, transfer, and detection of heat are the central topics that we will discuss from a physical, modeling, and clinical perspective.

37.2 Physical Background of Intracoronary Thermography

37.2.1 Production of Heat

High metabolism of macrophages or other exothermal processes related to macrophages, such as matrix breakdown by matrix metalloproteinases (MMP), are the most likely sources of heat production in diseased coronary arteries.

Macrophage heat generation can be measured *in vitro* using microcalorimetry. Heat production values of mouse macrophages were shown to be 0.8–6.5 pW per cell, its value decreasing upon increasing cell density.[4] Heat production values of alveolar rabbit macrophages grown in a monolayer were 19 pW per cell.[5] Adding 20% rabbit serum to the growth medium increased the heat production to 27 pW per cell. Heat production values of phagocytosing mouse macrophages were measured at approximately 20 pW per cell. When, for example, lipopolysaccharide-bounded high-density polyethylene particles were phagocytosed, these values could increase up to 92 pW per cell, implying an increase of up to five times the basal value.

There are several other mechanisms that could be responsible for heat production in vulnerable plaques. MMP-related breakdown of extracellular matrix in atherosclerotic plaques might also be an exothermal process. Toutouzas et al.[6] demonstrated that patients with acute coronary syndromes show increased MMP-9 concentration, which was well correlated with temperature differences between the atherosclerotic plaque and the normal vessel wall. Krams et al.[7] correlated temperature heterogeneity to regions of increased MMP-9 activity. Degradation of lipids or upregulation of uncoupling proteins might be involved in heat generation in vulnerable plaques as well.

37.2.2 Transfer of Heat

Heat transfer between two objects occurs only when the objects are at different temperatures and it occurs either by conduction, convection, radiation, or by a combination of these depending on the media involved.

Conduction is the transfer of heat through a solid or fluid medium due to a temperature gradient by the exchange of molecular kinetic energy during collisions of molecules. Thermal parameters governing this process are the specific heat capacity and heat conductivity. Materials can be divided into conductors, for example metals, known for their ability to transfer heat fast, and insulators, for example air and most plastics, which have low heat conductivity values. Fatty tissue is also known for its insulating properties[8]; so lipid present in the lipid core of a vulnerable plaque is likely to behave as an insulating material and may obstruct the transfer of heat produced by macrophages to its neighborhood.

Convection is the transfer of heat due to bulk motion of the medium. Heat produced by the vulnerable plaque will be transferred through the wall by means of conduction as described above. At the lumen wall, the heat will be transported by the flow of blood. Curvature of the vessel and presence of a measuring device like a thermographic catheter will influence the velocity distribution, resulting in changes of temperature distribution. Increase in local flow velocity will decrease the temperatures at the lumen wall.[9]

Every object radiates electromagnetic waves, the amount and the wavelength depending on the temperature and its emissivity. Heat transfer by radiation does not require a medium. The radiation in its turn may again be absorbed in a medium, which results in local heating. This type of heat transfer is usually neglected in intracoronary thermography. Whether it is correct to neglect radiation depends on, among others, the dimensions of the heat source, with a tendency for larger sources to have a higher radiative emission, and on the emissivity of the heat source.

37.2.3 Detection of Heat

A temperature increase, the result of heat production, can be measured by a variety of methods, using electrical, optical, magnetical, or other types of sensors. Electrical methods include resistance, thermistor, and thermocouple. Some of these have already been applied in the catheters that are currently used for intracoronary thermography.[10–13] These methods can be divided into either contact or noncontact methods.

Heat detection by a contact method occurs by conduction of heat from the luminal wall to the sensor. When the sensor does not approach the wall close enough, however, the sensor area is exposed to surrounding influences and this will affect the measurement.

Noncontact thermographic methods require a material, which will absorb the electromagnetic radiation emitted by the medium readily. The choice of this material is wavelength dependent. Many materials used for this purpose are sensitive for absorption in the infrared region. Blood is highly absorptive for most electromagnetic waves of these wavelengths, which makes *in vivo* detection difficult, since it requires flushing of the artery. Microwave detection could be a better option for *in vivo* temperature measurements, since microwaves have greater penetration depth for these wavelengths.

37.3 Modeling Temperature in Coronary Arteries

Many studies applied numerical simulations to model intracoronary thermography,[9,14–17] and this technique can be applied to answer a number of questions with respect to this technique. Some of these questions will be discussed in the subsequent section.

37.3.1 Cooling Effect of Blood Flow

A simplified geometry was created representing the lumen wall and the tissue surrounding it, which is depicted in Figure 37.1. In this tissue, a heat source was embedded and a heat source production value was assigned. Through the lumen, blood flow could be simulated. The cooling effect of blood flow can be studied by comparing the images of the temperature profiles at the lumen wall when no flow exists in the lumen and when a flow is simulated (Figure 37.1). It is clear that the flow influences both the maximal temperature at the lumen wall and the shape of the profile. The maximal temperature of 40.7°C under no flow conditions in the lumen is reduced to 38.8°C when a flow exists in the lumen. The profile when no flow exists in the lumen covers a greater part of the lumen than when a flow exists. In addition, the maximal temperature at the lumen when a flow exists has moved to a location further downstream. The profile is smeared out in the flow direction when a flow exists. The influence of flow can be calculated using the flow influence factor (FIF).[9] This factor is the ratio between the maximal temperature difference at the lumen wall when no flow exists in the lumen and the maximal temperature difference at the lumen wall when a flow does exist in the lumen. It can be interpreted by how much the measured temperature differences are underestimated when they are measured in the presence of blood flow. In this case the FIF equals 2.0. We can conclude that blood flow has a strong impact by reducing the maximal lumen wall temperature greatly and by

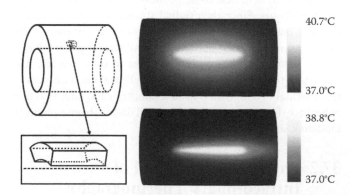

FIGURE 37.1 Simplified geometry of a vessel (top left panel). In the tissue a heat source is embedded (bottom left panel). The influence of flow on the temperature distribution at the vessel wall is shown in the right panel. Top: no flow, bottom: flow. Note the different temperature scales.

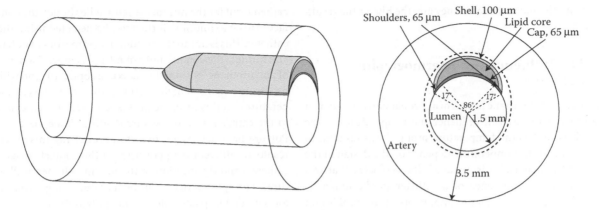

FIGURE 37.2 Schematic representation of the 3D geometry (left) and a cross-sectional view (right).

limiting the region where this maximal temperature can be measured.

37.3.2 Influence of Plaque Composition

Even though vulnerable plaques are characterized by distinct features, they can show great variation with respect to their composition. These variations include the location of the macrophages, the lipid density of the lipid core, and the location of the vasa vasorum with respect to the plaque. These differences in composition may influence the temperatures at the lumen wall. The simulations[16] were performed on a schematic representation of the lumen wall in which a realistically shaped plaque was embedded (Figure 37.2) and the location of the heat source in the plaque was varied.

The temperature distributions at the lumen wall are given when the heat source was assigned to the different regions and when a flow was present in the lumen (Figure 37.3). From top to bottom, the heat source was assigned to the cap, the shoulders, the shell, and the mixed region. The scales belonging to the individual profiles range from 37.0°C to the maximal temperature at the lumen wall, which differed for the different heat source

regions. The resulting profile of the simulation in which the shoulder region was assigned to be the heat source clearly showed the shape of the shoulder region of the plaque. The profiles belonging to the shell and the mixed region were longest in both the direction of the blood flow and the circumferential direction.

The temperature profiles at the centerline, the line of symmetry in the direction of blood flow, of the profiles are given in the left panel of Figure 37.3. These centerlines represent pullbacks along these lines that could have been obtained with a catheter used for intracoronary thermography. The pullbacks obtained when the cap, shell, and mixed regions were assigned to be the heat source were of identical appearance, with the mixed region showing the highest T_{max}, whereas the pullback for the shoulder region differed strongly. In the pullback for the shoulder region, two peaks could clearly be distinguished, representing the proximal and distal shoulder regions. In addition, the centerline pullback of the shoulder region did not reveal the maximal lumen temperature, which is 37.08°C. This maximal temperature was located at the parts of the shoulder region that were parallel to the blood flow direction. From the above observations, we can conclude that the position of the heat-producing macrophages strongly influences the resulting lumen temperature: the closer to

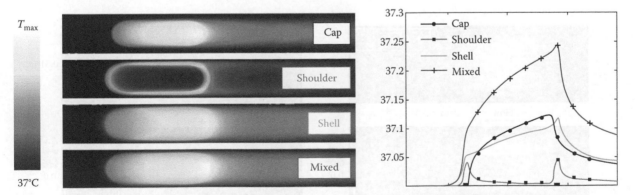

FIGURE 37.3 Temperature distribution at the vessel wall with blood flow (left panel). From top to bottom: a heat source present in the cap, the shoulder, the shell, and the mixed region. The right panel shows the temperature distribution one would have found after performing a pullback along the centerline.

the lumen and the more clustered they are, the higher the resulting temperature.

37.3.3 The Influence the Thermographic Catheter

Introducing a catheter in the lumen of a coronary artery to detect temperature differences will influence the temperatures at the lumen wall, as well as the temperature profile.[15] The amount to which it will affect the temperature depends on the design of the catheter and the material it is made of. The geometry that was created to study the influence of a catheter in the lumen is depicted in Figure 37.4. In Figure 37.5, temperature profiles on the lumen wall are shown when no catheter is present in the lumen (top left and right), when a 1–mm-diameter nitinol catheter contacts the lumen wall (bottom left) and when a 1-mm-diameter polyurethane catheter (bottom left) contacts the lumen wall. The maximal lumen wall temperature is 37.12°C when no catheter lies in the lumen, whereas these values increase to 37.51°C for the polyurethane catheter and to 37.14°C for the nitinol catheter. The maximal temperature lies on the contact surface of the catheter and the lumen for the polyurethane

catheter, but for the nitinol catheter, it lies beside the contact surface. This maximum will thus not be detected using the nitinol catheter. This is due to the conductive properties of polyurethane and nitinol. The conductivity of nitinol is much higher than that of polyurethane. Nitinol thus acts as a conductor and will remove heat from regions of higher temperature to regions of lower temperatures. In this case, the heat is removed from the heat source to the catheter and to the surrounding tissue via the catheter. Polyurethane acts as an insulator, preventing this process. This results in different temperatures on the contact surfaces of the catheters and the location of the maximal lumen wall temperatures. For optimal detection temperature differences, and thus for vulnerable plaque detection, polyurethane, or any other insulator, is a better choice than a conducting material for intracoronary thermography catheters.

37.4 Clinical Applications

Further development of intracoronary thermography could be a very important subject in the field of vulnerable plaque detection. However, great variations exist in the temperature differences that have been reported in clinical studies. This makes it difficult to interpret the results and nearly impossible to validate the method. We have covered the basic principles of intracoronary thermography, splitting it up into three components: heat generation, transfer, and detection. Using the principles of physics, literature study, and numerical simulations, these three aspects were clarified. The clinical studies will be discussed from this perspective.

37.4.1 Heat Generation

The data regarding the heat generation that have been published show great variation. The last *in vivo* studies reported temperature differences that converged to values between 0°C and 0.3°C.[13,18–22] This makes it plausible that these—relatively small—temperature differences are most likely to be expected *in vivo*. These temperature differences are hard to reconcile with the heat production values of macrophages that can be found in the literature.[9] It is therefore necessary to quantify more accurately the heat-generating processes in atherosclerotic plaques. However,

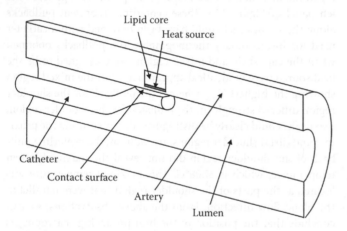

FIGURE 37.4 Simplified geometry of a vessel. In the tissue, a heat source is embedded and a catheter is modeled in lumen.

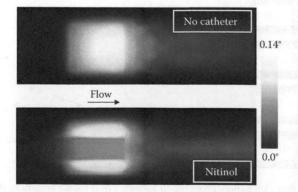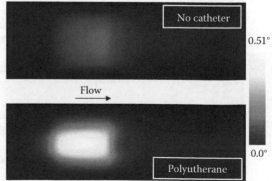

FIGURE 37.5 Influence of the presence of a thermographic catheter and its material on the temperature distribution. Top left and right: no catheter; bottom left: a nitinol catheter; bottom right: a polyurethane catheter. All with flow. Note the different temperature scales.

the experimental procedure to do so is a real challenge. To determine the amount of heat that can be generated by plaques, one could isolate fresh atherosclerotic plaques and study them using calorimetric methods. After this, one could isolate different cells, such as macrophages and foam cells, and/or other substances from the atherosclerotic plaque and study these cells calorimetrically to determine the exact origin of the heat generation. When a relationship between temperature increases in vulnerable plaques and plaque composition has been determined, vulnerability of a plaque could be related to temperature measurements.

37.4.2 Heat Transfer

Reported results from *ex vivo* studies[3,11] on temperature differences should be interpreted with great care. When tissue is removed from a living organism, it will cool down when it is not kept at body temperature. Temperature measurements on this tissue can be interpreted as differences in heat production, but they could represent different cooling rates as well. This has to do with the transfer of heat, which is only possible when temperatures differ. Furthermore, metabolic processes—and thus heat production—could be compromised by removing tissue from the organism.

Since the temperature differences that have been reported most recently are small, it appears mandatory to exclude external thermal influences during temperature measurements, such as those from heart muscle heat production and cooling of blood in lungs, or to compensate for them. Evidence for such mechanisms may be derived from several studies, since flow of blood influences the lumen wall temperature measurements.[18,23,24] Furthermore, the location and organization of heat sources are of influence.[9] For these reasons, it is recommended that intracoronary temperature measurements are combined with both flow measurements and a vessel wall imaging.

37.4.3 Heat Detection

The electrical sensors, thermistor, and thermocouple are currently applied in intracoronary thermography catheters. These sensors are very sensitive to thermal changes of the object, but will also be affected by the surroundings of the object. Therefore, the sensors need to be well insulated to optimize the detection of local hot spots. A thermistor needs a current to achieve accurate readings. However, using a current will heat the thermistor, which will influence the measurement. The catheters that are currently used use electrical sensors. These all have in common that the electrical signal is transported from the sensor to the measuring device using metal wires. Metals are known to be very good conductors; so the flow of heat from a hot spot through these wires may have a significantly lower read-out. This thermal leakage should be minimized. The thermographic measuring devices that are used in clinical studies are discussed. The perspective from the previous sections will be used to discuss the results of these clinical trials.

The heat detection by intracoronary thermographic devices is highly influenced by the material they are made from. The hydrofoil catheter that was used in the studies by Stefanadis et al.[10,25] and Toutouzas et al.[26] is made of a single thermistor embedded in a polyurethane housing. From the results in the previous section it can be concluded that polyurethane is a very good material for vulnerable plaque detection. The spatial sampling of the lumen wall by this catheter type is very low, since only one thermistor is embedded. Although the sensitivity of this device may be high, its practical utility may be somewhat restricted regarding the necessity of repeated pullbacks at different angular positions, due to the presence of only one thermistor in the catheter. The catheters that were used for intracoronary thermography that were made of nitinol are those used by Verheye et al.,[11,23] Diamantopoulos et al.,[24] Krams et al.,[7] and Naghavi et al.[12] From the previous section we know that this material enhances the conductive heat transfer from the heat source to its surroundings, thereby lowering the temperatures on the contact surface.

Other methods to measure temperature exist that may prove to be advantageous regarding this matter. Optical equipment, using fiber technology, might be a more appropriate choice for temperature measurement, since optical fibers are bad heat conductors. Combined with heat-sensitive fluorescent coatings or liquid crystal coatings this might give a whole new direction to the field of intracoronary thermographic measurements. Another method to measure temperatures is using magnetic resonance,[27] but so far its resolution is not sufficiently high. A great advantage is the noninvasiveness of this method.

All catheters that are currently used for intracoronary thermography only sparsely sample the vascular wall. The hydrofoil thermistor design measures the temperature at one location, leaving most of the circumference of the vascular wall unexplored. The thermistor catheter and the thermocouple basket catheter measuring at four and five locations,[11,12] respectively, still leave over 2 mm of the vascular wall in the circumferential direction undetermined. If we assume that macrophage clusters are the heat sources, they may be focal and have sizes smaller than 1 mm. For that reason, temperature increases on the vascular wall may be missed and a higher spatial resolution is needed, although the homogeneity of the temperature measurements in the temperatures reported in the latest publications on this subject counteract this argument.

37.5 Summary

Thermography for coronary plaque detection is still in development. A lot of research has been done, but many questions are still unanswered. Temperature differences exist in coronary arteries *in vivo*, but it is unclear what causes them. Is it really a temperature difference due to inflammation that has been measured? And what is the contribution of noise and artifacts? External thermal influences must be investigated. Influences of catheter design must be studied intensively. Changes in temperature due to source differences and flow must be known. All of these, and probably even more, need to be studied in order to more extensively validate the technique and to be able to decide on the usefulness of the intracoronary thermography for vulnerable plaque detection.

References

1. Falk E, Shah PK, and Fuster V. Coronary plaque disruption. *Circulation* 1995; 92(3): 657–671.

2. Schaar JA, Muller JE, Falk E, et al. Terminology for high-risk and vulnerable coronary artery plaques. Report of a meeting on the vulnerable plaque, June 17 and 18, 2003, Santorini, Greece. *Eur Heart J.* 2004; 25(12): 1077–1082.

3. Casscells W, Hathorn B, David M, et al. Thermal detection of cellular infiltrates in living atherosclerotic plaques: Possible implications for plaque rupture and thrombosis. *Lancet.* 1996; 347(9013): 1447–1451.

4. Loike JD, Silverstein SC, and Sturtevant JM. Application of differential scanning microcalorimetry to the study of cellular processes: Heat production and glucose oxidation of murine macrophages. *Proc. Natl. Acad. Sci. U.S.A.* 1981; 78(10): 5958–5962.

5. Thoren SA, Monti M, and Holma B. Heat conduction microcalorimetry of overall metabolism in rabbit alveolar macrophages in monolayers and in suspensions. *Biochim. Biophys. Acta.* 1990; 1033(3): 305–310.

6. Toutouzas K, Stefanadis C, Tsiamis E, et al. The temperature of atherosclerotic plaques is correlated with matrix metalloproteinases concentration in patients with acute coronary syndromes. *J. Am. College Cardiol..* 2001; 37(2): 356a.

7. Krams R, Verheye S, van Damme LCA, et al. *In vivo* temperature heterogeneity is associated with plaque regions of increased MMP-9 activity. *Eur Heart J.* 2005; 26(20): 2200–2205.

8. Duck FA. *Physical Properties of Tissue. A Comprehensive Reference Book.* London: Academic Press Limited; 26: 2200-5.

9. ten Have AG, Gijsen FJ, Wentzel JJ, et al. Temperature distribution in atherosclerotic coronary arteries: Influence of plaque geometry and flow (a numerical study). *Phys. Med. Biol.* 2004; 49(19): 4447–4462.

10. Stefanadis C, Diamantopoulos L, Vlachopoulos C, et al. Thermal heterogeneity within human atherosclerotic coronary arteries detected in vivo: A new method of detection by application of a special thermography catheter. *Circulation.* 1999; 99(15): 1965–1971.

11. Verheye S, De Meyer GR, Van Langenhove G, et al. *In vivo* temperature heterogeneity of atherosclerotic plaques is determined by plaque composition. *Circulation.* 2002; 105(13): 1596–1601.

12. Naghavi M, Madjid M, Gul K, et al. Thermography basket catheter: *in vivo* measurement of the temperature of atherosclerotic plaques for detection of vulnerable plaques. *Catheter Cardiovasc. Interv.* 2003; 59(1): 52–59.

13. Wainstein M, Costa M, Ribeiro J, et al. Vulnerable plaque detection by temperature heterogeneity measured with a guidewire system: Clinical, intravascular ultrasound and histopathologic correlates. *J. Invasive Cardiol.* 2007; 19(2): 49–54.

14. Kim T and Ley O. Numerical analysis of the cooling effect of blood over inflamed atherosclerotic plaque. *J Biomech. Eng.* 2008; 130(3): 031013.

15. ten Have AG, Draaijers EB, Gijsen FJ, et al. Influence of catheter design on lumen wall temperature distribution in intracoronary thermography. *J. Biomech.* 2007; 40(2): 281–288.

16. ten Have AG, Gijsen FJ, Wentzel JJ, et al. A numerical study on the influence of vulnerable plaque composition on intravascular thermography measurements. *Phys. Med. Biol.* 2006; 51(22): 5875–5887.

17. Lilledahl MB, Larsen EL, and Svaasand LO. An analytic and numerical study of intravascular thermography of vulnerable plaque. *Phys. Med. Biol.* 2007; 52(4): 961–979.

18. Stefanadis C, Toutouzas K, Tsiamis E, et al. Thermal heterogeneity in stable human coronary atherosclerotic plaques is underestimated in vivo: The "cooling effect" of blood flow. *J. Am. Coll. Cardiol.* 2003; 41(3): 403–408.

19. Verheye S, Van Langenhove G, Diamantopoulos L, et al. Temperature heterogeneity is nearly absent in angiographically normal or mild atherosclerotic coronary segments: Interim results from a safety study. *Am. J. Cardiol.* 2002; 90(6A): 24h.

20. Schmermund A, Rodermann J, and Erbel R. Intracoronary thermography. *Herz.* 2003; 28(6): 505–512.

21. Webster M, Stewart J, Ruygrok P, et al. Intracoronary thermography with a multiple thermocouple catheter: Initial human experience. *Am. J. Cardiol.* 2002; 90(6A): 24h.

22. Toutouzas K, Synetos A, Stefanadi E, et al. Correlation between morphologic characteristics and local temperature differences in culprit lesions of patients with symptomatic coronary artery disease. *J. Am. Coll. Cardiol.* 2007; 49(23): 2264–2271.

23. Verheye S, De Meyer GR, Krams R, et al. Intravascular thermography: Immediate functional and morphological vascular findings. *Eur. Heart J.*2004; 25(2): 158–165.

24. Diamantopoulos L, Liu X, De Scheerder I, et al. The effect of reduced blood-flow on the coronary wall temperature. Are significant lesions suitable for intravascular thermography? *Eur. Heart J.* 2003; 24(19): 1788–1795.

25. Stefanadis C and Toutouzas P. *In vivo* local thermography of coronary artery atherosclerotic plaques in humans. *Ann. Intern. Med.* 1998; 129(12): 1079–1080.

26. Toutouzas K, Vaina S, Tsiamis E, et al. Detection of increased temperature of the culprit lesion after recent myocardial infarction: The favorable effect of statins. *Am. Heart J.* 2004; 148(5): 783–788.

27. Wlodarczyk W, Hentschel M, Wust P, et al. Comparison of four magnetic resonance methods for mapping small temperature changes. *Phys. Med. Biol.* 1999; 44(2): 607–624.

Schlieren Imaging: Optical Techniques to Visualize Thermal Interactions with Biological Tissues

Rudolf Verdaasdonk

38.1 Introduction

During surgical procedures, thermal energy is commonly used to create hemostasis by heating the tissue above coagulation temperatures. During the denaturation of proteins and collagen, microvessels shrink and the passage of blood is blocked. The surgeon can resect diseased tissues from the body with minimal loss of blood. Electrosurgery or diathermia is the standard instrument in the operating room (OR) for surgical incisions or resections. Additionally, various instruments are able to heat tissue locally like lasers, RF needles, and ultrasonically vibrating tips. Each instrument has its unique characteristics and optimal settings depending on the application. Either tissue can be evaporated at a small spot without thermal effect in the direct environment or a large volume is coagulated before resection of, for example, a tumor.

It is important to understand the behavior and characteristics of these instruments for a safe and controlled application during the treatment of patients. Especially, the thermal effects in biological tissue need to be studied (McKenzie, 1990). Basically, the thermal effect in tissue is the resultant of the amount of energy (ΔE) deposited in a particular volume during a particular time (Δt) (van Gemert and Welch, 1989)

$$\Delta T = \frac{\Delta E}{\Delta V} \Delta t, \tag{38.1}$$

where ΔV is the volume heated and ΔT is the temperature change.

To study thermal effects in biological tissue, a thermocamera or an array of thermocouples are commonly being used (Torres et al., 1990).

Thermocouples can be put invasively underneath the tissue surface to monitor the temperature in time at one position. They are composed of two thin metal wires of which the ends make contact creating a voltage that is temperature dependent and can be calculated to a calibrated temperature. Thermocouples provide highly accurate temperature readings. Although the thermocouples can be as small as 100 μm, they have to be used invasively. Placing the thermocouples will disturb the environment. There will also be interference with the method by which the thermal energy is delivered in the tissue. In case of electrosurgery or RF energy, currents will be induced in the metal wires of the thermocouple and heating of the wire itself results in an error in the temperature reading. During laser irradiation, the thermocouple can absorb laser light directly when positioned in the field of irradiation. It is possible to correct for these errors by delivering the energy in an intermittent way so that the thermocouples can relax and reflect the actual temperature of the environment. Thermocouples provide a temperature reading at one location only, which is a major limitation of thermocouples for the study of the temperature distribution during tissue heating.

For imaging of temperature fields, infrared (IR) cameras are the most suitable instruments. They have been used for many years and the techniques have been greatly improved as to detectors and cooling techniques. Smart software packages and libraries make it possible to correct for material characteristics and emissivity to obtain calibrated temperatures. However, only temperatures at a surface can be visualized, which is a major limitation of the IR camera when studying thermal effects in biological tissues (Torres et al., 1990). The data at the surface are only marginally useful and can be used to extrapolate to the temperature distribution underneath the surface. For temperature distributions inside the tissue, thermocouples will need to be employed.

To overcome some of the limitations of common temperature measure techniques, this section will describe an optical method based on Schlieren techniques that has been developed and applied by the author for real-time deep-tissue visualization of thermal dynamics in phantom tissues with a high temporal and a spatial resolution in combination with calibrated temperature measurements (Verdaasdonk, 1995; Verdaasdonk et al., 2006). This setup has shown to be useful to obtain a better understanding of medical instruments based on thermal energy and is a great tool to educate medical professionals.

38.2 Schlieren Techniques

Schlieren imaging is an optical method used to visualize density changes in media based on spatial filtering resulting in an enormous contrast enhancement (Schwartz et al., 1956). The method is used in a broad range of applications such as fluid dynamics, ballistics, aerodynamics, and ultrasonic wave analysis image processing (Settles, 1985). Each of these interactions produces a density distribution specific to the respective mechanism. The application discussed in this chapter is primarily aimed at temperature distributions (Agrawal et al., 1998).

The basic setup for the Schlieren method consists of a so-called optical processor and is shown in Figure 38.1.

(Figure 38.1a) From a light source, a parallel beam is created by positioning the source in the focal point of a collimator. This source can either be a laser whose beam is expanded or, in this case, white light emitted from a very small point source. A position between the collimator and imaging lens can be used as the object plane where interaction with an object takes place. The imaging lens will focus the parallel beam in its focal point on the optical axis (when using monochromatic laser light, the Fourier transform of the object, representing the spatial frequency spectrum of the object, is formed in the focal plane).

(Figure 38.1b) The imaging lens will produce an image of the object at the image plane. This plane is located at the position prescribed by the lens formula:

$$\frac{1}{f} = \frac{1}{i} + \frac{1}{o}, \tag{38.2}$$

where f is the focal length, o is the object distance, and i is the image distance.

(Figure 38.1c) Due to variations in the refraction index or irregularities in the medium in the object plane (e.g., temperature gradient), some rays of the parallel beam will be deflected and they

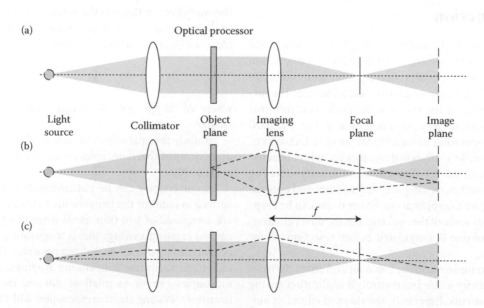

FIGURE 38.1 Setup optical processor.

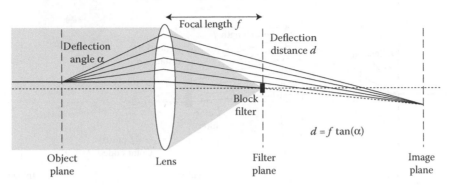

Spatial filtering with block filter

FIGURE 38.2 Schlieren setup with a block filter.

will cross the focal plane at a particular distance d from the optical axis. Nondistorted rays will be focused on the optical axis.

By inserting a mask or a filter in the focal plane of the imaging lens (Figure 38. 2), it is possible to block out the nondeflected rays, preventing them from reaching the image plane, or to earmark rays crossing the plane at certain positions. This process of modifying the object information in the focal or transform plane is known as spatial filtering. The lens in fact performs a Fourier transform of the acquired data and the filter acts as a transfer function. By blocking the rays crossing the optical axis, only refracted and diffracted rays will pass the transform plane (high band pass filter) and reconstruct an image at the image plane (inverse transformation). This results in an enormous contrast enhancement of the image due to the subtraction of the background light.

The result of such filtering is shown in Figure 38.3. Using a block filter, an image is obtained from all the deflected light but it does not contain information on the degree of deflection.

Depending on the deflection angle α and the focal length of the transform lens, a ray will pass the filter plane at a deflection distance d from the optical axis (Figure 38.2):

$$d = f \tan(\alpha). \tag{38.3}$$

The information on the degree of deflection can be preserved by color coding the rays coming through the filter plane using a color filter (Figure 38.4). This filter consists of concentric rings of discrete color bands separated by small black rings. The center of the filter is a black dot blocking the background light. Adjacent to the black dot, going away from the center, the colors shift gradually from blue to red, the "rainbow filter" (Howes, 1984; Greenberg et al., 1995).

Rays passing the filter plane will be color coded depending on the deflection distance d and will be reconstructed to an image at the image plane (Figure 38.5). The generated color image will show, position dependent, the degree of deflection in the object plane. From each color, the deflection angle can be determined, which is related to the variation in the refractive index of the medium in the object plane. The color image can be interpreted as a thermal image when the relation between refractive index and temperature gradient is known. The black rings in the filter will result in black lines in the image separating the discrete colors giving an impression of "isotherms" (Figure 38.6).

The position of the imaging lens, the filter, and the CCD camera are chosen depending on the desired magnification according to the lens formula. The diameter of the filter determines the dynamical range of temperatures that can be visualized. Using an x–y microtranslator, the filter can be optimally aligned on the optical axis. The CCD camera is positioned in the image plane.

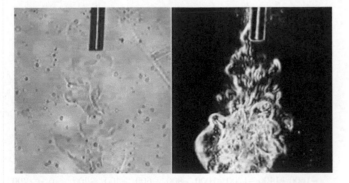

FIGURE 38.3 Effect of spatial filtering using a block filter. Left: normal view of turbulent hot water in front of a fiber after exposure to a holmium laser pulse. Right: after subtraction of background light the heated environment becomes clearly visible. (Adapted from Verdaasdonk, R. M., et al. *J. Biomed. Opt.* 11(4): 041110, 2006.)

FIGURE 38.4 Black and white representation of the "rainbow" color filter. (Adapted from Verdaasdonk, R. M., et al. *J. Biomed. Opt.* 11(4): 041110, 2006.)

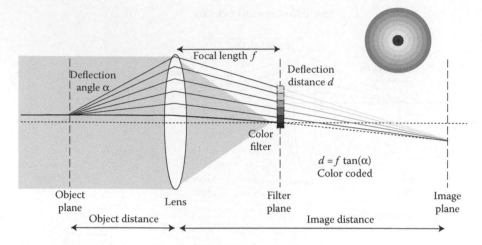

FIGURE 38.5 Scheme of spatial filtering using a color filter. (Adapted from Verdaasdonk, R. M., et al. *J. Biomed. Opt.* 11(4): 041110, 2006.)

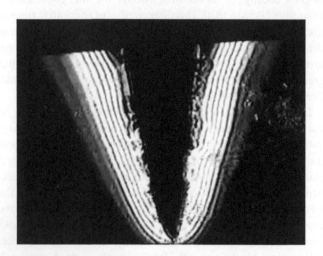

FIGURE 38.6 Example of a color-coded image showing the thermal zone around an ablation crater during laser irradiation. The black lines can be interpreted as isotherms. (Adapted from Verdaasdonk, R. M., et al. *J. Biomed. Opt.* 11(4): 041110, 2006.)

38.3 Deflection of Rays by a Medium

Rays can be deflected in a transparent medium due to variation in the refraction index. This variation results from the following:

a. *Inhomogeneity of the medium*: The medium might be a mixture of more compounds with a different density or a liquid with a solvent that is not uniformly distributed.
b. Local stresses can be induced by shockwaves traveling through the medium changing the local density. In transparent solids, external forces will induce local density variation in the structure of the material.
c. Media, such as gases and liquids, usually expand gradually due to a temperature rise so that the density decreases and hence the refractive index (Settles, 1981, 1985). In biological tissues, the basic medium is water, which will act as the optically active medium.

The deflection of rays due to the variation in the refractive index of water is used to visualize temperature gradients in tissues. Figure 38.7 shows the relation between temperature and the refractive index of water (Schiebener et al., 1990). The refractive index decreases with increasing water temperature although the degree is minute.

When a ray of light passes at a shallow angle through water above a hot surface, it will be deflected due to diffraction in the layers of water with varying refractive index due to the temperature gradient as depicted in Figure 38.8a and b.

The gradient in the water can be considered as infinite thin layers with increasing density (=refractive index). The light ray will be refracted on going from one layer to the other according to Snell's law of refraction given as

$$n_1 \cdot \sin(\theta_1) = n_2 \cdot \sin(\theta_2), \tag{38.4}$$

where n_1 is the refractive index of layer 1 and n_2 of layer 2, θ_1; is the entry angle and θ_2 the exit angle.

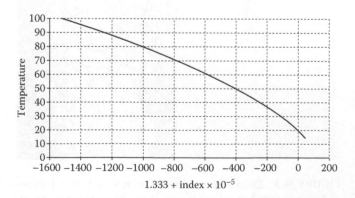

FIGURE 38.7 Refractive index of water in relation to temperature. (Adapted from Verdaasdonk, R. M., et al. *J. Biomed. Opt.* 11(4): 041110, 2006.)

(a)   (b)

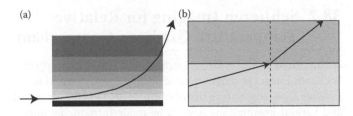

FIGURE 38.8 (a) Diffraction path of the ray through layers with increasing reflective index above a hot surface. (b) Close-up of diffraction of the ray at the border between infinite thin layers. (Adapted from Verdaasdonk, R. M., et al. *J. Biomed. Opt.* 11(4): 041110, 2006.)

38.4 Interpretation of the Color Images as Temperature Images

The deflection of the rays is due to the presence of a temperature gradient. For a larger local gradient $\Delta T/dx$ perpendicular to the direction of the parallel light beam, the deflection angle will be larger. The steep temperature gradient is usually present when the heat source is turned on. The temperature will usually drop exponentially with increasing distance from the source. In this phase, the color Schlieren technique will show colorful images. As soon as the heat source is turned off, the temperature gradient becomes less steep and the colors in the image fade and merge representing the low angle deflection range of the color filter (Figure 38.9). The series of bars on the left-hand side show the buildup of the temperature gradient at steps of 0.4 s after onset. The right-hand bars show the collapse of the gradient after the source has been shut off also at steps of 0.4 s.

When there is no temperature gradient, the image will show a uniform color despite the temperature of the medium being relatively higher than the environment.

Since the deflection of the ray is a resultant of a series of stepwise deflections traveling through the medium, it is not possible to ascribe a temperature to a particular color. The ray has traveled (in the z-direction) through various temperature gradients

depending on position z. So the total deflection will be the summation of

$$\text{Angle}\,\alpha = \Sigma\Delta\alpha\left(z, \text{grad}\,\frac{dT}{dx}\right). \qquad (38.5)$$

By approximation, one could state that the largest deflection results from position z, where the gradient was largest. With the knowledge of the symmetry of the temperature field one could make an attempt to ascribe temperatures to the colors; however, the error can be large. If there is knowledge of the temperature field, in case of a relatively simple symmetry like a uniform hot plate, hot point, or cylindrical source, ray-tracing could be used to calculate the path of the ray through a static temperature field (distribution) (Settles, 1985; Greenberg et al., 1995). However, usually, the temperature field is dynamic and finite element methods have to be introduced additionally to attempt a reliable calculation. This exercise will go beyond the goal of this method of visualization. The Schlieren imaging method for this application is intended to provide an intuitive imaging technique for a relative comparison between various conditions.

As an example, the "calibration" results will be shown for some symmetric and static temperature distributions. The following parameters will be of influence: (1) the angle of deflection, (2) the focal length of imaging lens, (3) the object distance, (4) the image distance, and (5) the symmetry of the refractive index distribution.

The most simple temperature distributions are unidirectional or axial symmetric.

38.5 Unidirectional Temperature Gradient

Above a heated flat surface (Figure 38.10) the temperature distribution can be considered as a unidirectional or a 1-D situation. An example of the calibration curve for a unidirectional temperature distribution is given in Figure 38.11.

38.6 Axially Symmetric Temperature Gradient

When the heat source is a "hot spot" in a cylindrical or spherical geometry, it can be considered as a two-dimensional situation with an axial symmetry for the temperature gradient (Figure 38.12).

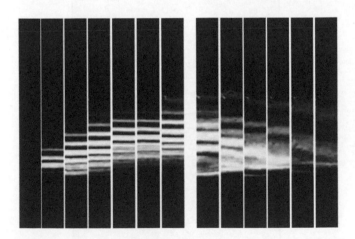

FIGURE 38.9 Growth (left columns) and decline (right columns) of a temperature gradient above a surface heated for several seconds. Each column represents a step of 0.4 s.

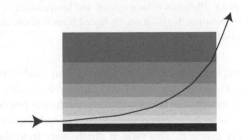

FIGURE 38.10 Deflection of the ray in a medium with a temperature gradient above a hot surface. (Adapted from Verdaasdonk, R. M., et al. *J. Biomed. Opt.* 11(4): 041110, 2006.)

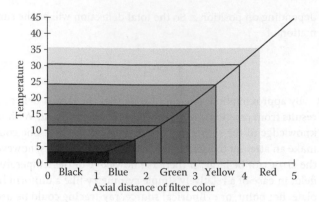

FIGURE 38.11 Relation between color and temperature in the color image of temperature distribution. (Adapted from Verdaasdonk, R. M., et al. *J. Biomed. Opt.* 11(4): 041110, 2006.)

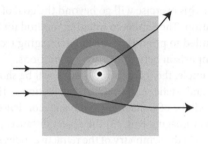

FIGURE 38.12 Deflection of rays in a medium with a radial temperature gradient. (Adapted from Verdaasdonk, R. M., et al. *J. Biomed. Opt.* 11(4): 041110, 2006.)

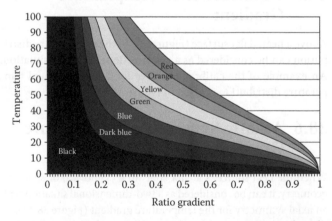

FIGURE 38.13 Relation between color and temperature in the color image of temperature distribution. (Adapted from Verdaasdonk, R. M., et al. *J. Biomed. Opt.* 11(4): 041110, 2006.)

An example of the calibration curve for an axially symmetric temperature distribution is given in Figure 38.13. This graph is tricky to interpret. The temperature at a particular position in an image has to be determined from the color and the ratio of the temperature gradient. This ratio is the distance from the axis of symmetry (usually the highest temperature) divided by the distance over which the temperature gradient extends [from the axis of symmetry to ambient temperature (black)].

38.7 Schlieren Imaging for Relative Temperature Gradient Comparison

As discussed in the previous paragraph, a calibration of color Schlieren images to derive reliable absolute temperatures is not possible. The temperature distribution needs to be symmetrical and various assumptions have to be made introducing uncertainties and errors. Additionally, major calculations have to be performed to derive the results.

The Schlieren imaging method is however perfect for relative comparison in a situation where the (unknown) geometry stays the same but parameters of the heat source are varied, for example, comparing laser wavelength A to wavelength B or laser setting A to setting B.

38.8 Simulation of Interaction of Thermal Energy Applications in Tissue

The Schlieren technique cannot be applied during the "real" laser-tissue interaction in (living) biological tissues. To visualize the thermal effects, the tissue has to be transparent. Therefore, a model is generally used to qualify the mechanism of action. A transparent polyacrylamide gel is applied, which is assumed to have similar thermal properties as biological tissues. To some extent, it also simulates the mechanical properties of soft tissues.

For electrosurgery and RF applications, the gel is made from saline to mimic the electrical conductivity properties. For laser, depending on the wavelength, a dye can be added to mimic the effective absorption properties.

Schematic of testing conditions in the object plane is depicted in Figure 38.14.

The gel (medium 2) can be in an air or a water environment (medium 1). To simulate the *in vivo* situation as close as possible, a slab of the original tissue can be put on top of the gel. The

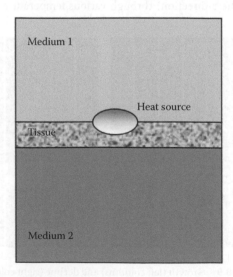

FIGURE 38.14 Schematic of testing conditions for interactions of thermal energy with artificial and real tissues.

original tissue effect will take place in the slab and heat will diffuse or be conducted to the gel giving an indirect impression of the thermal effects taking place inside the tissue.

In the tissue or gel, the heat transfer can only take place by heat conduction. In case of a liquid environment, convection is also involved, which is a very effective way of transferring heat. An air environment, on the other hand, will act as an isolator.

38.9 Color Schlieren Setup for Laser Research

To study the thermal interaction of continuous wave (CW) and pulsed laser systems with biological tissues, a rainbow Schlieren setup was used as illustrated in Figure 38.15.

A white light source is coupled into a ball-shaped distal tip fiber. The light emitted from the fiber is focused due to the spherical fiber end and subsequently divergences. The focal point of the collimating lens coincides with the focus of the fiber. The diameter of the lens is matched with the divergence of the beam so that all the light is collimated. A rectangular tank filled with

water is positioned in the object plane. The walls are perpendicular to the parallel beam to prevent any optical distortion due to refraction. A fiber with or without a probe is submerged in the water along with tissue or model tissue within the field of view (Figure 38.16). Additional filters can be used to filter out the scattered light from the primary laser wavelength.

38.10 High-Speed Schlieren Imaging

The Schlieren setup can also be used in combination with high-speed imaging depending on the camera type. By using a video camera with a high-speed shutter, it is possible to obtain a time resolution down to 100 μs, however, with a relatively slow frame rate. Specialized high-speed cameras are able to shoot many thousands of frames per second. However, the light from a cw white light source might not be sufficient. Instead, a flashed white light source can be used, which is synchronized with the video camera. A flash light can also be used to "freeze" images with illumination times in the microsecond range (van Leeuwen et al., 1991) or even nanosecond range (Vogel et al., 2006). The

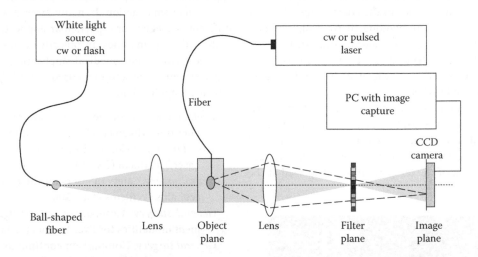

FIGURE 38.15 Rainbow Schlieren setup to study laser–tissue interaction.

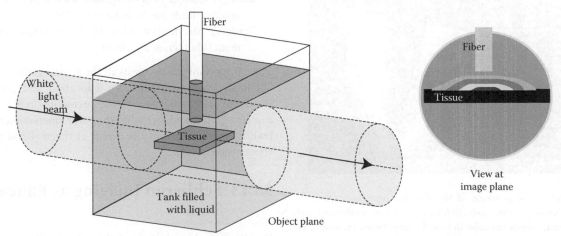

FIGURE 38.16 Close-up of the object plane where a collimated beam passes along the object of investigation with a schematic of the resulting image in the image plane.

millisecond time range, however, is far sufficient to capture thermal phenomenon using the color-coded technique.

38.11 Color Schlieren Imaging Combined with Absolute Temperature Measurements

The color imaging provides a real-time false color image of temperature gradients present in the transparent medium, which gives an impression of a three-dimensional (3-D) thermal distribution. However, assigning absolute temperatures to the colors in the images is not possible as discussed in the previous paragraphs. The images, however, are very useful for relative comparison of settings and the dynamics of the temperature distribution. To complement the color Schlieren images, absolute temperature measurements have been incorporated in the setup. Up to six miniature thermocouples can be positioned in the field of view. The readings are acquired with a PC and converted to absolute temperatures. Using specialized video mixing software, the positions and temperatures are added to the video frame in real time. The numbered dots indicate the position of thermocouples in the color image while the actual temperature of the corresponding thermocouple is presented at the bottom of the image. By choosing strategic positions for the thermocouples in the temperature field, the colors in the Schlieren image can be considered "calibrated" in real time; however, the geometry has to be taken into account.

Figure 38.17 shows a representative video frame of an experiment mimicking the cutting of soft tissue using a 2 μm cw laser. An array of 5 × 100 μm diameter thermocouples (numbered dots) is positioned alongside the position where the "tissue" will be cut. The readout of the temperatures is displayed at the bottom of the frame.

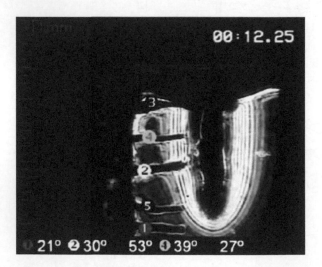

FIGURE 38.17 Video frame of the thermal imaging experiment: (upper right corner) a time code, (left rim) ruler with millimeter scale, and (lower rim) numbered color dots with temperature reading corresponding to the numbered thermocouples (thin black lines) in the image. (Adapted from Verdaasdonk, R. M., et al. *J. Biomed. Opt.* 11(4): 041110, 2006.)

In Figure 38.18, a series of frames is presented during the formation of an ablation crater in the gel over a time period of 25 s. It shows the ablation of the tissue (gel) with the dynamics of the thermal effects along the canal wall. After 12 s the laser exposure stops and the canal wall cools off. The temperatures of the various thermocouples are shown at the bottom of each frame. The complete temperature registration of the five thermocouples is presented in Figure 38.19. The curves show some interference noise during the acquisition that can happen easily with thin thermocouples that are difficult to shield from external (electromagnetic) interference.

38.12 Applications of "Thermal Imaging" Using the Color Schlieren Technique

Since we developed the color Schlieren imaging technique, we have used it for many studies of laser–tissue interaction with various lasers. After the introduction of a new medical laser application, the setup was used to get a better understanding of the mechanism of action. In many cases, it was used to compare different strategies for a particular application. In cooperation with medical companies an objective comparison of methods was performed. For medical applications in the University Medical center, the setup was used to find the safe and optimal setting for application.

Urology: comparison Benign Prostate Hypertrophy (BPH) fiber tips (Molenaar et al., 1994; van Swol et al., 1994)

Vascular surgery: Comparison various laser for EndoVenous Laser Treatment (EVLT) (Disselhoff et al., 2008)

Neuro surgery: Safety of lasers for the treatment of hydrocephalus (Willems et al., 2001)

General surgery: Comparison interstitial thermal treatment modalities for liver tumors (de Jager et al., 2005)

General surgery: Comparison continuous 2 μm laser with other surgical lasers (de Boorder et al., 2007)

Urology: Comparison treatment modalities for partial nephrectomie (de Boorder et al., 2008)

Dentistry: Mechanism of Erbium laser for root canal treatment (Blanken et al., 2009)

An important advantage of all the studies is the video footage obtained, which is perfect for education and presentations. Instead of explaining laser–tissue interaction to physicians with complex graphs and formulas, the color video sequences show a realistic presentation of the extent of thermal and mechanical effect inside tissue.

38.13 Schlieren Imaging as Educational Tool

Despite several limitations of the color Schlieren technique to study thermal interactions with tissues as discussed above, the most important advantage is its value for education (Verdaasdonk

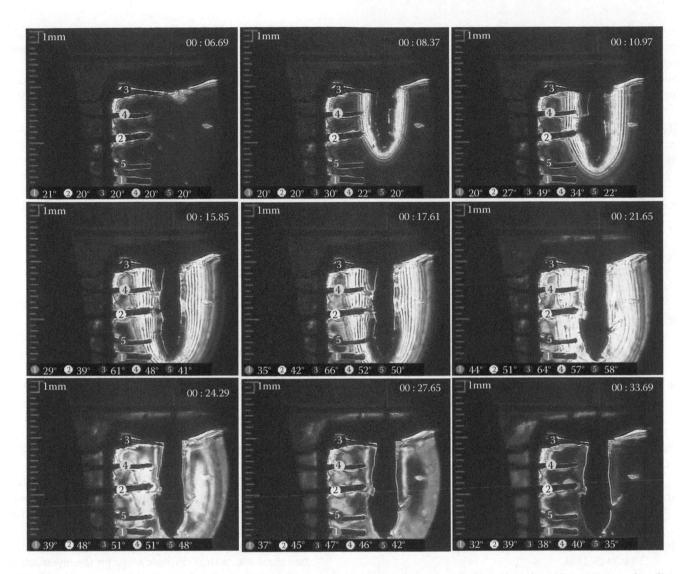

FIGURE 38.18 Series of frames showing the dynamics of the temperature distribution during the formation of an ablation crater in the gel over a time period of 25 s.

et al., 2006). Obtained movie clips of the dynamics during thermal interactions with tissues can be used to explain in real-time what the 3-D temperature evolution is without the need for extensive background information. Although, the tissue interactions

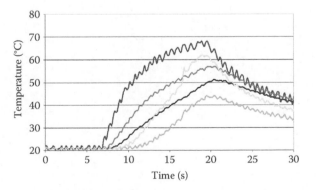

FIGURE 38.19 Temperature registration of the five thermocouples in the experiment depicted in Figure 38.18.

cannot be simulated in detail, the essence can be mimicked to show a good representation of reality. The imaging has proven to be very effective for education and training of students, scientists, and physicians. Even professionals in this area of research have shown their appreciation for this imaging technique as it shows the basic essentials of laser–tissue interactions.

38.14 Summary

This chapter describes an imaging method that visualizes temperature gradients in transparent phantoms tissue using color Schlieren techniques. This method enables the study of the dynamics of thermal interactions within (model) tissues and provides an instrument to perform a relative comparison for various treatment modalities. A calibration of the color images to absolute temperatures is not possible or practical. Therefore, additional thermocouple measurements can be incorporated in the imaging technique to obtain calibrated temperatures. This imaging method has been used in many research projects to investigate the mechanism of new medical instruments for providing safe and optimal settings for application in the clinic. The color Schlieren technique has proven to be a great education tool for researchers, surgeons, nurses, and students.

References

Agrawal, A. K., et al. Three-dimensional rainbow Schlieren tomography of a temperature field in gas flows. *Appl. Opt.* 37.3: 479–485, 1998.

Blanken, J., et al. Laser induced explosive vapor and cavitation resulting in effective irrigation of the root canal. Part 1: A visualization study. *Lasers Surg. Med.* 41: 514–519, 2009.

de Boorder, T., et al. Use of the 2-mu m Cw laser as addition and/or alternative for the Nd:YAG in urology. In: *Photonic Therapeutics and Diagnostics III*, Kollias, N. et al., (eds), SPIE, Vol. 6424, 2007.

de Boorder, T., et al. Comparing different treatment modalities for partial nephrectomies without ischemic period: Laser, hydro-jet and RF. In: *Photonic Therapeutics and Diagnostics IV*, Kollias, N., et al. (eds), SPIE, Vol. 6842, 2008.

de Jager, A. A., et al. Comparison of laser- and RF-based interstitial coagulation systems for the treatment of liver tumors (Invited Paper). In: *Thermal Treatment of Tissue: Energy Delivery and Assessment III*, Ryan, T., (eds), SPIE, Vol. 5698, 2005.

Disselhoff, B. C., et al. Endovenous laser ablation: an experimental study on the mechanism of action. *Phlebology* 23.2: 69–76, 2008.

Greenberg, P. S., Klimek, R. B., and Buchele, D. R. Quantitative rainbow Schlieren deflectometry. *Appl. Opt.* 34.19: 3810–3825, 1995.

Howes, W. L. Rainbow Schlieren and its applications. *Appl. Opt.* 23.14: 2449, 1984.

McKenzie, A. L. Physics of thermal processes in laser-tissue interaction. *Phys. Med. Biol.* 35.9: 1175–1209, 1990.

Molenaar, D. G., et al. Evaluation of laser prostatectomy devices by thermal imaging. In: *Medical Applications of Lasers II*, Bown, SG., et al. (eds), SPIE, Vol. 2327, 1994.

Schiebener, P., et al. Refractive index of water and steam as function of wavelength, temperature and density. *J. Phys. Chem. Ref. Data* 19.3: 677–717, 1990.

Schwartz, D. S., et al. Schlieren and shadow-graph techniques; some possible applications to the medical and biological sciences. *Med. Biol. Illus.* 6.1: 23–33, 1956.

Settles, G. S. Color Schlieren optics—a review of techniques and applications. *International Symposium on Flow Visualization*, Bochum, West Germany, September 9–12, 1980. West Germany, Ruhr-Universitaet Bochum, pp. 187–197, 1981.

Settles, G. S. Colour-coding Schlieren techniques for the optical study of heat and fluid flow. *Int. J. Heat Fluid Flow* 6.1: 3–15, 1985.

Torres, J. H., et al. Limitations of a thermal camera in measuring surface temperature of laser-irradiated tissues. *Lasers Surg. Med.* 10.6: 510–523, 1990.

van Gemert, M. J. and A. J. Welch. Time constants in thermal laser medicine. *Lasers Surg. Med.* 9.4: 405–421, 1989.

van Leeuwen, T. G., et al. Noncontact tissue ablation by holmium: YSGG laser pulses in blood. *Lasers Surg. Med.* 11.1: 26–34, 1991.

van Swol, C. F. P., et al. Physical evaluation of laser prostatectomy devices. In: *Lasers in Urology*, Watson, GM., et al. (eds), SPIE, Vol. 2129, 25–33, 1994.

Verdaasdonk, R. M. Imaging laser-induced thermal fields and effects. In: *Laser-Tissue Interaction VI*, Jacques, SL., (eds), SPIE, Vol. 2391, 165–175, 1995.

Verdaasdonk, R. M., et al. Imaging techniques for research and education of thermal and mechanical interactions of lasers with biological and model tissues. *J. Biomed. Opt.* 11.4: 041110, 2006.

Vogel, A., et al. Sensitive high-resolution white-light Schlieren technique with a large dynamic range for the investigation of ablation dynamics. *Opt. Lett.* 31.12: 1812–1814, 2006.

Willems, P. W., et al. Contact laser-assisted neuroendoscopy can be performed safely by using pretreated "black" fibre tips: Experimental data. *Lasers Surg. Med.* 28.4: 324–329, 2001.

39

Helium Ion Microscopy

John A. Notte and
Daniel S. Pickard

39.1 Introduction

The helium ion microscope (HIM) is a newly developed technology which offers high-resolution imaging with unique contrast mechanisms and a depth of field superior to competitive technologies. The range of applications for the HIM is presently being explored, but biological samples present unique challenges that can be addressed with this new microscope.

The HIM (Figure 39.1) is closely related to the scanning electron microscope (SEM) and the traditional gallium focused ion beam (Ga-FIB).[1,2] All these technologies use a well-focused charged particle beam to interrogate a single location on a specimen and measure the excitations that result—such as the ejection of particles or photons. The location of impingement is moved from point-to-point on the specimen in a raster pattern (Figure 39.2). The corresponding points in the image are assigned a gray level, which indicates the abundance of the detected excitations. The contrast in the image is caused by some interesting variation in the production of these excitations. As with all microscopies, there is an implied assumption that the pixel-to-pixel contrast variations are well-correlated characteristics that are of interest to the microscopist. In comparison with the SEM, the HIM offers higher resolution because of its smaller probe size (~0.35 nm), as well as a more favorable sample interaction. In comparison with the Ga-FIB, the helium ion beam produces

relatively little damage to the sample; thus, high-magnification images do not destroy the specimen.

39.2 Technology of the HIM

The development of the HIM was pursued over the last four decades with significant progress achieved by research groups around the globe. These groups included the Max Plank Institute,[3] the University of Lyon,[4] the IBM Corporation,[5] the Cornell University,[6] the JEOL Corporation,[7] Taiwan's Academia Sinica,[8] the University of Chicago,[9] the Oregon Graduate Center,[10] and the ALIS Corporation,[11] At present, the only HIM that is commercially available is the ORION™ product offered by the Carl Zeiss Corporation. To the extent possible, this chapter will describe the generic version of the HIM, although the example images will necessarily be produced from the singularly existing product.

The critical technology that enables the HIM is the helium ion source. It relies upon the creation of helium ions within a volume which is just a few angstroms in size. The concept of an atom-sized source of ions was first demonstrated in 1955 by Erwin Müller and his student researchers at the Pennsylvania State University.[12] In the subsequent 50 years, the helium ion source was optimized to produce ions at a faster and steadier rate, and with higher source longevity. It was only when the

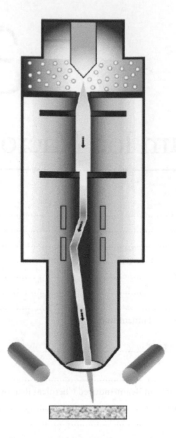

FIGURE 39.1 A diagram of the HIM showing the ion source, the ion optical column, the sample, and the detectors.

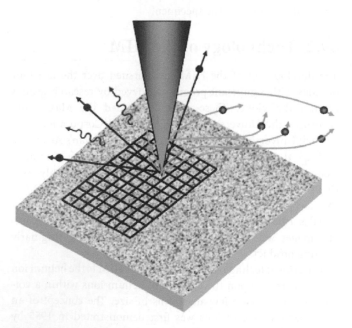

FIGURE 39.2 As the helium ion beam strikes the sample, numerous particles and excitations are produced. The detection of these on a pixel-by-pixel basis is the principle behind image formation. Graphically overlaid on the sample are the regions corresponding to the pixels in the final image.

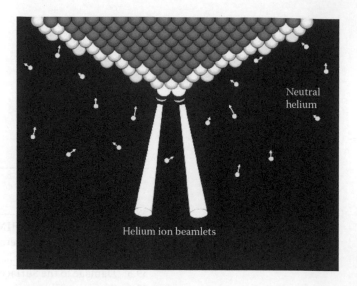

FIGURE 39.3 The helium ion source showing the atoms comprising the apex of the needle. (For simplicity of this diagram, only two terminal atoms are shown.) The most protruding atoms have sufficiently large electric fields so that the neutral helium atoms become ionized to produce three gradually diverging ion beamlets.

helium ion source became easily operable (an achievement of the ALIS Corporation in 2005) that the helium ion source became commercially viable.

The helium ion source consists of a pointed metal needle with a strictly controlled geometry at the apex. Indeed, the apex of the needle tapers gradually until it terminates in exactly three atoms (Figure 39.3). The needle is biased positively by a voltage of about +15 kV, thereby producing a large electric field in the vicinity of the apex. Immediately adjacent to the three terminal atoms, the electric field can be as large as 40 V/nm. At this enormous field strength most atomic bonds are broken, and even electrons can be extracted from otherwise neutral atoms via quantum mechanical tunneling.[13]

It is in the presence of this applied voltage that a small quantity of helium gas is admitted into the source region, which is maintained at cryogenic temperatures. The helium atoms are neutral, but they are polarized in the presence of the electric field and are drawn to the apex of the needle. When a helium atom approaches one of the three terminal atoms, one of its electrons can tunnel out, leaving the positively charged helium ion to be accelerated away from the positively biased apex. As the helium atoms become ionized, each of them is accelerated away from the needle, forming three beamlets of singly ionized helium atoms. One beamlet is produced for each of the three terminal atoms and a simple phosphor can be employed to observe the emission pattern (Figure 39.4). The production rate per atom is commonly as large as 10^8 s^{-1}—corresponding to a beamlet current on the order of 100 pA. And since all the ions are created at a potential of 15 kV (in this example), their ultimate energy when they strike a grounded specimen is 15 keV.

For each beamlet, all the ions are produced in the same region (about 0.3 nm wide) and follow the same trajectory with a very

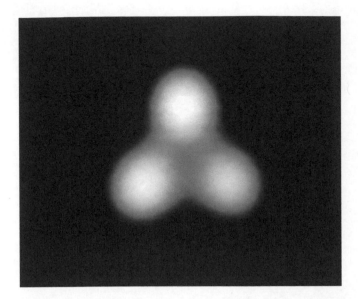

FIGURE 39.4 The emission pattern from the helium ion source. Each of the bright spots corresponds to a single atom on the apex of the needle.

narrow angular divergence (about 0.5°). For the helium ion source, this brightness (current per area per solid angle) is significantly higher than other conventional electron and ion sources. Further, the ions are nearly monochromatic—having an energy spread of less than 1 eV, while having an average energy of 15 keV in this example. It is well recognized within the discipline of charged particle microscopy that the combination of high brightness and narrow energy spread is critical to forming a beam that can be focused to a small probe size.[14] It is the unique properties of the helium ion beam (high brightness and low energy spread) that makes it possible to produce a probe size as small as 0.35 nm.

Of the three diverging beamlets, only one of them is selected to pass down the ion-optical column (as shown in Figure 39.1). The selected beamlet is then passed through an electrostatic lens that controls the beamlet's divergence. Once collimated, the ion

beam is shaped by passing it through an aperture. Thereafter, the beam is doubly deflected and passed through a second electrostatic lens. The second electrostatic lens serves to bring the beamlet to a finely focused probe—nominally at the surface of the specimen. Compared to the SEM, the helium probe size is not strongly limited by diffraction—an artifact that limits the ultimate probe size of the SEM. Without significantly suffering from diffraction, the helium beam can be focused to a probe size as small as 0.35 nm. Although not suitable for treatment in this text, this small probe size is a direct result of the high brightness and the narrow energy spread of the helium ion source.

39.3 Beam Specimen Interaction

Since the beam is focused and strikes the specimen, it interacts with it in a manner that is quite different from a SEM or a Ga-FIB. To better understand and visualize the differences, Monte Carlo computer models[15,16] are used to simulate individual trajectories of ions and electrons. The computer models emulate the well-established physics that concerns how the charged particles interact with the electrons and nuclei that comprise the specimen. By performing many hundreds of these simulations, it is possible to gain a new level of understanding of the typical ion's trajectory and the manner in which it interacts with the specimen.

In the Ga-FIB, the massive gallium atoms (mass = 69 amu) interact strongly with the electrons within the specimen, but owing to their much larger mass, the gallium ions continue on their trajectory with only minor deflections. The gallium atoms will also strike the nuclei within the specimen, and here they can impart significant energy and are deflected to larger angles. The sample atoms are commonly given enough energy to continue with their own cascade of collisions. Thereby, the volume of excitation produced by a gallium beam can be relatively broad (middle part of Figure 39.5). In a SEM, as the electrons (mass = 5.5×10^{-4} amu) strike the specimen, they interact strongly with the other electrons comprising the specimen. In these collisions, the energy loss and angular deflections can be

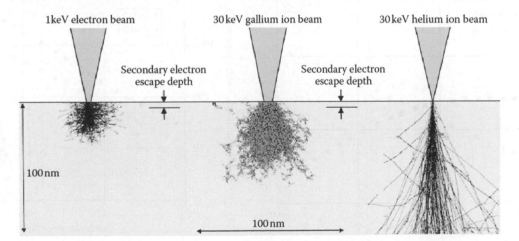

FIGURE 39.5 The sample interaction volumes for incident beams: SEM, Ga-FIB, and HIM. In these computer simulations, red and blue mark the trajectory of the incident ions and electrons, and green shows the location of the displaced atoms of the specimen.

significant since the two participants are of equal mass. Therefore, the electron beam quickly diverges from its original trajectory once it enters the specimen. This is shown in the computer simulations present in the left part of Figure 39.5.

In contrast with these unfavorable extremes, the intermediate mass of the helium atom (mass = 4 amu) gives it a unique advantage. The helium ion interacts strongly with electrons in the specimen, but these collisions induce only small angular deflections and energy losses; hence, the helium atom continues forward on its trajectory (right part of Figure 39.5). The helium ion has a low probability of scattering from a nucleus, and when it does, it transfers relatively little energy to it, since virtually any other atom is much more massive. Indeed, the relatively low mass of the helium ion causes it to occasionally scatter back out of the specimen. Excepting these cases of backscattering, the helium beam remains comparatively collimated as it penetrates into the specimen. Overall, the helium ion interacts with the specimen in a uniquely different manner, and therefore it can provide a means of imaging and analysis which is distinctly different from the SEM or the Ga-FIB.

39.4 Beam Penetration

The helium ion penetrates into the sample for some depth until its interactions with the electrons and nuclei deprive it of its kinetic energy and it comes to rest. The rate at which energy is lost to the specimen is termed stopping power. The stopping power varies with the energy of the ion as well as with the composition of the sample. At higher energies, the stopping power is primarily electronic (Figure 39.6) and therefore the scattering is primarily from electrons. But as the energy is gradually reduced, the scattering becomes increasingly nuclear and will therefore result in more large angle collisions. The overall shape of the trajectory (Figure 39.7) can be seen to be initially quite smooth, which terminates in an abrupt jumble.

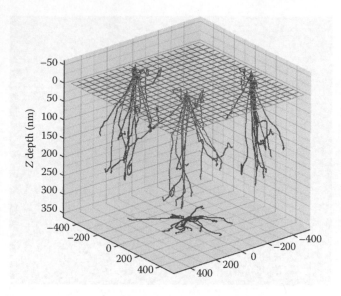

FIGURE 39.7 The trajectory of 25 representative helium ions (in red) shows the general shape of their trajectories. The "shadow" projections on the *XZ*, *YZ*, and *XZ* planes are shown in dark gray.

The penetration depth of the helium ion depends greatly upon the initial energy of the helium ion and the composition of the specimen. The ORION instrument can produce beam energies from 10 to 35 keV and this will correspond to penetration depths from 200 nm to 1.5 μm in typical materials (Figure 39.8).

39.5 Damage to the Surface and Subsurface

Helium ions can directly damage a specimen, as can any charged particle beam. However, compared with the gallium ions, the helium ions produce relatively low damage rates. For this reason,

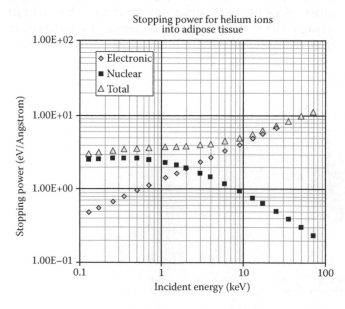

FIGURE 39.6 The stopping power of helium ions is shown to be primarily due to electronic scattering for energies above 2 keV.

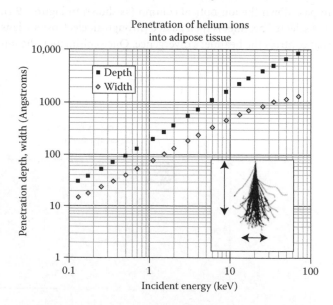

FIGURE 39.8 The average penetration depth and width is shown to vary nearly linearly with the initial beam energy.

helium ions are not a suitable substitute for the milling applications, which is normally fulfilled with a gallium beam. Among the various forms of specimen damage, the helium ions can be implanted into the specimen. For many materials, the helium atoms (usually neutralized once they enter the specimen) diffuse out of the specimen at the same rate in which they enter. The rate difference between helium implantation and helium escapement varies from one material to another, and varies strongly with specimen temperature. In some crystalline materials (silicon is one example), the helium escapement at room temperature is so slow that subsurface helium bubbles can be formed.[17] For most biological specimens, the helium atoms can escape before they can accumulate.

During nuclear scattering events, the helium atoms can transfer momentum to the atoms of the specimen, causing them to be displaced from their initial locations. The displacement is typically a matter of nanometers. For a single ion, the rate of displacement is quite small—perhaps 100 displacements occur along its trajectory. Since the typical sample interaction volume contains 10^9 atoms, this represents a displacement rate of 1 part in 10 million per incident ion. When this displacement occurs near the specimen surface and causes atoms to be ejected, it is termed sputtering. The rate of sputtering is small—about 1 sputtered atom from the specimen for every 100 incident helium ions.

39.6 Specimen Charging

From the ion source to the specimen, the helium atoms are in a singly charged state. However, as the ions penetrate into the sample, they extract an available electron from the specimen and continue along their trajectory. The large ionization energy of helium (25 eV) means that they are quite effective in stripping electrons from other atoms. Throughout the remainder of the trajectory, the helium atom/ion will re-lose its electron, and regain another one repeatedly for an indefinite number of times.[18] Thus, there is no net charge transport except at the surface which receives a net positive charge. This positive surface charging is increased by the ejection of 2–6 electrons scattering from each incident helium ion. Compared to a SEM, which can induce positive or negative specimen charging, the HIM produces only positive charging, and hence this effect can be mitigated through the use of an electron flood gun. The mitigation of charging artifacts is discussed in a later section.

39.7 Detectable Particles

A broad range of particles is generated as the helium ions penetrate into the specimen. The generation of these particles and their ultimate detection is critical to the image formation of the HIM. The information conveyed in an image is expressed solely through some characteristics of these particles such as their abundance, angle of emission, energy, polarization, charge state, and so on. It has even been demonstrated that multiple detectors can simultaneously be used to generate two independent images from a single raster scan (Figure 39.9). These different images

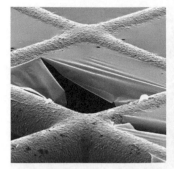

FIGURE 39.9 Images can be generated simultaneously from different detectors. The left image was generated from SEs and exhibits primarily topographic information. The right image was generated from the backscattered helium and exhibits primarily material information. The specimen consists of a copper grid with an overlying carbon film just 10 nm thick.

can provide different contrast mechanisms such as topographic contrast and material contrast.

39.7.1 Secondary Electrons

Of all the generated particles, secondary electrons (SEs) are the most abundant. These are electrons that are ejected from the specimen due to collisions initiated by the incident ion. They typically have an energy of less than 3 eV. A given material will have an SE yield (average electrons ejected per incident ion), which is characteristic of that sample. The SE yield ranges from a low of about 2 (for carbon) to a high of about 7 (for platinum), although the variation of the SE yield is not necessarily a simple function of the atomic number, since many other factors come in to play. The images that are produced from the abundance of the SEs will show distinctly different gray levels for different materials (Figure 39.10). It is experimentally observed that for a given material, the SE yield increases steadily with an increase in the incident ion's energy.

SEs are created all along the trajectory of the incident ion within the specimen; however, only those created very near the surface have a reasonable probability of escape. The typical SE escape depth can be just a few nanometers. For these reasons, the collected SEs provide the most surface-specific information. And since the ion beam diverges relatively slowly as it enters the sample, the SEs that can escape are produced very close to the point of incidence. For these reasons, the highest resolution HIM images rely upon the detection of SEs. The image shown in Figure 39.11 is at extremely high magnification. The image spans a field of view of 200 nm (square), and it was acquired in a 1024 × 1024 pixel image. Correspondingly, each pixel represents a region of the sample that is just 0.2 nm (square). A statistical analysis of the sharp edges in the image indicates that the edge width is just 0.21 nm.

As the helium ion beam impinges upon sample surfaces that are not orthogonal to the beam, a larger number of SEs are produced near the surface and hence can escape and be detected.

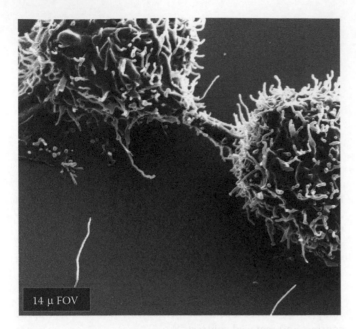

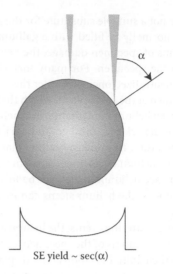

FIGURE 39.10 This image of Chinese hamster ovary cells shows very strong contrast from the different materials. The field of view is 14 μm. (Sample provided by NMI Stuttgart.)

FIGURE 39.12 As the beam strikes at a glancing angle, many more SEs are produced compared to normal incidence.

39.7.2 Backscattered Helium

As discussed earlier, the helium atom can scatter from nuclei as well as from electrons. When this occurs, the helium atom can be deflected by large angles—even to the point where the helium atom is ejected from the specimen. In this case, the ejected helium atom (neutral or ionized) is "backscattered." The backscattering can occur near the surface or deeper, and can be the

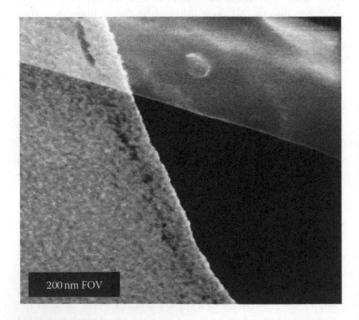

FIGURE 39.11 The images that are based upon the detection of SEs provide the highest resolution. This asbestos crystal overlying a carbon film reveals the truly small probe size of the HIM.

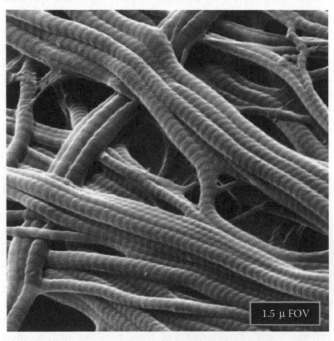

A spherical surface will show characteristic bright edges due to this effect (Figure 39.12). For this reason, SE images provide clear topographic information—making them easy to interpret as three-dimensional structures. For example, Figure 39.13 shows a complex weave of collagen fibers, but their shape and arrangement is readily understood.

FIGURE 39.13 The complex weave of collagen fibers is shown at a high-magnification [1 μm field-of-view (FOV)] SE image. Note how the edges are brighter, making the three-dimensional nature of the specimen easy to recognize. (Sample provided by NMI Stuttgart.)

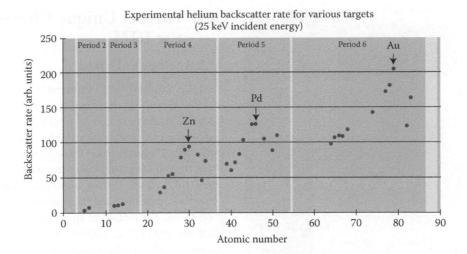

FIGURE 39.14   Experimentally measured backscatter rate. Note the oscillations in the backscatter rate which recur for rows 4, 5, and 6 of the periodic table.

result of a single large angle scattering event, or a sequence of smaller angle scattering events. The backscatter probability is relatively small, with a probability typically between 0.1 and 0.001. The probability tends to steadily decrease for higher incident beam energy. The backscatter probability also increases for targets with larger atomic numbers—although the trend is not monotonic (Figure 39.14).

A detector that is used for backscattered helium is commonly a microchannel plate (MCP), which is annular in its shape and coaxial with the incident beam. The images that are generated from the abundance of detected backscattered helium atoms tend to show primarily the atomic number contrast—with heavier elements revealed as bright in the images. When using the MCP detector, the backscatter detection rate is nearly independent of the angle of incidence; so unlike the SE images, the backscatter images tend to exhibit minimal topographic information. A representative backscatter image is shown in Figure 39.15.

39.7.3 Transmitted Ions

In the special case in which the specimen is thinned (20–200 nm), a transmission detector can be configured to detect the helium ions that pass through the specimen while suffering small or large angular deflections. For a sufficiently thin specimen, and a detector configured to collect only ions that have been significantly scattered (darkfield), the brighter regions will correspond to higher atomic number or higher atomic density. This imaging mode is closely related to the well-established technique of scanning transmission electron microscopy (STEM). Prethinned samples are commonly used in order to minimize sample damage, or to produce images that convey the thickness-averaged information (instead of surface-emphasized information). The transmission image, shown in Figure 39.16, is taken in "brightfield" mode, so that the lighter regions correspond to fewer scattering events.

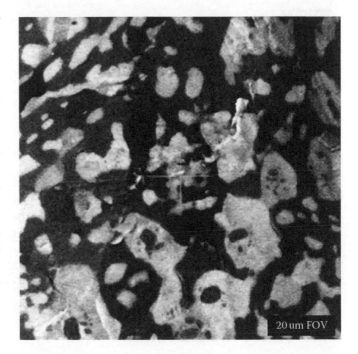

FIGURE 39.15   This is an image of a lead–tin eutectic using the MCP detector. Because of its larger atomic number, the lead ($Z = 82$) has a higher scatter rate and hence appears as a lighter shade of gray than the tin ($Z = 50$).

39.7.4 Photons

Although only briefly investigated, it has been observed that photons are also produced as a part of the ion–specimen interaction. Most materials seem to produce a low quantity of photons, and their origin is not fully understood. It is likely that some very high energy photons (~20 eV) are produced as the helium ion strikes the sample and returns to its neutral state. In addition, there are likely to be several lower energy (<1 eV) photons arising from the further relaxation of the excited helium atom to its ground state. Both these types of photons are likely to be produced just within

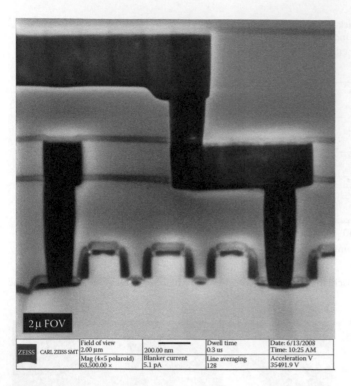

ZEISS CARL ZEISS SMT	Field of view 2.00 μm		Dwell time 0.3 us	Date: 6/13/2008 Time: 10:25 AM
	Mag (4×5 polaroid) 63,500.00 ×	200.00 nm Blanker current 5.1 pA	Line averaging 128	Acceleration V 35491.9 V

FIGURE 39.16 A thin slice (~100 nm) of the active region of a semiconductor device is imaged with the transmission detector. The contrast here indicates the thickness-averaged materials properties.

the surface of the specimen, and the probability of their escape is indicative of the optical transparency of that particular specimen. In addition to these mechanisms, there are certainly some materials that fluoresce in response to the incident ion. The wavelength of the ion-induced fluorescence will be a characteristic of the specimen (not the helium atom's energy spectrum) and hence can be used as a means of imaging the distribution of fluorescing molecules.[19] The incorporation of optical detectors with specific wavelength filters can be used to image the distribution of distinctly different fluorescing molecules.

39.7.5 Other Detectable Particles

Due to the infancy of this technology, there are likely to be other detectable particles or other excitations that have not yet been detected. For example, it is likely that there will be some number of x-rays produced, although there is some uncertainty regarding their rate of production.[20,21] The production of auger electrons are also expected at a low rate. In each of these cases, energy spectra could be used for elemental identification. It is also expected that the helium ions will produce some number of low-energy sputtered particles from the specimen—as well as higher energy recoiled atoms. Although there is not presently a detector to measure these particles, it is expected that such a detector could be used to analyze the mass of the sputtered atoms and hence deduce the specimen composition. Among the sputtered particles, the production of hydrogen atoms may provide a unique ability to "see" their distribution.

39.8 Further Unique Characteristics of the HIM

39.8.1 Depth of Field

The HIM is distinguished from the SEM in terms of its long depth of field (sometimes described as depth of focus). With a traditional SEM, it is difficult to attain good focus throughout the image when the specimen exhibits height variations. However, the helium ion beam strikes the sample with a very tightly focused beam that converges and diverges with a very narrow angle (as small as 0.3 mrad). Correspondingly, features that are nearer to or further from the ideal focal plane will appear in a relatively good focus. For biological specimens, this is uniquely valuable since the presence of important detailed features in the foreground as well as in the background is likely. Figure 39.17 demonstrates the long depth of field for the HIM. The features at the top of the image are several microns further from the features at the bottom of the image, but they are both well focused.

39.8.2 Charging Mitigation

For the SEM, charging arises from an imbalance between the rate of arriving electrons, and the rate of ejected electrons. For a SEM, this imbalance can produce a net-negative or a net-positive charging. The excess charge creates an induced voltage (hundreds of volts are possible) on the surface which will adversely affect the detector's ability to collect SEs. This results

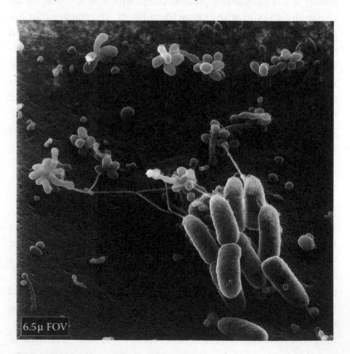

FIGURE 39.17 A colony of pseudomonas bacteria infecting lung fibroblast cells (sample is not metal coated). Note that the features in the foreground and background are both in sharp focus. (From Dan Pickard, National University of Singapore. Accepted for publication in the Proceedings of the EIPBN Conference. Copyright 2009, American Vacuum Society. With permission.)

in images that are dark, or streaked in white, or exhibit distortions. In contrast to this, the HIM has positive ions arriving and negative electrons being ejected (predominantly). For this reason, the HIM induces distinctly positive charging—a problem that can be countered through the use of an electron flood gun, a feature that is integrated into the ORION HIM. The flooding of electrons can be alternated with the normal sequences of scanning and acquisition. This technique produced the images in Figures 39.18 and 39.19. An alternative way of avoiding charging artifacts is to use a detected particle that is not affected, such as the much more energetic backscattered helium atoms. With an energy of 10–20 keV, these particles are not strongly affected by voltages up to a few kilovolts. Finally, it is also possible to operate the HIM with extremely low beam currents—as low as a few femtoamps. At these low rates of charging, the accumulated charge can be conducted away before an appreciable voltage develops. With these various methods of mitigating charging artifacts, it is commonly possible to use the HIM to image biological specimens without resorting to a metal-coating process—a process that obscures the finer surface details.

39.9 Materials Modification

While the HIM was developed for the purpose of imaging, some researchers have developed non-imaging applications. Some of these applications may be well suited for biological applications.

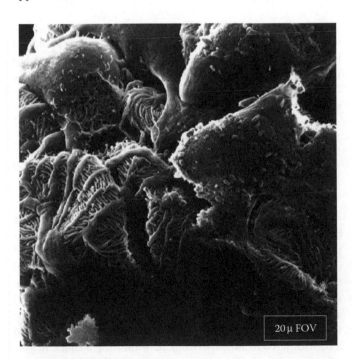

FIGURE 39.18 This image shows the filter-like structure of a rat kidney with a 20-μm FOV. Although the specimen is not metal coated, there is no evidence of charging artifacts. (From Dan Pickard, National University of Singapore. Accepted for publication in the Proceedings of the EIPBN Conference. Copyright 2009, American Vacuum Society. With permission.)

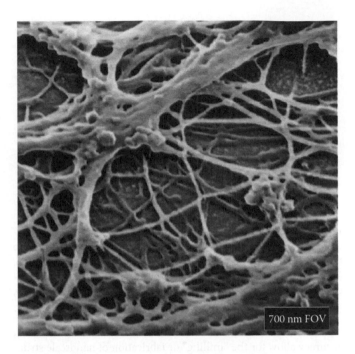

FIGURE 39.19 This image shows protein fibers with a 700-nm FOV. Although this specimen is not metal coated, there is no evidence of charging artifacts. (From Dan Pickard, National University of Singapore. Accepted for publication in the Proceedings of the EIPBN Conference. Copyright 2009, American Vacuum Society. With permission.)

39.9.1 Pattern Generation with a Helium Beam

The small focused probe size of the helium beam, and the ability to position it precisely, has attracted researchers who are interested in creating nanostructures. These can be created by standard lithographic techniques: A resist is uniformly applied to a substrate, and the beam is positioned in a controlled fashion so that it alters the chemical properties of the resist. After the resist is developed, and the unexposed material is dissolved, the patterned region persists. This technique is commonly used in the semiconductor industry for making fine structures (Figure 39.20), but it can also be used for creating microfluidic channels or for patterning biologically relevant molecules.

The helium ion beam has also been used for introducing stress in a thin silicon nitride membrane.[22] Depending on the beam energy, the helium ions will introduce stress near the top or the bottom surface. Correspondingly, the free standing membrane will bend up or down in response (Figure 39.21).

39.9.2 Helium Beam-Induced Chemical Reactions

There is a well-established science of using electron and gallium beams for inducing chemical reactions.[23] These methods involve the delivery of a "precursor gas" to the surface of the samples, and then using the incident beam to alter the gas molecules so that they become chemically active. In some cases, the beam-activated molecule can cause a chemical etching. In other cases, the beam-activated molecule can be made to adhere to the substrate. In

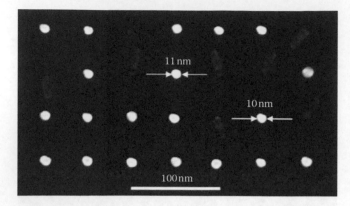

FIGURE 39.20 These pillars (viewed top-down in a SEM) were patterned in hydrogen silsesquioxane (HSQ) resist using helium ions. Note that some of the pillars are lying down—indicating their high aspect ratio. (From Dan Pickard, National University of Singapore. Accepted for publication in the Proceedings of the EIPBN Conference. Copyright 2009, American Vacuum Society. With permission.)

either case, the small probe size and the subnanometer positioning accuracy allow for the "milling" or fabrication of nanoscale structures. While these capabilities are highly refined for a gallium beam or a SEM, they have not yet been explored for a helium beam. It is expected that the quality of the helium-generated nanostructures could be quite good since the helium can diffuse out of the structure more readily than gallium. An example of a tall tungsten pillar, which was grown with a helium ion beam, is shown in Figure 39.22. The precursor gas is $W(CO)_6$.

39.10 Backscattered Helium Energy Spectroscopy

As discussed earlier, the helium ion (mass = M_1) can scatter from nuclei as well as from electrons. When this occurs, the helium

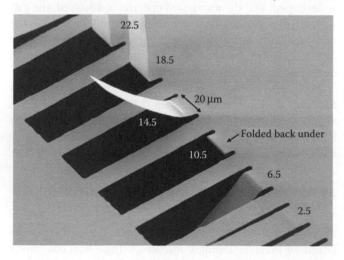

FIGURE 39.21 The silicon nitride membrane is bent up or down depending on the indicated beam energy. (From Dan Pickard, National University of Singapore. Accepted for publication in the Proceedings of the EIPBN Conference. Copyright 2009, American Vacuum Society. With permission.)

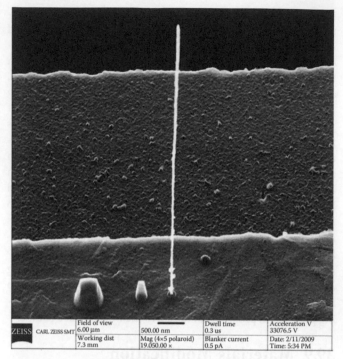

FIGURE 39.22 This tall tungsten pillar (45 nm wide and 7.9 μm tall) was grown by fixing the helium beam in one location, while delivering a tungsten precursor gas. (From Dan Pickard, National University of Singapore. Accepted for publication in the Proceedings of the EIPBN Conference. Copyright 2009, American Vacuum Society. With permission.)

ion will be deflected by an angle θ, from its original path, and have its initial energy E_0 reduced to E_1. The nuclei, (mass = M_2) will be knocked forward by an angle φ and with an energy E_2. Conservation of momentum and energy dictates the relationship given by Equation 39.1:

$$E_1 = E_0 \left(\frac{\mp\sqrt{M_2^2 - M_1^2 \sin^2\theta} + M_1 \cos\theta}{M_1 + M_2} \right)^2. \quad (39.1)$$

Therefore, a measurement of the backscatter energy E_1 and the backscatter angle θ allows the mass of the target atom to be computed. This general technique of ion scattering spectroscopy[24] (ISS) has long been used by scientists to determine the elemental identities and their depth distributions. The physics of ion scattering at the intermediate energies of the ORION (10–35 keV) are not well understood, being affected strongly by multiple scattering events. The ORION microscope has now incorporated a backscattered helium spectrometer—allowing a single instrument to perform high-resolution imaging and ISS. Several experimentally obtained spectra (Figure 39.23) show that in some cases, elements of similar mass (copper and nickel) can be readily discerned. Other applications of the detector allow the precise thickness measurement of a thin film with subnanometer resolution.

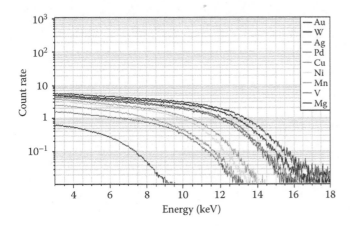

FIGURE 39.23 The spectra for nine elemental samples are shown superimposed. The vertical axis is the count rate for each corresponding energy bin (horizontal axis).

39.11 Summary

Recent improvements in the helium ion source have made the first commercial HIM practical. While it is related to the contemporary SEM or Ga-FIB, this new technology provides higher-resolution images due to the superior focusing properties of the helium ions. The helium ions also interact with the specimen differently from electrons or gallium ions, and therefore produce images with distinctly different contrast mechanisms. The HIM also has unique advantages such as charge mitigation, and a long depth of field. Aside from its imaging capabilities, the microscope is finding new applications for elemental and structural analysis, as well as nanofabrication.

References

1. L. Reimer, 1998, *Scanning Electron Microscopy*, 2nd edition, New York, NY: Springer.
2. J. Goldstein, D. E. Newbury, D. C. Joy, C. Lyman, P. E. Echlin, E. Lifshin, L. Sawyer, and J. Michael, 2003, *Scanning Electron Microscopy and X-Ray Microanalysis*, 3rd edition, London, UK: Kluwer Academic/Plenum Publishers.
3. S. Kalb itzer and A. Knoblauch, 1998, High brightness source for ion and electron beams, *Rev. Sci. Instrum.*, 69(2), 1026–1031.
4. V. T. Binh, 1988, *In situ* fabrication and regeneration of microtips for scanning tunneling microscopy, *J. Microsc.*, 152(2), 355–361.
5. H.-W. Fink, 1986, Mono-atomic tips for scanning tunneling microscopy, *IBM J. Res. Develop.*, 30(5), 460–465.
6. P. R. Schwoebel and G. R. Hanson, 1984, Beam current stability from localized emission sites in a field ion source, *J. Vac. Sci. Technol. B*, 3(1), 214–219.
7. T. Sakata, K. Kumagai, M. Naitou, I. Watanabe, Y. Ohhashi, O. Hosoda, Y. Kokubo, and K. Tanaka, 1992, Helium field ion source for application in a 100 keV focused ion beam system, *J. Vac. Sci. Technol. B*, 10(6), 2842–2845.
8. H.-S. Kuo, I.-S. Hwang, T.-Y. Fu, J.-Y. Wu, C.-C. Chang, and T. T. Tsong, 2004, Preparation and characterization of single-atom tips, *Nano Lett.*, 4(12), 2379–2382.
9. W. H. Escovitz, T. R. Fox, and R. Levi-Setti, 1975, Scanning transmission ion microscope with a field ion source, *Proc. Natl. Acad. Sci. USA*, 72(5),1826–1828.
10. J. H. Orloff and L.W. Swanson, 1975, Study of field-ionization source for microprobe applications, *J. Vac. Sci. Technol.*, 12(6), 1209–1213.
11. B. Ward, J. Notte, and N. P. Economou, 2006, Helium ion microscope: A new tool for nanoscale microscopy and metrology, *J. Vac. Sci. Technol. B*, 24, 2871–2874.
12. A. J. Melmed, 1995, Recollections of Erwin Muller's laboratory: The development of FIM (from 1951 to 1956), *Appl. Surf. Sci.*, 94, 17–25.
13. J. R. Oppenheimer, 1928, Three notes on the quantum theory of aperiodic effects, *Phys. Rev.*, 31, 66–81.
14. V. N. Tondare, 2005, Quest for a high brightness monochromatic noble gas ion source, *J. Vac. Sci. Technol. A*, 23(6), 1498–1507.
15. D. Drouin, A. R. Couture, D. Joly, X. Tastet, V. Aimez, and R. Gauvin, 2007, CASINO V2.42—A fast and easy-to-use modeling tool for scanning electron microscopy and micro-analysis users, *Scanning*, 29(3), 92–101.
16. J. F. Ziegler, J. P. Biersack, and U. Littmark, 1984, *The Stopping and Range of Ions in Solids*, Vol. 1 of series "Stopping and Ranges of Ions in Matter," Pergamon Press, New York. For the updated version, see SRIM Version 2006.02 (www. SRIM.com)
17. R. Livengood, S. Tan, Y.l Greenzweig, R. Hallstein, J. Notte, and S. McVey, 2010, Sub-surface damage from helium ions as a function of dose, dose rate, and beam energy, *J. Vac. Sci. Technol. B*, January, 2010.
18. T. M. Buck and J. M. Poate, 1974, Ion scattering for analysis of surfaces and surface layers, *J. Vac. Sci. Technol.*, 11(1), 289–296.
19. C. M. MacRae and N. C. Wilson, 2008, Luminescence database I—minerals and materials, *Microsc. Microanal.*, 14(02), 184–204.
20. L. Giannuzzi, 2005, Particle induced X-ray analysis using focused ion beams, *Scanning*, 23, 165–169.
21. D. C. Joy, H. M. Meyer III, M. Bolorizadeh, Y. Lin, and D. Newbury, 2007, On the production of X-rays by low energy ion beams, *Scanning*, 29, 1–4.
22. W. J. Arora, S. Sijbrandij, L. Stern, J. Notte, H. I. Smith, and G. Barbastathis, 2007, Membrane folding by helium ion implantation for three-dimensional device fabrication, *J. Vac. Sci. Technol. B*, 25(6), 2184–2187.
23. I. Utke, P. Hoffman, and J. Melngailis, 2008, Gas assisted focused electron beam and ion beam processing and fabrication, *J. Vac. Sci. Technol. B*, 26(4), 1197–1276.
24. J. Wayne Rabalais, 2003, *Principles and Applications of Ion Scattering and Spectrometry*, Wiley-Interscience, Hoboken, NJ.

40

Electron Microscopy: SEM/TEM

Fernando Nieto,
Juan Jiménez Millán,
Gleydes Gambogi Parreira,
Hélio Chiarini-Garcia,
and Rossana Correa
Netto Melo

40.1 Introduction

A microscope can be defined as an instrument that uses one or several lenses to form an enlarged (magnified) image. The term microscope comes from the Greek "mikros" = small and "skopos" = to look at. Microscopes can be classified according to the type of electromagnetic wave employed, depending on whether or not this wave is transmitted through the specimen. In transmission microscopes, the electromagnetic wave passed through the specimen is differentially refracted and absorbed. The most common types of transmission microscopes are transmitting light microscopes, in which visible spectrum or selected wavelengths pass through the specimen, and transmission electron microscopes (TEMs, Figure 40.1a) where the source of illumination is an electron beam.[1-6] Electron beams can also be passed over the surface of the specimen, thus causing energy changes in the sample. These changes are detected and analyzed to give an image of the specimen. This type of microscope is called scanning electron microscope (SEM, Figure 40.1b).[2,4,7] The optical paths of the illumination beam in light microscopes and TEMs are nearly identical. Both types of microscopes use a condenser lens to converge the beam onto the sample. The beam penetrates the sample and the objective lens forms a magnified image, which is projected to the viewing plane. SEM is nearly identical to TEM with respect to the illumination source and the condensing of the beam onto the sample. However, significant features distinguish SEM from TEM. Before contacting the sample, the SEM beam is deflected by coils that move the beam in scan pattern. Then a final lens (which is also called objective lens) condenses the beam to a fine spot on the specimen surface.

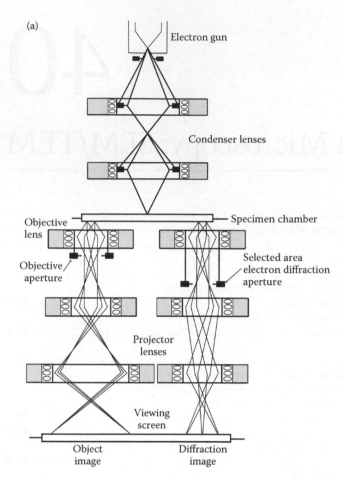

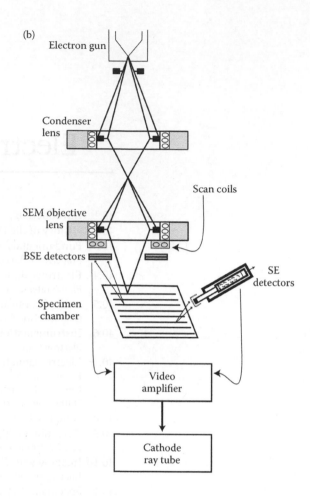

FIGURE 40.1 Schematic of EM imaging: (a) TEM and (b) SEM.

The signals produced by the effect of the beam on the sample are interpreted by specialized signal detectors.

Electron microscopy (EM), as it is understood today, is not a single technique but a diversity of different ones that offer unique possibilities to gain insights into morphology, structure, and composition of a specimen. The observable samples include biological specimens as well as inorganic and organic materials whose characterization need various imaging and spectroscopic methods acopled to the basic EM instrumentation.[2,5]

40.2 History of the Development of the Instrumentation

Hans Busch theoretically showed in 1927 that electron beams can be focused in an inhomogeneous magnetic field.[1–8] He also predicted that the focal length of such a magnetic electron lens could be changed continuously by varying the coil current. In 1931, Ernst Ruska and Max Knoll confirmed this theory by constructing a magnetic lens of the type that has been used since then in all magnetic electron microscopes leading to the construction of the first transmitted electron microscope instrument. The main limitation of their microscope was that electrons were unable to pass through thick specimens. Thus it was impossible

to utilize the instrument to its full capacity until the diamond knife and ultramicrotome were invented in 1951. In 1938, Manfred von Ardenne (1907–1997) constructed a scanning transmission electron microscope (STEM) by adding scan coils to a TEM. Vladimir Kosmo Zworykin (1889–1982), J. Hillier, and R. L. Zinder developed the first SEM in 1942 without using the transmitted electron signal. Charles Oatley and his PhD students of the University of Cambridge provided many improvements incorporated to subsequent SEM models that finally in 1964 resulted in the first commercial SEM by Cambridge Instruments. In 1986, E. Ruska (together with G. Binning and H. Rohrer, who developed the scanning tunneling microscope) obtained the Nobel Prize for their groundbreaking work on electron imaging.

40.3 Fundamentals of EM

40.3.1 Properties of Electron Beams

EM is based on the use of a stable electron beam that interacts with the matter.[3] Electrons are elementary particles with negative charge. These particles were discovered by J. J. Thompson in 1897 (Nobel Prize 1906), who deduced that the cathode rays consisted of negatively charged particles (corpuscles) that were constituents of the atom and over 1000 times smaller than a hydrogen atom.

The possibility of developing a microscope that utilized electrons as its illumination source began in 1924 when De Broglie (Nobel Prize 1929) postulated the wave–particle dualism according to which all moving matter has wave properties, with the wavelength λ being related to the momentum *p* by Equation 40.1:

$$\lambda = \frac{h}{p} = \frac{h}{mv},$$ (40.1)

where *h* is the Planck's constant = 6.626×10^{-34} Js, *m* is the mass, and *v* is the velocity).

It means that accelerated electrons also act as waves. The wavelength of moving electrons can be calculated from this equation considering their energy *E*. The energy of accelerated electrons is equal to their kinetic energy, given by Equation 40.2:

$$E = \frac{m_0 v^2}{2} = eV,$$ (40.2)

where *V* is the acceleration voltage, *e* is the elementary charge 1.602×10^{-19} C, m_0 is the rest mass of the electron 9.109×10^{-31} kg, and *v* is the velocity of the electron.

These equations can be combined to calculate the wavelength of an electron with certain energy from Equation 40.4, after substitution of Equation 40.3:

$$p = m_0 v = \sqrt{2m_0 eV},$$ (40.3)

$$\lambda = \frac{h}{\sqrt{2m_0 eV}} \approx \frac{1.22}{\sqrt{V}} \text{nm}.$$ (40.4)

At the acceleration voltages used in TEM, relativistic effects have to be taken into account according to the following equation:

$$\lambda = \frac{h}{\sqrt{2m_0 eV \left[1 + (eV / (2m_0 / c^2)) \right]}},$$ (40.5)

where m_0 is the rest mass of an electron = 9.109×10^{-31} kg and *c* is the speed of light in vacuum = 2.998×10^8 m/s.

Resolution in a microscope, defined as the ability to distinguish two separate items from one another, is related to wavelength of illumination (see Chapter 27 for explanation of "Abbe principle").

Table 40.1 shows several electron wavelengths at some acceleration voltages used in TEM.

Electron waves in beams can be either coherent or incoherent. Waves that have the same wavelength and are in phase with each

TABLE 40.1 Electron Wavelengths at Some Acceleration Voltages used in TEM

V_{acc}/kV	B/pm	A/pm
100	3.86	3.70
200	2.73	2.51
300	2.23	1.97
400	1.93	1.64
1000	1.22	0.87

V_{acc}, accelerating voltage; A, nonrelativistic wavelength; B, relativistic wavelength.

other are designated as coherent. In contrast, beams comprising waves that have different wavelengths or are not in phase are called incoherent. Given that electron waves interact with matter, to reduce the complexity of signals produced by these interactions, it is an essential prerequisite to use coherent beams.

Electrons accelerated to a selected energy have the same wavelength. The generation of a highly monochromatic and coherent electron beam is an important challenge in the design of modern electron microscopes. After interacting with a specimen, electron waves can form either incoherent or coherent beams which interact with each other producing either constructive or destructive interferences that can lead extinguish waves.

40.4 Electron–Matter Interactions

When an electron encounters a material, different interactions occur producing a multitude of signals (Figure 40.2), which can be classified into elastic and inelastic interactions.

In the elastic interactions, no energy is transferred from the electron to the sample and the electron leaving the sample still conserves its original energy ($E_{el} = E_0$). This is the case of the electron passing the sample without any interaction contributing to the direct beam (Figure 40.2). In thin samples, these signals are exploited in TEM and electron diffraction (ED) methods whereas in thick specimens, backscattered electrons (BSE) are the main elastic signals studied.[1,4]

In the inelastic interactions, an amount of energy is transferred from the incident electrons to the sample. Figure 40.3 shows the electron energy spectrum of several signals produced during electron–matter interaction. Low-energy peaks, such as the large secondary electron (SE) peak, correspond to inelastic interaction whereas high-energy peaks, BSE distribution, correspond to cases where only a negligible amount of energy is lost. Signals produced by inelastic electron–matter interactions (Figure 40.2) are predominantly utilized in analytical EM.[8-12]

The volume of interaction is controlled by energy loss through inelastic interactions and electron loss or backscattering through essentially elastic interactions. Factors controlling the electron penetration depth and the interaction volume are the angle of incidence, the current magnitude and the accelerating voltage of the beam, and the average atomic number (*Z*) of the sample.[8,11] The resulting excitation volume whose penetration generally ranges from 1 to 5 μm is indicated by the jug-shaped region (Figure 40.2).

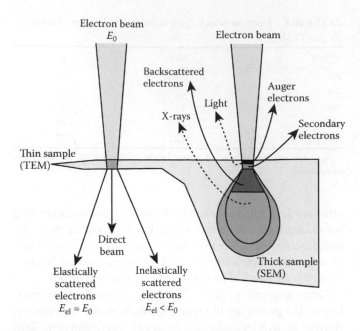

FIGURE 40.2　Diagram of electron-matter interaction.

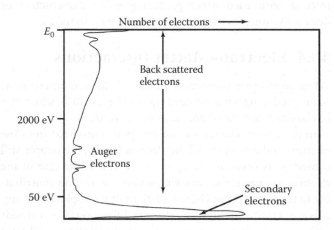

FIGURE 40.3　Electron spectrum.

40.4.1 Elastic Interactions

40.4.1.1 Incoherent Scattering, the BSE Signal

When an electron penetrates into the electron cloud of an atom, it is attracted by the positive potential of the nucleus deflecting its path toward the core. The Coulomb force F is defined in Equation 40.6[2,10,12]:

$$F = \frac{Q_1 Q_2}{4\pi\varepsilon_0 r^2},\qquad(40.6)$$

with r being the distance between the charges Q_1 and Q_2 and ε_0 the dielectric constant. The closer the electron comes to the nucleus, the larger is F and consequently the scattering angle (Figure 40.4). Scattering angles range up to 180°, but average about 5°.

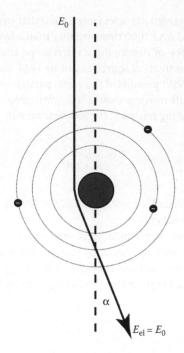

FIGURE 40.4　Representative illustration of incoherent scattering.

40.4.1.2 Backscattered Electrons

In some cases, even complete backscattering can occur in an individual interaction (Figure 40.5a). In thick samples, many incident electrons undergo a series of such elastic events that cause them to be scattered back out of the specimen (Figure 40.5b). BSEs are high-energy primary electrons that suffer large angle (>90°) scattering and re-emerge from the specimen. The energy of BSE depends on the number of interactions that they have undergone before escaping the sample.

The force F with which an atom attracts an electron is stronger for atoms containing more positives charges (higher atomic number, Z). Therefore, the fraction of beam electrons backscattered from a sample (η) depends strongly on the sample's average atomic number, Z (Figure 40.6).

40.4.1.3 Coherent Scattering, the ED Signal

When electrons are scattered by atoms in a regular array, collective elastic scattering phenomenon, known as ED, occurs. All atoms in such a regular arrangement act as scattering centers that can deflect the incoming electron from its direct path. Since the spacing between the scattering centers is regular now, interference of the scattered electron in certain directions happens. This occurs either constructively (reinforcement at certain scattering angles generating diffracted beams) or destructively (extinguishing of beams) which gives rise to a diffraction pattern (Figures 40.2 and 40.13). The scattering event can be described as a reflection of the beams at planes of atoms according to the Bragg law, which gives the relation between interplanar distance d and diffraction angle Θ, as described by Equation 40.7[5,7,12]:

$$\eta\lambda = 2d\sin(\theta).\qquad(40.7)$$

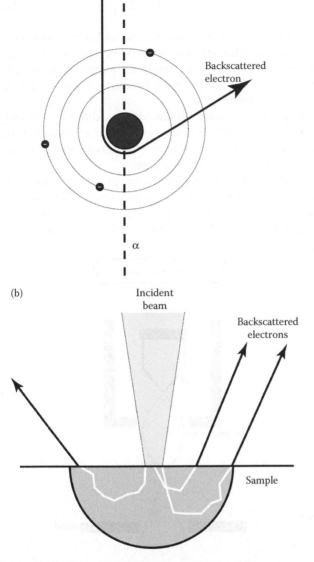

(a)

Backscattered
electron

α

(b) Incident
beam

Backscattered
electrons

Sample

FIGURE 40.5 (a) Generation of individual BSE and (b) generation of BSE in thick samples.

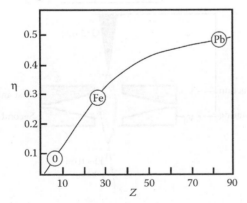

FIGURE 40.6 Relation between the fraction of electrons backscattered from a sample (η) and the atomic number (Z).

Since the wavelength λ of the electrons is known, interplanar distances can be calculated from ED patterns.

40.4.2 Inelastic Interactions

Most electrons of the incident beam follow complicated trajectories through the sample material, losing energy as they interact with the specimen atoms producing a number of interactions.[3] Some of the most significant effects are shown in Figure 40.2. Several interaction effects due to electron bombardment emerge from the sample and some, such as sample heating, stay within the sample.

40.4.3 Secondary Electrons

SEs are produced by inelastic interactions of high-energy incident electrons with valence electrons of atoms in the specimen causing the ejection of the electrons from the atoms (Figure 40.7) which can move towards the sample surface through elastic and inelastic collisions until it reaches the surface, escaping if its energy exceeds the surface work function, E_w. The strongest region in the electron energy spectrum is due to SEs that are defined as those emitted with energies less than 50 eV (Figure 40.3).

The mean free path length of SEs in many materials is ~1 nm (10 Å). Although electrons are generated in the whole region excited by the incident beam, only those electrons that originate less than 1 nm deep in the sample are able to escape, giving rise to a small volume production. Therefore, the resolution using SE is effectively the same as the electron beam size. The shallow depth of production of detected SEs makes them very sensitive to topography and they are used for SEM.

The fraction of SEs produced, δ, is the average number of SE produced per primary electron, and is typically in the range 0.1–10.

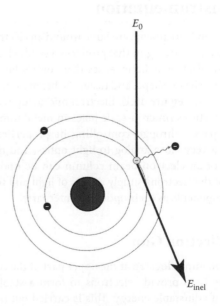

FIGURE 40.7 Diagram of SE generation.

40.5 Other Significant Inelastic Signals

40.5.1 Characteristic X-rays

When an electron from an inner atomic shell is displaced by colliding with a primary electron, it produces a vacancy in that electron shell. In order to re-establish the proper balance in its orbitals, an electron from an outer shell of the atom may fall into the inner shell and replace the spot vacated by the displaced electron. In doing so, this falling electron loses energy and this energy is referred to as x-rays.[1-8] The characteristic x-ray of a given element can be detected and identified; therefore, information about the chemical composition of different points of the sample can be obtained.

40.5.2 Auger Electrons

Auger electrons are produced when an outer shell electron fills the hole vacated by an inner shell electron that is displaced by a primary or BSE.[2,3] The excess energy released by this process may be carried away by an Auger electron. Because of their low energies, Auger electrons are emitted only from near the surface. The energy of an Auger electron can be characteristic of the type of element from which it was released and thus Auger electron spectroscopy can be performed to obtain chemical analysis of the specimen surface.

40.5.3 Cathodluminescence

The interaction of the primary beam with the specimen in certain materials releases excess energy in the form of photons.[2,3] These photons of visible light energy can be detected and counted forming an image using the signal of light emitted.

40.6 Instrumentation

Electron beam instruments are built around an electron column containing an electron gun that produces a stable electron beam and a set of electromagnetic lenses that control beam current, beam size and beam shape, and raster the beam (in the SEM and STEM cases, see Figure 40.1). Electron microscopes also have a series of apertures (micron-scale holes in metal film) by which the beam passes through controlling its properties. Electron optics are a very close analog to light optics, and most of the principles of an electron beam column can be understood by thinking of the electrons simply as rays of light and the electron optical components as their optical counterparts.

40.6.1 Electron Gun

The electron gun is located at the upper part of the column and its purpose is to provide electrons to form a stable beam of electrons of adjustable energy. This is carried out by allowing electrons to escape from a cathode material. The total energy required for a material to give up electrons is defined by Equation 40.8:

$$E = E_w + E_f, \qquad (40.8)$$

where E is the total amount of energy needed to remove an electron to infinity from the lowest free energy state, E_f is the highest free energy state of an electron in the material (which must be achieved), and E_w is the work function or work required to achieve the difference.

Electron beams can be produced by two different electron gun types (Figure 40.8): thermoionic guns (electron emission through heating) and field emission guns (electron emission through the application of an extraction voltage).

In the thermionic sources (Figure 40.8a), electrons are produced by heating a conductive material to the point where the outer orbital electrons gain sufficient energy to overcome the work function barrier and escape. Most of the thermoionic electron

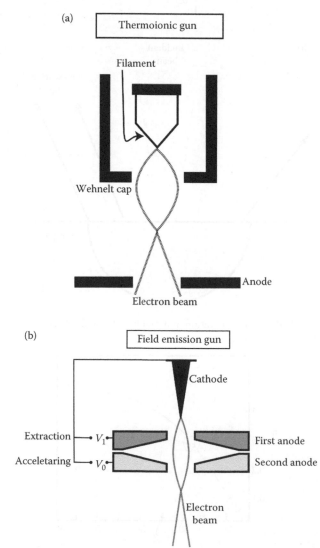

FIGURE 40.8 Electron generation: (a) thermal electron gun and (b) field-emission electron production.

guns have a triode configuration consisting of a cathode, a Wehnelt cap, and an anode.

The cathode is a thin filament (about 0.1 mm) wire bent into a "V" to localize emission at the tip, yielding a coherent source of electrons emitted from a small area, which in many cases is not perfectly circular. There are two main types of thermionic sources: tungsten metal filaments and LaB_6 crystals. These two types of sources require vacuums of $\sim10^{-5}$ and $\sim10^{-7}$ Torr, respectively.

High-vacuum conditions for the electron column and the specimen chamber are required in order to:

- Avoid damaging the filament. The volume around the electron gun must be kept free of gas molecules, especially oxygen, which will greatly shorten the filament life.
- Avoid arcing between the filament and the anode. There is very high voltage between these two components and stray air or gas molecules can cause electrical arcing between them.
- Avoid volatilizations produced by collisions between electrons of the beam and stray molecules. These collisions can result in spreading or diffusing of the electron beam or, more seriously, can result in volatilization event if the molecule is organic in nature (e.g., vacuum oil). Volatilizations can severely contaminate the microscope column, especially apertures, and degrade the image.

Tungsten filaments resist high temperatures without melting or evaporating, but they have a very high operating temperature (2700 K), which decreases their lifetime due to thermal evaporation of the cathode material. The electron flux from a tungsten filament is minimal up to a temperature of approximately 2500 K. Above 2500 K, the electron flux increases essentially in an exponential mode with increasing temperature, until the filament melts at about 3100 K. However, in practice, the electron emission reaches a plateau termed as saturation. Proper saturation is achieved at the edge of the plateau; higher emission currents serve only to reduce filament life. LaB_6 cathodes yield higher currents at lower cathode temperatures than tungsten, exhibiting 10 times the brightness and more than 10 times the service life of tungsten cathodes. Moreover, its emission region is smaller and more circular than that of tungsten filaments, which improves the final resolution of the electron microscope. However, LaB_6 is reactive at the high temperatures needed for electron emission.

The cathode is surrounded by a slightly negative biased Wehnelt cap (200–300 V) to localize at one spot the cloud of primary electrons emitted from the heated filament. The electrons emitted from the cathode–Wehnelt assembly are drawn away by the anode plate, which is a circular plate with a hole in its center. A voltage potential between the cathode and anode plates is used to accelerate the electrons down the column (accelerating voltage), condensing and roughly focusing the beam of primary electrons.

In the field emission guns (Figure 40.8b), the cathode consists of a sharp metal tip (usually Tungsten) with a radius of less than 100 nm. An extraction voltage (V_1) is established between the first anode and the tip and an electric field concentrated at the tip produces electron emission accelerated by the accelerating voltage (V_0) between the tip and the second anode.

There are two types of field emission guns (FEG): cold and thermally assisted. Both types require that the tip remains free of contaminants and oxide, and thus they require ultra-high-vacuum conditions (10^{-10} to 10^{-11} Torr). In the cold FEG, the electric field emission gun produced by the extraction voltage lowers the work function barrier and allows electrons to directly tunnel through it, thus facilitating emission. The cold FEGs must periodically heat their tip to be polished of absorbed gas molecules. The thermally assisted FEG (Schottky field emitter) uses heat and chemistry (nitride coating) in addition to voltage to overcome the potential barrier level.

40.7 Electromagnetic Lens

Electromagnetic lenses are made of a coil of copper wires inside several iron pole pieces (Figure 40.9). An electric current through the coils creates a magnetic field in the bore of the pole pieces. The rotationally symmetric magnetic field is strong close to the bore and becomes weaker in the center of the gap. Thus, when an electron beam passes through an electromagnetic lens, electrons close to the center are less strongly deflected than those passing the lens far from the axis. The overall effect is that

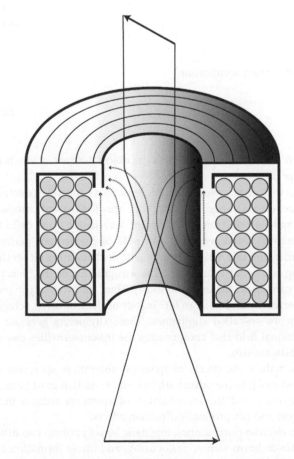

FIGURE 40.9 Construction of the electron microscope lens.

a beam of parallel electrons is focused into a spot. In a magnetic field, an electron experiences the Lorentz force *F*, as expressed in Equation 40.9:

$$F = -e(E + v \times B).$$ (40.9)

Or in magnitude only:

$$|F| = evB \sin(v \cdot B),$$ (40.10)

where *E* is the strength of the electric field, *B* is the strength of the magnetic field, and *e/v* is the charge/velocity of electrons.

The focusing effect of a magnetic lens therefore increases with the magnetic field *B*, which can be controlled via the current flowing through the coils. As it is described by the vector product, the resulting force *F* is perpendicular to *v* and *B*. This leads to a helical trajectory of the electrons and to the magnetic rotation of the image (Figure 40.9).

Electromagnetic lenses influence electrons in a similar way as convex glass lenses do with light. Thus, very similar diagrams can be drawn to describe the respective ray paths. Consequently, the imaginary line through the centers of the lenses in an electron microscope is called optical axis as well. Furthermore, the lens equation of light optics is also valid in electron optics, and the magnification is defined by Equation 40.11:

$$\frac{1}{u} + \frac{1}{v} = \frac{1}{f},$$ (40.11)

yielding the magnification

$$M = \frac{v}{u},$$ (40.12)

where *f* is the focal length, *u* is the object distance, and *v* is the image distance.

As with glass lenses, magnetic lenses have spherical (electrons are deflected stronger the more they are off-axis) and chromatic aberrations (electrons of different wavelengths are deflected differently). Moreover, the iron pole pieces are not perfectly circular, and this makes the magnetic field deviating rather than being rotational symmetric. The astigmatism of the objective lens can distort the image seriously. Thus, the astigmatism must be corrected, and this can fortunately be done by using octupole elements, so-called stigmators. These stigmators generate an additional field that compensates the inhomogeneities causing the astigmatism.

To reduce the effects of spherical aberration, apertures are introduced into the beam path (see Figure 40.1). It must be taken in account that the introduction of apertures reduces beam current and can produce diffraction effects.

In electron microscopes, magnetic lenses perform two different tasks: beam formation/focusing, and image formation and magnification.

40.7.1 Beam Formation and Focusing (Condenser Lenses in TEM and SEM and Objective Lens in SEM)

Condenser lenses are located at the upper part of the column (Figures 40.1a and b) ensuring that the electron beam is symmetrical and centered and controlling the beam intensity by focusing the electron beam onto a lower aperture. In the TEM, they focus the illuminating beam on the specimen.

In the SEM, the objective lens, located at the base of the electron column (Figure 40.1b) focuses the electron beam onto the sample and controls its final size and position. An objective aperture of variable size, located under the lens, controls aberrations, resolution, and probe current. Hosted within it are scanning coils, which raster the beam across the sample for textural imaging by varying the current through their magnets.

40.7.2 Image Formation and Magnification (Objective, Diffraction, Intermediate, and Projective Lenses in TEM)

The TEM has a complex electromagnetic lenses group as imaging system under the sample chamber. The objective lens focuses the beam after passing through the specimen and forms an intermediate image. Electrons scattered in the same direction are focused in the back focal plane forming a diffraction pattern. Electrons coming from the same point are focused in the image plane. An objective aperture, situated below the objective lens, allows the selection of only a few beams to form the images. Eye piece, intermediate and projector lenses magnify the intermediate image to form the final image.

40.8 Sample Chamber

Samples involved in the TEM and SEM techniques have different characteristics. TEM samples are usually small and must be as thin as possible, whereas SEM samples are far more variable ranging from small specimens, similar to those observed in the TEM, to large pieces of specimens. The sample chambers of the two instruments reflect this difference. TEM has a sample chamber just large enough to permit observation of specimens included in a ring or grid of a few mm. In contrast, SEM has a sample chamber that permits large and complex movements of the sample stage; for example, the SEM can accommodate samples up to 200 mm in diameter and ~80 mm high. Moreover, SEM sample chambers are designed to accommodate a wide set of signal detectors that allow the developing of their imaging and analytical facilities.

40.9 Detection of Signals

40.9.1 BSE Detection

The BSE signal is mainly used in SEM instruments. BSEs are detected using diodes, acting as semiconductors, located at the

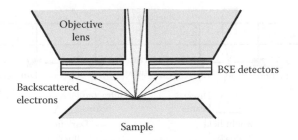

FIGURE 40.10 Diagram of BSE detector position.

base of the objective lens (Figure 40.10). When BSEs, imparting energy, strike these semiconductor chips, a current flows. The size of the current (signal) is proportional to the number of electrons hitting the semiconductor. The BSE detectors are sensitive to light and cannot be used if a sample illumination device is on at the chamber.

The BSE detector systems commonly consist of either two semicircular plates or four semiconductor plates mounted at the base of the bottom of the electron column.

40.9.2 SE Detection

Owing to the low energy of the SEs (Figure 40.2) emitted from the sample in all directions, in a first stage of the detection process, they have to be drawn from the sample surface toward the detector. They are gathered by a charged collector grid, which can be biased from -50 to +300 V (Figure 40.11). SEs are then detected using a scintillator–photomultiplier detector. The amplified electrical signal is sent to further electrical amplifiers, which increase the electrical signal and thus increase the brightness.

40.9.3 Low-Vacuum Mode

SEM has the added capability to also work with a relatively poor vacuum in the specimen chamber. Air (or another gas) is admitted into the sample chamber to neutralize the build-up of the negative electrostatic charge on the sample. In "low-vacuum (LV) mode," the sample chamber pressure, typically 20–30 Pa, is maintained at a specifiable value by a separate extra pump with a large foreline trap. The electron column is separated from the chamber by a vacuum orifice (aperture) that permits the column and objective

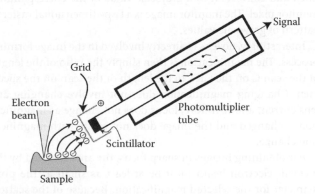

FIGURE 40.11 Configuration of SE detector.

lens apertures to be maintained at a higher vacuum pressure. LV mode allows the study of samples with poor or no electrical conductivity. Normally, these materials would be either coated with an electrically conductive coating, which takes time and hides the true surface, or studied at low voltages, which do not give backscatter or x-ray information. Additionally, samples containing volatile substances (water) can often be studied directly.

An environmental SEM utilizes relatively low-vacuum pressures (up to 50 Torr ~6700 Pa) not only to neutralize charges, but also to provide signal amplification. The electron column is gradually pressured by a series of pumps and apertures.

A positively charged detector electrode is placed at the base of the objective lens at the top of the sample chamber (Figure 40.12). Electrons emitted from the sample are attracted to the positive electrode, undergoing acceleration. As they travel through the gaseous environment, collisions occur between electrons and gas particles, producing more electrons and ionization of the gas molecules and effectively amplifying the original SE signal (electron cascade). This form of signal amplification is identical to that which occurs in a gas-flow x-ray detector. Positively charged gas ions are attracted to the negatively biased specimen and offset charging effects.

40.10 Imaging with EM

40.10.1 Imaging in TEM

Image formation in the TEM is carried out by the objective lens, which physically performs: (a) A Fourier transform (FT, Fourier analysis) that creates the diffraction pattern of the object in the back focal plane of the objective lens and (b) an inverse FT (Fourier synthesis) that causes the interference of the diffracted beams to go back to a real space image in the image plane.

Therefore, the selection of the focusing plane and the beams allowed to pass through the aperture of the objective lens play critical roles in the TEM image formation.

40.10.1.1 Diffraction Patterns

The FT of periodic structures gives rise to sharp spots at well-defined positions in the resulting diffraction pattern. Each

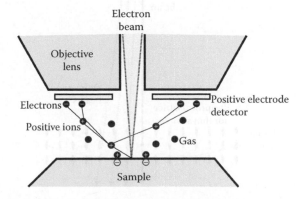

FIGURE 40.12 Diagram of the electron detection in an environmental SEM.

set of parallel lattice planes of the periodic structure is represented by spots which have a distance of $1/d$ (d is the interplanar spacing) from the origin and which are perpendicular to the reflecting set of lattice planes. The diffraction can be described in reciprocal space by the Ewald sphere construction (Figure 40.13). A sphere with radius $1/\lambda$ is drawn through the origin of the reciprocal lattice. For each reciprocal lattice point that is located on the Ewald sphere surface, the Bragg condition is satisfied and diffraction arises. Such diffraction patterns provide useful information about the nature of the periodic structure.

40.10.1.2 Contrast Images

The focus in the image plane of the electrons coming from the same point of the object produces images whose points have different contrast. Basic contrast mechanisms in TEM are controlled by the local scattering power of the sample. The interaction of electrons with heavy atoms having a high charge (Q) is stronger than with light atoms; hence, areas in which heavy atoms are localized appear with darker contrast than those with light atoms (mass contrast). In thick areas, more electron scattering events occur and thus thick areas appear darker than thin areas (thickness contrast). In particular, this mass-thickness contrast is important in bright (BF) and dark field (DF) imaging.

In the BF mode of the TEM (Figure 40.14), the aperture placed in the back focal plane of the objective lens allows only the direct beam to pass through. In this case, the image results from a weakening of the direct beam by its interaction with the sample and therefore the mass-thickness contrast contributes to image formation: thick areas and areas enriched in heavy atoms appear with darker contrast. If a specimen with a crystalline structure is

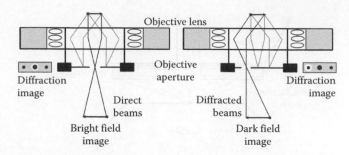

FIGURE 40.14 Representation of the BF and DF imaging in the TEM.

appropriately oriented, many electrons will be strongly scattered to contribute to the reflections in the diffraction pattern, and only a few electrons pass without interactions; therefore, this specimen appears with dark contrast in the BF image (diffraction contrast). In real specimens, all contrast mechanisms, namely mass-thickness and diffraction contrast, occur simultaneously, making interpretation of TEM images difficult sometimes. In DF images (Figure 40.14), the direct beam is blocked by the objective lens aperture while one or more diffracted beams are allowed to pass through, resulting in an image where the specimen is in general weakly illuminated and the surrounding area that does not contain the sample is dark.

40.10.1.3 Lattice Images

The lattice images are formed by the interference of various diffracted beams. A larger objective aperture has to be selected to obtain this type of images, allowing many beams, including, or not, the direct beam, to pass. If the resolution of the microscope is sufficiently high and a suitable sample is oriented along, then high-resolution TEM (HRTEM) images, showing crystalline periodicity, are obtained.

40.10.2 Imaging in SEM

In the SEM images, there is a conjugate correspondence between the rastering pattern of the specimen and the rastering pattern used to produce the image on the monitor. The signal produced by the rastering is collected by the detector and subsequently processed to generate the image (Figure 40.1b). That processing takes the intensity of the signal coming from a pixel on the specimen and converts it to a grayscale value of the corresponding monitor pixel. The monitor image is a two-dimensional rastered pattern of grayscale values.

Interestingly, no lens is directly involved in the image forming process. The magnification is then simply the ratio of the length of the scan C on the CRT to the length of the scan on the specimen. Changing magnification does not involve changing any lens current and therefore, focus does not change as magnification is changed and the image does not rotate with magnification change.

For obtaining images in sharp focus, the area sampled by the incident electron beam must be at least as small as the pixel diameter for the selected magnification. Because of the scattering of the incident electrons, the signals produced are actually

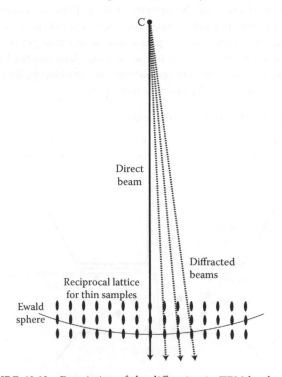

FIGURE 40.13 Description of the diffraction in TEM by the construction of the Ewald sphere.

generated from a larger area (perhaps 1.5–2 times) than the beam diameter. Therefore, the beam must be focused to a diameter between 0.5 and 0.7 times the diameter of the magnification.

Contrast C involves the signals produced by the detector for two points A and B on the sample, which depend on the number of signal electrons emitted from the sample, the number of these electrons reaching the detector, and its efficiency to record them. The signal produced by the detector can be modified to change the appearance of the image; however, such a change does not alter its information content.

40.10.2.1 Imaging with BSE

BSE provide valuable information because of their sensitivity to atomic number variations. However, it must be taken into account that BSE are produced from the entire upper half of the interaction volume (Figure 40.2) and, therefore, the spatial resolution of BSE images is poor (usually ~1 μm; at best ~0.1 μm).

The BSE signal is mainly used to obtain information on the specimen composition (average Z). SEM uses paired detectors to receive BSE (Figure 40.15). Addition of signals received by each detector highlights the compositional characteristics of the sample, thus providing a composition image (COMPO mode).

40.10.2.2 Imaging with SEs

SE imaging provides morphological information about the specimen. The topographical aspects of a SE image depend on how many electrons actually reach the detector. Given that, SE yield depends on the angle of tilt of the specimen relative to the primary electron beam, φ; when the incident electron beam intersects the tilted edges of topographically high portions of a sample at lower angles, it puts more energy into the volume of SE production, producing more SEs and generating a larger signal (Figure 40.16). Therefore, places where the beam strikes the surface at an angle have larger volumes of electron escape than the perpendicular incidence ($\varphi = 0$). These effects cause tilted surfaces to appear brighter than flat surfaces, and edges and ridges to be markedly highlighted, in images formed with SEs.

Faces oriented toward the detector also generate more SEs. SEs that are prevented from reaching the detector do not contribute to the final image. Therefore, faces oriented towards

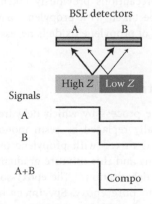

FIGURE 40.15 Generation of a composition image with paired BSE detectors in SEM.

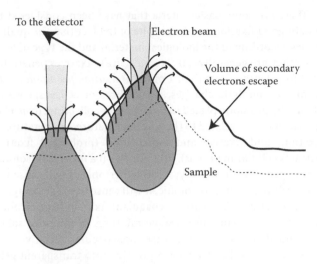

FIGURE 40.16 SEM configuration.

the detector will be brighter, whereas those in the opposite direction will be dark.

40.11 Specimen Preparation for TEM

Specimen processing for EM represents considerable effort to follow a rigid procedure that influences the quality of the final electron micrographs. The biological preparation begins with a living hydrated tissue and ends with the tissue being water-free which is preserved in a static state within a plastic resin matrix. Basically, the plastic resin mixture permeates the tissue, replacing all water within the cell and making the cell firm enough for sections to be cut. To get to this point, specimen processing should begin with careful planning and proceed with meticulous attention to details following this rigid sequence: (i) *primary chemical fixation*, (ii) *washing*, (iii) *secondary chemical fixation*, (iv) *dehydration*, (v) *use of transitional solvents*, (vi) *infiltration*, (vii) *embedding*, and (viii) *microtomy*.[13,14]

40.11.1 Fixation

Conceptually, the purpose of fixation in EM is to preserve the structure of the biological specimen as closer as possible to the living state. This means that fixation must protect the sample against destructive autolytic processes and additionally protect the specimen during the other processing steps and subsequent exposure to the electron beam. Basically, fixatives form crosslinks, not only between their reactive groups and the reactive groups of the specimen but also between different reactive groups within the specimen.[13,14]

Optimal fixation requires special care regarding the preparation of the solutions. Buffering system is crucial to maintain the cellular physiologic pH, for example, 7.2–7.4 for most mammalian tissues. Commonly used buffers are phosphate and cacodylate. Just as important as the selection of the buffer is the osmolarity of the buffering system, since it may induce shrinkage or swelling of tissue due to osmotic effects.[13,14]

There are some basic criteria that have been considered to obtain good fixation results in spite of the fact that the quality varies according to the biological material and the type of fixative. They are generally: (a) *specimen size*—fixatives penetrates well to a depth of less than 1 mm; (b) *fixation promptness*—to minimize autolytic changes, tissue slices or cell suspensions should be promptly placed in fixative after collecting; (c) *period of fixation*—it is related to the specimen nature, temperature of fixation, and fixative and buffer used (prolonged fixation extracts cellular materials); (d) *tonicity*—if a fixative is isotonic it may penetrate too slowly; if it is hypertonic, it will cause tissue shrinkage; if it is hypotonic, it will cause tissue swelling; (e) *the specificity of the fixative*—coagulant fixation (e.g., alcohol) transforms proteins into opaque mixtures of granular solids suspended in fluid. In this case, noncoagulant fixatives are indicated since they transform proteins into transparent gels, stabilizing them without much structural distortion; and (f) *method of fixation*—the best preservation is obtained by perfusing fixative through the blood vascular system of a previously anesthetized animal (*in vivo*). A less effective method but still useful in several cases is the sample immersion in fixative after removal from the body (*in vitro*). This method requires rapid removal and immediate immersion in the liquid fixative with as little mechanical damage as possible. Since it is absolutely critical that the specimen does not dry even slightly at this stage, it may be stored temporally in phosphate-buffered saline, if no fixative is immediately available.[13] As mentioned before, the specificity of the fixative is another critical issue in the fixation procedure. The choice of a proper fixative will depend on the purpose of the study.

The most common and best method of fixation for EM involves the use of glutaraldehyde as a primary fixation followed by secondary exposure (postfixation) to osmium tetroxide. *Glutaraldehyde* ($C_5H_8O_2$) is a dialdehyde and the most effective of the aldehyde fixatives for preserving fine structures. It stabilizes cell structures, prevents distortion during processing, and increases the permeability of tissue to embedding media. No other fixative surpasses glutaraldehyde in its ability to cross-link proteins although it also reacts with lipids, carbohydrates, and nucleic acids. It penetrates very slowly into the tissue (1 mm per hour), which calls the attention to the first criterion mentioned above. *Osmium tetroxide* acts as both a fixative and an electron-dense stain. It penetrates often slower than glutaraldehyde but reacts quickly with lipids, proteins, and lipoproteins. It does not react with ribonucleic acid (RNA) or deoxyribonucleic acid (DNA) and, in low concentrations, does not preserve microtubules. In its reduced state, this fixative renders tissue black, permitting it to be more easily seen during processing and cutting.[13,14]

There are some others fixatives such as paraformaldehyde, acrolein, potassium permanganate, and certain electron-dense stains such as uranium salts and ruthenium tetroxide, etc. Just for a general description, *acrolein* associated with glutaraldehyde gives a good preservation to microtubules; the high contrast of *potassium permanganate* ($KMnO_4$) is perfect to preserve cell membranes. It also preserves DNA and prevents its clumping during dehydration.[14] Again, a better choice will depend on the purpose of study.

40.11.2 Washing

In order to eliminate any free unreacted fixative that remains within the tissue, the specimen has to pass through several rinses in the same buffer vehicle used in the primary fixation step. If this step is not taken carefully, the micrographs will show many small black dots as a background and could generate a false analysis. The procedure includes one or two 10-min-washes after primary fixation with glutaraldehyde.[13]

40.11.3 Dehydration

Water is a highly polar molecule that is, by far, the major component of virtually all cells. Tissues have to be infiltrated by the hydrophobic embedding media, which makes dehydration a very important step during the specimen processing. It consists of replacing the water in cells with solvent agents as ethanol or acetone, by using a graded series of dehydration agents. Usually it begins with 30% or 50% ethanol or acetone followed by the secondary fixation with 70%, 85%, 95% and absolute ethanol or acetone during determined periods of time in order to eliminate the small amount of water remaining in the specimen. There are two cautions required during the dehydration process: (a) keeping the vials with dehydrant sealed tightly to avoid the absorption of water from the air and (b) replacing one dehydrant with another as quickly as possible to avoid drying of the tissue.[13]

40.11.4 Use of Transitional Solvents

This step is basically taken not only to further dehydrate the tissue but also to speed up the infiltration with the plastic embedding media that require a highly miscible solvent to it. Thus, another intermediary solvent, propylene oxide or acetone, being the first the standard solvent used, will replace the alcohol. Like absolute alcohol or ethanol, it is a highly hygroscopic, and the same precautions previously taken with these substances should be taken with propylene oxide. Usually more than one change of propylene oxide is necessary to replace the alcohol.[13]

40.11.5 Infiltration

Infiltration is the process by which dehydrants or transition fluids are gradually replaced by resin monomers. Due to its viscosity, epoxy is mixed with propylene oxide under determined proportions and this mixture gradually penetrates into the sample after dehydration.[13] The epoxy-solvent proportions is increased up to pure epoxy. Specimens in 100% resin are then transferred into molds or capsules containing the resin and are finally placed in an oven where the epoxy components polymerize to form a solid block.[13] The infiltration schedule may vary depending on the specimen.

40.11.6 Embedding

The next step is the embedding process, by which the liquid embedding media must permeate the specimen and harden to form a solid matrix. In common practice, this is accomplished by using epoxy monomers that harden with time and under certain curing conditions. Usually, an epoxy embedding media, Epon 812, has been used along the years for EM. The Epon embedding media is prepared by thoroughly mixing the resin (Epon 812 or substitute), the hardeners (DDSA—dodecenyl succinic anhydride and NMA—nadic methyl anhydride), and the accelerator (BDMA—bensyldimethylamine). The hardness of the Epon mixture will be reached by varying the proportions of the relative amounts of DDSA and NMA, the former producing softer blocks and the latter harder blocks.[13] The most adequate firmness will depend on the manufacturer's recommendations and the used proportions are commonly available in the literature.

40.11.7 Microtomy

This procedure involves cutting the specimen into extremely thin slices or sections for adequate viewing on the TEM. It requires practice and attention to details such as the quality of the knives and specimen support grids, and cleanliness of all reagents and glassware. Generally, specimen microtomy for EM is performed in two steps: semithin sectioning followed by ultramicrotomy. The former is useful to indicate the quality of the sample preparation and it helps to locate areas of interest for example, lesion, specific cell type, and so on. Semithin sections are cut with a glass knife and their thickness ranges between 0.5 and 2 μm .[13] These sections are usually stained with toluidine blue for evaluation at light microscopy. As the desired area of the block is selected, the specimens are subjected to ultramicrotomy, which is done using a diamond knife. The section thickness is determined by the interference colors of the sections floating on the surface of the diamond knife trough, varying between gray (around 60 nm and useful for high-resolution), gold (90–150 nm and useful for low magnification) and silver (60–90 nm, for general purposes).[14] Thin sections are then placed on copper meshed grids and stained with uranyl citrate followed by lead citrate, the standard method. However, uranyl acetate can be used as an additional step during the regular processing and, in this case, the grids are stained just with lead citrate.[13]

40.12 EM Applications to Biological and Clinical Studies

EM is a powerful tool for understanding the complexity of cellular mechanisms involved in numerous diseases as well to allow accurate diagnosis. Some examples highlighting the use of EM for biological and clinical investigations are discussed in the following.

40.12.1 Immune Cells Associated to Inflammatory and Allergic Diseases

The ultrastructure of cells from the immune system has been extensively investigated in order to obtain insights into the mechanisms of inflammatory and allergic diseases. EM techniques have allowed understanding the mechanisms of secretion of cells such as human eosinophils, cells typically associated with host defense against pathogens and with inflammatory and allergic diseases such as asthma.[15] Figure 40.17 shows an electron micrograph of a human eosinophil from a patient with hypereosinophilic syndrome (HES), which is characterized by increased numbers of activated eosinophils in the blood and tissues. TEM has identified in these patients a significant increase of large transport vesicles within blood eosinophils in parallel to the presence of altered secretory granules, indicating the occurrence of a secretion process termed as piecemeal degranulation.[16] This process which involves vesicular trafficking is important to understand how citotoxic proteins are released from eosinophils in affected tissues of patients with HES.[16]

40.12.2 Infectious Diseases

EM has also facilitated the study of infectious diseases, for example, helping to understand the function of inflammatory macrophages, the first line of defense in protecting the host from invading microorganisms. Activated macrophages exert critical activities in immunity to parasites, playing a pivotal role in the mechanism for halting diseases such as infections caused by the parasite *Trypanosoma cruzi*. This parasite is the causal agent of Chagas' disease, a significant public health issue which is still a major cause of morbidity and mortality in Latin America.

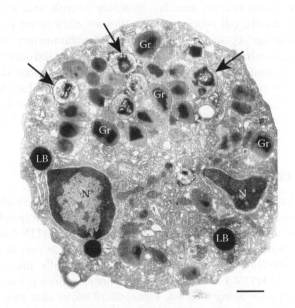

FIGURE 40.17 Ultrastructure of a blood eosinophil from a hypereosinophilic patient. Specific granules (Gr) with typical crystalloid cores are observed together with enlarged granules with disassembled cores (arrows). N, nucleus; LB, lipid bodies. Scale bar, 800 nm.

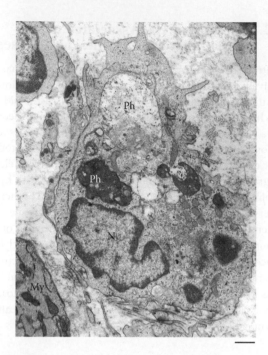

FIGURE 40.18 Ultrastructural features of a rat inflammatory macrophage in the heart following infection with *T. cruzi* (12 days). This cell exhibits cytoplasm that is voluminous, rich in organelles, and includes phagolysosomes (Ph) with degenerating material. N, nucleus; My, myocardium. Scale bar, 500 nm.

Figure 40.18 shows the ultrastructure of an inflammatory macrophage from the heart of a rat infected with *T. cruzi*. These activated macrophages exhibit a significant increase in their organelle apparatus, mainly Golgi complex, rough endoplasmic reticulum, polysomes, lysosomes, and vesicles. Phagolysosomes with varying sizes and electron densities containing amorphous or granular materials, cell debris, and parasites are often found in the macrophage cytoplasm. The nucleus of inflammatory macrophages is irregular in outline, more euchromatic and occasionally multinucleated. Moreover, these cells are characterized by a striking increase in surface projections and pseudopodia. These morphological characteristics of activation observed only by TEM have been recognized as an accurate indication of high phagocytic and microbicidal activities of macrophages.[17]

40.12.3 Biopsies of Virus-Infected Skin

EM has proved to be an important tool in studying dermal alterations caused by the infection with the human T-cell lymphotropic virus type 1 (HTLV-1).[18] The association of HTLV-1 with infectious dermatitis and other diseases is due in part to the immunodeficiency induced by this virus. Basically, the main target of HTLV-1 is the CD4 + T lymphocyte, which is immortalized after being infected. Figure 40.19 shows skin biopsies from normal (Figure 40.19a) and HTLV-1-infected individuals with dermatological manifestations of chronic dermal infection (Figure 40.19b).[19,20] While normal skin (Figure 40.19a) shows typical structural components such as cell junctions (desmosome

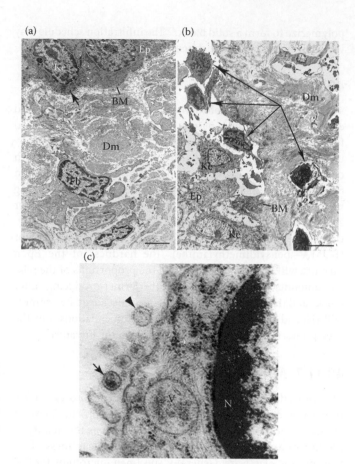

FIGURE 40.19 Ultrastructural features of a normal and virus-infected skin. In (a), typical epithelial structures such as basal membrane (BM), desmosome and hemidesmosome (arrow) are observed in addition to fibroblasts (Fb). In (b), dermal papillae from a virus-infected skin show a destroyed basal membrane (BM), with disruption of hemidesmosome and desmosome between keratinocytes (Ke). Several mononuclear cells (arrows) are infiltrating the epidermis (Ep) and dermis (Dm) in response to the infection with the HTLV-1 virus; Figure (c) shows HTLV-1 particles with ultrastructural morphological of likely mature (arrow) and immature viral (arrowhead) blossomed from the fibroblast cell membrane. V, vacuole, N, nucleus. Scale bar, 500 nm (a) 6 μm, (b) 12 μm, and (c) 120 nm.

and hemidesmosome), basal membrane, and connective tissue with fibroblast and collagen fibers, ultrastructural alterations are observed in the case of chronic dermal infection (Figure 40.19b). EM reveals intercellular spaces due to the disruption of desmosomes between keratinocytes, condensed nucleus, and vacuolated cytoplasm. Disruption of hemidesmosomes between epidermis and dermis and an infiltration of mononuclear cells into the dermal papillae are also observed (Figure 40.19b).[19,20]

Because of their very small size, viruses can only be observed by EM. Structures with diameters of approximately 100 nm and ultrastructural characteristics of the HTLV-1 particles are observed near or emerging from cells such as fibroblasts (Figure 40.19c). This virus continues to multiply in the compartments of the infected skin leading to a local chronic inflammatory reaction which attempts to control the infection.

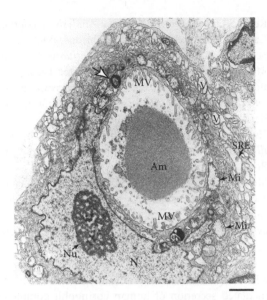

FIGURE 40.20 Ultrastrutural features of an adenocarcinoma cell—tumor cells showing a cytoplasm and nucleus with an abnormal vesicle aspect. A pleiomorphic nucleus with highly dispersed chromatin is pushed to the cytoplasm periphery by a large vesicle containing an amorphous electron dense material (Am) merged by numerous microvilli (MV). In addition to the numerous small vesicles (V), the tumor cells cytoplasm displays vacuolated mitochondria (Mi), dilated smooth endoplasmic reticulum (SER), and a dense lamellar body (white arrow). N, nucleus; Nu, Nucleolus. Scale bar, 2 μm.

40.13 Cancer Diagnosis Using EM

EM is a powerful adjunct to routine light microscopy when applied to the diagnosis of human tumors.[21] In many cases, accurate diagnosis of the origin of tumor depends on EM. For example, some tumors submitted for analysis are classified as adenocarcinomas because they reveal glandular structures by light microscopy, but the site of origin can only be determined by EM.[21] Metastatic adenocarcinoma from an unknown primary site requires extensive roentgenographic and endoscopic studies for localization of the site of tumor origin. Given these difficulties, rapid ultrastructural analysis allows the accurate diagnosis of the cell of tumor origin during metastatic adenocarcinomas. By providing magnification high enough to identify numerous intracytoplasmic elements, such as the presence of filaments, EM can identify typical cell structures such as lamellar (surfactant) bodies, which are typical of alveolar cell carcinoma of the lung or presence of apical terminal webs and diagnostic of gut epithelial cells (Figure 40.20).[22] This is important because it can provide an adequate therapeutic strategy for the patient.

40.14 EM Imaging Artifacts

Ideally, electron micrographs are of little scientific value if the information obtained from them is not of good quality. Mistakes made in an EM laboratory during specimen preparation can reflect in distortions named as *artifacts*. Some of the artifacts commonly observed on electron micrographs are

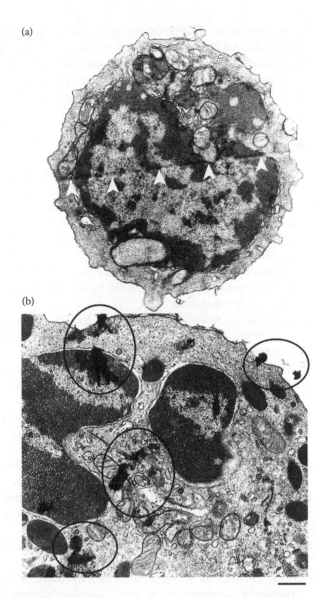

FIGURE 40.21 Artifacts commonly observed on electron micrographs. In (a), knife marks, which appear as perpendicular lines crossing the whole cell section (arrowheads) are observed on a blood cell (lymphocyte). In (b), staining artifacts appears as crystals or dense particles (circles) on the surface of a section from an eosinophil. Scale bar, (a) 1.2 μm and (b) 0.5 μm.

discussed below. It is important to emphasize that the interpretation skills come with experience and with a reasonable background in the fundamentals of cell structure and tissue architecture.

Generally, a common sign of poor fixation is the swelling of organelles. Mitochondria cristae often become peripherally positioned, turning to a rarified matrix or swollen appearance. On the other hand, inadequate removal or replacement of either the water or the fluids usually results in holes in the tissue. The sectioning step also requires special care. Knife marks or scrape marks on the section, which appears as lines or tears perpendicular to the knife edge are common findings on electron micrographs (Figure 40.21a) if the knives are not properly used or if they are defective or dirty in certain areas. The most common and problematic artifacts are those related to the staining of the specimens, which may take a variety of forms on the electron micrographs. Lead precipitate may appear as fine grains ("pepper"), crystals (Figure 40.21b) or as dense spherical particles.[13]

References

1. Buseck, P.R. (ed.), *Minerals and Reactions at the Atomic Scale: Transmission Electron Microscopy*, Reviews in Mineralogy 27. Mineralogical Society of America 1992.
2. Bozolla, J.J. and Russell, L.D., *Electron Microscopy: Principles and Techniques for Biologists*, Second edition, Boston: Jones and Bartlett Publishers, 1998.
3. Egerton, R.F., *Physical Principles of Electron Microscopy: An Introduction to TEM, SEM, and AEM*, Second edition, New York, NY: Springer, 2008.
4. Goldstein, J.I., Newbury, D.E., Echlin, P., Joy, D.C., Fiori, C., and Lifshin, E., *Scanning Electron Microscopy and X-ray Microanalysis*, New York, NY: Plenum Press, 1981.
5. Williams, D.B. and Carter, C.B., *Transmission Electron Microscopy: A Textbook for Materials Science*, Second edition, Springer, New York, 2009.
6. De Graef, M., *Introduction to Conventional Transmission Electron Microscopy*, Cambridge, MA: Cambridge University Press, 2003.
7. Goldstein, J.I. and Yakowitz, H. (eds), *Practical Scanning Electron Microscopy*, New York, NY: Plenum Press, 1975.
8. Lloyd, G.E., Atomic number and crystallographic contrast images with the SEM: A review of backscattered electron techniques, *Mineral. Mag.*, 51: 3–19.
9. Putnis, A., *Introduction to Mineral Sciences*, Cambridge, MA: Cambridge University Press, 1992.
10. Potts, P.J., Bowles, J.F.W., Reed, S.J.B., and Cave, M.R., *Microprobe Techniques in the Earth Sciences*, Chapman & Hall, 1995.
11. Reed, S.J.B., *Electron Microprobe Analysis*, Second edition, Cambridge University Press, Cambridge, 1993.
12. Scott, V.D. and Love, G. (eds), *Quantitative Electron-Probe Microanalysis*, Ellis Horwood, Chichester, 1983.
13. Bozzola, J.J. and Russel, L.D., *Electron Microscopy: Principles and Techiniques for Biologists*. Second edition, Jones and Barttlet Publishers, Sudbury, MA, 1998.
14. Hunter, E, *Practical Electron Microscopy: A Beginner's Illustrated Guide*, Second edition, Cambridge University Press, New York, 1993.
15. Melo, R.C.N., Spencer, L.A., Dvorak, A.M., and Weller, P.F., Mechanisms of eosinophil secretion: Large vesiculotubular carriers mediate transport and release of granule-derived cytokines and other proteins. *J. Leukocyte. Biol.*, 2008, *83*: 229–236.
16. Melo, R.C.N., Spencer, L.A., Perez, S.A., Neves, J.S., Bafford, S.P., Morgan, E.S., Dvorak, A.M., and Weller, P.F., Vesicle-mediated secretion of human eosinophil granule-derived major basic protein (MBP). *Lab. Invest.*, 2009, *89*: 769–781.
17. Melo, R.C.N., Acute heart inflammation: Ultrastructural and functional aspects of macrophages elicited by infection. *J. Cell Mol. Med.*, 2009, *13*: 279–294.
18. Bittencourt, A.L., Vertical transmission of HTLV-I/II: A review. *Rev. Inst. Med. Trop. Sao Paulo*, 1998, *40*(4): 245–251.
19. D'Agostino, D.M., Zotti, L., Ferro, T., Cavallori, I., et al., *Virus Research*, 2001, *78*: 35–43.
20. Smith, N., Eady, R.A.J., and Spittle, M.F., HTLV-1-associated lymphoma presenting in the skin. *Br. J. Dermatol.*, 2006, *127 I40*: 37–38.
21. Dvorak, A.M. and Monahan, R.A., *Diagnostic Ultrastructural Pathology I: A Text-Atlas Illustrating the Correlative Clinical Ultrastructural Pathologic Approach to Diagnosis*, CRC Press, Boca Raton, FL, 1992.
22. Dvorak, A.M. and Monahan, R.A., Metastatic adenocarcinoma of unknown primary site. Diagnostic electron microscopy to determine the site tumor origin. *Arch. Pathol. Lab. Med.*, 1982, *106*: 21–24.

VI

Physics of Accessory Medicine

IV

Physics of Accessory Medicine

41

Lab-on-a-Chip

Shalini Prasad,
Yamini Yadav,
Vindhya Kunduru,
Manish Bothara, and
Sriram Muthukumar

41.1 Need for Lab-on-a-Chip Devices

Lab-on-a-chip (LOC) is a miniaturized, portable device that integrates different laboratory functions such as real-time polymerase chain reaction (PCR), biochemical assays, immunoassay, dielectrophoresis, and bioseparations on a single device. These novel LOC systems have various applications in biotechnology, medicine, clinical diagnostics, chemical engineering, and pharmaceutics. LOC devices have several advantages in terms of low-volume consumption, thereby indicating less wastage of reagents/chemical solutions and requiring less sample bodily fluid for clinical diagnostics. Moreover, nowadays, there is also an ongoing need for active continuous time monitoring systems capable of rapid analysis with substantial accuracy for detecting biomolecules. LOC devices have the ability to carry out fast analysis and have a very short response time in the range of a few seconds to minutes. In addition, compactness of the system and small volumes parallelization allows for better process control of the complete reaction sequence, that is, mixing, heating, and separation in the microfluidic chambers. These high-throughput LOC systems, due to their miniaturized size, have lower fabrication and disposal costs.

Additionally, monitoring human health for early detection of disease conditions or health disorders is vital for maintaining a healthy life. Many cellular and subcellular structures such as tissues, microorganisms, organelles, cell receptors, enzymes, antibodies, and nucleic acids help determine the physiological state of a disease condition. In addition, analysis of the food and the environment for pertubants such as pesticides and river water contaminants in the form of harmful pathogens has also become invaluable for health diagnosis. A "real-time" LOC biosensor helps detect a number of analytes of interest in a near-continuous manner and plays an important role in accomplishing effective data generation and data processing, thereby supporting real-time decision-making and rapid manipulation. The critical features in a microfluidic system include the ability to characterize the flow in a microfluidic channel and the associated electrokinetic effects that can be leveraged for detection.

41.2 Microfluidics

The flow of a fluid through a microfluidic channel can be characterized by the Reynolds number. The fluid can be actuated by electro-osmotic pumping or pressure-driven flow. Laminar flow occurs at Reynolds numbers (Re) below ~2000:

$$Re = \frac{\upsilon \rho l}{\eta}, \tag{41.1}$$

where υ is the velocity, ρ is the density, η is the viscosity, and l is the channel diameter.

Typically, Re is much less than 100, often less than 1.0, due to smaller size of microchannels. In this Reynolds number regime, flow is completely laminar and no turbulence occurs. The transition to turbulent flow generally occurs in the range of Reynolds number 2000. Laminar flow provides a means by which molecules can be transported in a relatively predictable manner through microchannels. Therefore, the fluid actuation through microchannels can be achieved by electro-osmotic pumping and pressure-driven flow. In electro-osmotic pumping, electrical potential is applied across the electrolyte-filled channel between specific points to generate electro-osmotic effect. The walls of a microchannel have an electric charge which creates an electric double layer of counterions at the walls. When an electric field is applied across the channel, the ions in the double layer move toward the electrode of opposite polarity. The potential is applied between specific points on a microfabricated device to control

FIGURE 41.1 Schematic representation of electro-osmotic pumping, where the fluid flow is controlled by simply applying the potential between two specific points.

the direction of flow, as shown in Figure 41.1. The electro-osmotic flow velocity (v_{eo}) is given by Equation 41.2[1]:

$$v_{eo} = \frac{\varepsilon\xi}{4\pi\eta}E, \tag{41.2}$$

where ε is the solution's dielectric constant, ζ is the potential arising from the charge on the channel wall, η is the solution viscosity, and E is the electric field in V cm^{-1}.

As opposed to voltage potential, pressure-driven flow requires valves to control the flow direction. In pressure-driven flow, the fluid is pumped through the device via positive displacement pumps, such as syringe pumps. For pressure-driven laminar flow, the fluid velocity at the walls is zero. Hence applications that use microfluidics in LOC devices focus on optimizing the device geometry to achieve laminar flow where the biomolecules of analytes of interest can be subjected to electrokinetic phenomenon.

41.3 Potential Laboratory Functions

In the field of microfluidics, the long-term goal is to create an integrated, portable clinical diagnostic device for point-of-care applications. The revolutionizing potential of LOC systems, based on microfluidics technology, is widely recognized in life sciences and in industrial process control. Current technology has enabled the integration of microfluidic components, micro-optical devices, and control electronics in a complete, compact system. The purpose of the microfluidic LOC device is to manipulate, process, and deliver a solution to the detector which can collect, record, or transmit data. These microfluidic LOC devices typically consist of a monolithic material that is patterned with microscale channels and features such as mixers, reactors, valves, injectors, and separators that can assume the role of equivalent macroscopic laboratory equipment.

Peter B. Howell et al. designed a microfluidic mixer, consisting of a rectangular channel with grooves placed on the top and bottom of the channel.[2] Figure 41.2 shows the schematic diagram demonstrating the channel layout of the mixer with the grooves. Compared to standard microfluidic mixer with a low rate of diffusion, the novel mixer increased the driving force. It also formed advection patterns and a pair of vertically stacked vortices positioned side by side.

Cindy K. Harnett et al. designed a microfluidic mixer driven by induced charge electro-osmosis by creating microvortices within a fluidic channel by application of alternating current

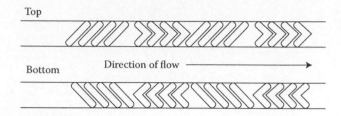

FIGURE 41.2 Demonstrating rectangular channel with grooves that help mixing of two liquids.

(AC) electric fields.[3] On the other hand, recently, Suk Tai Chang et al. designed novel, actively controlled microfluidic–electronic chips for pumping or mixing of the fluid in the microfluidic channel.[4] In these mixers, the principle of microfluidic pumping and mixing was guided by electronic components. The miniature diodes embedded into the microchannel walls rectify the voltage induced between their electrodes. This resulted in an electro-osmotic flow near the diode surface that was utilized for mixing.

Similar to a micromixer, a microreactor is one of the essential components of LOC devices. Typical microreactors are continuous lateral flow microfluidic channels for the chemical reaction of two or more chemical agents. Harry L. T. Lee et al. fabricated microbioreactor arrays with integrated mixers and fluid injectors for high-throughput experimentation with pH and dissolved oxygen controls.[5] The experiment included fermentation of *Escherichia coli* bacteria under controlled conditions. *E. coli* fermentations to cell densities greater than 13 g-dcw L^{-1} were achieved. Moreover, Masaya Tokoro et al. developed a new technique for microchemical reactors where the reactions take place inside the droplets.[6] The experiment helped prevent contamination and evaporation of immiscible liquid medium and similar problems that become apparent when handling microdroplets. Figure 41.3 demonstrates microchemical reactors where the reactions take place inside the droplets.

Kazunori Iida et al. built a microreactor for living anionic polymerization for LOC applications.[7] Conducting these reactions in a batch reactor results in uncontrolled heat generation with potentially dangerous rises in pressure; however, by using a microreactor, the reactions were safely maintained at a controlled temperature at all points in the reactor. On the other

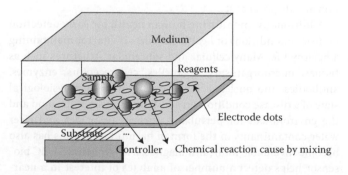

FIGURE 41.3 Schematic diagram representing concept of a microchemical reactor.

hand, Yong Lu et al. fabricated the microfibrous reactors for small-scale hydrogen production.[8] The high efficiency hydrogen was produced by decomposition of ammonia at moderate temperatures. This hydrogen-producing chemical reactor had various advantages in terms of large void volume, entirely open structure, high heat/mass transfer, high permeability, good thermal stability, and unique form factors. Konstantin Seibel et al. built a programmable planar micropump for LOC applications, which consists of an electro-osmotic micropump combined with a mass flow sensor in a closed control loop.[9] The electro-osmotic micropump was designed with microchannels of SU-8 and polyacrylamide gel electrodes with an on-chip mass flow sensor and external control circuitry for flow rates between 0 and 30 nL min⁻¹. Similarly, Seung-Mo Ha et al. presented a disposable thermo-pneumatic micropump with microscale check valves for easy incorporation in a fully integrated biochip.[10]

Typically, for biological applications, these micropumps are made of biocompatible materials such as glass and polydimethylsiloxane (PDMS). The pumping system was simulated by different electrical signal parameters such as electrical power, voltage pulse frequency, and duty ratio on the flow rate. The maximum flow rate up to 1000 μL min⁻¹ was achieved.

41.4 Methods of Detection

Diagnostic LOC-based chemical/biosensors are probably the largest area of research in the field of portable detection in liquid media. Over the last few years, many new ideas and technologies have been proposed in literature. In this section, comparative studies of these detection mechanisms have been discussed. Each detection mechanism differs significantly in the design of the device; the fabrication methodology adopted for such a design also differ based on the area of application. However, each mechanism has the same goals in mind that include reducing the sensing elements to be equivalent to the size of the target species, improving the sensitivity, reducing the reagent volumes to offset the costs of the reagents, moving toward a real-time system to acquire the results and simultaneously obtain a lower detection limit, and miniaturizing the entire system to improve portability. Biosensor devices can be broadly classified based on the basic principles of detection into two classes: labeled detection and label-free detection. Label-free detection is classified based on detection mechanism, namely, electrical detection, optical detection, and mechanical detection where fluorescent labels are not used for the purpose of detection. The most widely used label-free detection mechanism is the use of electrical parameters for detection. Electrical biosensors can be further classified based on the electrical measurement, which includes voltammetric, amperometric/coulometric, potentiometric, and impedance. Chunxiong Luo et al. also developed a highly specific immunoassay system for goat antihuman IgG using gold nanoparticles and microfluidic techniques.[11] The gold nanoparticles were coated with protein antigens in the presence of their corresponding antibodies to microfluidic channel surface. The effects of time accumulation, flow velocity, and concentration of antibodies to the red light

absorption percentage (RAP) of deposition were investigated, and by controlling the reaction time and flow velocity, a dynamic range of three orders of magnitude and a detection sensitivity of 10 ng mL⁻¹ of goat antihuman IgG were achieved. Another example of optical detection is miniaturized optical detection system for sensing in microchannels, based on integrated optical waveguide technology.[12] The absorption and fluorescence measurement schemes allowed detection limits to be 10 nM and 1 μM, and showed potential integration into an LOC system. Frank B. Myers et al. have identified various emerging detection paradigms involving nanoengineered materials and highlighted microfluidic diagnostic systems that demonstrate practical integration of sample preparation, analyte enrichment, and optical detection. On the other hand, Mehdi Javanmard et al. have demonstrated the use of rapid electrical detection techniques for quantification of target protein biomarkers using protein functionalized microchannels.[13] Detection of anti-hCG antibody at a concentration of 1 ng mL⁻¹ and at a dynamic range of three orders of magnitude was achieved in less than 1 h. The multiplex high-throughput LOC device can also be used to probe different protein biomarkers in human serum for cancer detection. XingJiu Huang et al. were able to detect pathogens such as *E. coli* using the electrical mechanism.[14] Electrical determination of *E. coli* O157:H7 was performed and investigated by antibody-modified individual SnO$_2$ nanowire (TONW) coupled with microfluidic chip. Also, Vindhya Kunduru et al. used electrical characterization to demonstrate protein detection using a platform-based assembly of microelectrode arrays (MEAs).[15] Using soft lithography methods, the sensing device was incorporated into a PDMS encasement engineered to support the base chip for further interrogation during the protein detection process. The biochip in Figure 41.4 demonstrates protein biosensor.

In addition, Vinu Lalgudi Venkatraman et al. designed an IrOx nanowires-based biosensor for health monitoring system.[16] Detection of two inflammatory proteins C-reactive protein (CRP) and myeloperoxidase (MPO), which are biomarkers of cardiovascular diseases, was demonstrated. This device was

FIGURE 41.4 Optical micrograph of the LOC device used for the formation of "microbridges." The device comprises of a base MEA encapsulated by a polymeric package.

based on electrical detection of protein biomarkers wherein an immunoassay was built onto the IrOx nanowires that in turn undergoes specific electrical parameter perturbations during each binding event associated with the immunoassay. The optical image demonstrates real-time sensing circuits embedded with microfluidic system in Figure 41.5.

Yeolhob Lee et al. incorporated a mechanical-based microbiosensor for detection of biomaterials, that is, for detecting the human insulin–antihuman insulin binding protein, the poly T-sequence DNA, the K20-Thiol DNA, and the K40-Thiol DNA in air.[17] The microsensor consists of a microcantilever actuated by piezoelectric transducer (PZT) film, designed to act as both a sensor and an actuator.

41.5 Mechanisms of Manufacturing

The most common techniques for manufacturing LOC device are by using soft lithography and photo lithography. Soft lithography suggests a new conceptual approach in nanomanufacturing that refers to a family of techniques for fabricating or replicating structures using "elastomeric stamps, molds, and conformable photomasks." Soft lithography includes techniques such as microcontact printing, microtransfer molding, micromolding in capillaries, microreplica molding, solvent-assisted microcontact molding, and near-field phase-shift lithography. These techniques are called soft lithography because all these techniques use soft, flexible, and elastomeric polymer material called PDMS. PDMS "stamps" with patterned relief are stamped onto the substrate to generate features. The stamps can be prepared by casting PDMS against masters patterned by conventional lithographic techniques, as well as against other masters of interest. Figure 41.6 represents the process sequence of soft lithography.

Many groups have implemented soft lithography to drive the pumps, valves, and electrodes used to manipulate fluids in microfluidic devices. John Paul Urbanski et al. developed a programmable and scalable control of discrete fluid samples in a PDMS microfluidic system using multiphase flows with an aim to achieve a new level of scalability and flexibility for LOC

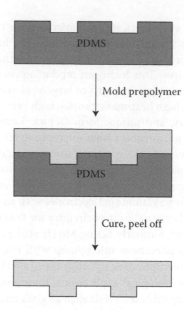

FIGURE 41.6 Schematic representation of process sequence using the soft lithography technique.

experiments.[18] Using soft lithography fabrication, world-to-chip connections can also be integrated directly into the masters, enabling devices with robust, well-aligned fluidic ports directly after molding.[19] Yu Hongbin et al. suggested a novel method for fabricating PDMS-based microchannel fabrication method for LOC application.[20] In this method, the microstructure was prepared by transferring the desired microchannel pattern into one PDMS substrate via soft lithography and bonding it to a spin-coated PDMS membrane with the oxygen plasma activated technology. Christopher J. Easley et al., with the help of soft lithography, on a PDMS–glass hybrid microdevice fabricated PDMS membrane valves for diaphragm pumping in order to couple infrared-mediated DNA amplification with electrophoretic separation of the products in a single device.[21] Schematic three-dimensional (3D) representation of microchip used for PCR is shown in Figure 41.7.

Although photolithography has been used for fabrication of integrated circuits to create a path for flow of electrons, similarly this technique can also produce the path for the flow of fluids. Functional microfluidics elements such as microchannels, micropumps, microvalues, bioreactors, and so on, are also intregrated to form LOC systems. Photolithography generally uses a pre-fabricated photomask or reticule as a master from which the final pattern is derived. In this method, radiation-sensitive polymeric materials called resists are used to produce microchannels in the plastic substrates. Figure 41.8 demonstrates schematic representation of the photolithography process.

In photolithography, the process sequence consists of five different steps. The step-by-step procedure consists of spin coat, exposure, development etch, and strip. In spin coat process, a thin layer of photoresist is coated onto base substrate and prebake to remove casting solvent. Later, the resist is exposed to an optical light of specific wavelength through a mask. The exposed

FIGURE 41.5 IrOx biosensor integrated with microfluidic channels and a chip with external pin layout for real-time analysis.

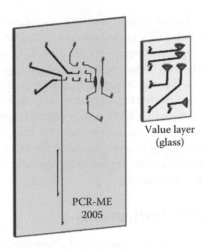

FIGURE 41.7 Schematic 3D representation of microchip used for PCR. (Adapted from Easley, C. J., Karlinsey, J. M., and Landers, J. P., *Lab Chip* 6: 601–610, 2006.)

resist film is then developed typically by immersion in a developer solvent to generate 3D relief images. Exposure may render the resist film more soluble in the developer, thereby producing a positive-tone image of the mask. Conversely, it may become less soluble upon exposure, resulting in the generation of a negative-tone image. The resist image is transferred onto the

substrate, and the resist film that remains after the development functions as a protective mask which is later etched. The resist film must "resist" the etchant and protect the underlying substrate while the bare areas get etched. The remaining resist film is finally stripped, leaving an image of the desired structures in the substrate. The process is repeated many times to fabricate complex LOC devices. Section 41.6 describes clinical applications of LOC devices for diagnosis of harmful diseases.

41.6 Clinical Applications

Analogous to the miniaturization of computer chips, LOC devices undergoes reduction in their size which offers the benefit of shorter reaction speeds with faster analysis times. Multiple active components on the microfluidic-based LOC chip offer parallel operation modes, thereby resulting in multiplexed analyses and higher throughput. These LOC devices can be cost effectively mass produced with a promise of higher analysis rates and better efficiency owing to the compactness and better process control; lower analyte consumption saves the cost of expensive reagents and is environment friendly during disposal. The most common potential clinical applications on an LOC device includes microfluidic dispenser, concentration gradient generator, electrophoretic separator, PCR chip for DNA amplification, quantitative DNA sensor chip, flow cytometry-based immunoassay for bacteria such as *E. coli* detection, real-time PCR detection, blood sample preparation for cellular analysis, DNA microarray, and protein microarray.

The development of a disposable plastic lab-on-a-biochip incorporates smart passive microfluidics with embedded on-chip power sources and integrated biosensor array for applications in clinical diagnostics and point-of-care testing. Chong H. Ahna et al. had discussed various handheld analyzers capable of multiparameter detection of clinically relevant parameters to detect the signals from the cartridge-type disposable biochip.[22] Figure 41.9 illustrates disposable smart LOC for point-of-care clinical diagnostics.

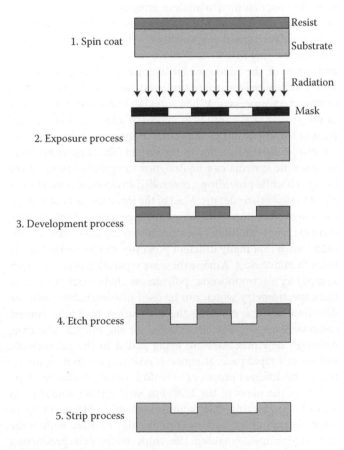

FIGURE 41.8 Schematic representation of photolithography process.

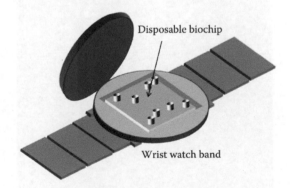

FIGURE 41.9 Microfluidic portable assay in a wristwatch configuration for disease diagnostics. The disposable wristwatch configuration comprises microfluidic devices with multiple inlet and outlet ports. Sample handling and analysis is performed within the wristwatch configuration.

Kewal K. Jain et al. explained in brief about postgenomic applications of LOC devices and microarrays.[23] The author further discussed in detail about biochip tools as aids to cancer management, diagnosis of infections, protein biomarkers and protein biochips, and finally on protein biochips for point-of-care diagnosis. DNA amplification by the PCR is a widely used tool in routine DNA assays. Jing Wang et al. constructed a pneumatically driven disposable microfluidic cassette for DNA amplification and detection.[24] The microfluidic device consists of a PCR thermal cycler, an incubation chamber to label PCR amplicons with upconverting phosphor (UPT) reporter particles, conduits, temperature-activated, normally closed hydrogel valves, and a lateral flow strip. On the other hand, Thomas Ming-Hung Lee et al. developed silicon-/glass-based devices for performing PCR target amplification and sequence-specific electrochemical (EC) detection.[25] M. Bothara et al. developed a "point-of-care" device for early disease diagnosis through protein biomarker characterization. The microfluidic-based protein biosensor is shown in Figure 41.10. The detection principle lies in the formation of an electrical double layer and its perturbations caused by proteins trapped in a nanoporous alumina membrane over a MEA platform measured 20 mm × 20 mm.

Electrophoretic separations are the most fundamental applications of an LOC device. On electro-osmotic pumping, the charged analytes is treated by electrophoretic separation based on the ratio of their charge to hydrated radius:

$$v_{ep} = \frac{(5/3)q}{\pi r \eta}, \qquad (41.3)$$

where v_{ep} is the electrophoretic velocity of the analyte, q is the charge, r is the hydrated radius, and η is the solution viscosity.

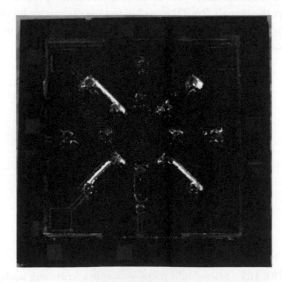

FIGURE 41.10 Integration of microfabrication and soft lithography to generate a lateral flow immunoassay for early disease diagnostics through detection of disease biomarkers using electrical detection mechanisms.

Jeong-Gun Lee et al. designed and microfabricated microchip for one-step DNA extraction and real-time PCR in one chamber for rapid pathogen identification.[27] A single pulse of 40 s lysed pathogens including *E. coli* and gram-positive bacterial cells, as well as the hepatitis B virus mixed with human serum, was demonstrated. Pathogen detection system was developed by integrating 808 nm laser and carboxyl-terminated magnetic beads. In the current section, we have discussed the clinical diagnostic application, while in the following section, we will discuss about the future indication and trend toward application of LOC systems.

41.7 Future Indications

The field of microfluidics combines the fabrication methods of micro- and nanotechnology with knowledge about the behavior of fluids on the fundamental microscopic level to give rise to very powerful techniques in controlling and measuring chemical reactions and physical and biological processes on the micro- and nanoscale.

In the LOC, a miniaturized fluidic system, measurements or sample manipulations that otherwise require considerable human involvement and sizeable laboratory equipment and space are performed on a chip that would fit in the palm of a hand. The development of the LOC also bring about possibilities of handling very small volumes, in the picoliter range, bringing about the opportunity to analyze samples that were previously beyond our reach. In addition, it has proven to have the capacity to increase both speed and sensitivity. These qualities combined with the fact that this is a tool on the same scale as the single cell and many of the fundamental biological processes make the LOC a well-suited means for the investigation and manipulation of these very processes. When studying biological systems, one of the greater challenges is to approach and examine the vital parts of the organisms or molecules in their natural states without disturbing the function of the system. The characteristics of microfluidic systems can be designed to resemble those of the living cell while providing a controlled environment for the scientist to study their details. Most of the intricate parts of biological systems function in the very size range of these tools making them excellent candidates for studying biological systems. It has been shown that many different processes can be carried out in microfluidic devices. Among these are separation processes such as capillary electrophoresis, polymerase chain reactions (PCR), mass spectrometry which can be used in conjunction with the detection devices, enabling the integration of fluidic control components, and transport systems for the microfluidic chip. Although new processes are being added to the microfluidic toolbox at a rapid pace, attention is now turning to the integration of the different processes to build a useful lab-on-the-chip.

Taking the ideas of the LOC one step further would be to instead of analyzing micro- and nanoscale objects trying to assemble molecules and macromolecular entities within the chip environment making the chips microscale production facilities or factory-on-a-chip devices.

Assembling molecules and other biological entities on a chip could enable us to construct complex molecular structures in a bottom-up high-throughput process. These chips could utilize novel techniques such as molecular motors as a "conveyor belt" transporting the assembled parts through the factory.

References

1. Figeys, D. and Pinto, D., Lab-on-a-chip: A revolution in biological and medical sciences, *Analytical Chemistry* 72: 330–335, 2000.
2. Howell, P. B., Jr., Mott, D. R., Fertig, S., Kaplan, C. R., Golden, J. P., Oran, E. S., and Ligler, F. S., A microfluidic mixer with grooves placed on the top and bottom of the channe, *Lab Chip* 5: 524–530, 2005.
3. Harnett, C. K., Templeton, J., Dunphy-Guzman, K. A., Senousy, Y. M., and Kanouff, M. P., Model based design of a microfluidic mixer driven by induced charge electroosmosis, *Lab Chip* 8: 565–572, 2008.
4. Chang, S. T., Beaumont, E., Petsev, D. N., and Velev, O. D., Remotely powered distributed microfluidic pumps and mixers based on miniature diodes, *Lab Chip* 8: 117–124, 2008.
5. Lee, H. L. T., Boccazzi, P., Ram, R. J., and Sinskey, A. J., Microbioreactor arrays with integrated mixers and fluid injectors for highthroughput experimentation with pH and dissolved oxygen control, *Lab Chip* 6: 1229–1235, 2006.
6. Tokoro, M., Katayama, T., Taniguchi, T., Torii, T., and Higuchi, T., PCR using electrostatic micromanipulation, in *SICE 2002. Proceedings of the 41st SICE Annual Conference*, 2002, pp. 954–956, Vol. 952.
7. Iida, K., Chastek, T. Q., Beers, K. L., Cavicchi, K. A., Chun, J., and Fasolka, M. J., Living anionic polymerization using a microfluidic reactor, *Lab Chip* 9: 339–345, 2009.
8. Lu, Y., Wang, H., Liu, Y., Xue, Q., Chen, L., and He, M., Novel microfibrous composite bed reactor: High efficiency H_2 production from NH_3 with potential for portable fuel cell power supplies, *Lab Chip* 7: 133–140, 2007.
9. Seibel, K., Scholer, L., Schafer, H., and Bohm, M., A programmable planar electroosmotic micropump for lab-on-a-chip applications, *Journal of Micromechanics and Microengineering* 18: 025008, 2008.
10. Ha, S.-M., Cho, W., and Ahn, Y., Disposable thermo-pneumatic micropump for bio lab-on-a-chip application, *Microelectronic Engineering* 86(4–6): 1337–1339, 2009.
11. Luo, C., Fu, Q., Li, H., Xu, L., Sun, M., Ouyang, Q., Chen, Y., and Ji, H., PDMS microfluidic device for optical detection of protein immunoassay using gold nanoparticles, *Lab Chip* 5: 726–729, 2005.
12. Malica, L. and Kirkb, A. G., Integrated miniaturized optical detection platform for fluorescence and absorption spectroscopy, *Sensors and Actuators A: Physical* 135: 515–524, 2007.
13. Javanmard, M., Talasaz, A. H., Nemat-Gorgani, M., Pease, F., Ronaghi, M., and Davis, R. W., Electrical detection of protein biomarkers using bioactivated microfluidic channels, *Lab Chip* 9: 1429–1434, 2009.
14. XingJiu, H. and YingYing, Z., Electrical determination of *E. coli* O157:H7 using tin-oxide nanowire coupled with microfluidic chip, *Sensors Journal, IEEE* 6: 1376–1377, 2006.
15. Kunduru, V. and Prasad, S., Electrokinetic formation of "microbridges" for protein biomarkers as sensors, *Journal of the Association for Laboratory Automation* 12: 311–317, 2008.
16. Zhang, F., Ulrich, B., Reddy, R. K., Venkatraman, V. L., Prasad, S., Vu, T. Q., and Hsu, S.-T., Fabrication of submicron IrO_2 nanowire array biosensor platform by conventional complementary metal–oxide–semiconductor process *Japanese Journal of Applied Physics* 47: 1147–1151, 2008.
17. Lee, Y., Lim, G., and Moon, W., A self-excited micro cantilever biosensor actuated by PZT using the mass micro balancing technique, *Sensors and Actuators A: Physical* 130–131: 105–110, 2006.
18. Urbanski, J. P., Thies, W., Rhodes, C., Amarasinghe, S., and Thorsen, T., Digital microfluidics using soft lithography, *Lab Chip* 6: 96–104, 2006.
19. Desai, S. P., Freeman, D. M., and Voldman, J., Plastic masters—rigid templates for soft lithography, *Lab Chip* 9: 2009.
20. Hongbin, Y., Guangya, Z., Siong, C. F., Shouhua, W., and Feiwen, L., Novel polydimethylsiloxane (PDMS) based microchannel fabrication method for lab-on-a-chip application, *Sensors and Actuators B: Chemical* 137: 754–761, 2009.
21. Easley, C. J., Karlinsey, J. M., and Landers, J. P., On-chip pressure injection for integration of infrared-mediated DNA amplification with electrophoretic separation, *Lab Chip* 6: 601–610, 2006.
22. Ahn, C. H., Choi, J.-W., Beaucage, G., Nevin, J. H., Lee, J.-B., Puntambekar, A., and Lee, J. Y., Disposable smart lab on a chip for point-of-care clinical diagnostics, *Proceedings of the IEEE* 92: 154–173, 2004.
23. Jain, K. K., Post-genomic applications of lab-on-a-chip and microarrays, *Trends in Biotechnology* 20: 184–185, 2002.
24. Wang, J., Chen, Z., Corstjens, P. L. A. M., Mauka, M. G., and Bau, H. H., A disposable microfluidic cassette for DNA amplification and detection, *Lab Chip* 6: 46–53, 2006.
25. Lee, T. M.-H., Carles, M. C., and Hsing, I.-M., Microfabricated PCR-electrochemical device for simultaneous DNA amplification and detection, *Lab Chip* 3: 100–105, 2003.
26. Bothara, M., Venkatraman, V., Reddy, R., Barrett, T., Carruthers, J., and Prasad, S., Nanomonitors: Electrical immunoassays for protein biomarker profiling, *Nanomed*, 3(4): 423–436, 2009.
27. Lee, J.-G., Cheong, K. H., Huh, N., Kim, S., Choi, J.-W., and Ko, C., Microchip-based one step DNA extraction and real-time PCR in one chamber for rapid pathogen identification, *Lab Chip* 6: 886–895, 2006.

The Biophysics of DNA Microarrays

Cynthia Gibas

42.1 Introduction

The deoxyribonucleic acid (DNA) polymer is the primary carrier of biological information in living cells. An individual DNA unit, referred to as a nucleoside, is made up of a 2-deoxyribose molecule attached to one of four types of cyclic compounds (known collectively as bases). The four bases commonly found in DNA are the following: the purines adenine and guanine (which are heterocyclic compounds made up of two fused rings, one having five and one having six members) and the pyrimidines cytosine and thymine (which are six-membered ring compounds). The backbone of a DNA strand is made up of alternating sugar and phosphate molecules, with the sugar being the 2-deoxyribose component of the nucleoside. The complete monomer unit of base, sugar, and phosphate is referred to as a nucleotide. DNA, as found in living cells, consists of two polynucleotide strands, which coil around each other in the familiar double helix. The sugar–phosphate backbone of the DNA is constant and unvarying, but the base attached to each sugar molecule can vary arbitrarily.

The property of DNA that allows it to carry, store, and replicate biological information with high fidelity is that each base, adenine (A), guanine (G), cytosine (C), or thymine (T), pairs uniquely with only one other base. In a standard double-helical DNA structure, A pairs with T, and G pairs with C (Figure 42.1). If the sequence of a DNA strand reads 5′-ATTCTAGTC-3′, the complementary strand that binds to it must read 5′-GACTAGAAT-3′.

The labels 5′ and 3′ refer to the ends of the DNA strand, and the terminology derives from the structure of the backbone. The 5′ end of the DNA is the end of the DNA with a terminal phosphate group, whereas the 3′ end has a terminal hydroxyl group. In the DNA double helix, the two strands are reverse complements of each other, so that the two strands above are paired as shown in Equation 42.1:

$$5′\text{-ATTCTAGTC-}3′$$
$$3′\text{-ATTCTAGTC-}5′ \tag{42.1}$$

Every part of the DNA in the cell carries with it a unique reverse complement. When the two strands unwind, the molecular machinery of the cell can build a new perfect copy of each strand using the opposite strand as a template. The property of complementarity also makes DNA an excellent target for molecular assays. For any strand of DNA above a certain length, a unique molecular probe can be designed, which will bind, or hybridize, only to its complementary target. If the probe is labeled using a radioactive or fluorescent tag, the presence of the target can be confirmed. The complementarity of DNA also allows for a chain reaction of amplification. In the presence of appropriate amounts of enzymes, free nucleotides, and short oligonucleotides called primers, the polymerase chain reaction will faithfully reproduce a sequence of DNA along with its complement, making massive quantities of a particular sequence available for use in molecular biology experiments.

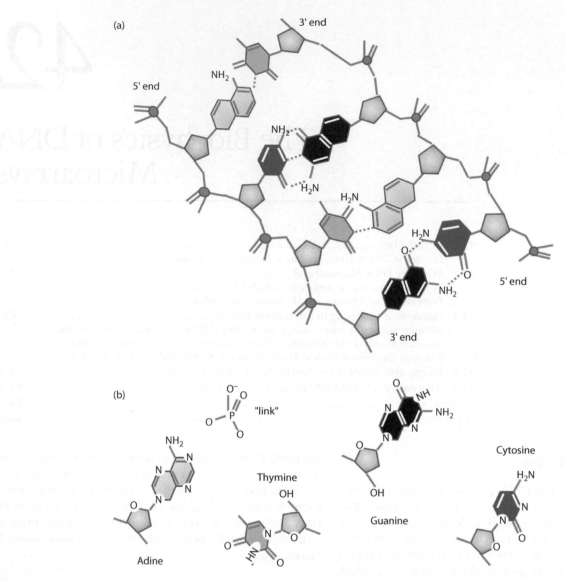

FIGURE 42.1 Molecular structure of DNA. (a) Phosphate-deoxyribose backbone DNA structure and (b) base pair chemistry: adenine (A), cytosine (C), guanine (G), and thymine (T) groups.

In this chapter, we will describe the DNA microarray experiment, along with the physical properties of DNA that play a role in design and interpretation of microarrays, and some applications of this important technology in biology and medicine.

42.1.1 Location of DNA in Living Systems

In eukaryotic cells, DNA is found sequestered in the cell's nucleus, usually in the form of multiple linear chromosomes. Chromosomes are made up of the genomic DNA, compacted around histone proteins and other molecular components. The function of chromatin formation is to pack a very long sequence of DNA into a very small space, while organizing it in such a way that it can be efficiently made available when a gene needs to be activated, or expressed, as we will describe below. Taken together, nuclear DNA specific to an organism is called its genome. In many eukaryotic cells, organelles, such as mitochondria and chloroplasts, have been integrated into the cell to perform energy conversion functions. These organelles also have small amounts of independently replicating DNA.

In prokaryotic cells, DNA is usually found as one or more circular molecules. These prokaryotic "chromosomes" are also referred to as genophores; this is the more correct term, as they are compacted without the other chromatin components that makes up a large proportion of the chromosome. The genophore or genophores are usually contained in a distinct but irregular area of the cell's cytoplasm called the nucleoid. Prokaryotic chromosomes are compacted via supercoiling and without chromatin. Prokaryotic cells often contain additional replicating units called plasmids, which may carry genetic information that is used in the cell along with the genomic information. Plasmids are self-replicating bits of "naked" DNA—not packaged within a cell—that can horizontally transfer from cell to cell and even species to species.

42.1.2 Biological Role of DNA

Three major biological processes are involved in the maintenance and expression of an organism's genetic information. Replication is the most fundamental of these processes; replication is the one that ensures that when a cell reproduces itself by division, a complete copy of the genomic DNA is transmitted to each new cell. Replication involves the wholesale duplication of the genome. The chromosomes are unwound and a complete new complementary strand is created for each of the two strands of the double helix. Expression of genetic information does not involve a wholesale duplication of the genome. Instead, within the larger genomic DNA, there are active regions called genes. Some parts of genes are only signals to the cellular machinery, but many are *transcribed* into RNA as part of their function. mRNA molecules encode proteins, the amino acid polymers that catalyze life's chemistry and form cellular structures; this is one of their many functions. When a gene produces an mRNA, this process is called transcription, and the mRNA is the transcript. The gene is said to be *expressed*. *Translation* is the final process of functional expression of a gene's information. In that process, a protein complex called the ribosome binds to the mRNA transcript. Amino acids are lined up and linked together by the ribosome, in the order specified by the gene's DNA. The molecular details of these processes are complex and differ between eukaryotic and prokaryotic systems. These details can be found in any good molecular biology text.

At various points in the processes of replication, transcription, and translation, molecular errors can occur, meaning that the genetic information is not transmitted with absolute fidelity. If this happens in an ordinary somatic cell, that individual cell may go haywire, resulting in cell death or disease. However, if a mutation occurs in a germ line cell, that mutation may eventually become fixed in the sequence. This results in a gradual process of sequence change, which underlies the biological process of evolution. Selective pressures on functional genes can act either to fix favorable mutations in genes or to purge unfavorable mutations. These pressures result in maintenance of a functional, recognizable gene sequence even in the face of a slow and constant rate of error in replication. Specific DNA sequences that are maintained in this way, instead of drifting to randomness, are said to be *conserved*. Conservation of functional pattern is the property of DNA that allows us to recognize particular functional genes and to probe for specific sequences using molecular diagnostic tests such as *microarrays*.

42.2 What are DNA Microarrays?

DNA microarray methods have been in use in biology and medicine since approximately 1995.[1] In a microarray experiment, a precision printing instrument is used to deposit or directly synthesize many thousands of individual small spots of material on a solid surface. In the DNA microarray experiment, each of these spots consists of picomoles of a distinct oligonucleotide molecule. Each of these molecules is referred to as a *probe*. A mixture of nucleotide molecules of unknown composition (referred to collectively as the *target*) is labeled, usually using a fluorescent molecule, and applied to the surface of the microarray. The system is allowed to equilibrate for several hours, so that each labeled nucleotide molecule in the unknown finds its complementary probe on the surface. The unknown mixture is then washed away under conditions that leave target molecules bound to each individual microarray spot. The intensity of the fluorescence signal from each individual spot is measured using a scanning device, such as a confocal laser scanner, resulting in an array of spot images of varying intensity (Figure 42.2).

The historical antecedent of the microarray experiment is the Southern blot,[2] a technique that has been widely used in molecular biology laboratories since its invention in the mid-1980s. In the Southern blot experiment, DNA is attached to a substrate, usually by adsorption out of an agarose gel (where the DNA in the sample has been separated by size) onto a filter membrane. A labeled probe of known sequence is then hybridized to the blotting membrane to show where the molecule of interest can be found in the gel. The microarray technique simply inverts this experiment, separating many known probes onto the surface and using them to pull specific labeled targets out of a mixture. This change in experimental perspective allows the investigator to identify thousands of different molecules in the same experiment. DNA microarrays can be used to detect either DNA or RNA, although because of its instability in solution, RNA is

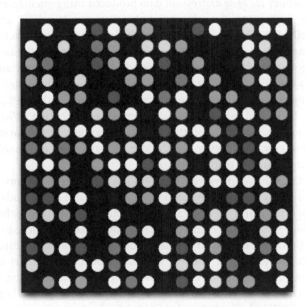

FIGURE 42.2 Schematic of an image from a two-color DNA microarray, in which two samples being compared are labeled with either red or green fluorescent labels. Dark gray spots represent signal predominantly from the green fluorescence channel. Light gray spots represent signal predominantly from the red fluorescence channel.

usually transcribed immediately into copy DNA (cDNA) after sampling and the cDNA is used as the target.

In a nutshell, the interpretation of microarray data is quite straightforward. The intensity of the fluorescence signal at a particular spot on a microarray should be roughly indicative of the amount of labeled material that has bound to that spot. If the binding behavior of each oligonucleotide probe/molecular target pair can be assumed to be similar, then one should be able to take the fluorescence intensity at a particular spot as indicative of the quantity of that particular target present in the mixture. In a gene expression experiment, the intensity of the signal would correspond in some way to the amount of an individual transcript present in the target mixture. However, over a decade of research has shown that microarrays are not consistently able to detect the target quantitatively. In many contexts, results of DNA microarray experiments are reported as ratios between two measurements taken under different conditions (such as from a healthy specimen and a diseased specimen). We will discuss our interpretation of the microarray data, and the role the physical properties of DNA play in the microarray response, in more detail below.

42.2.1 Types of Microarrays

The exact details of the microarray experiment differ depending on the array's manufacturer, the experimental goals for which the array is intended, and the individual researcher performing the experiments. However, there are many commonalities among DNA microarray platforms, and the types of experimental platforms available and actively in use fall into a few common categories. A survey of the Gene Expression Omnibus,[3] a public repository for gene expression data produced using microarrays and similar experiments, shows that between March 12 and April 12, 2009, data sets were deposited that had been collected using spotted oligonucleotide microarrays, *in situ* synthesized oligonucleotide microarrays, spotted cDNA microarrays, and arrayed oligonucleotide beads, as well as non-array-based technologies such as real-time polymerase chain reaction (PCR) (RT-PCR) and massively parallel signature sequencing. While gene expression is not the only application of microarray technology, this survey readily identifies the most commonly used microarray platform types.

cDNA microarrays were used in some of the earliest reported microarray gene expression assays[1] and are still in use today. This type of array is fabricated by mechanically spotting either full-length cDNAs, or shorter PCR products corresponding to a portion of the cDNA, onto a glass slide surface. Prior to spotting, PCR products to be used in fabrication of the array must be individually amplified. The PCR reaction begins with a region of the target DNA and two primers, short oligonucleotides that bind near the 3′ end of each strand of DNA defining the sequence region of interest. The duplex is un-annealed, and a polymerase extends each primer in the 3′ direction, producing the complement of each strand. Subsequent rounds of reaction start with twice as many template molecules, and result in selective production of large quantities of the sequence between the two primers. From a

bioinformatics perspective, optimal design of uniform, biophysically well-behaved cDNA microarrays is complicated, because in addition to the constraints that are needed to produce an amplicon that will identify its target uniquely, the process is constrained by the need to pick optimal primers for the PCR reaction, and frequently these different constraints come into conflict. cDNA microarrays therefore contain probes of widely varying length and biophysical properties, and it has been shown that the behavior of the long PCR products on the surface of these arrays is not uniform or consistent enough to allow for quantitative analysis of the results.[4,5]

Oligonucleotide microarrays use probes of roughly equal length and sequence composition to approximate uniform melting and hybridization behavior. There are two main formats for oligonucleotide arrays: short-oligo arrays using probes approximately 25 nucleotides in length, with 11–20 probes per target, and long-oligo arrays using probes approximately 50–70 nucleotides in length, with one or at most a few probes per target. 70-mers have been shown to have sensitivity comparable to cDNA probes[6] and are useful in applications where detection of low-copy-number transcripts is required; sensitivity decreases twofold between 60-mers and 35-mers, and 10-fold between 35-mers and 25-mers.[7] Shorter oligomers, on the other hand, are necessary to obtain the specificity needed for single nucleotide polymorphism (SNP) detection, but can rapidly lose their uniqueness, especially when used as probes against larger genomes where gene duplication has been significant. Each of these array formats can be produced by either mechanical spotting or *in situ* synthesis, although the large number of probes required for the short-oligo format generally demands that those arrays be produced using a high-density photolithographic process such as that pioneered by Affymetrix. Probes for short-oligo microarrays are picked and optimized in a computerized design process, and it is during this process that physical models of DNA hybridization can be most usefully applied, as we will see below.

Array experiments are performed in both one-color and two-color formats. In the two-color format, two samples, differing in one variable of interest, are labeled, each with a different fluorescent dye. The array is scanned using two different excitation wavelengths, and fluorescence data are collected in two detector channels simultaneously. The fluorescent labels commonly used in these experiments are Cy5 and Cy3, which fluoresce in the red and green wavelengths, respectively. The measurement of interest on two-color microarrays is the signal intensity ratio between the two detection channels, after correction for the different labeling efficiencies and quantum yields of the two dyes. In an expression experiment, genes that have changed in expression level significantly between two samples show up as having a statistically significant fold change from one condition or treatment to the other. The use of the two-color procedure arose in part because early microarrays were inconsistent in fabrication and often the only way to eliminate nuisance variables from the data analysis was to contrast samples of interest on the same array. However, the use of the two-color protocol has persisted, even though many manufacturers now produce arrays that are highly consistent, and even labs that produce their own arrays in-house are able to produce a relatively consistent product. One-color protocols have always been used in processing of high-density short-oligonucleotide microarrays, and have lately become more common with all types of arrays, although because the use of a one-color protocol requires twice as many microarrays as the use of a two-color protocol for the same number of samples, the use of one-color protocols is not universal. In theory, the signal from one-color array experiments can be interpreted as corresponding directly to the concentration of the target in solution, and this has resulted in an intense interest in theoretical models of probe affinity versus signal intensity for short-oligonucleotide arrays. Signal intensity is related to concentration either by using an affinity-based model to process the data and infer a result, or in some cases by direct comparison with a control concentration series of spiked-in targets, which is packaged with the microarray by some manufacturers (e.g., Agilent).

42.2.2 Applications of DNA Microarrays

DNA microarrays have broad applications in biology and medicine.[8] Analysis of gene expression is perhaps the best known of these applications, and over 200,000 individual samples have been assayed using microarrays and deposited in the Gene Expression Omnibus. However, many other applications of microarrays are growing in popularity. With the availability of the Affymetrix SNP arrays for the human genome, which are designed to allow the researcher to detect SNPs in individuals within a population, SNP-based studies have become increasingly popular. Some applications of SNP arrays include (i) whole genome association studies in which large numbers of individuals having different phenotypes are assayed to identify SNP variants that associate with a particular phenotype, (ii) SNP profiling in which individuals are assayed for incidence of particular disease-associated SNPs, and (iii) population biology applications in which particular SNPs are mapped to different human lineages. Exon junction and tiling arrays are two types of arrays designed to probe the structure of the gene rather than its expression. Eukaryotic genes are complex in structure and often contain an unknown number of expressed exons. Exons are coding components of the gene that are separated by untranslated regions called introns, and exons may be translated or left untranslated depending on biological conditions. Assaying transcription with arrays that contain many probes for each exon and for regions thought to be potential exons can help elucidate the structure of the gene, as well as showing which forms of the gene product are produced under different conditions. Comparative genomic hybridization (CGH) arrays and GeneID arrays are two different approaches to identification of genomic differences. GeneID arrays contain a reduced number of probes and are designed to look for specific known sequence variants or presence/absence patterns, often as a diagnostic for pathogen contamination. CGH arrays cover the entire genome with probes, and are designed to assay genomic content variation between closely related organisms. One common application of CGH arrays is to detect gene copy number variation between different individuals, cell cultures, or closely related strains or species.

42.2.3 Automated Methods for Probe and Target Selection and Microarray Data Analysis

Microarray design and analysis are, by nature, bioinformatics-intensive processes. Entire books could be and have been written about the analytical procedures that accompany a microarray experiment.[9,10] The major analytical steps in a microarray experiment (Figure 42.3) ideally begin before the microarray is printed, with planning of the technical specifications of the experiment in which the microarray will be used, followed by selection of probes and optimization of array chemistry. Following hybridization of the experimental sample to the microarray, and washing and drying of the slide under controlled conditions to stabilize the sample for scanning, computer analysis is involved in processing the scanned image to determine signal intensities at each spot, and subsequently in data normalization, in identification of statistically significant changes among sample types, and in classification of the response patterns of different targets across samples.

The physics of DNA hybridization is most relevant to the microarray experiment at two of these stages: first, in the design stage, biophysical models of hybridization can be used to predict properties of candidate probes and their targets. These properties can be used as criteria in the probe selection process. Second, in the normalization phase, physical models can inform analysis by giving insight into innate sequence properties or per-probe thermodynamics that affect the overall result. Additionally, physical models are currently being used in various ways to infer the absolute concentration of the target from microarray signal.

42.3 Biophysical Modeling in Microarray Design

Established computational prediction of duplex thermodynamics, which can predict the properties of an oligonucleotide molecule or duplex, and equilibrium models of hybridization,

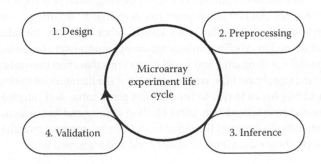

FIGURE 42.3 The life cycle of a microarray experiment includes several analytical steps, from design of the probes and the experimental strategy, to preprocessing of the data to obtain accurate intensity values, to inference of biological significance from the preprocessed data, to validation against corroborating experiments.

which can be simulated to estimate the endpoint behavior of the hybridization reaction, are widely available. DNA and RNA in the microarray experimental context exist in a fundamentally non-native state. Probe molecules, whether PCR products or synthetic oligonucleotides, are sequences that have been removed from their natural context and their behavior can be modeled using methods and parameters developed for small oligonucleotide molecules. Duplexes formed on the microarray can denature relatively easily unless the probe–target pair is designed with a high enough melting temperature, and the single-stranded DNA of the probe or the target can also form a stable internal structure, which may in some cases be stable enough to prevent duplex formation.

Traditionally, DNA microarrays have been designed based on the observable or computable physical properties of the molecular sequences involved. Early design pipelines[11,12] relied on standardizing GC content across the probe set as a proxy for thermodynamic consistency and used simple complementarity cutoffs to eliminate hybridization and internal structure formation. These evolved into more sophisticated design pipelines[13,14] that incorporated biophysics-based design parameters, such as melting temperature computed using a nearest-neighbor model of duplex formation free energy and internal probe structure predicted using a dynamic programming-based 2D optimal structure search with a scoring system based on the thermodynamics of base pairing. Many early probe design methods were greedy algorithms that simply identified all probes meeting particular parameter cutoffs. Recently published methods have attempted to integrate thermodynamic properties into a quantitative model that can be optimized across the entire array.[15] Each of these procedures requires the use of thermodynamic models of folded or paired DNA molecules, which can be accomplished using methods described in Section 42.3.1.

A summary of features in several published probe design software packages developed in the last several years is included in Table 42.1. Design parameters such as ΔG of duplex formation, melting temperature (T_m), and the process of screening for secondary structure (unimolecular structure) in the probe molecule require use of the molecular models described in Sections 42.3.2 and following.

Since the development of the first DNA microarrays in the mid-1990s, the implicit assumption has been that by avoiding cross-hybridization of probes with targets other than the intended target, and by avoiding obvious self-complementarity in the probe, most if not all binding and association events that affect the quality of the signal have been circumvented. But the literature on nucleic acid biophysics is rich in information about intra- and intermolecular interactions that affect binding affinity and hence signal intensity. To interpret the signal from DNA microarrays quantitatively, these effects must be understood and accounted for.

42.3.1 Modeling Macromolecular Structure and Binding of DNA

In the microarray hybridization milieu, two types of reactions with heterogeneous rate constants are in competition (Figure 42.4). The desired hybridization process is a bimolecular interaction between a probe and its intended target. The probe and the target encounter each other via a diffusive process and then hybridize in a two-part reaction. In a multiplex milieu like the microarray, many targets may compete for a particular probe via partial complementarity, and may associate and dissociate from multiple sites before finding the intended binding partner. The degree of complementarity will control the stability of the duplex and hence the rate of dissociation of the target from the probe. In a small number of cases, target molecules may form significant duplexes with each other, as well. These interactions constitute the multiplex component of the model. The multistate component of the model describes the tendency of the individual molecules to exist in multiple stable states, some of which may not be available for bimolecular interaction due to their internal stability. Unimolecular structure formation in either the probe or the target is therefore a competing process in the overall hybridization reaction, where the relevant rate constants are dependent on the length of the molecule and complexity of the structure. Many computational modeling methods exist that describe the behavior of short DNA oligonucleotides in solution, and these can be used in various ways to model the behavior of probes and targets on a DNA microarray.

42.3.2 The Nearest-Neighbor Model

At least for short oligonucleotides (≤40 nucleotides in length), the thermodynamics of hybridization can be accurately predicted using a nearest-neighbor model. For a DNA duplex at a given temperature, the free energy of duplex formation can be expressed as a function of the nearest-neighbor free energy parameters appropriate for that temperature, including values for hybridization initiation, a symmetry correction applicable to self-complementary duplexes, base pair stacking, and an energy penalty that applies only if the terminal base pair is an AT:

$$\Delta G_{total} = \Delta G_{init} + \Delta G_{sym} + \Delta G_{stack} + \Delta G_{term}. \quad (42.2)$$

Thermodynamic parameters for the nearest-neighbor model have been published by several research groups, and available software packages use different sets of values, not all of which give accurate results. A unified parameter set comprising values of ΔH, ΔS, and ΔG for each of these components, as well as modifications for internal mismatches, hairpin loops, and other common structures, is now widely accepted and used in nucleic acid modeling software such as UnaFold.[16,17] These parameters have been reported by SantaLucia and Hicks,[18] and the article is an excellent resource for understanding further details and applications of the nearest-neighbor model.

42.3.3 Two-State Models of Hybridization

The nearest-neighbor model can be used to compute duplex melting temperature for two oligonucleotide sequences based on a simple two-state model of hybridization, in which the molecules are assumed to be either in an unfolded, single-stranded

TABLE 42.1 Commonly Used Criteria for Probe Design and Selection in Freely Available Oligonucleotide Probe Design Tools[49]

Design Tool	Sequence Similarity Search	% Identity	Contiguous Identity	Target–Probe Mismatch Position	Forward/Reverse Strand Match	%GC	ΔG	T_m	T_m Range	NSH	Probe secondary structure	Dimer	Hairpin
ArrayOligoSelector[50]	BLAST	?	No	No	No	Yes	?	NN (?)	No	Yes	SW	Yes	?
GoArrays[51]	BLAST, $w=7$	Yes	Yes	No	?	No	?	NN; SL98	Yes	No	MFOLD	?	?
OligoArray[52,13]	BLAST, $w=7$	No	?	No	No	Yes	Yes	NN; SL98	Yes	Yes	MFOLD	Yes	Yes
OliCheck[53]	BLAST	No	Yes	Yes	Yes	?	?	Yes (?)	?	No	?	?	?
Oligodb[54]	BLAST	No	No	No	No	Yes	No	NN; melting	?	No	MFOLD	Yes	Yes
OligoDesign[55]	BLAST, $w=9$	Yes	No	No	No	No	No	NN; SL98	No	No	Nussinov	Yes	?
OligoPicker[11]	BLAST, $w=8$	Yes	Yes	No	No	No	No	GC; Schildkraut	Yes	No	BLAST	Yes	Yes
OligoWiz[56]	BLAST	Yes	Yes	No	?	No	Yes	NN (?)	Yes	No	Yes (unknown)	Yes	?
Oliz[57]	BLAST	Yes	Yes	No	No	Yes	No	Yes (?)	Yes	No	?	No	No
Osprey[58]	BLAST	?	Yes	No	Yes (?)	No	Yes	NN; SL98	Yes	Yes	MFOLD	Yes	?
Picky[59]	Suffix array	Yes	Yes	No	Yes	Yes	?	NN; SL96	Yes	Yes	Yes (unknown)	Yes	Yes
PRIMEGENS[60]	BLAST	?	No	?	?	Primer3	?	Breslauer	No	?	Primer3	Yes	?
PROBESEL[61]	Suffix tree	Yes	No	No	?	No	No	NN; SL98	No	No	No	No	No
ProbeSelect[12]	Suffix array	No	Yes	No	Yes	Yes	Yes	NN; SL98	No	No	No	No	?
Promide[62]	Suffix array	No	No	No	?	No	No	NN; SL98	Yes	No	Yes (unknown)	Yes	No
ROSO[14]	BLAST, $w=7$	Yes	No	No	?	Yes	Yes	NN; SL98	Yes	No	Yes (unknown)	Yes	Yes
YODA[63]	SeqMatch, $w=4$	Yes	Yes	No	?	Yes	?	NN; SL98	Yes	No	Yes (unknown)	Yes	?

Notes: ?: Not known from information published.

w: word size; NSH: Nonspecific hybridization; NN: nearest-neighbor two-state model.

SL98[64], SL96[65], Schildkraut[66], Breslauer[67], melting[68], Nussinov[69], Primer3[70], BLAST[71], MFOLD[72], SW[73].

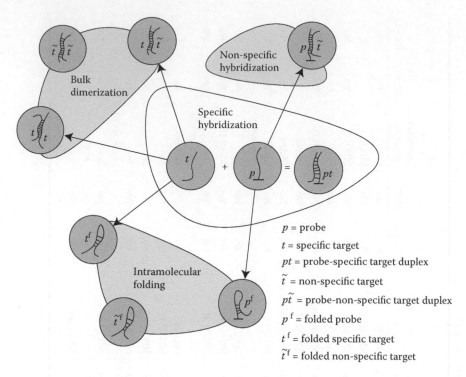

p = probe

t = specific target

pt = probe-specific target duplex

$\tilde{t}$ = non-specific target

$p\tilde{t}$ = probe-non-specific target duplex

$p^{\,f}$ = folded probe

$t^{\,f}$ = folded specific target

$\tilde{t}^{\,f}$ = folded non-specific target

FIGURE 42.4 Possible interactions among molecules in the microarray hybridization milieu. (Courtesy of V. Gantovnik, UNCC.)

state, or paired in a DNA duplex. Construction of microarrays using probes and targets having a highly uniform melting temperature allows the user of the microarray to assume that most targets will be associated with their respective probes below a particular experimental temperature, and completely or mostly dissociated at higher temperatures. The following formula, where R is the gas constant, C_T is the total strand concentration (M), and x is equal to 4 for self-complementary and 1 for non-self-complementary duplexes, gives the two-state T_m of a duplex:

$$T_m = \Delta H^\circ \times 1000 / \left(\Delta S^\circ + R \times \ln\left(\frac{C_T}{x} \right) \right) - 273.15. \qquad (42.3)$$

The two-state model formula for T_m has been frequently used in microarray design protocols. The recent advent of numerical models of coupled multistate equilibria, however, is likely to result in more accurate molecular simulation.

42.3.4 Models of Unimolecular Structure Formation

As mentioned in the introduction to this section, many interactions, both inter- and intramolecular, are possible in the microarray context. Single-stranded nucleic acids in solution form a unimolecular structure, and in the microarray context, the oligonucleotide is immediately available for formation of self-interactions, while it may take some time to diffuse through the reaction chamber and find a partner for duplex formation.

Most secondary structures of single-stranded nucleic acids include hairpins composed of duplex stems and a terminal loop of unpaired nucleotides. Stable secondary structures in either

the probe or the target nucleic acid can make the complement inaccessible to intermolecular Watson–Crick base pairing. Fragmenting the target can reduce the secondary structure effect but the extent of fragmentation is difficult to control and success is highly sequence dependent. A hybridization temperature sufficiently high to melt out internal structure is a possible strategy, but finding a consistent temperature for a large set of probe/target sets may be extremely challenging. Even small amounts of structure can have dramatic effects on hybridization. In one experiment, the T_m for 50% dissociation of an 8 bp loop in a 23-mer was 70°C, higher than the standard hybridization temperature in most microarray experiments. In another experiment contrasting the performance of a modified nucleoside and an unmodified probe, the use of an unmodified probe displaced only 30% of the target secondary structure when presented with 600 equivalents of a complementary nonamer.[19] Therefore secondary structure formation, especially in the probe molecule, is now generally considered as a factor when designing oligonucleotide microarrays, and the equilibrium between unimolecular and bimolecular structure must be modeled in order to improve the prediction of hybridization thermodynamics.

The standard computational procedure for predicting the folded state of a single-stranded nucleic acid molecule, for either DNA or RNA (which requires somewhat different energy parameters, but is otherwise similarly modeled), is a dynamic programming-based optimal structure search. The optimal structure is considered to be the most probable structure, or in chemical terms the structure with the lowest total free energy. Nearest-neighbor free energies are used to parameterize the model, and the lowest-energy structures are rewarded in the search. Dynamic programming can be used to identify multiple low-energy structures that may be nearly equally likely. Programs such as ViennaRNA[20] and SFold[21] use stochastic methods to evaluate the ensemble of likely structures and provide an estimate of accessibility for each nucleotide in a sequence. Accessibility, or the tendency to be unpaired and available for hybridization, can be evaluated across the entire molecule to identify regions of sequence in a target that may be most accessible to the microarray probe. Microarray design applications that do not take secondary structure into consideration may produce probe sets where up to 30% of probes are designed to hybridize to inaccessible regions in the target.[22]

42.3.5 Multistate Equilibrium Models of Hybridization

A common assumption in microarray experiments is that the probe is present in sufficiently high concentrations to drive the reaction to completion, even when the target is present in quite small quantities and competitors for binding are present. Limited secondary structures are assumed to present ineffective competitive problems, despite the fact that unimolecular reactions often have larger forward rate constants than bimolecular reactions. Effective reaction volumes, electrostatic and surface effects, and limits on diffusion are seldom mentioned, and rarely modeled, for arrays. However, these individual behaviors have been studied and shown to have profound effects on the degree to which a reaction has proceeded, in solution studies of oligonucleotide behavior. The interpretation of a result can change dramatically if any of these effects are present.

While some of these physical factors have not yet been addressed in detail, the problems of multistate and multiplex equilibria have widely accepted solutions. In 2004, Dimitrov and Zuker[16] published a multistate statistical mechanical model for duplex formation, by which equilibrium concentrations of the unimolecular and bimolecular species in a pairwise hybridization reaction could be predicted. This approach was integrated with an optimal structure search and is incorporated in the DINAmelt[17] webserver and UnaFold software. Around the same time, the program OMP[23] from DNA Software, an outgrowth of extensive work by Santalucia and coworkers, became available, with capabilities for prediction of equilibrium fraction bound for multiple interacting species and multiple states.

However, the assumption that microarray signal intensities are observed at equilibrium is likely not correct, since it has been demonstrated that times as long as 66 h[24] are required for a microarray experiment to reach equilibrium, and yet hybridization times this long are rarely used in practice. Simulation of a kinetic model would be required to allow prediction of the state of the hybridization reaction at intermediate time points.

42.3.6 Kinetic Models of Hybridization

Analyses of microarray results generally make global assumptions about both the thermodynamic and kinetic behaviors of the species present. For example, one cannot know *a priori*

the concentrations of targets, since the goal of most experiments is to determine those values, so modeling of concentration-dependent behavior in the experiment is somewhat uncertain.

Several attempts have been made to develop comprehensive models of hybridization kinetics between immobilized probes and free targets. Early approaches were based on a receptor–ligand model and included two hybridization mechanisms: direct hybridization from the bulk to the immobilized probes (direct hybridization) and nonspecific adsorption of the targets on the surface followed by two-dimensional (2D) diffusion toward the probes (indirect hybridization). Several assumptions are made in this model. First, DNA hybridization is assumed to be irreversible; dissociation does not play a role in the model. Second, the number of available probes is assumed to be constant throughout the hybridization, and is independent of the reaction rate. Because the model assumes that hybridization is an irreversible process, it does not allow comparison of specific and nonspecific kinetics.

The model developed by Erickson et al.[25] is more comprehensive. They also used two hybridization mechanisms: formation of DNA hybrids by direct hybridization of bulk target and immobilized probes, and nonspecific adsorption of the target on the surface followed by 2D diffusion towards the probes. The model assumes that the probes are well spaced and do not interact with each other. Zhang et al.[26] introduced competitive hybridization of perfectly matched and mismatched duplexes. They did not consider diffusion effects and treated the DNA array like a perfectly stirred reactor. Bishop et al.[27] used a model similar to the one developed by Zhang et al. In this model, perfect and incorrect hybridizations have the same association rate constant, and differ by their dissociation rate constants. Bishop et al. used a finite element software package Femlab that simulates the hybridization reaction and the DNA diffusive transport in the DNA array. In their computer simulations, one type of immobilized probe was considered, and two types of competing targets were included: a matched target and a mismatched target. Horne et al.[28] developed a general analytical description of the equilibrium and reaction kinetics of DNA multiplex hybridization. In this approach, multiplex hybridization is considered to be a competitive reaction process. They discovered complications associated with competition and cross-hybridization between two and three strands in the same reaction, and anticipated added complexities associated with competition between multiple-strand interactions in multiplex mixes. It was assumed that there are no diffusional barriers to the reaction process, and hairpin formation was ignored for simplicity. Many of these models have used approximate rates, treated all reactions as having uniform rates, or have ignored the competing folded states of the probe and the target in order to simplify the model, and therefore these models are not likely to be fully representative of the complex hybridization equilibrium on microarrays.

The complete reaction kinetics of a system of molecules interacting on a microarray can be formulated as a system of differential equations amenable to simulation. Consider a multicomponent system wherein a set of N probes p_i, $1 \leq i \leq N_p$, is designed to hybridize with set of N targets, t_j, $1 \leq j \leq N_t$. Every probe and target strand may form either a perfect match duplex or a degenerate structure: a nonperfect match hybrid duplex or a unimolecular secondary structure. Since the model is restricted to microarray hybridization, we assume that probes cannot interact with each other and cannot form $p_i p_j$ duplexes. Therefore, for a system of probes p_i and targets t_j we have the following system of chemical reactions expressed by Equations 42.4 through 42.7. Equation 42.4 describes the equilibrium between each target and surface-bound probe in the system:

$$p_i + t_j \underset{k^d_{p_i t_j}}{\overset{k^a_{p_i t_j}}{\rightleftharpoons}} p_i t_j. \tag{42.4}$$

Equation 42.5 describes interactions between targets in solution:

$$t_i + t_j \underset{k^d_{t_i t_j}}{\overset{k^a_{t_i t_j}}{\rightleftharpoons}} t_i t_j. \tag{42.5}$$

Equations 42.6 and 42.7 describe the unimolecular folding of each probe and target species:

$$p_i \underset{k^d_{p_i^f}}{\overset{k^a_{p_i^f}}{\rightleftharpoons}} p_i^f \tag{42.6}$$

$$t_i \underset{k^d_{t_i^f}}{\overset{k^a_{t_i^f}}{\rightleftharpoons}} t_i^f. \tag{42.7}$$

Within the bulk phase, transport of the targets is described using the traditional convection–diffusion equation shown in Equation 42.8:

$$\frac{\partial C^o_{t_j}}{\partial \tau} + v \cdot \nabla C^o_{t_j} = D_j \nabla^2 C^o_{t_j}, \tag{42.8}$$

where v is the solution-phase velocity, $C^o_{t_j}$ are the concentration of targets in solution, and D_j are the 3D diffusion coefficients. The corresponding boundary conditions, applied at all reacting surfaces, are given by

$$D_j \nabla C^o_{t_j} \cdot \vec{n} = \frac{\partial C_{t_j}}{\partial \tau}, \tag{42.9}$$

where n is the normal to the surface. Along nonreacting surfaces, this equation reduces to a zero-flux boundary condition. Other boundary conditions are geometry and situation specific. The mass conservation in the chemical reactions leads to the fundamental system of nonlinear differential equations shown in Equations 42.10 through 42.15:

$$\frac{\partial C_{p_i}}{\partial \tau} = -\sum_{j=1}^{N_t} (k^a C_{p_i} C_{t_j} = k^d_{p_i t_j} C_{p_i t_j}) - (k^a_{p_i^f} C_{p_i} - k^d_{p_i^f} C_{p_i^f}), \tag{42.10}$$

$$\frac{\partial C_{t_j}}{\partial \tau} = -\sum_{i=1}^{N_p} (k^a C_{p_i} C_{t_j} - k^d_{p_i t_j} C_{p_i t_j})$$

$$- \sum_{i=1}^{N_t} (\delta_{ij} - 1)(k^a C_{t_i} C_{t_j} - k^d_{t_i t_j} C_{t_i t_j}) - (k^a_{t_j^f} C_{t_j} - k^f_{t_j^f} C_{t_j^f}) \tag{42.11}$$

$$\frac{\partial C_{t_i t_j}}{d\tau} = k^a C_{t_i} C_{t_j} - k^d_{t_i t_j} C_{t_i t_j} \tag{42.12}$$

$$\frac{\partial C_{p_i t_j}}{\partial \tau} = k^a C_{p_i} C_{t_j} - k^d_{p_i t_j} C_{p_i t_j} \tag{42.13}$$

$$\frac{\partial C_{p_i^f}}{\partial \tau} = k^a_{p_i^f} C_{p_i} - k^d_{p_i^f} C_{p_i^f} \tag{42.14}$$

$$\frac{\partial C_{t_j^f}}{\partial \tau} = k^a_{t_j^f} C_{t_j} - k^d_{t_j^f} C_{t_j^f} \tag{42.15}$$

where $C_{p_i}(\tau)$, $C_{t_j}(\tau)$, $C_{p_i t_j}(\tau)$, $C_{t_i t_j}(\tau)$, $C_{p_i^f}(\tau)$, and $C_{t_i^f}(\tau)$ are the concentrations of the probe, target, probe–target duplex, target–target duplex, folded probe and folded target species, respectively, as a function of time τ, and δ_{ij} is the Kronecker delta-function. The initial conditions at time $\tau = 0$ are $(C_{p_i}, C_{t_j}, C_{p_i t_j}, C_{t_i t_j}, C_{p_i^f}, C_{t_i^f}) = (C^o_{p_i}, 0,0,0,0,0)$.

This formulation represents a complete general model applicable to most heterogeneous hybridization situations occurring in a microarray hybridization reaction. It is assumed that the diffusion coefficients, kinetic rate constants, and their dependence on temperature, salt concentration, and probe density are known in advance or can be determined experimentally.

The reasons for the use of simplified models are straightforward; first, the required information to parameterize a kinetic model of hybridization is not available. Reaction-rate equations require quantitative parameters referred to as rate constants. The equilibrium rate constant is a function of the association rate constant (k_a) and the dissociation rate constant (k_d). These values must be known or estimated in order to obtain quantitative results. Both association and dissociation rate constants depend on the DNA sequences involved, the ionic strength of the medium, and the reaction temperature. However, the dependence of dissociation rate constant on these factors is much stronger. Based on experimental observations reported in the literature, the range of association rate constants is from 10^5 to 10^6 M^{-1} s^{-1}, and the typical range for the dissociation rate constants is from 10^{-5} to 10^{-3} s^{-1}.[29–31]

Our survey of the literature shows that there is a lack of knowledge of precise parameters for bimolecular reactions for the oligonucleotides with length above 30, and for unimolecular folding reactions in general. Using approximate values of reaction rate constants results in an unrealistically short time to

TABLE 42.2   Reaction Rate Coefficients

Number of Base Pairs	Association Rate Constant, 10^3 (M^{-1} s^{-1})	Dissociation Rate Constant, 10^{-3} (s^{-1})
<10	24	20
10–20	120	0.21
20–30	390	0.01

equilibrium in model simulations, on the order of seconds to minutes.[27,28] While there is a great disparity in published rate constants for oligonucleotides, some parameters for different size classes, obtained by Henry et al.,[29] are shown in Table 42.2. While exact values disagree, it is clear that association rate constant values increase with increasing sequence length, whereas the dissociation rate constant values decrease. Thus, in longer nucleotide chains, hybridization becomes very stable, with the higher binding rate and slower dissociation rate. When paired molecules in the duplex have a mismatched base pair, the binding rate is decreased and the dissociation rate is increased, and consequently both the hybridization efficiency and association constant are decreased, more so with an increasing number of mismatches. However, without an improved database of rate information for surface hybridization kinetics, implementation of a full kinetic model to simulate hybridization curves and predict non-equilibrium concentrations at arbitrary time points is not likely to be successful.

42.4   Biophysical Issues in Interpretation of DNA Microarrays

The utility of thermodynamic and kinetic models of hybridization, as described above, is that they can potentially be used to establish a connection between observed signal intensity at a probe, I_p, and the concentration of the corresponding target, C_t, in the microarray experiment. As mentioned earlier in this chapter, many microarray results are still interpreted relative to a reference, with the fold change in the sample of interest relative to a control sample being the result. Inference of absolute target concentrations from microarray data has proven to be an elusive goal. In the last decade, statistical and physical modeling approaches have been used to examine the relationship of signal intensity to known solution concentration in control microarray experiments. Different observed and computed properties of the probe or the duplex have been fitted to binding isotherm models that describe concentration response. A solution adsorption isotherm relates the amount of a molecule adsorbed on a surface to the concentration of the molecule in solution.[32] Langmuir isotherm models have frequently been used in analysis of microarray data, using Equation 42.16:

$$\theta = \frac{\alpha \cdot C}{1 + \alpha \cdot C}. \tag{42.16}$$

where θ is the percentage coverage of the surface, C is the concentration, and α is a constant. The Langmuir equation is derived from the equilibrium between empty surface sites $[S*]$, particles $[P]$, and filled particle sites $[SP]$:

$$K = \frac{[SP]}{[S*][P]}. \tag{42.17}$$

The number of filled surface sites is proportional to θ, the percentage coverage of the surface; the number of unfilled sites is proportional to $1 - \theta$, and the number of particles or molecules available for binding is proportional to the concentration C. The constant α is expressed as

$$\alpha = \frac{\theta}{1 - \theta \cdot C}, \tag{42.18}$$

which rearranges to Equation 42.18 above. Data can be fitted to the Langmuir isotherm, expressed as

$$\Gamma = \frac{\Gamma_{max} K C}{1 + K C}, \tag{42.19}$$

where K is the Langmuir equilibrium constant, C is the aqueous concentration of the adsorbate, Γ is the amount of solute adsorbed, and Γ_{max} is the maximum amount of solute that can be absorbed as the aqueous concentration increases, using either linear or nonlinear regression methods.

Both statistical and physical approaches have been used in attempting to connect solution adsorption behavior to observed microarray intensities. In 2004, Hekstra et al.[33] used a Langmuir isotherm model parameterized with physical properties of the probe to predict absolute mRNA concentrations in Affymetrix Gene Chip array experiments. This model assumes the concentration of mRNA to be linearly related to the resulting fluorescence signal intensity I. These authors attempted to predict the Langmuir model parameters from the content of the probe sequence, allowing them to then invert the model and predict solution concentrations of target with some accuracy.

Subsequent to this key study, there have been several efforts to exploit the relationship between the Langmuir model hybridization rate constant, K, and the free energy of hybridization as predicted by the nearest-neighbor model, by assuming that the relationship between the two will follow the relationship:

$$K = e^{-\Delta G/RT}. \tag{42.20}$$

One can apply such an approach to extract free energy parameters directly from microarray data, as in a recent study by Hooyberghs et al.[34] and in older studies by Held et al.[35] In both of these cases, nearest-neighbor free energy parameters were estimated from a fit of microarray data (from a gold-standard experiment where concentrations are known) to the Langmuir model, and correspondence between solution and microarray free energies was shown to be imperfect. Target concentration prediction based on Langmuir models has so far proven to be elusive. Li et al. recently developed a Langmuir-based approach incorporating per-probe association and dissociation rate constants.[36] As we have discussed above, association and dissociation rate data for oligonucleotides, either in solution or on surfaces, are relatively rare in the literature, especially for nucleotides longer than 20–30 base pairs. In the Li model, rate constants computed using the NN model are incorporated into the underlying Langmuir model, in an attempt to capture the effects of per-probe kinetics in the model. However, the model as described is only applicable to probes having no competing unimolecular structure formation, so that predictions of concentration will only be accurate for the subset of the array that demonstrates two-state behavior.

In a recent study, carried out using a model microarray with controlled target sample composition, we have attempted to explain the response of each probe in the experiment by applying current state-of-the-art solution models as a predictor of surface behavior.[37] This model makes use of probe percent bound (PPB) obtained from a simulation of the hybridization equilibrium of all probes and targets in the experiment, as a predictor of the signal intensity generated from each probe, following Equation 42.21:

$$I_j = B_0 + B_1 < F_B >_j + \varepsilon_j, \qquad (42.21)$$

where I_j is the signal intensity from the probe at target concentration j. F_B is the PPB of the probe at target concentration j. B_0 and B_1 are free parameters and ε_j is an error term. The responses are typical of the Langmuir isotherm model and show that the model captures the physical chemistry of hybridization with $R^2 \geq 0.97$. The clear result of this experiment is that PPB computed using a detailed, multistate, and multiplex simulation of the hybridization equilibrium captures the variations in signal intensity according to target concentrations with high accuracy in a model microarray system. As more detailed and physically realistic models of hybridization are developed, it is likely that model-based recovery of absolute concentrations of target from observed microarray signal will become more reliable.

42.5 Alternatives to Standard Microarray Platforms

Microarray technologies continue to evolve rapidly, in part driven by the goal of improving the quantitative response of the assay. This is often approached by development of assay conditions designed to be solution-like, because our understanding of the behavior of nucleic acids in solution is developed well beyond our understanding of surface behavior.

While most reported microarray data sets are still collected using one of the planar substrate platforms described in this chapter, a growing number of experiments are conducted using the solution bead array technology developed by Illumina. In this array platform, individual micron-sized beads are functionalized with probe oligonucleotides. The bead includes an identifying label that allows the reader to recognize which probe sequence is present. The beads self-assemble on the end of an optical fiber etched with wells sized to hold an individual bead. Complementary targets present in solution bind to the beads, and the amount of target present is measured based on fluorescence signal, as it is in standard microarrays.[38] One claim associated with this array platform is that the bead substrate results in more solution-like hybridization behavior between the probe and the target. Improving fabrication techniques have led to recent reports of nanoscale fiber arrays, where densities of 4.6×10^6 array elements/mm^2 are possible.[39]

Label-free detection methods are increasingly under investigation, both for DNA microarrays and array-like proteomic platforms. Surface plasmon resonance methods enable quantitative measurement of DNA binding, as well as kinetics, on a 2D surface,[40] and this technology also shows promise for measurement of DNA–protein and protein–protein interactions.[41] Electrical detection of hybridization has been accomplished using electrical impedance spectroscopy[42] and by electrochemical detection of label-mediated chemical reactions on a nanoscale gold electrode array.[43]

Lab-on-chip approaches to studies of nucleic acids, proteins, and other molecular species are increasingly being merged with a microarray style, array-of-probes approach.[44] Microfluidic platforms generally consist of micro- or even nanoscale channels through which very small volumes of sample can be delivered precisely to the point of interaction. Microarray-style lab-on-chip approaches usually consist of multiple fluidic paths with low potential for cross-contamination and cross-hybridization. The benefits of such approaches for microarray-type studies are generally thought to be reduction of the required sample volume, reduced hybridization time due to improved mixing and other factors, and separation of the probe from the solid support. However, there are substantial surface adsorption effects involved in microfluidic systems due to the high surface-to-volume ratio of the microscale channel structures.[45] These effects have been extensively studied and various fluidics platforms for transcript detection via hybridization or PCR have been proposed.[46–48] Despite these efforts, there is not as yet any microfluidics-based microarray platform that has matched the rate of adoption seen with Affymetrix-type short-oligo, long-oligo, or even cDNA microarrays.

Acknowledgments

The author would like to thank Vladimir Gantovnik and Raad Gharaibeh for assistance with the preparation of some of the figures, equations and tables used in this chapter.

References

1. Schena M., Shalon D., Davis R.W., and Brown P.O., 1995. Quantitative monitoring of gene expression patterns with a

complementary DNA microarray. *Science*, 270(5235): 467–470.

2. Southern, E.M., 1975. Detection of specific sequences among DNA fragments separated by gel electrophoresis, *J. Mol. Biol.*, 98: 503–517.

3. Sayers E.W., Barrett T., Benson D.A., Bryant S.H., Canese K., Chetvernin V., Church D.M., et al., 2009. Database resources of the National Center for Biotechnology Information. *Nucleic Acids Res.*, 37(Database issue): D5–15 (Epub October 21, 2008).

4. Yue H., Eastman P.S., Wang B.B., Minor J., Doctolero M.H., Nuttall R.L., Stack R., et al., 2001. An evaluation of the performance of cDNA microarrays for detecting changes in global mRNA expression. *Nucleic Acids Res.*, 29(8): E41–1.

5. Shi S.J., Scheffer A., Bjeldanes E., Reynolds M.A., and Arnold L.J., 2001. DNA exhibits multi-stranded binding recognition on glass microarrays. *Nucleic Acids Res.*, 29(20): 4251–4256.

6. Wang H.Y., Malek R.L., Kwitek A.E., Greene A.S., Luu T.V., Behbahani B., Frank B., Quackenbush J., and Lee N.H., 2003. Assessing unmodified 70-mer oligonucleotide probe performance on glass-slide microarrays. *Genome Biol.*, 4(1): R5 (Epub January 6, 2003).

7. Relógio A., Schwager C., Richter A., Ansorge W., and Valcárcel J., 2002. Optimization of oligonucleotide-based DNA microarrays. *Nucleic Acids Res.*, 30(11): e51.

8. Stoughton R.B., 2005. Applications of DNA microarrays in biology. *Annu. Rev. Biochem.*, 74: 53–82.

9. Schena M., 2003. *Microarray Analysis*, Wiley, Hoboken, NJ.

10. Draghici S., 2003. *Data Analysis Tools for DNA Microarrays*, CRC Press LLC, Boca Raton, FL.

11. Wang X., 2003. Seed B: Selection of oligonucleotide probes for protein coding sequences. *Bioinformatics*, 19(7): 796–802.

12. Li F. and Stormo G.D., 2001. Selection of optimal DNA oligos for gene expression arrays. *Bioinformatics*, 17(11): 1067–1076.

13. Rouillard J.M., Zuker M., and Gulari E., 2003. OligoArray 2.0: Design of oligonucleotide probes for DNA microarrays using a thermodynamic approach. *Nucleic Acids Res.*, 31(12): 3057–3062.

14. Reymond N., Charles H., Duret L., Calevro F., Beslon G., and Fayard J.M., 2004. ROSO: Optimizing oligonucleotide probes for microarrays. *Bioinformatics*, 20(2): 271–273.

15. Leparc G.G., Tüchler T., Striedner G., Bayer K., Sykacek P., Hofacker I.L., and Kreil D.P., 2009. Model-based probe set optimization for high-performance microarrays. *Nucleic Acids Res.*, 37(3): e18 (Epub December 22, 2008).

16. Dimitrov R.A. and Zuker M., 2004. Prediction of hybridization and melting for double-stranded nucleic acids. *Biophys. J.*, 87(1): 215–226.

17. Markham N.R. and Zuker M., 2005. DINAMelt web server for nucleic acid melting prediction. *Nucleic Acids Res.*, 33(Web Server issue): W577–W581.

18. SantaLucia J. Jr. and Hicks D., 2004. The thermodynamics of DNA structural motifs. *Annu. Rev. Biophys. Biomol. Struct.*, 33: 415–440.

19. Nguyen H.K. and Southern E.M., 2000. Minimising the secondary structure of DNA targets by incorporation of a modified deoxynucleoside: Implications for nucleic acid analysis by hybridisation. *Nucleic Acids Res.*, 28(20): 3904–3909.

20. Wuchty S., Fontana W., Hofacker I.L., and Schuster P., 1999. Complete suboptimal folding of RNA and the stability of secondary structures. *Biopolymers*, 49(2): 145–165.

21. Ding Y., Chan C.Y., and Lawrence C.E., 2005. RNA secondary structure prediction by centroids in a Boltzmann weighted ensemble. *RNA*, 11(8): 1157–1166.

22. Ratushna V.G., Weller J.W., and Gibas C.J., 2005. Secondary structure in the target as a confounding factor in synthetic oligomer microarray design. *BMC Genomics*, 6(1): 31.

23. SantaLucia J., Jr, 2007. Physical principles and visual-OMP software for optimal PCR design. *Methods Mol. Biol.*, 402: 3–34.

24. Sartor M., Schwanekamp J., Halbleib D., Mohamed I., Karyala S., Medvedovic M., and Tomlinson C.R., 2004. Microarray results improve significantly as hybridization approaches equilibrium. *Biotechniques*, 36(5): 790–796.

25. Erickson D., Li D., and Krull U.J., 2003. Modeling of DNA hybridization kinetics for spatially resolved biochips. *Anal. Biochem.*, 317: 186–200.

26. Zhang Y., Hammer D.A., and Graves D.J., 2005. Competitive hybridization kinetics reveals unexpected behavior patterns. *Biophys. J.*, 89: 2950–2959.

27. Bishop J., Blair S., and Chagovetz A.M., 2006. A competitive kinetic model of nucleic acid surface hybridization in the presence of point mutants. *Biophys. J.*, 90: 831–840.

28. Horne M.T., Fish D.J., and Benight A.S., 2006. Statistical thermodynamics and kinetics of DNA multiplex hybridization reactions. *Biophys. J.*, 4133–4153.

29. Henry M.R., Wilkins Stevens P., Sun J., and Kelso D.M., 1999. Real-time measurements of DNA hybridization on microparticles with fluorescence resonance energy transfer. *Anal. Biochem.*, 276: 204–214.

30. Okahata Y., Kawase M., Niikura K., Ohtake F., Furusawa H., and Ebara Y., 1998. Kinetic Measurements of DNA hybridization on an oligonucleotide-immobilized 27-MHz quartz crystal microbalance. *Anal. Chem.*, 70: 1288–1296.

31. Tsoi P.Y., Zhang X., Sui S., and Yang M., 2003. Effects of DNA mismatches on binding affinity and kinetics of polymerase–DNA complexes as revealed by surface plasmon resonance biosensor. *Analyst*, 128: 1169–1174.

32. Burden C.J., Pittelkow Y.E., and Wilson S.R., 2004. Statistical analysis of adsorption models for oligonucleotide microarrays. *Stat. Appl. Genet. Mol. Biol.*, 3: Article 35,.

33. Hekstra D., Taussig A.R., Magnasco M., and Naef F. 2004. Absolute mRNA concentrations from sequence-specific calibration of oligonucleotide arrays. *Nucleic Acids Res.*, 31: 1962–1968.

34. Hooyberghs J., Van Hummelen P., and Carlon E., 2009. The effects of mismatches on hybridization in DNA microarrays: Determination of nearest neighbor parameters. *Nucleic Acids Res,* 37(7): e53-1-11.

35. Held G.A., Grinstein G., and Tu Y., 2003. Modeling of DNA microarray data by using physical properties of hybridization. *Proc. Natl. Acad. Sci. USA,* 100(13): 7575–7580.

36. Li S., Pozhitkov A., and Brouwer M., 2008. A competitive hybridization model predicts probe signal intensity on high density DNA microarrays. *Nucleic Acids Res.,* 36(20): 6585–6591 (Epub October 18, 2008).

37. Gharaibeh R.Z., Newton J.S., Weller J.W., and Gibas C. J., 2010. Application of equilibrium models of solution hybridization to microarray design and analysis. *BMC Bioinformatics,* in review.

38. Kukol A., Li P., Estrela P., Ko-Ferrigno P., and Migliorato P., 2008. Label-free electrical detection of DNA hybridization for the example of influenza virus gene sequences. *Anal Biochem.,* 374(1): 143–153 (Epub October 30, 2007).

39. Tam J.M., Song L., and Walt D.R., 2009. DNA detection on ultrahigh-density optical fiber-based nanoarrays. *Biosens. Bioelectron.,* 24(8): 2488–2493 (Epub December 30, 2008).

40. Gao Y., Wolf L.K., and Georgiadis R.M., 2006. Secondary structure effects on DNA hybridization kinetics: A solution versus surface comparison. *Nucleic Acids Res.,* 34(11): 3370–3377. Print 2006.

41. Nedelkov D., 2007. Development of surface plasmon resonance mass spectrometry array platform. *Anal. Chem.,* 79(15): 5987–5990 (Epub June 23, 2007).

42. Li X., Lee J.S., and Kraatz H.B., 2006. Electrochemical detection of single-nucleotide mismatches using an electrode microarray. *Anal. Chem.,* 78(17): 6096–6101.

43. Elsholz B., Wörl R., Blohm L., Albers J., Feucht H., Grunwald T., Jürgen B., Schweder T., and Hintsche R., 2006. Automated detection and quantitation of bacterial RNA by using electrical microarrays. *Anal. Chem.,* 78(14): 4794–4802.

44. Situma C., Hashimoto M., and Soper S.A., 2006. Merging microfluidics with microarray-based bioassays. *Biomol. Eng.,* 23(5): 213–231 (Epub July 7, 2006). Review.

45. Seemann R., Brinkmann M., Kramer E.J., Lange F.F., and Lipowsky R., 2005. Wetting morphologies at microstructured surfaces. *Proc. Natl. Acad. Sci. USA,* 102(6): 1848–1852 (Epub January 27, 2005).

46. Erickson D., Liu X., Krull U., and Li D., 2004. Electrokinetically controlled DNA hybridization microfluidic chip enabling rapid target analysis. *Anal Chem.,* 76(24): 7269–7277.

47. Keramas G., Perozziello G., Geschke O., and Christensen C.B., 2004. Development of a multiplex microarray microsystem. *Lab Chip.,* 4(2): 152–158 (Epub January 16, 2004).

48. Lenigk R., Liu R.H., Athavale M., Chen Z., Ganser D., Yang J., Rauch C., et al., 2002. Plastic biochannel hybridization devices: A new concept for microfluidic DNA arrays. *Anal. Biochem.,* 311(1): 40–49.

49. Kreil D.P., Russell R.R., and Russell S., 2006. Microarray oligonucleotide probes. *Methods in Enzymology,* 410: 73–98.

50. Bozdech Z., Zhu J., Joachimiak M.P., Cohen FE, Pulliam B., and DeRisi J.L., 2003. Expression profiling of the schizont and trophozoite stages of *Plasmodium falciparum* with a long-oligonucleotide microarray. *Genome Bioliogy,* 4(2): R9.

51. Rimour S., Hill D., Militon C., and Peyret P., 2005. GoArrays: Highly dynamic and efficient microarray probe design. *Bioinformatics,* 21(7): 1094–1103.

52. Rouillard J.M., Herbert C.J., and Zuker M., 2002. Oligoarray: Genome-scale oligonucleotide design for microarrays. *Bioinformatics,* 18(3): 486–487.

53. Charbonnier Y., Gettler B., Francois P., Bento M., Renzoni A., Vaudaux P., Schlegel W., and Schrenzel J., 2005. A generic approach for the design of whole-genome oligoarrays, validated for genomotyping, deletion mapping and gene expression analysis on Staphylococcus aureus. *BMC Genomics,* 6: 95.

54. Mrowka R., Schuchhardt J., and Gille C., 2002. Oligodb—interactive design of oligo DNA for transcription profiling of human genes. *Bioinformatics,* 18(12): 1686–1687.

55. Tolstrup N., Nielsen P.S., Kolberg J.G., Frankel A.M., Vissing H., and Kauppinen S., 2003. OligoDesign: Optimal design of LNA (locked nucleic acid) oligonucleotide capture probes for gene expression profiling. *Nucleic Acids Res.,* 31(13): 3758–3762.

56. Nielsen H.B., Wernersson R., and Knudsen S., 2003. Design of oligonucleotides for microarrays and perspectives for design of multi-transcriptome arrays. *Nucleic Acids Res.,* 31(13): 3491–3496.

57. Chen H. and Sharp B., 2002. Oliz, a suite of Perl scripts that assist in the design of microarrays using 50mer oligonucleotides from the 3′ untranslated region. *BMC Bioinformatics,* 3(1): 27.

58. Gordon P.M.K. and Sensen C.W., 2004. Osprey: A comprehensive tool employing novel methods for the design of oligonucleotides for DNA sequencing and microarrays. *Nucleic Acids Res.,* 32(17): e133-1–9.

59. Chou H.-H., Hsia A.-P., Mooney D.L., and Schnable P.S., 2004. Picky: oligo microarray design for large genomes. *Bioinformatics,* 20(17): 2893–2902.

60. Xu D., Li G., Wu L., Zhou J., and Xu Y., 2002. PRIMEGENS: Robust and efficient design of gene-specific probes for microarray analysis. *Bioinformatics,* 18(11): 1432–1437.

61. Kaderali L. and Schliep A., 2002. Selecting signature oligonucleotides to identify organisms using DNA arrays. *Bioinformatics,* 18(10): 1340–1349.

62. Rahmann S., 2002. Rapid Large-Scale Oligonucleotide Selection for Microarrays. In: *First IEEE Computer Society Bioinformatics Conference (CSB): 2002.* IEEE Press.

63. Nordberg E.K., 2005. YODA: selecting signature oligonucleotides. *Bioinformatics,* 21(8): 1365–1370.

64. SantaLucia J., Jr, 1998. A unified view of polymer, dumbbell, and oligonucleotide DNA nearest-neighbor thermodynamics.

Proceedings of the National Academy of Sciences USA, 95(4): 1460–1465.

65. SantaLucia J., Jr., Allawi H.T., and Seneviratne P.A., 1996. Improved nearest-neighbor parameters for predicting DNA duplex stability. *Biochemistry*, 35(11): 3555–3562.

66. Schildkraut C., 1965. Dependence of the melting temperature of DNA on salt concentration. *Biopolymers*, 3(2): 195–208.

67. Breslauer K.J., Frank R., Blocker H., and Marky L.A., 1986. Predicting DNA duplex stability from the base sequence. *Proceedings of the National Academy of Sciences USA*, 83(11): 3746–3750.

68. Le Novere N., 2001. MELTING, computing the melting temperature of nucleic acid duplex. *Bioinformatics*, 17(12): 1226–1227.

69. Nussinov R., Shapiro B., Le S.Y., and Maizel J.V., Jr, 1990. Speeding up the dynamic algorithm for planar RNA folding. *Math. Biosci.*, 100(1): 33–47.

70. Rozen S., and Skaletsky H., 2000. Primer3 on the WWW for general users and for biologist programmers. Methods in *Molecular Biology*, 132: 365–386.

71. Altschul S.F., Gish W., Miller W., Myers E.W., and Lipman D.J., 1990. Basic local alignment search tool. *J. Mol. Biol.*, 215(3): 403–410.

72. Zuker M., 2003. Mfold web server for nucleic acid folding and hybridization prediction. *Nucleic Acids Res.*, 31(13): 3406–3415.

73. Smith T.F., Waterman M.S., and Fitch W.M., 1981. Comparative biosequence metrics. *J. Mol. Evol.*, 18(1): 38–46.

43

Nuclear Medicine

Robert Splinter

43.1 Introduction

The branch of medicine that uses the radioactive and nuclear properties of matter as a diagnostic and therapeutic tool is often referred to as nuclear medicine.[1-3] All matter is constructed of atoms, which are composed of protons (positively charged), neutrons (no charge, neutral), and electrons (negatively charged with valence 1, or -1.6×10^{-19} C).[4-7] An electron has 0.0005 times the mass of the proton, 1.6×10^{-27} kg, and a neutron has about the same mass as a proton.[1,5] This chapter will describe the atomic characteristics and the various kinds of nuclear radiation that can be generated. Additionally, the concept of radioisotope is described. Other nuclear activity discussed ranges from the concept of radioactive decay to radioactive date establishment. The radioactive applications in medicine are described along with means of detection. A brief overview is also presented describing radiation units and radioactive exposure.

Radiation is associated with both electromagnetic radiation as well as charged particles. Antoine Henri Becquerel (1852–1908) discovered natural radioactivity in 1896.[8] Rutherford identified the radioactive properties of helium nuclei in the form of α particles in the year 1909. The Bohr (Niels Henrik David Bohr, 1885–1962) model for the atom eventually led to the understanding of these radioactive isotopes in 1913, improving on the earlier work by Ernest Rutherford (1871–1937).[5,9] The electron-cloud model around the nucleus is often also referred to as the Rutherford–Bohr model. The α particle is one of several atoms with an unstable nucleus, called radionuclides or radioisotopes (radioactive isotopes).[1-4,10] Isotopes are atoms with similar chemical properties as atoms with the same number of protons (which means they have the same atomic number) but with a different number of neutrons. Since the neutrons have a non-negligible mass, isotopes have different mass numbers (A) but relatively similar chemical

behavior. Radionuclides generally originate from the same chemical element (as listed in the periodic table). Selected isotopes are unstable and decay to lighter isotopes or stable elements. The notation for the atom has the mass number on the top left of the element and the atomic number on the bottom left; for example for lithium, Li, this becomes, with $A = 6$ and $Z = 3$, $^{6}_{3}$Li.[1-5,7]

A student of Rutherford, Hans Geiger, together with Erich Regener, discovered the scintillation of α particles impacting on a zinc sulfide screen, making the impact of the α particle visible.[3,4,11] Geiger developed and implemented the mechanism of measuring the energetic impact of the charged helium nucleus, the Geiger counter.

Nuclear medicine uses isotopes or radionuclides to perform testing of metabolic activity by means of selective uptake of the radiopharmaceuticals and hence image the emission of the decay radiation.

Another notable name in this spectrum is Wilhem Conrad Röntgen (1845–1923).[12] Röntgen's name is associated with the discovery of x-ray, the foundation of one of today's primary medical imaging applications. Although x-ray is also used in nuclear medicine, this is not the primary focus of this chapter. Marie Curie (1867–1934), on the other hand, made a significant contribution to the advancement of nuclear medicine with the discovery of the isotopes Polonium and Radium.[13] Unfortunately, Madame Curie died of the effects of her nuclear decay studies from radiation exposure-induced leukemia.

43.1.1 Physical Characteristics of Nuclides

Nuclides generally decay to more stable elements, often in stages, and the energetic loss is expressed either in electromagnetic radiation or the emission of physical particles. The emitted radiation is classified into four (4) categories based on the manner of decay and the energetic content.[1-4,6,7,11] The high-energy photon emission (electromagnetic radiation) is called γ rays, electron emission is called β radiation (negatively charged -1 "electron charge"), and the release of a much heavier composite particle constructed of two (2) protons and two neutrons is classified as α particle decay (positively charged +2 "electron charge"). The fourth category is neutron emission; but this type of radiation is less useful since it is hard to detect and indices little radioactive effects due to its neutral charge. All radioactive decay itself has ionizing properties, meaning it will remove electrons from the orbit around the nucleus of atoms, making this a mechanism that changes the chemical and physical behavior of the molecules that contain the ionized atoms. Due to the nature in which x-ray radiation is generated, x-ray has high energy content and is also considered ionizing radiation as well as short-wavelength ultraviolet electromagnetic radiation. Generally, wavelengths of electromagnetic radiations longer than 300 nm are considered nonionizing radiation (visible light, radio and television waves, etc.).

One prime example of a radioactive isotope is Uranium-235, which is the unstable isotope for Uranium-238.

43.1.2 Physical Characteristics of Different Types of Radiation

α radiation as heavy particle emission ($^{4}_{2}\alpha^{++}$) cannot travel far in free space and will extinguish within only a few centimeters in air and a few micrometers in tissue, approximately less than a cell layer. The energy content of α particles is 4–8 MeV.

β radiation ($^{0}_{-1}\beta^{-}$) is much lighter and has a broad energy spectrum, depending on the mode of generation and the material used to generate the free electron. β radiation travels a few meters in air and several millimeters in tissue.

γ radiation ($^{0}_{0}\gamma$), due to its electromagnetic character, is free of mass and will travel several meters in air and requires aluminum or lead to attenuate or block the emission. The energy range of γ radiation is 1–100 MeV.

43.1.2.1 Physical Characteristics of γ Radiation

γ radiation is part of the short-wavelength spectrum of electromagnetic radiation. As such the γ radiation can be classified based on wavelength and the associated energy. The basic formulation of the dependence of the energy (E) content of electromagnetic radiation is a linear relationship with the frequency (υ) or reciprocal relationship with the wavelength (λ) of the radiation. The energy in turn is directly related to the isotope source producing the radiation, as expressed by

$$E = h\upsilon = h\frac{c}{\lambda}. \tag{43.1}$$

43.1.2.2 Physical Characteristics of Particulate Radiation

Both α and β radiation represent particulate radiation. The energy content of a particle in motion with velocity v and mass m is described by the kinetic energy. The kinetic energy is defined by

$$E = \frac{1}{2}mv^2. \tag{43.2}$$

In a head-on collision with an atom, the α particle or β particle will release electrons during an inelastic collision. In the case of an inelastic collision, the kinetic energy needs to match or exceed the electric Coulomb force holding the electron in the orbit around the nucleus. This condition is given by

$$\frac{1}{2}mv^2 = k\frac{q_1 q_2}{d}, \tag{43.3}$$

where d is the distance from the nucleus the particle comes to rest, $q_1 = ne$ and $q_2 = Ze$ are the charges of the colliding particle and the nucleus, respectively, and $k = 9 \times 10^9$ Nm²/C² is the Coulomb constant. The number of electron charges ($e = 1.60 \times 10^{-19}$C) of the colliding particle will be $n = -1$ for the β particle and $n = 2$ for the α particle, respectively, and Z is the atomic number of the atom. The negative sign for the β particle indicates attraction, but for the calculation of the distance, the absolute value of the Coulomb force is used.

43.2 Nuclear Medicine

The use of radioactive materials in both treatment and diagnostic modalities of medicine falls in the field of nuclear medicine. One imaging application of nuclear medicine is the detection of decaying radionucleotides that are introduced in the biological medium. Measuring the radioactive decay is performed by scintillation counters and is called scintigraphy.[1-3,6,7,11,14] An example of a configuration of a coded mask used to provide a filter system to assist in calculating the source distribution in the biological sample or patient is shown in Figure 43.1.

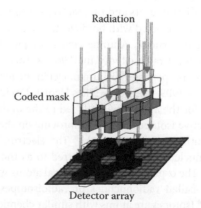

FIGURE 43.1 Representation of a coded mask casting a shadow pattern on the radiation detector. The raster can be used to perform a mechanical filter to assist in the calculation of the distribution of sources in the field of view.

The topic of radionucleotides describes the use of radioactive isotopes that are selectively absorbed in specific organs depending on the metabolic functioning of the organ. The selective absorption and subsequent emission of radioactive radiation thus paints a picture of the biological function of the organ and is considered a form of functional imaging.

43.2.1 Radioactivity and Radionucleotides

Most chemical elements have more than one isotope. Isotopes are chemically identical to the primary atom but have a different atomic weight. The nucleus of the atom has the same number of protons, but a different number of neutrons. The number of protons in the nucleus is identified by the atom number, Z; the atomic weight A is the sum of all protons and neutrons in the nucleus.[5,9–11,14,15]

The atomic number is placed on the left top side of the notation for the element, while the atomic weight is placed on the left bottom: for example, $^{131}_{53}J$ for iodine. Iodine has 53 protons and 131 - 53 = 78 neutrons.

While some isotopes are chemically and energetically stable, many isotopes are unstable and will disintegrate to a lower atomic weight within a specific time frame. The degeneration to a lower chemical binding state releases energy in the form of radiation.

43.2.1.1 Radioactive Decay

The unstable isotopes will decay to a lower chemical energy state over a time frame that is determined by the chemical forces and the product of the decay. The time frame over which half the amount of isotopes reduces to the product isotope, which itself may or may not be stable, is called the half-life.

Under the decay, three major types of radiation can be identified: α, β, and γ decay radiation. Additional neutron and neutrino radiation also takes place, but has no direct use in medicine since this carries no charge, and is for various other reasons difficult to detect.

The energy of the isotopes is determined by the atomic weight and the fact that mass and energy are correlated by the Einstein principle $E = mc^2$. Here m equals the atomic mass and c is the speed of light in vacuum.

43.2.1.2 α Decay

α decay is identified by emission of a helium nucleus: two protons and two neutrons. α decay is electropositive with charge two, that is, positive charge equivalence of two electron charges. The velocity of the α particle will depend on the total energy difference between the primary isotope and the product element, minus the energy in the α particle itself; the remaining energy is used as kinetic energy for the particle.

Since the α particle is positively charged it will have an ionizing effect when interacting with other atoms and molecules.

In general, due to the ionizing effect α radiation is not preferred for nuclear medicine applications, since the damage caused by the decaying radiation does not justify the potential benefits.

43.2.1.3 β Decay

Two kinds of β particle radiation can be distinguished: electron and positron. Electron emission makes the atomic number Z increase by one, while positron emission will make the atomic number decrease by one.

In general, the electric charge again presents a dilemma for nuclear medicine applications. However, the positron emission is used for specialized imaging using the positron-emission-tomography technique (PET imaging) as described in Chapter 16.

43.2.1.4 γ Decay

γ radiation involves high energy photons only; the atomic number and atomic weight of the isotope do not change during this decay process. The energy quantum of the photon is characteristic of the isotope used and has a relatively narrow bandwidth.

Since γ radiation does not carry a charge, the interaction with tissue molecules is minimal. γ radiation is omnipresent in our common life resulting from solar radiation and cosmic radiation interacting with atoms and molecules in the upper atmosphere.

43.2.2 Radioactive Labeling

Radioactive labeling describes the use of isotopes to mark and identify the mechanism of action of a specific metabolic process.[1–3,11,14,16] Most frequently, radioactive labeling is used to monitor the drug interaction of a prototype pharmaceutical. The radiopharmaceutical concentration distribution obtained by various imaging techniques will establish the targeting capability and the potential peripheral deposition, hence the efficacy of the drug. Radioactive labeling provides concentration tracking in DNA down to 0.1 pg (pictogram) per cell.

43.2.3 Scintigraphy

By injecting or ingesting specific isotopes that are isotopes of regular atoms that are a common element in one or more particular metabolic activities, the metabolic activity of the organ can be measured through measuring the absorbed amount of the isotope. Sometimes atoms need to be chosen to create isotopes that are not a normal part of the cellular or organ's metabolic activity, but they are close enough in chemical appearance to fake belonging in the cellular digestive process, and can thus be detected as being absorbed for the particular purpose their real cousins perform. This second option is critical if no stable isotopes can be produced from the indigenous atoms or molecules. Most often the radioactive isotopes are geared to get incorporated in a molecular strain that is recognized by the cells.

As mentioned before, the majority of isotopes in question will produce γ radiation, since α or β radiation will produce harmful biological effects.

Specialized scintillation counters will ionize under the influence of γ radiation and will excite or release electrons from a sensitive material designed to have this particular capability. For instance, the sodium–iodine crystal will fluoresce under γ

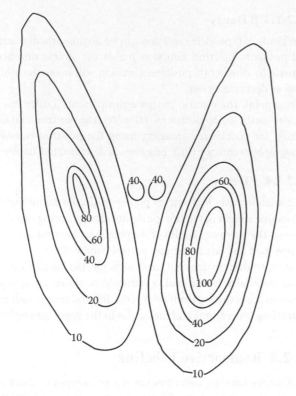

FIGURE 43.2 Illustration of the distribution of radiation emitted from a thyroid gland obtained through the administration of radioactive iodine. The radiation pattern is a direct indication of the local activity.

radiation from the iodine isotope; this fluorescence is captured by a photomultiplier tube and a current is generated under the influence of γ radiation. Other materials are CSI, bismuth germ, and xenon gas at 25 atm. Figure 43.2 illustrates the radiation measured with a charged coupled device (CCD) scintillation counter emitted from a thyroid that has been treated with a radioactive iodine isotope.

To ensure a good correlation between the point of emission of the γ radiation and the location of detection in the scintillation counter, a collimator is placed in the scintillation counter to allow only γ rays leaving the organ in a trajectory that is a direct line of connection to the face of the detector in order to be counted and localized (Figure 43.1). This way diffuse radiation is eliminated from detection. A collimator is a block of lead with holes drilled all the way through in such a way that a focal point can be established (Figure 43.3).

The scintillation counter is scanned over the surface of the organ and the counted values for each location are stored with respect to location and displayed for analysis of the organ function.

Other methods of γ ray detection use semiconductor materials arranged in an array to form an instantaneous image. This type of device is called a γ camera. The γ camera works in a similar fashion as the CCD camera, only with γ sensitive material doped.

The γ camera is significantly faster than the scintillation counter and can be used to examine dynamic processes, instead of merely performing static scans.

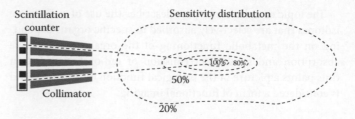

FIGURE 43.3 Scintillation counter with a lead collimator block facing to the right. The tubular holes drilled all the way through are arranged in a way that the focal point can be established. The focusing sensitivity is displayed by the hypothetical intensity response profile on the right-hand side.

43.2.4 Measurement and Classification of Radiation Activity

All nuclear radiation is measured in number of decay interactions per unit of time.[1–4,7,10,11,14] Several units of decay are commonly used, Becquerel [Bq], one (1) disintegration per second, and the curie [Ci], 3.7×10^{10} transformations per second. The curie is derived from the decay of 1 g of radium. Radiation energies are typically measured in units of electron volt, keV or MeV.

43.2.5 Radioactive Dating

The rate of decay (dN/dt) is directly proportional to the number of isotopes present at one time (N) as expressed in Equation 43.1, yielding the quantity of radioactivity (A_q) expressed by solving:

$$\frac{dN}{dt} = -\lambda N = A_q, \qquad (43.4)$$

where λ is the decay constant, which is particular to the isotope in question.

Solving Equation 43.4 for time gives the equation of radioactive decay:

$$A_q(t) = A_{q0} \exp(-\lambda t), \qquad (43.5)$$

where A_{q0} is the activity of the isotope at the start of the time period in which the decay is measured.

Equation 43.5 can be rewritten in terms of the number of isotopes [$N(t)$] as

$$N(t) = N_0 \exp\left(\frac{-t}{\tau}\right), \qquad (43.6)$$

where N_0 represents the number of decays at the start of the time frame in which the decay is monitored, and τ is defined as the half-life of the isotope.

43.2.5.1 Radioactive Half-Life

The half-life indicates the time lapse in which the activity decays by one half.[4,15,17,18] The half-life is often used as a measure to gauge the emission level of the decay process. This measure translates into the following statement: the longer the half-life, the higher the energy and the more reactive the decay radiation. Each nucleotide has a unique half-life, which can be used to specifically identify the isotope.

Other applications of the measure of the half-life are a means of determining the age of a component with a specific organic isotope, carbon dating. The radioactive isotope ^{14}C is produced in the upper atmosphere by cosmic radiation interaction with nitrogen, ^{14}N. The carbon-14 is incorporated in all carbon-based life forms as if it were regular ^{12}C. Generally, in a living organism, only 1.1% of all bound carbon is carbon-14. The organic isotope does not get replenished when the object is dead or outside the influence of the source that generates the isotope, and as such the decay of a specific mass will reveal the remaining fraction of the isotope against the standard fraction of the isotope present under natural circumstances. The naturally occurring carbon isotope ^{14}C ecays with a half-life of 5730 years.

As per definition, at $t = 0$, Equation 43.7 holds true:

$$N(t) = N_0. \tag{43.7}$$

At the half-life time, $t = \tau$, the nucleotide count has dropped to the expression:

$$N(t) = \frac{1}{2}N_0. \tag{43.8}$$

Substitution of Equation 43.6 into Equation 43.8 yields

$$\frac{1}{2} = \exp\left(\frac{-t}{\tau}\right). \tag{43.9}$$

Solving for τ requires taking the natural logarithm on both sides of Equation 43.9, as shown below:

$$\ln\left(\frac{1}{2}\right) = \frac{-t}{\tau} = -\lambda\tau. \tag{43.10}$$

Solving Equation 43.10 gives the correlation between the half-life and the decay constant as

$$\frac{\ln(2)}{\lambda} = \tau = \frac{0.693}{\lambda}. \tag{43.11}$$

Since the decay constant is known for a certain isotope, the half-life and as such the age of the sample can easily be determined from the measured decay rate.

The following are a few examples of decay of specific isotopes: Carbon-14, C-14, which decay under β emission with an energy

TABLE 43.1 Examples of the Half-Life Time Frame of Various Isotopes

Isotope	Half-Life
Tc-99 m	6.0 hours
I-131	8.05 days
Co-60	5.26 years
Sr-90	28.1 years
Pu-239	24,400 years
U-238	4,150,000,000 years

of 0.157 MeV or 157 keV; cesium, Cs-137, which decays under γ emission with an energy of 662 keV; and Americium-241, Am-241, which emits α particle with an energy of 5.485 MeV. Several half-life times are shown in Table 43.1.

In addition to the radioactive half-life, the body also removes isotopes in the metabolic process. The biological half-life (τ_b) needs to be combined with the radioactive half-life to determine the effective half-life (τ_e) that can be used to measure the accumulated isotope retention and the associated decay as follows:

$$\frac{1}{\tau_e} = \frac{1}{\tau_b} + \frac{1}{\tau}. \tag{43.12}$$

43.2.6 Radiation Detection

Several devices are available for the detection of radiation.[2,3,10,11,14,16,17]

43.2.6.1 Geiger Counter

The Geiger counter relies on the fact that ionizing radiation does just that; it ionizes seed molecules in a vacuum chamber. The charged unit is accelerated in an electric field applied between a cathode and an anode. The ion moving toward the positive cathode in turn produces a cascade of ionization, resulting in an effective current that is converted into an electron current when impacting the cathode. The current induces a voltage when passing through a resistor. The voltage is used to create a click in a speaker and makes the voltmeter dial deflect with each disintegration. The audible cracks from the speaker and the deflection of the dial are used to indicate the number of particulate or photonic disintegrations and hence the strength of the radiation.

43.2.6.2 Scintillation Counter

The scintillation counter uses the conversion of radiation energy (particle or electromagnetic) to a photon that is channeled through an avalanche multiplier that amplifies the photon for detection. This technique is called a photomultiplier. Typical gains for photomultipliers are in the order of 10^7. Closing the photomultiplier with a 50 Ω load will result in a voltage of several mV, which can be measured by a high-resistance voltmeter.

43.2.6.3 Semiconductor Detectors

Semiconductor detectors use the p–n junction effect to generate electron–hole pairs under the influence of external energy. The transistor principle is used to apply an electric field at the p–n junction to prevent the electron–hole pair to recombine, and instead creates a current. Electronic integration of the current yields the total energy delivered over the integration time.

43.2.6.4 Cloud Chamber

A cloud chamber is a sealed chamber filled with a saturated gas cloud that condenses under the influence of a passing particle. A cloud chamber will thus only detect the α and β particle. The motion of the charge will act as a current seen electronically. Application of an external homogeneous magnetic field will apply a force on the charge and divert the charge according to the sign of the charge (positive or negative for the α and β particle, respectively) and the right-hand rule. The cloud chamber thus is primarily a visual aid with little quantitative information.

43.2.6.5 Bubble Chamber

The bubble chamber relies on the same principle as the cloud chamber, only it uses a liquid in a sealed container that is just at its boiling point.

43.2.6.6 Photographic Emulsion

The use of a photographic emulsion provides a qualitative measure based on the oxidation of the silver chloride, providing visible tracks after developing the film. Additionally, this system can also detect γ rays, by priding a darkening of the film and as such the amount of darkening of the film can be used as a quantitative indication of the radiation exposure. When used in one particular radiation environment only, this can be used as a continuous dosimeter counter and is done so for radiation monitoring in hospital rooms with x-ray.

43.2.6.7 γ Camera

The γ camera collects the γ radiation emitted by isotopes using scintigraphy. Scintigraphy uses a sodium–iodide crystal primarily doped with thallium, which causes light flashes when electrons are released from the iodine. The photons are amplified by a photomultiplier tube. The γ camera is the result of the work of Robert Hofstadter and Hal Anger. Sometimes the γ camera is also called the anger camera. The γ camera is sensitive to the energy level of the γ radiation by the voltage produced in the photomultiplier. The γ camera is in fact capable of distinguishing the signals from various isotopes used simultaneously. The photomultiplier uses a grid imaging method that will give spatial resolution of the isotope distribution.

43.2.7 Units of Radiation Exposure

The impact of radiation can be quantified in various ways.[1–3,14] The exposure to radiation can be in Röntgen (R), or in SI units:

TABLE 43.2 Quality Factors Used by the Nuclear Regulatory Commission (NRC)

Radiation	Q
γ and β radiation	1
α particles	20
Neutrons of unknown energy	10
High-energy protons	10

C/kg. The absorbed amount of radiation measures the amount of energy absorbed per unit mass. This quantity applies to all types of ionizing radiation and all types of materials. In this measure, one (1) Röntgen of γ or x-ray radiation is approximately equal to one rad in human tissue. The absorbed dose and the absorbed dose rate are expressed in rad and in SI as gray (Gy), respectively. The dose equivalent and the dose equivalent rate apply to human exposure and are related to the risk of a radiation dose calculated by multiplying the absorbed dose by a "quality factor." The quality factor depends on the type of radiation and the tissue exposed. This measure as such becomes independent of the type of radiation and is measured in rem or sievert (Sv). Exposure to one (1) rem poses the same risk for any type of ionizing radiation, α, β, γ, *x-ray, or neutron*. Table 43.2 shows the quality factor for various types of radiation.

43.2.8 Natural Background Radiation

Depending on one's location (residential or rural) and altitude, on average a person is exposed to a certain amount of radiation (natural background radiation). Table 43.3 lists the natural background radiation.

In contrast, a chest x-ray with modern equipment exposes the patient to approximately 5–10 mrem, whereas a full CT-scan exposes the patient to approximately 100–200 mrem.

43.3 Clinical Applications of Nuclear Medicine

Nuclear medicine used isotopes and radiation exposure for various diagnostic and therapeutic applications. Select examples of clinical radiation exposure are listed in Table 43.4.

TABLE 43.3 Average Exposure to Various Sources of Natural Background Radiation

Radiation Source	Radiation Exposure (mrem/year)
Cosmic radiation	30
Terrestrial radionuclides (e.g., uranium-235, nuclear weapon testing)	30
Radionuclides naturally occurring in the body	40
Inhaled radionuclides (primarily from radon, but also from drinking water)	200
Total	300

TABLE 43.4 Selected Examples of Radiation Exposure in Diagnostic Nuclear Medicine

Diagnostic Application	Radiation (mrem)
Lung perfusion (Tc-99 m)	50
Thyroid uptake (I-131)	70
Chest x-ray	5–10
Full CT-scan	100–200

Source: From NCRP No. 124.

TABLE 43.5 Acute Effects of Whole Body Exposure on Man

Absorbed Dose (rad)	Biological Effect
5	Initial changes to blood observed
25	Temporary changes to the blood
50	No significant impact
100	Likely recovery
450	Lethal dose 50/30
600	Fatal in weeks
1200	Fatal in days
10,000	Fatal in several hours

43.3.1 Therapeutic Nuclear Medicine

The most common use of therapeutic nuclear medicine is in the following three areas: the treatment of thyroid problems (e.g., hyperthyroidism), polycythemia (abnormal red cell and blood increase), and prostate cancer. Iodine isotopes are commonly used (I-131) in the treatment of both thyroid cancer as well as an overactive thyroid gland.

In the treatment of prostate cancer, specific isotopes are injected to target the diseased tissue.

The prostate therapy is part of a technique called radioimmunotherapy (RIT). In RIT, the radioisotopes are engineered to attach to antibodies with a specific affinity for selected groups of cells. The respective antibodies in turn will incorporate themselves in the cancer cells with the radionuclides attached. The radionuclides are now fully deployed to destroy the cancer cells locally.

Another application in RIT is in breast cancer treatment; however still investigational, it may deliver a promising new method to treat metastatic cancer of the breast with minimized dosage. The breast cancer treatment has the isotope yttrium-90 attached to the m-170 monoclonal antibody, which is digested by the cancer cells and selectively destroys the growth.

One common use of radionuclides is brachytherapy. In brachytherapy, isotopes are placed close to cancerous tissue, either externally or by surgical implant, and the time exposure provides a measure of the exposure dosage clinically relevant for the treatment and destruction of the cancer. Cobalt-60 is another common isotope used to primarily produce γ radiation. When focused, this creates an ablative beam referred to as the γ knife. The use of focused radiation is called teletherapy.

43.3.2 Diagnostic Nuclear Medicine

The use of radiopharmaceuticals for imaging evolved in the 1970s. Specific diagnostic applications are in the liver, brain, spleen, thyroid, and gastrointestinal track. The use of isotopes for metabolic imaging is one of the examples of diagnostic nuclear medicine. The radionuclide iodine-131 is incorporated by the thyroid and the emitted γ radiation is a direct indication of the local metabolic activity. Additionally, the use of isotopes in PET imaging has gained significant popularity. For PET or SPECT (single-photon emission computed tomography) imaging, nucleotides are incorporated in specific substances that will target a designated biological process, for instance.

Water with deuterium isotopes can be used to monitor blood flow rate noninvasively on a small scale. Cellular carbohydrate consumption can be diagnosed by including the isotope fluorodeoxyglucose (FDG), which acts as glucose in the cellular metabolism. Also protein synthesis can be imaged with the use of radiopharmaceutical fatty acids.

In most cases, the emitted γ radiation is collected with a specifically designed γ camera. For instance, technetium (Tc-99) isotopes under various radiopharmaceutical names depending on the antibody attachment (e.g., Pertechnetate®, Pulmonite®, Microlite®, Glucoscan®, Verluma®, and Ceretec®) are generally used for a multitude of imaging applications, such as the bladder, breast, ectopic gastric mucosa, heart, liver, parathyroid glands, renal, salivary glands, spleen, thyroid, as well as tear ducts.

SPECT in particular uses γ-emitting short-lived isotopes such as technetium-99, indium-111, iodine-131, or thallium-205.

43.3.3 Acute Effects of Whole Body Exposure on Man

Table 43.5 lists a range of examples of the clinical consequences of exposure to increasing amounts of radiation. Note that radiation exposure is cumulative, and the exposure over time adds up to result in the effects listed for the total exposure. However, the expression may be delayed due to the fact that the body does show a certain form of adaptation. Generally, there is a difference between the exposure and the absorbed dose, also based on the radiation source.

References

1. Greenberg D and Leslie WD. *Nuclear Medicine*. Austin, TX: Landes Bioscience; 2001.
2. Ell P and Gambhir S. *Nuclear Medicine in Clinical Diagnosis and Treatment*. New York: Churchill Livingstone; 2004.
3. Schiepers C and Allen-Auerbach M. *Diagnostic Nuclear Medicine*, 2nd edition. Basel: Birkhäuser; 2005.
4. Knoll GF. *Radiation Detection and Measurement*, 3rd edition. Hoboken, NJ: Wiley; 2000.
5. Bohr N. On the constitution of atoms and molecules. *Philos. Mag.* 1913; 26: 1–25.
6. Audi G and Wapstra AH. The 1995 update to the atomic mass evaluation. *Nucl. Phys.* 1995; A595: 409–480.

7. *CRC Handbook of Chemistry and Physics.* pp. 1–40, 11–14. Boca Raton, FL: CRC Press; 2008.

8. Becquerel AH. Sur les radiations émises par phosphorescence. *Comptes Rendus* 1896; 122: 420–421.

9. Rutherford E. *Radioactivity.* Cambridge: Cambridge University Press; 1904.

10. Krubsack AJ, Wilson SD, Lawson TL, Kneeland JB, Thorsen MK, Collier BD, Hellman RS, and Isitman AT. Prospective comparison of radionuclide computed tomographic, sonographic, and magnetic resonance localization of parathyroid tumors. *Surgery* 1989; 106: 639–646.

11. Eckerman K and Endo A. *MIRD: Radionuclide Data and Decay Schemes.* Reston, VA: The Society of Nuclear Medicine; 2008.

12. Röntgen WC. Über eine neue Art von Strahlen. *Sitzungsberichte der Würzburger Physikalischen-Medicinischen Gesellschaft* 1895: 132–141.

13. Curie M and Curie P. Sur une nouvelle substance radioactive, contenue dans la pechblende. *Comptes Rendus de Seances de l'academie de Sciences* 1898; 127: 175–178.

14. Mas J. *A Patient's Guide to Nuclear Medicine Procedures: English-Spanish.* Reston, VA: Society of Nuclear Medicine; 2008.

15. Bertin C and Bourg ACM. Radon-222 and chloride as natural tracers of the infiltration of river water into an alluvial aquifer in which there is significant river/groundwater mixing. *Environ. Sci. Technol.* 1994; 28(5): 794–798.

16. Diggles L, Mena I, and Khalkhali I. Technical aspects of prone dependent-breast scintimammography. *J. Nucl. Med. Technol.* 1994; 22: 165–170.

17. Jull AJT. *Radiocarbon.* Department of Geosciences, University of Arizona, http://www.radiocarbon.org/

18. Arnold JR and Libby WF. Age determinations by radiocarbon content: Checks with samples of known age. *Science* 1949; 110: 678–680.

VII

Physics of Bioengineering

Biophysics in Regenerative Medicine

Amanda W. Lund,
George E. Plopper, and
David T. Corr

44.1 Native Tissue Structure and Function

Natural and engineered tissues are a collection of cells arranged within a structural scaffold of insoluble extracellular matrix (ECM) proteins and soluble signaling molecules that provide biochemical and mechanical cues directing function. This function is therefore dependent on the spatiotemporal resolution of matrix proteins, cells, and signaling molecules, and their interactions with one another. While the ECM is ubiquitous throughout development, its composition, ratio, geometrical arrangement, and mechanical properties are important distinguishing features between tissue types.[1,2] Cells interact with the ECM in their microenvironment to direct specialization and tissue-specific behavior. This interaction occurs in a bidirectional, reciprocal manner wherein cells respond to the cues present in their environment by initiating signaling programs of genetic and phenotypic transformation.[3] This transformation in turn allows the cell to exert physical and biochemical influence on its microenvironment leading to cell/matrix cooperativity during development, homeostasis, and repair.

Adhesion to the ECM regulates cell function. All normal healthy cells, with the exception of circulating blood cells, are contact dependent and therefore require physical association with matrix proteins for survival. In response to the formation of matrix adhesions, cells assemble intracellular machinery that link cellular signaling and behavior to mechanical and biochemical changes in their microenvironment.[3] Tissue geometry, determined by the spatial organization of ECM components, presents both a template and an instructive cue for tissue morphogenesis.[4] Through changes in geometry, long-range tissue interactions can be controlled by coordinating cell behavior in distal areas of the tissue.[4,5] This coordination is achieved by gradients of soluble factors and mechanical stresses that restrict and define localized activity: changes in solute.

Cell–cell interactions also play an important role in tissue homeostasis and repair. Cells form adhesive structures between each other that have both intracellular and extracellular components, and when disrupted induce a series of signals that propagate through the cell population. For example, osteocytes within mature bone use these junctions to sense and coordinate responses to microfracture, and endothelial cells form tight junctions to present a selective barrier to solute diffusion and cell migration. Taken together, these mechanisms coordinate to maintain tissue function and integrity during development, homeostasis, and repair.[6]

44.2 *In Vivo* Regeneration and Repair

Regeneration is the recovery of both the structure and function of injured and diseased tissues. Some amphibians have the extraordinary ability to regenerate limbs following amputation, and embryos (including human) share this capacity to heal without scarring. Most organisms respond to severe injury through the induction of a spontaneous repair process, forming a contractile fibrotic clot that ultimately forms a scar. Embryos and select amphibians uniquely respond to this same type of injury with conserved mechanisms of tissue synthesis that result in the near "perfect" healing of the wounded tissue and no scarring. During embryonic healing in *Drosophila melanogaster*, a cable of actin filaments is created in the cells surrounding the wound that steadily contracts and pulls the injury together like a drawstring to interlock and seal the wound.[7] This closure is dependent on the extension of small filopodia and the necessity of filopodia-based tissue development is conserved across species including mice, worms, and humans.[8] Embryos are in the business of building tissue. Therefore, within this context, a synthesis-based response to injury makes sense.[9] When disrupted, embryos may borrow mechanisms of synthesis from adjacent developing tissues. This unique property is restricted, however, to injuries incurred prior to the third trimester; thereafter, healing undergoes a transition from regeneration to repair and scarring.[10]

Repair, the mechanism through which all adult mammals respond to injury, is a spontaneous process of wound closure that results from the contraction of the stroma (i.e., loose connective tissue), scar formation, and healing of the underlying tissue. During adult wound healing, a blood clot is formed at the site of injury to stem the flow of blood; fibroblasts then secrete a matrix of support proteins that fill the defect. This matrix contracts, leading to scar formation as cells eventually seal the wound.

Most mature epithelial tissues contain three principal layers: epithelial cells, a basement membrane, and an underlying connective layer (stroma). Excision of the cells and the basement membrane results in spontaneous regeneration much like that seen in the embryo. However, the stroma is unable to replicate this process, and scar formation in the stroma is a key inhibitor of functional tissue regeneration in adults.[10–13]

Scar formation may be an adaptive response to challenges faced by mature tissues. Adult wound healing may reflect the need for a defense mechanism against infection, as scar formation is induced by inflammation. Embryos prior to the third trimester have an immature immune system and are unable to mount a significant inflammatory reaction. Mice lacking macrophages and neutrophils do not scar and heal as quickly as normal newborns.[14–16] Modulation of the inflammation response, therefore, stands to play a significant role in the inhibition of scar formation and the induction of regenerative healing within adult tissues. Understanding the mechanisms of embryonic regeneration and applying them as design principles in engineered materials is a critical goal for the field of regenerative medicine. Ideally, this type of tissue growth could be induced without compromising the immune system of the recipient.

44.3 Natural and Synthetic Scaffolds in Regenerative Medicine

44.3.1 Materials

Cells, scaffolds, and signaling molecules are often referred to as the tissue engineering triad.[17] The ideal graft would include a role for each component in a controlled and predictable manner. Scaffolds, which are composed primarily of polymeric biomaterials, provide mechanical stability and act as a substrate for cell adhesion, growth, and differentiation. The enormous variety of methods available to tune material and cell function complicates the search for optimal regenerative medicine scaffolds.[17]

Three-dimensional (3D) scaffolds are composed of a variety of different materials that are able to guide cell function, differentiation, and morphogenesis. The simplest consist of ceramics, glass, and/or synthetic polymers. However, to achieve more physiological regenerative effects, a number of ECM proteins and polysaccharides[18] have also been used as scaffolds, including collagen I,[19–21] gelatin,[22,23] fibrin,[24,25] chitosan,[26–28] hyaluronan,[29,30] agarose,[31,20] and alginate.[32,33] In addition to these natural polymers, a variety of synthetic polymers have been developed as bases for scaffolds. For example, polyethylene glycol (PEG),[34] poly(lactic-*co*-glycolide) (PGLA),[35,36] poly(lactic acid) (PLA),[37,38] and poly(glycolic acid) (PGA)[39,40] are widely used in regenerative medicine. These polymers allow for specific control over bulk properties, surface chemistry, and sourcing that offer advantages over their natural counterparts. These materials are biodegradable, biocompatible, and bioactive and therefore present a significant advantage in their ability to induce regeneration responses when implanted *in vivo*.

An additional advantage of these proteins, polysaccharides, and polymers is their ability to be solubilized and polymerized into a hydrogel, allowing for the direct encapsulation of living cells. The major limitation of acellular substitutes is that they do not self-renew and thus have a limited useful life span. In cases where healing is delayed or halted, host cell infiltration is prevented or lacking a vascular supply, incorporation of living cells into the engineered construct is especially important for stimulating tissue formation. The rapid generation of quality tissue is needed to prevent morbidity in such cases. For example, the reader is referred to several recent reviews for bone,[41–43] cardiac,[44,45] tendon,[46,47] nerve,[48,49] cartilage,[50,51] and fat[17,52,53] tissues.

44.3.2 Biodegradability

Beyond providing structural support within the void space of the defect, the implanted material should be readily degradable to promote host integration. There are two types of biodegradable polymers: (1) natural materials such as polysaccharides (starch, alginate, chitan/chitosan, and hyaluronic derivatives) and proteins (collagen, fibrin, and silk) and (2) synthetic polymers.[54] Each type has important advantages and disadvantages.

Natural polymers are degraded through cell-mediated secretion of proteases via mechanisms consistent with those in native tissue. Ideally the degradation rate should equal the rate of new matrix deposition by implanted and host cells.[17] Engineering biodegradability into synthetic polymers likewise provides control over the useful life of the scaffold and release of bioactive growth factors and protein fragments within a diseased or injured tissue. Synthetic polymers produced under controlled conditions result in predictable and highly reproducible material properties, that is, tensile strength, elastic modulus, and degradation rate. The degradation kinetics of biomaterials can be controlled by a variety of tissues and handling properties including chemical composition, hydrophilicity, structure, processing history, molar mass, polydispersity, environmental conditions, stresses and strains, and porosity.[40,55]

The most commonly used biodegradable synthetic polymers are poly-α-hydroxy esters including PLA, PGA, and PLGA.[56] These polymers are hydrolytically degraded through de-esterification and their degradation products are removed by natural pathways. For example, PGA degradation products are treated as metabolites and PLA is cleared through the tricarboxylic acid cycle.[54] The degradation rate of each of these polymers is greatly affected by treatment conditions and the host environment, but in general take from 12 to greater than 24 months to completely degrade. However, rapid bulk degradation of these polymers can occur, resulting in the premature failure of implants, largely due to losses in mechanical integrity. This abrupt release of degradation products can lead to a strong inflammatory response that is deleterious to the implant.[40,57] Polymers that do not exhibit bulk erosion properties but instead degrade selectively at the polymer water interface are advantageous in applications where the controlled release of factors is important (e.g., as a drug delivery vehicle). These surface-eroding polymers have several advantages including the retention of mechanical integrity, minimization of toxic degradation products, and enhanced tissue ingrowth.[58]

44.3.3 Biocompatibility

Materials used for the purposes of tissue engineering must have the capacity to support the growth, differentiation, and integration of both endogenous and exogenous cells.[17] There are seven requirements for biocompatibility with respect to the implantation of biomaterials. Biomaterials must be (1) noninflammatory, (2) immunologically inert, (3) noncytotoxic, (4) able to resist physical stresses within the site of implantation, (5) structurally adaptable, (6) affordable, and (7) easy to sterilize during manufacture.[59,60] The biocompatibility of the starting construct, the toxicity of its degradation products, and its tendency to illicit an immune response are critical design features for any engineered tissue.

44.3.4 Bioactivity

Beyond being able to coexist with host tissue and degrade in a balance with regenerative events, many scaffolds are engineered to induce specific cellular behaviors. ECM-based scaffolds have inherent bioactivity that, when correctly understood and applied,

can specifically direct cell function and phenotype upon encapsulation. The bioactivity of these ECM proteins comes from specific sequence and structural motifs that guide the interactions between cells and their extracellular environments. Using these motifs, engineers have designed bioactivity into synthetic polymers in a controlled and systematic way to drive specific cell function and morphogenesis. Adhesion to ECM proteins within the native tissue can induce the correct positioning,[61] polarization,[62,63] and differentiation[64,65] of cells within their microenvironment. Using these proteins alone or in combination can result in successful engineering of cell and tissue function. However, the heterogeneity present in isolated sources of these proteins does present a problem in terms of engineering simplicity and consistency into tissue constructs.

Fragments of ECM proteins are also used in so-called "smart" biomaterials that provide a means for locally directing cell function through engineering the cell/scaffold interface. For example, the tripeptide adhesive ligand, RGD, is widely applied in conjunction with hydrogel carriers to guide cellular adhesion, proliferation, and differentiation of preosteoblast cells.[66] Increasing ligand density promotes enhanced proliferation and may provide a method of controlling *ex vivo* expansion and the development of transplanted cells. PEG hydrogels containing RGD and laminin sequence insertions induce enhanced lumen and apical pole formation with proper basolateral polarization in MDCK cells.[67] The proteoglycan hyaluronic acid (HA) is also widely used as a base component of hydrogel culture. HA hydrogels promote proliferation, branching morphogenesis, and tubule differentiation of epithelial cells for the purposes of renal tissue engineering.[68] These observed effects are both molecular weight-and concentration-dependent.

The two-way interaction between a cell and its matrix is crucial for the simultaneous reorganization of the ECM and induction of cell behavior required during development.

44.4 Mechanical Characterization of Biomaterials

The emerging field of biomedical engineering exists at the intersection of three classical areas of science: biology, medicine, and engineering. Biomechanics is a specific discipline within biomedical engineering that addresses the mechanics of living organisms. That is, it focuses on how living organisms produce force and respond to forces and displacements. To study biomechanics requires a base knowledge in theoretical and applied mechanics (structural, fluid, and thermal mechanics), combined with a similar understanding in biology and physiology. Occasionally, the field of biomechanics is associated solely with human and animal movement science. However, this is a very narrow view, as the study of mechanics in biological organisms occurs at a wide variety of levels (e.g., structures, tissues, and organisms), and extends well beyond applications in kinesiology. As a result, a number of subfields have been created, all within the traditionally defined area of biomechanics, to more accurately identify the specific area of study, such as molecular-, cellular-, tissue-, organ- or multi-tissue system-, joint-, and whole body-biomechanics. While there are thermal, fluid, and structural disciplines within each of these areas, we will focus the following discussion on those applications relating to structural mechanics.

Structural biomechanics combines the basic principles of classic structural mechanics, with the necessary biology and/or physiology appropriate for the structure, tissue, or system studied. To begin, from a structural mechanics perspective, one must consider three factors: (1) the geometry of the specimen or structure, (2) the loading conditions applied, and (3) the basic principles of mechanics, such as the enforcement of the appropriate physical conservation laws (e.g., conservation of momentum, conservation of mass, and conservation of energy). Enforcement of these conservation principles yields vector equations that incorporate the magnitude and direction of the loadings with their point of application on the geometry of the specimen or structure. Since these relations are determined based solely on the specimen geometry and the applied loads, they do not include any information or assumptions of the material, or materials, that comprises the specimen. These relations arise from the application of basic vector mechanics and can be derived from first principles. This process is followed to generate the vector equations that govern a variety of loading conditions, including the loads and energies generated through mechanical (structural), fluid, or thermal means.

44.4.1 Mechanical Behavior of Material

44.4.1.1 Stress

With the loading established, and equilibrium enforced, the internal forces and moments that develop within the specimen or structure can be determined. These values can then be normalized by the area over which they act to get a load distribution per unit area, or *stress*. Stresses represent the intensity of the force experienced, or "felt," by the material under the given loading condition, and are given in units of pressure (psi or Pa). If one considers an arbitrary solid subjected to loading, within the solid every differential area, dA, is subjected to a differential force vector, $d\mathbf{F}$. This differential force vector can be resolved into a component dF_n acting normal (perpendicular) to the area dA and a component dF_t acting tangent to dA. The normal stress (σ) is defined as the normal component per unit area (also see Chapters 15 through 18),

$$\sigma = \frac{dF_n}{dA},$$ (44.1)

and the shear stress (τ) is defined as the tangential component of force per unit area,

$$\tau = \frac{dF_t}{dA}.$$ (44.2)

Typical convention is that normal stresses that are tensile, in which the forces "pull" on dA, are considered positive, and

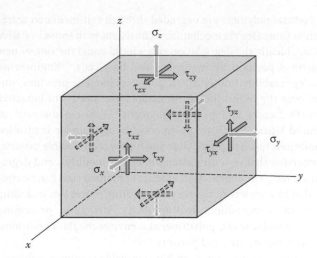

FIGURE 44.1 General 3D state of stress, as shown on a cubic differential volume.

compressive stresses that "push" on dA are negative. These differential stresses define the stress on a differential area, and thus represent the stress at a point in the material. However, stresses can vary considerably from point to point within a given cross section. Therefore, it is often advantageous and convenient to determine the average stresses acting on a material. This is done by integrating the differential definitions of stress (Equations 44.1 and 44.2) over the entire cross-sectional area to gain average values of normal stress (σ_{ave}),

$$\sigma_{ave} = \frac{P}{A},$$ (44.3)

in which P represents the total normal force acting on the cross-sectional area, A; and the average shear stress (τ_{ave}),

$$\tau_{ave} = \frac{V}{A},$$ (44.4)

where V represents the total shearing force acting tangent to the cross section.

By extending these ideas to a differential volume element, the generalized state of stress in a solid can be represented on a unit cube, as illustrated in Figure 44.1. Each face of the cube has three stresses acting on it: a single normal stress acting perpendicular to the plane and a pair of orthogonal shear stresses to define the stresses acting within the plane. Through the enforcing of equilibrium, it is easily shown that the following equalities must hold.

$$\tau_{xy} = \tau_{yx}, \quad \tau_{yz} = \tau_{zy}, \quad \tau_{xz} = \tau_{zx}.$$ (44.5)

Thus, the generalized state of stress is defined by six terms: three normal stresses and three shear stresses (e.g., σ_x, σ_y, σ_z, τ_{xy}, τ_{yz}, τ_{zx}). While this example utilized a Cartesian coordinate system, a similar process can be followed for other orthogonal coordinate systems.

44.4.1.2 Strains

In a similar fashion, *strains* are determined by normalizing the displacements in the specimen or structure by their undeformed size (length or angle). Considering a solid of original length Δx that deforms an amount $\Delta\delta$ along its length (the x-axis) when loaded, the axial strain, or normal strain (ε) along the x-axis, is defined as

$$\varepsilon_x = \lim_{\Delta x \to 0} \frac{\Delta\delta}{\Delta x} = \frac{d\delta}{dx}, \qquad (44.6)$$

where the limit as Δx approaches zero indicates the elongation (or compression) of an infinitesimal length dx, or the derivative of axial deformation with respect to the axial direction. As with the derivative definition of stress, Equation 44.6 defines the strain at a point, and as seen in stress, the axial strain may vary considerably from point to point. Therefore, at times it is convenient to define an average normal strain (ε_{ave}) as the change in length over the original length,

$$\varepsilon_{ave} = \frac{L_f - L}{L} = \frac{\delta}{L}, \qquad (44.7)$$

where L_f is the final length, L is the original undeformed length, and δ is the total change in length. Normal strains that elongate the solid are considered positive and those that contract or compress are considered negative.

Whereas axial (normal) strains describe the normalized length change along an axis, shearing strains measure the change in angle between two, originally perpendicular, axes. Thus, the shearing strain (γ) in an x–y plane in which the x and y directions are perpendicular (angle $= \pi/2$ rad) in the undeformed state is defined as

$$\gamma_{xy} = \frac{\pi}{2} - \theta, \qquad (44.8)$$

where θ is the angle between the x and y axes in the deformed state. As such, shearing strains act to decrease (positive shearing strain) or increase (negative shearing strain) the angle between two perpendicular axes, and therefore represent a change in *shape* rather than a change in *size*. Thus, for the general state of 3D strain, as shown in Figure 44.2, the differential volume ($\Delta V = \Delta x \Delta y \Delta z$) is a cube with all right angles in its undeformed state. When deformed due to applied loading, the change in the size of each side of the cube is given by the axial strains (ε),

whereas the change in angles between the cube sides is reflected in the shearing strains (γ).

Strain measurements describe the deformations of the specimen or structure and are typically described as either elastic or plastic in nature. Purely elastic strains can be completely recovered once the loading is removed, whereas plastic, or inelastic, strains result in permanent deformation. Since the determination of stresses and strains already accounts for specimen geometry, their values are not influenced by the size of the specimen. As a result, stresses and strains can be directly compared among different specimens over a variety of materials and sizes.

44.4.1.3 Constitutive Equations

Once the governing loading equations are established, the contributions of the material itself must be considered. To account for material behavior, constitutive equations are constructed. Unlike equations derived from first principles that apply similarly to all materials at virtually all levels, constitutive equations are based on relationships observed in experiments, and the precise relationship is specific to the given material. These equations contain a constant (or constants) intrinsic to that specific material. Thus, the observed response is characterized by a behavior specific to the material, which relates to the value of the intrinsic material-specific constant, or material property.

Material properties have been determined to describe a vast number of materials over a wide range of applications. Materials can be characterized by their behavior using structural material properties (e.g., Young's modulus, ultimate strength, yield stress, shear modulus, and Poisson's ratio),[69,70] fluid and thermal material properties (e.g., viscosity, coefficient of thermal expansion, specific heat capacity, thermal conductivity, specific enthalpy, melting point, and thermal resistance),[71–73] electrical material properties (e.g., resistivity, permittivity, and conductivity),[74,75] as well as physical material properties such as density. These properties depend solely on the material and are not affected by the loading or the geometry of the specimen. Although a wide variety of constitutive relations have been established, leading to the identification of numerous material properties, we will focus the discussion that follows on those relating to applications in structural mechanics.

One structural example, the relationship between stress and strain, over the linear region, has been well characterized by a material constant—Young's elastic modulus. This value is intrinsic to the material and does not depend on specimen geometry. Thus, a small steel orthopedic implant will have the same Young's modulus as a steel strut from a building or bridge, provided the grade of the material (e.g., ASTM 316L stainless steel) and its treatment (e.g., annealed, anodized, heat treated, and cold rolled) are the same.

Young's modulus, as well as a number of additional important structural material properties, can be determined by plotting the experimental stress data versus the corresponding strain data from any number of different mechanical tests. One such test, an elongation to failure, is shown qualitatively for a brittle material in Figure 44.3. In this experiment, the specimen is stretched to failure, typically at a constant rate in displacement (or strain)

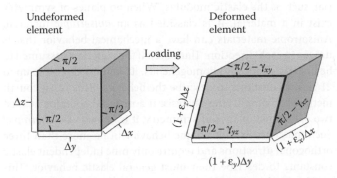

FIGURE 44.2 General 3D state of strain, illustrating the normal strains and shearing strains that develop with loading, and their effect on changing the size and shape of the volume, respectively.

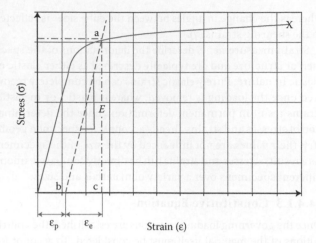

FIGURE 44.3 Graphical representation of typical stress–strain curve for a brittle material stretched to failure. The slope of the linearly elastic portion of the curve gives Young's elastic modulus, *E*.

control, and the load and elongation are recorded. These data are converted into stress and strain, respectively, and plotted as shown in Figure 44.3. From such a plot, the stress and strain values corresponding to the point of failure (denoted by an "X" in Figure 44.3) can be readily obtained, providing the ultimate strength and failure strain, respectively. The early portion of the curve shows a linear relationship between stress and strain, such that the slope is constant. In this area, the material behaves elastically, and any deformation that occurs with loading is recovered once the load is removed. The slope of this portion, *E*, is the Young's modulus of the material. Once the material is strained beyond the linearly elastic region, an increase in strain will result in permanent, plastic deformation. For example, if loaded to point "a" (Figure 44.3) the total amount of strain in the material is $\varepsilon_p + \varepsilon_e$, and the specific energy in the material is equal to the area under the stress–strain curve up to point "a." If completely unloaded from point "a," the curve will follow the same linear slope (*E*) to the point of zero load "b." While the stress has been returned to zero, a significant amount of permanent strain exists. ε_e represents the amount of elastic strain recovered by unloading, and ε_p is the permanent, plastic strain. With unloading, the specific energy associated with elastic deformation is recovered (area of triangle a–b–c), and the remaining area under the curve represents the amount of energy required to plastically deform the material. Additionally, by calculating the area under the entire stress–strain curve, from the onset of strain to the point of failure, the specific energy required to fail the material, or the material's toughness, can be determined. Thus, a variety of important material properties and characteristics can be determined from these fairly simple experimental data.

44.5 Structural and Mechanical Properties

In addition to those properties intrinsic to each material (i.e., *material properties*), similar groups of properties define the

mechanical response of structures and multiple material systems. The former, termed *structural properties*, are similar to material properties; however the influence of specimen geometry is retained. As such, in addition to the material composition, structural properties are specific to the size and architectural arrangement of the specimen or structure. This provides insight into the behavior of the structure, but does not allow for direct comparison between specimens of different sizes. For example, the stiffness of a steel orthopedic implant can be orders of magnitude lower than the stiffness of a steel structural support beam, even if the specimens are made of identical steels, due to the vastly different geometries and architectures. Thus, a structural property (e.g., stiffness) is determined both by the specimen geometry and arrangement, as well as the intrinsic properties (e.g., elastic modulus) of the material, or materials, that comprises it. For multiple material systems, such as composite materials, or materials with multiple levels of structural organization, it is common to calculate structural properties and then normalize to the specimen geometry. These approaches require a specific distinction, *mechanical properties*. Because such a specimen is heterogeneous in composition or organization, these properties cannot be attributed to a single material over all ranges of scale.

Materials can behave in a very complex and multidirectional fashion. However, this general behavior is often greatly simplified with a number of assumptions. First, a material is considered *linearly elastic* if there exists a linear relationship between stress and strain. Linear elasticity is a valid assumption for a wide range of metals and other materials, for strains below the point of yield and permanent deformation. Second, analyses can be further simplified if the strains in the material are sufficiently small, such that small angle approximations are valid. Third, it is also common to assume that the material is *homogeneous*, or spatially uniform, without localized voids or impurities. In its most general form, strain is defined by a fully populated second-order tensor, requiring nine unique terms to define the strain at a given point. However, if the material is assumed homogeneous, the tensor becomes symmetric, and the number of unique components required to describe a general state of strain reduces to six.

In a similar way, the planes of symmetry in a material reduce the complexity of the intrinsic properties that describe the material, such as the elastic modulus. When no planes of symmetry exist in a material, it is classified as an *anisotropic* material. Anisotropic materials can have a mechanical behavior that is unique in each direction. Thus, even if the material is assumed to be linearly elastic and homogeneous, it would still require up to 21 elastic constants to describe the behavior. This is the fourth method for simplifying analysis: if a material contains at least two orthogonal planes of symmetry, it is known as *orthotropic*. Such materials exhibit unique behavior in no more than three orthogonal directions and require only nine independent elastic constants to describe their most general elastic behavior. This can be further simplified for *isotropic* materials, which have planes of symmetry in every direction, and thus their material behavior is identical in all directions. The elastic behavior of an isotropic material can be completely described by just two elastic constants. These four approaches can thus greatly reduce the complexity of mechanical analysis.

44.6 Biological Tissue and Biomaterial Behavior

Many of the assumptions used to simplify general material behavior are no longer valid when considering biological tissue or tissue-engineered biomaterials. For example, while the small strain approximation is typically valid for metals, ceramics, and hard biologic tissues such as bone, it is a poor approximation for materials that experience larger strains (greater than 1% strain)[76] and is inappropriate for compliant polymers, elastomers, and biological soft tissues (e.g., cartilage, muscle, tendon, ligament, skin, vasculature, and connective tissue) that experience strains well above 1%. Some of these can even achieve greater than 100% strain before failing. For such materials, finite strain formulations should be employed to more accurately predict the mechanical response. Finite strain treatment of soft tissues was pioneered by Y. C. Fung[77,78] and has since been applied in a variety of soft tissue biomechanical applications.[79-88] The violation of the small strain assumption leads to a similar violation of the assumption of linear elasticity. Linear elasticity is also violated when a material exhibits a stiffening or softening response, such that the mechanical property of the material is different at different loads. Many collagenous soft tissues (e.g., skin, tendon, ligament, and muscle) when strained in tension exhibit a strain-stiffening behavior at low loads, followed by a linear region, and material softening prior to failure. This nonlinear behavior is often attributed to the progressive recruitment of slack or misaligned collagen fibers during the toe-in (stiffening) region and the progressive failure of stretched collagen fibers during the softening region,[89,90] as illustrated in Figure 44.4.

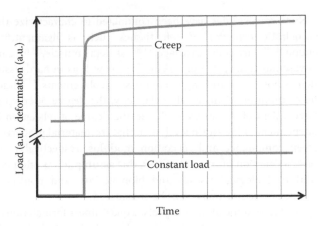

FIGURE 44.5 Graphical illustration of viscoelastic material creep, showing increasing deformation (or strain) with time while subjected to a constant applied load.

Most biological materials and tissues also exhibit *viscoelastic* behavior, a combination of both elastic and viscous behaviors. Such materials display a time-dependent mechanical behavior, such as creep, stress relaxation, and hysteresis. *Creep* is the progressive strain or deformation of a material with time, while subjected to a constant controlled load or stress, as illustrated in Figure 44.5, whereas *stress relaxation* describes the gradual decrease in stress in the material with time, when subjected to a constant strain, as illustrated in Figure 44.6. For cyclical loading protocols, a material exhibits *hysteresis* if the relationship between stress and strain (i.e., the shape of the curve on a stress–strain diagram) during loading is different than that observed in the unloading phase. In all of these behaviors—creep, stress relaxation, and hysteresis—the relation between stress and strain is not unique, but rather it changes in time. This time dependency of the stress–strain relationship is a distinguishing trait of viscoelastic materials and further illustrates the challenges in characterizing their behavior.

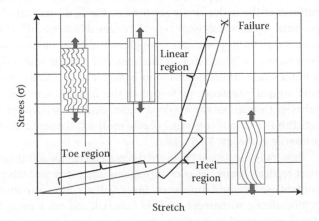

FIGURE 44.4 Typical nonlinear elastic response of a fiber-based material, such as collagen-based biological soft tissues like tendons and ligaments. The distinct nonlinear strain-stiffening behavior in the toe-in region is attributed to the progressive recruitment of slacked fibers. Once all fibers in the cross section are recruited, the material displays a linear response.

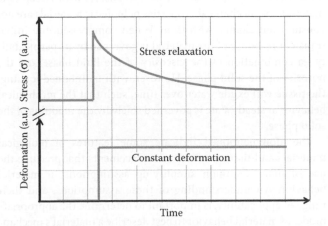

FIGURE 44.6 Graphical illustration of material stress relaxation, showing decreasing stress values with time while subjected to a constant applied deformation.

Numerous studies have been conducted to characterize the viscoelastic behavior of soft tissues such as ligament,[91–97] tendon,[98–100] articular cartilage,[101,102] skeletal muscle,[103,104] cardiovascular tissue,[105,106] and skin,[107–109] as well as in hard tissue such as bone[110–112] and engineered tissue replacements.[113] Under the assumption of linear viscoelasticity, the degree of creep is independent of the stress level, and the degree of relaxation is independent of the strain level, such that the material properties (creep compliance and relaxation modulus, respectively) are solely a function of time. However, this assumption is often not valid in biological tissues and biomaterials, and nonlinear aspects of viscoelasticity must be considered.

To address some of these aspects, a quasi-linear formulation of general viscoelasticity (QLV) was developed by Y. C. Fung.[114] QLV greatly simplifies generalized viscoelasticity, yet provides the flexibility to incorporate many of the nonlinear behaviors displayed by biologic soft tissues. While it has shown great applicability in soft biological tissues and biomaterials,[92,93,101,103,106,114–117] there are ranges of stress, strain, or strain rate, in which the QLV model fails to accurately describe the material behavior. One such example is in the application to strain-rate-sensitive materials. Materials that are *strain rate sensitive* exhibit different responses (stresses) to loading depending on the rate at which they are strained. Many biologic materials display a significantly different material response when evaluated at different rates of strain. For example, both bone and ligament display an increased tensile strength and elastic modulus when evaluated at higher rates of strain. As a result, alternatives to QLV are often employed for such nonlinear viscoelastic applications.[85,118,119]

An additional complication that many biomaterials introduce to the typical simplifying assumptions of material behavior is the presence of more than one material phase (e.g., solid, liquid, and gas). These multiphase materials exhibit a more complex mechanical behavior due to the individual responses and interactions of the different material phases. For example, articular cartilage is a biological tissue that is well characterized as a biphasic material—a liquid phase (water) contained within a solid matrix of collagen and proteoglycans. As with viscoelastic material, the mechanical properties of biphasic materials are not constant but change with time. When compressed, the initial transient response of the material is governed by its permeability—a combination of the viscosity of the fluid phase and the porosity of the solid phase. However, with maintained loading, the tissue will exude water over time, such that the mechanical behavior at steady state is governed solely by the modulus of the solid phase.

The aforementioned indicate some features of biological materials and tissue-engineered replacements that preclude the use of many common simplifying assumptions in material behavior. An understanding of these assumptions, and their range of applicability, is important to determine the appropriate model of material behavior to best describe a material's mechanical response. The aim is to find the model that simplifies the analysis as much as possible, while retaining all of the flexibility necessary to accurately describe the material's behavior.

44.7 Tissue Engineering: Identification of Target Function/Properties

One of the principal aims of functional tissue engineering is to produce a material capable of restoring the normal function of the tissue it seeks to replace. In particular, the native tissue's biochemical and biomechanical response throughout physiologically relevant ranges of chemical stresses and force provides an important functional target in the design of engineered replacements. Thus, a critical first step in this process is to characterize the behavior of the healthy, uninjured tissue to establish the desired "target" performance. Many of the first studies aimed at characterizing tissue mechanical behavior and properties focused primarily on the tissue stiffness or elastic modulus, Poisson's ratio, and failure properties such as failure stress, failure strain, and energy to failure.[120–124] While these properties adequately describe the general behavior of many traditional engineering materials, they often do not sufficiently capture the characteristic behavior of biologic materials, particularly soft- and high-strain materials. This is due in part to the fact that the behavior of biologic tissues rarely conforms to the most commonly made simplifying assumptions of material behavior discussed above.

Furthermore, the linear stiffness and failure loads typically occur well outside the normal physiological range of most soft tissues. These properties thus cannot provide a suitable metric for evaluating a replacement material's ability to restore normal function in soft tissues. Some recent studies of soft tissues under low loads have examined the strain-stiffening behavior throughout the physiologically relevant range of motion in ligaments,[95,125] tendons,[126] skeletal muscles,[104] and skin.[109] Characterizing this behavior is crucial, as it will provide the target nonlinear material behavior that an engineered replacement material should achieve for successful restoration of normal function.

Evaluation of tissue-engineered replacements is similarly impacted by this approach. If failure strength and high-load stiffness of the native tissue are used as the "gold standard," a potential replacement tissue could be deemed "successful" without concern for its properties throughout the *entire* range of use. For example, a "successful" replacement for a tissue with a significant strain-stiffening or toe-in behavior at low loads could have similar resistance to failure as the native tissue; yet its behavior throughout the low-load region could be wildly different. This is a serious concern, because most soft tissues function primarily in the low-load region.

Thus, fully functional engineered replacements for soft tissues must replicate tissue behavior throughout the entire physiologic range of use. Some exciting recent investigations utilize variations in bioreactor environment, scaffold material, and cell seeding to "tune" engineered tissue replacements to respond appropriately within the proper load range. For example, David Butler and colleagues have employed a multidisciplinary approach to tune functional tissue-engineered tendon repairs to match the full strain-stiffening and linear behavior observed in uninjured rabbit patellar tendon.[126–128] Such "tuning" approaches show great promise for creating constructs or replacement tissues precisely engineered to replicate the native tissues' behavior over the entire physiologically relevant range of motion.

44.8 Modeling/Informatics in Tissue Engineering

Developing methods of efficiently closing the tissue engineering loop, from biological hypothesis to successful *in vivo* application, require rigorous quantitative methods of analyzing the spatiotemporal function of tissues. Modeling functional, temporal, and spatial networks is a powerful means for testing these hypotheses and improving the throughput of material optimization. Models inherently provide quantitative analysis that provides feedback loops to inform engineering design. From the single-cell level up to the level of developing tissue and organization dynamics, creating models that quantitatively capture the principles guiding each tier of decision making is an emerging field that promises to accelerate tissue engineering efforts.

Cell signaling occurs through a complex and nonlinear network of protein interactions and genetic regulation that, even in two-dimensional (2D) culture conditions, becomes far too complex for traditional methods of data analysis. Because a relatively small number of signaling molecules can interact to form thousands of distinct signaling pathways, the combinatorial complexity of these networks requires advanced methods of data mining.[129] Computational modeling presents an approach to understanding network dynamics that defines function and dysfunction within these systems while also providing predictive capabilities. These methods are well reviewed[129-132] and provide insight into the dynamics of interactions between soluble cues, signaling pathways, and neighboring cells.[133-136]

3D imaging has historically been a powerful tool for cell biologists and engineers, providing spatial information including the location of specific structures within cells and tissues. Critical to the throughput and effectiveness of imaging studies, however, is the development of methods for quantitative and automated analysis of multispectral images over time [so-called five-dimemsional (5D) analysis]. Automated tracking of cell lineages during proliferation and migration from a series of image sequences provides a wealth of information such as cell positioning, shape, motility, and ancestry, as well as temporal relationships that would be difficult to quantify by hand.[82,137] For example, analysis of the neural stem cell (SC) niche using automated image analysis permits the identification of specifically labeled cells and quantifies cell–cell interactions including proximity to specific structures.[83,138] Inhibition of specific interactions within this model results in a quantitative shift in cell positioning and proliferation.

Image analysis and quantitative modeling permit us to step up from subcellular localization and microenvironment interactions to examine cell-based features in the context of global tissue organization and development. The coordination of interaction between cells and their matrix defines tissue structure and function. Therefore, the spatiotemporal quantification of these interactions should provide insight into population dynamics and the cooperativity within developing tissues. Several groups have begun to apply quantitative methods in order to understand differentiation, morphogenesis, and remodeling. Models have been developed that predict the sprouting location of branching tubules in 3D type I collagen culture based on the geometry of the starting structure.[5] Another approach, based on graph theory, permits quantitative understanding of network dynamics and is applied in fields as diverse as airplane trafficking, celebrity networks, and cellular metabolism.[139,140,162] This idea, when applied to tissue structure, enables the extraction of parameters that define function and dysfunction in terms of the structural organization of cells to aid pathology diagnoses.[141-144] These cell graphs have also recently been applied to compacting type I collagen hydrogels, with encapsulated SCs, resulting in the extraction of cellular patterns that govern the spatiotemporal progression of tissue organization in the early stages of mesenchymal stem cell (MSC) culture, matrix remodeling, and differentiation.[163]

These emerging fields of research all provide quantitative metrics, or design principles, that can be used to more efficiently engineer the cell/biomaterial interface for consistent and controlled cell fate determination. These methods provide an important and quantifiable link between starting materials and end products that can be compared across various culture systems in a systems biology "loop." Understanding the early structural and biochemical events that can predict long-term functional outputs should increase the throughput, cost-effectiveness, and consistency of biomaterial development and characterization. The rigor inherent in quantitative approaches has great promise to further enhance biomaterial translation into the clinic for the purposes of directing tissue function and regeneration.

44.9 Role of SCs in Regenerative Medicine

SCs have the unique ability to both self-renew and differentiate along a variety of lineages making them attractive for application in tissue engineering and regenerative medicine. The ultimate goal for the use of SCs in regenerative medicine is to create a self-renewing source for properly differentiated cells. But maintaining both regeneration and controlled differentiation of SCs in engineered niches remains a significant barrier to the widespread use of SC-based regenerative therapies: most SCs fail to maintain these behaviors once they are removed from their native niche *in vivo*. Two tiers of SC differentiation potential define the extent to which an SC population can recreate mature tissue within the organism.[145,146] *Pluripotent* SCs are capable of differentiating into a greater number of different cell lineages than *multipotent* SCs. Embryonic stem cells (ESCs) are regarded as the "gold standard" for regenerative therapies because they are the most pluripotent SCs and can therefore replicate all tissues in the adult human body. Adult MSCs, found within mature tissue, are an attractive source of SCs due to their ease of isolation and multipotency.[147,148] Only fertilized eggs and the cells derived from the first few divisions are *totipotent*, in that they can form the entire body and the placenta that supports it during gestation. No SC is totipotent.

44.10 Summary

This summary is brief and is far from an exhaustive treatment of solid mechanics and the mechanical behavior of materials. For a more complete development of these concepts, the reader should consult introductory texts in solids mechanics[69,70,156] as well as reference material concerning advanced mechanics of materials[157,158] and those that describe a continuum mechanics approach.[159-161] The continuum approach becomes particularly important when the simplifying assumptions of infinitesimal deformations (strains) and material planes of symmetry (i.e., isotropic and orthotropic) are no longer valid. As such, the continuum approach is often employed in soft tissues and biomaterials that exhibit large strain and/or anisotropic material behavior, as well as in advanced biomechanical models that incorporate mechanobiology, injury, and tissue development, growth, and remodeling. Humphrey provides a very elegant and comprehensive review of these aspects of continuum biomechanics.[149]

References

1. Robert, L. Matrix biology: Past, present and future. *Pathol. Biol. (Paris)* 49(4), 279–283, 2001.
2. Piez, K. A. History of extracellular matrix: A personal view. *Matrix Biol.* 16(3), 85–92, 1997.
3. Xu, R., Boudreau, A., and Bissell, M. J. Tissue architecture and function: Dynamic reciprocity via extra- and intra-cellular matrices. *Cancer Metastasis Rev.* 28(1–2): 167–176, 2009.
4. Nelson, C. M. Geometric control of tissue morphogenesis. *Biochim. Biophys. Acta.* 1793(5): 903–910, 2009.
5. Nelson, C. M., Vanduijn, M. M., Inman, J. L., Fletcher, D. A., and Bissell, M. J. Tissue geometry determines sites of mammary branching morphogenesis in organotypic cultures. *Science* 314(5797), 298–300, 2006.
6. Shields, J. D., Fleury, M. E., Yong, C., Tomei, A. A., Randolph, G. J., and Swartz, M. A. Autologous chemotaxis as a mechanism of tumor cell homing to lymphatics via interstitial flow and autocrine CCR7 signaling. *Cancer Cell* 11(6), 526–538, 2007.
7. Martin, P. and Lewis, J. Actin cables and epidermal movement in embryonic wound healing. *Nature* 360(6400), 179–183, 1992.
8. Wood, W., Jacinto, A., Grose, R., Woolner, S., Gale, J., Wilson, C., and Martin, P. Wound healing recapitulates morphogenesis in Drosophila embryos. *Nat. Cell Biol.* 4(11), 907–912, 2002.
9. Wadman, M. Scar prevention: The healing touch. *Nature* 436(7054), 1079–1080, 2005.
10. Yannas, I. V. Similarities and differences between induced organ regeneration in adults and early foetal regeneration. *J. R. Soc. Interface* 2(5), 403–417, 2005.
11. Yannas, I. V., Burke, J. F., Orgill, D. P., and Skrabut, E. M. Wound tissue can utilize a polymeric template to synthesize a functional extension of skin. *Science* 215(4529), 174–176, 1982.
12. Yannas, I. V., Burke, J. F., Warpehoski, M., Stasikelis, P., Skrabut, E. M., Orgill, D., and Giard, D. J. Prompt, long-term functional replacement of skin. *Trans. Am. Soc. Artif. Intern. Organs* 27, 19–23, 1981.
13. Yannas, I. V., Lee, E., Orgill, D. P., Skrabut, E. M., and Murphy, G. F. Synthesis and characterization of a model extracellular matrix that induces partial regeneration of adult mammalian skin. *Proc. Natl. Acad. Sci. USA* 86(3), 933–937, 1989.
14. Dovi, J. V., He, L. K., and DiPietro, L. A. Accelerated wound closure in neutrophil-depleted mice. *J. Leukoc. Biol.* 73(4), 448–455, 2003.
15. Martin, P., D'Souza, D., Martin, J., Grose, R., Cooper, L., Maki, R., and McKercher, S. R. Wound healing in the PU.1 null mouse-tissue repair is not dependent on inflammatory cells. *Curr. Biol.* 13(13), 1122–1128, 2003.
16. Ashcroft, G. S., Yang, X., Glick, A. B., Weinstein, M., Letterio, J. L., Mizel, D. E., et al. Mice lacking Smad3 show accelerated wound healing and an impaired local inflammatory response. *Nat. Cell Biol.* 1(5), 260–266, 1999.
17. Chan, B. P. and Leong, K. W. Scaffolding in tissue engineering: General approaches and tissue-specific considerations. *Eur. Spine J.* 17(Suppl. 4), 467–479, 2008.
18. Prabaharan, M. and Mano, J. F. Stimuli-responsive hydrogels based on polysaccharides incorporated with thermo-responsive polymers as novel biomaterials. *Macromol. Biosci.* 6(12), 991–1008, 2006.
19. Kemp, P. D. Tissue engineering and cell-populated collagen matrices. *Methods Mol. Biol.* 522, 1–8, 2009.
20. Lund, A. W., Bush, J. A., Plopper, G. E., and Stegemann, J. P. Osteogenic differentiation of mesenchymal stem cells in defined protein beads. *J. Biomed. Mater. Res. B Appl. Biomater.* 87(1): 213–221, 2008.
21. Cen, L., Liu, W., Cui, L., Zhang, W., and Cao, Y. Collagen tissue engineering: Development of novel biomaterials and applications. *Pediatr. Res.* 63(5), 492–496, 2008.
22. Li, X., Xie, J., Yuan, X., and Xia, Y. Coating electrospun poly(epsilon-caprolactone) fibers with gelatin and calcium phosphate and their use as biomimetic scaffolds for bone tissue engineering. *Langmuir* 24(24), 14145–14150, 2008.
23. Young, S., Wong, M., Tabata, Y., and Mikos, A. G. Gelatin as a delivery vehicle for the controlled release of bioactive molecules. *J. Control. Release* 109(1–3), 256–274, 2005.
24. Rowe, S. L. and Stegemann, J. P. Interpenetrating collagen-fibrin composite matrices with varying protein contents and ratios. *Biomacromolecules* 7(11), 2942–2948, 2006.
25. Gorodetsky, R. The use of fibrin based matrices and fibrin microbeads (FMB) for cell based tissue regeneration. *Expert. Opin. Biol. Ther.* 8(12), 1831–1846, 2008.
26. Prabaharan, M. and Mano, J. F. Chitosan-based particles as controlled drug delivery systems. *Drug Deliv.* 12(1), 41–57, 2005.
27. Wu, L., Li, H., Li, S., Li, X., Yuan, X., Li, X., and Zhang, Y. Composite fibrous membranes of PLGA and chitosan prepared by coelectrospinning and coaxial electrospinning. *J. Biomed. Mater. Res. A.* 2009. E-published.

28. Yang, K. C., Wu, C. C., Lin, F. H., Qi, Z., Kuo, T. F., Cheng, Y. H., Chen, M. P., and Sumi, S. Chitosan/gelatin hydrogel as immunoisolative matrix for injectable bioartificial pancreas. *Xenotransplantation* 15(6), 407–416, 2008.

29. Niemeyer, P., Krause, U., Kasten, P., Kreuz, P. C., Henle, P., Sudkam, N. P., and Mehlhorn, A. Mesenchymal stem cell-based HLA-independent cell therapy for tissue engineering of bone and cartilage. *Curr. Stem Cell Res. Ther.* 1(1), 21–27, 2006.

30. Zhang, H., Wei, Y. T., Tsang, K. S., Sun, C. R., Li, J., Huang, H., Cui, F. Z., and An, Y. H. Implantation of neural stem cells embedded in hyaluronic acid and collagen composite conduit promotes regeneration in a rabbit facial nerve injury model. *J. Transl. Med.* 6, 67, 2008.

31. Batorsky, A., Liao, J., Lund, A. W., Plopper, G. E., and Stegemann, J. P. Encapsulation of adult human mesenchymal stem cells within collagen-agarose microenvironments. *Biotechnol. Bioeng.* 92(4), 492–500, 2005.

32. Matsuno, T., Hashimoto, Y., Adachi, S., Omata, K., Yoshitaka, Y., Ozeki, Y., Umezu, Y., Tabata, Y., Nakamura, M., and Satoh, T. Preparation of injectable 3D-formed beta-tricalcium phosphate bead/alginate composite for bone tissue engineering. *Dent. Mater. J.* 27(6), 827–834, 2008.

33. Duggal, S., Fronsdal, K. B., Szoke, K., Shahdadfar, A., Melvik, J. E., and Brinchmann, J. E. Phenotype and gene expression of human mesenchymal stem cells in alginate scaffolds. *Tissue Eng. Part A* 15(7): 1763–1773, 2009.

34. Kraehenbuehl, T. P., Zammaretti, P., Van der, V, Schoenmakers, R. G., Lutolf, M. P., Jaconi, M. E., and Hubbell, J. A. Three-dimensional extracellular matrix-directed cardioprogenitor differentiation: Systematic modulation of a synthetic cell-responsive PEG-hydrogel. *Biomaterials* 29(18), 2757–2766, 2008.

35. Lim, T. Y., Poh, C. K., and Wang, W. Poly (lactic-*co*-glycolic acid) as a controlled release delivery device. *J. Mater. Sci. Mater. Med.* 20(8): 1669–1675, 2009.

36. Bible, E., Chau, D. Y., Alexander, M. R., Price, J., Shakesheff, K. M., and Modo, M. The support of neural stem cells transplanted into stroke-induced brain cavities by PLGA particles. *Biomaterials* 30(16): 2985–2994, 2009.

37. Li, J., Zhu, B., Shao, Y., Liu, X., Yang, X., and Yu, Q. Construction of anticoagulant poly(lactic acid) films via surface covalent graft of heparin-carrying microcapsules. *Colloids Surf. B Biointerfaces* 70(1), 15–19, 2009.

38. Prabaharan, M., Rodriguez-Perez, M. A., de Saja, J. A., and Mano, J. F. Preparation and characterization of poly(L-lactic acid)-chitosan hybrid scaffolds with drug release capability. *J. Biomed. Mater. Res. B Appl. Biomater.* 81(2), 427–434, 2007.

39. Zhou, X. Z., Leung, V. Y., Dong, Q. R., Cheung, K. M., Chan, D., and Lu, W. W. Mesenchymal stem cell-based repair of articular cartilage with polyglycolic acid-hydroxyapatite biphasic scaffold. *Int. J. Artif. Organs* 31(6), 480–489, 2008.

40. Martin, C., Winet, H., and Bao, J. Y. Acidity near eroding polylactide-polyglycolide *in vitro* and *in vivo* in rabbit tibial bone chambers. *Biomaterials* 17(24), 2373–2380, 1996.

41. Bonzani, I. C., George, J. H., and Stevens, M. M. Novel materials for bone and cartilage regeneration. *Curr. Opin. Chem. Biol.* 10(6), 568–575, 2006.

42. Bonzani, I. C., Adhikari, R., Houshyar, S., Mayadunne, R., Gunatillake, P., and Stevens, M. M. Synthesis of two-component injectable polyurethanes for bone tissue engineering. *Biomaterials* 28(3), 423–433, 2007.

43. Stevens, B., Yang, Y., Mohandas, A., Stucker, B., and Nguyen, K. T. A review of materials, fabrication methods, and strategies used to enhance bone regeneration in engineered bone tissues. *J. Biomed. Mater. Res. B Appl. Biomater.* 85(2), 573–582, 2008.

44. Stegemann, J. P., Hong, H., and Nerem, R. M. Mechanical, biochemical, and extracellular matrix effects on vascular smooth muscle cell phenotype. *J. Appl. Physiol.* 98(6), 2321–2327, 2005.

45. Niklason, L. E. and Langer, R. S. Advances in tissue engineering of blood vessels and other tissues. *Transpl. Immunol.* 5(4), 303–306, 1997.

46. Liu, Y., Ramanath, H. S., and Wang, D. A. Tendon tissue engineering using scaffold enhancing strategies. *Trends Biotechnol.* 26(4), 201–209, 2008.

47. Rotini, R., Fini, M., Giavaresi, G., Marinelli, A., Guerra, E., Antonioli, D., Castagna, A., and Giardino, R. New perspectives in rotator cuff tendon regeneration: Review of tissue engineered therapies. *Chir. Organi. Mov.* 91(2), 87–92, 2008.

48. An, Y., Tsang, K. K., and Zhang, H. Potential of stem cell based therapy and tissue engineering in the regeneration of the central nervous system. *Biomed. Mater.* 1(2), R38–R44, 2006.

49. Schmidt, C. E. and Leach, J. B. Neural tissue engineering: Strategies for repair and regeneration. *Annu. Rev. Biomed. Eng.* 5, 293–347, 2003.

50. Stoddart, M. J., Grad, S., Eglin, D., and Alini, M. Cells and biomaterials in cartilage tissue engineering. *Regen. Med.* 4(1), 81–98, 2009.

51. Kerker, J. T., Leo, A. J., and Sgaglione, N. A. Cartilage repair: Synthetics and scaffolds: Basic science, surgical techniques, and clinical outcomes. *Sports Med. Arthrosc.* 16(4), 208–216, 2008.

52. Stosich, M. S., Moioli, E. K., Wu, J. K., Lee, C. H., Rohde, C., Yousref, A. M., Ascherman, J., Diraddo, R., Marion, N. W., and Mao, J. J. Bioengineering strategies to generate vascularized soft tissue grafts with sustained shape. *Methods* 47(2), 116–121, 2009.

53. Morgan, S. M., Ainsworth, B. J., Kanczler, J. M., Babister, J. C., Chaudhuri, J. B., and Oreffo, R. O. Formation of a human-derived fat tissue layer in P(DL)LGA hollow fibre scaffolds for adipocyte tissue engineering. *Biomaterials* 30(10), 1910–1917, 2009.

54. Rezwan, K., Chen, Q. Z., Blaker, J. J., and Boccaccini, A. R. Biodegradable and bioactive porous polymer/inorganic composite scaffolds for bone tissue engineering. *Biomaterials* 27(18), 3413–3431, 2006.

55. Heidemann, W., Jeschkeit, S., Ruffieux, K., Fischer, J. H., Wagner, M., Kruger, G., Wintermantel, E., and Gerlach, K. L. Degradation of poly(D,L)lactide implants with or without addition of calcium phosphates *in vivo*. *Biomaterials* 22(17), 2371–2381, 2001.

56. Lu, L. and Mikos, A. G. Biodegradable polymers. *Sci. Med. (Phila)* 6(1), 6–7, 1999.

57. Bergsma, E. J., Rozema, F. R., Bos, R. R., and de Bruijn, W. C. Foreign body reactions to resorbable poly(L-lactide) bone plates and screws used for the fixation of unstable zygomatic fractures. *J. Oral Maxillofac. Surg.* 51(6), 666–670, 1993.

58. Zelikin, A. N., Lynn, D. M., Farhadi, J., Martin, I., Shastri, V., and Langer, R. Erodible conducting polymers for potential biomedical applications. *Angew. Chem. Int. Ed. Engl.* 41(1), 141–144, 2002.

59. Lemons, J. E. and Lucas, L. C. Properties of biomaterials. *J. Arthroplasty* 1(2), 143–147, 1986.

60. Stynes, G., Kiroff, G. K., Morrison, W. A., and Kirkland, M. A. Tissue compatibility of biomaterials: Benefits and problems of skin biointegration. *ANZ. J. Surg.* 78(8), 654–659, 2008.

61. Tanentzapf, G., Devenport, D., Godt, D., and Brown, N. H. Integrin-dependent anchoring of a stem-cell niche. *Nat. Cell Biol.* 9(12), 1413–1418, 2007.

62. Wang, F., Weaver, V. M., Petersen, O. W., Larabell, C. A., Dedhar, S., Briand, P., Lupu, R., and Bissell, M. J. Reciprocal interactions between beta1-integrin and epidermal growth factor receptor in three-dimensional basement membrane breast cultures: A different perspective in epithelial biology. *Proc. Natl. Acad. Sci. USA* 95(25), 14821–14826, 1998.

63. Weaver, V. M., Lelievre, S., Lakins, J. N., Chrenek, M. A., Jones, J. C., Giancotti, F., Werb, Z., and Bissell, M. J. beta4 integrin-dependent formation of polarized three-dimensional architecture confers resistance to apoptosis in normal and malignant mammary epithelium. *Cancer Cell* 2(3), 205–216, 2002.

64. Salasznyk, R. M., Williams, W. A., Boskey, A., Batorsky, A., and Plopper, G. E. Adhesion to vitronectin and collagen I promotes osteogenic differentiation of human mesenchymal stem cells. *J. Biomed. Biotechnol.* 2004(1), 24–34, 2004.

65. Lukashev, M. E. and Werb, Z. ECM signalling: Orchestrating cell behaviour and misbehaviour. *Trends Cell Biol.* 8(11), 437–441, 1998.

66. Alsberg, E., Anderson, K. W., Albeiruti, A., Franceschi, R. T., and Mooney, D. J. Cell-interactive alginate hydrogels for bone tissue engineering. *J. Dent. Res.* 80(11), 2025–2029, 2001.

67. Chung, I. M., Enemchukwu, N. O., Khaja, S. D., Murthy, N., Mantalaris, A., and Garcia, A. J. Bioadhesive hydrogel microenvironments to modulate epithelial morphogenesis. *Biomaterials* 29(17), 2637–2645, 2008.

68. Rosines, E., Schmidt, H. J., and Nigam, S. K. The effect of hyaluronic acid size and concentration on branching morphogenesis and tubule differentiation in developing kidney culture systems: Potential applications to engineering of renal tissues. *Biomaterials* 28(32), 4806–4817, 2007.

69. Beer, F., Johnston, R. E. J., and DeWolf, J. *Mechanics of Materials*, 4th Edition. New York: McGraw-Hill, 2005.

70. Hibbeler, R. *Mechanics of Materials*, 7th Edition. Englewood Cliffs, NJ: Prentice Hall, 2007.

71. Roberson, J. A. and Crowe, C. T. *Engineering Fluid Mechanics*, 4th Edition. Boston, MA: Houghton Mifflin, 1990.

72. Incropera, F. P. and DeWitt, D. P. *Introduction to Heat Transfer*, 2nd Edition. New York: Wiley, 1990.

73. Moran, M. J. and Shapiro, H. N. *Fundamentals of Engineering Thermodynamics*. New York: Wiley, 1988.

74. Plonus, M. A. *Applied Electromagnetics*. New York: McGraw-Hill, 1978.

75. Halliday, D. and Resnick, R. *Fundamentals of Physics*, 3rd Edition. New York: Wiley, 1988.

76. Rees, D. *Basic Engineering Plasticity: An Introduction with Engineering and Manufacturing Applications*. Illustrated edition. New York: Butterworth-Heinemann, 2006.

77. Fung, Y. C. Elasticity of soft tissues in simple elongation. *Am. J. Physiol.* 213(6), 1532–1544, 1967.

78. Fung, Y. C. Biorheology of soft tissues. *Biorheology* 10(2), 139–155, 1973.

79. Humphrey, J. D. and Yin, F. C. A new constitutive formulation for characterizing the mechanical behavior of soft tissues. *Biophys. J.* 52(4), 563–570, 1987.

80. Holmes, M. H. and Mow, V. C. The nonlinear characteristics of soft gels and hydrated connective tissues in ultrafiltration. *J. Biomech.* 23(11), 1145–1156, 1990.

81. Kwan, M. K., Lai, W. M., and Mow, V. C. A finite deformation theory for cartilage and other soft hydrated connective tissues-I. Equilibrium results. *J. Biomech.* 23(2), 145–155, 1990.

82. Kwan, M. K., Lin, T. H., and Woo, S. L. On the viscoelastic properties of the anteromedial bundle of the anterior cruciate ligament. *J. Biomech.* 26(4–5), 447–452, 1993.

83. Best, T. M., McElhaney, J. H., Garrett, W. E., Jr., and Myers, B. S. Axial strain measurements in skeletal muscle at various strain rates. *J. Biomech. Eng.* 117(3) 262–265, 1995.

84. Mendis, K. K., Stalnaker, R. L., and Advani, S. H. A constitutive relationship for large deformation finite element modeling of brain tissue. *J. Biomech. Eng.* 117(3), 279–285, 1995.

85. Johnson, G. A, Livesay, G. A., Woo, S. L., and Rajagopal, K. R. A single integral finite strain viscoelastic model of ligaments and tendons. *J. Biomech. Eng.* 118(2), 221–226, 1996.

86. Kohles, S. S., Thielke, R. J., Vanderby, R. Jr., Finite elasticity formulations for evaluation of ligamentous tissue. *Biomed. Mater. Eng.* 7(6), 387–390, 1997.

87. Simon, B. R., Kaufmann, M. V., McAfee, M. A., Baldwin, A. L., Wilson, L. M. Identification and determination of material properties for porohyperelastic analysis of large arteries. *J. Biomech. Eng.* 120(2), 188–194, 1998.

88. Secomb, T. W. and El-Kareh, A. W. A theoretical model for the elastic properties of very soft tissues. *Biorheology* 38(4), 305–317, 2001.

89. Frisen, M., Magi, M., Sonnerup, I., and Viidik, A. Rheological analysis of soft collagenous tissue. Part I: Theoretical considerations. *J. Biomech.* 2(1), 13–20, 1969.

90. Frisen, M., Magi, M., Sonnerup, L., and Viidik, A. Rheological analysis of soft collagenous tissue. Part II: Experimental evaluations and verifications. *J. Biomech.* 2(1), 21–28, 1969.

91. Haut, R. C. and Little, R. W. Rheological properties of canine anterior cruciate ligaments. *J. Biomech.* 2(3), 289–298, 1969.

92. Woo, S. L., Gomez, M. A., and Akeson, W. H. The time and history-dependent viscoelastic properties of the canine medical collateral ligament. *J. Biomech. Eng.* 103(4), 293–298, 1981.

93. Thornton, G. M., Oliynyk, A., Frank, C. B., and Shrive, N. G. Ligament creep cannot be predicted from stress relaxation at low stress: A biomechanical study of the rabbit medial collateral ligament. *J. Orthop. Res.* 15, 652–656, 1997.

94. Thornton, G. M., Leask, G. P., Shrive, N. G., and Frank, C. B. Early medial collateral ligament scars have inferior creep behaviour. *J. Orthop. Res.* 18(2), 238–246, 2000.

95. Thornton, G. M., Shrive, N. G., and Frank, C. B. Ligament creep recruits fibres at low stresses and can lead to modulus-reducing fibre damage at higher creep stresses: A study in rabbit medial collateral ligament model. *J. Orthop. Res.* 20(5), 967–974, 2002.

96. Panjabi, M. M., Moy, P., Oxland, T. R., and Cholewicki, J. Subfailure injury affects the relaxation behavior of rabbit ACL. *Clin. Biomech. (Bristol, Avon)* 14(1), 24–31, 1999.

97. Provenzano, P., Lakes, R., Keenan, T., and Vanderby, R., Jr., Nonlinear ligament viscoelasticity. *Ann. Biomed. Eng.* 29(10), 908–914, 2001.

98. Atkinson, T. S., Ewers, B. J., and Haut, R. C. The tensile and stress relaxation responses of human patellar tendon varies with specimen cross-sectional area. *J. Biomech.* 32(9), 907–914, 1999.

99. Haut, R. C. and Little, R. W. A constitutive equation for collagen fibers. *J. Biomech.* 5(5), 423–430, 1972.

100. Screen, H. R. C. Stress relaxation of tendon fascicles at the microscale. *Proceedings of the International Symposium of Ligament & Tendon*, February 10, 2007, San Diego, CA; 24.

101. Woo, S. L., Simon, B. R., Kuei, S. C., and Akeson, W. H. Quasi-linear viscoelastic properties of normal articular cartilage. *J. Biomech. Eng.* 102(2), 85–90, 1980.

102. June, R. K., Ly, S., and Fyhrie, D. P. Cartilage stress-relaxation proceeds slower at higher compressive strains. *Arch. Biochem. Biophys.* 483(1), 75–80, 2009.

103. Best, T. M., McElhaney, J., Garrett, W. E., Jr., and Myers, B. S. Characterization of the passive responses of live skeletal muscle using the quasi-linear theory of viscoelasticity. *J. Biomech.* 27(4), 413–419, 1994.

104. Corr, D. T., Leverson, G. E., Vanderby, R., Jr., and Best T. M. A nonlinear rheological assessment of muscle recovery from eccentric stretch injury. *Med. Sci. Sports Exercise* 35(9), 1581–1588, 2003.

105. Rousseau, E. P., Sauren, A. A., van Hout, M. C., and van Steenhoven, A. A. Elastic and viscoelastic material behaviour of fresh and glutaraldehyde-treated porcine aortic valve tissue. *J. Biomech.* 16(5), 339–348, 1983.

106. Sauren, A. A., van Hout, M. C., van Steenhoven, A. A., Veldpaus, F. E., and Janssen, J. D. The mechanical properties of porcine aortic valve tissues. *J. Biomech.* 16(5), 327–337, 1983.

107. Eshel, H. and Lanir, Y. Effects of strain level and proteoglycan depletion on preconditioning and viscoelastic responses of rat dorsal skin. *Ann. Biomed. Eng.* 29(2), 164–172, 2001.

108. Ruvolo, E. C., Jr., Stamatas, G. N., and Kollias, N. Skin viscoelasticity displays site- and age-dependent angular anisotropy. *Skin Pharmacol. Physiol.* 20(6), 313–321, 2007.

109. Corr, D. T., Gallant-Behm, C. L., Shrive, N. G., and Hart, D. A. Biomechanical behavior of scar tissue and uninjured skin in a porcine model. *Wound Repair Regen.* 17(2), 250–259, 2009.

110. Lakes, R. S., Katz, J. L., and Sternstein, S. S. Viscoelastic properties of wet cortical bone-I. Torsional and biaxial studies. *J. Biomech.* 12(9), 657–678, 1979.

111. Lakes, R. S. and Katz, J. L. Viscoelastic properties of wet cortical bone-III. A non-linear constitutive equation. *J. Biomech.* 12(9), 689–698, 1979.

112. Yeni, Y. N., Shaffer, R. R., Baker, K. C., Dong, X. N., Grimm, M. J., Les, C.,M., et al. The effect of yield damage on the viscoelastic properties of cortical bone tissue as measured by dynamic mechanical analysis. *J. Biomed. Mater. Res. A* 82(3), 530–537, 2007.

113. Moutos, F. T. and Guilak, F. Composite scaffolds for cartilage tissue engineering. *Biorheology* 45(3–4), 501–512, 2008.

114. Fung, Y. C. Stress strain history relations of soft tissues in simple elongation. In: Y. C. Fung, N. Perrone, and M. Anliker (eds), *Biomechanics: Its Foundations and Objectives.* Englewood Cliffs, NJ: Prentice Hall, 1972.

115. Woo, S. L. Mechanical properties of tendons and ligaments. I. Quasi-static and nonlinear viscoelastic properties. *Biorheology* 19(3), 385–396, 1982.

116. Carew, E. O., Talman, E. A., Boughner, D. R., and Vesely I. Quasi-linear viscoelastic theory applied to internal shearing of porcine aortic valve leaflets. *J. Biomech. Eng.* 121(4), 386–392, 1999.

117. Carew, E. O, Barber, J. E., and Vesely, I. Role of preconditioning and recovery time in repeated testing of aortic valve tissues: Validation through quasilinear viscoelastic theory. *Ann. Biomed. Eng.* 28(9), 1093–1100, 2000.

118. Provenzano, P. P., Lakes, R. S., Corr, D. T., and Vanderby, R., Jr. Application of nonlinear viscoelastic models to describe ligament behavior. *Biomech. Model Mechanobiol.* 1(1), 45–57, 2002.

119. Lakes, R. S. and Vanderby R. Interrelation of creep and relaxation: A modeling approach for ligaments. *J. Biomech. Eng.* 121(6), 612–615, 1999.

120. Attarian, D. E., McCrackin, H. J., DeVito, D. P., McElhaney, J. H., and Garrett, W. E., Jr. Biomechanical characteristics of human ankle ligaments. *Foot Ankle* 6(2), 54–58, 1985.

121. Nikolaou, P. K., Macdonald, B. L., Glisson, R. R., Seaber, A. V., and Garrett, W. E., Jr. Biomechanical and histological evaluation of muscle after controlled strain injury. *Am. J. Sports Med.* 15(1), 9–14, 1987.

122. Siegler, S., Block, J., and Schneck, C. D. The mechanical characteristics of the collateral ligaments of the human ankle joint. *Foot Ankle* 8(5), 234–242, 1988.

123. Taylor, D. C., Dalton, J. D., Jr., Seaber, A. V., and Garrett, W. E., Jr. Experimental muscle strain injury. Early functional and structural deficits and the increased risk for reinjury. *Am. J. Sports Med.* 21(2), 190–194, 1993.

124. Vogel, H. G. Tensile strength of skin wounds in rats after treatment with corticosteroids. *Acta Endocrinol. (Copenh)* 64(2), 295–303, 1970.

125. Hurschler, C., Provenzano, P. P., and Vanderby, R., Jr. Scanning electron microscopic characterization of healing and normal rat ligament microstructure under slack and loaded conditions. *Connect Tissue Res.* 44(2), 59–68, 2003.

126. Butler, D. L., Juncosa-Melvin, N., Boivin G. P., Galloway, M. T., Shearn, J. T., Gooch, C., et al. Functional tissue engineering for tendon repair: A multidisciplinary strategy using mesenchymal stem cells, bioscaffolds, and mechanical stimulation. *J. Orthop. Res.* 26(1), 1–9, 2008.

127. Juncosa, N., West, J. R., Galloway, M. T., Boivin, G. P., and Butler, D. L. *In vivo* forces used to develop design parameters for tissue engineered implants for rabbit patellar tendon repair. *J. Biomech.* 36(4), 483–488, 2003.

128. Juncosa-Melvin, N., Shearn, J. T., Boivin, G. P., Gooch, C., Galloway, M. T., West, J. R., et al. Effects of mechanical stimulation on the biomechanics and histology of stem cell-collagen sponge constructs for rabbit patellar tendon repair. *Tissue Eng.* 12(8), 2291–2300, 2006.

129. Zisch, A. H., Schenk, U., Schense, J. C., Sakiyama-Elbert, S. E., and Hubbell, J. A. Covalently conjugated VEGF-fibrin matrices for endothelialization. *J. Control. Release* 72(1–3), 101–113, 2001.

130. Zisch, A. H., Lutolf, M. P., Ehrbar, M., Raeber, G. P., Rizzi, S. C., Davies, N., et al. Cell-demanded release of VEGF from synthetic, biointeractive cell ingrowth matrices for vascularized tissue growth. *FASEB J.* 17(15), 2260–2262, 2003.

131. Seliktar, D., Zisch, A. H., Lutolf, M. P., Wrana, J. L., and Hubbell, J. A. MMP-2 sensitive, VEGF-bearing bioactive hydrogels for promotion of vascular healing. *J. Biomed. Mater. Res. A* 68(4), 704–716, 2004.

132. Park, Y., Lutolf, M. P., Hubbell, J. A., Hunziker, E. B., and Wong, M. Bovine primary chondrocyte culture in synthetic matrix metalloproteinase-sensitive poly(ethylene glycol)-based hydrogels as a scaffold for cartilage repair. *Tissue Eng.* 10(3–4), 515–522, 2004.

133. Adelow, C., Segura, T., Hubbell, J. A., and Frey, P. The effect of enzymatically degradable poly(ethylene glycol) hydrogels on smooth muscle cell phenotype. *Biomaterials* 29(3), 314–326, 2008.

134. Blinov, M. L., Ruebenacker, O., and Moraru, I. I. Complexity and modularity of intracellular networks: A systematic approach for modelling and simulation. *IET. Syst. Biol.* 2(5), 363–368, 2008.

135. Rangamani, P. and Iyengar, R. Modelling spatio-temporal interactions within the cell. *J. Biosci.* 32(1), 157–167, 2007.

136. Karlebach, G. and Shamir, R. Modelling and analysis of gene regulatory networks. *Nat. Rev. Mol. Cell Biol.* 9(10), 770–780, 2008.

137. Aldridge, B. B., Burke, J. M., Lauffenburger, D. A., and Sorger, P. K. Physicochemical modelling of cell signalling pathways. *Nat. Cell Biol.* 8(11), 1195–1203, 2006.

138. Giurumescu, C. A., Sternberg, P. W., and Asthagiri, A. R. Intercellular coupling amplifies fate segregation during Caenorhabditis elegans vulval development. *Proc. Natl. Acad. Sci. USA* 103(5), 1331–1336, 2006.

139. Kodiha, M., Brown, C. M., and Stochaj, U. Analysis of signaling events by combining high-throughput screening technology with computer-based image analysis. *Sci. Signal.* 1(37), 12, 2008.

140. Albeck, J. G., Burke, J. M., Aldridge, B. B., Zhang, M., Lauffenburger, D. A., and Sorger, P. K. Quantitative analysis of pathways controlling extrinsic apoptosis in single cells. *Mol. Cell* 30(1), 11–25, 2008.

141. Lazzara, M. J. and Lauffenburger, D. A. Quantitative modeling perspectives on the ErbB system of cell regulatory processes. *Exp. Cell Res.* 315(4), 717–725, 2009.

142. Chen, Y., Ladi, E., Herzmark, P., Robey, E., and Roysam, B. Automated 5-D analysis of cell migration and interaction in the thymic cortex from time-lapse sequences of 3-D multi-channel multi-photon images. *J. Immunol. Methods* 340(1), 65–80, 2009.

143. Shen, Q., Wang, Y., Kokovay, E., Lin, G., Chuang, S. M., Goderie, S. K., Roysam, B., and Temple, S. Adult SVZ stem cells lie in a vascular niche: a quantitative analysis of niche cell–cell interactions. *Cell Stem. Cell* 3(3), 289–300, 2008.

144. Barabasi, A. L. and Oltvai, Z. N. Network biology: Understanding the cell's functional organization. *Nat. Rev. Genet.* 5(2), 101–113, 2004.

145. Barabasi, A. L. Network medicine—from obesity to the "diseasome." *N. Engl. J. Med.* 357(4), 404–407, 2007.

146. Bilgin, C., Demir, C., Nagi, C., and Yener, B. Cell-graph mining for breast tissue modeling and classification. *Conf. Proc. IEEE Eng. Med. Biol. Soc.* 5311–5314, 2007.

147. Demir, C., Gultekin, S. H., and Yener, B. Learning the topological properties of brain tumors. *IEEE/ACM. Trans. Comput. Biol. Bioinform.* 2(3), 262–270, 2005.

148. Demir, C., Gultekin, S. H., and Yener, B. Augmented cell-graphs for automated cancer diagnosis. *Bioinformatics* 21(Suppl. 2), ii7–ii12, 2005.

149. Humphrey, J. D. Continuum biomechanics of soft biological tissues. *Proc. R. Soc. Lond. A* 459, 3–46, 2003.

150. Tengvall, P., Lundstrom, I., Sjoqvist, L., Elwing, H., and Bjursten, L. M. Titanium-hydrogen peroxide interaction: Model studies of the influence of the inflammatory response on titanium implants. *Biomaterials* 10(3), 166–175, 1989.

151. Suzuki, R., Muyco, J., McKittrick, J., and Frangos, J. A. Reactive oxygen species inhibited by titanium oxide coatings. *J. Biomed. Mater. Res. A* 66(2), 396–402, 2003.

152. Suzuki, R. and Frangos, J. A. Inhibition of inflammatory species by titanium surfaces. *Clin. Orthop. Relat. Res.* 372, 280–289, 2000.

153. Rushing, G. D., Goretsky, M. J., Gustin, T., Morales, M., Kelly, R. E., Jr., and Nuss, D. When it is not an infection: Metal allergy after the Nuss procedure for repair of pectus excavatum. *J. Pediatr. Surg.* 42(1), 93–97, 2007.

154. Jansen, J. A., van der Waerden, J. P., and de Groot, K. Epithelial reaction to percutaneous implant materials: *In vitro* and *in vivo* experiments. *J. Invest. Surg.* 2(1), 29–49, 1989.

155. Hoiby, N., Espersen, F., Fomsgaard, A., Giwercman, B., Jensen, E. T., Johansen, H. K., Koch, C., Kronborg, G., Pedersen, S. S., and Pressler, T. Biofilm, foreign bodies and chronic infections. *Ugeskr. Laeger* 156(41), 5998–6005, 1994.

156. Timoshenko, S. and MacCullough, G. H. *Elements of Strength of Materials*, 3rd Edition New York: D. Van Nostrand Company, 1951.

157. Den Hartog, J. P. *Advanced Strength of Materials*, New York: McGraw-Hill, 1952.

158. Cook, R. D. and Young, W. C. *Advanced Mechanics of Materials*. New York: Macmillan, 1985.

159. Fung, Y. C. *First Course in Continuum Mechanics*. Englewood Cliffs, NJ: Prentice Hall, 1977.

160. Lai, M. W., Rubin, D., and Krempl, E. *Introduction to Continuum Mechanics*, 3rd Edition. Oxford: Pergamon Press, 1993.

161. Malvern, L. E. *Introduction to the Mechanics of a Continuous Medium*. Englewood Cliffs, NJ: Pretence Hall, 1969.

162. Gunduz, C., Yener, B., and Gultekin, S. H. The cell graphs of cancer. *Bioinformatics* 20(Suppl. 1), i145–i151, 2004.

163. Lund, A. W., Bilgin, C. C., Hasan, M. A., McKeen, L. M., Stegemann, T. P., Yener, B., Zaki, M. J., and Plopper, G. E. Quantification of spatial parameters in 3D cellular constructs using graph theory, *J Biomed Biotech*. 1–16, 2009.

Additional Reading

Banerjee, C., McCabe, L. R., Choi, J. Y., Hiebert, S. W., Stein, J. L., Stein, G. S., and Lian, J. B. Runt homology domain proteins in osteoblast differentiation: AML3/CBFA1 is a major component of a bone-specific complex. *J. Cell Biochem.* 66(1), 1–8,1997.

Barry, F. P. and Murphy, J. M. Mesenchymal stem cells: Clinical applications and biological characterization. *Int. J. Biochem. Cell Biol.* 36(4), 568–584, 2004.

Becker, A. J., Mcculloch, E. A., and Till, J. E. Cytological demonstration of the clonal nature of spleen colonies derived from transplanted mouse marrow cells. *Nature* 197, 452–454, 1963.

Boyer, L. A., Lee, T. I., Cole, M. F., Johnstone, S. E., Levine, S. S., Zucker, J. P., et al. Core transcriptional regulatory circuitry in human embryonic stem cells. *Cell* 122(6), 947–956, 2005.

Chambers, I., Colby, D., Robertson, M., Nichols, J., Lee, S., Tweedie, S., and Smith, A. Functional expression cloning of Nanog, a pluripotency sustaining factor in embryonic stem cells. *Cell* 113(5), 643–655, 2003.

de Crombrugghe, B., Lefebvre, V., Behringer, R. R., Bi, W., Murakami, S., and Huang, W. Transcriptional mechanisms of chondrocyte differentiation. *Matrix Biol.* 19(5), 389–394, 2000.

Ducy, P., Zhang, R., Geoffroy, V., Ridall, A. L., and Karsenty, G. Osf2/Cbfa1: A transcriptional activator of osteoblast differentiation. *Cell* 89(5), 747–754, 1997.

Engler, A. J., Sen, S., Sweeney, H. L., and Discher, D. E. Matrix elasticity directs stem cell lineage specification. *Cell* 126(4), 677–689, 2006.

Gregory, C. A., Ylostalo, J., and Prockop, D. J. Adult bone marrow stem/progenitor cells (MSCs) are preconditioned by microenvironmental "niches" in culture: A two-stage hypothesis for regulation of MSC fate. *Sci. STKE.* (294), e37, 2005.

Jaiswal, N., Haynesworth, S. E., Caplan, A. I., and Bruder, S. P. Osteogenic differentiation of purified, culture-expanded human mesenchymal stem cells *in vitro*. *J. Cell Biochem.* 64(2), 295–312, 1997.

Owen, M. and Friedenstein, A. J. Stromal stem cells: Marrow-derived osteogenic precursors. *Ciba Found. Symp.* 136, 42–60, 1988.

Perry, R. L. and Rudnick, M. A. Molecular mechanisms regulating myogenic determination and differentiation. *Front Biosci.* 5, D750–D767, 2000.

Poliard, A., Nifuji, A., Lamblin, D., Plee, E., Forest, C., and Kellermann, O. Controlled conversion of an immortalized mesodermal progenitor cell towards osteogenic, chondrogenic, or adipogenic pathways. *J. Cell Biol.* 130(6), 1461–1472, 1995.

Rosen, E. D. and Spiegelman, B. M. Molecular regulation of adipogenesis. *Annu. Rev. Cell Dev. Biol.* 16, 145–171, 2000.

Siminovitch, L., Mcculloch, E. A., and Till, J. E. The distribution of colony-forming cells among spleen colonies. *J. Cell Physiol.* 62, 327–336, 1963.

Index